Routledge Handbook of Sustainable Heritage

This handbook presents cutting-edge and global insights on sustainable heritage, engaging with ideas such as data science in heritage, climate change and environmental challenges, indigenous heritage, contested heritage and resilience. It does so across a diverse range of global heritage sites.

Organized into six themed parts, the handbook offers cross-disciplinary perspectives on the latest theory, research and practice. Thirty-five chapters offer insights from leading scholars and practitioners in the field as well as early career researchers. This book fills a lacuna in the literature by offering scientific approaches to sustainable heritage, as well as multicultural perspectives by exploring sustainable heritage in a range of different geographical contexts and scales. The themes covered revolve around heritage values and heritage risk; participatory approaches to heritage; dissonant heritage; socio-environmental challenges to heritage; sustainable heritage-led transformation and new cross-disciplinary methods for heritage research.

This book will be an invaluable resource for students and scholars in heritage studies, archaeology, museum studies, cultural studies, architecture, landscape, urban design, planning, geography and tourism.

Kalliopi Fouseki, Professor in Sustainable Heritage Management, UCL Institute for Sustainable Heritage, UK.

May Cassar, Director of the UCL Institute for Sustainable Heritage, UK.

Guillaume Dreyfuss, Director of Research at the Architecture Project LTD, Malta.

Kelvin Ang Kah Eng, Director in Conservation Management of the Urban Development Authority of Singapore.

Routledge Handbook of Sustainable Heritage

Edited by
Kalliopi Fouseki, May Cassar,
Guillaume Dreyfuss and
Kelvin Ang Kah Eng

LONDON AND NEW YORK

Cover image: ©Henk Leerink

First published 2023
by Routledge
4 Park Square, Milton Park, Abingdon, Oxon OX14 4RN

and by Routledge
605 Third Avenue, New York, NY 10158

Routledge is an imprint of the Taylor & Francis Group, an informa business

© 2023 selection and editorial matter, Kalliopi Fouseki, May Cassar, Guillaume Dreyfuss and Kelvin Ang Kah Eng; individual chapters, the contributors

The right of Kalliopi Fouseki, May Cassar, Guillaume Dreyfuss and Kelvin Ang Kah Eng to be identified as the authors of the editorial material, and of the authors for their individual chapters, has been asserted in accordance with sections 77 and 78 of the Copyright, Designs and Patents Act 1988.

All rights reserved. No part of this book may be reprinted or reproduced or utilised in any form or by any electronic, mechanical, or other means, now known or hereafter invented, including photocopying and recording, or in any information storage or retrieval system, without permission in writing from the publishers.

Trademark notice: Product or corporate names may be trademarks or registered trademarks, and are used only for identification and explanation without intent to infringe.

British Library Cataloguing-in-Publication Data
A catalogue record for this book is available from the British Library

Library of Congress Cataloging-in-Publication Data
A catalog record has been requested for this book

ISBN: 978-0-367-48274-9 (hbk)
ISBN: 978-1-032-27699-1 (pbk)
ISBN: 978-1-003-03895-5 (ebk)

DOI: 10.4324/9781003038955

Typeset in Bembo
by codeMantra

Contents

List of figures ix
List of tables xii
Notes on contributors xiii

Introduction: sustainability for heritage and heritage for sustainability 1
KALLIOPI FOUSEKI, GUILLAUME DREYFUSS, KELVIN KAH ENG ANG
AND MAY CASSAR

PART I
Heritage values and risk 15

1 **Values and sustaining heritage** 17
   RANDALL MASON

2 **Heritage and change management** 30
   TORGRIM SNEVE GUTTORMSEN AND JOAR SKREDE

3 **Combining theory and practice: incorporating value in risk assessment at heritage sites** 44
   KATHERINE CURRAN, JUNNAN BAO, HEATHER OAKLEY, ARGYRO GILI AND
   KALLIOPI FOUSEKI

4 **Heritage values and heritage management frameworks in Nigeria** 58
   JOY ACHEYINI EDEOJA, KALLIOPI FOUSEKI AND ALEJANDRA ALBUERNE

5 **Evaluating the management plan of Bali Cultural Landscape from the local community's perspective** 72
   DIANA RAHMAN AND KALLIOPI FOUSEKI

PART II
Participatory heritage — 85

6 Managing participatory heritage for enhancing social well-being — 87
   EIRINI GALLOU AND KALLIOPI FOUSEKI

7 Social sustainability and witnessing difficult heritage — 104
   ROSS WILSON

8 Citizen science in sustainable heritage conservation — 115
   ROSIE BRIGHAM AND JOSEP GRAU-BOVÉ

9 Community-centred sustainable heritage management: reality and challenges in practice — 126
   SUJEONG LEE

10 Heritage conservation as a social process: assessing social impacts of participatory cultural heritage conservation — 138
   ELIA QUIJANO QUIÑONES AND KALLIOPI FOUSEKI

11 Sustainable heritage through a sustainable community — 154
   CHIHEI SUZUKI

PART III
Dissonant and 'Pacific' heritage — 167

12 Developing international cultural relations through the negotiation of cultural property disputes: a sustainability perspective — 169
   MARIA SHEHADE

13 Is world heritage politically sustainable? — 181
   KRISTEN BARRETT-CASEY, LORIKA HISARI AND KALLIOPI FOUSEKI

14 Political ruptures and the cultural heritage of Iraq — 196
   RENÉ TEIJGELER AND MEHIYAR KATHEM

15 Najaf, Iraq: developing a sustainable approach to threatened heritage — 213
   CAROLINE A. SANDES AND ALI N. ATTIYAH

16 Sustaining cultural heritage in post-conflict Syria: the case of Aleppo — 226
   HIBA ALKHALAF

17 Heritage and peacebuilding: challenges, possibilities and
    sustainable practices   241
    FERAS HAMMAMI, DAVID C. HARVEY, DANIEL LAVEN AND DIANA WALTERS

PART IV
Environment, heritage and society   261

18 Sustainable heritage and climate change   263
    ALESSANDRA BONAZZA

19 Environmental design strategies for heritage   272
    ROSA SCHIANO-PHAN AND TALIA QUESADA CAMPAÑA

20 Energy efficiency in historic buildings   290
    GUSTAF LEIJONHUFVUD, PETRA ERIKSSON AND TOR BROSTRÖM

21 Balancing heritage values, thermal comfort and energy efficiency
    in world heritage sites: the case of Mexico City   305
    KRISANGELLA SOFÍA MURILLO CAMACHO, KALLIOPI FOUSEKI AND
    HECTOR ALTAMIRANO

22 Food heritage as a catalyst for environmental sustainability:
    reflections on the cultural value imbued by citizens to food
    and its role in supporting scientific debate about food security   323
    GEORGIOS ALEXOPOULOS, SUZANNE KAPELARI AND THEANO MOUSSOURI

23 The search for virtue: sustainability and systemic protection
    of agricultural heritage   348
    ISABELLE ANATOLE-GABRIEL

PART V
Sustainable heritage-led transformation   363

24 A boundaries approach to urban heritage: the case of
    Egyptian antiquity   365
    AMR EL-HUSSEINY AND KALLIOPI FOUSEKI

25 Sustaining heritage places: crossroads between urban imaginaries,
    heritage use and sustainability   381
    TU-CHUNG LIU AND KALLIOPI FOUSEKI

26 FORT ST ANGELO is not a billboard: image-driven media
    and the resilience of the project   392
    ERICA GIUSTA AND GUILLAUME DREYFUSS

Contents

27 Integrating urban conservation into urban planning   405
ARTHUR PARKINSON AND MARK SCOTT

28 Foregrounding ethics in conservation in Singapore: issues, questions, and framework   421
KANG SHUA YEO AND JEFFREY KOK HUI CHAN

29 Urban values-centred regeneration in the perspective of the circular economy model: an overview of the key issues   438
LUIGI FUSCO GIRARD AND FRANCESCA NOCCA

PART VI
Inter-temporal and inter-spatial, dynamic heritage research methods   457

30 Using system dynamics in heritage research   459
KALLIOPI FOUSEKI, KATHERINE CURRAN AND JOSEP GRAU-BOVÉ

31 Port city resilience: piloting a socio-spatial method for understanding, comparing and representing linked maritime heritage   471
CAROLA HEIN, LUCIJA AŽMAN MOMIRSKI AND YVONNE VAN MIL

32 Heritage data science   484
SCOTT ALLAN ORR

33 Capturing heritage significance: a critical analysis of economics-based methods   497
MINA DRAGOUNI

34 A rambling field role for the heritage practitioner: a means to come to more socially sustainable heritage (re-) development projects   511
GISÈLE GANTOIS

35 Teaching futures literacy for the heritage sector   527
CORNELIUS HOLTORF

Index   543

Figures

2.1	The tomb cover, which will hide the tomb until Margrethe's death	34
2.2	The artist Bjørn Nørgaard's sepulchral monument, which will be the tomb of Queen Margrethe II of Denmark when she dies	34
2.3	The UNESCO toxic industrial heritage landscape of Røros and the Circumference	37
3.1	Unglazed waterglass paintings by E. M. Ward in Commons' Corridor	50
5.1	The rice field and the irrigation system at the Subak Pakerisan Watershed, part of the World Heritage Site of Bali Cultural Landscape	73
5.2	Subak attributes from the World Heritage Committee and the local community's perspectives	75
6.1a	Typical stages or phases, linking project planning with delivery and evaluation process of heritage projects from a project management perspective	91
6.1b	Heritage life cycle, as a typical 'linear' process of heritage management to support operational analysis: following Burra charter values-based approach framework. Dissemination can be considered to include celebrations, interpretations activities and conservation to include conservation planning, maintenance and documentation activities and risk assessment tasks	91
6.2	Key factors for participatory projects to develop and flourish from participants' perspectives – at individual and community levels. Expectations from both sides (communities/institutions) stemming from interest and expected benefits	97
10.1	Conceptual framework for assessing social impacts of heritage conservation	142
11.1	Correlation between conservation and sustainability	163
15.1	Imam Ali Shrine, Najaf, September 2016	213
16.1	A map of Syria showing the location of the city of Aleppo in the North	230
17.1	A World War I military cemetery overlooking the popular 'Obstanserseehütte' mountain station along the Obstansersee Lake	246
17.2	'Peace Park, Nairobi'	249

17.3	Al-Shuhada Street, centrally located in the Old Town of Hebron. The Israeli military welded shut the street-facing doors of all the homes and shops, and forcibly displaced most of its residents and shopkeepers	252
19.1	Comparison between conventional photo, HDR photo and false colour luminance plot of historic courtyard in Lahore, Pakistan	281
19.2	Golden Lane Estate	283
19.3	Atares Castle, Havana, Cuba	285
20.1	Visby is a world heritage town located on the island of Gotland in the Baltic Sea. Gotland will be a Swedish pilot region for a 100% renewable energy system	291
21.1	The Zócalo in the historic centre of Mexico City	310
21.2	The buildings and their architectural styles	312
21.3	Reinforcing causal loop created by the authors	314
21.4	Works in walls due to damp presence	317
23.1	The Corton Hill in the World Heritage property Les Climats du vignoble de Bourgogne, Burgundy, France. © Michel Joly	357
24.1	The Egyptian public 'dwelling in the boundaries' of urban heritage sites in the context of historic Cairo. The boundary marks the distinction between the early 16th century complex of Sultan Al-Ghouri, a symbolic representation of Egyptian history and identity, and the public everyday life of street vendors selling mainly undergarments to passers-by	365
24.2	The mosque of Al-Saleh Talaʾiʿ in Cairo, one of the first interventions of the Comité de Conservation des Monuments de l'Art Arabe in 1919 in the process of establishing modern heritage preservation practices in Egypt. Their approach included clearing all structures adjacent to the mosque's walls, labelling them as 'parasite structures', and introducing the idea of a fence or boundary to surround the mosque and separate it from its urban surroundings. The later layers of practice added two more boundaries surrounding the original one resulting in its current state, surrounded by three layers of boundaries and interspaces	369
24.3	Within the preference of the Temple of Khnum in the old city of Esna in Upper Egypt, the only building in a maintained condition, other than the temple isolated in its pit, is the green dome of Ibrahim Maghazi Mosque, an early 20th-century structure that was recently rebuilt except for its minaret. Not only is the mosque rebuilt by local funding and craftsmanship amidst the decaying area but it is also the only exception to the ban of all construction work in the area surrounding the temple along with the ongoing restoration work of a few houses funded by the United States≈Agency for International Development. The mosque is only one example of many other mosques being rebuilt and renovated by local families in Esna but with no attention to the structures' material, historic significance or other ancient registers of the past	372
24.4	Bab Al-Muzayinīn – the current main entrance of Al-Azhar Mosque, one of the most significant Egyptian national symbols dating back to the Fatimid 10th century and the formal religious institution in Egypt. In its recent renovation works, completed in 2018, a significantly higher boundary replaced the existing one, limiting the accessibility	

	for the public and rendering the historic mosque out of reach. Also, the mosque's World Heritage Site status seems ambiguous in local media as Al-Ahram, one of Egypt's leading official newspapers, states that it has lost its status after the 1996 renovation work while other sources claim it is still maintained. In any case, the recent works have not been addressed by the World Heritage Committee and were executed in a state of silence	373
24.5	The Aga Khan Historic Cairo project is celebrated as the most successful community-oriented conservation project by Egyptian practitioners. Despite the massive component of community work involved in the project, the status of urban heritage objects and their spatial manifestations remain unaltered – the Eastern side of the Fatimid wall (the main heritage component of the first phase of the project) remains beyond reach to its neighbouring community of Al-Darb Al-Ahmar, surrounded by boundaries complemented with barbed wire resembling high-security borders. The absence of the idea of a 'public space' in the modern sense and the specific sociopolitical context produces forms of participation and community components in heritage practices that are fundamentally different from the global conceptions of such concepts	376
27.1	Nyhavn, Copenhagen – an example of restoration of built heritage for regeneration. However, the area is now described as a tourist enclave and entertainment strip for visitors	409
28.1	Tan Si Chong Su's entrance hall before the 1998–2001 works, circa 1980	426
28.2	The brutalist architecture of Golden Mile Complex	429
31.1	The datawheel methodology	476
31.2	First draft for comparative geospatial mapping methodology, with the case study of London, Hamburg and Rotterdam	479
32.1	A framework for heritage data science, combining data, thinking, and the heritage domain	487
34.1	Spatial Narratives – Entering the Memory Palace – Noeveren – Wherever the feet go the eyes will follow soon. Clay model representing the experienced space of a former claypit as part of the cuesta in Noeveren along the river Rupel	517
35.1	Will the heritage preserved from the past be a valuable legacy in the future?	532

Tables

3.1	List of interviewees	49
16.1	To the left, An attempt to map the main stakeholders at the city of Aleppo, including the local, national, private, professional and international stakeholders as well as the local civil society. To the right, mapping stakeholders using the interest/power tool to put them into groups that each play different roles within the reconstruction process, each group requires different engagement techniques	236
21.1	Interview number and building status	313
21.2	Coding groups	313
22.1	Definitions	324
22.2	The 15 botanical garden partners of the BigPicnic project	329
22.3	Safeguarding intangible cultural heritage and sustainable development at the national level (UNESCO, 2018a, Chapter VI, Paragraphs 170–197)	335
27.1	A typology of conservation-planning policy approaches. While each column represents a chronological shift, the different approaches tend to co-exist and overlap, rather than representing 'clean breaks'	416
30.1	Example extrapolated from the detailed matrix of cause and effect variables. This formed the basis for developing the causal loop diagram	461
32.1	Objectives assessed by heritage data science thinking against the priorities of stakeholders	486
35.1	A necessarily flawed sketch of some changing practices and perceptions of the values and uses of heritage in European society over the past few centuries. The point is not accuracy in the details but to illustrate change over time	528
35.2	Some didactic methods and techniques for teaching futures literacy in the heritage sector	536

Contributors

Alejandra Albuerne is Lecturer at the Institute for Sustainable Heritage. With a background in structural engineering, built heritage and international development, her work explores the role of heritage in sustainable development, including situations of conflict and disaster management. Alejandra holds a PhD from the University of Oxford and has lectured at the University of Oxford and the School of Architecture of Madrid. She has over a decade of industrial experience in leading built environment firms, such as Arup and Ecosistema Urbano, and international development, including Engineers Without Borders and Architecture Sans Frontieres.

Georgios Alexopoulos is Assistant Professor (Postdoc position) at the University of Innsbruck and Adjunct Lecturer at the Hellenic Open University (Greece). He studied archaeology at the University of Athens and received his MA and PhD degrees from the Institute of Archaeology at University College London (UCL). George has worked at UCL as Teaching Fellow coordinating modules in the MA programmes for Heritage and Museum Studies and has also taught at UCL Qatar and the Open University of Cyprus. He has participated in research projects (regional, national and international) in the UK and Greece and was most recently employed as Research Associate at UCL for the EU-funded BigPicnic project. His research interests encompass cultural heritage and museums studies, the management of intangible and living religious heritage, food heritage, sustainability in heritage and museum practice.

Hiba Alkhalaf is an Architect with academic and professional experience in Architectural Conservation and Heritage Management. She holds a BArch from the University of Damascus and MSc in Architectural Conservation and PhD in Architecture from the University of Edinburgh. The interdisciplinary nature of her research bridges architecture, heritage conservation and sustainable urban development, that is, by connecting the tangible elements of heritage (buildings and sites), their meanings and values (to people, stakeholders and community) and the function. Hiba has been actively involved in various projects in the MENA region focusing on the role of heritage in conflict area in peacebuilding and reconciliation, achieving sustainable heritage management and empowering local community. She had developed tools that are people driven to assess and manage heritage, such as sense of place digital tool. Her research had led to develop several capacity-building training programmes, and she worked closely with heritage professionals in Syria, Tunisia, Libya and the UK.

Hector Altamirano is an Associate Professor at the Bartlett School of Environment, Energy and Resources, Institute for Environmental Design and Engineering. He is a Building Scientist interested in the health impact of buildings and has a wide-ranging experience in the indoor environment and operational performance of buildings, the development of mould and humidity conditions, in-situ monitoring, field surveys and experimental set-up of buildings and their components. The outcome of Dr Altamirano-Medina's work on the Humidity in Dwellings project (2005–2007) for DCLG was incorporated into Appendix A of the ADF 2010 and now forms part of the Building Regulations for England and Wales. Dr Altamirano-Medina has led a number of Building Performance Evaluation (BPE) projects funded by the Technology Strategy Board (TSB, now Innovate UK) that have investigated the post-construction and in-use performance of new residential developments. He has also participated in the Facilitation, Learning and Sharing (FLASH) programme, which benefited thousands of small- and medium-sized businesses involved in low-carbon building and retrofit projects.

Isabelle Anatole-Gabriel is an International Civil Servant and a Researcher in International Heritage and Museology Studies. She is currently the Head of the Europe and North America Unit at the World Heritage Centre (UNESCO) and Lecturer at Ecole du Louvre and the University of Poitiers (France). She is Associate Member of the research laboratory *Heritages: Culture/s, Patrimoine/s, Création/s* at CY Cergy Paris University-Centre national de la recherche scientifique (CNRS). She also teaches regularly in several universities abroad. Editor-in-Chief of the former UNESCO journal *MUSEUM International*, she is the Author of books on the intellectual and political history of international heritage and heritage value of agricultural landscapes.

Ali N. Attiyah is an Assistant Professor at the Faculty of Engineering, University of Kufa. He graduated from Civil Engineering Department in 1986, and he completed his higher studies in Structural Engineering at the University of Baghdad. His interest in heritage started in 2003 when he worked as a consultant on the conservation of the Imam Ali Shrine at Najaf City. In 2009, he wrote a book titled the Spiritual Values of the Holy Shrines Architecture, where he tried to explore the intangible values affected the traditional design of the shrines. Later, he was appointed to be a Member of the National Committee to Inscribe Wadi Al-Salam Cemetery to the World Heritage List in 2013. He got training courses at the UNESCO Iraq Office on the protection and enhancement of tangible and intangible heritage. In 2019, he got a grant of 30,000 GBP to document the heritage buildings in Kufa City.

Junnan Bao is a Former Student at University College London and holds an MSc degree in Built Environment: Sustainable Heritage. Her research interests include risk assessment methodologies for heritage assets, with a particular focus on the application of heritage significance/values in risk assessment for heritage.

Kristen Barrett-Casey is an MSc Built Environment: Sustainable Heritage Student at the Institute for Sustainable Heritage, University College London. She is currently conducting research for the Council of Europe's European Routes of Industrial Heritage on the future of management at industrial heritage sites across Europe, and is interested in heritage management, heritage values and public engagement generally. Her previous and ongoing research focuses on post-conflict heritage management and sustainability in the MENA region, the politics of the past and the looting and trafficking of cultural property. Her most recent publication is "An Introduction to Political Sustainability" in Perspectives on Sustainability in Cultural Heritage Conservation.

Alessandra Bonazza is responsible for the Research Unit "Impact on Environment, Cultural Heritage and Health" at CNR-ISAC and Professor of Environmental Impact on Materials, Deterioration and Ageing at the University of Bologna. Principal investigator in several EC projects, she recently coordinated the study funded by DG-EAC, "Safeguarding Cultural Heritage from Natural and Man-Made Disasters", and the Interreg Central Europe ProteCHt2save. She is currently the Project Manager of the Interreg Central Europe STRENCH (STRENgthening resilience of Cultural Heritage at risk in a changing environment through proactive transnational cooperation) and Member of the "Copernicus Cultural Heritage Task Force". Her academic background is geology, though she has worked in the field of heritage science for her entire career moving from building material characterization to evaluation of pollution and climate impact on monuments and archaeological sites. Currently, her research activity mainly focuses on the following:
- Risk and vulnerability assessment of built heritage and cultural landscape exposed to climate extreme events.
- Protection of cultural heritage in facing climate change impact by damage modelling and risk mapping.

Rosie Brigham is a Software Engineer and Researcher at the Institute for Sustainable Heritage (ISH), University College London. She is a Founder of the Monument Monitor, a joint citizen science initiative between ISH and Historic Environment Scotland that uses visitor photographs to monitor remote heritage sites. Her PhD explores how artificial intelligence can improve conservation management processes, and her previous publications have investigated the reliability of citizen scientist provided data.

Tor Broström is a Professor in Conservation at the Department of Art History, Uppsala University. He has a Bachelor of Science in Engineering from Princeton University, a Licentiate of Technology from Chalmers Institute of Technology and a PhD from the Swedish Royal Institute of Technology. His research and teaching is about energy efficiency, indoor climate control and climate change adaptation of historic buildings.

May Cassar, Director of the UCL Institute for Sustainable Heritage, UK.

Talia Quesada Campaña is a Principal Architect at the Office of the Historian in Havana, Cuba, and holds an MSc in Architecture and Environmental Design from the University of Westminster in London. As part of her work in Old Havana, she has participated in several restoration and retrofit projects, as well as co-authoring the Energy efficiency's guide for buildings in the historic centre. In 2016, she became a Fellow at the Transsolar Academy in Transsolar, an engineering firm that develops and validates climate and energy concepts for the built environment. She was also a part-time Instructor for Environmental Design at the Design, History and Technology Department of the School of Architecture of the "José A. Echeverría" Polytechnic Institute. Her publications include "Conversion of Santo Domingo of Atares Castle: a bioclimatic evidence approach" in PLEA 2017.

Jeffrey Kok Hui Chan received his BArch from the Southern California Institute of Architecture (SCI-Arc), his MEd in Mind, Brain and Education from Harvard University and PhD in Architecture (Design Theory and Methods) from the University of California, Berkeley. His research focuses on design ethics in the context of large-scale socio-technical systems and urbanization.

Katherine Curran is an Associate Professor at the UCL Institute for Sustainable Heritage. Her research focuses on the degradation of heritage materials, with a particular focus on the conservation of plastics. She leads the ERC Starting Grant project "COMPLEX: The Degradation of Complex Modern Polymeric Objects in Heritage Collections: A System Dynamics Approach".

Mina Dragouni is a Postdoctoral Researcher at the Department of History & Archaeology, University of Patras; a Research Associate at the Department of Economic and Regional Development, Panteion University of Social & Political Sciences; and an Adjunct Lecturer in Cultural Economics and Management at the Open University of Cyprus. She holds an MSc in Cultural Studies and Entrepreneurship (University of Nottingham) and a PhD in Sustainable Heritage (UCL). She has participated in various projects related to heritage management and economics, whereas currently she is involved in the external evaluation of Elefsina European Capital of Culture 2023, measuring the title's impact on the city. Her research interests revolve around cultural heritage and politics, heterodox economics, tourism studies and community research. Her work has been published by leading journals in the field, including the *International Journal of Heritage Studies, Journal of Heritage Tourism, Journal of Sustainable Tourism* and *Economic Modelling*.

Guillaume Dreyfuss, Director of Research at the Architecture Project LTD in Malta.

Joy Acheyini Edeoja is a Research Student at University College London's Institute for Sustainable Heritage. She received a bachelor's degree in Building Technology from the University of Jos, Nigeria, and a master's degree in Energy and Environmental Management from the Glasgow Caledonian University and has been a Teacher and an Administrator. Her wider research interest is in areas related to heritage perception and participation among community members. She has also shown specific interest in the maintenance and preservation of heritage buildings. Currently, her PhD research is on the impact of education on heritage perception among young adults within communities in Nigeria.

Amr El-Husseiny holds a BArch and an MSc in History and Theories of Architecture and has been a Teaching Assistant and an Assistant Lecturer at the Department of Architecture, Faculty of Engineering, Cairo University, since 2012. Primarily a full-time Educator and Researcher in Architecture, Amr has also contributed through part-time and private practice in local and international architectural design projects. His fields of interest include photography, architectural design pedagogy and urban heritage, motivated by the notions of identification and representation. Amr is an Egyptian scholar funded by a joint scholarship programme between the Egyptian government and the British Council in Egypt to carry out his PhD at the Institute for Sustainable Heritage, The Bartlett.

Kelvin Ang Kah Eng, Director in Conservation Management of the Urban Development Authority of Singapore.

Petra Eriksson is a Lecturer in conservation at the Department of Art History, Uppsala University. She is a PhD Candidate in Conservation at Gothenburg University. Petra has a background in the applied field of building conservation, and her research is in the field of conservation in relation to energy efficiency.

Kalliopi Fouseki, Professor in Sustainable Heritage Management, UCL Institute for Sustainable Heritage, UK.

Luigi Fusco Girard is Emeritus Professor of Economics and Environmental Evaluation at the University of Naples Federico II, Italy; Director of the Laboratory of Research on Creative and Sustainable City, Lead Partner World Urban Campaign; Scientific Coordinator of the European Horizon 2020 Research and Innovation project "CLIC – Circular models Leveraging Investments in Cultural heritage adaptive reuse" (2017–2020); Associate Professor at Institute for Research on Innovation and Services for Development (IRISS) of Italian National Research Council (2010–today); Member of General Assembly of UN-HABITAT Universities (2015–today); Member of Steering Committee "World Urban Campaign", United Nations Centre for Human Settlements (UN-Habitat), Nairobi, as Leading Member (1 November 2009–today); Co-President of the "GUD Program Committee on Celebrating Our Urban Heritage" (Global Urban Development); President of ICOMOS ISCEC (International Scientific Committee on the Economics of Conservation); and Deputy Vice President of ICOMOS, Italy (2016–2019).

Eirini Gallou is a Senior Social Analyst at Historic England. Prior to this role, Eirini conducted her PhD at the UCL Institute for Sustainable Heritage. Her thesis was entitled "Exploring impacts of heritage participation: reciprocal links between local community well-being and sustainable heritage management in rural island context".

Gisèle Gantois is an Associate Professor affiliated with the Faculty and Department of Architecture and the RLICC, KU Leuven. As an Architect specialized in the conservation of monuments and sites, she focuses on community heritage as part of a social, cultural and ecological fabric. Her research is directed towards the development of methods and tools to uncover the less visible contemporary meanings of heritage with the aim to come to socially more accepted projects of reuse.

Argyro Gili is a Research Associate at the Institute for Sustainable Heritage at University College London. She holds an MSc degree in Applied Mathematics and Physical Sciences and also an MEng and a PhD in Electrical and Computer Engineering from the National Technical University of Athens in Greece. Her research interests include device and process modelling. Her current research focuses on mathematical modelling of the physicochemical processes involved in degradation of plastic artefacts in museum collections in relation to the environment within the framework of the European research project COMPLEX.

Erica Giusta is Director of Innovation at research-based architecture practice AP Valletta in Malta. She holds an MSc in Sustainable Architecture from Politecnico di Torino (Italy) and a Post-graduate Master in Media & Communications from the Sole 24 Ore Business School in Milan (Italy). At AP, she has contributed to and is currently involved in a number of heritage-related projects and competitions, both in Malta and overseas. Her research focuses on the relationship between new media, built environment and heritage. In 2017, she participated in the first International

Congress on Architectural Communication organized by ETSAM (Universidad Politecnica de Madrid) with the paper "Hysterical Symptoms of the Communication of Post-Truth Architecture", co-authored with Guillaume Dreyfuss. Erica also regularly contributes to design and architecture magazines such as *Il Giornale dell'Architettura*, *A10 New European Architecture* and *The Malta Artpaper*.

Josep Grau-Bové is a Lecturer in Heritage Science at the UCL Institute for Sustainable Heritage studying (1) new technologies for preventive conservation, including the development of tools for the analysis of environmental data; (2) transport phenomena in heritage materials and microenvironments. This often involves computational fluid dynamics, for example, of pollutants, rain and particulate matter; (3) citizen heritage science and crowdsourcing in historic sites and museums; and (4) data-driven approaches to the analysis of large collections, with the objective of informing conservation management and decision-making. He is also the Director of the MSc in Data Science.

Torgrim Sneve Guttormsen, PhD, is an Archaeologist and Research Professor at Norwegian Institute for Cultural Heritage Research. He has been partaking in and leading his institute's strategic research target areas during the period 2016–2020, which is coordinated under the programmes "Heritage Politics" and "Urban Heritage". His previous and ongoing research areas include public archaeology, theory of heritage, heritage politics and management, urban heritage, heritage routes, memorials and monument studies, contentious heritage and immigrant heritage. He has been the co-editor of *Heritage, Democracy and the Public – Nordic Approaches* (2016) and *Heritage as a Driver for Sustainable Cities – Deep Cities* (2020). Some of his main publications include "Archaeology as Conceptual Tool in Urban Planning" (in *Heritage as a Driver for Sustainable Cities*, 2020) and "Using Contestation to Elicit Values for Heritage Planning: The Case of the Urban Park at Ekeberg in Oslo, Norway" (in *Cultural Heritage,* Campelo et al. 2019).

Feras Hammami is Associate Professor of Conservation at the University of Gothenburg. His research concerns heritage, peace and conflict in cities. Feras conducted several empirical studies in Sweden, Palestine and Botswana where he explored the entanglement of heritage in urban change and the social responses to them, and contributed to the debates about the political instrumentality of heritage. In his current research projects, he explores heritage relations to notions of reconciliation, peacebuilding and gentrification. His latest publications include "Reviewing Neighborhood Sustainability Assessment Tools through Critical Heritage Studies" (2020); "Heritage Necropolitics and the Capture of Hebron: The Logic of Closure, Fear, Humiliation and Elimination" (2019); and "Heritage and Resistance: Irregularities, Temporalities and Cumulative Impact" (2018). He co-founded and is currently the Editor-in-Chief of AESOP's open-access journal *plaNext–Next Generation Planning.*

David C. Harvey is an Associate Professor in Critical Heritage Studies at Aarhus University, Denmark, and an Honorary Professor of Historical and Cultural Geography at the University of Exeter (UK). His work has focussed on the geographies of heritage, and he has contributed to some key heritage debates, including *processual* understandings of heritage, extending the temporal depth of heritage, the outlining of heritage–landscape and heritage–climate change relations and the opening-up of hidden memories through oral history. His recent works include *The Future of*

Heritage as Climates Change: Loss, Adaptation and Creativity (edited with Jim Perry, 2015); *Commemorative Spaces of the First World War: Historical Geography at the Centenary* (edited with James Wallis, 2018); and *Creating Heritage: Unrecognised Pasts and Rejected Futures* (edited with Tom Carter, Roy Jones and Iain Robertson, 2020). He is on the Editorial Board of *The International Journal of Heritage Studies* and co-edits a Berghahn Book Series *Exploration in Heritage Studies*.

Carola Hein is Professor and Chair, History of Architecture and Urban Planning at Delft University of Technology. Her (co-)edited books and monographs include *Urbanisation of the Sea (2020)*, *Adaptive Strategies for Water Heritage (2020)*, *The Routledge Planning History Handbook* (2018), *Port Cities: Dynamic Landscapes and Global Networks* (2011) and *Brussels: Perspectives on a European Capital* (2007). She has also published numerous articles in peer-reviewed journals, books and magazines.

Lorika Hisari is an Architect and a PhD Researcher at the Bartlett UCL, Institute for Sustainable Heritage, London, UK. She holds a degree of a Graduated Engineer of Architecture from the University of Pristina, Republic of Kosovo. She is experienced in post-war reconstruction, regeneration and development as a responsible Architect in multidisciplinary projects for international donor-aid programmes. She completed a postgraduate professional course at master's level on International Spatial Planning Practice and Public Realm at the University of Westminster, London, UK, where she was awarded a JCDecaux Prize for Urban Design, 2012. Since 2009, she is engaged in built heritage conservation, and her research interest is urban heritage and area-based conservation in diverse historic urban environments in conflict and post-conflict contexts. Her most recent publication is "Post-War Cultural Heritage Preservation in Kosovo: Rethinking the Implementation of Ahtisaari Plan Annex V" in the Special Issue of *Heritage*.

Cornelius Holtorf is Professor of Archaeology and holds a UNESCO Chair on Heritage Futures at Linnaeus University in Kalmar, Sweden, where he is also directing the Graduate School in Contract Archaeology (GRASCA). In his research, he is particularly interested in contemporary archaeology, heritage theory and heritage futures, with numerous international publications in these areas. He also enjoys sailing and travelling.

Suzanne Kapelari is University Professor in Biology and Environmental Education and Head of the Science, Geography, Computer Science and Mathematics Education Group at the Department of Subject-Specific Education at the University of Innsbruck, Austria.

She was and is a Coordinator and/or Partner of a couple of European-funded projects and has taught Biology and Environmental Education at high schools for a couple of years. Currently, she is Dean of the Faculty of Teacher Education at the University of Innsbruck.

Mehiyar Kathem is Research Associate and Coordinator for the Nahrein Network, a project based at University College London's History Department which works to strengthen Iraq's and its neighbours' cultural and heritage infrastructure (*www.ucl.ac.uk/nahrein*). Since 2017, he has worked to strengthen research collaborations and partnerships between UK and Iraqi universities and academics, particularly in the field of heritage and education. Dr Kathem completed his doctoral research at the

School of Oriental and African Studies (SOAS), where he researched the formation of Iraq's NGO sector and statebuilding after the Iraq War.

Daniel Laven is an Associate Professor of Human Geography and Head of Department, Dept. of Economics, Geography, Law and Tourism at Mid Sweden University. Daniel's research is conducted under the auspices of the university's European Tourism Research Institute (ETOUR) and explores how heritage can be used to encourage more sustainable environments. Daniel's recent works include *Heritage and Peacebuilding* (edited with Diana Walters and Peter Davis, 2017) and a series of studies that examine the intersection of tourism, heritage and peacebuilding in the Palestinian–Israeli context. Together with Jack Shepherd, Daniel recently authored a chapter titled "Military Occupations and Tourism", which will be included in the forthcoming *Routledge Handbook on Political Borders and Tourism* (edited by Dallen Timothy and Alon Gelbman). He is currently collaborating with Robert Manning and other former members of the University of Vermont's Park Studies Laboratory on the fourth edition of *Studies in Outdoor Recreation* (Oregon State University Press), which has been a foundational text in the field of outdoor recreation and nature-based tourism for more than 30 years.

Sujeong Lee has been a Research Staff in the Cultural Heritage Administration of Korea since 2009. She holds an MA in Art History from Dongguk University of Korea and a PhD in Conservation Studies from the University of York. She has led various heritage projects in revising Korean law and setting out new policy for heritage conservation and management. Her main research scope is conservation principles for designated heritage, criteria of value assessment and planning permission, and process of rational decision-making. Her interest extends from single monument and sites into historic city. Her previous projects include "Ethical Guidelines for Conservators" and "Revision of Historic City Law". Ethical guidelines, enacted in 2012 for the first time in Asia, have been adopted to the National Examination for Certificated Conservators and their daily practice. She has participated in revising Korean law for historic city and historic environment between 2018 and 2019.

Gustaf Leijonhufvud is a Researcher in Conservation at the Department of Art History, Uppsala University. He has a BSc in Building Conservation from Gotland University and a PhD in Conservation from Gothenburg University. His research is about the sustainable management of cultural heritage, with a focus on the interaction between science, policy and practice.

Tu-Chung Liu is a PhD Graduate from the Institute for Sustainable Heritage at University College London, UK, as well as the Cultural and Educational Director of Sunmake Cultures Foundation in Taiwan. He holds a Master of Architecture from National Chiao Tung University, Taiwan, and a Postgraduate Diploma of Cultural and Creative Industries from King's College London, UK. His research concerns include cultural imagination, heritage-led urban regeneration and sustainable development of the historic environment. In addition, he started his career as an Assistant Researcher for cultural policy analysis at National Centre of Traditional Arts, Taiwan, and the Bureau of Cultural Heritage, Ministry of Culture, Taiwan.

Randall Mason is Professor at the University of Pennsylvania's Weitzman School of Design, Senior Fellow at PennPraxis and former Chair of its Graduate Program in

Historic Preservation. His scholarly and professional work addresses conservation, planning and public space issues, working with organizations including the Getty Conservation Institute, William Penn Foundation, City of Philadelphia, National Park Service and the government of Rwanda. He is educated in geography, history and urban planning, with a PhD from Columbia University. He was recently named founding Faculty Director of the WSOD's Center for the Preservation of Civil Rights Sites.

Yvonne van Mil is Researcher in Spatial History and Cartographer, affiliated with the TU Delft. Her research topics include urban and spatial planning in the 19th and 20th centuries, focusing specifically on regional history and geo-spatial mapping. She was Co-author of several books including *Driven by Steel. From Hoogovens to Tata Steel 1918–2018* (2018) and contributed chapters to the *Atlas of the Dutch Urban Landscape* (2014).

Lucija Ažman Momirski, PhD, is an Associate Professor of Urban Design at the University of Ljubljana. Her research interests lie in the past, present and future spatial development of the port city regions. She was the project leader of the professional guidelines for the overall spatial organization of the Port of Koper, Slovenia, from 2007 to 2011, and published research papers, such as "The Resilience of the Port Cities of Trieste, Rijeka, and Koper" (2020) and "Shifts in Governance: Who Governs and What is Governed in the Port of Koper" (2020).

Theano Moussouri is an Associate Professor in Museum Studies at University College London (UCL). Theano holds a BA in Preschool Education from the University of Athens and an MA and a PhD in Museums Studies from the University of Leicester. Before joining UCL, she worked for the Science Museum in London as an Audience Researcher, was a Doctoral Researcher at the Institute for Learning Innovation in Annapolis and was Senior Research Fellow at the Centre for Museums and Gallery Education at the Museum Studies Department, University of Leicester. She worked on large exhibition as well as research project, including BigPicnic, where she was the UCL PI. Her research focuses on knowledge construction and the value and impact of cultural institutions on people's lives, using participatory research approaches. Her research interests in the area of food choice and sustainability include food and decision-making, food knowledge and culture and food heritage. Theano is currently a Member of the UCL Food, Metabolism and Society research domain and serves as Associate Editor for *Curator: The Museums Journal* and for *Museums and Society* and on the Editorial Board of the *Science Museum Group Journal*.

Krisangella Sofia Murillo Camacho is a Doctoral Researcher in Built Environment at Institute for Sustainable Heritage and holds an MSc in Advanced Architectural Studies from University College London. She is also a Lecturer at the National Autonomous University of Mexico. Sofia has worked in the private and public sectors and completed different architectural and urban projects.

Francesca Nocca graduated in November 2010 with summa cum laude in Architecture from the University of Naples Federico II, Italy. In 2017, she obtained the title of PhD in Architecture (specific curriculum in urban planning and evaluation) from the University of Naples Federico II, with a thesis entitled "Hybrid evaluation tools for operationalizing Historic Urban Landscape approach". From 2018 to 2020, she was a Postdoc Researcher at the Department of Architecture of the University of Naples

Federico II, within the PRIN 2015 Project entitled "Metropolitan cities: economic-territorial strategies, financial constraints and circular regeneration". Since February 2020, she has been an Adjunct Professor of Appraisal and Professional Practice at the Department of Architecture of the University of Naples Federico II. From 2014, she has been a full-time Researcher of the International Research Laboratory on the Creative and Sustainable City (Lead Partner World Urban Campaign, in collaboration with the UN-Habitat Urban Regeneration HUB – UN-Habitat Universities). She is currently a Member of ICOMOS (International Council of Monuments and Sites) and ISCEC (Italian Scientific Committee on the Economics of Conservation), as well as being a Member of the Research Centre for Appraisal and Land Economics (Ce.S.E.T.) and the Association of Environmental Analysts (AAA).

Heather Oakley is a Senior Project Leader at UK Parliament, managing conservation projects within the Palace of Westminster. She holds a BA in Architecture from the University of Sheffield and an MSc in Sustainable Heritage from UCL. Her present work focuses on sustainable conservation of historic stonework within a live environment.

Scott Allan Orr is a Lecturer in Heritage Data Science at UCL Institute for Sustainable Heritage. An engineer with broad interests, his research within heritage science primarily uses data-driven approaches to assess environmental impacts on the historic built environment, the use of non-destructive tools in building surveys and incorporating value and perception into scientific evaluations. His research emphasizes a holistic approach to considering the historic built environment in its context: an upcoming project with the University of Antwerp will evaluate synergies between urban greening and sustainable environments for historic building materials. He is the Deputy Programme Director of the MSc Data Science for Cultural Heritage, on which he teaches modules about heritage data visualization, heritage science and digital technologies for the historic built environment.

Arthur Parkinson is an Assistant Professor and Programme Director of the MSc Urban Design and Planning in the School of Architecture, Planning and Environmental Policy, University College Dublin, Ireland. He holds a BSc in Architectural Studies and an MSc in Urban Design from the University of Strathclyde and a PhD in Urban and Regional Planning from University College Dublin. Prior to entering academia, Arthur worked in architectural practice in both Ireland and Scotland, in the fields of both architectural conservation and urban design. He is a Member of the RIAI (Royal Institute of the Architects of Ireland) and an RIAI Grade 2 Conservation Architect. Arthur's research is focused principally on conservation planning, the role of competing values in shaping heritage policy and practice outcomes, and the interface with other aspects of planning.

Elia Quijano Quiñones is a PhD Student at the UCL Institute for Sustainable Heritage. She holds an MA in Principles of Conservation from the UCL Institute of Archaeology and a BA in Archaeology from the Autonomous University of Yucatán, México. She has participated in several archaeological heritage conservation projects conducted by the National Institute of Anthropology and History in Yucatán, México. Her research interests include the social impact of heritage conservation, participatory heritage conservation, the contribution of heritage conservation to the Sustainable Development Goals, social development, mixed methodologies in heritage studies and the role of heritage conservation in the social sustainability of rural contexts. Her PhD research project, entitled *Identifying social development through heritage conservation: the case of contemporary Maya communities in Yucatán, México*, explores the common ground between participatory heritage conservation and social development through the identification of indicators of social impact.

Diana Rahman is a PhD Graduate from the Institute for Sustainable Heritage, University College London. She holds an MA in World Heritage Studies from the University of Birmingham and a BSc in Architecture and Planning from Institut Teknologi Bandung. She has participated in various projects related to World Heritage Site management, cultural landscape, food heritage and community participation. Her previous and ongoing research comprises studies related to indigenous community, intangible heritage and values, Authorised Heritage Discourse and cultural ecology. She has also been involved in a transdisciplinary research project looking at the role of cultural knowledge for tackling global food insecurity. Currently, she carries out work at the UN Disaster Risk Reduction Office (New York office).

Caroline A. Sandes is an Independent Archaeological Researcher based in London, UK, with an MA from University College Dublin and a PhD from the Institute of Archaeology, University College London. Her research specializes in cultural heritage in (post-)conflict situations in the Middle East, most recently Iraq. She has also done fieldwork in Ireland, the UK, Australia, Germany and Lebanon. Her publications include a book, *Archaeology, Conservation and the City: post-war redevelopment in London, Berlin and Beirut* (2010), and numerous academic papers and chapters. She is also a Guest Lecturer for the Institute of Sustainable Heritage, UCL, among other institutions, and also works for ICOMOS-UK.

Rosa Schiano-Phan is a Reader at the University of Westminster in London, where she leads an MSc in Architecture and Environmental Design. She holds an MSc in Architecture, Energy and Sustainability from London Metropolitan University and a PhD in Energy and Environment from the Architectural Association, School of Architecture in London. Rosa has worked on projects for the environmental retrofit of heritage buildings, with specific attention to the application of natural ventilation and passive cooling strategies. Her recent research includes studies on the environmental performance of modernist architecture and the natural ventilation strategy for the refurbishment of the Grade II* building, Richmond House. Her main publications include *The Architecture of Natural Cooling* by Routledge (2020); "Spatial Delight and Environmental Performance of Modernist Architecture in London – Golden Lane Estate" in *Future Cities and Environment* (2018); and "The Mitigative Potential of Urban Environments and Their Microclimates" in *Buildings* (2015).

Mark Scott is Professor of Planning and Dean in the School of Architecture, Planning & Environmental Policy, University College Dublin. His research interests are focused theories and practices of spatial planning and governance, with a particular focus on rural planning, conservation planning and climate action. His recent co-authored or co-edited books include *Rural Places and Planning* (Policy Press, 2022), *The Routledge Companion to Rural Planning* (Routledge, 2019), *Renewing Urban Communities* (Routledge, 2017) and *Rural Planning & Development* (Routledge, 2017). He is an Editor of the international journal *Planning Theory & Practice* and is also on the Editorial Board of *Town Planning Review*.

Maria Shehade is an Adjunct Lecturer at the Cyprus University of Technology and a Research Associate at the CYENS Centre of Excellence. She holds a PhD and a master's degree in Cultural Heritage Management from University College London, a bachelor's degree in History and Archaeology from the National and Kapodistrian University of Athens and a certificate in Negotiation from the Institute of Leadership and Management, UK. Her doctoral research focused on the development of a novel strategic framework for the negotiation and resolution of cultural property disputes. She has worked as a Research Associate in several research projects on heritage management, museology, negotiation, cultural diplomacy and cultural policy. She has also taught courses in several universities including the University of Nicosia, Open University Cyprus and the Cyprus University of Technology. Her academic and research work focuses on cultural policy and management, international negotiations, antiquities restitution, cultural diplomacy and museology. She has received several scholarships and awards for her research from funding organizations, including the National Foundation of Scholarships of Greece and the Leventis Scholarship Foundation.

Kang Shua Yeo received his BArch, MArch and PhD in Architecture from the National University of Singapore (NUS). His research interest is primarily in the area of architectural conservation. He is a recipient of the UNESCO Asia-Pacific Awards for Cultural Heritage Conservation in 2005, 2010 and 2014.

Joar Skrede, PhD, is a Sociologist and Research Professor at Norwegian Institute for Cultural Heritage Research. He works primarily with heritage studies, urban studies and discourse studies, often in combination. Skrede has published several scientific articles, nationally and internationally, within topics such as the relations between experts and laypersons in heritage studies, textual, visual and material representations of heritage, heritage-led urban regeneration, place identity and sustainable development. Skrede is also engaged with critical realism as a philosophy of science. Some of his most influential publications include "Uses of Heritage and beyond: Heritage Studies viewed through the lens of Critical Discourse Analysis and Critical Realism" in *Journal of Social Archaeology* (2018) and "What's wrong with heritage experts? An interdisciplinary discussion of experts and expertise in heritage studies" in *International Journal of Heritage Studies* (2019).

Chihei Suzuki is a Historical Geographer and a Senior Specialist for World Heritage at the Agency for Cultural Affairs, Government of Japan. He studied historical geography at Kyoto University and its graduate school. Since 2005, he has been in charge of conservation of cultural landscapes in Japan, has engaged in the affairs of World Heritage conservation. He has gained a PhD degree in Regional Policy from Takasaki City University of Economy. He belongs to various academic societies including Japan ICOMOS. He is at the moment partaking in conserving the World Cultural Heritage sites in Japan, creating nomination dossiers for the World Heritage List, reviewing the national policy for World Heritage and participating in the discussion on the implementation of World Heritage Convention. His main publications include "Whose World Heritage?" in World Heritage no. 97 (2021); and the nomination dossier of *Hidden Christian Sites in the Nagasaki Region* (2017, co-author).

René Teijgeler has been a Cultural Advisor for over 30 years and specialized at first in Culture and Development and the last 15 years in the Protection of Cultural Heritage in times of conflict. As a heritage risk manager, he has worked in the conflict areas, especially Afghanistan, Iraq and Syria, and provided short-term advice in many other countries. He co-founded the NGO "Heritage for Peace", which focuses on Syria. René combines an academic background in sociology, social psychology, anthropology and conservation.

Diana Walters is a Heritage Practitioner, Facilitator and Researcher from the UK. She works in international development and peacebuilding in many countries including Albania, the former Yugoslavia, Kenya and Palestine. She has been associated with the NGO Cultural Heritage without Borders for many years. Diana's activity is focused on interpretation, inclusion, learning, interspiritual dialogue, peace and good neighbourliness, social justice and human rights. Her core belief is that heritage can lead to transformation of individuals and communities. With co-editors Daniel Laven and Peter Davis, Diana edited *Heritage and Peacebuilding* (2017, Boydell and Brewer). She frequently contributes to debates and seminars, deliberately opening up opportunities for dialogue and seeking to gently challenge herself and others. She is an Honorary Fellow at the University of Plymouth.

Ross Wilson is Professor of History and Director of Liberal Arts at the University of Nottingham. His research interests include public history, heritage studies, First World War history, New York history and the history of science. His work includes *Cultural Heritage of the Great War in Britain* (2013), *Language of the Past* (2016) and *Natural History: Heritage, Place and Politics* (2017).

Introduction

Sustainability for heritage and heritage for sustainability

Kalliopi Fouseki, Guillaume Dreyfuss, Kelvin Kah Eng Ang, and May Cassar

Introduction: conceptualizing 'sustainable heritage'

The idea of a *Handbook on Sustainable Heritage* has grown over the last decade and a half. Since the launch of the MSc in Sustainable Heritage in 2004 at the then Centre for Sustainable Heritage – and now Institute for Sustainable Heritage (ISH) at The Bartlett, University College London – the seeds for a handbook on this subject were planted. It was indeed fortunate to be invited by Routledge 15 years later to develop the first Handbook on Sustainable Heritage – at a time when research and teaching on sustainable heritage led by ISH had been strengthened by scholarly projects as well as novel investigations by doctoral and master students. The timeliness of the invitation coincided with the publication of the 17 Sustainable Development Goals (SDGs) in 2016 by the United Nations which, among others, makes specific reference to heritage in SDG 11: Sustainable Cities and Communities. Although 'heritage' is mentioned once under Target 11.4 which 'aims to strengthen efforts to protect and safeguard the world's cultural and natural heritage',[1] the inclusion of 'heritage' in the SDGs has triggered global work to provide evidence on the contribution of heritage to all SDGs (e.g. Pottgiesser et al, 2021).

The term 'sustainable heritage', at the time, reflected the innovative, cross-disciplinary expertise of the Director of ISH, Prof. May Cassar, in preventive conservation and environmental design and engineering under the broad umbrella of environmental sustainability. Emphasis was placed on understanding how heritage could take forward the principles of environmental, social and economic sustainability as envisioned in Our Common Future, also known as the Brundtland Report, published in October 1987 (World Commission on Environment and Development, 1987). This publication was in recognition of Gro Harlem Brundtland's, former Norwegian Prime Minister, role as Chair of the World Commission on Environment and Development. Climate change, the interaction of internal and external environments and their impact on immovable and movable heritage, adapting cultural heritage to a changing climate through sustainable environmental design and equipping heritage to contribute to environmental sustainability goals at national and global level, were all early drivers for the development of sustainable heritage at UCL. The approach to sustainability has been and continues to be holistic and inclusive, and this approach underpinned the design of the MSc in Sustainable Heritage from which flourished related research on heritage values, society and communities, as well as the economy. Together, they were at the heart of sustainable heritage teaching and research.

DOI: 10.4324/9781003038955-1

Undoubtedly, the term 'sustainable heritage' remains ambiguous. Indeed, it is not uncommon to be asked 'What is sustainable heritage'? We will endeavour to contribute a first, simple response to this complex question in this introduction, while the remaining chapters of the handbook unpick the nature of sustainable heritage in detail. To begin, we can contemplate the term 'sustainable heritage' in two distinct ways. The first represents a 'responsive approach' to sustainable heritage which hints that heritage is subject to risks, threats and deterioration, and as a result, it merits conservation and management. The second reflects a 'dynamic approach' to sustainable heritage which entails that heritage can potentially act as a catalyst for environmental, social and economic sustainable development. These two combined approaches illustrate that heritage is subject to change as much as it is a driver for sustainable change (Fouseki et al., 2020). If change is inherent in heritage, then 'sustainable heritage' may initially sound as an oxymoron between 'change' and the 'endeavour to sustain'. However, this oxymoron, we argue, connotes more the acceptance of change and the complex nature of heritage that fosters a constructive negotiation between 'accepting' and 'managing' change, rather than a tension between 'change' and 'maintenance'. More significantly, the term 'sustainable heritage' denotes that heritage is ultimately a social and cultural process (Smith, 2006), or in other words, a dynamic socio-cultural practice (Fouseki et al., 2020).

This argument aligns with our underpinning premise that heritage is a complex and dynamic system consisting of material and immaterial interconnected dimensions (Fouseki et al., 2020). Accordingly, we posit that the 'sustainability' and 'resilience' of heritage against environmental, social, economic or other threats as well as the impacts of heritage on communities, economies and the environment are concurrent processes that should, therefore, be examined together. Currently, the two processes are often disassociated in heritage practice and heritage research. This is somewhat apprehensible because a co-examination of the two processes requires collaborations of social scientists, humanity researchers and physical scientists or engineers, collaborations that are usually difficult to pursue due to lack of resources. Of course, resources are only one of the multiple factors contributing to lack of such collaborations. The *Mind the Gap* research project, funded by the Arts and Humanities Research Council, showcases a wide range of factors affecting the development and execution of cross-disciplinary and cross-sectoral heritage collaborative projects between universities and heritage institutions (Dillon et al., 2014). The project examined through a quantitative, attitudinal survey what enables and impedes effective collaboration between academic heritage researchers and users of heritage research evidence in practice. The study revealed a scepticism on behalf of heritage practitioners regarding the implementation of the impacts of the research with some respondents finding interdisciplinary collaboration challenging.

Despite the challenges of cross-disciplinary heritage research, it is widely acknowledged that studies on heritage is an interdisciplinary field (e.g. Labadi & Long, 2010, p. xi; Harisson, 2013, p. i). As Labrador and Silverman have rightly stressed, the 'field of cultural heritage is no longer simply a past-oriented discipline, dependent on the expertise of art and architectural historians, archaeologists, conservators, curators, and site and museum administrators' (2018, p. 1). To this end, their *Handbook on Public Heritage: Theory and Practice* aims to highlight potential contributions of disciplines that heritage studies have not engaged to a great extent such as 'development studies, political science, anthropology, management studies, human geography, ecology, psychology, sociology, cognitive studies, and education to heritage studies and management' (Labrador & Silverman, 2018, p. 2). What is still omitted from the aforementioned studies is a more forceful engagement of heritage studies and heritage management with physical sciences. We though concede that this engagement is a challenging task. Chapters 3, 8, 20, 21 and 30 in this *Handbook* constitute a first attempt towards this direction.

Structure

The chapters are organized into six parts/themes representing six topical and urgent areas of research and teaching in the field of sustainable heritage including Part I: Heritage values and risk; Part II: Participatory heritage; Part III: Dissonant and 'pacific' heritage; Part IV: Environment, heritage and society; Part V: Sustainable heritage-led transformation; and Part VI: Inter-temporal and inter-spatial, dynamic heritage research methods.

The six parts contain a few of the fundamental elements of 'sustainable heritage'. Values, the focal point of Part I, are the essence of heritage. Managing heritage is in effect about managing values (Avrami et al., 2019; see also Mason in this volume). There is already an extensive literature on the subject. However, the handbook takes a step further by examining how the abstract, subjective and complex construct of heritage values can be tangibly integrated into risk assessment strategies and tools.

The concept of values raises the critical question: why does heritage matter, and for whom. To answer this question, participatory approaches to managing and researching heritage are needed. Part II is dedicated to the participatory nature of heritage management which enables multivocality, negotiation and dialogue (see, for instance, Roued-Cunliffe & Copeland, 2017). For participatory approaches to heritage to be meaningful and sustainable beyond the lifetime of a heritage project, integrating the views of everyone involved is essential (e.g. Fouseki, 2010). The underpinning premise here is that without 'meaningful participation' neither heritage can be sustained in the long-term nor can sustainable development through heritage be accomplished. Part II unfolds this reciprocal, complex relationship through the application of novel methodological frameworks, which capture the dynamics of heritage participation and its impacts on heritage and those involved in the process.

We have already referred to the dissonant nature of heritage. There is no 'universally accepted set of values' attributed to heritage. Similarly, there is no 'universally agreed set of voices' expressed by the various communities and stakeholders involved in participatory heritage processes. If heritage management is about managing heritage values, then it is also about instigating and empowering dialogue and negotiation. The dissonant nature of heritage becomes even more perplexing in war and post-war contexts where heritage is often used as a divisive means. Part III deals specifically with heritage as the target of war and/or heritage as the medium for peacebuilding in post-war contexts. As the role of heritage in reconciliation and post-war peacebuilding is largely unexplored, this part yields an opportunity to introduce a focal study on 'pacific heritage', to use the term of Hammami et al. in this volume.

Part IV accentuates how interwoven environment, society and culture are. This is critical for the development of sustainable environmental policies and strategies that aim to either protect heritage from global environmental challenges (such as climate change) or to discover the opportunities within heritage to catalyse environmental and social sustainability.

Part V is dedicated to the shift towards heritage-led regeneration/transformation urban projects. This model intends ultimately to sustain built heritage during the economic and social 'regeneration' of urban areas under pressure. The chapters adopt a critical approach to current heritage-led regeneration models, advocating the need to embed values and communities into an integrated approach to urban conservation and urban planning – which is often not the case.

Part VI suggests new methodologies to study and research heritage that will enable students, scholars and practitioners to deconstruct 'heritage values' and the notion of 'change' in the past, present and future (inter-temporal) as well across scales, spaces and places (inter-spatial). The development and application of the proposed methods require synergy between the natural sciences (e.g. Piñar & Sterflinger, 2021), physical sciences (e.g. Dillon et al., 2014), social sciences and the humanities (e.g. Labrador & Silverman, 2018).

Before proceeding with a more thorough presentation of each part, it is important to highlight at this juncture that we strove to include chapters written by established heritage scholars and experienced heritage practitioners as well as early career researchers conducting their doctoral studies at ISH. While preparing reading lists for students or conducting research on heritage, the absence of voices – especially of female authors – on heritage matters in English literature is notable. This *Handbook* seeks to bridge this gap through the inclusion of 12 chapters associated with research at ISH, seven of which are based on international doctoral research work in Egypt, Indonesia, Kosovo, Mexico, Nigeria and Taiwan. The remaining chapters complete the diverse geographic representation with case studies from Belgium, Iraq, Japan, Korea, Malta, Netherlands, Norway, Singapore, Sweden, Syria, UK and the USA.

Heritage values and risk

As aforementioned in the introduction, heritage values encompass the critical question 'what to preserve, for whom and why'. Heritage management will be a sustainable process only if decisions on heritage management are informed by the values attached by all those linked to a heritage place or all those affected by a heritage decision. Thus, heritage values 'are attributed, not intrinsic; mutable, not static; multiple and often incommensurable or in conflict' (De la Torre, 2013, p. 155; see also Smith et al., 2017). In a way, the various values (social, economic, environmental, political, cultural) reflect the various sustainability dimensions (i.e. social sustainability, economic sustainability, environmental sustainability, cultural sustainability). Cultural sustainability, in particular, is less discussed as it only recently got recognized as a distinct pillar of sustainable development models (e.g. Auclair & Fairclough, 2015). Because of the pertinent role that heritage values have in heritage management and in unpacking the sustainability dimensions with which heritage as a system is linked, Part I of this *Handbook* is dedicated on this abstract notion. The chapters in Part I offer a fresh perspective by looking at 'change' as a value itself and also by investigating how to integrate the abstract and subjective concept of values into management tools related to risk assessment.

Part I begins with Mason's assertion that 'sustaining heritage means sustaining values' (Chapter 1). Mason's conceptual analysis reminds us that, since values are in constant flux and change, dynamic frameworks of value-centred conservation need to be developed. Current frameworks are static in that they rely on either the use of pre-existing value typologies or short-term assessments that can only provide a snapshot of the diverse values attached to heritage. To this end, Mason presents the use of a more dynamic valued-centred conservation framework reflecting to the urban heritage reality of Pennsylvania. Guttormsen and Skede in Chapter 2 utilize examples from Nordic countries in order not only to acknowledge the changeable and dynamic nature of heritage and values but also to contend that change and continuity is a value itself (see, also, Fouseki et al., 2020). This provocative proposition re-orientates the development of heritage strategies and policies of managing change (e.g. Roders & van Oers, 2013) towards heritage strategies and policies on creatively accepting and using change for sustainable heritage futures.

Someone could argue that approaching change as a value itself cannot be implemented in actual heritage practice. Indeed, the complexity of integrating values and change as a value into heritage management systems is evidenced by Curran et al. in Chapter 3. In this chapter, the question that is being discussed revolves around the integration of values into risk assessment models. Risk assessment reflects the 'responsive' approach to sustainable heritage as it aims to identify and ultimately protects heritage from risks and threats to which it is exposed. By integrating values into risk assessment, a concept that necessitates involvement of diverse stakeholders and acceptance of change may prove a challenging task. The chapter draws on interviews with experts in the field of heritage risk assessment on the role of value and on the development of two novel risk assessment approaches. The examples focus more on the impact of certain risks on values attributed by different stakeholders. Despite this focus, the chapter provides a starting basis to consider on how such an approach can provide a driver for participatory conservation practices which in effect can lead to social cohesion and other social outcomes, as outlined in Part II.

While the first three chapters in Part I employ case studies from the 'Western world', the following two chapters address the issue of heritage values, change and risk in the context of Nigeria and Indonesia. In both chapters, Kalliopi Fouseki with her doctoral students, Joy Edeoja and Diana Rahman, juxtaposes community values attributed to world heritage sites with the ones assigned by national and international authorities. While there is already extensive critique regarding the absence of community voices in the management of world heritage (e.g. Kato, 2006; Li et al., 2020), the thorough ethnographic work in the case of Indonesia and the historic, critical review of the case of Nigeria provide much-needed evidence to support this critique.

Participatory heritage

The need to assess the values of multiple stakeholders invites the adoption of participatory approaches to heritage. 'Participatory heritage could be thought as a space, a space in which individuals engage in cultural activities outside of formal institutions for the purpose of knowledge sharing and co-creating with others' (Roued-Cunliffe & Copeland, 2017, p. xiv). Participatory heritage can also be conceptualized as a social, dynamic process during which both those initiating the process (in our case heritage institutions) and those invited to participate in the process are affected. How and in what ways everyone is affected during and beyond the lifetime of a participatory heritage project may be hard to 'observe, recognize and measure' (Neal, 2015, p. 346).

It is for this reason that sophisticated methodological frameworks are needed in order to implement sustainable participatory projects the outcomes and impacts of which outlast the project itself. Kalliopi Fouseki with her doctoral student, Eirini Gallou (Chapter 6), proposes a theoretical and methodological framework for understanding and assessing the reciprocal impacts of participatory heritage projects on heritage institutions and heritage participants. However, such frameworks become more sensitive in the context of places dealing with traumatic and difficult histories (e.g. MacDonald, 2010). Wilson in Chapter 7 interprets the difficulties with which public engagement with such histories is associated. He argues that objects and sites linked to traumatic events require 'dialogue and engagement, they need a mode of witnessing that recognizes the inequalities and divisions that persist in societies' (Chapter 7, in this volume). Brigham and Grau-Bové present an alternative form of public participation, that of citizen science. Through citizen science, the non-specialist public transforms to co-researchers in heritage science research projects. Although citizen science is an established field (e.g. Bonney et al., 2014), it has not been applied in heritage research extensively and in a critical manner.

The remaining chapters in Part II are specific case study analyses. Lee (Chapter 8) extracts cases of decision-making in Korea's historic city of Gongju in order to address a lack of understanding on community and their needs in heritage as a process. Lee advocates for the need of a clearer definition of the term 'community' and the communication between residents and heritage authorities. Indeed, the term 'community' is another ambiguous term (e.g. Waterton & Smith, 2010). Questions such as who represents the community, how is a community specified and determined, what ties people into a distinct community group are only some of the several questions that heritage practitioners and scholars working on participatory heritage are faced.

Quinones and Fouseki deal with some of these questions by comparing participatory heritage projects in Yucatan, Mexico. Each project was characterized by different levels of participation. They propose a novel evaluation framework for assessing social impacts of participation in heritage conservation. Heritage conservation is approached as a social process. This approach shifts the emphasis of the evaluation of participatory projects beyond the lifetime of the project.

Part II is concluded by the case of Japan's Hidden Christian Sites in the Nagasaki Region, which were added to the UNESCO World Heritage List in 2018. Suzuki in Chapter 11 illustrates a shift of a managerial paradigm from participatory governance (i.e. where the private sectors participate in the management led by the public sector) towards active management by local residents (i.e. where management is run wholly by local residents' initiative).

Dissonant and 'pacific' heritage

One of the first books to address the dissonant nature of heritage was the volume *Dissonant heritage: the management of past as a resource in conflict* edited by Tunbridge and Ashworth in 1996. The authors highlight the conflicts and disharmonies emerging from contemporary uses of the past such as the commodification and marketization of heritage. Indeed, dissonance is inherent in heritage. This is even more so in the case of 'negative heritage' (Rico, 2008) or 'dark heritage' (Clarke et al., 2017). In Part III, we opt to focus on political sustainability and the role of heritage in conflict resolution and peacebuilding, especially in war and post-war contexts.

Part III illustrates that heritage is not always harmonious, peaceful or beautiful. Heritage can be dark, linked to atrocities, conflicts and traumas. Part III begins with a chapter by Shehade that revolves around contentious claims of repatriation and restitution of cultural heritage objects. Through the example of the Getty–Italy restitution agreement, the chapter unveils the potential of alternative dispute resolution methods – such as negotiation – in the economic development of the involved parties, the physical sustainability and preservation of the contested heritage, the development of tourism initiatives and the involvement of all interested stakeholders and the preconditions for achieving such benefits. Negotiation implies dialogue and mutual willingness to compromise.

However, how feasible is this type of negotiation in cases where power levels are imbalanced? The power imbalance and the degree to which political sustainability can be achieved in world heritage sites located in divided countries are examined by Hisari et al. in the case of Kosovo. In Chapter 3, Hisari et al. examine whether the World Heritage status can contribute to achieving long-term reconciliation in multi-ethnic post-conflict environments such as in the case of Kosovo. The chapter calls for a new framework of 'political sustainability', which can place greater emphasis on who is making conservation and management decisions at heritage sites, and what future political implications of this could be.

Part III continues dealing with the association of heritage with political sustainability in Chapter 14. Through a historic review of the Iraq case, Teijgeler and Kathem unfold how heritage, politics and cultural continuity have been linked over time. Such historic analysis enables to better comprehend the ways in which cultural heritage is used for political means. Sandes and Attiyah (Chapter 15) discuss post-conflict heritage reconstruction in Iraq. They focus on the historic shrine city of Najaf where insensitive redevelopment to accommodate pilgrims occurs, alongside problems associated with Iraq's years of conflict and occupation. The authors advocate for the inclusion of intangible heritage into the post-war development process, if a sustainable future is to be achieved. Alkhalaf (Chapter 16) also deals with the issue of post-conflict reconstruction and the role of heritage in this process in Syria. Alkhalaf argues that 'for post-conflict heritage reconstruction to be sustainable, a bottom-up, people-driven, participatory, and holistic approach is required' (Chapter 16 in this volume). The author presents various methods that utilize cultural heritage values as the basis for the recovery process, from a war-torn community to the post-conflict reconstruction.

Part III concludes with an optimistic tone as its last chapter looks at the role of heritage in peacebuilding. Hammani et al. in Chapter 17 introduce the term 'pacific heritage' and explain how heritage can contribute to peacebuilding when 'drawing on ideas of reconstructive learning, heritage from below, co-resistance, and disinheritance' (Hammani et al., in this volume).

Environment, heritage and society

The underpinning premise of Part IV is that heritage is inherently 'green'. The 'greenness' of heritage derives not only from the fact that by preserving and adaptively reusing existing heritage carbon dioxide emissions can be substantially reduced, but also because by looking at heritage lifestyles of the past we can learn how to live more sustainably in the present (e.g. Fouseki & Nicolau, 2018). Barthel-Bouchier (2016) offers a comprehensive analysis of environmental challenges – such as rising waters, desertification, deforestation and polar melting – to which heritage is exposed. There is also exhaustive research on the 'responsive' approach to heritage at risk of climatic changes (e.g. Brimblecombe et al., 2010; Sabbioni et al., 2010; Sesana et al. 2018; Orr & Cassar, 2020) and natural disasters more generally (e.g. MacKee et al., 2014; Nicu, 2017).

What this research lacks is integration of the cultural dimensions of sustainability into our understanding of the impact of environmental changes on heritage and the contribution of heritage itself to environmental sustainability. Therefore, in Part IV, contributors demonstrate new areas of critical research in relation to heritage and climate change (Bonazza in Chapter 18), improving energy performance of heritage buildings (Chapters 19–21) but also the role of food heritage and agricultural landscapes as catalyst for environmental sustainability (Chapters 22–23).

Part IV begins with the 'responsive approach' to sustainable heritage – that is heritage affected by environmental challenges. Bonazza in Chapter 18 summarizes the outcomes of a large number of EU-funded projects on the topic. Most research on climate change and heritage has focused on the impact of climate change on the material degradation of heritage. As much as there is still more research needed on this subject given the diversity of heritage materials, there is also an urgent need to understand communities' attitudes to risk assessment and preparedness towards extreme weather phenomena emerging from the climate change crisis. While we need more research of this type, there is growing socio-technical research in understanding how residents of old buildings negotiate their decisions between their need for thermal comfort, energy improvement and heritage conservation (e.g. Fouseki et al., 2020). To this end, Schiano-Phan in Chapter 19 outlines how environmental design can be a catalyst for protecting heritage from the ever-increasing challenges posed by climate changes. Shiano-Phan summarizes current environmental design strategies available to promote environmentally sustainable heritage by minimizing their energy demand while maintaining architectural and historic significance, occupants' comfort, health and well-being. To do this, Schiano-Phan argues for minimizing reliance on mechanical systems, which not only contribute to the buildings carbon footprint, but are often detrimental to the users' spatial experience and human comfort, detracting from the poetics and the spatial delight of the heritage site.

The principles outlined in this chapter are further articulated in greater detail in two geographically different case studies of different climates and different cultural systems – Sweden and Mexico. Leijonhufvud et al. in Chapter 20 illustrate in cases in Sweden a systematic and interdisciplinary approach for reaching a balance between energy conservation and building conservation in the historic building stock. Murillo-Camacho et al. in Chapter 21 offer an insight into the decision-making process of residents in the World Heritage city of Mexico on energy efficiency with the ultimate goal of improving thermal comfort and reducing energy consumption.

All contributors in Part IV implicitly or explicitly stress that heritage lies at the interface between environmental and cultural sustainability as well as that environmental sustainability cannot be separated from cultural sustainability, and vice versa. This claim is made more emphatic by Alexopoulos et al. whose chapter is based on an EU-funded food sustainability project entitled 'BigPicnic'. The authors argue that food heritage can be a catalyst for discussing global and local challenges to sustainability and for promoting food security. This can only be attained by converging cultural heritage and environmental sustainability inseparably and by considering food's tangible and intangible heritage dimensions and values along with its multi-sensory elements.

The same call for cultural and environmental sustainability is also made by the last author of Part IV, Anatole-Gabriel. The author relies on World Heritage agricultural landscapes and presents an empirical reflection on the manner and measure with which sustainability is applied to the protection processes of the different components of agricultural landscapes and to suggest replicability for other agricultural territories.

Sustainable heritage-led transformation

Part V raises issues of heritage and sustainability related to the global challenge of rapid urbanization. Although heritage has often been at the core of urban transformation processes, the 'heritage-driven transformation' models have been subject to critiques as they often lead to the 'elitization' of heritage and the creation of 'pockets of unequal geographies' (e.g. Labadi, 2016; Fouseki & Nicolau, 2018). A call for a better integration of urban planning and urban conservation has been made (Labadi & Logan, 2015), an integration that can better be facilitated if the 'deep meanings' – often invisible – of a historic urban environment are unpicked (Fouseki et al., 2020). A 'deep cities' approach (Fouseki et al. 2020) puts at fore the urgency for adding a distinct 'heritage pillar' into sustainable development models (see, also, Albert et al., 2017).

In Part V, we advance the aforementioned critical arguments by showcasing methods and theories that facilitate an inclusive model of heritage-driven transformation. This part begins with the presentation of innovative approaches regarding the understanding of the 'deep layers' and 'deep values' of urban heritage. The 2011 Historic Urban Landscape Recommendation of UNESCO has prompted to look at the 'deep heritage' of places across space and society. Given that an urban system is complex and dynamic, special methods and theories are needed in order to unpack the values that will inform future urban heritage transformation strategies. El Husseiny and Fouseki apply boundaries theory as a means for understanding the latent symbolic and cultural boundaries hidden behind the physical boundaries in historic urban areas of Egypt. Liu and Fouseki (Chapter 25) utilize theories of imagination to discuss how imagined pasts, presents and futures are being planned by city planners in Taipei, Taiwan. Dreyfuss and Giusta (Chapter 26) critically review the impacts of the extensive number of visual/digital images of cultural heritage on identifying the values of the historic urban environment in Malta.

The aforementioned critical chapters on unpacking the deep meanings of urban heritage are followed by chapters focusing on urban heritage conservation and planning. Parkinson and Scott put forward the premise that urban planning and urban conservation need to be better integrated and tie up with wider social values. For Parkinson and Scott, 'conservation as an element within urban planning also intersects with wider societal values'. The authors summarize promising shits of urban planning 'from a focus on individual buildings towards the historic urban landscape; from top-down expert approaches to participatory methods; and from viewing conservation as a 'cost' to promoting heritage assets as an economic opportunity' (Chapter 27 in this volume). However, they note that tensions between exploiting and protecting heritage through urban planning are still present.

Yeo Kang Shua and Jeffrey Kok Hui Chan continue the conversation on urban conservation as a social practice focusing on the ethical dimensions of urban conservation. They use examples from Singapore in order to argue that 'conservation issues are framed and evaluated presupposes the use of certain normative evaluations, which must be made more explicit or publicly justifiable in any responsible conservation practices' (Chapter 28 in this Handbook).

Fusco et al. in Chapter 29 provide a holistic approach to sustainable heritage-led regeneration by suggesting a circular economy model that can be implemented in the cities in order to achieve sustainable development. According to this model, urban heritage provides a resource that can be 're-used, recycled, recovered, regenerated and shared'. This chapter thus re-iterates the opening statement of Part IV – that heritage is intrinsically 'green'.

Inter-temporal and inter-spatial, dynamic heritage research methods

Part VI encompasses six chapters on novel methods for studying and researching heritage. Collectively, the chapters represent inter-temporal, complex and dynamic methods which are needed in order to study how heritage changes and contributes to change. As has already been stated, heritage is a dynamic and complex system (e.g. Fouseki & Bobrova, 2018). Accordingly, systemic and dynamic methods are essential for researching heritage. Fouseki et al. in Chapter 30 demonstrate the application of system dynamics in three, diverse heritage projects including a heritage studies/management project looking at peoples' attitudes towards energy efficiency and heritage values; a heritage science project looking at the material degradation of heritage plastics; and a heritage science project looking at the material degradation of historic archival collections. System dynamics can be particularly useful when the research question is that of an 'over time' question.

Hein et al. also pose a novel inter-temporal methodological framework for understanding historic port cities from a long-term perspective which they name 'waterwheel' (see Chapter 31). By using their methodology in Rotterdam, Hamburg and London, an understanding of resilience in the past can be achieved which can inform future city resilience. Hein et al. emphasize the opportunities provided by existing longitudinal data sets and digitized maps which allow the application of such methods.

The data sets themselves posit issues of sustainability. How can the growing data information be maintained, sustained and better used for heritage management? An introduction to this topical subject is provided by Orr who introduces the niche field of heritage data science. Scott defines heritage data science as 'a transdisciplinary field that employs data-driven approaches with critical reasoning within the heritage domain, in awareness of its unique and pressing challenges, to inform engagement with heritage and its interpretation and long-term management' (Chapter 32 in this volume).

The chapters following introduce new methods from social sciences which have huge potential for exploring heritage research questions. Dragouni reviews some of the most promising methods for understanding the economic impacts of heritage. Gantois describes the application of the anthropological method of 'interactive walking'. This method is both inter-temporal and inter-spatial. It is inter-temporal in the sense that it allows to understand the deep heritage and values of a heritage place. It is inter-spatial in that it enables participants to develop a narrative towards a heritage place.

Part VI is concluded with a 'future'-related chapter. Having articulated complex and dynamic, inter-temporal and inter-spatial methods that capacitate understanding of the past and present, Holtorf outlines the significance of futures thinking in the heritage sector. He specifically points out the need on developing methodologies for communicating and teaching futures thinking to heritage professionals. He thus introduces a series of training methods of teaching 'future literacy' to heritage professionals.

Conclusions

'Sustainable heritage' connotes the complexity with which the study and management of heritage are imbued. The adjective 'sustainable' literally implies something that is 'able to be maintained at a certain level or rate'[2]. According to this interpretation, 'sustainable heritage' connotes the ability of heritage to be maintained at a certain level or rate' through, for instance, heritage conservation and management. However, we use the term 'sustainable heritage' more broadly as it goes beyond the material sustainability of heritage by encompassing the ability of heritage to drive sustainable social, economic, environmental and political sustainability.

The chapters in the handbook open up new research directions in the field of sustainable heritage. They look at both the ways in which heritage is sustained and how heritage contributes to sustainability. In terms of sustaining heritage, the handbook identifies environmental (such as climate change), political (such as war) and social (such as urbanization) challenges as global threats to heritage that require new ways of researching and managing heritage to deal with the challenges. One way is the critical integration of values assessment into risk assessment methods. Such work enables better understanding of the impact of heritage conservation decisions on the diverse range of values attributed to heritage by all stakeholders. Adopting participatory methods that involve communities as co-researchers and co-managers can also facilitate the sustainability of heritage by creating close and reciprocal links between communities, research and heritage institutions. In the context of conflict and post-conflict areas where heritage has been the target of destruction, post-war reconstructions that embed intangible heritage sustain the preservation of tangible heritage by making it relevant to communities. When it comes to urbanization, built heritage and intangible heritage are both at risk as transformation happens at a rapid pace leading often to loss of heritage and dislocation of communities. Heritage-led urban regeneration has often been used as a means to balance heritage conservation with development. However, the lack of participation of communities and lack of recognition of wider heritage values (and not just those linked to built heritage) have unintended social consequences (i.e. social division, communities dislocation).

While heritage is at risk from global challenges, it can also be a catalyst for coping with these challenges. From an environmental point of view, all contributions to the handbook demonstrate that there is an inseparable link between environment, society and culture which, if treated as such, can provide a solid basis for environmental sustainability. From a political sustainability perspective, heritage can drive peacebuilding if intangible meanings are embedded into the post-war construction. From an urban sustainability lens, heritage can be a catalyst for social equality and cohesion as well as sustainable and inclusive economic growth, if a heritage-driven transformation project is participatory and fully encompassing all types of heritage. To understand the complexity of material change in heritage and its multiple dynamic social, economic and political impacts, inter-temporal and inter-spatial, cross-disciplinary methods need to be developed and applied.

We would like to conclude this introduction with a call for more cross-disciplinary research on 'sustainable heritage' – that is more integrated research which does not separate the 'responsive' and 'dynamic' approaches to sustainable heritage and wider cross-disciplinary research that merge together the interdisciplinary fields of heritage studies, heritage management and heritage science.

Notes

1 http://uis.unesco.org/en/topic/sustainable-development-goal-11-4
2 https://www.lexico.com/definition/sustainable

References

Albert, M. T., Bandarin, F., & Pereira Roders, A. P. (Eds.) (2017). *Going beyond- perceptions of sustainability in heritage studies no. 2*. Heidelberg: Springer.

Auclair, E., & Fairclough, G. (Eds.) (2015). *Theory and practice in heritage and sustainability: Between past and future*. London: Routledge.

Avrami, E., Macdonald, S., Mason, R., & Myers, D. (Eds.) (2019). *Values in heritage management: Emerging approaches and research directions*. Los Angeles, CA: The Getty Conservation Institute.

Barthel-Bouchier, D. (2016). *Cultural heritage and the challenge of sustainability*. London: Routledge.

Bonney, R., Shirk, J. L., Phillips, T. B., Wiggins, A., Ballard, H. L., Miller-Rushing, A. J., & Parrish, J. K. (2014). Next steps for citizen science. *Science*, *343*(6178), 1436–1437.

Brimblecombe, P., Grossi, C. M., & Harris, I. (2010). Climate change critical to cultural heritage. In H. Gökçekus, U. Türker & J. LaMoreaux (Eds.), *Survival and sustainability: Environmental earth sciences* (pp. 195–205). Berlin, Heidelberg: Springer.

Cassar, M., Briblecombe, P., Nixon, T., Price, C., Sabbioni, C., Saiz-Jimenez, C., & Van Balen, K. (2004). Sustainable solutions in the conservation and protection of historic monuments and archaeological remains: A critical assessment of European research needs. In C. Saiz-Jimenez (Ed.), *Air pollution and cultural heritage* (pp. 249–261), London: CRC Press.

Clarke, D., Cento Bull, A., & Deganutti, M. (2017). Soft power and dark heritage: Multiple potentialities. *International Journal of Cultural Policy*, *23*(6), 660–674.

De la Torre, M. (2013). Values and heritage conservation. *Heritage & Society*, *6*(2), 155–166.

Dillon, C., Bell, N., Fouseki, K., Laurenson, P., Thompson, A., & Strlič, M. (2014). Mind the gap: Rigour and relevance in collaborative heritage science research. *Heritage Science*, *2*(1), 1–22.

Fouseki, K. (2010). 'Community voices, curatorial choices': Community consultation for the 1807 exhibitions. *Museum and Society*, *8*(3), 180–192.

Fouseki, K., & Bobrova, Y. (2018). Understanding the change of heritage values over time and its impact on energy efficiency: Decision-making at residential historic buildings through system dynamics. In *The 3rd international conference on energy efficiency in historic buildings (EEHB2018), Visby, Sweden, September 26th to 27th, 2018* (pp. 11–21). Upsalla: Uppsala University.

Fouseki, K., Guttormsen, T. S., & Swensen, G. (Eds.) (2020). *Heritage and sustainable urban transformations: Deep cities*. London: Routledge.

Fouseki, K., Newton, D., Murillo Camacho, K. S., Nandi, S., & Koukou, T. (2020). Energy efficiency, thermal comfort, and heritage conservation in residential historic buildings as dynamic and systemic socio-cultural practices. *Atmosphere*, *11*(6), 604. https://doi.org/10.3390/atmos11060604.

Fouseki, K., & Nicolau, M. (2018). Urban heritage dynamics in 'heritage-led regeneration': Towards a sustainable lifestyles approach. *The Historic Environment: Policy & Practice*, *9*(3–4), 229–248.

Harrison, R. (2013). *Heritage: Critical approaches*. London: Routledge.

Kato, K. (2006). Community, connection and conservation: Intangible cultural values in natural heritage—the case of Shirakami-sanchi world heritage area. *International Journal of Heritage Studies*, *12*(5), 458–473.

Labadi, S. (2016). The impacts of culture and heritage-led development programmes: The cases of Liverpool (UK) and Lille (France). In S. Labadi & W. Logan (Eds.) *Urban heritage: Development and sustainability: International frameworks, national and local governance* (pp. 137–150). London: Routledge.

Labadi, S., & Logan, W. 2015. *Urban heritage, development and sustainability: International frameworks, national and local governance.* London: Routledge.

Labadi, S., & Long, C. (Eds.) (2010). *Heritage and globalisation.* London: Routledge.

Labrador, A. M., & Silberman, N. A. (Eds.) (2018). *The Oxford handbook of public heritage: Theory and practice.* Oxford: Oxford University Press.

Li, J., Krishnamurthy, S., Roders, A. P., & van Wesemael, P. (2020). State-of-the-practice: Assessing community participation within Chinese cultural World Heritage properties. *Habitat International, 96*, 102107.

Macdonald, S. (2010). *Difficult heritage: Negotiating the Nazi past in Nuremberg and beyond.* London: Routledge.

MacKee, J., Haugen Askland, H., & Askew, L. (2014), Recovering cultural built heritage after natural disasters: A resilience perspective. *International Journal of Disaster Resilience in the Built Environment, 5*(2), 202–212. https://doi.org/10.1108/IJDRBE-09-2012-0032.

Neal, C. (2015). Heritage and participation. In E. Waterton & S. Watson (Eds.), *The Palgrave handbook of contemporary heritage research* (pp. 346–365). London: Palgrave Macmillan.

Nicu, I. C. (2017). Natural hazards-a threat for immovable cultural heritage. A review. *International Journal of Conservation Science, 8*(3), 375–388.

Orr, S. A., & Cassar, M. (2020). Exposure indices of extreme wind-driven rain events for built heritage. *Atmosphere, 11*(2), 163. https://doi.org/10.3390/atmos11020163.

Piñar, G., & Sterflinger, K. (2021). Natural sciences at the service of art and cultural heritage: An interdisciplinary area in development and important challenges. *Microbial Biotechnology. 14*(3), 806–809. https://doi.org/10.1111/1751-7915.13766.

Pottgiesser, U., Fatorić, S., Hein, C. M., de Maaker, E., & Roders, A. P. (Eds.) (2021). *LDE heritage conference on heritage and the sustainable development goals: Proceedings.* Delft: TU Delft Open.

Rico, T. (2008). Negative heritage: The place of conflict in world heritage. *Conservation and Management of Archaeological Sites, 10*(4), 344–352.

Roders, A. P., & van Oers, R. (2013). *Managing change. Integrating impact assessments in heritage conservation.* Berlin: De Gruyter.

Roued-Cunliffe, H., & Copeland, A. (Eds.) (2017). *Participatory heritage.* London: Facet Publishing.

Sabbioni, C., Brimblecombe, P., & Cassar, M. (Eds.) (2010). *The atlas of climate change impact on European cultural heritage: Scientific analysis and management strategies* (No. 19). London: Anthem Press.

Sesana, E., Gagnon, A. S., Bertolin, C., & Hughes, J. (2018). Adapting cultural heritage to climate change risks: Perspectives of cultural heritage experts in Europe. *Geosciences, 8*(8), 305. https://doi.org/10.3390/geosciences8080305.

Smith, L. (2006). *Uses of heritage.* London: Routledge.

Smith, G. S., Messenger, P. M., & Soderland, H. A. (Eds) (2017). *Heritage values in contemporary society.* London: Routledge.

Tunbridge, J. E., & Ashworth, G. J. (1996). *Dissonant heritage: The management of the past as a resource in conflict.* Chichester: John Wiley & Sons.

Waterton, E., & Smith, L. (2010). The recognition and misrecognition of community heritage. *International Journal of Heritage Studies, 16*(1–2), 4–15.

World Commission on Environment and Development. (1987). *Report of the world commission on environment and development: Our common future.* Retrieved from: https://sustainabledevelopment.un.org/content/documents/5987our-common-future.pdf (Accessed on 11 May 2021).

Part I
Heritage values and risk

1 Values and sustaining heritage

Randall Mason

Sustaining heritage means sustaining values

What is meant by 'sustaining heritage'? Most people, most of the time, think of buildings, archaeological sites, monuments or other built environments, and sustaining such seems (mostly[1]) to be a good thing. 'Heritage', though, contains many possibilities as to what it means. The meaning, functions, uses and other so-called intangible aspects of heritage have gained status as core parts of any thoughtful definition of heritage (Smith, 2006). So 'sustaining heritage' raises great and important questions, because it sends us immediately to wrestle with fundamental issues: How do we define heritage? What are our goals as we talk about, manage, design and debate it? How do we make decisions, individually and collectively about sustaining heritage? This chapter aims to build a deeper understanding of what 'sustaining heritage' can be, and how to approach it, through the lens of values – or, more precisely, values-centered conservation[2] theory (VCC) and practice. The simple assertion I start and end with is that 'sustaining heritage means sustaining values'. Why are values a key locus for sustaining heritage (as opposed to questions about materiality or financing or some other particular aspect of heritage)? Values, as they are ascribed and evolve over time, connect inherited fabric (built heritage) and experiences thereof to contemporary society. In other words, they are the social infrastructure of heritage, they represent the utility of heritage in the present and future (Jones & Leech, 2015). Values connect fabric to contemporary society in feedback loops that, if they work well, make heritage places relevant and meaningful and useful not just relict and ruinous and in need of fixing.

To support this assertion, I review some of the principles behind (and uses and difficulties with) VCC.[3] VCC is a set of discernment and decision-making practices premised on a dynamic understanding of heritage as a social process (as opposed to an object-centered practice). What kinds of decisions? All kinds: which places qualify as heritage, why, how, and who is involved. In other words, decisions about listing, design, policy. It is an interpretive framework, not merely descriptive or documentary or for classification. It is becoming increasingly attuned to the functions of heritage in contemporary society and thus evolving beyond the kinds of valorization aimed strictly at the fixed, formal, material qualities of heritage structures.

I am keen to reinforce VCC's particular focus on decisions and its utility for supporting decision-making related to all aspects of heritage. As heritage decisions continue to become more complex, more contested and more deeply implicated with other civil society issues like politics, economic development, armed conflict, migration and ecological crises (I ask readers to stipulate this), professionals in heritage-related fields are compelled to join debates about sustainability and sustainable development (among other issues, such as social equity, racial justice, hunger, poverty, climate change, and more).

From my desk in Philadelphia

My interest in this topic is driven by years of writing about, practicing and working to evolve VCC and also by the decidedly unsustainable development and mismanagement of cultural heritage going on in my home city of Philadelphia.

Philadelphia is in many ways a quintessential post-industrial city. A colonial plan and massive growth between the American Civil War and World War II endowed the place with gridded neighborhood packed with masonry buildings – grand public buildings, myriad factories and infrastructure, endless rows of terraced worker and middle-class houses. Postwar suburbanization and urban decline – between the 1940s and the 1980s – grew the region but decimated many of the folks and buildings left in city neighborhoods. A central-city resurgence of an 'eds-meds-beds' (universities, hospitals, tourism) economy from the 1990s to the present brought prosperity to Center City, gentrification to many neighborhoods on the verges of the center and entrenched poverty to extensive, largely minority neighborhoods more far-flung. It is worth noting that Philadelphia has the highest poverty rate among large American cities – 23% live below the poverty income line (Shields, 2020). Philadelphia is thus a city of intense extremes, and calls for preservation play very differently in these cities within the city.

As I sit down to finish this essay, needless, profit-driven loss of old buildings seems to be prevailing in my city, heightening the urgency with which heritage professionals must adopt more effective ways of articulating and arguing the public benefits of conservation. This is the opposite of sustaining heritage. The rules governing preservation and demolition are too loose, oversight is lax, and the incentives to discard old buildings are many. The loss of heritage buildings to demolition – rising to fever pitch here in Philly (Herr-Cardillo & Fedeli, 2021; Saffron, 2021) – is not just a lost argument over the artistic qualities or historic associations of a particular property. It is also loss of memories and stories. Loss of feeling. Discounting of possible futures. A loss of complexity. A loss of embodied energy. Demolition is a loss of so many types of value, beyond bricks and dollars. Many will ascribe these failures to the insecurity of Philadelphians: we're always afraid to say 'no' to a new development scheme, lest the money flow up to New York or to the suburbs. I ascribe it more to impatience and short-sightedness.

Preservation is, after all, about so much more than buildings. Even if, professionally, we seem to want to retreat more deeply into buildings and technical solutions and scientific insights for the answers. The local debate over preservation or demolition erodes the potential value of preservation itself, in that political debates pushing back against preservation redefine 'what preservation is about' to the narrowest of bases of legal-regulatory decisions, which effectively denies the existence of any values outside what might be clearly protected in law.

The long Philadelphia preservation struggle in a nutshell: 19th- and early 20th-century row houses, churches and synagogues, factories, a rail station and other buildings and places of manifest heritage value are being destroyed in a long moment of profligate, short-sighted, selfish overdevelopment. Perfectly reusable places with long histories of adaptation and evolution are being sacrificed for profit. The state and municipality have proven both unable and unwilling to manage the longer-term costs and benefits of this trend (now 15 years and running). The result is un-sustaining and unsustainable in every sense: economic, environmental, social and cultural. And it's demoralizing to preservation advocates. The decisions about demolition/conservation abide only by profit motive and property ideology (which, ironically, one could argue is the heritage of all American cities), sacrificing future social benefits and cultural functions. The decisions processes lack transparency, make a joke of intergenerational and intragenerational equity, worsen economic inequity, incur myriad environmental costs, and unreflectively dispose of generations of the city's shared past. We (and future generations) all have to live with the tyranny of these individual decisions, as well as the lack of coordinated decisions for which the government is responsible, as if the idea of sustainability and sustainable development didn't exist.

Issues of cultural heritage management and sustainable development could not be further from these Philadelphia decisions, to the detriment of the public in every sense. The situation makes more urgent the project of this chapter and the larger volume: to plot ways of questioning and integrating cultural heritage and development processes so they benefit future publics. This work must be both professionally and politically minded – attending to refinement of professional practices and theories as well as to power and realpolitik. This shift in theories of heritage is widely acknowledged (see Smith, 2006; Wijesuriya et al., 2013; Desilvey, 2017; Avrami & Mason, 2019), though not a matter of consensus. VCC is well suited to (indeed, is designed for) raising and even bridging these distinct concerns.

Heritage functions as both archive and agent

My approach to all questions related to heritage and its conservation begins with the idea that 'heritage is the past made useful'. This simple but profound definition calls our attention to the dynamic nature of heritage, and to its ongoing functions, uses and meanings in contemporary society. Heritage consists not simply of a collection of old things, but also the functions performed or social processes related to those things. Societies don't just have heritage, they use it. It is not just a leftover from the past, it is constructed, activated and used in the present. One can organize the myriad possible uses and functions of heritage in contemporary society by marking out two kinds of functions modern heritage places are expected to perform. They must simultaneously function as an 'archive' (representing the past, materially or otherwise, in somewhat stable form) and as an 'agent for change' (using inherited forms, places and ideas to meet non-archival needs, like public space, housing, tourism experiences, social reform arguments, economic spaces, political symbols, and more). Archive and agent both are fraught and contested functions; there is tension between them, too. VCC constructs a framework within which these different functions and future can be accounted for and acknowledged if not neatly reconciled. VCC recognizes the multiple ways heritage places are valued by society (in other words, the different qualities these places possess), the dynamic social contexts of different places/countries, and challenges decision-makers to reckon with the complexity of values at stake in designing heritage places.

In one sense, the link of heritage conservation to sustainable development is obvious: conservation aims (through varied means) to selectively manage development in order to sustain 'heritage places'[4] through time and space (slowing it, stopping it, redesigning it as the case may be). Some conservation processes, one could say, are required of any conception of sustainable development. While it has been universally embraced, sustainable development, at another level, has an ambiguous relationship with culture and heritage. Elaborations of 'what is sustainability' or 'sustainable development' routinely specify economic, ecological and social aspects – rarely centering on 'culture' (Avrami, 2016). Indeed, several kinds of 'ambiguity' separate notions of heritage conservation and sustainable development – in terms of scale, responsibility, who sets priorities (Wijesuriya et al., 2013, 19).

Developments in VCC are consonant with the developing discourse on sustainability in at least two ways. First, both discourses aim to explore and reckon with the larger systems underlying specific phenomena (respectively, why and how certain heritage places get identified and preserved; how to balance ecological, economic and social outcomes of development). Systems thinking is a shared driver, though the theoretical and practical realization of systems thinking is far more advanced for sustainability. Second, the linkage of heritage conservation and sustainability (and its common deployment as 'sustainable development') goes hand-in-hand with recognition of broadening of values of heritage places and consequent need to engage with non-conservation stakeholders and processes in order for heritage conservation to gain relevance in society. These developments are consequent and intertwined; I don't ascribe cause to either, they are both reflections of changing societal processes.

The VCC framework helps build a conception of conservation that is equally rooted in an understanding of dynamic, contemporary society and in the history and materiality of inherited built environments. In light of this, 'sustaining heritage' takes on several potential aspects: the obvious 'material' interventions to repair and maintain integrity; the 'cultural practices' and 'social processes' of discerning, interpreting and otherwise using heritage in contemporary society; the more complex sustaining of 'meanings' (through the interplay of materiality and cultural meaning, a la the Ship of Theseus); 'integrating' conservation practices with development in its economic sense (rendering development more 'sustainable'); and in terms of 'financed', keeping heritage conservation alive as an enterprise. Spanning all these aspects, I tend to think of sustainable heritage as marked by increasing 'relevance' of heritage to contemporary social needs, including its 'resonance' with politics, and therefore a useful means to pursue and debate various social ends.

There is no one right or best answer, I would submit, to which of these aspects is most important; different disciplines and different politics favor one or another definition. Indeed, the varied notions of 'sustainable' that pertain to heritage signal that most decisions about heritage are design decisions[5]: they have no 'best' answer in the sense that scientific methods yield one best solution; rather, there are many, alternative solutions and methods contingent on any number of incommensurable factors, which should be judged in context instead of being abstracted from context. Science is used in some aspects of conservation, of course, but heritage decisions are fundamentally design decisions, not scientific ones.

VCC frameworks are built to deal with this sort of complexity – offering a framework for comparing (not solving for) the very different qualities informing decisions and outcomes depending on decisions about heritage places. No matter what 'sustaining heritage' means to you in a given moment, that decision can be well informed (or explained or analyzed) by VCC frameworks. As will be explained below, these frameworks are dynamic, contingent (on time, place, participation, resources available, etc.) and are continually adapted with consciousness of politics and power relationship.

The various meanings of sustainable heritage are all valid, all potentially significant, and all are (can be) pursued simultaneously. VCC gives us a framework for modeling how to weigh different notions of 'sustaining heritage' against one another.

Values-centered conservation in brief

This section is not meant to be a full account of VCC or its evolution. Rather, the goal here is highlighting key issues and pointing to essential works in the literature, where more detailed discussions can be found.

At base, VCC tries to account for the whole range of ways people might use, think about, enjoy, worship, or otherwise give a place importance or meaning or utility. 'Values' discourse connects equally to formal-material aspects of places and to more immaterial, fugitive aspects of places (stories, associations, functions, uses). VCC opens the deeper, broader questions about how cultures make meaning out of built environments.[6]

The VCC framework's centrality to modern heritage conservation practice is widely acknowledged, by proponents and critics alike (Avrami et al., 2000; de la Torre et al., 2002; Poulios, 2010; Walter, 2014). Emphasizing VCC is meant to create, as Marta de la Torre framed it, a lingua franca through which mutual understanding could be built (across professions and between professionals and publics), dialogue and collaboration can be enabled, and complex decisions could be taken on sustaining cultural heritage places. As elaborated by various national policy frameworks, by NGOs and QUANGOs like the Getty Conservation Institute and the UK's National Lottery Heritage Fund (formerly Heritage Lottery Fund), and others, VCC aims to be a comprehensive, holistic framework robust enough to address all aspects of heritage and its uses of heritage, and flexible enough to be dynamic over time, in different geographic settings, and in the face of politicization. The VCC also has to do the work of harnessing theories from a number of disciplines and professions to the practical challenges of conservation. It is not intended as a totalizing system or a set of norms, but rather as an enabling device and a set of adaptable protocols.

VCC complements and encompasses fabric-centered conservation. VCC can be used to address very traditional curatorial preservation situations and decisions, and it also contains very radical potentials for challenging the hegemony of fabric-centered practices and complicating preservation decisions by embracing social and societal considerations. VCC is also a pushback against expertise and checks the power claims of professions and other powerful institutions. This points to a main insight of this chapter: VCC suggests a framework for making decisions. How the framework is 'used', and by whom, is the decisive question whether one is interested in using conservation to advance sustainability, profit, equity, beauty, connoisseurship or any other end.

My appeal for continued use and development of VCC is based on a few linked stipulations and hypotheses about heritage and its conservation as social practices:

- Heritage is constructed, and in any place or time typically takes a variety of material and immaterial forms.
- We typically deal with heritage places (a synonym for properties, and emphasizing matters of material and fabric) but acknowledge the movable, fugitive qualities heritage sometime has.
- Better, more informed decisions about sustaining heritage places will result from understanding heritage places holistically (responding to the full range of a place's values).

- Holistic understanding increases the chance of achieving greater relevance to more stakeholders and stronger collaboration (on the principle that if we don't know about values and interests behind, we can't purposefully satisfy them).
- Possibilities for sustaining heritage can best be realized by considering the whole range of qualities, functions, meanings, associations of the place – in other words, its values.

Definitions and principles

As a conceptual model linking theory and practice, the values framework is a way of thinking and modeling how heritage is constructed and should be conserved. It is not perfect, and is criticized, but it is still useful. The focus here is on describing and evolving the framework and – more significantly – exploring how the framework/theory is used – to make heritage places and their conservation more sustainable.

The basics of the VCC framework are rooted in object-centered, art-historical epistemology. This understanding has evolved more recently, though – influenced greatly by the Burra Charter – to be more function- or process-centered. This evolution continues and is evident in practical applications of the framework[7] as well as in trends observing (and advocating) that a broader range of values are being incorporated into values frameworks.[8] VCC is often introduced as an elaboration on Riegl and his values typology, which is related to both objects (monuments) and the functions of those monuments in society (Riegl, 1903; Arrhenius, 2012). And while Riegl still presents a meaningful starting point (rooted in the same modern tendency to classify, created to implement a decision-making system as public policy), his famous essay is not the exclusive rationale for continuing to elaborate on notions of value and valuing as the core issue of making heritage and conservation decisions. (See below for a discussion of typologies.)

The simple working definition of heritage as 'the past made useful' opens important complexities about heritage's role in society, and how this changes across cultures, jurisdictions and over time, and of course with the changing formal and material character of heritage places themselves. In other words, we embrace a conception of heritage as a set of things and a series of processes, dynamically connected and feeding back in myriad ways.[9] Analytical tools have to respect the dynamism of heritage and its conservation as processes, so we have to seek dynamic concepts of heritage to evolve beyond traditional static curatorial concepts around which the conservation field has evolved (Fouseki, 2010). VCC, by helping practitioners and decision-makers deal with this dynamism, helps meet central challenges of sustaining heritage.

It is unabashedly a presentist framework – built to look backward and forward; built to understand and model all aspects of the heritage place holistically (material and immaterial, cultural and natural; artifact and resource; intention and outcome). In order to stay relevant to present concerns, to shifting ideas about the future and to changing interpretations of the past, VCC is designed to be applied iteratively. That is, the discernments, analyses and decisions made with VCC are time- and context-bound and need to be periodically revisited.

To be succinct, the main ideas in VCC are best conveyed in bullet-point statements[10]:

- 'Values', as applied to heritage, connote qualities or characteristics of a place … everything that might prompt us to preserve/curate the place;
- Values can relate to fabric to function/use or to meaning, or all three in combination;

- There are a variety of values types (see discussion of typologies, below);
- Values are many: any particular place has many, simultaneous values;
- Values are not inherent in heritage places. They are ascribed (by a number of actors, institutions and other stakeholders) and strongly influenced by broader societal contexts/dynamics and by cultural as well as other motivations (profit, political influence, etc.);
- Values are therefore modeled as temporally, spatially, societally dynamic – in tension with the core tradition of conservation to valorize the static qualities, material authenticity, continuity of use/meaning;
- Values conflict – not all can be realized/maximized at once; resolving conflicts is more often a matter of trading off or 'satisficing' (à la Simon);
- Values change (through time, in light of social and societal contexts, and according to perspective/viewer);
- In a period with ever more demands on heritage, and conflicts surrounding how heritage places are identified, it remains essential that value frameworks be reconsidered and adapted.

Value typologies have been an important part of the values framework. By organizing thought about values categorically, they inspire deeper thinking about valuing and capture important differences perceived in the qualities and functions of heritage places. The usual typologies are well known through publication and practice, though there is no one a priori best typology for all heritage places or decisions. The Burra Charter's framework of 'aesthetic, historic, scientific, social or spiritual value for past, present or future generations' provides as good a starting point as any (quoted in Mackay, 2019, p. 101; also see Australia ICOMOS, 2013 and de la Torre, 2002). Riegl (1903) is important but isn't the last word; his value categories get complex as he tries to align monuments' functions as artwork or 'heritage' with the different social motivations, professional discernments and issues of reception. Subsequent interpretations of Riegl's value categories have led heritage professionals – over decades and in contexts around the globe – to center on a fairly stable and very useful range of 'heritage values' core to the conservation enterprise: historical, artistic, aesthetic qualities that dominate our understanding of the cultural significance of heritage places.

The differences between value categories highlight incommensurable epistemological breaks (economic versus cultural values, for instance) just as they reveal professional consensus. Typologies are meant as starting points for more detailed discussions, or rather as a scaffolding to structure more elaborate work discerning values. However, they are often over-interpreted or applied in a way that reifies values as static characteristics. Critics are right that the use of these typologies tends toward making values static – but this resides more in the use of the framework than the design of the framework itself (Poulios, 2010; Walter, 2014).

The major issue in the evolution of value typologies has been how to introduce social values – roughly, those external to traditional notions of monuments and their conservation, which have increasing resonance with conservation as a publicly demanded process in contemporary society. Opening to non-heritage, social values in some measure make this operation particularly fraught: different actors get involved, difference and divergent interpretations of value abound, different paces of change in material and society, changing technologies, etc. Elaborations of 'social value' (Johnston, 1994; Jones & Leech, 2015) make explicit, important connections to present-day functions, supporting the 'agency' of heritage as distinct from its archiving functions. Thus, the use of 'heritage values' and 'social values' as two meta-categories has emerged as a way to accommodate progressive theories of heritage construction, intangible values, non-Western approaches as well as traditional, curatorial approaches (Wijesuriya et al., 2013) that create a close parallel in using 'essential' and 'instrumental' values. In simplest terms, heritage values center on the qualities the conservation has traditionally guarded and developed as a priori social goods (due to age, association, etc. – as ends in themselves), while social values acknowledge that heritage places are also used as a means to other ends (housing, tourism development, public space, etc.) and thus have value beyond their discernable connections to the past.

In a recent collection of essays published by the Getty Conservation Institute, several of the authors offer a variety of interpretations for extending notions of 'heritage values' and 'societal values' as VCC continues to evolve in practice, respond to theoretical discourse, and struggle to be implemented in different policy and cultural contexts (Avrami et al., 2019).

Shortcomings and critiques

To be sure, VCC is complicated and has shortcomings, especially as one tries to implement it in practice. Value analyses are often quite subjective. Stakeholder consultation is time-consuming, risky and often foregrounds conflict. Being diligent about the whole spectrum of values requires involving a lot of different methods and expertise. The consultation, negotiation, discernment and design thinking needed to understand trade-offs and make decisions about a place's values are complex; they are emphatically *not* simply matters of scientific method or expert aesthetic judgment (which, though complex in their own ways, have the quality of seeming definitive). The variety of different values at play are complicated and often incommensurable (comparing apples and oranges); there are abiding practical and intellectual questions about how (and whether) to relate heritage and social values in a single framework.

VCC emphasizes the political nature of heritage decisions, while asking seeking to sustain the expert roles of professionals. In practice, this tension often seems exhausting and perhaps the *least* sustainable aspect of heritage processes.

VCC still depends on experts contributing and having to share authority with other stakeholders and the public; this is often uncomfortable and unwelcome. This kind of intensely collaborative analysis and decision-making takes a lot of time/human/financial resources, and a tolerance for messiness. And even though the VCC is designed to be iterative and adaptable, there remains an unfortunate tendency in practice to fix and reify 'official' judgments of value.

The cost, trouble and risks of a more uncertain and complex process, depending on many agents, seem questionable. Especially to credentialed professionals or their clients wedded to a notion of scientific expertise determining heritage decisions. Contemporary economic and social values seem particularly vexing on the face: they are imposed on heritage professionals and institutions from 'outside' heritage discourse; they are difficult to measure and document; they are motivated by short-term, contemporary, extra-heritage (economic) concerns; they seem too deferential to non-heritage-experts and to stretch the notion of heritage professionals' purposes.

Notwithstanding these difficulties, VCC is appropriately complicated to inform the complex design decisions that practitioners, policy-makers, owners and citizens have to make every day – incorporating many different inputs and addressing many interests.

In the scholarly literature, some recent critiques have raised legitimate issues about VCC (Poulios, 2010; Walter, 2014). They see VCC reinforcing static, artifact-centered views of conservation, which in turn empowers professionals and marginalizes various publics. The critiques don't take issue with the framework so much as how it is used; they don't invalidate the ongoing contributions and adaptation of VCC work. Can VCC be used to reinforce static thinking? Yes, especially the values typologies if they are used as a normative framework (to look for certain values). Can it also be used to advance more progressive, process-centered thinking? The answer is also yes – with engagement-centered, politically aware methods of describing values more than categorizing them. No doubt values frameworks can be used in ways suited to traditional, curatorial preservation, thereby reinforcing the politics of expertise and expert discourse. As continually evolving, though, VCC can be directed at functions and uses, at intangible experiences of heritage places, at social and societal values. If it is used as a dynamic tool for framing, communicating around and facilitating decisions (by a variety of actors in the process), then VCC will work toward more relevant and sustained outcomes.

In concluding this short section on weaknesses and critiques of VCC, it should be affirmatively stated that more, and more penetrating, research will be very beneficial. Have VCC ideas and practices helped sustain heritage any *better* than traditional ideas and practices, for instance? Ethnographic, case study and other lines of research would be very welcome.

Conclusion: continuing to adapt values-centered conservation

About 15 years ago, I wrote an essay on values-centered heritage conservation, making an argument that it presents a useful tool for progressive conservation practice, particularly important at a time when churning social change demands more of historic preservation. I started it with: 'one should expect that the social changes of the last couple generations would move the field toward new paradigms in preservation' (Mason, 2006, p. 21). We are now (still) in such a time. As the urgency grows to reform and remake conservation practice, values-centered theory and practice remains a useful and robust framework.

How we use it remains the big question. I ended that essay with:

> The social contexts of historic preservation—the changing structure and tenor of contemporary culture, the influence of market thinking, and the particular societal forces and tensions we face at any given moment—demand new frameworks and practices for historic preservation. We in preservation tend to think incorrectly that being resistant to change is a virtue and part of preservation ideology. As society and cultural process get more complex, the means and ends of preservation, too, get more complex. While not a panacea, values-centered preservation is a way to organize this new paradigm.
>
> (Mason, 2006, p. 45)

I still believe this. Years of practice and teaching have shown the values framework (VCC) to be extremely useful as a heuristic, a guide for making design decisions and a tool for analyzing others' decisions. VCC enables broader conversations about the values of a place and the implications of a decision with non-conservation professionals, owners and other stakeholders. It informs practices of what I call 'engaged preservation' – wherein conservation per se responds directly to the social, economic and political issues facing a community as well as to its cultural, memorial and preservation assets (Mason, 2018). Engaged preservation models heritage conservation as a means to other (social) ends, not only as a cultural practice. VCC also proves helpful as a research framework, for post-facto analysis of a site's evolution and even as a framework for mapping the evolution of the field (Avrami & Mason, 2019). And in the end, an indicator of how well heritage – in any and all senses – is being sustained.

What about heritage needs to be better sustained? Financing? Political support? Materiality? Continuous processes of cultural interpretation and political engagement? Its usefulness is satisfying contemporary social needs? All of the above – heritage is best sustained by doing something, not being something. Above all, sustained community engagement is what's needed. Not simply in the form of consumption or passive bequest or option values, but driving toward more profound reworking of power structures: decentering, decolonizing, repairing crisis-inducing cultural breaks past and present.

I've invested a lot of work in constructively criticizing and adapting VCC. Acknowledging my own lack of critical distance from the topic, I see several issues and points of weakness needing to be addressed in order for use of values frameworks to continue evolving. The shared goal of these evolutionary changes is sustaining heritage in all of its senses – culturally, politically and of course materially.

The power relationships driving heritage decisions, and otherwise inscribed in them as historical traces, should be fully acknowledged and revealed. Critical heritage studies have contributed much to the field's awareness of power and intentional transparency. But there are persistent lags between theory, practice and policy – by their nature, these domains change at different rates. Who is visible? Who has power in heritage discourse? In management? These need to be more rigorously revealed as part of standard practice. And claiming that VCC is participatory doesn't make it so; meaningful participation in decisions must be continually pushed and demonstrated. Perhaps we could add political 'valence' to assessments of value in order to foreground power. Why? If heritage decisions ignore changing values, and the power valences of these values, actors who wield disproportionate power will become more disconnected from contemporary society and heritage would be a province of the powerful.

To better reflect power in values assessment, our typologies should keep changing, too – more to the point, they should reflect important values ascribed from outside the heritage domain, even if they often run counter to the self-defined curatorial goals of many heritage professionals and institutions. Three meta-groupings of 'heritage', 'social' and 'societal values' can organize our higher-level understandings of how heritage 'gets used' in contemporary society. Finer distinctions would still be drawn within these meta-categories. In a recent essay analyzing places of traumatic heritage, I suggested such a framework of heritage, social and societal values (Mason, 2019). It distinguished social values (the non-heritage qualities and functions of heritage places, including public space and economic aspects) from societal values as a set of qualities constructed in broader society and effectively brought into the site in question (especially symbolic and political qualities and uses).

Discourse and language matter. How we invoke the framework affects how the framework is used. Is 'values' the best term to use in parsing the motivations behind heritage conservation decisions? There is ceaseless confusion (in English) between the sense of values as used in VCC (a quality or characteristic of a heritage place) and as otherwise widely used as a synonym for 'ethics' or 'morals'. Even conveying the canonical values-as-qualities definition seems to work against a dynamic conception of values (for which David Throsby's notion of 'flows of value' is helpful (Throsby, 2000, p. 46)). In addition, some find it awkward to speak of 'values-based', favoring the less-apt 'value-based' as a modifier. Strategically, should the VCC discourse pivot toward 'qualities-based conservation'? Or even decision-centered conservation? This really gets to the point of VCC: centering on the ethical imperative to serve society and the public; getting away the sense of conservation concerned mostly with repair of architectural fabric.

These weak spots of VCC, and the difficulty of communicating them in practice, reinforce that the processes of discerning, articulating and analyzing values should be dynamic, engaging, iterative, even creative. Another means of addressing these complexities is being intentional and inclusive about designing specific modes of engagement. There are many publications from the design fields these days on genuinely participatory engagement methods (de la Pena et al., 2017). Specific to the heritage field, Kate Clark's book *Playing with the Past* (2019) provokes us to be serious and joyful in pursuit of meaningful engagement. Working with design students over the years has reinforced the insights to be gained from experimenting with graphics that capture more of the dynamism and complexity of values and valuing processes in using the framework. Strong, exploratory graphic design can organize a great deal of complexity and tether it to the quite different goals of engaging with multiple publics while resonating with professional discourse.

Because matters of power are at stake in producing and conserving and sustaining heritage. And given the colonial power relations and path dependence inscribed into many heritage places and institutions, 'sustained' heritage might have to yield to 'transformed'. In order to not simply reproduce decisions made by the powerful and reproduce discourse scripted by experts alone, other theories and frameworks need to be proposed, critiqued and brought to bear in practice. Cultural practices like heritage conservation get sustained by being iterative, by constituting an ongoing discourse. This is how VCC needs to continue evolving. How well will VCC sustain heritage in all its senses by continuing to adapt decision-making to new power relationships?

Notes

1 Controversies over Jim Crow and other colonialist monuments have complicated this.
2 I use international parlance in this chapter: 'conservation' refers to the broad field of managing heritage. In the US context, the term would be 'preservation'.
3 I use 'values-centered conservation' though it is synonymous with 'values-based conservation', which is more commonly used in professional literature.
4 The formulation 'heritage place' intentionally contains the complexities of material and processual, formal and intellectual, individual and social factors. In other words, the meaning of 'place' is not simply materialist.
5 Embracing Herbert Simon's broad definition of design: 'Everyone designs who devises courses of action aimed at changing existing conditions into preferred ones'. (Simon, 1996, p. 111).
6 My own understanding of VCC was greatly sharpened by working at the Getty Conservation Institute on a team led by Marta de la Torre in the late 1990s. I have since learned a great deal by applying these ideas in theory, in teaching and in practice – in collaboration with many colleagues, students, partners at GCI, Penn and elsewhere.
7 I'll use 'values framework' as a shorthand for the prevailing model, based on Burra Charter and GCI publications.

8 The history of this evolution is schematically mapped out in Avrami and Mason (2019); also see Walter's illuminating account (2014).
9 Laurajane Smith was among the first to elaborate the complex character and discourses of contemporary heritage phenomena, in *Uses of Heritage* (2006) and elsewhere.
10 This summary draws on Avrami et al. (2000); de la Torre (2002); Mason (2006); Avrami et al. (2019).

References

Arrhenius, T. (2012). *The fragile monument—On conservation and modernity*. London: Artifice.

Australia ICOMOS (2013). *Burra charter*. Retrieved from: https://australia.icomos.org/wp-content/uploads/The-Burra-Charter-2013-Adopted-31.10.2013.pdf (Accessed on 28 April 2021).

Avrami, E. (2016). Making historic preservation sustainable. *Journal of the American Planning Association*, 82(2), 104–112.

Avrami, E., Macdonald, S., Mason, R., & Myer, D. (Eds.). (2019). *Values in heritage management: Emerging approaches and research directions*. Los Angeles, CA: J. Paul Getty Trust/Getty Conservation Institute.

Avrami, E., & Mason, R. (2019). Mapping the issue of values. In E. Avrami, S. Macdonald, R. Mason, & D. Myer (Eds.), *Values in heritage management: Emerging approaches and research directions* (pp. 9–33). Los Angeles, CA: J. Paul Getty Trust/Getty Conservation Institute.

Avrami, E., Mason, R., & de la Torre, M. (2000). *Values and heritage conservation: Research report*. Los Angeles, CA: Getty Conservation Institute.

Clark, K. (2019). *Playing with the past: Exploring values in heritage practice*. New York: Bergahn Books.

de la Pena, D., Jones Allen, D., Hester, R. T., Hou, J., Lawson, L. J., & McNally, M. J. (Eds.). (2017). *Design as democracy: Techniques for collective creativity*. Washington, DC: Island Press.

de la Torre, M. (Ed). (2002). *Assessing the values of heritage*. Los Angeles, CA: Getty Conservation Institute.

Desilvey, C. (2017). *Curated decay*. Minneapolis: University of Minnesota Press.

Fouseki, K. (2010). Community voices, curatorial choices: Community consultation for the 1807 exhibitions. *Museum & Society*, 8(3), 180–192.

Herr-Cardillo, S., & Fedeli, D. (2021, February 15). Philly's plan to fight climate change has a glaring absence on demolitions and cultural heritage. *Philadelphia Inquirer*.

Johnston, C. (1994). *What is social value?: A discussion paper*. Canberra: Australian Heritage Commission.

Jones, S., & Leech, S. (2015). *Valuing the historic environment: A critical review of existing approaches to social value*. Retrieved from: Arts and Humanities Research Council website: https://ahrc.ukri.org/research/fundedthemesandprogrammes/culturalvalueproject/ (Accessed on 28 April 2021).

Mackay, R. (2019). Values-based management and the Burra Charter: 1979, 1999, 2013. In E. Avrami, S. Macdonald, R. Mason, & D. Myer (Eds.), *Values in heritage management: Emerging approaches and research directions* (pp. 110–126). Los Angeles, CA: J. Paul Getty Trust/Getty Conservation Institute.

Mason, R. (2006). Theoretical and practical arguments for values-centered preservation. *CRM: The Journal of Heritage Stewardship*, 3(2), 21–48.

Mason, R. (2018). Engaged preservation. *Journal of Architectural Education*, 72(2), 200–204.

Poulios, I. (2010). Moving beyond a values-based approach to heritage conservation. *Conservation and Management of Archaeological Sites*, 12(2), 170–185.

Riegl, A. (1903). Modern cult of monuments. In N. S. Price, M. K. Talley Jr, A. Melucco Vaccaro (Eds.), (1996). *Historical and philosophical issues in the conservation of cultural heritage* (pp. 69–83). Los Angeles, CA: Getty Conservation Institute.

Saffron, I. (2021, March 28). Mayor Kenney's lack of involvement in Somerset El closure isn't just a transit issue. It's a leadership failure. *Philadelphia Inquirer.*

Shields, M. (2020, December 16). The changing distribution of poverty in Philadelphia. *Economy League of Philadelphia.* Retrieved from: https://economyleague.org/providing-insight/leadingindicators/2020/12/16/phlpov19 (Accessed on 28 April 2021).

Simon, H. (1996). *Sciences of the artificial.* Cambridge: MIT Press.

Smith, L. (2006). *Uses of heritage.* London: Routledge.

Throsby, D. (2000). *Economics and culture.* Cambridge: Cambridge University Press.

Walter, N. (2014). From values to narrative: A new foundation for the conservation of historic buildings. *International Journal of Heritage Studies, 20*(6), 634–650.

Wijesuriya, G., Thompson, J., & Young, C. (Eds.). (2013). *Managing cultural world heritage. World heritage resource manual.* Paris: UNESCO. Retrieved from: http://whc.unesco.org/en/managing-cultural-world-heritage/ (Accessed on 28 April 2021).

2 Heritage and change management

Torgrim Sneve Guttormsen and Joar Skrede

Introduction

Preserving the 'frozen image' of a specific historical time, represented by intact cultural environments such as whole building complexes, quarters and park designs, has been a priority throughout the history of cultural heritage conservation. Italian Renaissance cities such as Florence, Venice and Pienza, the Gothic Quarter of the old city of Barcelona or the well-preserved medieval old town of Tallinn, Estonia, to mention a few, are all designated and protected because they are perceived as original, authentic and unique heritage places representing a distinct historical period (for a discussion of authenticity, see Viñas, 2002). For instance, in the heritage management history of the historic Røros Mining Town in Norway, the United Nations Educational, Scientific and Cultural Organization (UNESCO) stated that the priority was the care of a particular (idealized) historic cityscape that was to be preserved, and anything that could confuse that image was removed or redone to fit into this conservation strategy (Guttormsen & Fageraas, 2011). No doubt this type of heritage has created symbolic capital that is economically and socially sustainable (ibid).

On the other hand, the heritage management sector often specifies that cultural heritage that expresses historical fragmentation or relict that is temporal and undergoing transformation has less value as heritage (for an overview, see Guttormsen, 2020). This view corresponds to what is frequently referred to as the authorized heritage discourse (AHD; Smith, 2006), without denying the existence of local and diverse types of AHDs (Pendlebury, 2013; Skrede & Hølleland, 2018). There is now a burgeoning scholarly critique of viewing derelict landscapes from our not-so-distanced past as inferior to 'archaic' and 'aesthetic' ruins (e.g. Cowie & Heathcott, 2003; Edensor, 2005; High & Lewis, 2007). However, although our environments are constantly changing and most of the heritage that surrounds us has been affected by historical transformation, cultural heritage management practices are far more concerned about management and preventing change through preservation rather than valuing the transformative character of heritage and the principles of 'curating decay' (Pereira Roders & Hudson, 2012; Pereira Roders & Oers, 2013; DeSilvey, 2017). This springs from a conservation strategy wherein cultural heritage is perceived as static and immutable, as a closed container or glass showcase. This is sometimes also called a 'bell jar' protection strategy, where change is perceived as unwanted interventions against which heritage management must defend (Ruggles, 2012; for perspectives going beyond these dichotomies, see Taylor, 2015; Karlsson & Gustafsson, 2020; Gao & Jones, 2021). This chapter will explore the 'responsive approach' to heritage 'sustainability', which often implies 'freezing'

DOI: 10.4324/9781003038955-4

heritage in time through a provocative lens – that of historical transformation or change as heritage value in itself (see Chapter 5 in this handbook). It will use case studies from a Scandinavian context, including buildings, urban historic environments and cultural landscapes, to demonstrate how continuity and change (or discontinuity) could be studied and assessed for heritage significance. The chapter uses the approach of 'deep cities' (Fouseki et al., 2020; NIKU.no., 2020) to unpack how continuity in historic environments is sustained or vanishes over time.

Sustainability and sustainable development as concepts for change as heritage

Sociologist Anthony Giddens has questioned the usefulness of the concept of 'sustainable development'. He argues that the two verbs – to sustain and to develop – have somewhat contradictory meanings. In the linguistic sense, 'sustainability' implies continuity and balance, while 'development' implies dynamism and change (Giddens, 2011, pp. 60–61). He claims that this ambiguity may contribute to environmentalists being drawn to the sustainability side, whilst business people are drawn towards the (economic) development side (Andersen & Skrede, 2017, p. 583). Giddens also argues that 'development' can have at least two different meanings: development as economic growth, typically measured by growth in Gross Domestic Product (GDP), and development as the social process of taking people out of poverty (Giddens, 2011, p. 583). Giddens' contextual framing is environmental politics and the necessity of sustaining natural resources; however, if we broaden the scope to include tangible and intangible heritage, the discussion becomes different. Undoubtedly, heritage is also torn between different priorities – social, cultural, environmental, economic and other – but it is doubtful that heritage can always be sustained by putting a 'bell jar' on it. Identifying, valuing and sustaining heritage is a social and cultural process that involves communication and meaning-making (Smith, 2006, p. 2). DeSilvey reflects on these processes of change and transformation:

> If memory is understood not as something that is deposited within material containers for safekeeping but as something that is "ignited in dialogue between mind and matter," then it does not necessarily need to rely on a stable material form for its expression. In the interface between materiality and sociality, different agencies—discursive and practiced, textual and tactile—may contribute to the production of memory. Remembrance in this mode involves a willingness to accept the unsettling of our sense of ourselves as autonomous agents and to think instead about the work of assembling meaning as collaborating with an array of other materials, forces, and organisms.
>
> (DeSilvey, 2017, p. 14)

She adds that we 'must forget in order to remain present, forget in order not to die, forget in order to remain faithful', and that we should 'look beyond loss to conceive other ways of understanding and acknowledging material change' (DeSilvey, 2017, p. 5; see also DeSilvey & Harrison, 2020). Personal and collective memories are sensitive to historical transformation and become activated when societal changes and changes in our daily life routines make a past seem alien, like 'a foreign country' (Lowenthal, 1985). Changes are immaterialized in the present as something lost, absent and disappeared, as well as materialized in things, practices and places from the past that are decayed, abandoned and fragmented. A past can haunt us, like the ghostly post-Soviet heritage similar to a film-still from Andrei Tarkovsky's *The Stalker* (Wood, 2015).[1] The philosophical and theoretical universe from which perceptions of change as a heritage value spring has many sources, from Japanese *wabi-sabi* (侘寂) aesthetics and Derrida's 'hauntology' to Schumpeter's theory of 'creative destruction' and Foucault's concept of 'heterotopia' (for an overview, see Sterling, 2017; Guttormsen, 2020).

Furthering this argument, when going from 'valuating' to 'enabling' various conceptualizations of change as heritage, value would be an asset as a complex planning tool (for a processual approach to 'valuate' and 'enable', see DIVE analysis in Swensen & Berg, 2020; see also Guttormsen, 2020, p. 35). Dissolution, collage, palimpsest and stratigraphy are all concepts of change (see Table 3.2 in Guttormsen, 2020, p. 50). While dissolution, as a heritage concept with aesthetic qualities, is associated with decay and ruins and extracts the character of an abandoned place, collage resembles the metaphors of bricolage and brecciation as a heritage value, extracting the character of a place with a mixture of elements in which issues of temporality are collapsed or compressed into a collection of things or elements that are not necessarily interrelated but become 'fused together and form a whole' (Bartolini, 2014, p. 523). Palimpsest, on the other hand, is defined by the 'rewriting' of cultural expressions, the transparency of various traces from the past remaining visible, now mixed in with the new, where temporality is, metaphorically speaking, a kind of flattened time pushed up against the windshield of a speeding car. Places affected by palimpsest can also be identified as giving a feeling of visualized absence of the past, the negative imprint of something which is not here anymore, like the shadows of a demolished building on the walls of a neighbouring building. The concept of stratigraphy resembles palimpsest as it entails layering, but it defines more specific sequences and linear successions of time as a colourful sandwich of compact material history. Dissolution, collage, palimpsest and stratigraphy are concepts of change as heritage value that, each in its own way, are recognizable by their definition of heritage places as abandoned, mixed, rewritten or layered. Thus, if we do not – like Giddens – look upon continuity and balance as concepts in contrast to dynamism and change, we may open a space for creative exploration of what heritage is and what we can do with it. In nature management, 'conservation' is generally associated with the protection of natural resources, while 'preservation' is associated with the proper use of natural resources (National Park Service, 2020). This distinction corresponds to the difference between non-use and sustainable use, respectively, where sustainability does not equal putting a 'bell jar' on heritage.

In what follows, we will provide three examples of a form of heritage management that enables heritage to be both sustained *and* developed in different ways. The first example is the royal tombs at Roskilde Cathedral, Denmark. Almost 40 Danish kings and queens are buried in this cathedral. And yet, they are continually extending the burial chapels to provide space for new monarchs when they die, perfectly demonstrating that continuity and change are two sides of the same coin. Thereafter, we will provide an example of a ruin in Oslo, describing how it has been integrated into a new building structure in contradiction to the argument that ruins are 'spaces of waste that contain nothing, or nothing of value' (cf. Edensor, 2005, p. 7). Finally, we will delve into a more or less unplanned case of decay and discuss how it was dealt with and made more sustainable.

Three domains of change as a heritage value

We will take the theoretical and conceptual perspectives further by exemplifying three epistemologically and sociologically based organizational domains where change as a heritage value plays a significant role in the design of places. We are borrowing three familiar concepts from change theory (systems theory) – developmental, transitional and transformational change – to understand the relationships of change and sustainability in various heritage management practices (Marshak, 1993; Davis et al., 2006). We use change theory as an opportunity to theorize about what change entails as a strategic heritage approach to change management. In this way, we will also be able to define how change is an active social component in the organization of the practices that take place through the uses of heritage.

Valuing developmental change: the royal tombs at Roskilde Cathedral, Denmark

Developmental 'change builds on the past and leads to better performance over time, e.g., better teamwork' (Marshak, 1993, p. 8). Transferred to the organizing principles of heritage management, developmental change would characterize a form of heritage management that recognizes the need to make improvements to an existing situation. It is a change that shows continuity, but at the same time accepts change as part of the processes that define and develop heritage places over time. Developmental change involves a planned or intended change, usually taking place through institutional practices, wherein historical roots are linked to a utopian future.

An illustrative example of developmental change as a specific heritage strategy is Roskilde Cathedral (Danish: Roskilde Domkirke, a UNESCO World Heritage Site) in the city of Roskilde on the island of Zealand (Sjælland) in eastern Denmark (UNESCO, 1995). Since the cathedral was constructed during the 12th and 13th centuries, its architecture shows 800 years of European architectural styles and incorporates both Gothic and Romanesque architectural features. The cathedral is the official royal burial church of Danish monarchs from the early period of the Danish (and Danish-Norwegian) medieval kingdom to the present. As such, it has been significantly extended and altered over the centuries to accommodate a considerable number of burial chapels, and the many added chapels show different architectural styles.

Almost 40 Danish kings and queens have been buried in Roskilde Cathedral, dating back to Sweyn II in 1076 and possibly even to the great king Harald Bluetooth in 986. Changes have been made inside the cathedral, and new building complexes and burial chapels have been added to make room for the royal tombs. The grave practices also point to the future. Since 2003, the sculptor Bjørn Nørgaard has, on commission of Queen Margrethe II (born 1940), been designing the Queen's tomb in the cathedral, which is now ready and waiting (Figure 2.1). Named 'Sarkofag', or 'sarcophagus', the tomb features a transparent, rounded capsule in cast glass holding two supine figures mounted on pillars topped with silver elephant heads (Figure 2.2). St. Birgitta's Chapel in the cathedral has been restored to house the tomb of Queen Margrethe II, thus confirming how developmental changes manifested by the temporality of the place are legitimizing the linkage of the Danish royal family and kingdom from a mythical past to an infinite future.

Figure 2.1 The tomb cover, which will hide the tomb until Margrethe's death.
Source: Photo by Keld Navntoft. © The Royal Danish House.

Figure 2.2 The artist Bjørn Nørgaard's sepulchral monument, which will be the tomb of Queen Margrethe II of Denmark when she dies.
Source: Photo by Keld Navntoft. © The Royal Danish House.

The example from Roskilde Cathedral illustrates an organizational principle that enables intended historical continuity based on developmental change as a heritage value, where new forms of heritage within a common context are added with the prospect of shaping future heritage. The Roskilde example refers to a royal succession that points to the future based on various historical time expressions that have also changed over time. Since heritage is a cultural and social process, a 'bell jar' approach would not have worked in this case. One could, of course, embalm the bodies of the dead, but one can do nothing about the inevitable fact that future monarchs will also pass away. Thus, the Roskilde example demonstrates that to 'sustain' also implies to 'develop'. The memory of the royals must be continued within Roskilde Cathedral's changing conditions, which indicates that this heritage project, from a sustainable heritage management perspective, is not an attempt to freeze time. Rather, such a combination of sustainability and development can be an organic solution when dealing with tangible and intangible heritage. The memory of the royals is sustained *and* developed with the design of new intergenerational tombs.

Valuing transitional change: a ruin wall from a prison in downtown Oslo, Norway

Transitional change (involving, providing or consisting of the passage, movement or change from one state, condition, subject, place, etc., to another, see Marshak, 1993, p. 8) characterizes an organization of heritage management that recognizes the need to implement a completely new course of action. There are a number of examples of such a heritage strategy, especially in urban planning, such as through ruin conservation based on fragmented anecdotal evidence. Ruin conservation also largely includes situations where fragments are implemented in new building designs, as is often the case in 'deep cities'. An example of this type of ruin conservation in a Norwegian context is the so-called Tukthusmuren (prison wall) in Oslo.

The historic Tukthus quarter (Tukthuskvartalet) from the 18th century covered two blocks between Storgata and Torggata streets in central Oslo and was entirely reserved for the prison's activities. During the 1930s, the prison property was gradually separated and the buildings were demolished. When the main building was demolished in 1938, an era of about 200 years of use as a prison site was finally over. Only parts of the prison walls remained, fragmented parts of a large building complex which, with its ruin-like appearance, yielded an anecdotal storytelling materiality that concealed a monumental and dark past in the middle of the city. In 2000, some new buildings were completed, with the restored ruins of the prison wall as part of the new architecture. The wall was integrated into the new structures as part of a restaurant, as traces in the asphalt over a road and through a new large office building and police station. When observing the wall today, it becomes obvious that the new quarterly plan differs from the original quadrature of the prison complex (Guttormsen & Simon, 2018).

This example illustrates a planning situation that is common in today's society. That is the adaptive reuse of fragments of buildings and infrastructure from past societies found at archaeological excavations or above-ground buildings that have been partially demolished for new purposes. Ruins are, as waste in the circular economy, a reusable resource that may be both economically and environmentally sustainable. Contrary to the Danish case, the sustainable management of the ruins of Tukthuset is not about sustaining an intergenerational memory of monarchs. This was a vernacular site in need of being managed one way or another. One solution could have been to put a 'bell jar' on it, making it into a heritage site 'frozen' in time as an 'aestheticized' and 'romanticized' fragment of the past. Putting a 'bell jar' on every ruin would probably not be feasible, but we often use this strategy in cases where ruins are ascribed significant heritage values. One example is the enormous glass hall erected above the ruins of the medieval cathedral of Hamar in south-central Norway (Eriksen, 2007, p. 15), where 'the architect has enveloped the ruins in a large vacumous space' (Schnipper, 2015). However, most ruins are (more) vernacular, such as Tukthuset. In such cases, we can try to bring about a transitional change for the site, preserving the old, parallel to developing new structures *on* or in prolongation *of* the ruins. In this way, we could also potentially, by highlighting differences by means of palimpsest and stratigraphy, direct attention to fragments of heritage that would otherwise have remained unrecognized. Thus, a transitional approach to heritage management may, in such cases, be a sustainable management of heritage. We are sustaining the ruin *and* integrating it into something new. Such a transitional change both sustains and develops heritage, indicating that heritage is both 'deep' and contemporary. Whereas developmental change is defining changes within the same historical tracks by showing continuity, transitional change will find totally new tracks and create a new direction compared to previous uses and functions. The Tukthuset case, for instance, demonstrates a totally new function where the transition from being a prison (dark heritage) in the past becomes a newly designed restaurant (the dark heritage is hidden and purified by becoming something totally different with this new function).

Valuing transformational change: the Røros UNESCO industrial toxic mining landscape, Norway

Transformational change refers to the transfiguration from one state of being to a fundamentally different state of being. It connotes a marked change in form, nature or appearance; a process by which one figure, expression or function is converted into another one of similar value (Marshak, 1993, p. 8). Transformational change constitutes the third domain of change in terms of heritage values we believe express heritage management practices. Transformational change happens when a process or 'norm' creates itself as a result of the destruction of the existing mode and assumes its own shape. You can refine the conservation strategy as it develops, but it is usually random, unintentional and unplanned. Transformational change entails valuing heritage as a process and a transient character. Thus, heritage is like a verb, actively doing and in motion. This is heritage of incremental decay and decomposition caused by natural processes (e.g. climate change) and the absence of human intervention.

Transformation is usually considered a threat, as expressed by Kaslegard (2011) in a heritage management report about heritage and climate change: 'Even though ruins and buildings are exposed to the same types of physical decay, ruins can be even more vulnerable because they lack a protective roof' (p. 24). Management strategies that value decay are, therefore, normally not talked about or considered acceptable, but they do occur to a large extent, especially under conservation conditions where it is difficult to maintain the status quo due to the scale, type and importance of the heritage, or where

Figure 2.3 The UNESCO toxic industrial heritage landscape of Røros and the Circumference. Source: Photo by Tommy Gildseth [CC BY-SA 4.0]). Retrieved from: https://no.wikipedia.org/wiki/Christianus_Sextus.

the location poses a risk of destruction that is beyond human control. Typical places are shady industrial ruins in abandoned areas and of large scale making their conservation through restoration a difficult task. In Norway, there is cultural heritage from the 20th-century mining operations on Svalbard, the mining operations at Røros and the Circumference from the 17th to 20th centuries, with the latter protected as a UNESCO World Heritage Site and both constituting heritage values where change is associated with decay, decomposition and toxic waste (UNESCO, 2010, see Figure 2.3).

Other examples of built heritage on Danish shores are gradually disappearing due to coastal erosion and flooding, and the choice is either to relocate the heritage or to let it go and appreciate the transformational value of 'ruinification' as a process of a vanishing heritage (Wienberg, 2014). Equally relevant examples are the remote and outdated places and landscapes bearing Soviet heritage in a rapidly transforming post-socialist society (Litchfield, 2014), forgotten locations such as an outdoor cinema in the Sinai Peninsula, Egypt (Kikkas, 2020), and urban public places in New York City that are no longer in use (Sollis, 2013). Elements of decay are evident when the material heritage is kept as it is, or more precisely by not doing any preservation and instead valuing decay, putrefaction and fragmentation as a (post-) preservation practice (DeSilvey, 2017).

The dimension of transformational change as heritage value is also underestimated from a sustainability perspective. For instance, Sollis (2013) seeks a discussion about what is being preserved and what is gradually disappearing, about what represents our anxieties about and hopes for the sustainability of our society. In such a discussion, heritage becomes an educational tool that nourishes the intellectual capital that promotes socially and culturally sustainable societies. When transformational change is valued, sustainable development is not necessarily a 'tool', but rather a way of coming to terms with and accepting natural decay. From derelict sites, there will arise new structures and activities that are organic and of social and cultural relevance. Nature and humans will reclaim wasteland and develop new functions and practices. In the literature on sustainability, a distinction between 'strong' and 'weak' sustainability has emerged (Neumayer, 2010). Proponents of the former argue that renewable resources must not be drawn down faster than they can be renewed. This corresponds to viewing heritage as a non-renewable resource (Brattli, 2009, p. 24). The latter approach, however, allows that certain resources can be depleted as long as they can be substituted by others. These somewhat different understandings of sustainability have been referred to as the non-substitutable and the substitutability paradigms, respectively (Neumayer, 2010, pp. 20–21; Skrede & Berg, 2019, p. 87). If we view heritage as a renewable and constantly changing resource, its nature would be transformational and substitutional. No society has surrounded itself with as many heritage sites and objects as modern Western societies (Holtorf, 1998, p. 2). That does not mean, however, that we should demolish buildings or remove ruins if there are good reasons to preserve them; rather, we should value change as intrinsic to heritage. Heritage management should 'actively and responsibly' renew the past in our time (Holtorf, 1998, p. 5). This includes considering how we can understand the needs of future generations that are not necessarily concurrent with our own (Taylor, 2013, p. 5).

Conclusions

In this chapter, we have advanced three main points. First, although for some heritage management professionals the best form of conservation would be a 'bell jar' approach, protecting heritage from change based on, for example, the appearance of authenticity and a specific time period considered valuable, it turns out that, in practice, even a 'bell jar' strategy involves effecting significant changes to maintain the status quo. This critique has a hermeneutic dimension. It questions the extent to which conservation practices represent a historical transformation that strengthens the 'true authentic' place. It further questions whether conservation efforts should be seen as the idealization of a place that is far removed from historical reality, which is continually changing and being shaped by historical transformation, not least during continuous 'heritagization' processes. Viewed from this social constructionist perspective, heritage sites appear authentic not because they are inherently authentic but because they are constructed as such in terms of points of view, beliefs, perspectives or powers (see also Guttormsen & Fageraas, 2011, pp. 447–448).

Second, we address the need for a new theoretical and methodological reservoir to approach heritage under change as a prosperous value. In nature heritage management, it is accepted that nature can both be a system (ecosystem) and consist of values that are constantly changing. But when it comes to cultural heritage, there is more opposition to such a view. We have provided insight into key theoretical approaches to management practices and thereby put forth systematic ways of organizing the uses of heritage, which is analytically illustrated by three types of change – developmental, transitional and transformational – which in turn opens up for exploring the premises and consequences for very different approaches in the heritage management's strategies to change management. Thus, we have shown that although heritage management may associate change with threats and the negative effects caused by a lack of conservation measures, we are surrounded with a vast amount of heritage that is continuously under change. We have shown that heritage management practices have various ways of handling change as a heritage value, even when change is not articulated or desired.

This brings us to our third point, and main conclusion, which is that heritage characterized by change (e.g. fragmented anecdotal traces) is a heritage value with significant potential as a sustainable resource. It could be argued that the focus on change as a heritage value in planning would involve the absence of holistic preservation strategies that opens for greater fragmentation and more reduction in the value of historic environments, which in turn give a green light to speculative entrepreneurs where short-term economic profit takes precedence over long-term sustainable strategies (Guttormsen, 2020, p. 51). In handling change as a value, one enters a political field where caution regarding why and how to protect must be clearly articulated. To reject conservation ideologies as reactionary or to refuse to consider their potential would be a poor utilization of the grand diversity of heritage resources that could benefit society. We have argued that the terms 'sustain' and 'develop' are not necessarily contradictory. Sustainability as a concept does not have to imply preserving status quo, as sustainable 'development' involves managing change in ways in which heritage is used as a resource for future generations. In as much as heritage is a social and cultural process, it is dynamic and constantly developing. We need to deal with heritage as something organic, not as a static entity. A sustainable use of heritage can involve developmental, transitional and transformative approaches, sustaining the material and immaterial remnants of the past, in parallel to managing and developing heritage on behalf of future generations.

We will conclude by outlining new areas for the next decade of research in the field of sustainable heritage, all of which emphasize change as a heritage value. First of all, with ever-increasing urban population growth, valuing change as a heritage value will have a tremendous impact on urban planning and design. It will benefit society to use heritage as part of future planning. Such an approach allows cities to grow and include parts of the past into city planning instead of sealing off the city (the 'bell jar' protection) from the past. Considering how much materiality is affected by change and the types of heritage that are removed because they are considered to be of little value, it is unjustifiable from a sustainability perspective to ignore change as a heritage value. In addition, a people-centred approach and the use of digital tools to create good places based on change as a heritage value will be essential in well-developed change management.[2] In planning new research, it will be vital to develop and implement theories and methods that will yield new input for preservation ideology.

Another field of research that will become increasingly relevant is how to deal with heritage from war turmoil and falling regimes, such as the heritage from the aforementioned post-Soviet societies, which is often associated with decay, petrification, absence and transformation. Just as with the large industrial heritage landscapes that were exemplified earlier, a number of challenges arise with regard to curating decay because the changes are so sweeping and extensive from one society to another. At the same time, the willingness to preserve can stand in contradiction to differing attitudes about the (painful) heritage of falling regimes.

A characteristic of heritage places and landscapes that are decayed, abandoned and fragmented is that they appeal to aesthetic and emotional experiences associated with poetic reflection on the changes of life and the perishable, to that which is transient and associated with clues from the past, death and the afterlife. Emotional and affective aspects of (vanishing) heritage are significant elements of social and psychological life (cf. Smith, 2020). Memorization relates to both the 'permanence and absence' of landscapes (Castañeda López & Vela Cossío, 2020), and heritage is as much about 'feelings' as about 'material structures'. What has been referred to as the 'emotionalization' of culture and social life (Bondi, 2005; Bille & Simonsen, 2019) is, therefore, highly relevant for heritages studies, and the already established interest in the affective and emotional dimension of heritage should be continued.

In the majority of academic literature on sustainable development, there are primarily three forms of sustainability: environmental, social and economic. However, there are scholars who have argued the benefits of including a fourth dimension of sustainability, namely 'cultural' (e.g. Hawkes, 2001; Dessein et al., 2015; Skrede, 2016; Throsby, 2017; Skrede & Berg, 2019). Undoubtedly, heritage serves as an economic asset and plays a role in environmental policies. Heritage also brings about social well-being, belonging and place identity; however, without being inscribed into a fourfold sustainability model, heritage may run the risk of being subsumed by other priorities (Skrede & Berg, 2019, p. 87). We have tried to demonstrate how cultural considerations tie in with debates on sustainable development; however, there is still a need to shed light on this topic in heritage and change management.

Notes

1 *The Stalker* is a 1979 Soviet science fiction art film directed by Andrei Tarkovsky. The film represents

> a post-Soviet and post-industrial genre of photography which seeks to document sites of decay and abandonment [...] Stalker's characters move through poetically-charged landscapes of industrial ruins and resurgent nature, with Tarkovsky's camera often lingering on abandoned artifacts and apparently contaminated landscapes. This imagery simultaneously inaugurated and transcended a so-called "ruin porn" aesthetic that similarly features industrial ruins, abandonment, architectural decay, and urban failure.
>
> (Madson, 2019)

2 See www.deepcities.eu for information on the newly begun research project 'Curating Sustainable URBAn Transformations through HERItage' [CURBATHERI], running between 2020 and 2022 (NIKU.no. 2020).

References

Andersen, B., & Skrede, J. (2017). Planning for a sustainable Oslo. The challenge of turning urban theory into practice. *Local Environment. The International Journal of Justice Sustainability*, 22(5), 581–594. https://doi.org/10.1080/13549839.2016.1236783.

Bartolini, N. (2014). Critical urban heritage: From palimpsest to brecciation. *International Journal of Heritage Studies*, 20(5), 519–533. https://doi.org/10.1080/13527258.2013.794855.

Bille, M., & Simonsen, K. (2019). Atmospheric practices: On affecting and being affected. *Space and Culture*, 1–15. Advanced online publication. https://doi.org/10.1177/1206331218819711.

Bondi, L. (2005). Making connections and thinking through emotions: Between geography and psychotherapy. *Transactions of the Institute of British Geographers*, 30(4), 433–448. https://doi.org/10.1111/j.1475-5661.2005.00183.x.

Brattli, T. (2009). Managing the archaeological world cultural heritage: Consensus or rhetoric? *Norwegian Archaeological Review*, 42(1), 24–39. https://doi.org/10.1080/00293650902904505.

Castañeda López, C., & Vela Cossío, F. (2020). The imprint of the Spanish tobacco industry on the urban landscape: Permanences and absences of an industrial memory. *Industrial Archaeology Review*, 42(1), 16–28. https://doi.org/10.1080/03090728.2020.1725286.

Cowie, J., & Heathcott, J. (2003). Introduction. The meanings of deindustrialization. In J. Cowie, & J. Heathcott (Eds.), *Beyond the ruins: The meanings of deindustrialization* (pp. 1–15). Ithaca, NY: Cornell University Press.

Davis, J., Galvarin, M., Williams, M., & Johnson, C. T. (2006). *Change theory and historical model matrix*. Tempe: School of Advanced Studies, University of Phoenix. Retrieved from: https://www.academia.edu/7843194/Change_Theory_and_Historical_Model_Matrix (Accessed on 5 May 2021).

DeSilvey, C. (2017). *Curated decay: Heritage beyond saving*. Minneapolis: The University of Minnesota University Press.

DeSilvey, C., & Harrison, R. (2020). Anticipating loss: Rethinking endangerment in heritage futures. *International Journal of Heritage Studies*, 26(1), 1–7. https://doi.org/10.1080/13527258.2019.1644530.

Dessein, J., Soini, K., Fairclough, G., & Horlings, L. (Eds.) (2015). *Culture in, for and as sustainable development. Conclusions from the COST action IS1007. Investigating cultural sustainability*. Finland: University of Jyväskylä.

Edensor, T. (2005). *Industrial ruins: Spaces, aesthetics, materiality*. New York: Berg Publishers.

Eriksen, A. (2007). The murmur of ruins. *Ethnologia Europaea*, 36(1), 5–20.

Fouseki, K., Guttormsen, T. S., & Swensen, G. (2020). Heritage and sustainable urban transformations: "deep cities" approach. In K. Fouseki, T. S. Guttormsen & G. Swensen (Eds.), *Heritage as a driver for sustainable cities – deep cities* (pp. 1–15). London & New York: Routledge.

Gao, Q., & Jones, S. (2021). Authenticity and heritage conservation: Seeking common complexities beyond the 'Eastern' and 'Western' dichotomy. *International Journal of Heritage Studies*, 27(1), 90–106. https://doi.org/10.1080/13527258.2020.1793377.

Giddens, A. (2011). *The politics of climate change*. 2nd ed. Cambridge: Polity Press.

Guttormsen, T. S. (2020). Archaeology as conceptual tool in urban planning. In K. Fouseki, T. S. Guttormsen & G. Swensen (Eds.), *Heritage as a driver for sustainable cities – deep cities* (pp. 35–54). London and New York: Routledge.

Guttormsen, T. S., & Fageraas, K. (2011). The social production of 'attractive authenticity' at the World Heritage Site of Røros, Norway. *International Journal of Heritage Studies*, 17(5), 442–462. https://doi.org/10.1080/13527258.2011.571270.

Guttormsen, T. S., & Simon, V. K. (2018). *Har ruiner en verdi i dagens bybilde?* forskning.no, 9 May 2018. Retrieved from: https://blogg.forskning.no/kulturminnebloggen/har-ruiner-en-verdi-i-dagens-bybilde/1091646 (Accessed on 5 May 2021).

Hawkes, J. (2001). *The fourth pillar of sustainability: Culture's essential role in public planning*. Retrieved from: http://www.culturaldevelopment.net.au/community/Downloads/HawkesJon(2001)TheFourthPillarOfSustainability.pdf (Accessed on 5 May 2021).

High, S., & Lewis, D. W. (2007). *Corporate wasteland: The landscape and memory of deindustrialization*. Ithaca, NY: Cornell University Press.

Holtorf, C. (1998). Is the past a non-renewable resource? *World Archaeology Congress (WAC) Inter-Congress on the destruction and conservation of cultural property*. Retrieved from: https://learning.hccs.edu/faculty/brett.furth/anth2302/additional-readings/holtorf-1998 (Accessed on 5 May 2021).

Karlsson, H., & Gustafsson, A. (2020). Staging antiquity: A comparison of five Greek cultural heritage sites and the construction of their authenticity. *Journal of Heritage Management*, 5(1), 7–23. https://doi.org/10.1177/2455929620919257.

Kaslegard, A. S. (2011). *Climate change and cultural heritage in the Nordic countries*. TemaNord 2010:599. Copenhagen: Nordic Council of Ministers. Retrieved from: https://www.diva-portal.org/smash/get/diva2:700575/FULLTEXT01.pdf (Accessed on 5 May 2021).

Kikkas, K. (2020). *End of the world cinema*. Retrieved from: https://www.kaupokikkas.com/end-of-the-world-cinema (Accessed on 6 September 2020).

Litchfield, R. (2014, 10 July). Soviet ghosts: An empire in decay - in pictures. *The Guardian*. Retrieved from: https://www.theguardian.com/culture/gallery/2014/jul/10/soviet-ghosts-an-empire-in-decay-in-pictures (Accessed on 5 May 2021).

Lowenthal, D. (1985). *The past is a foreign country*. Cambridge: Cambridge University Press.

Madson, R. (2019, 30 April). Post-stalker: Notes on post-industrial environments and aesthetics. *Streka Mag*. Retrieved from: https://strelkamag.com/en/article/post-stalker-notes-on-post-industrial-environments-and-aesthetics (Accessed on 5 May 2021).

Marshak, R. J. (1993). Managing the metaphors of change. *Organizational Dynamics*, 22(1), 44–56. https://doi.org/10.1016/0090–2616(93)90081-B.

National Park Service. (2020). *Conservation vs preservation and the National Park Service*. Retrieved from: https://www.nps.gov/teachers/classrooms/conservation-preservation-and-the-national-park-service.htm (Accessed on 5 May 2021).

Neumayer, E. (2010). *Weak versus strong sustainability: Exploring the limits of two opposing paradigms*. Cheltenham: Edward Elgar.

NIKU.no. (2020). *CURBATHERI – Curating sustainable urban transformations through heritage*. Retrieved from: https://curbatheri.niku.no/ (Accessed 27 September 2020).

Pendlebury, J. (2013). Conservation values, the authorised heritage discourse and the conservation-planning assemblage. *International Journal of Heritage Studies*, 19(7), 709–727. https://doi.org/10.1080/13527258.2012.700282.

Pereira Rodes, A., & Hudson, J. (2012). Change management and cultural heritage. In E. Finch (Ed.), *Facilities change management* (pp. 175–190). West Sussex: Blackwell.

Pereira Roders, A., & Oers, R.v (2013). Managing change: Integrating impact assessments in heritage conservation. In: M. T. Albert, R. Bernecker & R. Britta (Eds.), *Understanding heritage: Perspectives in heritage studies* (pp. 89–104). Bonn: Walter de Gruyter GmbH.

Ruggles, D. F. (2012). Introduction: The social and urban scale of heritage. In D. F. Ruggles (Ed.), *On location. Heritage cities and sites* (pp. 1–14). New York: Springer.

Schnipper, E. (2015, 26 October). Kjell Lund. Cathedral of Hamar, protecting the cathedral. *Rooms of Possibilities*. Retrieved from: http://www.roomofpossibilities.com/index.php/2015/10/26/protecting-the-cathedral-of-hamar/ (Accessed on 27 September 2020).

Skrede, J. (2016). What may culture contribute to urban sustainability? Critical reflections on the uses of culture in urban development in Oslo and beyond. *Journal of Urbanism. International Research on Placemaking and Urban Sustainability*, 9(4), 408–425. https://doi.org/10.1080/17549175.2015.1074603.

Skrede, J., & Berg, S. K. (2019). Cultural heritage and sustainable development: The case of urban densification. *The Historic Environment: Policy & Practice*, 10(1), 83–102. https://doi.org/10.1080/17567505.2019.1558027.

Skrede, J., & Hølleland, H. (2018). Uses of heritage and beyond: Heritage studies viewed through the lens of critical discourse analysis and critical realism. *Journal of Social Archaeology*, 18(1), 77–96. https://doi.org/10.1177/1469605317749290.

Smith, L. (2006). *Uses of heritage*. London: Routledge.

Smith, L. (2020). *Emotional heritage: Visitor engagement at museums and heritage sites*. London and New York: Routledge.

Sollis, J. (2013). *Stages of decay*. Munich: Presrel.

Sterling, C. (2017, 6 November). *On heritage and hauntology*. Retrieved from: http://www.carmah.berlin/reflections/on-heritage-and-hauntology/ (Accessed on 27 September 2020).

Swensen, G., & Berg, S. K. (2020). The 'garden city' in the green infrastructure of the future: Learning from the past. *Landscape Research*, 45(7), 802–818. https://doi.org/10.1080/01426397.2020.1798365.

Taylor, J. (2013). Intergenerational justice: A useful perspective for heritage conservation. CeROArt [Online], HS | 2013, I. Brajer (Ed.), *Conservation: Cultures and Connections*. Retrieved from: https://journals.openedition.org/ceroart/3510 (Accessed on 27 September 2020).

Taylor, J. (2015). Embodiment unbound: Moving beyond divisions in the understanding and practice of heritage conservation. *Studies in Conservation*, *60*(1), 65–77. https://doi.org/10.1179/2047058413Y.0000000122.

Throsby, D. (2017). Culturally sustainable development: Theoretical concept or practical policy instrument? *International Journal of Cultural Policy*, *23*(2), 133–147. https://doi.org/10.1080/10286632.2017.1280788.

UNESCO. (1995). *Roskilde Cathedral*. UNESCO. Retrieved from: https://whc.unesco.org/en/list/695/ (Accessed on 27 September 2020).

UNESCO. (2010). *Røros Mining Town and the circumference*. UNESCO. Retrieved from: https://whc.unesco.org/en/list/55/ (Accessed on 27 September 2020).

Viñas, S. M. (2002). Contemporary theory of conservation. *Studies in Conservation*, *47*(1), 3. https://doi.org/10.1179/sic.2002.47.Supplement-1.25.

Wienberg, J. (2014). Four churches and a lighthouse—preservation, 'creative dismantling' or destruction. *Danish Journal of Archaeology*, *3*(1), 68–75. https://doi.org/10.1080/21662282.2013.910366.

Wood, P. (2015). Selective amnesia and spectral recollection in the bloodlands. In R. Richardson (Ed.), *Walking inside out: Contemporary British psychogeography* (pp. 89–102). London: Rowman & Littlefield.

3 Combining theory and practice
Incorporating value in risk assessment at heritage sites

Katherine Curran, Junnan Bao, Heather Oakley,
Argyro Gili and Kalliopi Fouseki

Introduction

This chapter investigates how the complex concept of 'value' can best be incorporated into the practical heritage management exercise of risk assessment. Many risk assessment methods estimate 'loss of value' (e.g. Waller, 1994; Michalski & Pedersoli Jr, 2016); however, this is a non-trivial exercise and risks oversimplifying a complex idea by forcing users to quantify value numerically. On the other hand, attempts to exclude the concept of value completely from a risk assessment exercise produce a less thoughtful exercise and are ultimately doomed, as any judgement will inevitably include some form of value-based decision, however poorly it is articulated or framed.

How can value be incorporated into risk assessment in a way that reflects the real complexity of the concept, without making practical risk assessment overly theoretical and unfeasibly time-consuming? This dilemma is inherent in many aspects of heritage management where decisions related to condition assessment or environmental control, while fundamentally practical, nevertheless require engagement with the concept of value. Resolving this requires engagement with a holistic definition of sustainable heritage management. A risk assessment exercise may appear to focus on the more tangible aspects of sustainability such as balancing the risk of material deterioration (physical sustainability) against the need to reduce carbon emissions (environmental sustainability) or budget constraints (economic sustainability). However, it is essential to consider how it can ensure social and cultural sustainability by a thoughtful incorporation of the concept of value into the risk assessment exercise.

This chapter includes interviews with experienced practitioners in the field of risk assessment, which explore their views on the role of value in the risk assessment process and the challenges they perceive with its inclusion. The chapter also describes two case study risk assessment approaches, which incorporate the concept of heritage value more explicitly than is usually seen. It will outline the approaches and discuss the benefits, challenges and limitations associated with them. The chapter therefore explores how this fundamental but complex concept can be incorporated into a risk assessment exercise in a way that is thoughtful, critical, holistic but also practical. This has wider implications for the field, as it speaks to the need to marry the theoretical and the practical to achieve true sustainable heritage management practice. It is worth noting here that the 'value' refers to all values (social, economic, cultural, etc.) attributed to heritage.

DOI: 10.4324/9781003038955-5

Literature context

Risk assessment

Risk assessment is a widely used heritage management tool that is used to identify priorities and allocate resources. Put simply, risk assessment involves the identification of hazards (such as light exposure, theft or flooding) that may affect a heritage asset, followed by a systematic assessment of both the probability that each hazard will have an impact on the site and the severity of the impact associated with it.

Risk = Probability X Severity

There are multiple risk assessment methods that have been designed for a heritage context. Two prominent methods are those by Robert Waller and Stefan Michalski. Waller's Cultural Property Risk Analysis Model (CPRAM; Waller, 1994, 2003, 2019) approach involves assigning a score between 0 and 1 (i.e. a ratio scale) to each of three factors: probability (or extent for long-term hazards such as high relative humidity), fraction susceptible (FS) and loss of value (LV). These three scores are then multiplied by each other to give an overall magnitude of risk. Further development of the CPRAM method has led to subdivision of these three risk variables to four, five or more variables (Waller, 2019). Michalski's ABC method (Michalski, 2004; Michalski & Pedersoli Jr, 2016) is similar and involves assigning scores of 1–5 for three parameters, A: frequency or rate, which is similar to probability, B: loss of value and C: items affected. The three scores are then added together to give a risk score. In both methods, the hazards are aligned with the agents of deterioration developed by the Canadian Conservation Institute (2017) and include both long-term hazards such as fluctuating relative humidity and catastrophic events such as fires. Both methods have been adopted internationally. While both have been developed for museum collections, they have found broader application; for example, a recent UNESCO publication on the application of risk assessment to World Heritage Sites used both the CPRAM and ABC methods (Paolini et al., 2012).

Together with the work of Jonathan Ashley-Smith (1999), the methods of Waller and Michalski have had a significant impact on the field. However, other approaches have also been explored. For example, Brokerhof and Bülow's (2016) 'QuiskScan' method aims to provide a quicker alternative to the CPRAM and ABC models. Rather than numerical scores, vulnerability to a hazard and value are ranked as 'high', 'medium' or 'low'. This enables a colour coding of the collection according to the 'vulnerable value' of the different units to each of the hazards. Subsequently, exposure to the hazard is assessed as 'high', 'medium' or 'low' to identify the most significant risk. Romão, Paupério and Pereira (2016) provide the user with a flow chart to assess risks to built cultural heritage through the expected level of damage to the asset, the extent to which that damage is repairable, an estimation of the loss of value and the likelihood of the hazard occurring.

A key benefit of risk assessment within heritage management is that it enables different hazards to be compared systematically and transparently (Henderson & Waller, 2016). This can provide an evidence base with which to identify the most vulnerable parts of a museum collection (Bratasz et al., 2018) or to make a case for the need for new facilities (Waller & Lee, 2018). Garside et al. (2018) showed that a risk assessment process can enable a team to explore multiple options to tackle a single issue. They note that a risk assessment approach 'allows complex issues to be presented in a simple, readily understandable manner to non-specialist stakeholders, particularly those in other departments and at higher levels of management' (p. S99). Other research identified that taking a risk assessment approach to decision-making allowed conservators to gain greater leverage and demonstrate their value to other colleagues (Rogerson & Garside, 2017). Colour-coded outcomes such as those produced by Brokerhof and Bülow (2016) and Garside, Bradford and Hamlyn (2018) are valuable tools to communicate with relevant stakeholders such as management, funders, maintenance staff or curatorial colleagues.

However, risk assessment does entail significant challenges. One is the time-consuming nature of the process, with one practitioner citing six meetings of one and a half hours each, alongside additional time spent collecting data as the time it took to assess the most heterogeneous collection at a museum (Alice Cannon, personal communication). Another issue is the subjectivity of the process. Subjectivity plays a role at all stages of a risk assessment, including the choice of hazards to focus on, the assessment of severity and in addressing the role of uncertainty (Morgan & Henrion, 1990). Cannon et al. (2017) found that different stakeholders prioritised different risks based on their area of expertise, their professional experience or simply their cultural background. The role of cultural context in the perception of risk has been explored in detail by Douglas and Wildavsky (1983).

It is worth stating in the context of this chapter that the concept of 'risk' within heritage has not received the level of scrutiny to date that it deserves (Rico, 2015, p. 150). It is a widely applied approach at both a national and an international level, via schemes such as the Historic England Heritage at Risk Register (Historic England, 2020) or the ICOMOS Heritage at Risk Programme (International Council on Monuments and Sites, 2020). It is beyond the scope of this chapter to provide an in-depth critique of this topic, however it is worth bearing in mind that the use of risk assessment is a methodological choice that comes with limitations such as an emphasis on negative consequences and threats (Aven, 2014, p. 20). Approaches to risk assessment and management depend on approaches and perceptions of risk. In the context of heritage, Rico (2015, p. 153) argues that heritage professionals tend to perceive risk in terms of 'undesirable change'. Such an approach fosters quantitative and scientific ways to measure risk and risk probability. However, a values-based approach to heritage and risk implies that heritage may be 'at risk' or 'not at risk' depending on which values are taken into consideration as being vulnerable and how these are affected by specific threats (Rico, 2015, p. 156). Moreover, as Rico rightly states, the ways in which the changing context, within which material decay occurs, is culturally valued should be a key informant of the 'at risk' condition (Rico, 2015, p. 156). Indeed, it could be argued that 'risk' and 'value' are interdependent concepts as the 'perception of risk and endangerment is a fundamental element in the production of heritage value' (DeSilvey & Harisson, 2020, p. 1). Consequently, 'loss aversion' is centrally positioned in heritage management practices (Holtorf, 2015).

Value

The importance of the concept of value in heritage conservation has been central to heritage management since at least the 1960s. The first scholar who explicitly discusses the idea of 'value' in relation to art and archaeological objects was the Austrian art historian, Alois Riegl (Zerner, 1976). In his paper 'The Modern Cult of Monuments, its Nature and Development' originally published in 1903, Riegl distinguishes a range of values including value as a monument, artistic value, value as commemoration or remembrance, historical value, art historical value, present day value, newness value, antique value, functional or use value (Zerner, 1976, p. 186). A similar attempt to distinguish between the 'historic' and 'aesthetic' values of 'historic monuments' occurs in the Venice Charter for the Conservation and Restoration of Monuments and Sites (International Council on Monuments and Sites, 1964). The charter states that restoration aims to 'preserve and reveal the aesthetic and historic value of the monument' based on 'original material and authentic documents' (Article 9). The Venice charter introduced a Western-centric, material-focused approach to originality and authenticity that proved not to be applicable in the global context (Goetcheus & Mitchell, 2014). As early as in 1979, the Australia ICOMOS Burra Charter for Places of Cultural Significance enhanced the focus on value by including not just aesthetic, historic and scientific but also social values as part of the overall cultural significance of a heritage place (Waterton et al., 2006). Having gone through several revisions since then, the latest charter makes special reference to spiritual values. It reflects the long-standing work, dialogue and collaboration between heritage managers and the Aboriginal communities who over the years have undertaken an active role in decision-making processes surrounding heritage management issues. Assessing heritage values and preparing a statement of significance is highlighted as the first step in heritage management processes at international (e.g. Burra Charter) or national (e.g. Historic England Conservation Principles (2008)) level. The importance of value as the intrinsic reason for conservation is also widely discussed in the academic literature. Avrami, Mason and de la Torre (2000) state that 'The ultimate aim of conservation is not to conserve material for its own sake but, rather, to maintain (and shape) the values embodied by the heritage' (p. 7).

In effect what the emphasis on values dictates is that it is essential to identify the groups or individuals who have an interest in a heritage site (stakeholders). Consequently, values are attributed by people, and as such, they are mainly 'extrinsic'. As de la Torre (2013) explains values are 'attributed, not intrinsic; mutable, not static; multiple and often incommensurable or in conflict' (p. 155), and further suggests that heritage conservation is largely dependent on 'cultural contexts, societal trends, political and economic forces' (p. 159). On this note, there is still a debate in literature regarding the extent to which values can be both intrinsic (that is that a heritage object or place has values even if they are not acknowledged by people) and extrinsic, or are always extrinsic. Indeed, as shown in our recent, European research on the different approaches and methods utilised by different disciplines to deconstruct and assess heritage values (Fouseki et al., 2020), there is a wide array of conceptualisations and methodological approaches among disciplines and sectors.

While the importance of taking heritage values into account in conservation decision-making has been discussed extensively, the assessment of values continues to be challenging. Value can encompass multiple attributes such as aesthetic, historic or economic value and is often approached through the concept of a value typology (Fredheim & Khalaf, 2016), especially by heritage institutions. On the contrary, academic research on heritage values tends to adopt qualitative methods instead of using existing typologies (Fouseki & Sakka, 2013; Fouseki et al., 2020). To conclude this section, heritage values are the starting point in heritage conservation and heritage management. They reflect the voices, aspirations, emotions and identities of a wide range of groups and individuals. As such, they are often conflicting and divergent (Fouseki & Alexopoulos, 2016) making heritage management and conservation a process of negotiation of the different values attached to heritage. Negotiation and 'stakeholder' participation are therefore critical in heritage management, and as such, appropriate time should be allocated for effectively assessing the cultural significance of a heritage site.

Risk assessment and value

Value is thus a vital aspect of sustainable heritage management and a concept that is multifaceted, subjective and variable over time. Value is also key to the process of risk assessment. It is interesting to note that the BS ISO standard 31000:2018 *Risk Management – Guidelines* (2018), while relevant to any type of risk assessment including fire safety or health hazards, puts a primary emphasis on value: 'The purpose of risk management is the creation and protection of value' (p. 2).

Many existing risk assessment methods include the concept of value explicitly. As has been discussed, both the CPRAM and ABC methods from Waller and Michalski use 'loss of value' as one of their three key factors when calculating the 'magnitude of risk', while others use concepts such as 'vulnerable value' (Brokerhof & Bülow, 2016). These methods sometimes include recommended approaches to the assessment of value. For example, Michalski and Pedersoli (2016) recommend the use of a 'value pie' through which subgroups of the heritage site under assessment can be assigned a relative value. This will draw on a value typology, and Michalski states that 'The primary guidance to quantify the relative value of items for risk management purposes will be the organization's mandate and the judgements of stakeholders' (p. 45). For the QuiskScan approach, Brokerhof and Bülow (2016) recommend using the criteria from the Cultural Heritage Agency of the Netherlands (2014) which includes both attributes such as condition or rarity and a value typology including historic value and artistic value.

The assessment of value is not used uncritically within the field. Michalski and Pedersoli (2016, p. 45) ask directly 'Isn't quantification of heritage value meaningless?' They advise that an assessment of value within a risk assessment is not an absolute measurement but an effort to quantify 'as best we can, the shared feeling that some things are more important than others' (p. 45). Cannon et al. (2017) explore the challenges of estimating value within risk assessment in more detail. She notes that participants in risk assessments in Museums Victoria, Australia (using Waller's CPRAM method), found it 'frustrating to try to reduce complex and nuanced feelings into numbers' (p. 3). She also states that participants felt that they were 'committing to a level of precision they could not defend' (p. 4). The paper is titled Towards the 'collectively subjective' and reflects a sense that the exercise is not an effort to eliminate subjectivity but to arrive at a shared understanding.

Some of the literature provides recommendations on how to approach the assessment of loss of value. The value of consulting a wide range of stakeholders has been mentioned by some authors (e.g. Michalski & Pedersoli Jr, 2016; Garside et al., 2018). The value of a moderated discussion and the need for pragmatism, e.g. by taking the mode or median of varying individual value assessments, are also mentioned (Brokerhof & Bülow, 2016). However, there is significant scope for a more in-depth study of this topic. Incorporating value into a practical heritage management exercise requires a balance between respecting the inherent complexity of value while recognising that heritage risk assessments need to be done within reasonable timescales. The goal of this chapter is to reflect further on this challenge by exploring the opinions of experienced practitioners within the field of heritage risk assessment and through the study of two case studies.

Methodology

Interviews

Semi-structured interviews were conducted with heritage professionals selected according to the relevance of their published works to the topic of risk assessment and value. The interviews were conducted either face-to-face or via Skype between 9 July and 8 August 2019. Transcripts of the interviews were studied, and the key conclusions drawn from the interviews are discussed below. The list of interviewees is shown in Table 3.1.

Chapel of the Nativity of the Virgin case study

This case study looked at the Chapel of the Nativity of the Virgin at Fort St Angelo, Malta, a site managed by Heritage Malta. The Chapel is currently closed to visitors, due to concerns about the impact of the external environment on the fragile interior. In November 2018, a group of students from the UCL Institute for Sustainable Heritage visited the site to assess the risks of opening the Chapel to visitors. A values assessment was conducted through a review of historical sources, visual appraisal of the Chapel and discussion with key stakeholders. The values were divided into tangible values, such as vernacular limestone construction or ceremonial artefacts, and intangible values such as religious and social values. Temperature and relative humidity data from inside and outside the Chapel were studied, and condition assessments of the interior, façade and collection items of the Chapel were performed. The visitor experience was evaluated through analysis of comments from the visitor book.

The risk assessment process was adapted from the British Library method (Garside et al., 2018). Relevant hazards were identified through the environmental and condition assessments, and the impact and likelihood of these on the tangible and intangible values of the Chapel were given a score from one to five and colour coded green, orange and

Table 3.1 List of interviewees

Interviewee	Organisation
Agnes Brokerhof	Cultural Heritage Agency of the Netherlands
Alice Cannon and Helen Privett (interviewed together)	Museums Victoria
Paul Garside	British Library
Cordelia Rogerson	British Library
Xavier Romão	International Scientific Committee on Risk Preparedness (ICORP) of ICOMOS
Robert Waller	Protect Heritage Corp.

red, indicating low, medium and high risks. Two scenarios were considered, the Chapel remaining closed and the Chapel being opened to visitors. Discussions among the group of four students assigned to work on the risk assessment were recorded. Three of this group were interviewed together after the risk assessment was complete using similar questions to those used for the other interviewees. The recordings of the discussions and the transcript of the interview were studied, and the conclusions drawn are discussed below.

Palace of Westminster case study

This study considers wall paintings in the Palace of Westminster, executed in the unusual waterglass technique. Paintings in the Peers' Corridor are glazed, whereas those in Commons' Corridor are unglazed (Figure 3.1).

A question was posed by the Curator's Office as to whether the paintings should all be glazed, all unglazed or left as they are. This study involved an interview with Caroline Babington, the Collections Care Manager in the Curator's Office of UK Parliament on 2 July 2019. This interview, in combination with a literature review, was used to understand the significance of the paintings. A values map was created which defined the key values associated with the paintings as historical, symbolic, evidential and aesthetic value. Past condition surveys were studied, and a new condition survey of four paintings was carried out. A condition survey of the built fabric surrounding the paintings was undertaken. Monitoring of temperature and relative humidity in the Peers' and Commons' Corridors and externally was performed, along with measurements of light intensity and visitor observations.

The risk assessment process studied the impact of hazards deemed relevant to the paintings from the condition and environmental analysis. It considered three relevant scenarios: (i) maintaining the status quo, (ii) glazing all of the paintings and (iii) removal of the glazing in the Peers' Corridor. The framework was adapted from the British Library's risk assessment method (Garside et al., 2018). The impact and likelihood of each of the hazards were scored from one to five which gave a risk score of low, medium, high or housekeeping with a colour coding system. The impact of the hazards on each of the values (historical, symbolic, evidential and aesthetic) was considered

Figure 3.1 Unglazed waterglass paintings by E. M. Ward in Commons' Corridor.
Source: Photo by by Heather Oakley.

separately and repeated for each scenario. The risk assessment was first conducted individually by one of the co-authors (Heather Oakley) and then repeated by the researcher along with the curator.

Analysis

Interviews

The interviews demonstrate a clear consensus from experienced practitioners in the field of risk assessment that the concept of value is critical to the practice of risk assessment. When asked *'Do you think the consideration of values in heritage risk assessment is important and why?'*, respondents described it as 'core', 'critical', 'fundamental' or 'important'. Alice Cannon states that: *'you don't know what you're trying to preserve if you haven't thought about that [value] first'*. The British Library risk assessment template does not explicitly incorporate value; however, interviews with British Library staff demonstrated that the concept is still implicitly part of the process. In this case, it is incorporated as part of 'risk appetite' which defines the extent to which the Library is willing to accept risks for different types of items. Risk appetites are lowest for objects considered as 'treasures', a well-established categorisation at the British Library, which involves a judgement based on the values associated with that object.

There is also consensus on the fact that incorporating the concept of value into a risk assessment is challenging. Several interviewees point to the fact that values change over time and will be perceived differently by different stakeholders. Xavier Romão states that *'we could not value quantitatively cultural heritage in 99.9% of the cases'*. Most participants point to the importance of working with as broad a range of stakeholders as possible. Agnes Brokerhof recommends the use of a group interview process with stakeholders, which could include experts but also wider stakeholder groups such as different ethnic or religious communities. Cordelia Rogerson, Paul Garside, Alice Cannon and Helen Privett concur on the importance of consulting a wide range of stakeholders, although from an institutional point of view, wider community perspectives are often mediated through curators. Robert Waller recommends *'as open as possible a brainstorming exercise'* which could include heritage managers, conservators, exhibition developers, educators, schools and artists. The importance of communication, networks and negotiation came through very strongly in the interviews. A crucial point that emerged from the interviews was that attempts to reduce or eliminate subjectivity are hopeless. With the concept of value intrinsically subjective, it is not possible to make assessments of value (and thus the risk assessment process) fully objective. What can be aimed for instead is to make it as comprehensive as possible by open discussions with as wide a range of stakeholders as possible. As Waller puts it: *'this is the idea of recognizing that our challenge isn't to reduce subjectivity but it's to make it as comprehensively collective and respectful of all perspectives'*.

There is an inherent tension in risk assessment. Heritage value is often deeply felt, with a strong emotional or cultural context. Undertaking a risk assessment requires the practitioner to incorporate these feelings into a systematic process. There is an inevitable discomfort involved in applying an analytic-deliberative approach to something as instinctive and emotional as value (Henderson & Waller, 2016). Waller describes risk assessment as the process of making *'aspirational goals operational'*. Ultimately, sustainable heritage management requires practical decisions to be made within budget and to a deadline. At some point, the nuance and complexity of heritage value need to be operationalised, and as Cannon points out, *'it may be an imperfect understanding or description* [of value], *but you at least have to attempt, even if you could only get part way, that's better than no way'*. A risk assessment may also mean that difficult decisions need to be made. Waller pointed to a specific case in which there was a direct clash between the ways in which an indigenous community perceived the value of a heritage collection, compared with the western curatorial point of view. He says that we ultimately need to decide: *'how should we as the combined society allocate our resources to protecting both of those interests?'* He believes that this requires decisions that are effectively political, involving the coming together of different, small groups of people, willing to make trade-offs and to compromise.

Accepting the reality of simplifying value is important within the risk assessment process; however, this needs to be done with care. Waller flags up the risk of *'technocratic overreach'* whereby a process such as risk assessment, which involves numbers, can be seen as more objective, or more absolute than it really is. Several of the interviewees raised the issue of using numbers in risk assessment. Cannon states in relation to using numerical scores, *'it's just kind of alien when meaning and value is something you express in words'*. Brokerhof points out that value should be considered as relative rather than absolute and that it's *'categories more than real numbers'*.

It is tempting to avoid numerical scores altogether and to consider the numbers within a risk assessment as categories, similar to the QuiskScan approach (Brokerhof & Bülow, 2016) of using 'high', 'medium' and 'low'. This would certainly ease practitioners' discomfort. However, Waller sounds a note of caution here. He warns that the use of categories can lead to judgements based on intuition, rather than more effortful evidence-based decisions (Henderson & Waller, 2016). He believes that the use of categories absolves the risk assessor of the responsibility of making difficult judgements and reduces the value of the available information.

Perhaps the most useful way to view the numerical scores of a risk assessment is to consider them as falling somewhere between the use of numbers to quantify properties such as length or mass and the use of categories such as 'high' and 'low'. A 'loss of value' score of 0.2 does not represent an objectively defined quantity, in the way that 0.2 metres or 0.2 kilograms does. However, there can be a value in using numbers rather than categories because of the way in which it makes the risk assessor think. As Rob Waller puts it, *'these are not accurate numbers, these are useful numbers'*.

In reality, a risk assessment appears deceptively simple. Often, there is a simple matrix to use, a small number of scores to assign and an easily understandable colour-coded outcome. But this conceals the complexity of the issues considered, the data that need to be collected and the time involved in building consensus. Rogerson advises practitioners to consider risk assessment as a funnel. She says: *'what you need to pop out the other end has to be an easily understood answer, but it still contains all of that complexity from the top of the funnel'*. Easily understood outcomes are critical if a risk assessment is going to be an effective heritage management tool. Several interviewees pointed out that a benefit of risk assessment is that it uses concepts and a language that will be familiar within finance, insurance and management, enabling decisions about conservation and value to be communicated to other, possibly powerful stakeholders. As Brokerhof puts it, *'you have to turn something passionate, ideological, meaningful into hard business talk'*. Given the apparently simple appearance of risk assessment, it is very important that its complexity is not underestimated. Three interviewees highlighted the importance of experience and training and the benefits of developing methods over time. Also highlighted was the fact that a risk assessment will always be done in context and will change from organisation to organisation but also within an organisation, depending on what type of decision it is being used for (Bao, 2019).

This chapter will now briefly explore two case studies in which a risk assessment process was applied to aid decision-making.

Chapel of the Nativity of the Virgin case study

During the follow-up interview with the team, many of the same issues arose that are mentioned above. In this case, due to the limitations of time involved in the student project, the consultation with stakeholders was limited to a small number of academics, Heritage Malta staff and a member of the parish. The team clearly identified this as a limitation of their process, highlighting the absence of a community voice as a weakness. In more practical terms, when considering the process followed, some previously raised points arose again. The importance of compromise and negotiation was identified, and a discomfort with using numerical scores was evident. The team said that they had chosen the British Library matrix because other systems had a more numeric scale. They also valued the way in which that method involved presenting the findings with colour coding. They also mentioned the risk that the person with the '*loudest voice*' would dominate and the value of an effective moderator. The importance of the group dynamics during a risk assessment has been raised by Cannon et al. (2017) who discusses the pros and cons of 'consensus' and 'dictator' models.

The approach of dividing the values of the site into 'tangible' and 'intangible' had mixed success. One of the team members felt that it would have been more effective to break these categories down further: '*I think if you really, really want to do the appropriate risk assessment, you have to have tables for all values – historical value, religious value, social value, economic value, and then average them*'. They pointed to the fact that a certain type of physical damage might affect different values differently. For example, damage to the interior surfaces of the Chapel walls was felt to have an impact on its religious value as it negatively affected the appearance of the Chapel, but not on its historic value, as the surfaces were recently painted. An approach in which the value is broken down into sub-categories is presented in the next case study.

A further issue with the tangible/intangible approach that was not raised by the team in their interview but was evident in the recordings of their discussions was that considering the intangible consequences of material change is conceptually challenging but critical to heritage risk assessment. The team were reasonably comfortable considering the impact of air pollution on limestone weathering, something that affected the architectural value of the Chapel. The challenge arose when the impact of such hazards on values such as religious value or social value was considered. While the physical impact of a hazard such as high relative humidity on the corrosion of metallic religious artefacts can be understood with the correct data, translating this into an impact on religious value is much more challenging. Is value lost incrementally as a surface is weathered? Or is a loss of value threshold reached when a specific feature is no longer identifiable? The team understandably struggled with translations like these. It is worth considering whether the impact part of current risk assessment needs to be broken down into physical impact and intangible impact or whether the relationship between the physical and intangible aspects of the heritage in question needs to be mapped and discussed prior to a risk assessment exercise.

Palace of Westminster case study

The risk assessment process at the Palace of Westminster enabled the differences between different scenarios to be more clearly understood. It could be seen that the decision between glazing the paintings or removing all the glazing boiled down to making a decision between the risks of damage to unglazed paintings from hazards such as fluctuating relative humidity and the risks to the appearance of the paintings due to the negative visual impact of the glazing. The visual impact of the glazing affected the aesthetic and symbolic values the most but was felt to have little impact on the evidential and historical values. The hazards that had high impacts on the values in the scenario in which the paintings were glazed were considered to be less permanent and therefore less damaging as the glazing could theoretically be removed at a later date; however, the compromise would be to accept the detrimental effect on the aesthetic and symbolic values. This shows one of the ways in which the risk assessment process facilitates decision-making. A decision still needs to be made, and some risk to the values accepted; however, the risk assessment process enables the consequences of a conservation decision to the values of a heritage artefact to be understood and mapped and provides the evidence base for the final decision.

The way in which the loss of value was broken down to look at the impact of different hazards on specific values, e.g. aesthetic and evidential, was felt to be helpful. While the risk matrix was large, the process in fact seemed faster than if the impact of a hazard on the value was assessed in a broader way. As the researcher (author Heather Oakley) says, *'there were more decisions, but they were simpler decisions'*. The question *'What is the potential impact of light on the aesthetic value of these paintings?'* is more refined and thus easier to answer than *'How could light lead to loss of value in these paintings?'* or *'What is the potential impact of light on these paintings?'* The outcomes of the risk assessment were similar when it was done by the researcher alone, compared to when it was done together with the curator. This consistency was felt to be partially due to the method of separating the values, as more precise questions were less likely to be interpreted differently by different assessors. However, given that the values are intrinsically interlinked, it was sometimes difficult to discern a difference between the impact of a hazard on one value or another, although there were some hazards that posed different levels of threat to each value. It proved valuable to refer frequently to the values map and easiest to consider one hazard at a time, across the effect on values. The colours and conditional formatting of the document also assisted with the comprehensibility and speed of the process (Oakley, 2019).

Other recommendations from this study are similar to those which have already been discussed. A broader range of stakeholders would have been helpful; suggestions include the conservation architect, staff from visitor services or facilities management. The importance of having good quality data was also highlighted. This study involved the collection of environmental and condition data and historical research as part of the research, and having this data available was critical for the risk assessment. The benefits of a risk assessment approach for prioritising and for interacting with other departments within an organisation were also pointed out, particularly given that risk assessment is an approach used in many fields.

Conclusions

Ultimately, successful sustainable heritage management involves balancing all aspects of sustainability, environmental, economic, physical, social and cultural. Risk assessment is a tool that combines all of these aspects and is thus a useful way to study the tensions involved in balancing them. From this study, it seems that a level of discomfort needs to be accepted if a practical heritage management exercise is going to engage with a nuanced concept such as value. By engaging with as wide a range of stakeholders as possible and by maintaining an awareness of the risk of 'technocratic overreach', risk assessment can engage with the concept of value in a way that recognises its complexity. However, at some point, this will need to be simplified in order to work through the risk assessment process. The authors feel that this simplification is worth it, for the benefits that risk assessment can bring in terms of transparent decision-making and communication with stakeholders. This is what is required if we are to operationalise the theoretical.

What is clear is that risk assessment is not as simple as it looks. A large amount of data, about value but also about condition, environment, organisational strategy, etc., is needed before the exercise can be done. There is a limit to how quickly a risk assessment can be done, if it is to be at all meaningful or even not misleading. In addition, we recommend that a mapping takes place prior to risk assessment that links physical change to loss of value. Is value lost incrementally, e.g. as a surface erodes, or is it lost once a threshold is reached, e.g. once a key feature is unrecognisable? We also warn against the misinterpretation of numerical scoring systems, as scores for concepts like 'loss of value' are not objectively defined quantities. There is a risk that the method's perceived objectivity is overestimated.

Some interviewees directly flagged up areas for future research in this area. Robert Waller proposed more study of the nature of decision-making within risk assessment and of the nature of time horizons for risk assessment. We advise further research into the role of non-heritage experts in risk assessment practice. With such clear consensus on the importance of wide stakeholder groups, the difference between direct engagement with non-heritage experts versus mediated engagement through curators warrants further study.

References

Ashley-Smith, J. (1999). *Risk assessment for object conservation*. London: Butterworth Heinemann.

Australia International Council on Monuments and Sites. (1979). *The Australia ICOMOS guidelines for the conservation of places of cultural significance ("Burra Charter")*. (pp. 1–2). Burra: Australia ICOMOS.

Aven, T. (2014). *Risk, surprises and black swans: Fundamental ideas and concepts in risk assessment and risk management*. Abingdon: Routledge.

Avrami, E., Mason, R., & de la Torre, M. (2000). *Values and heritage conservation*. Los Angeles, CA: The Getty Conservation Institute.

Bao, J. (2019). The application of heritage significance in risk assessment for cultural heritage: Process, challenges, and suggestions. (Unpublished master's thesis). UCL, London.

Bratasz, L., White, T., Butts, S., Sease, C., Utrup, C., Boardman, R., & Simon, S. (2018). Toward sustainable collections management in the Yale Peabody museum: Risk assessment, climate management, and energy efficiency. *Bulletin of the Peabody Museum of Natural History*, 59(2), 249–268. https://doi.org/10.3374/014.059.0206.

Brokerhof, A. W., & Bülow, A. E. (2016). The QuiskScan - a quick risk scan to identify value and hazards in a collection. *Journal of the Institute of Conservation*, 39(1), 18–28. https://doi.org/10.1080/19455224.2016.1152280.

British Standards Institute. (2018). *BS ISO 31000 : 2018 risk management — guidelines*. London: British Standards Institute.

Canadian Conservation Institute. (2017). *Agents of deterioration*. Retrieved from: https://www.canada.ca/en/conservation-institute/services/agents-deterioration.html. (Accessed on 21 April 2020).

Cannon, A., Privett, H., McCubbin, M., & Waller, R. (2017). Towards the "collectively subjective": The challenges of estimating loss in value during collection risk assessments. In J. Bridgland (Ed.), *ICOM-CC 18th Triennial Conference Preprints, Copenhagen, 4–8 September 2017*, Paris: International Council of Museums. Retrieved from: https://www.icom-cc-publications-online.org/1661/Towards-the-collectively-subjective--The-challenges-of-estimating-loss-in-value-during-collection-risk-assessments (Accessed on 21 April 2020).

Cultural Heritage Agency of the Netherlands. (2014). Steps in the valuation process. In A. Versloot (Ed.), *Assessing museum collections: Collection valuation in six steps* (pp. 11–53). Amersfoort: Cultural Heritage Agency of the Netherlands.

de la Torre, M. (2013). Values and heritage conservation. *Heritage & Society, 6*(2), 155–166. https://doi.org/10.1179/2159032x13z.00000000011.

DeSilvey, C., & Harrison, R. (2020). Anticipating loss: Rethinking endangerment in heritage futures. *International Journal of Heritage Studies, 26*(1), 1–7. https://doi.org/10.1080/13527258.2019.1644530.

Douglas, M., & Wildavsky, A. (1983). Conclusion : Risk is a collective construct. In *Risk and culture: An essay on the selection of technological and environmental dangers* (pp. 186–198). Berkeley: University of California Press.

Fouseki, K., & Alexopoulos, G. (2016). Gender exclusion and local values versus universal cultural heritage significance: The Avaton debate on the monastic community of Mount Athos. *European Journal of Post-Classical Archaeologies, 6*, 235–254.

Fouseki, K., & Sakka, N. (2013). Valuing an ancient palaestra in the centre of Athens: The public, the experts, and Aristotle. *Conservation and Management of Archaeological Sites, 15*(1), 30–44.

Fouseki, K., Taylor, J., Diaz-Andreu, M., van dre Linde, S., & Pereira-Roders, A. (2020). Locating heritage value. In S. Saville & G. Hoskins (Eds.), *Locating value: Theory, application and critique* (pp. 37–50), London: Routledge.

Fredheim, L. H., & Khalaf, M. (2016). The significance of values: Heritage value typologies re-examined. *International Journal of Heritage Studies, 22*(6), 466–481. https://doi.org/10.1080/13527258.2016.1171247.

Garside, P., Bradford, K., & Hamlyn, S. (2018). The use of risk management to support preventive conservation. *Studies in Conservation, 63*(1), 94–100. https://doi.org/10.1080/00393630.2018.1479935.

Goetcheus, C., & Mitchell, N. (2014). The Venice Charter and cultural landscapes: Evolution of heritage concepts and conservation over time. *Change Over Time, 4*(2), 338–357.

Henderson, J., & Waller, R. (2016). Effective preservation decision strategies. *Studies in Conservation, 61*(6), 308–323. https://doi.org/10.1179/2047058415Y.0000000019.

Historic England. (2008). *Conservation principles: Policies and guidance for the sustainable management of the historic environment*. London: Historic England.

Historic England. (2020). *Search the heritage at risk register*. Retrieved from: https://historicengland.org.uk/advice/heritage-at-risk/search-register/ (Accessed on 21 April 2021).

Holtorf, C. (2015). Averting loss aversion in cultural heritage. *International Journal of Heritage Studies, 21*(4), 405–421. https://doi.org/10.1080/13527258.2014.938766.

International Council on Monuments and Sites. (1964). International Charter for the conservation and restoration of monuments and sites (The Venice Charter 1964). *IInd international congress of architects and technicians of historic monuments* (pp. 1–4). Venice: ICOMOS.

International Council on Monuments and Sites. (2020). *Heritage@Risk Reports*. Retrieved from: https://www.icomos.org/en/what-we-do/risk-preparedness/heritage-at-risk-reports (Accessed on 21 April 2021).

Michalski, S. (2004). *Risk assessment of collections : Simple scales that add up*. Canada: Canadian Conservation Institute.

Michalski, S., & Pedersoli Jr, J. L. (2016). *The ABC method: A risk management approach to the preservation of cultural heritage*. Canadian Conservation Institute. Retrieved from: http://canada.pch.gc.ca/DAMAssetPub/DAM-PCH2-Museology-PreservConserv/STAGING/texte-text/risk_Manual_2016_1486742306045_eng.pdf. (Accessed on 21 April 2021).

Morgan, M. G., & Henrion, M. (1990). *Uncertainty: A guide to dealing with uncertainty in quantitative risk and policy analysis*. Cambridge: Cambridge University Press.

Oakley, H. (2019). *To glaze or not to glaze: The use of values-based risk assessment to inform decision-making in the conservation of wall paintings*. (Unpublished master's thesis). UCL, London.

Paolini, A., Vafadari, A., Cesaro, G., Santana Quintero, M., Van Balen, K., Vileikis, O., & Fakhoury, L. (2012). *Risk management at heritage sites: A case study of the Petra world heritage site*. Paris: UNESCO and KU Leuven.

Rico, T. (2015). Heritage at risk: The authority and autonomy of a dominant preservation framework. In K. Lafrenz Samuels & T. Rico (Eds.), *Heritage keywords: Rhetoric and redescription in cultural heritage* (pp. 147–162). Colorado: University Press of Colorado. https://doi.org/10.5876/9781607323846.c009.

Rogerson, C., & Garside, P. (2017). Increasing the profile and influence of conservation—An unexpected benefit of risk assessments. *Journal of the Institute of Conservation*, 40(1), 34–48. https://doi.org/10.1080/19455224.2016.1214848.

Romão, X., Paupério, E., & Pereira, N. (2016). A framework for the simplified risk analysis of cultural heritage assets, *Journal of Cultural Heritage*, 20, 696–708. https://doi.org/10.1016/j.culher.2016.05.007.

Waller, R. (1994). Conservation risk assessment: A strategy for managing resources for preventive conservation, *Studies in Conservation*, 39(2), 12–16. https://doi.org/10.1179/sic.1994.39.Supplement-2.12.

Waller, R. (2003). *Cultural property risk analysis model: Development and application to preventive conservation at the Canadian museum of nature*. Göteborg, Sweden: Acta Universitatis Gothoburgensis.

Waller, R. (2019). Collection risk assessment. In L. Elkin & C. A. Norris (Eds.), *Preventive conservation: Collection storage* (pp. 59–90). New York: Society for the Preservation of Natural History; American Institute for Conservation of Historic and Artistic Works; Smithsonian Institution; The George Washington University Museum Studies Program.

Waller, R., & Lee, K. (2018). Impact of facility renewal deferment on risk to Royal British Columbia Museum, Canada, collections. *Studies in Conservation*. 63(1), 288–292. https://doi.org/10.1080/00393630.2018.1504518.

Waterton, E., Smith, L., & Campbell, G. (2006). The utility of discourse analysis to heritage studies: The Burra Charter and social inclusion. *International Journal of Heritage Studies*, 12(4), 339–355.

Zerner, H. (1976). Alois Riegl: Art, value, and historicism. *Daedalus*, 105(1), 177–188.

4 Heritage values and heritage management frameworks in Nigeria

Joy Acheyini Edeoja, Kalliopi Fouseki and Alejandra Albuerne

Introduction

The World Heritage Convention (WHC) in 1972 broadened the focus of what heritage is. Prior to the Convention, heritage largely connoted tangible objects such as monuments, collections and sites. The Convention introduced other forms of heritage, including festivals, dresses and food (UNESCO, 1972). The management of heritage has become an interesting field globally because of what heritage contributes to the social life of people. These contributions include, among others, cultural diversity, tourism, community well-being, a sense of place and sustainable development (Mason, 2008). Participatory heritage management has been touted by many as the answer to the negligence and abandonment of heritage, but is it just about engagement? Indeed, there is rich literature on community engagement and heritage (Crooke, 2010; Waterton & Smith, 2010; Watson & Waterton, 2010), but very few resources detail the democratic involvement and representation of communities when it comes to decision-making and decision-taking or how to be group conscious in engaging the diverse groups found within the community (Fouseki, 2010; Smith & Fouseki, 2011; Dragouni & Fouseki, 2017). Understanding the multiple meanings attached to heritage by various groups is key to an all-inclusive heritage management approach. All-inclusivity can only be accomplished through collaborative efforts that will be beneficial to both people and heritage, thereby contributing to sustainable development; getting involved increases attachment and a sense of ownership (Court & Wijesuriya, 2015; Jaafar et al., 2015; Gallou & Fouseki, 2018).

Though without a universally agreed definition, a commonly employed definition of sustainable development (SD) is one offered by the Brundtland Report which defines it as the ability to meet the needs of current generations without compromising the ability of subsequent generations to meet their needs (Brundtland, 1987; Landorf, 2009). Defined as a valuable inheritance from the past, heritage is expected to contribute socially, economically and politically to the lives of community members (Lowenthal, 2005). However, key issues that affect community contributions relate to the meanings and values attributed to heritage (Mason, 2002; Smith, 2008; McClelland et al., 2013). Value is a social and cultural construct that pertains to the perceptions of people. Attaining sustainability requires that the diverse views held by community members should be considered. The concept of a value-based heritage management approach is therefore people-centred (Mason, 2002; Gibson & Pendlebury, 2009; Court & Wijesuriya, 2015; Dragouni et al., 2018). Value, which is defined as the degree of significance, is comparably affected by government legislations and politics which created a top-down approach to heritage management in the past (Boyd, 1996; Court & Wijesuriya, 2015).

DOI: 10.4324/9781003038955-6

Heritage management is often defined as a process in which experts evaluate the meaning and value of heritage through the development of policies aimed at its sustainability (Mason, 2002; Smith et al., 2003). Given that values are diverse, conflict and dissonance are inevitable. Therefore, heritage management also involves the management of conflicts related to the values and significance of heritage (Smith et al., 2003; Chirikure et al., 2010). Smith et al. (2003) in their work among the Waanyi women (Australian aborigines) have, for instance, demonstrated how empowering women to take control over a heritage project was more meaningful than the project itself. The process of empowering and relinquishing control is certainly a complex and dynamic process. It is though fundamental for 'generating a sense of ownership and belonging as well as of constructing identities' (Fouseki, 2010, p. 181).

In Nigeria, the geographical focus of this chapter, there is a growing number of heritage sites that are no longer valued by indigenous communities. We assume that this may be due to the introduction of legislations which empowered governments to make decisions over heritage, thereby alienating communities from the management of heritage and, subsequently, leading to abandonment and neglect. Indeed, when the relationship between people and heritage weakens, there is a disconnection which can lead to conflicting arguments about heritage values among community members as well as between heritage experts and the communities (Smith & Waterton, 2012). A sustainable management framework, therefore, goes beyond ownership and custodianship. It should also include the ability to harness the conflicting views and values held by various stakeholders. Indeed, it could be stated that sustainable heritage management involves, among other abilities, the ability to enable dialogue and negotiation among those groups who have an interest or lack of interest in a heritage place.

As Liwieratos points out (2017, p. 16):

> The management task is basically to answer such questions as: what do we perceive as cultural heritage, with what criteria do we define cultural heritage, how do we want to conserve it … what is the role of the public.

This statement has been echoed by Harrison (2010) and other scholars who have pointed out that often heritage under the management of the national government is controlled by some few experts who become the determining factor on what heritage value should be, thereby alienating indigenous communities. To contribute to sustainable development, every stakeholder's (government's, expert's, and community members') views and inputs are relevant (Smith, 2008; de la Torre, 2013; Fouseki et al., 2020). Through the case of Nigeria, we will demonstrate how indigenous communities in Nigeria are alienated from heritage management practices and contrast this current alienation with practices in the past when indigenous communities were the main custodians of their heritage.

Indigenous communities in Nigeria

The United Nations (UN) special rapporteur of the Sub-commission on the Prevention of Discrimination and Protection of Minorities defined the generic term indigenous communities as follows:

> Those which, having a historical continuity with pre-invasion and pre-colonial societies that developed on their territories, consider themselves distinct from other

sectors of the societies now prevailing on those territories, or parts of them. They form at present non-dominant sectors of society and are determined to preserve, develop and transmit to future generations their ancestral territories, and their ethnic identity, as the basis of their continued existence as peoples, in accordance with their own cultural patterns, social institutions and legal system.
(as cited by José R. Martínez Cobo in Food and Agriculture Organization of the United Nations (FAO), 2008, p. 355)

Similarly, an indigenous community in this chapter will be considered as the community with varied ethnic identities and localities; whose ancestors were the initial inhabitants or natives of the place (Chirikure & Pwiti, 2008). It is interesting to note that most African and non-African post-colonial nations fit the definition of José R. Martínez Cobo in FAO (2008). These nationalities, particularly in Africa, comprise communities with numerous dialects and varied ethnicities which collectively represent a repository of indigenous knowledge, norms, values, crafts and practices passed from generation to generation.

It is generally accepted that there are various views and values among multi-ethnic communities. These views are essentially aimed at connecting people with heritage (Silberman, 2013). However, there are a couple of considerations: Firstly, do indigenous communities have the ability to distinguish between what is currently existing that should be protected or what to discard? Secondly, what forms of engagement exist between heritage, local communities and heritage professionals? (Harrison, 2010).

In post-colonial communities, values attributed to heritage are shifting from what they were during the pre-colonial era. In a bid to meet the needs of present generations, there is an ongoing transformation that is giving rise to a heritage hybridization or a disconnection and/or abandonment. For instance, in Nigeria, the spiritual value attributed to heritage places, such as markets, during the pre-colonial era diminished in post-colonial times due to the advent of missionaries (Chukwuemeka et al., 2019). Before the advent of missionaries, some markets were places connected with indigenous deities where a day of the week (a day of no visitation or transaction) was usually devoted to such markets. Most of these marketplaces today have changed from having a dedicated sacred day within the week into a daily marketplace where transactions take place at all times. The advent of Christianity and formal education by missionaries introduced a shift in the relationship with indigenous deities; relationships were seen as paganistic (Baharvand, 2016). The market example is a testament to heritage evolving in order to meet the needs of current generations.

Contemporary challenges and implications in Nigeria (A case study)

Values, management and evolution

It is essential to assess and periodically review the values placed by community groups on heritage instead of imposing a restrictive copy-paste management system. As Taylor (2004, p. 419) opines, heritage management should be informed by local conditions:

International organizations such as United Nations Educational, Scientific and Cultural Organization (UNESCO), International Council for Monuments and Sites (ICOMOS), International Council of Museums (ICOM), and International Centre for the Conservation and Restoration of Monuments (ICCROM) are powerful ones. They lay down international standards for professional practice - world best practice - in the cultural heritage field as well as influencing thinking in those fields in less direct ways. But they also stand accused of imposing a common stamp on culture across the world.

Having a global outlook on management will be a good foundation. In addition, a local perspective that considers the peculiarity of each site and region in the management plan will achieve a better people-centred approach (Taylor, 2004). For example, many indigenous communities in Africa live below poverty line. Over 70% of the world's poorest people reside in Africa with an estimated 400 million living below the poverty line (Ezekwesili, 2020). Inevitably, the management and appropriation of heritage acquire economic values. What previously was traditional has become valorized by communities, not just for its social and cultural values, but also for its economic benefits. Attention is therefore shifting from the management and conservation of heritage to how heritage can be produced and used profitably (Probst, 2013). Though economic values are necessary for the resilience and well-being of communities, they should though align with other values. A holistic heritage management approach will benefit both heritage and communities. Providing a holistic approach to heritage management requires addressing the following questions:

i   What drives values attributed to heritage by certain community?
ii Who decides what is or is not heritage?
iii How involved is the indigenous community in the process of heritagization?

In countries, such as Nigeria, the complexity of what should be defined as heritage is more pronounced due to the presence of what Quist (2001, p. 297) called the 'triple cultural heritage' which is the heritage resulting from the three dominant religions, namely tradition (African), Christianity and Islam. Value perceptions among community members are, therefore, not a constant feature but may change over time as a result of proselytization or migration from one belief to the other. This has, thus, affected individual heritage management in some part while producing an evolved heritage that meets the needs of current generation elsewhere. Confirming the fact that values are expressions of emotions or beliefs and are influenced by circumstances. Evolution can result from changes within the society, such as those in the demography or in the function of a place. In many instances, the factors that influence the values of a place can easily be identified, while at other times, it results from more subtle factors and can only be detected over longer periods of time (de la Torre, 2013).

The uniqueness of heritage and the justification for its management are achieved through the value community members attributed to it (Cassar & Fouseki, 2014). A sense of ownership is critical to its existence. For instance, the Osun-Osogbo grove (one of the two World Heritage Sites (WHS) in Nigeria), a sacred ground, is regarded by its devotees as heritage and they attribute value to this landmark, but to those who hold different viewpoints or religion, it is paganistic and should not be in existence. What then happens when a religious ground is dispossessed of its sacredness? Evolution to meet the needs of the current generation becomes crucial in the absence of stakeholders who view such heritage with significance. This issue is of particular importance, especially in cases where heritage is dispossessed of whatever it is that made it a heritage. Indigenous heritage, such as the Osun-Osogbo grove, for instance, may transform into a global or national 'piece of the past' which is disconnected from the present.

There is an ongoing evolution of indigenous heritage and involving community members in its management will increase the level of awareness and appreciation. This may, consequently, reduce abandonment or destruction of icons and monuments whenever there are religious or political uprisings (iconoclasm), a common occurrence within communities in Nigeria. Indeed, many icons and monuments have been lost among indigenous communities as a result of religious or ethnic uprisings (Jaafar et al., 2015; Sabri & Olagoke, 2019). The younger generation, whose voices are often absent when it comes to decision-making and management of heritage, are also sadly the ones usually employed by the older generation to carry out such acts of destructions. Giving community members the opportunity to be part of heritage management will increase the desire to protect and also reduce the replacement of heritage buildings with new development (Said et al., 2013).

The framework for managing indigenous heritage in Nigeria

This section focuses on the form of heritage management framework developed by Nigerian indigenous communities. Nigeria, a former British colony, gained its independence in 1960 (Adedimeji, 2009). Prior to the colonial era, works of arts and valuables linked to indigenous communities were archived in the palaces of Kings and shrines (Shyllon, 2005). This type of heritage included artefacts such as carved bronzes and sculptures (that were often totemic representations of deities) and terracotta. These ancient archives could be regarded as the 'museums' of each community (Shyllon, 2005; Adedimeji, 2009). Therefore, the assumption that heritage management in Africa came with colonialism is not accurate, since the colonial masters encountered many heritage sites and artefacts indicating an existence of a form of heritage management (Ndoro, 2003). Currently, artefacts, such as the Bronzes, believed to have derived from the 9th-century Nri (Eastern) and Benin (Southern) kingdoms of Nigeria are all examples of the evidence of a form of heritage management prior to colonization (Arowolo, 2010). Firstly, the majority of these artefacts were totemic representations of community's deities and rulers. Secondly, there were some forms of sacredness attached to the objects because most of these communities believed in these deities and rulers. Some of these totemic objects were regarded as progenitors and protected through taboos and restrictions. Taboos such as the one that forbids women from interacting directly with the deities or the one that restricts community members from going to their farms, markets or stream on sacred days are but a few examples (Baharvand, 2016). The traditional African is guided by whatever deity they worship through the shrine, a place dedicated to ancestral spirits, which is protected by custodians who guide the people through these taboos and restrictions (Ndoro, 2003).

Prior to government involvement, the two World Heritage Sites, Sukur Cultural Landscape (SCL) and the Osun-Osogbo grove – both belief-related heritage sites – were managed traditionally through taboos and restrictions such as those that prohibit hunting or fishing within the grove (UNESCO, 2016). During the pre-colonial era, indigenous communities interacted with these places through the custodianship of the sites; thus, there existed an interrelationship and a commitment.

Colonialism and government involvement led to the separation of the people from the heritage. Colonial government, alongside missionaries and Islamic jihad, were factors that may have contributed to distancing indigenous communities from the custodianship of indigenous heritage. With the arrival of missionaries, who came with colonialism, some heritage objects, places and ceremonies were termed paganistic. While the proselytization of community members was ongoing, some of the condemned artefacts were being looted, ending up in Europe and other parts of the world, while others were burnt (Eze-Uzomaka, 2014; Adewumi, 2015). The looting of Nigeria's artefacts in the 19th century resulted in the colonial government drafting relevant legislations; one of such is the Antiquities Act in 1953 aimed at preventing antiquity theft (Shyllon, 1996; Adewumi, 2015).

Nigeria, like most sub-Saharan Africa, sometimes has been critiqued for being out of touch with local communities in regard to heritage management. The management of heritage in colonial Nigeria was the responsibility of colonialists. Upon the arrival of independence, post-colonial heritage experts simply adopted what was handed down from colonialism (Munjeri, 2003; Ndoro, 2003; Chirikure et al., 2010).

Currently, heritage management is domiciled in the Ministry of Culture as the National Commission for Museums and Monuments (NCMM). The NCMM has the oversight of listed and declared heritage nationally and has branches in all the states of the federation (Onyima, 2016; NCMM, 2020). For instance, the management of the two World Heritage Sites is carried out through the so-called Traditional-Modern System (TMS), a combined effort of the local, state, federal government, local people, non-governmental organizations, and heritage professionals (UNESCO, 2016). This is relevant because the recognition and return of power that goes with traditional custodianship to host communities in the African context have been advocated as a means of effective heritage management (Chirikure et al., 2010). The organizational structure is such that the government manages the heritage by providing technical and legal directions, while communities, as joint managers, provide traditional know-how and guidance through taboos and restrictions. However, reports of failure or inability of some stakeholders to involve community members in the decision-making process can be observed in current management plans (UNESCO, 2016). The aim of the TMS among indigenous communities should be to direct and connect people to their heritage, thereby establishing a strong community with social well-being of the people at its heart.

Participation as a democratic process includes a proper representation of community members in the management team. A situation where the community is under-represented should be termed undemocratic. Nevertheless, it should be acknowledged that a participatory approach to heritage management is imbued with several challenges in Nigeria, as described below in detail.

Challenges

Challenge 1: Diversity and belief

Located in Western Africa, Nigeria is made up of an estimated 500 indigenous dialects and 250 ethnicities, along with a tripartite belief demography which is majorly traditional, Christianity and Islamic (Quist, 2001; Onyima, 2016). These beliefs, together with the dynamism due to ethnicity, have produced heritage with a more fluid-like and dynamic life cycle that has set it apart from the feudal and static forms found in Europe (Breen, 2014). This dynamism is evident in the diversity found across the country.

Diversity connotes varied heritage, but does heritage exist, if it is not accepted? Diversity may become an obstacle to participation, but there are situations where the barrier has been overcome for the common good of heritage and communities. An indicative example of this is the Osun-Osogbo sacred grove, a living monument whose survival depends on the interaction between devotees and the grove. Prior to the late 1970s, it used to be an exclusive heritage for the devotees (worshippers) of the indigenous community. However, it has since gone through several progressions in order to meet the needs of the present generation. According to Probst (2009), the initiation of an Austrian artist named Susanne Wenger who settled in Osogbo in 1958 and became a devoted worshipper of the Osun deity marked the beginning of series of transformation or hybridization of the grove. He stated that:

> In the 1970s and '80s the reservation tied in with the debate on the foreignness of the grove's images on the one hand and the debate on the foreignness of Wenger as the artist who created and prompted these works on the other. Both debates went together, doubling the feature of difference and thus creating a double hybridity, as it were. And as if this is not enough, there was—and still is—also the hybridity of the site as such, simultaneously an active Yoruba ritual site and a Western sculpture garden where Nigerian school children and American and European tourists alike get guided tours, during which they learn about traditional Yoruba ritual and religion.
>
> (Probst, 2009, p. 25)

The grove, as a totemic representation of deities and a heritage from the past, will only exist if there are devotees. Hybridity, transformation and/or evolution are, therefore, based on the ability of heritage to meet the needs of different generations.

The globe is increasingly becoming an urban centre, thereby rendering it inclusive as new traditions are transforming heritage to one that meets the needs of present generations. The ongoing proselytization among community members and globalization are few of the factors that had transformed some sacred heritage sites, practices into hybrids or the abandonment of some old heritage and the creation of new forms. This also proves that cultural traditions are not static but fluid adjusting to times and situations, consistent with the clichéd phrase 'culture is dynamic' (Ndoro, 2003, p. 84; Fouseki & Sakka, 2013).

Challenge 2: Globalization and Western ideas

Participatory heritage management is a collaborative effort of every group within the community. At the same time, there are specific forms of cultural heritage or knowledge prioritized by certain communities, thus excluding other forms of heritage and knowledge. Among indigenous communities in Nigeria, maintaining a balance between Western ideas and what is obtained traditionally is a major threat to the management of heritage currently. Western ideas are competing with the cultural way of life of communities, thereby replacing, creating hybrids, or eroding other forms of indigenous knowledge (Arowolo, 2010). For example, the homogenization and standardization of official language such as the English language (though relevant for interaction) have threatened some indigenous languages. What is the value attributed to indigenous languages in Nigeria by community members? This may have unintended social consequences. Inability to communicate in one's language means that you are cut off from some people within the family and communities (Bale, 2010; Nwegbu et al., 2011). For instance, a situation where grandparents can only communicate in the indigenous dialect and grandchildren can communicate only in English language puts pressure on family relationships. Such a situation robs a community of the expected benefits of heritage and pressurizes or may ultimately end relationships. From a previous study, the heritage language distress started in Nigeria in the 1950s when the so-called elites, armed with some learning in English language as a result of colonialism, established themselves as leaders of the nation but never saw anything good or constructive in the traditions and cultures of the people (Shyllon, 1996). Consequently, intangible cultural heritage, such as languages, dressings, ceremonies and places, was abandoned and replaced with the culture of the colonialist.

According to research by the Linguistics Association of Nigeria (LAN), about 400 indigenous languages in Nigeria are under threat, with 152 languages on the way to extinction (Ohiri-Aniche, 2014). Accordingly, in decades to come, it is expected that some intangible heritage such as languages and festivals may either have evolved beyond their present form, become a hybrid or even extinct. Although this in a way is part of the dynamism of heritage, evolution towards extinction would be a cultural deprivation for future generations.

Challenge 3: Lack of documentation

A common phenomenon among indigenous communities in Nigeria is the lack of documentation of cultural heritage which is only managed traditionally through oral transmission. Even now, after the establishment of Western education, proper documentation of indigenous cultural heritage in many African countries is still lagging (Shyllon, 2005; Chanda, 2018). Documenting oral traditions, cultural skills and craftsmanship will ensure a smooth transmission from generation to generation and answer some questions such as: Why are we here? Why do we look different from others? Why are we doing what we are doing? Those vested with this knowledge will not be around forever. This paucity of documentation is best understood through Chanda (2018, p. 35) who posits that:

> This is because modern learning, and scholars in general, have frequently underplayed, trivialized or simply ignored indigenous knowledge, labelling it as being primitive and lacking scientific proof. Many elderly men and women die, taking with them the rich knowledge acquired through many years of traditional learning and practice. Therefore, it is usually said that the demise or death of one old African man is like setting ablaze a big university library.

In the Sukur Cultural Landscape (WHS), a weakness mentioned by the management plan is the low level of documentation due to limited funds and expertise (UNESCO, 2016). Documentation is necessary in the transmission of heritage to the next generation. According to Fouseki (2010, p. 181), as communities transform in nearly 'chameleon-like fashion', documentation will preserve the different phases of a heritage through time.

Challenge 4: Conflicting values between experts and communities

Sabri and Olagoke (2019), in their study on religious heritage buildings in Nigeria, argued that heritage management gets entangled in politics and social struggles when stakeholders have conflicting views. Indigenous communities often attach less importance to methodical value but attribute greater value to experience, belief or the symbolic nature of indigenous heritage (Boyd, 1996). This can be exemplified in the Robben Island experience. While the residents of Robben Island regard the quarry site where Nelson Mandela worked and his old prison quarters as the real place of value, to the site staff, however, Robben Island is the cell where Nelson Mandela lived. The prisoners attributed little value to the prison quarters but called the limestone quarry, where they worked, 'the University' (Munjeri, 2003, p. 75). This example demonstrates that heritage is ultimately all about values.

Indigenous communities are the creators of indigenous heritage; therefore, heritage experts' endeavour to understand what heritage means to indigenous communities should be the first step in heritage management. The inability of some heritage experts to involve community members in decision-making processes can possibly cut off communities from interacting with heritage (Arowolo, 2010; Chirikure et al., 2010). Subsequently, causing heritage to be viewed as government proprietorship and community members, remote observers instead of custodians. Heritage management should therefore take a value-based approach that is democratic in participation as well as representative in decision-making (Mason, 2002; Taylor, 2004; McClelland et al., 2013). This is necessary because:

> Many of heritage's data are social factors where personal and collective memory inhere and cannot be defined scientifically. What has occurred, where and when in history, can be studied objectively (there may be differences of opinion on verification of facts or whether something occurred), but it is the who was involved and the why they did things the way they did that fascinate people. Much of this can be, and is, open to interpretation.
>
> (Taylor, 2004, p. 421)

Article 29 of the United Nations Declaration on the Rights of Indigenous Peoples (UNDRIP) (UNESCO, 2014) states that indigenous peoples have the:

> Right to the conservation and protection of the environment and the productive capacity of their lands or territories and resources. This includes the right to maintain and strengthen their distinctive spiritual relationship with their traditionally owned or otherwise occupied and used lands, territories, waters and coastal seas and other resources.

Management of heritage is first and foremost a right, then the duty of the custodians as well as government agencies. The role of government agencies is to encourage people–heritage interaction through policy provision.

Challenge 5: Economic benefits

The economic benefit is one additional dimension of the multi-faceted values of heritage. The Osun-Osogbo sacred grove (WHS), as an example, depicts a heritage site with an outlined guideline in accordance with UNESCO Outstanding Universal Value (OUV) for its management. An annual festival organized by the community has drawn global focus to this heritage where, in recent times, attendance has increased with attendees from within Nigeria and the diaspora (Adedeji & Fadamiro, 2018). Although this has proved beneficial in multiple ways, it has not been without its constraints. The fragility of the grove, resources and management were a few considerations mentioned in the ICOMOS monitoring report (UNESCO, 2015; Adedeji & Fadamiro, 2018). Though heritage management may involve some form of commodification in its awareness campaign, however, understanding the capability of the heritage should be central. Therefore, if the need for economic benefits outweighs this capability, the heritage may become pressurized and wear out eventually. Commodification of heritage contributes to social and economic status of a community but understanding the capability of a heritage resource should be the centrality of heritage management.

Conclusion

This chapter aimed to illustrate some of the challenges and obstacles in implementing participatory heritage management. It did so by focusing on the case of Nigeria where a shift in the management of heritage has been noted between the pre- and post-colonial era. We argued that while in the pre-colonial era, indigenous communities were the main custodians of their heritage, in the colonial era, a top-down, hierarchical governance system was imposed which led to the alienation of indigenous communities from their heritage. And yet, there is a growing emphasis on the need to foster the democratic management of indigenous heritage. To achieve this in Nigeria, five main challenges need to be overcome including diversity of beliefs, lack of documentation, predominance of Western views in Nigerian heritage systems, balancing economic benefits with other values such as sacred value, and negotiating opposing values between experts and communities.

One of the ways to pioneer and manage indigenous heritage could be to get the education sector to introduce and facilitate heritage awareness and appreciation in the classroom, with the development of new curricula. Heritage differs from one community to another; the content of such curricula should be tailored uniquely or customized to the needs of communities. It is hereby recommended that narratives, such as myths and legends, surrounding national and local heritage become part of classroom activities. Activities may include ways in which heritage can benefit young people economically, trainings such as entrepreneurship and extracurricular activities, for instance, debates, competitions that are non-faith-based (most communities in Nigeria are diverse when it comes to faith and religion, with religious crises being common occurrences). These benefits should take into account the social impact of the classroom on the community's well-being and cohesion. Young people should be able to learn more about heritage through artefacts and objects related to indigenous communities and other communities too since they are not constrained to be educated only within their localities.

The appreciation of heritage among community members through awareness creation should be encouraged by allowing indigenous communities to participate in decision-making. More research is needed in order to understand the engagement of younger people with culture and heritage as well as the role of language as heritage, for the creation of sustainable communities. Further research on participatory heritage management is required in the context of multi-ethnic and multi-belief communities where there is diversity of voices and values. Collaborating with communities through the presence of localized development groups, non-governmental bodies as well as the government in many multi-ethnic states in Africa will enhance such participatory research.

References

Adedeji, J. A., & Fadamiro, J. A. (2018). Urbanisation forces on the landscapes and the changing value-systems of Osun sacred grove UNESCO site, Osogbo, Nigeria. *Landscape Research*, 43(6), 798–816. https://doi.org/10.1080/01426397.2018.1459525.

Adedimeji, M. A. (2009). Globalization and the survival of the Nigerian cultural and linguistic heritage: The American paradigm. Retrieved from: https://www.researchgate.net/publication/267557630 (Accessed on 5 August 2019).

Adewumi, A. A. (2015). *Return and restitution of cultural property in African states under the 1970 UNESCO and 1995 UNIDROIT conventions*. (Doctoral dissertation). University of Ibadan. Retrieved from: http://hdl.handle.net/123456789/152 (Accessed on 7 October 2019).

Arowolo, D. (2010). The effects of Western civilisation and culture on Africa. *Afro Asian Journal of Social Sciences*, 1(1), 1–13.

Baharvand, P. A. (2016). The role of British missionaries in the rejection of Igbo religion and culture. In Chimamanda Ngozi Adichie's Purple Hibiscus. *Journal of Novel Applied Sciences*, 5(2), 43–51.

Bale, J. (2010). International comparative perspectives on heritage language education policy research. *Annual Review of Applied Linguistics*, 30, 42–65. https://doi.org/10.1017/S0267190510000024.

Boyd, W. E. (1996). The significance of significance in cultural heritage studies: A role for cultural analogues in applied geography teaching. *Journal of Geography in Higher Education*, 20(3), 295–304. https://doi.org/10.1080/03098269608709374.

Breen, C. (2014). World heritage sites, culture and sustainable communities in Africa. In S. Makuvaza (Ed.), *The management of cultural world heritage sites and development in Africa history, nomination processes and representation on the world heritage list* (pp. 83–91). New York: Springer.

Brundtland, G. (1987). Our common future—call for action. *Environmental Conservation*, 14(4), 291–294. Retrieved from: www.jstor.org/stable/44518052. (Accessed on 15 August 2020).

Cassar, M., & Fouseki, K. (2014). My historic environment. *The Historic Environment: Policy & Practice*, 5(2), 226–228. https://doi.org/10.1179/1756750514Z.00000000057.

Chanda, K. C. (2018). *Invoking co-management as a tool in the conservation and sustainable development of World Heritage properties in Africa*. Publication in World Heritage for Sustainable Development in Africa. Paris: UNESCO.

Chirikure, S., Manyanga, M., Ndoro, W., & Pwiti, G. (2010). Unfulfilled promises? Heritage management and community participation at some of Africa's cultural heritage sites. *International Journal of Heritage Studies*, 16(1–2), 30–44. https://doi.org/10.1080/13527250903441739.

Chirikure, S., & Pwiti, G. (2008). Community involvement in archaeology and cultural heritage management: An assessment from case studies in southern Africa and elsewhere. *Journal of Current Anthropology*, 49(3), 467–485.

Chukwuemeka, V., Gantois, G., Scheerlinck, K., Schoonjans, Y., Onyegiri, I., Guttormsen, T. S., Fouseki, K., & Swensen, G. (2019). Embodying local identity as heritage in transition. The case study of Onitsha markets, Nigeria. In K. Fouseki, T. S. Guttormsen, & G. Swensen (Eds.), *Heritage and sustainable urban transformations: Deep cities* (pp. 55–66). London: Routledge.

Court, S., & Wijesuriya, G. (2015). *Guidance notes- People-centred approaches to the conservation of cultural heritage: Living heritage* (pp. 1–9). Rome: ICCROM. Retrieved from: https://www.iccrom.org/sites/default/files/PCA_Annexe-2.pdf# (Accessed on 22 July 2020).

Crooke, E. (2010). The politics of community heritage: Motivations, authority and control. *International Journal of Heritage Studies*, 16(1–2), 16–29. https://doi.org/10.1080/13527250903441705.

de la Torre, M. (2013). Values and heritage conservation. *Journal of Heritage & Society*, 6(2), 155–166.

Dragouni, M., & Fouseki, K. (2017). Drivers of community participation in heritage tourism planning: An empirical investigation. *Journal of Heritage Tourism, 13*(3), 237–256. https://doi.org/10.1080/1743873X.2017.1310214.

Dragouni, M., Fouseki, K., & Georgantzis, N. (2018). Community participation in heritage tourism planning: Is it too much to ask? *Journal of Sustainable Tourism, 26*(5), 759–781. https://doi.org/10.1080/09669582.2017.1404606.

Ezekwesili, O. (2020, April 16). Opinions China must pay reparations to Africa for its Coronavirus failures. *The Washington Post.* Retrieved from: https://www.msn.com/en-gb/news/in-depth/opinions-china-must-pay-reparations-to-africa-for-its-coronavirus-failures/ar-BB12Lq6y?ocid=spartanntp. (Accessed on 16 April 2020).

Eze-Uzomaka, P. (2014). Archaeology and heritage legislation: A comparative study. *Research on Humanities and Social Sciences, 4*(8), 135–145.

FAO (2008). The concept of indigenous peoples. Retrieved from: http://www.fao.org/3/i3144e/I3144e18.pdf. (Accessed on 20 January 2020).

Fouseki, K. (2010). Community voices, curatorial choices: Community consultation for the 1807 exhibitions. *Museum and Society, 8*(3), 180–192.

Fouseki, K., Guttormsen, T. S., & Swensen, G. (2020). Heritage and sustainable urban transformations. A deep cities approach. In K. Fouseki, T. S. Guttormsen, & G. Swensen (Eds.), *Heritage and sustainable urban transformations: Deep cities* (pp. 1–15). London: Routledge.

Fouseki, K., & Sakka, N. (2013). Valuing an ancient palaestra in the centre of Athens: The public, the experts, and Aristotle. *Conservation and Management of Archaeological Sites, 15*(1), 30–44.

Gallou, E., & Fouseki, K. (2018). Heritage within landscapes as catalyst for socio-economic development: Locating social impacts for rural communities outside of museum walls. *Proceedings of safeguarding the values of European cultural heritage* (pp. 107–125). Athens, Greece: ICOMOS.

Gibson, L., & Pendlebury, J. (2009). Introduction: Valuing historic environments. In L. Gibson, & J. Pendlebury (Eds.), *Valuing historic environments* (pp. 51–63). Surrey: Ashgate.

Harrison, R. (2010). What is heritage? In R. Harrison (Ed.), *Understanding the politics of heritage* (pp. 5–42). Manchester: Manchester University Press.

Jaafar, M., Noor, S. M., & Rasoolimanesh, S. M. (2015). Perception of young local residents towards sustainable conservation programmes: A case study of the Leggong world cultural heritage site. *Journal of Tourism Management, 48,* 154–163.

Landorf, C. (2009). Managing for sustainable tourism: A review of six cultural world heritage sites. *Journal of Sustainable Tourism, 17*(1), 53–70. https://doi.org/10.1080/09669580802159719.

Liwieratos, K. (2007). *The competitive advantage strategy in cultural heritage management: The case-study of the Mani area in the southern Peloponnese.* (Doctoral dissertation), University of London. Retrieved from: ProQuest dissertations open access: https://ethos.bl.uk/OrderDetails.do?uin=uk.bl.ethos.498705 (Accessed on 14 June 2020).

Lowenthal, D. (2005). Natural and cultural heritage. *International Journal of Heritage Studies, 11*(1), 81–92.

Mason, R. (2002). Assessing values in conservation planning: Methodological issues and choices present. In E. Avrami, R. Mason, & M. de la Torre (Eds.), *Values and heritage conservation research report* (pp. 5–30.). Los Angeles, CA: The Getty Conservation Institute.

Mason, R. (2008). Be interested and beware: Joining economic valuation and heritage conservation. *International Journal of Heritage Studies, 14*(4), 303–318. https://doi.org/10.1080/13527250802155810.

McClelland, A., Peel, D., Hayes, C. L., & Montgomery, I. (2013). A values-based approach to heritage planning: Raising awareness of the dark side of destruction and conservation. *Town Planning Review, 84*(5), 583–603.

Munjeri, D. (2003). Anchoring African cultural and natural heritage: The significance of local community awareness in the context of capacity-building. In E. de Merode, R. Smeets, & C. Westrik (Eds.), World Heritage Papers 13. *A conference organized by the Netherlands National Commission for UNESCO, in Collaboration with the Netherlands Ministry of Education, Culture and*

Science, 22–24 May (pp. 75–80). France: UNESCO. Retrieved from: whc.unesco.org/documents/publi_wh_papers_13_en.pdf (Accessed on 27 November 2019).

NCMM (2020). National Commissions for Museums and Monuments. Retrieved from: https://ncmm.gov.ng/ (Accessed on 20 September 2019).

Ndoro, W. (2003). Traditional and customary heritage systems: Nostalgia or reality? The implications of managing heritage sites in Africa. In E. de Merode, R. Smeets, & C. Westrik (Eds.), World Heritage Papers 13. *A conference organized by the Netherlands National Commission for UNESCO, in Collaboration with the Netherlands Ministry of Education, Culture and Science,* 22–24 May (pp. 81–84). France: UNESCO. Retrieved from: whc.unesco.org/documents/publi_wh_papers_13_en.pdf (Accessed on 27 November 2019).

Nwegbu, M. U., Eze, C. C., & Asogwa, B. E. (2011). Globalization of cultural heritage: Issues, impacts, and inevitable challenges for Nigeria. *Library Philosophy and Practice,* Paper 674. Retrieved from: http://digitalcommons.unl.edu/libphilprac/674 (Accessed on 29 April 2021).

Ohiri-Aniche, C. (2014, February 26). More than 400 Nigerian indigenous languages are endangered. *Vanguard Newspapers.* Retrieved from: https://www.vanguardngr.com/2014/02/400-nigerian-indigenous-languages- endangered/ (Accessed on 27 July 2020).

Onyima, B. N. (2016). Nigerian cultural heritage: Preservation, challenges, and prospects. *African Journals Online, 12.* Retrieved from: https://www.ajol.info/index.php/og/article/view/141270 (Accessed on 16 June 2019).

Probst, P. (2009). Yoruba heritage as project. Reauthenticating the Osun grove in Osogbo, Nigeria. *African Arts, 42*(4), 24–37.

Probst, P. (2013). Preserving heritage and the values of exchange: Lessons from Nigeria. *History Compass, 11*(12), 1035–1046. https://doi.org/10.1111/hic3.12102.

Quist, H. O. (2001). Cultural issues in secondary education development in West Africa: Away from colonial survivals, towards neo-colonial influences? *Journal of Comparative Education, 37*(3), 297–314.

Sabri, R., & Olagoke, O. A. (2019.) Predicaments in the management of religious heritage buildings and sites in Nigeria. *Conservation and Management of Archaeological Sites, 21*(1), 45–65. https://doi.org/10.1080/13505033.2019.1596521.

Said, S. Y., Aksah, H., & Ismail, E. D. (2013). Heritage conservation and regeneration of historic areas in Malaysia. *Procedia - Social and Behavioural Sciences, 105,* 418–428.

Shyllon, F. (1996). Cultural heritage legislation and management in Nigeria. Retrieved from: https://www.cambridge.org/core (Accessed on 7 September 2020).

Shyllon, F. (2005). The poverty of documentary heritage management in Nigeria. *International Journal of Cultural Property, 9*(1), 23–48. https://doi.org/10.1017/S0940739100770937.

Silberman, N. A. (2013). Heritage interpretation as public discourse: Towards a new paradigm. In M. Albert, R. Bernecker & B, Rudolph (Eds.), *Understanding heritage: Perspectives in Heritage Studies (pp. 21-34).* Berlin, Boston: De Gruyter. https://doi.org/10.1515/9783110308389.21.

Smith, L. (2008). *Uses of heritage.* London: Routledge.

Smith, L., & Fouseki, K. (2011). *The role of museums as places of social justice: Community consultation and the 1807 bicentenary.* London: Routledge.

Smith, L., Morgan, A., & van der Meer, A. (2003). Community driven research in cultural heritage management: The Waanyi women's history project. *International Journal of Heritage Studies, 9*(1), 65–80. https://doi.org/10.1080/1352725022000056631.

Smith, L., & Waterton, E. (2012). Constrained by common sense: The authorised heritage discourse in contemporary debates. In J. Carman, R. Skeates, & C. McDavid (Eds.), *The Oxford handbook of public archaeology* (pp. 153–171). Oxford: Oxford University Press.

Taylor, K. (2004). Cultural heritage management: A possible role for charters and principles in Asia. *International Journal of Heritage Studies, 10*(5), 417–433. https://doi.org/10.1080/1352725042000299045.

UNESCO (1972). Convention concerning the protection of the world cultural and natural heritage. Retrieved from: https://whc.unesco.org/archive/convention-en.pdf (Accessed on 23 July 2019).

UNESCO (2014). United Nations Declaration on the Rights of Indigenous Peoples (UNDRIP). Retrieved from: https://www.un.org/esa/socdev/unpfii/documents/DRIPS_en.pdf (Accessed on 24 October 2019).

UNESCO (2015). Report on the ICOMOS reactive monitoring mission to Osun-Osogbo Sacred Grove, 25–30 October 2015. Retrieved from: http://whc.unesco.org/en/documents/141546/ (Accessed on 2 June 2020).

UNESCO (2016). Sukur world heritage site management plan 2017–2021. Retrieved from: https://whc.unesco.org/en/list/938/documents (Accessed on 05 May 2020).

Waterton, E., & Smith, L. (2010). The recognition and misrecognition of community heritage. *International Journal of Heritage Studies*, 16(1–2), 4–15. https://doi.org/10.1080/13527250903441671.

Watson, S., & Waterton, E. (2010). Heritage and community engagement. *International Journal of Heritage Studies*, 16(1–2), 1–3. https://doi.org/10.1080/13527250903441655.

5 Evaluating the management plan of Bali Cultural Landscape from the local community's perspective

Diana Rahman and Kalliopi Fouseki

Introduction

The aim of a World Heritage Site management plan is to facilitate sustaining the Outstanding Universal Value (OUV) of the site. However, the implementation of those management plans is often challenging and complex. Several issues related to the implementation of World Heritage Site management plans are discussed in great length by many scholars, for instance, related to tourism and infrastructure development on the site, the collaboration between stakeholders, and the effectiveness of the management strategy (Jones & Shaw, 2012; Santos & Zobler, 2012; Fibiger, 2015). Many scholars also have engaged with issues related to local communities of World Heritage Sites, including local disengagement from conservation practices, displacement of indigenous community's residence, and contrasting attitudes between local communities and site managers towards heritage management (Maikhuri et al., 2001; Chirikure et al., 2010; Suntikul & Jachna, 2013; Zhang et al., 2014).

The discrepancy between the perceptions of local communities and site managers of World Heritage Site and OUVs is a recurrent issue in heritage management (Okech, 2010; Yan, 2015). Similarly, local communities and site managers often have different management approaches for World Heritage Sites (Chirikure et al., 2010). It has been demonstrated that conflicts that led to ineffective management strategies happen when this discrepancy is disregarded (Goh, 2015), but positive outcomes might also appear as long as dialogues are encouraged (Landorf, 2009).

As a cultural World Heritage Site, the Bali Cultural Landscape or widely known as the Subak landscape was also inscribed without the absence of numerous management problems. The site (Figure 5.1), which consists of a network of rice terraces and water temples surrounding the traditional farming system called Subak, was inscribed in 2012. Since then, the World Heritage Committee (WHC) highlighted the worrying problems of land conversion and threats to the traditional rice production system. The Statement of Conservation reports (hereafter SOC reports) during 2015–2019 also repeatedly indicated the unsolved problems regarding insufficient management strategies and the lack of local community involvement. The lack of community involvement in the management of World Heritage Sites is not of surprise. Several scholars who have attempted to investigate what drives or prohibits community engagement in heritage practices have demonstrated that several factors, including gender, education, social impacts, and access to resources, can affect the degree of community participation (Dragouni & Fouseki, 2018; Gallou & Fouseki, 2019). Other studies also pointed out the issue of awkward engagement and tourism development as part of management problems in Bali Cultural Landscape (Salamanca et al., 2015; MacRae, 2017).

Figure 5.1 The rice field and the irrigation system at the Subak Pakerisan Watershed, part of the World Heritage Site of Bali Cultural Landscape.
Source: Photo by Diana Rahman, May 2019.

As an understanding of the local community's perception towards World Heritage Site and its management are crucial for the management of the World Heritage Site, there is a growing need to investigate this topic as it is still under-researched. Moreover, by understanding the local community's perception towards a World Heritage Site and its management plan, the effectiveness of conservation strategies could be increased and ways of improving their engagements could be established (Silva & Chapagain, 2014). That being said, this chapter attempts to analyse the perception of the local community of the Bali Cultural Landscape towards both the World Heritage Site and its management plan. It aims to evaluate the effectiveness of the management plan through the local community's perspective as well as uncover any discrepancy that leads to inconsistent implementation of the management plan. Ultimately, this chapter offers an investigation of the relationship between stakeholders' perceptions and different conservation approaches, providing new insights into the ways in which a changing, living cultural landscape is being valued and approached by the local community that directly uses the landscape.

Methods of the study

The case study analysis is based on empirical data collected through ethnography fieldwork and examination of the management plan of the Bali Cultural Landscape and SOC reports. The study compares the written version and the local community's version of the World Heritage Site's values as well as compares the management plan and its implementation. By doing so, this study expects to broaden our understanding of the management problems within the Bali Cultural Landscape, especially in relation to stakeholders' discrepancy.

Ethnography was employed because it allows the researcher to immerse in the everyday life of a local community and engage in cultural activities to become less of an 'outsider' (Malinowski, 1922). By doing so, as Creswell (2013) mentioned, impacts of research activities on the local community's behaviour can be minimized. Creswell (2013) also asserted that ethnography is useful to explore beliefs, meanings, and behaviours of a cultural group, which is particularly beneficial to understanding heritage management issues from the local community's perspective. Scholars have argued on the appropriate duration of ethnography work, and there is a tendency to assume that ethnography requires a long-term fieldwork duration. However, Fetterman (2004) and Madden (2017) persisted that ethnography can be conducted over a shorter period of typical 12–18 months. It is the 'intensive excursion' into the lives of the cultural group and the aim to reveal what matters to them that separate ethnography with other data collection methods (Pink & Morgan, 2013, p. 352). Nevertheless, there might be a direct relationship between the duration of the fieldwork and the quality of data that are collected (Fetterman, 2004).

This study uses critical discourse analysis (CDA) to analyse information gathered from the management plan and SOC reports of the Bali Cultural Landscape. As Fairclough (1995, p. 7) mentioned, CDA is useful to investigate 'how texts work within socio-cultural practices'. It is considered the most appropriate data analysis method as this study attempts to examine non-obvious relationships between the official documents and management practices at the local community level. Alongside ethnography study, the use of CDA will help untangle any discrepancy and conflict related to the management of the landscape.

Comparing the Outstanding Universal Value and the local community's version of values

The Statement of Outstanding Universal Value (SOUV) of World Heritage Sites summarizes the significance of a World Heritage Site and the management requirements surrounding the site that are usually proposed by advisory bodies (International Council on Monuments and Sites (ICOMOS), International Union for Conservation of Nature (IUCN), and/or International Centre for the Study of the Preservation and Restoration of Cultural Property (ICCROM)) and agreed by the WHC. SOUV contains necessary information and acts as a future reference in managing World Heritage Sites. Although it is rarely considered a problem, SOUVs and the nomination dossiers might have a few significant differences regarding its contents. The differences between the nomination dossiers and the SOUV seem trivial as the latter is a much shorter document than the former. However, considering that monitoring and evaluation, management recommendations, and intervention of the WHC to World Heritage Sites use SOUV as its reference point, those differences would bring significant implications on the management strategies of World Heritage Sites.

The Bali Cultural Landscape was inscribed to the World Heritage List under the criteria (iii), (v), and (vi). It is particularly inscribed as a physical manifestation of the Tri Hita Karana (three causes of goodness), a Balinese philosophy which encourages a harmonious relationship between individuals and the realms of the spirit (parahyangan), individuals and the human world (pawongan), as well as individuals and nature (palemahan). As a World Heritage Site, the Bali Cultural Landscape is substantially valued for its OUVs. Thus, its protection and management strategies are also developed around these values. However, the fieldwork observations revealed that the local community assigned different values to the site and the Subak system that are not acknowledged on the SOUV. In addition to symbolic and religious values, the Bali Cultural Landscape is also considered as having economic and social values.

The difference between the local community's and the WHC's version of values occurs because they conceptualize the Bali Cultural Landscape differently. As a World Heritage Site, the Bali Cultural Landscape is officially considered a physical manifestation of the Subak system and the Tri Hita Karana philosophy. However, the local community perceives the Subak system as the integration between social components, irrigation techniques, and religious rituals that make the rice production works. Thus, the local community interprets the significance of the site as arising from all those components and the relationship between them.

It is observed that there is also a disparity between the local community and the WHC in determining the attributes of the Bali Cultural Landscape. Attributes of World Heritage Sites could be understood as any physical elements, tangible aspects, intangible aspects or processes that convey the OUVs of the site (Marco, 2013). Determining attributes is a crucial part of the nomination process as they affect the management strategy and the site's conservation approach. However, as seen in Figure 5.2, there is a significant difference between the WHC and the local community's version regarding elements that could demonstrate the significance of the Bali Cultural Landscape. While the WHC recognized many tangible attributes, the local community of the Bali Cultural Landscape raised the importance of including intangible aspects as the relevant attributes of the site.

The findings suggest a correlation between the different versions of attributes and how the WHC and the local community interpret values of Bali Cultural Landscape. As attributes are determined in order to be able to develop appropriate measures of protecting values of the site, having different versions of values means that there are different versions of attributes that should be protected. More importantly, this also means that there would be different versions of appropriate management approaches that are developed to sustain those values. This is how the discrepancy between the local community, the WHC, and other stakeholders emerged.

Figure 5.2 Subak attributes from the World Heritage Committee and the local community's perspectives.
Source: Created by Diana Rahman & Kalliopi Fouseki.

For the local community of the Bali Cultural Landscape, it is evident that other values are as important as the OUVs. Depending on their roles within the traditional village and Subak organization, farmers and the local community might also have different interpretations of which Subak components are the most valuable. However, it is crucial to acknowledge that intangible components are important parts of the Subak system and of Balinese culture in general. Unfortunately, as many intangible attributes are not recorded in the SOUV, they are not a priority of the current management plan.

Although the World Heritage Site's management plans are developed primarily to protect the OUVs of the site, this study argues that disregarding the protection of other values would be a catastrophe for the sustainability of the site. First of all, this would create a prolonged tension among stakeholders as they have different interpretations of the site's significance and what is considered worthy to be protected for the future. Secondly, it disarticulates the local community's perspectives. This would not only hinder a possibility of fruitful collaborations between stakeholders but could also become an obstacle to community participation.

Concepts and implementations of Bali Cultural Landscape's management strategies

Initially, the Governing Assembly was the responsible committee that oversees the implementation of the management plan of the Bali Cultural Landscape. The Governing Assembly consists of representatives of all Subak organizations, villages, and relevant government and non-government organizations and has a primary role in coordinating stakeholders, implementing the management plan, and conducting monitoring and evaluation activities. As the effectiveness of the Governing Assembly was criticized by the WHC, the Indonesian government established a replacement committee called the Coordination Forum in 2014. This substitution primarily aimed to remove negative connotations of the Governing Assembly as a hierarchical top-down institution (Salamanca et al., 2015). It also attempts to offer a better platform for the local community and stakeholders who previously had limited involvement in the Governing Assembly.

The Coordination Forum has successfully conducted routine meetings and discussions, yet negative impressions towards this new committee remain. The local community of Subak Pakerisan Watershed, an area within the boundaries of the Bali Cultural Landscape, particularly criticized the lack of support and proactiveness from the Coordination Forum to farmers and issues related to rice production. Farmers often need to manage Subak-related problems by themselves with minimal support from site managers or the government. For instance, they have been repairing the leaks from the irrigation canals using their own resources and trying to integrate innovative farming techniques with limited funding. Disappointment and scepticism towards the government are increasing as farmers are disadvantaged by inadequate management measures.

As an adaptive management system, the management plan of the Bali Cultural Landscape is designed to be able to respond to uncertain situations by incorporating a monitoring and evaluation stage. This stage is a requirement for all adaptive management frameworks as it improves the management strategies by developing a better system that suits different situations (National Research Council, 2004). As it is widely understood, an adaptive management framework differs from trial and error as it involves the identification of management alternatives, management consequences, and recognition of key uncertainties (National Research Council, 2004).

Unfortunately, the management plan of the Bali Cultural Landscape lacks these components. Site managers did not fully comprehend the adaptive management framework as many of them were not involved in the development of the management plan in the first place. Alongside insufficient resources, inappropriate strategies, and poor designs, the adaptive management framework has been poorly interpreted and executed. This is reflected in the prolonged problem of unsuccessful collaborations between governments, academia, and NGOs, as well as the absence of updated and detailed management actions. It can also be seen from the lack of a clear methodology and system for conducting monitoring and evaluation activities as well as a strategy on how it could be incorporated into the management plan of the Bali Cultural Landscape. As mentioned by a respondent, monitoring and evaluation activities are seen as a formality rather than a tool to evaluate the management plan.

The recent SOC reports indicated that financial incentives remain a management priority for the Bali Cultural Landscape in order to help farmers with the high cost of rice production and rituals. The government has also provided incentives to heads of farmers (Pekasehs) who manage Subak organization as well as financial assistance to conduct religious ceremonies. Tax reduction for all Subak members within the World Heritage Site's boundary in Gianyar Regency is also an ongoing programme. However, Pekasehs and farmers contradict the effectiveness of these incentives. According to them, farmers receive 50% of land tax reduction or 150,000 rupiah ($10), which is less than 5% of the total cost of rice production. Pekasehs receive financial incentives of less than $30 a month, which is insufficient to compensate for the time that they have spent in dealing with additional tasks and issues on the site. These additional tasks which include mediating between farmers and the different levels of government, attending the Coordination Forum, welcoming international visitors, and prohibiting land conversion were only added after the inscription.

SOC reports indicated that the WHC requests site managers to carry out heritage impact assessment (HIA) and environmental impact assessment (EIA) before making irreversible decisions on the Bali Cultural Landscape. In response to this, the national government highlighted that the existing national laws have already managed activities related to the use of natural and cultural environments. This instrument is argued to be sufficient to protect the OUV of the site. However, as rice fields are legally owned by individuals, controlling private uses of rice fields is an impossible job for both the government and site managers. The government also does not have control over several other components of the Bali Cultural Landscape, such as the water temples and the water palace, as it was managed by *Puri* (the Balinese Palace) and customary villages. Thus, the use of EIA and HIA or the national laws alone would be inadequate to address the issue of irreversible change or development; alternative strategies that can correspond with the control and ownership system of Balinese culture are needed.

The SOC reports of the Bali Cultural Landscape, submitted both by the Indonesian government and the WHC, currently incorporate fewer discussions related to the protection of intangible aspects of the Subak landscape. Through these reports, it can be seen that the current priority of the WHC is the protection of the rice terraces and tangible components of the Subak system, such as water temples and the irrigation canals. As discussed, the SOC reports indicate that the site's conservation priority was determined by the WHC's interpretation of the site's significance, which is in contrast to the local community's concerns about the protection of intangible aspects of the Subak landscape.

Local community's perspectives towards the World Heritage status and the site's management

Not all farmers in Subak Pakerisan Watershed understand the reason behind the inscription of their rice fields into the World Heritage List. Most of them were not involved in the nomination process, let alone receive adequate information regarding the consequences and benefits of the World Heritage status. Although the nomination dossier of the Bali Cultural Landscape indicated that community consultation was extensively done, farmers and the local community of Subak Pakerisan Watershed felt they were only informed about the outcome of the process. Therefore, it is plausible that prior consultations were not inclusive and only involved several representatives of farmers and the local community.

Six years after the inscription of Bali Cultural Landscape, farmers and the local community continue to expect that the World Heritage status would bring significant financial benefits to the area. Farmers have hoped that in addition to the government's assistance in infrastructure development and farming incentives, the World Heritage status would significantly bring financial improvement that could improve their social and economic condition. The local community of Subak Pakerisan Watershed has an ongoing assumption that UNESCO shares the Indonesian governments' responsibilities to deliver economic improvements and agricultural supports to the area within the boundary of the World Heritage Site.

As the nomination brought some restrictions which affect how rice fields are utilized, the local community anticipates that there would be compensation for their willingness to support the conservation projects. The World Heritage status might facilitate farmers' access to agricultural support, but the restrictions that come alongside the status are considered more problematic for farmers and their families. For instance, as land conversion is prohibited, several families have difficulties to build a place to live. Others did not appreciate the decrease of the land price as it affects their source of income. Since the tangible benefits of the World Heritage status have not been felt, farmers have started to ignore the governments' advice and disengage with the conservation projects. This is in line with a respondent's statement:

> In short, they said there will be prosperity for farmers... So, we supported it, right? Because we were happy that subak will be tax free and so on. But in reality, there isn't anything from UNESCO that can be enjoyed by subak (community)... and there is a new rule too: land conversion is now forbidden. I was the one who is actually disappointed! It means we are trapped... because subak exists to give new life to its owners. So, if you (farmers) want to build something, I will not forbid it. The most important point is that every time we meet, we could smile to each other and there's no bad relationship between us.
>
> (R22)

In many Balinese families, rice fields are usually inherited to the children both as a symbolization of the relationship to the ancestors and as an economic asset. While working in the rice production process, owners of the rice fields must maintain the sanctity and the 'spirit' of the rice field by conducting rituals before and after an activity. Some farmers illustrated that rice fields are considered a backyard of the family, and it is the family's responsibility to ensure it passing on to the future generation. However, as rice fields have a primary function to support a family's life, it is considered reasonable to sell or convert rice fields, particularly when rice production is no longer sufficient as the main livelihood. Pekasehs, who are supposed to assist the site managers by overseeing activities which threaten the sustainability of the landscape, understand the complexities of this issue and become reluctant to prohibit land conversion.

That being said, it is evident that rice fields possess cultural and economic significance for the local community, especially farmers. Although the nomination dossier acknowledges the cultural value of rice fields, it failed to comprehensively discuss the economic value of rice fields, not only for rice production but also as a land investment. Arguably, the exclusion of this aspect in the discussion led to an incomplete understanding of the local community's perspective of the significance of the site as well as the needs for land conversion. Although the SOC reports indicated that there is a change in the way the local community values Subak landscape, this chapter argues that it was the nomination dossier that did not have the complete records of the local community's existing perspective.

The local community of the Bali Cultural Landscape raised some discussions related to the importance of modernization of farming tools in reducing physical labours. Modern equipment such as machine tractor and harvesting machine raise the cost of the whole agricultural processes but help to reduce the problem of a shortage of farmers and also enable farmers to find additional jobs. They also claimed that several modifications in Subak rituals have helped farmers cut the cost of the practices. The simplification of offerings and ceremonies has been done to enable cheaper rituals, so that farmers could continue their practices without compromising their economic condition. This, for instance, includes using more accessible food and materials. Interestingly, these adjustments also contribute to help farmers adapting to climate change as simpler rituals often use less of natural resources. As Balinese rituals are vulnerable to environmental change, those strategies are crucial to maintaining the continuity of traditional rituals without being heavily dependent on natural resources. Although the change of social and cultural activities is included on the list of factors that could degrade the OUV of World Heritage Sites, it is essential to note that in the Bali Cultural Landscape, these changes are needed to protect the sustainability of rice production and Balinese culture.

This case study highlights that issues of the Bali Cultural Landscape management might be triggered by unmet expectations of the local community and misunderstanding of the local community's perspective. It also shows that an inclusive consultation prior to the World Heritage Site nomination process is crucial not only to collect comprehensive knowledge regarding the significance of the site but also to provide necessary communication mechanisms for the local community to understand the consequences of the World Heritage Site inscription. By doing that, a prior consultation would also narrow the discrepancy between stakeholders and enable dialogues that minimize the possibility of future conflicts in the management practices.

Arguably, a community consultation approach will also narrow the gap that exists between the practice of managing the cultural landscape and the efforts of protecting intangible heritage and indigenous practices. The ability of site managers to document the correct values and attributes of a World Heritage Site is crucial to ensure that the protection of both tangible and intangible heritage is aligned and not competing with each other.

Stakeholders' discrepancy and its implications

In addition to the problem of insufficient management strategies, another challenge of managing the Bali Cultural Landscape can be seen in an inharmonious relationship between site managers and the local community. As the ethnography fieldwork successfully discovered, farmers of Subak Pakerisan Watershed are sceptical towards the World Heritage Site programme and site managers, primarily because they have not significantly benefited from the inscription for their livelihood. MacRae (2017) highlighted that a contradiction between local realities of the Bali Cultural Landscape and the global vision is evident. He observed that farmers' concerns differ from what was written in the World Heritage Site's official documents as they are interested more in the costs and the benefits of the listing rather than the management of the landscape. Alongside several other scholars, MacRae (2017) agreed that there is a gap between the ideal World Heritage conservation approach and local practices (Ndlovu, 2011; Alivizatou, 2012; Ginzarly et al., 2019).

Many studies raised a discussion about the stakeholders' discrepancy in perceiving World Heritage Site values as well as tensions between stakeholders in the management of heritage sites (Chirikure et al., 2010; Okech, 2010; Yan, 2015). From observations of the local community's perceptions related to both Subak values and the management strategies, this study demonstrates that incorporating local values into the SOUV strongly increases the chance of an effective management plan of World Heritage Site. The gaps between the universal and local values would be narrowed down by incorporating the local community's version of heritage values. Instead of focusing on the protection of the OUV that might not strongly benefit the local community, the protection of local values would reduce local community's scepticism towards the World Heritage Site programme and increase their involvement in the management of the site.

In the case of the Bali Cultural Landscape, the local community acknowledges more component values and attributes of the Bali Cultural Landscape than the WHC. Intangible values and attributes dominate the local community's version of the significance because the Balinese culture believes that all objects have a 'soul' that makes them alive. Although this does not mean that the protection of tangible values and attributes such as rice terraces, water temples, and water palaces is less important, neglecting the protection of intangible values and attributes of Subak would be considered an eradication of the core of the Bali Cultural Landscape. Some areas in Bali have already experienced this loss. To cut the cost of rice production, some farmers diminished Subak rituals and disengaged with Subak organization. Although rice fields and rice production are protected, there is no more integration between the farming system and the religious practices. In other words, the Subak system and cultural and religious significance of the rice field are no longer present.

It is to be noted that the local community's understanding of Subak values and attributes affects their attitude towards the conservation strategy that was developed by the WHC and the national government. Arguably, they have developed their ideal conservation approach that corresponds to their interpretation of Subak values. On the other hand, the WHC and the Indonesian government, who value the Bali Cultural Landscape from its OUV, also have developed a different understanding of the ideal conservation approach as they do not interpret the site in the same way as the local community. As an intermediate stakeholder, the governments might understand the local community's perspective; however, it is also their responsibility to fulfil the mandates that are given by the WHC.

Inconsistencies between the SOC reports, the management plan, and the actual condition of Subak Pakerisan Watershed can be seen as an implication of both an ineffective collaboration between stakeholders and a discrepancy in interpreting the significance of the Bali Cultural Landscape. In the long run, the absence of active evaluations of the management plan and dialogues between stakeholders would trigger a dissonant condition in the Bali Cultural Landscape which could lead to the destruction of not just the World Heritage Site, but more importantly, the degradation of the local community's skills to sustain traditional Balinese practices.

This study's findings indicated that the management plan of World Heritage Site could contribute to shaping and defining essential attributes for the site. Thus, an inappropriate management plan would not only affect the way future generations interpret the significance of the Bali Cultural Landscape, but also affect how the Subak system and traditional practices are implemented. The absence of one or some tangible and intangible components of the Subak system would modify the local community's perception towards the site, their relationship with the site, and more importantly, the traditional practices relating to the whole rice production system.

Conclusions

This study has illustrated that different interpretations regarding the significance of the Bali Cultural Landscape are present. In addition to the site's OUVs, the local community recognized other values that are not listed on the SOUV, such as the economic value of the rice field and values associated with farmers' relationships. As a result of this discrepancy, the WHC and the local community developed different lists of attributes that have influenced their conservation approach and priorities. As written on the SOUV, tangible aspects of the Subak landscape, including rice fields and water temples, have become the main focus of the WHC recommendations. On the other hand, intangible aspects of the Subak landscape, such as the relationship between farmers, farmers' livelihood, and religious rituals are the management priority for the local community of the Bali Cultural Landscape. This finding demonstrates that stakeholders' attitudes towards the management plan of World Heritage Site are directly linked to their interpretations towards the significance of the site.

The chapter has also illuminated the ineffectiveness of the management plan of Bali Cultural Landscape as well as its inconsistent implementation, particularly from the local community's perspective. It is evident that management-related problems remain to be seen at Bali Cultural Landscape. The unmet expectations of the local community regarding the benefits of the World Heritage status have triggered inharmonious relationships between them and the site managers. Moreover, it is illustrated that the current management plan did not contribute to maintaining the ability of the Subak system to adapt to extreme environmental changes as it did not acknowledge the importance of change and adaptation in Balinese context. The management strategy failed to consider traditional Balinese values and different interests between stakeholders, forcing the local community to have uncomfortable participation in the site management.

The inadequacy of the current management plan of the Bali Cultural Landscape can be observed from several points. As an adaptive management framework, the management plan still lacks monitoring and evaluation strategies. There is no substantial evidence that shows that the management plan has been evaluated and updated from its initial version. With the presence of various management problems in the Bali Cultural

Landscape, the absence of evaluation stages indicates that the management measures have not been sufficient. There are inadequate adaptive elements within the management plan as it seems more like a trial and error strategy than an adaptive framework. In addition, the management plan seems to unsuccessfully address problems that are raised by the WHC and the local community. It could be argued that many proposed strategies, such as financial incentives, were developed by the government to fulfil the WHC's recommendation for SOC reports rather than to solve the actual problems. It is observed that thorough consultation with the local community was rarely made prior to the development of management strategies. Consequently, many management actions are misdirected and often fail to resolve the intended problems.

The way ahead

This case study exemplifies how crucial is for site managers to conduct a periodic evaluation of World Heritage Sites' management plan to ensure that adequate and relevant protection measures are always present. Inclusive dialogue with the local community of World Heritage Sites would not only improve the effectiveness of the management plan but also narrow any discrepancy that are present among stakeholders, which might hinder the implementation of the management strategies. It is important to ensure that this dialogue is initiated from the beginning of the nomination process as it is crucial to collect local community's perspective on the significance of World Heritage Site. Arguably, prolonged tensions between local communities and site managers as well as the local community's disengagement from the management of World Heritage Sites could be avoided by making the World Heritage Site framework more relevant for them.

The case study of this chapter offered, we hope, a new insight into the implications of lack of assessment of the wide array of values attached to a cultural landscape which is constantly subject to change. Previous research has highlighted the significance of community participation in heritage management and the need to assess the values of all stakeholders directly or indirectly related to a site. With this case study, we also illuminated the practical implications for the sustainable future of a heritage site if the values of all stakeholders are not taken into account.

Therefore, we would like to conclude this chapter by stressing that participatory heritage management should not be an option but a standard practice. Many reports published by the World Heritage Centre have indeed underlined that the participation of local communities, such as in determining heritage values and engaging with the management actions, is vital in managing these universally significant sites (World Heritage Centre, 2002; Munjeri, 2004; Mitchell et al., 2009). However, looking at this case study, it is clear that the practice on the ground could actually be far from the ideal standard envisioned by the World Heritage Site framework and those working at the Paris headquarter. The need for the World Heritage Site framework to acknowledge and address this issue is therefore of paramount importance in making sure this practice is indeed implemented at the core of managing every World Heritage Site.

References

Alivizatou, M. (2012). Debating heritage authenticity: Kastom and development at the Vanuatu Cultural Centre. *International Journal of Heritage Studies*, *18*(2), 124–143. https://doi.org/10.10 80/13527258.2011.602981.

Chirikure, S., Manyanga, M., Ndoro, W., & Pwiti, G. (2010). Unfulfilled promises? Heritage management and community participation at some of Africa's cultural heritage sites. *International Journal of Heritage Studies*, 16(1–2), 30–44. https://doi.org/10.1080/13527250903441739.

Creswell, J. W. (2013). Five qualitative approach to inquiry. In Creswel, J. W. (Eds), *Qualitative inquiry and research design: Choosing among five approaches* (3rd ed., pp. 53–84). Thousand Oaks, CA: SAGE Publications.

Dragouni, M., & Fouseki, K. (2018). Drivers of community participation in heritage tourism planning: An empirical investigation. *Journal of Heritage Tourism*, 13(3), 237–256. https://doi.org/10.1080/1743873X.2017.1310214.

Fairclough, N. (1995). *Critical discourse analysis: The critical study of language*. New York: Norman Publishing.

Fetterman, D. M. (2004). Ethnography. In Bickman L., & Rog D. J. (Eds), *The SAGE encyclopedia of social science research methods* (pp. 329–332). Thousand Oaks, CA: SAGE Publications. https://doi.org/10.4135/9781412950589.

Fibiger, T. (2015). Heritage erasure and heritage transformation: How heritage is created by destruction in Bahrain. *International Journal of Heritage Studies*, 21(4), 390–404. https://doi.org/10.1080/13527258.2014.930064.

Gallou, E., & Fouseki, K. (2019). Applying social impact assessment (SIA) principles in assessing contribution of cultural heritage to social sustainability in rural landscapes. *Journal of Cultural Heritage Management and Sustainable Development*, 9(3), 352–375. https://doi.org/10.1108/JCHMSD-05-2018-0037.

Ginzarly, M., Farah, J., & Teller, J. (2019). Claiming a role for controversies in the framing of local heritage values. *Habitat International*, 88, 1–10. https://doi.org/10.1016/j.habitatint.2019.05.001.

Goh, H. M. (2015). UNESCO World Heritage Site of Lenggong Valley, Malaysia: A review of its contemporary heritage management. *Conservation and Management of Archaeological Sites*, 17(2), 143–158. https://doi.org/10.1080/13505033.2015.1124180.

Jones, R., & Shaw, B. (2012). Thinking locally, acting globally? Stakeholder conflicts over UNESCO world heritage inscription in Western Australia. *Journal of Heritage Tourism*, 7(1), 83–96. https://doi.org/10.1080/1743873X.2011.632482.

Landorf, C. (2009). A framework for sustainable heritage management: A study of UK industrial heritage sites. *International Journal of Heritage Studies*, 15(6), 494–510. https://doi.org/10.1080/13527250903210795.

MacRae, G. (2017). Universal heritage meets local livelihoods: 'awkward engagements' at the world cultural heritage listing in Bali. *International Journal of Heritage Studies*, 23(9), 846–859. https://doi.org/10.1080/13527258.2017.1339107.

Madden, R. (2017). *Being ethnographic: A guide to the theory and practice of ethnography*. London: SAGE.

Maikhuri, R. K., Nautiyal, S., Rao, K. S., & Saxena, K. G. (2001). Conservation policy–people conflicts: A case study from Nanda Devi Biosphere Reserve (a World Heritage Site), India. *Forest Policy and Economics*, 2(3–4), 355–365. https://doi.org/10.1016/S1389-9341(01)00037-5.

Malinowski, B. (1922). *Argonauts of the western pacific: An account of native enterprise and adventure in the archipelagoes of melanesian New Guinea*. London: Routledge and Kegan.

Marco, L. De. (2013). *Managing world heritage properties: The role of statements of outstanding universal value and attributes*. Florence. Retrieved from: https://whc.unesco.org/document/125026 (Accessed on 30 September 2020).

Mitchell, N., Rössler, M., & Tricaud, P.-M. (2009). *World Heritage Paper 26: World heritage cultural landscapes: A handbook for conservation and management*. Paris: UNESCO World Heritage Centre. Retrieved from: http://whc.unesco.org (Accessed on 2 October 2020).

Munjeri, D. (2004). Anchoring African cultural and natural heritage: The significance of local community awareness in the context of capacity-building. In U. World Heritage Centre (Ed.), *World heritage paper 13-linking universal and local values: Managing a sustainable future for world heritage* (pp. 75–80). Paris: World Heritage Centre.

National Research Council. (2004). *Adaptive management for water resources planning*. Washington, DC: The National Academic Press.

Ndlovu, N. (2011). Management versus preservation: Archaeological heritage management in a transforming South Africa. *Conservation and Management of Archaeological Sites*, 13(2–3), 123–133. https://doi.org/10.1179/175355211X13179154165944.

Okech, R. N. (2010). Socio-cultural impacts of tourism on World heritage sites: Communities' perspective of Lamu (Kenya) and Zanzibar Islands. *Asia Pacific Journal of Tourism Research*, 15(3), 339–351. https://doi.org/10.1080/10941665.2010.503624.

Pink, S., & Morgan, J. (2013). Short-term ethnography: Intense routes to knowing. *Interaction*, 36(3), 351–361. https://doi.org/10.2307/symbinte.36.3.351.

Salamanca, A. M., Nugroho, A., Osbeck, M., Bharwani, S., & Dwisasanti, N. (2015). *Managing a living cultural landscape: Bali's subaks and the UNESCO World Heritage Site*. Bangkok: Stokholm Environment Institute.

Santos, C. A., & Zobler, K. A. (2012). The bridge on the river Elbe: World heritage in a modern city. *Annals of Tourism Research*, 39(1), 484–486. https://doi.org/10.1016/j.annals.2011.06.005.

Silva, K. D., & Chapagain, N. K. (2014). *Asian heritage management*. New York: Routledge. https://doi.org/10.4324/9780203066591.

Suntikul, W., & Jachna, T. (2013). Contestation and negotiation of heritage conservation in Luang Prabang, Laos. *Tourism Management*, 38, 57–68. https://doi.org/10.1016/j.tourman.2013.02.005.

World Heritage Centre, U. (2002). *World heritage papers 7- Cultural landscapes: The challenges of conservation 2002*. Ferrara. Retrieved from: http://whc.unesco.org/venice20027 (Accessed on 2 October 2020).

Yan, H. (2015). World Heritage as discourse: Knowledge, discipline and dissonance in Fujian Tulou sites. *International Journal of Heritage Studies*, 21(1), 65–80. https://doi.org/10.1080/13527258.2014.894930.

Zhang, C., Fyall, A., & Zheng, Y. (2014). Heritage and tourism conflict within world heritage sites in China: A longitudinal study. *Current Issues in Tourism*, 18(2), 1–27. https://doi.org/10.1080/13683500.2014.912204.

Part II
Participatory heritage

6 Managing participatory heritage for enhancing social well-being

Eirini Gallou and Kalliopi Fouseki

Introduction: Participatory heritage management approaches for social and heritage sustainability through reciprocity

This chapter discusses the role of **participatory approaches to heritage management** and their potential for effective safeguarding of heritage resources while increasing **social well-being** among local communities involved in the process. The theme of participatory approaches provides an opportunity to reflect upon the extensive literature on *participation or engagement* in heritage and theoretical conceptualisation of heritage as commons (González, 2015; Gould, 2017) or as a public resource, with the relevant implications those have for theory, policy and practice. It also allows us to link the discussion with recent approaches for assessing social well-being benefits from heritage, a research area that has received extensive focus lately in heritage studies and the policy sphere and is intrinsically linked to the role of heritage for enhancing socially sustainable development (Duxbury, 2007; Duxbury and Jeannotte, 2011).

Social well-being has been identified by the World Health Organization (1948) as a central component of individuals' overall health. The term has been conceptualised and operationalised in many different ways: within social sciences, social well-being encompasses behaviours that reflect community and organisational participation, social support, active community or high levels of civic membership, or concentration or development of 'social capital' and social cohesion (Andrews & Withey, 1976; Coleman, 1988; Putnam, 2000). The concept therefore is critical in establishing the links between heritage engagement, wider social and citizen engagement as well as psycho-social aspects of individual health.

While participatory approaches have been widely explored in adjacent fields (mainly within museum studies) during the last two decades, their relevance for heritage management/institutions engaging in the management of built and intangible heritage resources has not been fully explored. This chapter will set the base for this exploration by discussing key differences in types of engagement in heritage between the spheres of project management and urban planning.

The challenges posed by lack of inefficient or non-inclusive public engagement in heritage management have been exposed in critical heritage studies (see Winter, 2013, for the definition) with a significant body of work qualitatively exploring case studies or critically engaging with policy implications, within the umbrella of humanities research. These are building upon the seminal work and influential critique on heritage practice by Smith (2004, 2006) (and later Waterton, 2010; Waterton & Smith, 2010, among others). These are mostly focusing at exposing unequal power dynamics, describing politics

DOI: 10.4324/9781003038955-9

and relationships between stakeholders engaged in heritage management across various parts of the world, while quite rarely engage with evaluating project outcomes (societal or broader) as part of their core focus. While this forms a valuable groundwork to start from, it also suggests a gap in methodological approaches to study positive aspects and benefits of participatory approaches or the difference they may make in establishing that lost 'power balance'. A significant body of researchers developing studies around the so-called authorised heritage discourse (AHD); Smith, 2006; Smith & Waterton, 2012; Akagawa, 2018) that has shaped theoretical advancement in the field of heritage studies has been fairly critical to institutions' role and policy bodies' position in reproducing the power gap between decision-makers and the public, exposing potentially negative implications in the way heritage is perceived and pointing towards low expectations on future engagement. This critique, in a sense, had as a result a distancing of practice-based studies (e.g. studies using evidence from public programmes and evaluations) that can be important to shed light on formulation of best professional and applied approaches and bridge the gap between theory and practice in the field. This chapter aims to advocate for bottom-up, policy-relevant and action research approaches that can inform practice to cover this gap and especially shape further evidence on the role of heritage for social sustainability.

The chapter also addresses one key dilemma in the heritage studies work on social value and social outcomes (of heritage projects): the focus on issue of management contrasted to that of heritage governance (which is considered to include planning and sustainable tourism management) and its implication for studies on participation and its added value. This is discussed here, as part of engaging with the AHD but also due to the mere link with the fundamental issue of benefits or dividends being linked to few types of stakeholders versus wider societal benefits, equally distributed among all members of society. Heritage is being considered as a source of 'capital', both monetary and social, with potentially exclusionary character: this could be attributed to the fact that preservation regulation and production processes have been dominated by experts, as opposed to locals, or by few communities 'holding' certain types of knowledge and power within a social system. Through comparing different disciplinary lenses, used in studies of participatory heritage management and approaches to it, one can study societal benefits ranging from global to community and individual level. It will argue that there is a gap in studies and tools for analysing (the evolution and outcomes of) social interactions that can enable us look not only at social well-being benefits for communities (as static outputs in time), but also at interactions and relationships between heritage institutions and communities and their indirect effects on heritage protection. The proposed theoretical stance is based on the social psychology concept of reciprocity and the adjacent concept of social capital that enables accounting for 'social resources' established, maintained and shared between institutions and communities.

The chapter argues that a systemic conceptualisation of institutions and communities as interacting and mutually dependant social actors is necessary development in this quest for benefits mapping – asking the questions: benefits for whom and through which (social) practices are they produced? It proposes a novel theoretical lens (that of reciprocity) that can shape conceptually and methodologically a framework for further studies, while linking participation with social and institutional impacts stemming from processes of engagement while enabling the introduction of dynamic analysis (across time and multiple actors).

The proposed novel theoretical framework can allow us to critically develop further the AHD (looking at heritage as social practice) and expand research horizons in the field. It can allow us to link heritage studies with research on social sustainability and local development through a new theoretical approach of reciprocity and suggesting relevant methodological paths to analyse actor's relationships and heritage contribution to social well-being outcomes and sustainable management approaches (Gallou, 2020).

Participation in heritage: links with social sustainability and sustainable management

Looking at existing research on heritage participation, a researcher would find multiple relevant works under variant disciplinary traditions which look into equally varied forms and levels of engagement: work on levels of civic engagement, participation in political decisions, or planning within urban policy studies (Healey, 1997; McGuirk, 2001; McAreavey, 2006, among others) and in heritage field (Pendlebury, 2013), pointing to the responsibilities and commitment of public institutions and authority gradients and discussing its levels and types (see, for instance, Ladder of participation by Arnstein, 1969). A body of critical work under development studies, looks into instrumentalisation of engagement, normative and tokenistic approaches to community participation that has been linked with the role of authorities (White, 1996). Such work lies next to researchers that connect public participation to empowerment (or *the transformational role of engagement*), linked to local development (Pretty & Ward, 2001; Pretty, 2003; Hickey & Mohan, 2005 among others) and indirectly, linked to the agenda for achieving sustainable development goals. Works that consider cultural heritage as public resource or part of 'common pool resources' (Ostrom 1990, see discussion in Dragouni, 2017) emphasise the increasingly high responsibilities of institutions for allowing and encouraging open access and wider engagement with cultural heritage management. Such works can be very helpful, as they provide a critical lens into the analysis of what types of activities and processes exist across countries and what level of engagement is achieved on the ground with historic sites and cities. These are the basic steps when one is considering how these activities and processes can lead to various expected outcomes and inform approaches to study *outcomes of participation* (as a form of capacity building for the parties involved, from managers to community members).

The chapter will focus on approaches for studying 'active forms of participation'(as defined in Gallou, 2020): considering multiple forms from engagement in volunteering to, informal or community-led initiatives, engagement in decisions made around management of heritage resources and spatial planning. This aims to cover engagement in (re)-production of cultural resources as opposed to a more passive consumption through visiting or using such resources (see, for instance, O'Brien & Oakley, 2015; Crossick & Kaszysnka, 2016).

This type of engagement in heritage is considered an enabler for one of the less explored pillars of sustainability, the social one: supporting social capital and social well-being goals, which have become especially important for the challenges faced mainly by isolated rural or socially fragmented communities. The role of heritage as enabler for social sustainability had been conceptualised through a human rights approach (Bonnici, 2009) to heritage management: this was expressed through the urges of the 2005 European Framework Convention (Council of Europe, 2005) that debated on the

Value of Cultural Heritage for society, prioritising democratic participation for both interested, marginalised and local communities to foster social cohesion and mutual respect (Logan, 2012, p. 236). Therond (2009) connects these emerging communities with the academic discourse on creating new states of balance and 'shared responsibility' between them and the respective functions of institutional experts in the processes of heritage management, which is the direction the chapter will adopt to study participatory approaches.

If we accept that heritage is an enabler for social sustainability, then a greater focus is needed in managing heritage, as a (socially defined) process (rather than a cultural product or a statically defined material resource), a part of culture, which, at the same time, functions to support and transmit culture (Ashworth et al., 2005, p. 7; Smith, 2006; Dicks, 2007; Harrison, 2013; Auclair & Fairclough, 2015).

By expanding this assumption to heritage institutions, researchers in the field need to study them as key social actors supporting local development initiatives but also as having social responsibility towards public heritage management, collaborating with communities through participatory processes to ensure benefits for all. Enabling participation in heritage has been traditionally considered 'good practice' in conservation, with international bodies like World Heritage Centre and the International Centre for the Study of the Preservation and Restoration of Cultural Property (ICCROM; see Wijesuriya & Thompson, 2016) advocating for 'people-centred approaches' as the future for heritage to engage with development goals. It has been praised for its potential to counteract conflicts and contribute to sustainable heritage management (Radice & Labadi, 2010; Spiridon & Sandu, 2015). However, its contribution to institutional goals through achieving instrumental and institutional benefits has only been indirectly and rather sporadically discussed (Clark, 2006; Scott, 2006).

The increasingly recognised potential of participation in heritage to affect human well-being (mental, psychological, social as well as physical health aspects of it, as in Chatterjee & Camic 2015) by policy-driven research reflects a certain emphasis on instrumental and more easily measurable benefits (such as job and economic growth with a recent turn to individual well-being via life satisfaction) compared to softer (and usually harder to quantify) set of benefits mainly being recognised and advocated for by civic society associations and bodies advocating for equal access (such as active citizenship, right to decision-making, social capital and social support) that indirectly support community and societal well-being. Many of the latter are gained through the process of engagement itself instead of being indirect outcomes of finalised capital work projects (as heritage tends to be reduced to sometimes) for example (Gallou & Fouseki, 2018).

The chapter advocates for the need of bottom-up, multi-actor and relational approaches that can shed light on social well-being benefits through looking at participation as a process rather than looking at the outputs or outcomes of engagement for individuals (and subsequently assuming the collective outcomes as being a sum of individual ones) alone. We assume that benefits from the process are mutually beneficial for both institutions and communities as they can maintain longer-term, sustainable interactions for the protection of heritage resources across time, exchange knowledge and shape networks in place that are key prerequisites for mere protective actions to take place.

We recognise that the various potentials of participation to provide both instrumental and intrinsic benefits (Gallou & Fouseki, 2018) can lead to tension and ineffective applications of it (not to mention that political manipulation of participation at governance level can easily result in tokenistic approaches) (White, 1996; Monno & Khakee, 2012). This makes the urge for applied research 'unpacking how participation is employed and what constitutes best participatory approaches in terms of both process and outcomes, vital for avoiding such ineffective applications and reducing risk of tokenism. The novel theoretical framework of reciprocity presented here can be conceptually and methodologically employed to allow for relational analysis, position the focus on institutional benefits next to societal well-being ones and enable truly examining heritage as social practice (as per AHD). Most critically, it enables locating social outcomes as distributed across various actors and stakeholders involved in management and planning (conceptualised as social exchanges) and can therefore enable us to answer the key question of 'who benefits'(Waterton, 2005; Gallou & Fouseki, 2019).

Defining heritage management and sustainable approaches to it: the role of community participation

In search of a definition of what falls under 'heritage management' and what are impacts to institutional practices (related to management approaches), we can draw upon Pendlebury's theorisation of 'heritage conservation–planning', perceived as a social entity under constant iteration, an 'assemblage' (DeLanda, 2006 cited in Pendlebury, 2013). This suggests that participatory approaches to heritage management can be viewed in different stages of the 'heritage protection life cycle' and transcend management to reach the planning sphere: from identification of new heritage, to mapping and identifying attributes, conserving, interpreting and sharing its values and significance but also legislating and applying planning frameworks to ensure its long-term safeguarding and regulatory frameworks around access to it (Figures 6.1a and 6.1b).

In that sense, studies on participation cannot just look at typical engagement forms (like volunteering in conservation of material resources) but on a broader set of types of engagement. It is, however, obvious that the processes of engagement in all the upper stages of heritage protection life cycle may differ a lot as the stakeholders involved

Figure 6.1a Typical stages or phases, linking project planning with delivery and evaluation process of heritage projects from a project management perspective.
Source: Adapted from Gallou, 2020, p. 443 and recreated by Kalliopi Fouseki.

Figure 6.1b Heritage life cycle, as a typical 'linear' process of heritage management to support operational analysis: following Burra charter values-based approach framework. Dissemination can be considered to include celebrations, interpretations activities and conservation to include conservation planning, maintenance and documentation activities and risk assessment tasks.
Source: Adapted from Gallou, 2020, p. 443 and recreated by Kalliopi Fouseki.

in each of those stages differ and national frameworks of protection and planning can decisively define who has (legal) rights to engage and how they can affect decisions related to the protection of material heritage resources. The involvement of voluntary or community sector has also had a long involvement in the delivery of conservation outcomes (Mac Donald, 2011), encouraged in certain national contexts like the UK more than others, where non-profit organisations that represent social interests and may include local residents (Fox et al., 2005) are also often involved in delivering conservation outcomes.

Academics and institutions (e.g. Sullivan, 1998; Demas, 2002; Mason, 2002; Mason & Avrami, 2002; de la Torre et al., 2005) have previously recognised socio-political aspects of 'heritage process': focusing on the right of multiple stakeholders to participate in conservation, or management of urban heritage (Mac Donald, 2011) becomes key for allowing for a value-based approach (Mason & Avrami, 2002, p. 25) to be operationalised. It is no chance that existing work on equalising access has focused on shared values: as values are key in re-establishing the dominant discourse each time (as defined in AHD) participation obtains a key role in assisting in values sharing, exchange versus perceptions and interactions between actors with different value sets (Gallou & Fouseki, 2019).

The subject of ownership of heritage resources (legal versus moral; Byrne, 1995) lies anyway at the heart of the AHD, questioning for the rights of heritage managers (as 'experts') to manage and decide alone, and requires acknowledging the rights of local populations ('non-experts') to decide over their past (Sørensen & Carman, 2009). Therond (2009) calls for shared responsibility between them and functions of institutional experts and participatory approaches lie in the heart of enabling any equal 'footing' in decision-making in order to share not only responsibilities but also benefits stemming from using and communicating heritage values.

However, those innovative and emerging conceptualisations of heritage, communities and new models of management come with a set of challenges regarding their link with achieving sustainable management practice. Existing institutional guidelines on sustainable heritage management (see, e.g., English Heritage, 2008) tend to show a strong concern on sustainability of material dimensions for future generations, but rather ignoring the importance of actors in achieving that through collaboration or commonly agreed and sanctioned frameworks for action. A relevant definition of sustainable heritage management framework based on the UK context is given by Landorf (2011) who recognises that while conflicts may exist between potential heritage (as a physical resource that needs protection) and societal benefits from heritage management and certain ideological lenses may prevail, the intersection between the two is worth exploring.

An assessment of wider socio-economic instrumental benefits stemming from heritage, recognised by Waterton (2010), Watson and Waterton (2010) and Pendlebury (2013) and lately discussed by Berger et al. (2020), can be beneficial in those lines to make the link: such studies identify areas of contribution of heritage to sustainable development agenda like social capital and local development-related outcomes (with social capital considered a driver for entrepreneurship as well). Other policy reports in the UK identify key benefits for participants engaging with heritage like cohesion, empowerment, ability to act as group but also close and wider relationships formation, mentioned (e.g. in BOP, 2011; Murzyn-Kupisz & Djiazek, 2013), but few practically focused studies exist to provide a strong empirical basis for those claims (Gallou, 2020). A body of studies uses well-being indicators like life satisfaction and subjective well-being, accompanied by a set of research looking into specific health benefits for participants that derive from the nature of their experience (physical and emotional connection or reminiscence for example; e.g. Konlaan et al, 2000; Staricoff et al., 2001; Staricoff, 2006; Bygren et al., 2009; Clift et al., 2009; Camic & Chatterjee, 2013; Napier et al., 2014; Gallou & Fouseki, 2018, 2019).

Less clarity occurs in what can be termed sustainable conservation and the role of communities and people centred approaches to achieve it, with some benefits in the realm of place-making (Delconte et al., 2016; Beck & Brookes, 2018; Corcoran et al., 2018) and sense of place (Hawke, 2011) considered of a reciprocal nature and synergistic between institutions and communities (Gallou, 2020, p. 320).

UNESCO'S (2010) aim to delineate the 'sustainable management of cultural heritage' at the service of development can help us conclude on a key definition of the concept and ways to assess the contribution to sustainable communities. UNESCO underlines at least two important dimensions of it: that of longevity of the material aspects and that of economic, environmental and social viability. Social viability of cultural heritage is thought as obtained through participation of local populations in its management, interpretation and dissemination and is essential to ensure collective and individual pride in heritage, interest and involvement in its protection. This is the aspect of sustainable management that we need to focus on and acknowledges that this needs to be integrated in evaluation frameworks for sustainability, next to measurements of longevity of material aspects. A starting point is monitoring continuous and recurring engagement in heritage and relationships between different communities and institutions.

There is also the key issue of gap between practice and official discourses that needs to be addressed: it was only lately that the discussion on cultural impact evaluation (as impact evaluation applicable in wider range of cultural projects and programmes) has been critically reviewing the distance between official discourses and those that correspond to cultural practitioners, showing how the gap between the two is being enlarged in through the process of instrumentalisation of policy (Newsinger & Green 2016, who discuss the topic through the construction of the concept of 'cultural value', suggest that it is construed 'through the epistemologies and methodologies of cultural evaluation', see also Banks, 2015). O'Hagan (2016) reflects on the discrepancy between institutional objectives for the cultural sector and societal benefits as experienced by communities. Such researchers underline the importance of approaching the subject of social impacts from the perspective of practitioners and communities as key for covering the gap in the way societal benefits are described and considered by researchers but also policy-makers in the future: the gap around processes of engagement (as compared to outcomes and measurable outputs) is key in reshaping this discourse. This is another fact that points to the need for understanding synergies, reciprocal links but also diversions in objectives between institutions and communities to achieve social well-being and the development of methodologies to study them.

Discussion of existing approaches: gaps and the need for a reciprocal theoretical framework

Given the multiplicity of types of heritage (tangible, intangible and intangible aspects of tangible) and types of engagement (level and degree of involvement by beneficiaries, type of engagement, process and activities involved, etc.), the options and approaches for researchers willing to study outcomes in the social sphere remain open but challenging. A shared attribute of existing studies revealed here is a gap on massive empirical evidence on outcomes or impacts of projects and interventions with a less critical lens on processes and interactions that led that undermines the role of institutional input for the production of (yielding) social outcomes. The effect on institutional processes, managerial approaches and behaviours needs to come more to the forefront as is the interdependency between communities and institutions (as social actors) and the systemic nature of the problem of participation.

To achieve this, the chapter suggests looking back into the theoretical basis of participatory management research and makes a proposition for a *reciprocal impact framework* as the way forward together with a systemic conceptualisation of communities and institutions as interdependent actors of the same (social) system (Avrami 2012; Gallou, 2020, p. 297). Conceptualising heritage as a social process of value deliberation and interpretation, and accepting that actors' actions are interlinked and one affects another, one needs to take a closer look into sociological theories that seek to explain the development of long-term relationships, partnership and links between actors, in this case communities and institutions, or the formation and sustenance of social capital (its various types bonding, bridging and linking) as outcome of the process or its exchange with other forms of capital (see Delconte et al., 2016). Moreover, being able to map changes and understand the dynamic nature of those links/relationships across time, allows researchers to understand their indirect impact and potentially causal effects: relational resources can act as enablers of achieving heritage conservation outcomes.

The concept of reciprocity, as defined within anthropological and social theory, consists of this intangible aspect of social capital that makes relationships work and sustains the connections that bind societies together: ' a compelling obligation that reflects the normative standards that sustain [social] exchange' (Misztal, 1996, as in Fu, 2004, p. 20). When the concept is applied in the study of the relationships between institutions and communities it can allow us to frame theoretically an understanding of impacts for all actors involved in participatory projects in a systemic way. This is possible through observing social (and potentially human, physical or financial resource) exchanges across time and considering actors' socio-political interactions: such interactions may imply expected rewards or links with powerful external actors that can support the missions of heritage protection bodies or groups, especially evident at planning-level decisions (Manzo & Perkins, 2006). In terms of social impacts from engagement, studying reciprocity in behaviours ties closely with studying trust towards public institutions and their representatives as cultivated through collaboration or exchanges within project structures. As trust forms key aspect of social capital, the ramifications for societal well-being can also be studied (e.g. lack of or presence of cohesion, degree of collaboration with public bodies from specific groups) but also, in return, their effect for achieving heritage planning objectives.

Social capital, trust, but also *reciprocity* between social actors are highly interlinked concepts and therefore relevant to the discussion on effects of engagement (as social activity) on attitudes to heritage institutions. The concept of *reciprocity* allows us to suggest an analytical framework (see Gallou, 2020, p. 107) that employs rich literature on social networks analysis and social capital mapping (in its various types and forms). It is key to recognise that it allows not only for quantitative mapping of networks but also for qualitative analysis of the nature of the relationship developed through social interactions.

How would one conceptualise participation and its study under such a framework?

i Participation as a social interaction involving exchanges and social network creation between participants, associations and professionals, sustaining social capital. In this way, participation can improve the social well-being of individuals (direct impact) and community groups (knock-on impact, as they may support social relationships but also affect individual actions for continuous engagement and support of participatory projects. Due to their nature their potential for maximising social well-being benefits is high).
ii Participation as a social interaction involving exchanges of (tangible and intangible) resources between local communities and institutions. Through these exchanges, participation can improve the way communities view institutions and their work, increase support in their future ventures and produce reciprocal behaviours.
iii However when actors influence each other and participants distribute impacts within their networks, not only relationships are developed in the process of engagement, but also positive outcomes for individuals are gained.

It is impossible to ignore that heritage studies need to account for social (and wider wellbeing) outcomes emerging from the significant spatial attributes and engagement with historic places as well; a spatial interaction, which has been seen to increase one's 'sense of place'. Through engagement with place, communities can contribute to safeguarding heritage material aspects but also contribute to place-making and decision making at planning and governance level too (Gallou, 2020, p. 108).

While the first two concepts look more in depth in stakeholder relationships, the third one hints to methods for research from disciplines like environmental and social psychology and the study of behaviour change due to interaction with certain physical contexts. It can provide insights for ways of managing physical resources that form part of wider environmental resources. It needs therefore to be weaved (rather indirectly) into the sociological lens for its outcomes to be interpreted. Well-being benefits from participation in heritage at community level can be conceptualised as stemming from social and spatial experiences. Institutions need to understand the interaction between social capital and sense of place (attachment and identity, dependency to place) and the effect of personal factors like interpersonal relationships and individual perceptions and experiences in shaping behaviours towards heritage and place, in order to grasp the complexity of those interdependencies.

Between processes, modalities of participation and outcomes: a systemic perspective

The notion of time and the notion of multi-actor framework are key from an analytical perspective to enable understanding of process and outcomes of participatory projects. Gallou (2020) has inductively explored such an approach, by analysing interviews and mapping behaviours of stakeholders across a time period, as a set of collaborative/participatory projects evolved. This can be helpful to provide typological clarity on forms of participation [like Arnstein's (1969) and followers' variations by Cohen & Uphoff,

1980; Choguill, 1996; Agarwal, 2001] and link them to commonly observed outcomes across places.

This also allows to study expected and non-expected but observed outcomes, as they are emerging through processes. In other words, it allows for the development of social impact frameworks, for identifying effectiveness of certain approaches versus others for achieving certain objectives (e.g. inclusion of elderly, disabled or incomers in heritage activities).

To allow more systematic studies to evolve, the heritage sector needs to employ new and effective tools – for example applying a systemic level analysis that looks into ecosystem of stakeholders and identifies exogenous and endogenous factors that affect their interactions (as a social system) can be useful start. Applied tools that can support causal link analysis and systems conceptualization, like causal loops diagrams (Fouseki & Bobrova, 2018; Gallou & Fouseki, 2018; Fouseki et al., 2020; Gallou, 2020, p. 138), but also theories of change development and stakeholder mapping approaches based on actor-network theory applications (see, for example, Ong, 2014), provide to the researcher the freedom to follow up the distinct processes and role of actors in management processes in the course of time. More complex research designs could look into developing models to perform factor analysis to identify key drivers and barriers for engagement for example, look into depth in impact evaluation framework based on the tripartite structure of participation described above, and start collecting pre- and post-intervention data on social attitudes to formulate experimental data sets or (if adopting a case study approach) isolate and develop control cases in a robust way. Impact evaluation studies have more to offer here. Causal effect analysis allows to locate inter-links between social and institutional impacts (how social capital development could lead to further changes and finally affect institutions in the course of time) but also indirect connections between impacts and behaviours towards heritage safeguarding.

Discussion: Expectations and roles for communities and institutions for achieving reciprocal links and collaborative management

Given the theoretical base presented above, achievement of reciprocal links between institutions and communities (and the individuals within both groups of actors) is key for achieving long-term relationship, enhanced collaborative management of heritage resources based on consensus and deliberation. This can support capacity building of local gatekeepers (especially crucial in isolated contexts rich in heritage resources, see Hewlett & Edwards, 2013), whereas the development of external links with new social actors can support local well-being (Uphoff, 1992).

Key for reciprocal behaviour between experts, their institutions and community groups are the amount, direction and quality of mutual exchanges of knowledge and resources. This allows for trust development and legitimisation of contribution to both sides (expert knowledge coming from local groups bound to their land and scientific support coming from conservation professionals) through collaborations based on mutual trust and empowerment.

In order to understand how an effective collaborative heritage management model (see also Aas et al., 2005) can be established, one needs to expose individual-, community- and institutional-level motivations and perceived ability to engage in such management approaches.

Figure 6.2 shows the different levels (individual, community and institutional levels) of expectations from the other side to engage in a reciprocal relationship as developed through the analysis for the doctoral research by Gallou (2020). These can define how

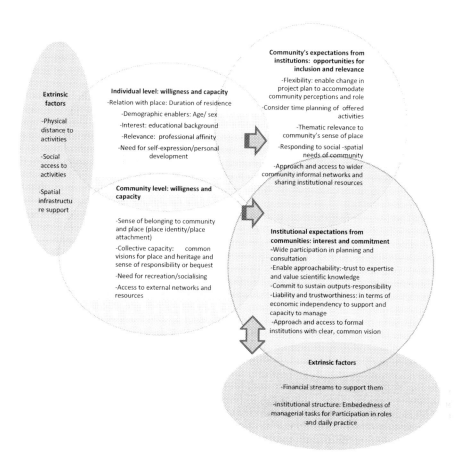

Figure 6.2 Key factors for participatory projects to develop and flourish from participants' perspectives – at individual and community levels. Expectations from both sides (communities/institutions) stemming from interest and expected benefits.
Source: Gallou (2020, p. 328).

effective participation may look like, in terms of achieving positive social and institutional expectations and goals that matter for the different groups engaged (experts, community members, etc.). The distance between community and institutional expectations help us identify gap areas for achieving reciprocal links (Figure 6.2).

The analysis of behaviours of agents at a systemic level confirmed the interdependencies between managers', planners' agency as well as the role of (less dynamic) factors like institutional structures in place, for enabling reciprocal and sustained links between institutions and local communities. Quality indicators for inclusivity, the effect of knowledge exchange channels and the quality of information, were identified as key factors in developing trust (Collier & Berman, 2002). The use of the framework enables understanding which individual and community level factors (like willingness to participate and capacity to engage) enable participation to *happen* at the first place (Figure 6.2).

Conclusions

Identifying typologies or modalities of participation as levels of engagement and processes as discussed above is, firstly, important for heritage studies in order to assist the development of theories and frameworks of engagement in the context of consensus (Fouseki, 2015) and collaborative value identification. Further developments will benefit from analytical focus on relevant levels of power to influence the projects and forms of capital invested in this process, as well as function of participation (as per White, 1966; see Gallou & Fouseki, 2019; Gallou, 2020 for a proposed framework for analysis based on processes or activities, benefits as expected outcomes including social well-being outcomes, and beneficiaries, that can be useful for evaluating the effectiveness of variant participatory projects). Secondly, such analysis will enable process-based research that would greatly benefit the field and facilitate institutional applications of academic knowledge.

The wider definitions of heritage that may differ across cultures but also across institutional and political contexts as well as the levels of study (local, territorial, regional or national focus) undoubtedly lead to a multiplicity of approaches to study participation and its benefits. The need for stakeholder collaboration and participatory management approaches remains, however, commonly sought ideal across international case study literature on management but also relevant to governance models for heritage.

The imminent value of the proposed theoretical concept of reciprocity and methodological approaches that can be derived from its adoption lies in the ability of the *framework to study reciprocal relationships and benefits* between communities and institutions through participation. Its novelty lies in incorporating three distinct but *interrelated perspectives on studying 'participation' and 'collaborative management' approaches* that have not been considered together until now: social interaction at the individual, community and institution levels but also spatial interaction with place and active engagement in planning and place-making that can all be associated with form of social (and in many cases other forms of) capital benefits. The systemic approach allows for dynamic exploration of the relationship and how behaviours shape outcomes, while revealing causal links in the process (as perceived by participants) – it therefore offers much more than a static observation of single actors' actions in a single time point.

This way, the approach enables bridging the gap between process-focused and outcome-focused approaches to accounting for the benefits of participation. Participatory management research relies a lot on operational-level analysis and data collection. Further research on the subject can advance practice and illuminate policy in two ways: (1) by focusing on engagement in planning decisions and place-making processes as part of a management-planning integrated framework. This can allow evidence collection on the contribution of heritage to education and social capital development and (2) by focusing on developing process evaluation methodologies through utilising systemic-level analytical methods. These allow to look into partnerships, interdependencies and cooperation instead of evaluation from single/leading institution's perspective alone. By employing effectively qualitative approaches, new research can inform us sufficiently about motivations, barriers and changes in actors' behaviours, around the evolving landscape of co-creative and collaborative management approaches in the heritage sector.

References

Aas, C., Ladkin, A., & Fletcher, J. (2005). Stakeholder collaboration and heritage management. *Annals of Tourism Research*, *32*(1), 28–48.

Agarwal, B. (2001). Participatory exclusions, community forestry, and gender: An analysis for South Asia and a conceptual framework. *World Development*, *29*(10), 1623–1648.

Akagawa, N. (2018). Authorized heritage discourse. *The encyclopedia of archaeological sciences*, 1–4. https://doi.org/10.1002/9781119188230.saseas0056.

Andrews, F. M., & Withey, S. B. (1976). *Social indicators of well-being*. New York: Plenum Press.

Arnstein, S. R. (1969). A ladder of citizen participation. *Journal of the American Institute of Planners*, *35*(4), 216–224.

Ashworth, G. J., Graham, B., & Tunbridge, J. E. (2005). The uses and abuses of heritage. *Heritage. Museums and galleries: An introductory reader* (pp. 26–37). London: Routledge.

Auclair, E., & Fairclough, G. (Eds.) (2015). *Theory and practice in heritage and sustainability: Between past and future*. London: Routledge.

Avrami, E. C. (2012). *A systems approach to historic preservation in an era of sustainability planning*. Retrieved from https://doi.org/doi:10.7282/T3RB73H0 (Accessed on 11 May 2021).

Avrami, E., Mason, R., & de la Torre, M. (Eds.) (2000), *Values and heritage conservation: Research report*. Los Angeles, CA: The Getty Conservation Institute.

Banks, J. A. (2015). *Cultural diversity and education: Foundations, curriculum, and teaching*. London: Routledge.

Beck, D., & Brooks, S. (2018, May). Social enterprise and the development of cultural heritage assets as catalysts for urban placemaking. In F. Calabrò, L. Della Spina, & C. Bevilacqua (Eds.), *International symposium on new metropolitan perspectives* (pp. 308–315). Cham: Springer.

Berger, S., Dicks, B., & Fontaine, M. (2020) 'Community': A useful concept in heritage studies?, *International Journal of Heritage Studies*, *26*(4), 325–351. https://doi.org/10.1080/13527258.2019.1630662.

Bonnici, U. M. (2009). The human right to cultural heritage–The Faro Convention's contribution to the recognition and safeguarding of this human right. In *Heritage and beyond* (pp. 53–58). Malta: Publishing editions, Council of Europe. Retrieved from: https://rm.coe.int/16806abdea (Accessed on 10 January 2020).

BOP Consulting (2011) for HLF. *Assessment of the social impact of volunteering in HLF-funded projects: Yr 3 Heritage Lottery Fund*. Retrieved from: https://www.heritagefund.org.uk/sites/default/files/media/research/social_impact_volunteering_2011.pdf (Accessed on 10 January 2020).

Bygren, L. O., Weissglas, G., Wikström, B. M., Konlaan, B. B., Grjibovski, A., Karlsson, A-B., Sven-Olof, A., & Sjöström, M. (2009). Cultural participation and health: A randomized controlled trial among medical care staff. *Psychosomatic Medicine*, *71*(4), 469–473.

Byrne, D. (1995). Buddhist stupa and Thai social practice. *World Archaeology*, *27*(2), 266–281.

Camic, P. M., & Chatterjee, H. J. (2013). Museums and art galleries as partners for public health interventions. *Perspectives in Public Health*, *133*(1), 66–71.

Chatterjee, H. J., & Camic, P. M. (2015). The health and well-being potential of museums and art galleries. *Arts & Health: An International Journal for Research, Policy and Practice*, *7*(3), 183–186. http://dx.doi.org/10.1080/17533015.2015.1065594.

Choguill, C. L. (1996). Toward sustainability of human settlements. *Habitat International*, *20*(3), v–viii.

Clark, K. (2006) *Capturing the public value of heritage: Proceedings of the London conference 24–25th January 2006*. London: English Heritage.

Clift, S., M. Camic, P., Chapman, B., Clayton, G., Daykin, N., Eades, G., & White, M. (2009). The state of arts and health in England. *Arts & Health*, *1*(1), 6–35.

Cohen, J. M., & Uphoff, N. T. (1980). Participation's place in rural development: Seeking clarity through specificity. *World Development*, *8*(3), 213–235.

Coleman, J. S. (1988). Social capital in the creation of human capital. *American Journal of Sociology*, 94, S95–120.

Collier, C., & Berman, G. (2002). *Community stakeholder involvement*. Retrieved from http://www.safegrounds.org/pdfs/W38_Safegrounds_Community_Stakeholder_final.pdf (Accessed on 03 May 2021).

Corcoran, R., Marshall, G., & Walsh, E. (2018). The psychological benefits of cooperative place-making: A mixed methods analyses of co-design workshops. *CoDesign*, 14(4), 314–328.

Council of Europe (2005). *Framework convention on the value of cultural heritage for society* (Faro Convention). Brussels: Council of Europe Treaty Series - No. 199. Retrieved from: https://rm.coe.int/CoERMPublicCommonSearchServices/DisplayDCTMContent?documentId=0900001680083746 (Accessed on 15 January 2020).

Crossick, G., & Kaszynska, P. (2016). *Understanding the value of arts & culture: The AHRC cultural value project*. Swindon: Arts and Humanities Research Council.

de la Torre, M., MacLean, M., Mason, M., & Myers, D. (Eds.) (2005), *Heritage values in site management: Four case studies*. Los Angeles, CA: The Getty Conservation Institute.

DeLanda, M. (2006). *A new philosophy of society: Assemblage theory and social complexity*. London & New York: Bloomsbury Publishing.

Delconte, J., Kline, C. S., & Scavo, C. (2016) The impacts of local arts agencies on community placemaking and heritage tourism. *Journal of Heritage Tourism*, 11(4), 324–335. https://doi.org/10.1080/1743873X.2015.1088019.

Demas, M. (2002) Planning for conservation and management of archaeological sites. In J. M. Teutonico & G. Palumbo (Eds.), *Management planning for archaeological sites: An international workshop, May 2000* (pp. 27–54). Los Angeles, CA: The Getty Conservation Institute.

Dicks, B. (2007). Review of the uses of heritage' by Laurajane Smith. *Museum and Society*, 5(1), 58–59.

Dragouni, M. (2017). Sustainable heritage tourism: Towards a community-led approach (Doctoral dissertation, UCL (University College London)). Retrieved January 5, 2020 from: https://www.semanticscholar.org/paper/Sustainable-heritage-tourism%3A-Towards-a-approach-Dragouni/3ebf99034ffdcf4993d30f59c2f38f205d57c2b2 (Accessed on 10 January 2020).

Duxbury, N. (2007). *Culture as a key dimension of sustainability: Exploring concepts, themes, and models*. Retrieved from: http://www.cultureandcommunities.ca/resources/publications.html. (Accessed on 10 January 2020).

Duxbury, N., & Jeannotte, M. S. (2011). Introduction: Culture and sustainable communities. *Culture and Local Governance*, 3(1–2), 1–10.

English Heritage (2008). *Conservation principles, policies and guidance*. London: English Heritage.

Fouseki, K. (2015). *Dispute management in heritage conservation: The case of in situ museums*. BAR International Series 2776. Oxford: British Archaeological Reports Ltd.

Fouseki, K., & Bobrova, Y. (2018). Understanding the change of heritage values over time and its impact on energy efficiency decision-making at residential historic buildings through system dynamics. In *The 3rd international conference on energy efficiency in historic buildings (EEHB2018), Visby, Sweden, September 26th to 27th, 2018*. (pp. 11–21). Upsalla: Uppsala University.

Fouseki, K., Newton, D., Murillo Camacho, K. S., Nandi, S., & Koukou, T. (2020). Energy efficiency, thermal comfort, and heritage conservation in residential historic buildings as dynamic and systemic socio-cultural practices. *Atmosphere*, 11(6), 604.

Fox, C., Brakarz, J., & Cruz, A. (2005). *Tripartite partnerships. Recognizing the third sector: Five case studies of urban revitalization in Latin America*. Washington, DC: Inter-American Development Bank, Steven Kennedy, SBK&A.

Fu, Q. (2004). Trust, social capital and organizational effectiveness (Master thesis). Virginia: Virginia Tech University, Retrieved from: https://vtechworks.lib.vt.edu/handle/10919/9926 (Accessed on 3 May 2021).

Gallou, E. (2020). *Exploring impacts of participation in heritage management: Reciprocal links between communities and heritage institutions in the case of Orkney islands* (Unpublished doctoral dissertation). London: UCL (University College London). Retrieved January 5, 2020 from: https://discovery.ucl.ac.uk/id/eprint/10089042/ (Accessed on 10 January 2020).

Gallou, E., & Fouseki, K. (2018, September). Heritage within landscapes as a catalyst for socio-economic development: Locating social impacts for rural communities outside of museum walls. In *Proceedings of safeguarding the values of European cultural heritage* (pp. 107–125). ICOMOS.

Gallou, E., & Fouseki, K. (2019). Applying Social Impact Assessment (SIA) principles in assessing contribution of cultural heritage to social sustainability in rural landscapes. *Journal of Cultural Heritage Management and Sustainable Development*, 9(3), 352–375. https://doi.org/10.1108/JCHMSD-05-2018-0037.

González, P. A. (2015). Conceptualizing cultural heritage as a common. In P. F. Biehl, D. C. Comer, C. Prescott, & H. A. Soderland (Eds.), *Identity and heritage* (pp. 27–35). Cham: Springer.

Gould, P. G. (2017). Considerations on governing heritage as a commons resource. In P. Gould, & K. Pyburn (Eds.), *Collision or collaboration. One world archaeology* (pp. 171–187). Springer: Cham. https://doi.org/10.1007/978[[sbn]]3[[sbn]]319[[sbn]]44515[[sbn]]1_12.

Harrison, R. (2013). *Heritage: Critical approaches*. London: Routledge.

Hawke, S. K. (2011). Local residents exploring heritage in the North Pennines of England: Sense of place and social sustainability. *International Journal of Heritage and Sustainable Development*, 1(1), 32–40.

Healey, P. (1997). *Collaborative planning: Shaping places in fragmented societies*. Hampshire & London: Macmillan International Higher Education.

Hewlett, D., & Edwards, J. (2013). Beyond prescription: Community engagement in the planning and management of national parks as tourist destinations. *Tourism Planning & Development*, 10(1), 45–63.

Hickey, S., & Mohan, G. (2005). Relocating participation within a radical politics of development. *Development and Change*, 36(2), 237–262.

Konlaan, B. B., Bygren, L. O., & Johansson, S. E. (2000). Visiting the cinema, concerts, museums or art exhibitions as determinant of survival: A Swedish fourteen-year cohort follow-up. *Scandinavian Journal of Public Health*, 28(3), 174–178.

Landorf, C. (2011). Evaluating social sustainability in historic urban environments. *International Journal of Heritage Studies*, 17(5), 463–477.

Logan, W. (2012). Cultural diversity, cultural heritage and human rights: Towards heritage management as human rights-based cultural practice. *International Journal of Heritage Studies*, 18(3), 231–244.

Macdonald, S. (2011). Leveraging heritage: public-private, and third-sector partnerships for the conservation of the historic urban environment. In: ICOMOS 17th General Assembly, 2011-11-27 / 2011-12-02, Paris, France. Retrieved from: http://openarchive.icomos.org/id/eprint/1303/1/IV-3-Article2_Macdonald.pdf (Accessed on 10 January 2020).

Manzo, L. C., & Perkins, D. D. (2006). Finding common ground: The importance of place attachment to community participation and planning. *Journal of Planning Literature*, 20(4), 335–350.

Mason, R., & Avrami, E. (2002). Heritage values and challenges of conservation planning. In J. M. Teutonico & G. Palumbo (Eds.), *Management planning for archaeological sites: An international workshop, May 2000* (pp. 13–26). Los Angeles, CA: The Getty Conservation Institute.

McAreavey, R. (2006). Getting close to the action: The micro-politics of rural development. *Sociologia Ruralis*, 46(2), 85–103.

McGuirk, P. M. (2001). Situating communicative planning theory: Context, power, and knowledge. *Environment and Planning A*, 33(2), 195–217.

Misztal, B. (1996). *Trust in modern societies: The search for the bases of social order*. Cambridge: Polity Press.

Monno, V., & Khakee, A. (2012). Tokenism or political activism? Some reflections on participatory planning. *International Planning Studies*, 17(1), 85–101.

Murzyn-Kupisz, M., & Działek, J. (2013) Cultural heritage in building and enhancing social capital, *Journal of Cultural Heritage Management and Sustainable Development*, *3*(1), 35–54.

Napier, A. D., Ancarno, C., Butler, B., Calabrese, J., Chater, A., Chatterjee, H., & Macdonald, A. (2014). Culture and health. *The Lancet*, *384*(9954), 1607–1639.

Newsinger, J., & Green, W. (2016). The infrapolitics of cultural value: Cultural policy, evaluation and the marginalisation of practitioner perspectives. *Journal of Cultural Economy*, *9*(4), 382–395.

O'Brien, D., & Oakley, K. (2015) *Cultural value and inequality: A critical literature review. A report commissioned by the Arts and Humanities research council's cultural value project. Documentation*. Swindon: Arts and Humanities Research Council. Wiltshire Publisher. Retrieved January 5, 2020 from: https://ahrc.ukri.org/documents/project-reports-and-reviews/cultural-value-and-inequality-a-critical-literature-review/ (Accessed on 10 January 2020).

O'Hagan, J. (2016). European statistics on cultural participation and their international comparability. *International Journal of Cultural Policy*, *22*(2), 291–303.

Ong, C. E. (2014). Rethinking (in) tangible heritage: Social constructionism and actor-network theory approaches. *Proceedings of XVIII ISA - facing and unequal world, challenges for global sociology*, Yokohama, Japan.

Ostrom, E. (1990). *Governing the commons: The evolution of institutions for collective action*. Cambridge: Cambridge University Press.

Pendlebury, J. (2013). Conservation values, the authorised heritage discourse and the conservation-planning assemblage. *International Journal of Heritage Studies*, *19*(7), 709–727.

Pretty, J. (2003). Social capital and the collective management of resources. *Science*, *302*(5652), 1912–1914.

Pretty, J., & Ward, H. (2001). Social capital and the environment. *World Development*, *29*(2), 209–227.

Putnam, R. D. (2000). *Bowling alone: The collapse and revival of American community*. New York: Simon & Schuster.

Scott, C. (2006). Museums: Impact and value. *Cultural Trends*, *15*(1), 45–75.

Smith, L. (2004). *Archaeological theory and the politics of cultural heritage*. London: Routledge.

Smith, L. (2006). *Uses of heritage*. London: Routledge.

Smith, L., & Waterton, E. (2012). Constrained by common sense: The authorized heritage discourse in contemporary debates. In R. Skeates, C. McDavid, & J. Carman (Eds.), *The Oxford handbook of public archaeology* (pp. 153–171). Oxford: Oxford University Press.

Sørensen, M. L. S., & Carman, J. (Eds.) (2009). *Heritage studies: Methods and approaches*. London; New York: Routledge.

Spiridon, P., & Sandu, I. (2015). Conservation of cultural heritage: From participation to collaboration. *ENCATC Journal of Cultural Management and Policy*, *5*(1), 43–52.

Staricoff, R. L. (2006). Arts in health: The value of evaluation. *The Journal of the Royal Society for the Promotion of Health*, *126*(3), 116–120.

Staricoff, R., Duncan, J., Wright, M., Loppert, S., & Scott, J. (2001). A study of the effects of visual and performing arts in healthcare. *Hospital Development*, *32*, 25–28.

Sullivan, S. (1998). A planning model for the management of archaeological sites. In M., De la Torre (Ed), *The conservation of archaeological sites in the Mediterranean Region. An International Conference Organized by the Getty Conservation Institute and the J. Paul Getty Museum, 6–12 May 1995* (pp. 15–26). Los Angeles, CA: Getty Conservation Institute.

Therond, D. (2009). *Heritage and beyond: Prospects opened up by the Faro convention*. Brussels: Council of Europe.

Radice, G., & Labadi, S. (2010). *Towards a UNESCO suite of indicators on culture and development (2009–2010) Working document*. Retrieved from: http://www.unesco.org/new/fileadmin/MULTIMEDIA/HQ/CLT/pdf/Conv2005_CDindicators_Literature.pdf (Accessed on 5 January 2020).

Uphoff, N. T. (1992). *Local institutions and participation for sustainable development.* London: Sustainable Agriculture Programme of the International Institute for Environment and Development.

Waterton, E. (2005). Whose sense of place? Reconciling archaeological perspectives with community values: Cultural landscapes in England. *International Journal of Heritage Studies, 11*(4), 309–325.

Waterton, E. (2010). *Politics, policy and the discourses of heritage in Britain.* Hampshire: Palgrave.

Waterton, E., & Smith, L. (2010). The recognition and misrecognition of community heritage. *International Journal of Heritage Studies, 16*(1–2), 5–15.

Watson, S., & Waterton, E. (2010). Heritage and community engagement. *International Journal of Heritage Studies, 16*(1–2), 1–3. https://doi.org/10.1080/13527250903441655.

White, S. C. (1996). Depoliticising development: The uses and abuses of participation. *Development in Practice, 6*(1), 6–15.

WHO (1948). *Preamble to the Constitution of the World Health Organization as adopted by the International Health Conference.* New York, 19–22 June, 1946; signed on 22 July 1946 by the representatives of 61 States and entered into force on 7 April 1948 (Official Records of the World Health Organization, no.2, p. 100).

Wijesuriya, G., Thompson, J., & Court, S. (2016). People-centred approaches. In G. Ghitty (Ed.), *Heritage, conservation and communities: Engagement, participation and capacity building* (pp. 34–50). Oxon, New York: Routledge.

Winter, T. (2013). Clarifying the critical in critical heritage studies. *International Journal of Heritage Studies, 19*(6), 532–545.

7 Social sustainability and witnessing difficult heritage

Ross Wilson

Introduction

This chapter demonstrates how studies of cultural heritage can take a leading role in debates regarding social sustainability. This is especially the case with aspects of 'difficult heritage', the tangible and intangible remnants of the traumatic events of the modern era. It is these objects and sites that require dialogue and engagement, they need a mode of witnessing that recognises the inequalities and divisions that persist in societies. This way of engaging and orientating individuals and communities with the past as a means of ensuring well-being, democracy and fostering a truly inclusive civil society is key for issues of sustainability. Whilst studies of sustainability have tended towards ecology, scholars within the arts and sciences, who are seeking to change the agenda away from purely environmental issues, have highlighted how an anthropocentric approach is needed to address the challenges that societies across the world will encounter in the present and the future. The problems of sustainability that the world faces must be addressed through the social factors that drive divisions within communities. By understanding the role of the witness in representing and acknowledging the past as a means of building an equitable present and future, heritage studies serve as a tool of sustainability.

Social sustainability and difficult heritage

The discussion of sustainability is often phrased within the context of the pressing issues faced by communities today. This framing is entirely necessary to mobilise action and awareness to deal with the issues of climate change and environmental collapse. However, the world is threatened by the events that have taken place over the course of the 20th century and not entirely by what occurs now. This threat is not solely the legacy of industrialisation and unrestrained capitalist growth, it is the consequences of the divisive, painful and difficult aspects of the past which leave unresolved debates, anger and suspicion. Whilst sustainability issues tend to be cast in the light of the present concern for future survival, there is a need to regard how past concerns shape current divisions. The disconnection between past, present and future within discussions of sustainability has meant that the notion of 'sustainable heritage' can sometimes focus on utilitarian responses to protect and preserve. This has inevitably obscured how sustainable heritage encompasses a concern for ensuring individuals and communities possess an ability to create and maintain equality and justice (Leichenko et al., 2010). This is essential as societies across the world are threatened by the divisions caused by the traumatic events of the modern era. To ensure the integrity of present and future societies, the legacy of the past must be addressed to build solutions to crises in democracy that threatens to dislocate people from one another (see Sachs, 1999; van der Leeuw, 2020). The representation of our individual and collective histories can be used to promote 'social sustainability' for our futures.

The movement away from a focus on addressing issues as imminent emergencies towards broader contextual studies follows the distinction drawn by scholars of sustainability that anthropocentric as opposed to eco-centric approaches are required to build resilient and tolerant communities (after Schlossberg & Zimmerman, 2003). This work has been particularly prominent within debates regarding sustainable urban planning where an awareness of how ethnicity, class and gender were at times forgotten within the broader environmental movement (Jarvis et al., 2001; Colantonio & Dixon, 2009; Manzi et al., 2010). The tendency to examine sustainability as an issue for the present and not one shaped by the past which is interlinked by issues of identity and power ensures that a commitment to be 'sustainable' could become another tool of control. To avoid such eventualities, notions of sustainability must be broadened to address the fractures that currently exist within our communities. The rise of populism, curtailment of democratic rights, the restrictions of movement across the globe and the inequalities which have been exacerbated by a global pandemic have demonstrated the need for 'social sustainability' to be regarded as a concern of paramount importance (Magis & Shinn, 2009). This focus on anthropocentric sustainability is sometimes confused with a drive to ensure jobs and opportunities are not lost in the campaign for an ecologically balanced future (see Nijkamp & Fritts, 1988; Elkington, 1997). However, social sustainability is more than just ensuring economic factors are included within environmental schemes. Such limited responses run the risk of accusations of 'green-washing', the proclamation of sustainable principles without the work to ensure their delivery. It has to be a critical awareness that it is the relationships between people that create sustainable communities in all senses of the term (Littig & Grießler, 2005).

Sustainability is not just a current environmental issue; it is a social issue and will only be addressed through a recognition of the factors that drive inequality and division. Studies of 'social sustainability' have sought to develop the debate with frameworks that seek to create societies based on equality and justice (see Cuthill, 2010). Magis and Shinn (2009, p. 38) forwarded a series of conditions that were intended to ensure that individuals and communities could participate in society and maintain engagement in the future: (i) well-being, (ii) equity, (iii) democratic government and (iv) democratic civil society. Social sustainability is less of an organised set of methods but rather a commitment to building principles of justice within schemes of sustainability (after Sen, 2013). It is this commitment to human relations as the basis of sustainability which connects to the place and value of heritage within society (see Landorff, 2011). It is within the spaces and through the objects of the past where this sense of social engagement and sustainability can be fostered and developed (Albert, 2015; Yadollahi, 2015). There could be a tendency in this respect to focus on the recognition of a 'shared heritage' and accentuate universal value of the past and a collective valuation of the historical environment as fundamentally positive and a benefit for communities (Smith, 2006; Labadi, 2012). This approach to heritage and social sustainability would be fundamentally flawed in its perspective as it would begin its assessment on the assumption of universal values and a pre-existing level of social equality (after Koning, 2001). Sustainability will not be achieved through an approach that examines the things we have in common to fashion a sense of civil society. It needs to be formed from a dialogue that concerns the 'difficult heritage' that communities possess. Sustainable communities can be formed but only within the context of public participation that addresses divisions and inequalities (after Macnaghten & Jacobs, 1997).

The discussion of 'difficult heritage' over the last two decades has focused on the way in which societies across the world attempt to address the legacy of war, genocide, repression and injustice (Logan and Reeves, 2009; Macdonald, 2009; Lee, 2019). The trauma of the modern era has created a tangible and intangible legacy across communities and society which have been neglected, obfuscated or repressed as the unyielding logic of late capitalism urges continuation and growth (after Jameson, 1990). Whilst some monuments and memorials are constructed to commemorate these moments of trauma, they are presented as 'reminders of the past' and 'lessons from history' (after Young, 1994). They do not address the structural inequalities, racism and prejudice that preceded the events, and which are still operating within contemporary society (see Hicks, 2020). The tendency to avoid painful or challenging aspects of national pasts is well-recognised within studies of memory, history and heritage as a means of maintaining notions of identity and concepts of stability. Whilst commemorative schemes may have proliferated to mark the events of the traumatic pasts, these processes can possess a tendency to obscure or historicise what are ongoing and recurrent issues. Indeed, the proliferation of memorials to mark the 20th century in particular has been critiqued as an attempt to reiterate a narrative of progress and development rather than reflect on the failings of civil societies to address these concerns (after Nora, 1989). The inability for individuals and groups to engage with recognising these aspects of the past will perhaps inevitably lead to the corruption and failure of civil society and democratic structures in the present and prevent a future community based on equality and justice. What is required is a means to respond to these pasts differently as a vital part of sustainability.

Witnessing the past, present and future

The 'difficult heritage' of the traumatic pasts of the modern era needs to be witnessed as part of our attempt to create places of equality and justice. Assessments of cultural heritage of conflict, genocide and repression all highlight the limitations of representational schemes to address inequalities and to respond to changes in society (Sørensen & Viejo-Rose, 2015; Echternkamp & Jaeger, 2019). The campaigns to remove memorials and monuments to the past demonstrate the limits of these material structures and their tendency towards the preservation of the *status quo*. The social, economic, political and cultural flux in which contemporary society exists entails that the manifestation of past ideals of stability will be an afront to any attempt to construct sustainable communities. Rather than distilling social memory into a memorial, exhibition or commemorative space, we require a way of engaging with these sites that builds towards principles of social sustainability (Magis & Shinn, 2009, p. 38). This is a recognition of the importance of witnessing the traumatic events of the modern era, not as an inactive process of acknowledgement but as a means of making these events part of our attempts to address the problems we face in the present. Witnessing is used here as a term to define the engagement society has with the past and its orientation to historic events. These are not points in history that may have been personally experienced, but they are moments that have structured contemporary society; they are events where the institutions, values and perspectives of individuals and communities have been formed from these aspects of the past. To ensure social sustainability, there is a need for us all to bear witness.

The perspective of the witness can be regarded here as an imaginative, emotional and intellectual tool for sustainability as it provides a point of reflection on the way communities are constructed (Wilson, 2016, 2017). To be a witness is to have an active commitment with and orientation towards the wider world, to observe, to reflect and to testify as to the occurrence of what has been seen, heard or experienced. The figure of the 'witness' is present within a variety of contexts across western societies, and it is frequently regarded as the bearer of truth. Within a legal context and within a religious connotation, the witness is one to whom the 'truth' has been revealed and it is this which establishes the privileged position of the witness and the duty to testify. However, the etymology of the word equates the term to personal knowledge rather than unassailable truth, derived as it is from the Old English 'witt' meaning conscience or understanding. Rather than the 'witness' possessing a singular truth that they are obliged to render up to their wider community, the role of the witness indicates the significance of active participation within the world (Ricœur, 2004, pp. 264–265). The 'witness' in this context is not an objective recorder but a figure whose perception is acknowledged to be specific in time, individual in scope and singular in experience. What is significant about the witness is their role towards the event; the manner in which their attitudes, ideas, values and identities are formed in relation to the action with which they have engaged (after Thrift, 2000, 2003). To act as a witness places moral, social and political obligations onto the individual through forming a connection between the individual and the event itself, as they are required to bear the burden of their witnessing and to testify to its significance.

Notions of witnessing have been explored within philosophical, media, theological and historical concepts for the purposes of social justice and equality (Felman & Laub, 1992; Oliver, 2001; Peters, 2001; Sontag, 2004; Frosh, 2006). Discussions of how social media enables witnesses across societies have demonstrated the significance of this observation in an era of 'fake news' (after Goolsby, 2010). It is within this context that witnessing for the purposes of social sustainability can be understood. For individuals and communities to find a means of engagement and understanding in order to build an inclusive and equitable civic identity for the future, there is a need to witness the past. This is the capacity of heritage studies to ensure communities are resilient and able to adapt to the challenges that will be encountered in the present and the future. The 'difficult heritage', which challenges the well-held notions of place and identity within the nation state, can serve as locations where well-being, empathy, democracy and civil society are generated. This is not brought about through the passive acknowledgement of the 'lessons of the past' but as a recognition that these events transform the individual and the community within the present. To witness is also not to essentialise our understanding or awareness of the past; the act of witnessing is always interpretative and flexible. Whilst the perspective of the witness may change, the moral responsibility of the witness does not alter; the purpose of the witness is to be knowledgeable and represent for others and the self (after Vivian, 2017). Witnessing can therefore provide an adaptable framework by ensuring that we address the historical divisions and injustices that shape our present. Studies of heritage are key in this respect, as they provide the locations where this act of witnessing can take place (after Simon, 2015).

To detail how this form of engagement with the past can address issues of social sustainability, the locations and places where difficult heritage is engaged with can be assessed for the acts of witnessing that are induced. Across the world, museums and heritage sites represent to visitors how acts of violence and brutality were inflicted in the past in order to address social issues in the present (see Aspel, 2015). However, the way this is undertaken is varied; different types of witnessing can be observed within exhibitions and displays for a range of objectives. As witnessing serves as a vital part of social sustainability, the use of these perspectives can be examined and assessed. This study has defined four types of witnessing that take place within heritage sites. These can be used to identify how concepts of well-being, empathy, democracy and civil society can be formed through our witnessing of the past to sustain our present and future (after Magis & Shinn, 2009, p. 38):

- Eyewitnesses – this focuses on establishing observation as a means of connection
- Expert witnesses – this emphasises knowledge to create awareness
- Character witnesses – this utilises an emotional connection to generate empathy
- Lay witnesses – this connects to the experience of everyday life

Using these concepts, the museums and heritage sites that represent the history of occupation and resistance in Europe can be assessed. These sites represent the 'difficult heritage' of the 20th century, where notions of citizenship, identity, democracy and empathy are negotiated for future and current generations (after Walsh, 2001).

Eyewitnesses

Over the last three decades, whether through audio–visual technology, narrative accounts or photography, museums have sought to create the perspective of the eyewitness (see de Jong, 2018). To be an eyewitness is to possess first-hand knowledge of the events: to have seen, heard or felt, and thereby become a vehicle for truth. This type of recognition appears to be prioritised within exhibitions and displays as it appears to bear greater weight than any other mode of representation. Whilst the visitors may not have engaged with the events themselves, they are placed within these sites as if they were able to observe them first-hand. Texts, photographs, video and even virtual reality (VR) have been used to attempt to place the audiences to these locations in the context of what happened to enable a corporeal or visceral engagement with the past. The effective nature of these exhibitions and displays is evident; individuals and groups are exposed to the trauma of the past as if they had lived through the events themselves. Museums and heritage sites that seek to create eyewitnesses are able to evoke deep emotional connection to the subject. This point of engagement can be found in museums that seek to engage audiences with histories of war and genocide in the 20th century such as Verzetsmuseum (Resistance Museum) in Amsterdam. In this display, visitors are taken back to Nazi-occupied Netherlands:

> A backdrop of recreated streets and wall-filling color photographs evokes the atmosphere of the war years. Authentic objects, photos and documents, film and sound fragments tell the history of people in wartime.
>
> (Verzetsmuseum, 2020)

The value of such displays is clear, as the events of the 20th century recede from 'living memory', as the generation that lived through those moments pass away, to make current generations 'eyewitnesses' ensures an important educational and cultural function. It maintains the remembrance of these moments in the present and thereby serves as a means of social sustainability. It fosters well-being and civic identity as the traumatic past is engaged with and visitors are asked to maintain this memory in the present.

However, there are limitations to the way in which these eyewitnesses are formed. The individual is always subject to the limitations of their visit, where time, information, knowledge and personal contexts can impinge upon their identity as eyewitnesses. The eyewitness will always be an observer who witnesses the experience of others. They may be called to speak of this knowledge, to testify their account but frequently museums or heritage sites do not request this application of knowledge. This limitation of the eyewitness stands in stark contrast to the proliferation of sites that use such techniques to return visitors to past eras as a means of gaining insight and understanding (after Speight, 2016, p. 24). Across the world, the role of the eyewitness is prioritised as the mode in which the previous century can be regarded by contemporary audiences. 'Seeing it for yourself' is vaunted as a means of acknowledgement and the dominant theory of knowledge acquisition (after Hooper-Greenhill, 1992). The eyewitness, which has been the predominant way in which modern exhibitions and displays seek to engage visitors, does not serve to ensure social sustainability. Indeed, as these institutions have asked individuals to become eyewitnesses to the past, we have allowed these events to become partial, disconnected from the present and possibly formulaic. With the role of the eyewitness, focused on experience and presence, there is little or no role for testimony and engagement. The past becomes a record which we have the responsibility to observe in a limited capacity but for which we cannot account for ourselves in relationship to it. Therefore, whilst the eyewitness perspective has dominated museum displays, it only provides a partial model for social sustainability.

Expert witness

In the confrontation with the traumatic history of the modern era, visitors to museums and heritage sites are presented with an array of data, facts and accounts to enable them to become 'experts' where they can observe and pass judgement on the events that are presented to them (after Dufresne-Tassé, 1995, p. 253). Turning visitors into expert witnesses is a mode of representation that insists upon the delivery of information as a priority (see Sodaro, 2018, p. 156). As individuals and groups process through these displays, the acquisition of data is presented as a means of 'learning from the past'. The learning objective within these displays is focused on ensuring that the visitor is in possession of the 'facts' and can then serve as an advocate for this truth (Golding, 2014). Such a perspective is utilised particularly in museums and heritage sites where divisive and dissonant histories are represented. The expert witness is able to consider the materials, assess the data, come to a considered opinion regarding the events and then make sure that these ideas are maintained within society (Weiser, 2017). Visitors become agents of knowledge in this respect as they are placed into positions of responsibility and power whilst asked to acknowledge what has happened through the acquisition of seemingly objective information.

The difficulty in this representation is the reduction of traumatic histories to the baseline of facts and figures and thereby an emotional distancing of the individual from that past. To be an expert is to cast a neutral, distanced perspective onto the events and only navigate their importance to the presence through that framework. The expert is also a figure that is lauded within specific cultural contexts for their position and function within society which ultimately excludes those without access to this perceived expertise. Installations, texts and objects that rely upon casting visitors in the role of the expert witness have proliferated across the world. For example, the Museo della Resistenza in Bologna, Italy, focuses on the scale of display within its account of the resistance against fascism:

> With photos, documentary material, also in original, period films, the theme of anti-fascism is proposed in its national and international aspects, with particular attention to the Spanish Civil War and the participation in this event of thousands of Italian volunteers, including many Bolognese, who would then play a fundamental role in the partisan struggle.
> (Museo della Resistenza, 2020)

The focus on the accumulation of knowledge to deal with traumatic histories is a common point of engagement. The role of the expert witness has been created at these sites to deal with the trauma of the past. In this manner, expert witnesses testify to the scale but perhaps not the reverberations of the traumatic events of the modern era.

Character witness

In the museums and heritage sites that recount aspects of difficult heritage, the use of oral, visual or written accounts of individuals who themselves were present during moments of brutality, horror and inhumanity is an important element in telling engaging accounts of the past (Bernard-Donals & Glejzer, 2001; Arrigoni & Galani, 2020). Visitors are frequently confronted with 'talking heads', narrations or videos that explain how an individual or community experienced these moments (Kavanagh, 2000, p. 141). This introduction of personal elements within the museum offers a learning model where visitors apprehend and engage with the immediate and emotive accounts of others. Instead of being made into eyewitnesses themselves, visitors are asked to regard the testimonies of others as a means of judging their experiences. In essence, visitors are asked to become character witnesses, to understand and engage with others and be guided by these voices as to their veracity, importance and their impact (Lehrer & Milton, 2001). The weight of responsibility is key here as visitors are asked to consider the lives of others, their experiences and their attitudes and regard themselves as arbiters of justice, able to assess and examine the evidence presented. The focus on the personal element, of an individual sharing their story with others, is used to add further emphasis on the history which is represented. This can be seen in the Latvijas Okupācijas muzejs (Museum of the Occupation of Latvia) in Riga. The collection of testimonies is used to establish connection and a sense of civic identity:

> These testimonies and the donated artefacts add personal experience to the history, through the life story of each person. These memories do not talk only of suffering, but of endurance, selflessness and resistance.
> (Latvijas Okupācijas muzejs, 2020)

This is the purpose of the character witness, to detail the human perspective as a means of establishing emotional and social connections. Using a range of media, textual, video, audio or even VR, the character witness is one which aims to establish empathy and to convince audiences of the importance of their recognition of another's history (after Gökçiğdem, 2019). The character witness offers a human and nuanced perspective on establishing a relationship to a past traumatic event. However, in the confrontation with the subjective experience of another, the power of the account and the importance in communicating the trauma of the past can be lost. In a world of competing narratives, expert opinion and 'fake news', the accounts of others can be obscured in the critical context of a mass media society. Where the stories of others are a constant point of engagement within social media, we are desensitised somewhat from the importance of another's story and its significance to our own lives. From the intimacies of life to communicating basic elements of humanity, our character witnesses are reliant upon a 'structure of feeling' which may be absent in the visitor (Williams, 1965, p. 64). As such, the impact of the trauma of the past may be lost in the narratives and stories that are present within exhibitions.

Lay witness

To be a lay witness is to speak from one's own perspective, to acknowledge the impact of an event or issue upon who you are and what you do. Over the last two decades, museums that detail the trauma of the 20th century have attempted to create exhibitions and displays that require individuals to assess how the past has shaped their own lives (see Wetherall et al., 2018). This type of witnessing within museums could be criticised as solipsistic but it links to modes of education and learning that require the individual to place themselves in the context of their study to enable comprehension (Hein, 2002). Visitors are asked to consider how the events they observe have impacted upon their lives, their families and their identities. For aspects of the recent past with a direct cultural or social connection to the visitor, this forms an immediate point of recognition. For those who may be distanced from the events, as a lay witness the individual is asked to regard how the histories they observe have shaped them and their world. The legacy of the wars, genocide and violence has left immutable marks on our political systems, our social orders and our sense of identity, morality and perspective. We are the products of the turmoil of the modern era. It is the role of the lay witness to recognise this connection and bear testimony to this effect. This can be seen with the displays in the Museu do Aljube Resistência e Liberdade in Lisbon, Portugal, which specifically asks visitors to take this perspective and connects the country's history of dictatorship with struggles for freedom across the world:

> It is a museum site and a historical museum that intends to fill a gap in the Portuguese museological fabric, projecting the valorization of the memory of the struggle against the dictatorship in the construction of an enlightened and responsible citizenship.
>
> (Museu do Aljube Resistência e Liberdade, 2020)

This fundamentally changes the way in which the history and legacy of the modern era is communicated to visitors. As we enter into the museum space, we are not asked to bear the burden of memory, but we are asked to be part of that history, to recognise it is part of our own sense of self. This can be done within an institution as part of a wider recognition within society. As we move further away from the 20th century and it ceases to be part of 'living memory', it is this mode of witnessing which will be essential. Our world has fundamentally shifted because of what occurred in the modern era. The lay witness, one who can testify to how this has altered their own perspective, is the most important carrier of the history of the past into the future. This is the capacity of heritage studies to address social sustainability. What we bear witness to is a past that has shaped us in order to create equality in the present and the future.

Conclusion

The legacy of war, colonialism, oppression and violence in the modern era is written into contemporary society. Therefore, there are points of engagement with the past that reference pain, trauma and unresolved issues that are divisive within democratic states. In response to these difficulties, governments, institutions and the wider population may resort to avoidance and obfuscation. However, exhibitions, displays and heritage sites provide a location where witnesses to the past can be formed who will testify to the abuse of power and recognise the suffering of others to create sustainable democracies and communities in the present. By regarding the point of engagement with these sites in the context of witnessing, we can begin to consider how these sites enable social sustainability. To be an eyewitness, expert witness, character witness and lay witness, we are placed in a relationship to the past that asks us to testify. This engagement demands that we orientate ourselves in relationship to the past as a means of addressing inequality in the present. By using the position of the 'witness', we can explore how difficult heritage can be mobilised for the benefit of individuals and communities within contemporary society.

References

Albert, M. (2015). Mission and vision of sustainability discourses in heritage studies. In M. Albert (Ed.), *Perceptions of sustainability in heritage studies* (Heritage Studies 4) (pp. 11–20). Berlin: Walter de Gruyter.

Arrigoni, G., & Galani, G. (2020). Recasting witnessing in museums: Digital interactive displays for dialogic remembering. *International Journal of Heritage Studies, 27*(2), 250–264.

Aspel, J. (2015). *Introducing peace museums.* London and New York: Routledge.

Bernard-Donals, M., & Glejzer, R. (2001). *Between witness and testimony: The Holocaust and the limits of representation.* Albany, NY: SUNY Press.

Colantonio, A., & Dixon, T. (2009). *Measuring socially sustainable urban regeneration in Europe.* Oxford: Oxford Brookes University.

Cuthill, M. (2010). Strengthening the 'social' in sustainable development: Developing a conceptual framework for social sustainability in a rapid urban growth region in Australia. *Sustainable Development, 18*(6), 362–373.

de Jong, S. (2018). *The witness as object: Video testimony in memorial museums.* Oxford: Berghahn.

Dufresne-Tassé, C. (1995). Andragogy (adult education) in the museum: A critical analysis and new formulation. In E. Hooper-Greenhill (Ed.), *Museum, media, message* (pp. 245–259). London and New York: Routledge.

Echternkamp, J., & Jaeger, S. (Eds.) (2019). *Views of violence. Representing the Second World War in German and European museums and memorials.* Oxford: Berghahn.

Elkington, J. (1997). *Cannibals with forks: The triple bottom line of 21st century business.* Oxford: Capstone.

Felman, S., & Laub, D. (1992). *Testimony: Crises of witnessing in literature, psychoanalysis and history.* London and New York: Routledge.

Frosh, P. (2006). Telling presences: Witnessing, mass media, and the imagined lives of strangers. *Critical Studies in Media Communication*, 23(4), 265–284.

Gökçiğdem, E. M. (2019). *Designing for empathy: Perspectives on the museum experience*. Lanham, MD: Rowman and Littlefield.

Golding, V. (2014). The elephant in the room. In A. Fromm, V. Golding, & P. Rekdal (Eds.), *Museums and truths* (pp. 3–20). Newcastle: Cambridge Scholars Press.

Goolsby, R. (2010). Social media as crisis platform: The future of community maps/crisis maps, *ACM Transactions on Intelligent Systems and Technology*, 1(1), Article 7.

Hein, G. E. (2002). *Learning in the museum*. London and New York: Routledge.

Hicks, J. (2020). *The Brutish museums: The Benin Bronzes, colonial violence and cultural restitution*. London: Pluto Press.

Hooper-Greenhill, E. (1992). *Museums and the shaping of knowledge*. London and New York: Routledge.

Jameson, F. (1990). *Postmodernism, or, the cultural logic of late capitalism*. Durham, NC: Duke University Press.

Jarvis, H., Pratt, A., & Cheng-Chong Wu, P. (2001). *The secret life of cities*. London: Routledge.

Kavanagh, G. (2000). *Dream spaces: Memory and the museum*. Leicester and London: Leicester University Press.

Koning, J. (2001). Social sustainability in a globalizing world: Context, theory and methodology explored. In H. J. van Rinsum & A. de Ruijter (Eds.), *More on MOST* (pp. 63–89). The Hague: National UNESCO Commission.

Labadi, S. (2012). *UNESCO, cultural heritage, and outstanding universal value: Value-based analyses of the world heritage and intangible cultural heritage conventions*. Lanham, MD: AltaMira Press.

Landorf, C. (2011). Evaluating social sustainability in historic urban environments. *International Journal of Heritage Studies*, 17(5), 463–477.

Latvijas Okupācijas muzejs (2020). Par muzeju. Retrieved from: http://www.okupacijasmuzejs.lv/lv/par-muzeju (Accessed on 29th March 2020).

Lee, H. K. (2019). *'Difficult heritage' in nation building: South Korea and post-conflict Japanese colonial occupation architecture*. New York: Palgrave Macmillan.

Lehrer, E., & Milton, C. E. (2001). Introduction: Witness to witnessing. Curating difficult knowledge: Violent pasts in public places. In E. Lehrer, C. Milton, & M. Patterson (Eds.), *Curating difficult knowledge: Violent pasts in public places* (pp. 1–22). New York: Macmillan.

Leichenko, R., O'Brien, K., & Solecki, W. (2010). Climate change and the global financial crisis: A case of double exposure. *Annual Association of American Geographers*, 100(4), 963–972.

Littig, B., & Grießler, E. (2005). Social sustainability: A catchword between political pragmatism and social theory. *International Journal of Sustainable Development*, 8(1–2), 65–79.

Logan, W., & Reeves, K. (2009). *Places of pain and shame: Dealing with 'difficult heritage'*. London and New York: Routledge.

Macdonald, S. (2009). *Difficult heritage: Negotiating the Nazi past in Nuremberg and beyond*. London and New York: Routledge.

Macnaghten, P., & Jacobs, M. (1997). Public identification with sustainable development: Investigating cultural barriers to participation. *Global Environmental Change*, 7(1), 5–24.

Magis, K., & Shinn, C. (2009). Emergent principles of social sustainability. In J. Dillard, V. Dujon, & M. C. King (Eds.), *Understanding the social dimension of sustainability* (pp. 15–44). New York: Routledge.

Manzi, T., Lucas, K., Lloyd-Jones, T., & Allen, J. (Eds.) (2010). *Social sustainability in urban areas. Communities, connectivity and the urban fabric*. London and Washington, DC: Earthscan.

Museo della Resistenza. (2020). L'antifascismo prima della Resistenza. Retrieved from: www.museodellaresistenzadibologna.it/tour-virtuale/mostre-permanenti2/l-antifascismo-prima2 (Accessed on 20th March 2020).

Museu do Aljube Resistência e Liberdade. (2020). Homepage. Retrieved from: https://www.museudoaljube.pt (Accessed on 12th March 2020).

Nijkamp, P., & Frits, S. (1988). Ecologically sustainable development: Key issues for strategic environmental management. *International Journal of Social Economics, 15*(3–4), 88–102.

Nora, P. (1989). Between memory and history: Les lieux de mémoire. *Representations, 26*, 7–24.

Oliver, K. (2001). *Witnessing: Beyond recognition*. Minneapolis: University of Minnesota Press.

Peters, J. D. (2001). Witnessing. *Media, Culture and Society, 23*(6), 707–723.

Ricœur, P. (2004). *Memory, history, forgetting*. Chicago, IL: The University of Chicago Press.

Sachs, I. (1999). Social sustainability and whole development: Exploring the dimensions of sustainable development. In E. Becker, & T. Jahn (Eds.), *Sustainability and the social sciences* (pp. 25–36). London: Zed Books.

Schlossberg, M., & Zimmerman, A. (2003). Developing statewide indices of environmental, economic and social sustainability: A look at Oregon and the Oregon Benchmarks. *Local Environment, 8*(6), 641–660.

Sen, A. (2013). The ends and means of sustainability. *Journal of Human Development and Capabilities, 14*(1), 6–20.

Simon, R. I. (2015). *A pedagogy of witnessing curatorial practice and the pursuit of social justice*. Albany, NY: SUNY Press.

Smith, L. (2006). *Uses of heritage*. London and New York: Routledge.

Sodaro, A. (2018). *Exhibiting atrocity: Memorial museums and the politics of past violence*. New Brunswick, NJ: Rutgers University Press.

Sontag, S. (2004). *Regarding the pain of others*. New York: Penguin.

Sørensen, M. L. S., & Viejo-Rose, D. (Eds.) (2015). *War and cultural heritage: Biographies of place*. Cambridge: Cambridge University Press.

Speight, C. (2016). Museums and higher education: A new specialist service. In: B. Cook, R. Reynolds, & C. Speight (Eds.), *Museums and design education: Looking to learn, learning to see* (pp. 11–27). London and New York: Routledge.

Thrift, N. (2000). Afterwords. *Environment and Planning D: Society and Space, 18*(2), 213–255.

Thrift, N. (2003). Performance and…. *Environment and Planning A, 35*(11), 2019–2024.

van der Leeuw, S. (2020). *Social sustainability, past and future undoing unintended consequences for the Earth's survival*. Cambridge: Cambridge University Press.

Verzetsmuseum. (2020). Nederland in de Tweede Wereldoorlog. Retrieved from: https://www.verzetsmuseum.org/museum/nl/exposities/nederland-in-de-tweede-wereldoorlog (Accessed on 20th March 2020).

Vivian, B. (2017). *Commonplace witnessing: Rhetorical invention, historical remembrance, and public culture*. Oxford: Oxford University Press.

Walsh, K. (2001). Collective amnesia and the mediation of painful pasts: The representation of France in the Second World War. *International Journal of Heritage Studies, 7*(1), 83–98.

Weiser, M. E. (2017). *Museum rhetoric: Building civic identity in national spaces*. University Park: Penn State University Press.

Wetherell, M., Smith, L., & Campbell, G. (2018). Introduction: Affective heritage practices. In L. Smith, M. Wetherell, & G. Campbell (Eds.), *Emotion, affective practices, and the past in the present* (pp. 1–21). Abingdon: Routledge.

Williams, R. (1965). *The long revolution*. Harmondsworth: Penguin.

Wilson, R. (2016). Witnessing and affect: Making new spaces to remember the Great War in Britain. In D. Drozdzewski, S. De Nardi, & E. Waterton (Eds.), *Memory, place and identity: Commemoration and remembrance of war and conflict* (pp. 221–235). London and New York: Routledge.

Wilson, R. (2017). Witnessing the Great War in Britain. In D. Harvey, & J. Wallis (Eds.), *Commemorating and remembering the First World War at its Centenary* (pp. 190–208). London: Routledge.

Yadollahi, S. (2015). Reflection on methodological approaches in assessing and implementing social sustainability in historic public spaces. In M.-T. Albert (Ed.), *Perceptions of sustainability in heritage studies* (pp. 159–172). Berlin: Walter de Gruyter GmbH & Co KG.

Young, J. (1994). *The texture of memory: Holocaust memorials and meaning*. New Haven, CT: Yale University Press.

8 Citizen science in sustainable heritage conservation

Rosie Brigham and Josep Grau-Bové

Finding a space in the scientific process

Citizen science (CS) can be defined as the involvement of non-experts in scientific research. It follows that, in order to examine CS, it is useful to start with the elements of the scientific process, and see where non-experts may fit in. The principal work of scientific discovery is often described like this: it involves conjecting hypotheses, drawing logical predictions and systematically carrying out empirical observations on them. A hypothesis is formed via inductive reasoning based on known observations; deductions drawn from these are then refined through the experimental findings. In a way, CS can be seen as the practice of including people not involved in traditional scientific discovery into one or more of these processes: hypothesis generation, prediction, empirical measurement and hypothesis refinement.

Experience shows that, in practice, CS projects tend to focus on only some of the activities outlined above. For example, it is much more common to involve non-experts in measurement than in hypothesis generation. As we shall see, this does not need to be the case. Heritage science has the potential to enable CS projects across all areas of the scientific process. Doing so is only a matter of identifying suitable opportunities.

It is important to note that the existence and the practice of CS can only exist in a world in which science is constructed as a preserve for academic institutions, professionals and industry (Haklay, 2013). Of course, the boundary between a 'scientist', a person actively engaged in a project involving a scientific method, and a 'citizen scientist' is unavoidably fuzzy. This chapter will not attempt to draw this boundary. A scientist in one area of expertise might be a citizen scientist in another area. This being said, terminology is important: it can change the way participants feel about themselves or their participation in an activity. Therefore, for a full exploration of this, we highly recommend reading Citizen Science Terminology Matters: Exploring Key Terms (Eitzel et al., 2017).

This chapter aims to be a practical introduction for the cultural heritage practitioner. Drawing from a variety of CS projects within heritage science and the wider field, we present some key issues that have to be addressed prior to conducting participatory research. As well as considering where research should include volunteers, researchers should consider who they will be targeting as well as giving foresight to issues surrounding data collection alongside managing a public engagement project. We hope this work assists in creating more participatory projects in heritage science, as there is a great potential for further research in this area.

Types of participation

A useful classifier of CS endeavours is the extent in which the participant is included into the scientific process. It can take many forms and varies greatly across projects; the crowd as a scientist, the crowd as a computer, the crowd as a reporter and the crowd as a sensor. Haklay's framework creates a useful way of dividing projects in regard to levels of participation and engagement within the scientific method. It consists of four different levels. At the highest level, participants are engaged across all areas of the research: in the problem definition, collection and analysis of data (Haklay, 2013). Such projects can be referred to as 'participatory science' or 'extreme citizen science'. In such projects, research teams such as the UCL Centre for Extreme Citizen Science (UCL ExCite) have pioneered using Bespoke Technology to assist with marginalised communities and indigenous groups on matters such as poaching and land management (Stevens et al., 2014). These projects focus on a 'bottom-up' practice, working with communities and interdisciplinary academics to ensure the research delivers the best outcome for all the participants within the project, be it researcher or contributor.

At the bottom level of the framework are projects wherein participants operate like sensors. Here the citizen scientist contributes only to a small section of the research process, adding little creative or considered input. This could be by submitting simple data or providing computing power. The utilisation of participants as distributed processors is not to be overlooked, shown by the success of projects such as Einstein@home (Abbott et al., 2009) and Foldit (Cooper et al., 2010).

Outlining the role of the participant within the scientific method should be of primary importance when designing a CS project. Will the project take on an active participatory model, or shall contribution be minimal? The majority of CS projects involve participants in one of two areas of the research process: data collection and analysis. This is not different in heritage science. CS within heritage projects often sits on the bottom two levels of Haklay's framework. Projects carried out in online settings often use the 'crowd' as a tool of analysis. Transcribe Bentham (Causer & Terras, 2014) and Old Weather (Romeo & Blaser, 2011) paved the way to use volunteers to transcribe historic documents. The Zooniverse and the Citizen Science Alliance are particularly popular portals for community-based analysis (CBA), hosting projects like Shakespeare's World, AnnoTATE and Criminal Characters (Zooniverse, 2020). There is distinct preference for CBA projects above other types of CS within galleries, libraries, archives, and museums. This may be because they hold considerable amounts of data and metadata in their collections that cannot be conveyed or analysed in full to audiences due to curatorial constraints. Thus, they successfully utilise the 'citizen historian', to engage and research their collections in a way traditional visiting does not allow (Ridge, 2013).

Projects which involve participants in both data collection and analysis are particularly popular within environmental conservation. Here, participants use sophisticated equipment to monitor air or water quality (Cohn, 2008), or simple tools, such as phone cameras to track plant growth (Johnson, 2016) or bird migration (Sullivan et al., 2009). Within cultural heritage schemes such as Micropasts (Bonacchi et al., 2014) and Rekrei (formerly Project Mosul), created and refined 3D models of heritage artefacts undertake both data collection and analysis (Vincent et al., 2015).

Recruiting a crowd

In addition to designing where volunteers fit into the research process, it is also crucial to identify who. Authors writing to recommend crowdsourcing for their particular field can frequently overlook the motivations of the crowd they want to gather (Bonney et al., 2016). Assuming participation without understanding the targeted audience could be detrimental to the project. Assumptions can often lead to bias, resulting in inefficient design and limited poor user reception.

In community-based monitoring (CBM) projects, participation is based mainly on the desire to contribute to scientific discovery over that of social factors, such as joining a community (Nov et al., 2014; Land-Zandstra et al., 2016; Lee et al., 2018); however, local objectives of raising awareness and filling data gaps have also proven to be an influence (Carlson & Cohen, 2018). Whilst a main objective behind Scottish Coastal Heritage at Risk Project (SCHARP) was to create a complete list of coastal heritage sites at risk across Scotland, it was also a highly useful tool to engage local communities. It is often found that the communication of tangible results is a stronger motivator than reward or recognition for contributing (Alender, 2016). CS researchers must therefore prioritise communicating their results clearly within an appropriate public space to allow participants to understand how their input has been used.

Seeking opportunity

Citizen science projects require a fertile soil. Their success largely depends on the existing interest of participants. We all have intellectual interests that do not find an expression as actions. We may be curious about monument conservation, but there are many barriers to contribute to a conservation project. We may be concerned with climate change, though can do little more than read about it. We all want to learn to code but cannot find the time. Projects thrive when they identify an existing motivation to contribute. No projects will succeed to convince participants to do something that they are not already interested in. Being alert to opportunity is as important as designing a robust method of participation. This is why, for example, there are many successful CS projects within the field of Ornithology; it is a serious leisure activity in which large groups of geographically distributed enthusiasts can carry out observations with little prior training. Projects such as eBird capitalise on this community, creating a platform in which bird watchers can contribute their sightings to a single repository, as well as creating a wider community. Bird watchers gain additional purpose from a hobby they already enjoy; their observations are taken seriously and contribute to important scientific research on a sometimes-global level.

The field of cultural heritage, as a popular sector for leisure activity and enthusiasts, holds great potential for both online and offline crowd-based researchers. CS can touch almost all streams of both cultural heritage and heritage science, from large-scale disaster management (Kumar, 2020) to object monitoring (Brigham et al., 2018) and curation (Ridge, 2013). In the 2018, Taking Part Survey, the main survey of cultural participation in England, 73% of the respondents had visited a heritage site in the past year (Department for Digital Culture Media and Sport, 2019). Similar to ornithology, this offers the heritage scientist an actively engaged community, who carries increasingly sophisticated technological equipment (smartphones) which can be utilised for research purposes.

This is the rationale behind the Monument Monitor project, which aims to assess the extent visitor photographs can assist with managing and observing change in remote heritage sites. Visitors to selected sites across Scotland are prompted to take and submit images of certain areas which managers require greater documentation for. The participants are a self-selecting group of enthused people who already demonstrate an active interest in heritage by their presence at a site. One of our findings is that the remoteness of the site does not correlate with the number of submissions. Submissions received from the remote Machrie Moor Standing Stone Circles on the Isle of Arran were ten times that received from Holyroodhouse Palace in the same month, even though the former had only a fraction of the visitor footfall. One possible explanation is that participants who make the effort to a remote site tend to be more interested and invested in monument. Both the Ghost Signs and Medieval Graffiti projects are further examples of this project methodology, in which volunteers documented historic advertising signs and medieval graffiti around the cities of Port Phillip and Norwich, respectively (Champion, 2015; Kyi et al., 2016). Here though, volunteers were actively recruited across an existing group of enthusiasts.

With the advent of the Internet, the opportunity to connect enthusiasts with scientific projects relevant to their interests increased exponentially. Consequently, the number of CS projects has increased dramatically, as has 'CS' as a field of scientific research (Tulloch et al., 2013; Follett & Strezov, 2015; Sauermann & Franzoni, 2015). From this, we see the rise of the 'digital historian' and the 'digital archaeologist'. Large data sets created in remote sensing and satellite archaeology (Parcak et al., 2017) lend themselves easily to CBA such as Fossil Finder (Wilson, 2018) and the Valley of the Khans project (Lin, 2010). In other projects, participants submit their own data; with increasing abilities of photogrammetry, community-based reconstructions that create digital facsimiles of heritage can also prove useful (Vincent et al., 2015). Online projects offer multiple advantages to the scientific researcher, the ability to either collect or organise a vast wealth of data whilst engaging audiences in scientific research.

This leads us into one of the final points of opportunity, science communication. CS projects have a double-positive effect as they can facilitate the explanation of scientific research as well as act as a form of public accountability. As such, CS projects satisfy research funders' requests for grant holders to conduct scientific outreach (Silvertown, 2009). Ortiz et al. point out that whilst there is a clear need for better science and technology outreach within cultural heritage, it is often beyond the reach of researchers' abilities, due to either time or budget constraints (Ortiz et al., 2018). However, CS and especially CBM projects directly engage the audience with a problem. Thus, such projects can have the potential to deliver a double benefit to heritage institutions in getting visitors to collect or analyse valuable data, as well as alerting them to any issues that they may have overlooked. It has been found that participants in community monitoring projects are often more likely to participate in local development (Conrad & Hilchey, 2011) and can assist with developing an awareness of cultural heritage values as well as different types of damage (Kyi et al., 2016).

Finding the correct methodology for your data

With an avenue of potential research found, it is important to apply the correct methodology for the project. This will vary according to the overall aims of the research. The development of new participatory methods is an area that would welcome further research. Not much is known about the benefits and drawbacks of different models of participation. Another unknown is how participants perceive and respond to different types of tasks. Above all, there is room for innovation and creating new types of participation.

In general, the more complex the task assigned to the citizen scientist, the smaller the pool of participants will be. This is not necessarily a drawback, especially if a project is focused on a small, local community. Projects which aim to recruit as many people as possible should be accessible with any potential barriers to participation being systematically erased (Ridge, 2014).

This is why platforms like Zooniverse are so exceptionally easy to use. At the time of writing, it was possible to contribute to scientific research from the main website landing page after only five mouse clicks, with about 20 seconds of time spent reading text. Anyone across the world with an Internet connection and web browser can easily contribute to scientific research. Their platform allows researchers to upload data and create projects in which volunteers can easily tag, annotate or transcribe with little training. It has been used for numerous heritage projects such as mapping 17th century weather data (Old Weather), identifying stories from ANZAC troops though WWI (Measuring the ANZACS) and identifying archaeological features in LiDAR images (Heritage Quest).

In contrast to this are projects such as the SCHARP run by the SCAPE Trust, a highly successful citizen heritage science project. In an attempt to fully record historical sites before being lost to coastal erosion, SCAPE created a standardised surveying methodology and recruited a network of volunteers across the country to assist in the documentation of coastal archaeology. Volunteers had to register and undertake basic training to participate in the study and when on site would record their findings through a bespoke mobile phone app as well as through comprehensive written surveys. This required a considerable time commitment from the volunteers. Training was crucial in this case due to the sometimes-dangerous nature of working on the coastline as well as the importance for comprehensive cataloguing. Over 500 people took part in various archaeology surveys, cataloguing 1,041 reports across 282 sites. It produced an excellent new body of data and shows that large-scale managed volunteer input can produce high-quality, detailed information (Hambly, 2017).

The Monument Monitor project sits between the two levels of engagement used by SCAPE and Zooniverse. Whilst barriers to entry are relatively low, it does require participants to be physically on the site. Instructions for participants are simpler than the documentation involved in the referenced SCHARP project; however, the participant still has more direction in their submissions than they may have in Mapping Historic Skies. In Monument Monitor, participants actively choose the subject of their photographs. In some monuments, they are asked to decide what could be considered as 'damage' as well as photographing the objects in an angle they deem appropriate. The submissions of the visitors are then assessed by site managers. This approach has led to a greater understanding of how visitors perceive damage at the sites. In one instance, images were frequently sent to report stone thefts. In fact, no such thing had occurred, rather a local sheep had pawed a hole in the ground to find shelter from the weather. This difference of participation engagement is reflected in the submission numbers. At the time of writing, Monument Monitor has over 4,200 submissions and SCHARP has just over 1,041. Both of these are well below Mapping Historic Skies which at the time of writing data from the project website indicates it has 47,887.

Data collection controls

In addition to developing a data collection method, it is important to apply measures to ensure the reliability of data. Data collected by citizen scientists are often held in low regard. In a study of biodiversity scientists, it was found that perceptions of the reliability of CS-collected data were very low; results were often treated with suspicion and thus rarely result in highly cited journal publications (Burgess et al., 2017). Haklay suggests that there is a cultural issue that leads to a suspicion of data generated by citizen scientists, a view that is based on an idea that 'science is best left to the scientists' (Haklay, 2013). In not knowing who is on the other end of the data point, an expert or an untrained amateur, it can feel like you are gambling the reliability of your results (Gura, 2013).

Measurement error is, however, measurable. Assessments of the reliability of CS data should be based on statistical evaluations of its quality. This has been assessed in several studies. In their review of different CBM projects in environmental monitoring, Conrad et al. found there to be little compensation for volunteer error in presented data (Conrad & Hilchey, 2011). Absent or problematic standardisation methods have also been raised as issues in CBA projects (Cohn, 2008; Dickinson et al., 2012) and CBM projects (Conrad & Hilchey, 2011). However, in an extensive review of community-based water monitoring projects, Carlson et al. demonstrate data interpretation bias was equal between professionals and citizen scientists (Carlson & Cohen, 2018). Similarly, systematic bias in aspects such as spatially and temporally non-random observations, weather, under detection of organisms, confusion of species and over-reporting of rare species was found in equal propensity in both professionally collected data sets and those collected by volunteers (Kosmala et al., 2016). With appropriate training, instructions or with clear procedures, citizen scientists can collect data with similar levels of accuracy as professional scientists.

The lack of awareness of the real credibility of data within the scientific community is likely a larger barrier than the competency of the participants (Carlson & Cohen, 2018). In order to improve this perception, projects must clearly show how data are collected and how potential inaccuracies are accounted for. In addition, CS data sets should be judged individually, according to the design and purpose and without presumptions of inferiority due to it being volunteer generated (Kosmala et al., 2016).

As mentioned, a common issue within many monitoring projects is bias. Bonney et al. (2009) noted that in ornithological surveys, certain species were often over-reported with others underreported, this being more exciting to report a rare bird, rather than to continually report common one (Bonney et al., 2009). McKinley et al. (2017) also found that more 'charismatic' species such as wolves, bears and certain birds got far more attention than other species. Within the context of Monument Monitor, a heritage site which has been popularised on TV may receive more visitors and thus more submissions. It is important that potential bias in any given data dimension should be considered and accounted for. Different project designs make erroneous data difficult to account for. In an ornithological CBA project such as eBird, confirmation of sightings by experts is very difficult as they are not physically with the volunteers.

A poorly laid out project has the potential to use up significant resources and result in poor data sets that can't inform decision making. Additionally, it can create the illusion that sufficient monitoring has been carried out leading to incorrect decisions being made (Tulloch et al., 2013). It is crucial to consider the data that the volunteers will generate early on and identify potential pitfalls and appropriate solutions to ensure the reliability of the collected data. In a CS project that collected records of local heritage, data analysts found that one submission could refer to the '14th century', another to '1356' and another to the 'middle ages'. Whilst all referencing to the same time period, the software processing the submissions struggled to accurately assimilate and categorise them (Blaise et al., 2018). Similarly, in a project that monitored mammal movement on the Appalachian Trail, the initial data gathered were found to be too imprecise to be of use. This was countered by asking the volunteers to produce clear numbers of animals, rather than ranges (Cohn, 2008).

There is a fine balance between asking volunteers to collect data that are highly specific in detail or data that are too generalised to be of any use. Consequently, it is important to develop a process with clear instructions and quality control measures and adapt them accordingly as data are collected. The methodology should be appropriate for the intended community of the project and should also allow for serendipitous discovery. This allows patterns or results to emerge which may not have been the original purpose of the investigation. Such was the case in the discovery of 'green pea' universes previously unknown to science, by Galaxy Zoo volunteers (Cardamone et al., 2009).

Do not relax

Once a project has been established, it is not necessarily a case of sitting back and letting the data flow in. CS projects are highly complex and should not be used as a way to get through a data set on the cheap, nor should there be any expectation of a task involving community engagement to involve minimal project management.

On the surface, CS programs deliver monetary gain; they recruit unpaid volunteers to collect data potentially saving costs that can be spent on other activities. Looking at six different projects on Zooniverse in 2015, Sauermann et al. calculated that the amount of contributions amounted around $1,554,500 dollars of unpaid labour costs (Sauermann & Franzoni, 2015). However, Zooniverse is one of the largest online CS portals, drawing over millions of daily observations, and represents an extreme case. For a smaller project such as Transcribe Bentham, the cost of creating software and recruiting volunteers to transcribe the documents over a six-month period was equivalent to hiring two full-time employees who could have completed the task in two weeks (Causer & Terras, 2014). Such aspects should be considered with projects that have a time-sensitive element which should be weighed against the potential benefits of engaging volunteers.

CS data sets come with additional costs of coordination, communication and data checking which must be accounted for. Without a careful level of planning and communication, the resultant data could prove worthless and thus could represent a considerable loss of resources. Thus, any costs saved on using volunteers for data collection or sorting must be considered alongside those of management, recruitment and sustaining participation.

Looking to the future

Citizen science projects can be far more complex than traditional scientific experiments. They have additional sources of uncertainty and involve additional decision-making steps. In order to be successful, a project needs to be engaging enough to capture the attention of a population of people, the participants need to be engaged enough to then collect data (possibly over a period of time) and finally the collected data, which has to be relevant, then need to be used and applied. This creates three main areas in which CS projects can fail: engagement, data collection and data use.

However, this complexity should not stop researchers using CS methods in the future. They offer multiple advantages to the scientific researcher such as the ability to collect a vast wealth of data that would otherwise be untenable alongside the ability to engage audiences in scientific research, and not only in primary data, but any associated metadata as well. Whilst submissions to Monument Monitor are useful to measure things such as groundwater flooding, each photograph is also equally useful to understand how such a project can succeed. Data can be extracted, for example, from the timestamp, image quality and submission method. By conducting CS, we also capture data about CS.

With a need for increasing levels of transparency and accountability, they have been known to help funding applications and make policy decisions (Gura, 2013). There is little doubt that CS can be, when used effectively, indistinguishable from traditional scientific investigation in the quality of output (McKinley et al., 2017). It provides a viable method to tackle some of the hugely complex and large-scale tasks facing researchers today.

This is especially true of researchers within the heritage sector and heritage science. With the future of the global economy looking increasingly unstable, there will no doubt be severe funding cuts implemented throughout the sector. Involving participants increases awareness of issues within heritage, from climate change to vandalism. One could also advocate for the higher neutrality of CS projects, in that they can compensate for the biases of academics; issues that matter to the scientist may not be the same as those that matter to participants. CS will give audiences increased agency over our shared heritage.

The huge complexities in such projects can be mitigated with time. Zooniverse has created a platform in which researchers can create their own CS project in a matter of minutes. Researchers should carry out their experiments in a similar manner, creating easily reproducible methodologies, creating any software open source and ensuring all results are made accessible within the public domain, thus adhering to the guiding principles of CS (Robinson et al., 2019). This will ensure that, similar to Zooniverse, experiments can be easily repeated for different purposes, with smaller levels of input time and time again. This type of streamlining can be enhanced by now established technology within artificial intelligence. Huge amounts of data collected from citizen scientists can be more easily processed and analysed, and any data analysed by citizen scientists can be verified in a similar manner (Bonney et al., 2009).

Fantastic work, some briefly referenced in this paper, is being done that establishes the effectiveness of citizen science within heritage and heritage science. However, as CS within the heritage sector grows, more coordination should be done to ensure the potential for global projects. Whilst Zooniverse engages global audiences to address localised projects, only eBird is really achieving CS observations at a global scale (Ellwood, et al., 2017). Heritage is at danger across the world, with 53 world heritage sites being classed 'at risk' globally, a citizen heritage science framework could be an affective measure to assist with accurate documentation and research. Such coordinating could come from ICOMOS, World Monuments Fund or UNESCO as international bodies that have a variety of international standards of heritage management and documentation. Global coordination would unlock many opportunities. For example, CS offers a good way to reduce the need of international air travel of experts to sites for scientific analysis, involving instead local expertise.

There is a great potential for CS within the cultural heritage sector, as the world attempts to recover from a global health and economic crisis, and participatory research could prove a useful tool to engage communities with their heritage in new ways and encourage directed investment. More research needs to be done to assess where participants can be involved in different stages of the scientific process, and what types of CS work best within heritage science. Creating frameworks from such research will allow a certain ease of reproduction, allowing more CS projects to be set up with fewer organisational costs. However, collaboration and sector-wide communication will be key to achieving this on a national or international level to ensure multiple projects do not work in silo towards the same goal.

References

Abbott, B. P., Abbott, R., Adhikari, R., Ajith, P., Allen, B., Allen, G., ... Anderson, D. P. (2009). Einstein@Home search for periodic gravitational waves in early S5 LIGO data. *Physical Review D - Particles, Fields, Gravitation and Cosmology, 80*(4), 042003. https://doi.org/10.1103/PhysRevD.80.042003.

Alender, B. (2016). Understanding volunteer motivations to participate in citizen science projects: A deeper look at water quality monitoring. *Journal of Science Communication, 15*(3). https://doi.org/10.22323/2.15030204.

Blaise, J.-Y., Dudek, I., & Saygi, G. (2018). Citizen contributions and minor heritage: Feedback on modeling and visualising an information mash-up. *Proceedings-2018 IEEE 5th International Conference on Data Science and Advanced Analytics, DSAA 2018*, 11–20. https://doi.org/10.1109/dsaa.2018.00011.

Bonacchi, C., Bevan, A., Pett, D., Keinan-Schoonbaert, A., Sparks, R., Wexler, J., & Wilkin, N. (2014). Crowd-sourced archaeological research: The Micropasts project. *Archaeology International, 17*, 61–68. https://doi.org/10.5334/ai.1705.

Bonney, R., Cooper, C. B., Dickinson, J., Kelling, S., Phillips, T., Rosenberg, K., & Shirk, J. (2009). Citizen science: A developing tool for expanding science knowledge and scientific literacy. *BioScience, 59*(11), 977–984. https://doi.org/10.1525/bio.2009.59.11.9.

Bonney, R., Cooper, C., & Ballard, H. (2016). The theory and practice of citizen science: Launching a new journal. *Citizen Science: Theory and Practice, 1*(1), 1. https://doi.org/10.5334/cstp.65.

Brigham, R., Grau-Bové, J., Rudnicka, A., Cassar, M., & Strlic, M. (2018). Crowdsourcing as an analytical method: Metrology of smartphone measurements in heritage science. *Angewandte Chemie - International Edition, 57*(25). https://doi.org/10.1002/anie.201801743.

Burgess, H. K., DeBey, L. B., Froehlich, H. E., Schmidt, N., Theobald, E. J., Ettinger, A. K., ... Parrish, J. K. (2017). The science of citizen science: Exploring barriers to use as a primary research tool. *Biological Conservation, 208*, 113–120. https://doi.org/10.1016/j.biocon.2016.05.014.

Cardamone, C., Schawinski, K., Sarzi, M., Bamford, S. P., Bennert, N., Urry, C. M., ... Vandenberg, J. (2009). Galaxy Zoo Green Peas: Discovery of a class of compact extremely star-forming galaxies. *Monthly Notices of the Royal Astronomical Society, 399*(3), 1191–1205. https://doi.org/10.1111/j.1365-2966.2009.15383.x.

Carlson, T., & Cohen, A. (2018). Linking community-based monitoring to water policy: Perceptions of citizen scientists. *Journal of Environmental Management, 219*, 168–177. https://doi.org/10.1016/j.jenvman.2018.04.077.

Causer, T., & Terras, M. (2014). "Many hands make light work. Many hands together make merry work": Transcribe Bentham and crowdsourcing manuscript collections. In M. Ridge (Ed.), *Crowdsourcing our cultural heritage* (pp. 57–68). London: Routledge.

Champion, M. (2015). *Medieval graffiti: The lost voices of England's churches*. 1st ed. London: Ebury Press.

Cohn, J. P. (2008). Citizen science : Can volunteers do real research? *BioScience*, *58*(3), 192–197. https://doi.org/10.1641/B580303.

Conrad, C. C., & Hilchey, K. G. (2011). A review of citizen science and community-based environmental monitoring: Issues and opportunities. *Environmental Monitoring and Assessment*, *176*(1–4), 273–291. https://doi.org/10.1007/s10661-010-1582-5.

Cooper, S., Khatib, F., Treuille, A., Barbero, J., Lee, J., Beenen, M., … Players, F. (2010). Predicting protein structures with a multiplayer online game. *Nature*, *466*, 756–760. https://doi.org/10.1038/nature09304.

Department for Digital Culture Media and Sport. (2019). *Taking Part Survey, 2017/19: Quarter 4 statistical release*. https://www.gov.uk/government/publications/taking-part-201920-heritage/heritage-taking-part-survey-201920#visiting-heritage-sites.

Dickinson, J. L., Shirk, J., Bonter, D., Bonney, R., Crain, R. L., Martin, J., … Purcell, K. (2012). The current state of citizen science as a tool for ecological research and public engagement. *Frontiers in Ecology and the Environment*, *10*, 291–297. https://doi.org/10.1890/110236.

Eitzel, M. V, Cappadonna, J. L., Santos-Lang, C., Duerr, R. E., Virapongse, A., West, S. E., … Jiang, Q. (2017). Citizen science terminology matters: Exploring key terms. *Citizen Science: Theory and Practice*, *2*(1), 1. https://doi.org/10.5334/cstp.96.

Ellwood, E. R., Crimmins, T. M., & Miller-Rushing, A. J. (2017). Citizen science and conservation: Recommendations for a rapidly moving field. *Biological Conservation*, *208*, 1–4. https://doi.org/10.1016/j.biocon.2016.10.014.

Follett, R., & Strezov, V. (2015). An analysis of citizen science based research: Usage and publication patterns. *PLoS ONE*, *10*(11), e0143687. https://doi.org/10.1371/journal.pone.0143687

Gura, T. (2013). Citizen science: Amateur experts. *Nature*, *496*, 259–261. https://doi.org/10.1038/nj7444-259a.

Haklay, M. (2013). Citizen science and volunteered geographic information: Overview and typology of participation. In D. Sui, S. Elwood, & M. Goodchild (Eds.), *Crowdsourcing geographic knowledge: Volunteered geographic information (VGI) in theory and practice*. 1st ed. Vol. 978940074 (pp. 105–122). Dodrecht, Heidelberg, New York, London: Springer. https://doi.org/10.1007/978-94-007-4587-2_7.

Hambly, J. (2017). *Scotland's coastal heritage at risk: Final evaluation report*. Retrieved from: http://scharp.co.uk/media/medialibrary/2017/09/SCHARP-HLF-EVALUATION_FINAL_web.pdf. (Accessed on 14 May 2021)..

Johnson, K. A. (2016). Real life science with dandelions and project budburst. *Journal of Microbiology & Biology Education*, *17*(1), 115–116. https://doi.org/10.1128/jmbe.v17i1.1064.

Kosmala, M., Wiggins, A., Swanson, A., & Simmons, B. (2016). Assessing data quality in citizen science. *Frontiers in Ecology and the Environment*, *14*, 551–560. https://doi.org/10.1002/fee.1436.

Kumar, P. (2020). Crowdsourcing to rescue cultural heritage during disasters: A case study of the 1966 Florence Flood. *International Journal of Disaster Risk Reduction*, *43*, 101371. https://doi.org/10.1016/j.ijdrr.2019.101371.

Kyi, C., Tse, N., & Khazam, S. (2016). The potential role of citizen conservation in re-shaping approaches to murals in an urban context. *Studies in Conservation*, *61*, 98–103. https://doi.org/10.1080/00393630.2016.1193690.

Land-Zandstra, A. M., Devilee, J. L. A., Snik, F., Buurmeijer, F., & Van Den Broek, J. M. (2016). Citizen science on a smartphone: Participants' motivations and learning. *Public Understanding of Science*, *25*(1), 45–60. https://doi.org/10.1177/0963662515602406.

Lee, T. K., Crowston, K., Harandi, M., Østerlund, C., & Miller, G. (2018). Appealing to different motivations in a message to recruit citizen scientists: Results of a field experiment. *Journal of Science Communication*, *17*(02), A1+. https://doi.org/10.22323/2.17010202.

Lin, A. Y.-M. (2010). The search for Genghis Khan: Using modern tools to hunt for an ancient past. *2010 IEEE Aerospace Conference*, 1–2. https://doi.org/10.1109/aero.2010.5447038.

McKinley, D. C., Miller-Rushing, A. J., Ballard, H. L., Bonney, R., Brown, H., Cook-Patton, S., … Soukup, M. A. (2017). Citizen science can improve conservation science, natural resource management, and environmental protection. *Biological Conservation, 208*, 15–28. https://doi.org/10.1016/j.biocon.2016.05.015.

Nov, O., Arazy, O., & Anderson, D. (2014). Scientists@Home: What drives the quantity and quality of online citizen science participation? *PLoS ONE, 9*(4), e90375. https://doi.org/10.1371/journal.pone.0090375.

Ortiz, P., Ortiz, R., Martín, J. M., Rodríguez-Griñolo, R., Vázquez, M. A., Gómez-Morón, M. A., … Becerra, J. (2018). The hidden face of cultural heritage: A science window for the dissemination of elementary knowledge of risk and vulnerability in cultural heritage. *Heritage Science, 6*. https://doi.org/10.1186/s40494-018-0224-z.

Parcak, S., Mumford, G., & Childs, C. (2017). Using open access satellite data alongside ground based remote sensing: An assessment, with case studies from Egypt's Delta. *Geosciences, 7*(4), 94. https://doi.org/10.3390/geosciences7040094.

Ridge, M. (2013). From tagging to theorizing: Deepening engagement with cultural heritage through crowdsourcing. *Curator: The Museum Journal, 56*(4), 435–450. https://doi.org/10.1111/cura.12046.

Ridge, M. (2014). Crowdsourcing our cultural heritage: Introduction. In M. Ridge (Ed.), *Crowdsourcing our cultural heritage* (1st ed., pp. 1–14). Routledge. https://doi.org/10.4324/9781315575162.

Robinson, L. D., Cawthray, J. L., West, S. E., Bonn, A., & Ansine, J. (2019). Ten principles of citizen science. In S. Hecker, M. Haklay, A. Bowser, Z. Makuch, J. Vogel, & A. Bonn (Eds.), *Citizen science: Innovation in open science, society and policy* (pp. 1–23). UCL Press. https://doi.org/10.2307/j.ctv550cf2.9.

Romeo, F., & Blaser, L. (2011). Bringing citizen scientists and historians together. In J. Trant & D. Bearman (Eds.), *Museums and the Web 2011: Proceedings*. http://conference.archimuse.com/mw2011/papers/bringing_citizen_scientists_historians_together.

Sauermann, H., & Franzoni, C. (2015). Crowd science user contribution patterns and their implications. *Proceedings of the National Academy of Sciences, 112*, 679–684. https://doi.org/10.1073/pnas.1408907112.

Silvertown, J. (2009). A new dawn for citizen science. *Trends in Ecology and Evolution, 24*(9), 467–471. https://doi.org/10.1016/j.tree.2009.03.017.

Stevens, M., Vitos, M., Altenbuchner, J., Conquest, G., Lewis, J., & Haklay, M. (2014). Taking participatory citizen science to extremes. *IEEE Pervasive Computing, 13*(04), 20–29. https://doi.org/10.1109/MPRV.2014.37.

Sullivan, B. L., Wood, C. L., Iliff, M. J., Bonney, R. E., Fink, D., & Kelling, S. (2009). eBird: A citizen-based bird observation network in the biological sciences. *Biological Conservation, 10*(142), 2282–2292. https://doi.org/10.1016/j.biocon.2009.05.006.

Tulloch, A. I. T., Possingham, H. P., Joseph, L. N., Szabo, J., & Martin, T. G. (2013). Realising the full potential of citizen science monitoring programs. *Biological Conservation, 165*(09), 128–138. https://doi.org/10.1016/j.biocon.2013.05.025.

Vincent, M. L., Gutierrez, M. F., Coughenour, C., Manuel, V., Bendicho, L.-M., Remondino, F., & Fritsch, D. (2015). Crowd-sourcing the 3D digital reconstructions of lost cultural heritage. *Digital Heritage*, 171–172. https://doi.org/10.1109/digitalheritage.2015.7413863.

Wilson, A. (2018). *Fossil Finder blog*. Retrieved from: https://www.zooniverse.org/projects/adrianevans/fossil-finder/about/results (Accessed on 14 May 2021).

Zooniverse. (2020). *Zooniverse.org*. Retrieved from: www.zooniverse.org (Accessed on 14 May 2021).

9 Community-centred sustainable heritage management

Reality and challenges in practice

Sujeong Lee

Introduction

Heritage is a social and cultural 'process' rather than a 'thing' (Smith, 2006). Heritage management, accordingly, not only encompasses understanding and protecting what is valued but it also uses those values as a resource to benefit people who appreciate, manage, merchandise and transmit these values. Hence, heritage management has been predominantly understood as a value-based process of identifying, sustaining and using heritage values.

Such process is a cyclic interaction between tangible and intangible elements. Tradition, thoughts, beliefs, aesthetic tastes or skills have produced tangible remains. Those remains from the past bring out memories, create new meanings and promote new activities within contemporary society. The tangible remains can be no more than a material object of no value without these intangible elements and vice versa. It is also a cyclic interaction between one generation and the next. The trace of life and spiritual activities of one generation can play a vital part of human history when the next generation perceives, interprets, values, commemorate and understands the meaning and value of it.

The way of practicing heritage has evolved during the past two centuries. During the 19th and early 20th centuries, heritage has been understood as material remains that serve as historical evidence to inform us about an important historical event or of a person which has imprinted on national or social identity. Heritage management was aimed at conserving material remains in a form tide to a certain moment of history. For the last decades, the concept of heritage has been extended. Now, heritage is a 'tool' to enjoy life and benefit society. Experience-based programmes, for instance, such as family barbeque dinners, organised by stone-age outdoor museums in Gongju, a historic city of Korea, have provided an opportunity to make fire as pre-historic people did. This allows the present generation to learn about the way of living of the period and also enjoy a special dining-out experience. Such heritage programmes can encourage local communities and visitors to develop their interest in the meaning of heritage and, at the same time, enrich their life experience. Rather than 'freezing time', the practice of heritage now aims at managing change of the material form and its meaning in a way that it can be both appreciated by the present generation and also delivered to the next generation. The practice of heritage also aims for the practice and process to be part of the development of a sustainable society.

However, the parties who operate the process are often limited to selected groups such as experts, politicians or officers of the heritage authority. Therefore, heritage has not yet been recognised by residents, visitors and local communities as a useful resource in sustainability.

This paper examines how local residents have been excluded from value-based processes in Gongju City – a city that has been designated as an important heritage city under Korean law. Located to the south of Seoul, Gongju was a capital of the Baekjae Kingdom (BC18~AD660).

Due to a lack of participation by residents in value identification as well as value management in particular, the heritage of the city has been and is still understood by residents as an obstacle in sustaining their lives. Three cases of managing historic sites within the historic centre of Gongju City displays the absence of local community in the process of heritage management. Although residents, who stand as a significant component of the local community, can play an important role and should be involved in heritage management, they have not been included in value-based decision-making. The chapter concludes with a set of practical recommendations that suggest how the community can be more involved in the process of managing heritage.

No community involvement in heritage management

International charters and principles, such as the 1987 Charter on the Conservation of Historic Towns and Urban Areas (the Washington Charter) and the 2011 Valletta Principles for the Safeguarding and Management of Historic Cities, Towns and Urban Areas, have highlighted the need for community involvement. However, there is no clear definition yet on who the community is (Neal & Walters, 2008).

One problem is that the literal meaning of community is too vague and wide (Waterton & Smith, 2010). Anyone or any group of people can be categorised as part of the 'community' if they are involved in some way when it comes to issues relating to heritage. An owner of a heritage building can be community. A person who visits a heritage site once in their lifetime can also be part of the community. The Valletta Principles list different stakeholders as 'elected authorities', 'municipal services', 'public administrations', 'experts', 'professional organisations', 'voluntary bodies', 'universities', 'residents', etc. The Australia ICOMOS Charter for Places of Cultural Significance, so-called the Burra Charter, stresses the importance of the participation of 'people' for whom 'the place has significant associations and meanings, or who have social, spiritual or other cultural responsibilities for the place' (2013, Article 12). No matter what we call or how we classify, it is clear that the different interests of each of these groups create difficulties in trying to define 'community'.

A legal framework therefore cannot secure community involvement without clarifying who the community is and what their interests are. Therefore, if we would like to include community in the heritage process by giving them a right to participate in a value-based heritage process, it is necessary for us to consider using a more legally appropriate terminology in laws and guidelines. As long as the word community is used without limiting it to mean a specific group, it would be difficult for parties in the process to mutually understand the word in the same meaning. Therefore, it would be more efficient for us to use a specific definition such as 'residents (ICOMOS, 1987, 2011)', 'property owners', 'neighbours', 'local activists', 'architects (Orbaşli, 2008)', 'visitors', 'local politicians', and 'local historians'.

Historically, heritage management in Korea has been led by national and local governments. They have the power to enact related laws and set out criteria to assess values, designate and protect what is valued in and also the specific ways of protection. Much influenced by the Western approach of heritage discourse, their policy perspective owes much to the 'authorised heritage discourse' which understands heritage as an object and is dominated by concepts of monumentality and aesthetics as Smith argued (2006). It means that the historical and artistic values of heritage have been at the centre of what is to be conserved. Such forms of heritage have been regarded as national assets rather than community assets. Accordingly, local residents and their interests have lost their seat at the table when it comes to decision making.

In particular, when the private right of a land owner is in the way of 'freezing' heritage for the purpose of its protection, planning decisions have been made on behalf of the nation or public by sacrificing the private right of the land owner to both living in their property (which so happens to contain a heritage site) and enjoying these heritage values in a living city.

The owner's right and interest in using and developing his land within a designated heritage area or its surroundings should instead give way to the protection of such a heritage site in order to preserve its historical value that is fixed to a time of the past. For the sake of the material preservation of the past, residents should remove their houses because their homes are on the top of archaeological remains.

The heritage authority would claim that such decisions were for the benefit of the public and future generations. Their action to preserve heritage as a physical object of monumental and artistic values may ensure the sustainability of the physical aspects of the heritage site. However, it is not possible to proceed to build a sustainable heritage practice in society if there is no involvement of related stakeholders such as site managers, residents, regular visitors from local community, and shop owners in the decision-making process.

The following chapters will examine three cases, which are all located in the historic city of Gongju. It will illustrate the way heritage discourse has been practiced in Korea, which has regarded heritage as an object to be protected and not as a process to be carried out.

Being 125km to the south of Seoul, Korea's capital city, Gongju City is a historic town of 120,000 residents. During the Three Kingdoms period (BC37~AD668), it was one of the three capital cities of the Baekjae Kingdom. After the Baekjae Kingdom was overthrown by the Silla Kingdom (BC57~AD935), it remained as provincial capital of Chungcheong Province, with military, trade and political functions. The capital of the province was moved again to Daejeon City during the Japanese Colonial Period (1910~1945). While its political function was removed, it remained as a touristic destination.

Divided by the Geumgang River, the present town has been separated into two, the old on the south and the new on the north. The former is full of monuments, historic buildings, old structures and archaeological remains that include a fortress, royal tombs, houses, religious sites and the disturbed remains of the old municipal hall. Ranging from the pre-historic era into the present, they have been carefully managed to manifest historical events and lifestyle. Gongsanseong Fortress and Songsanri Tomb have been designated as the Baekjae Historic Area by UNESCO in 2015.

A location of schools, public offices such as the city council and tax office, traditional markets, restaurants and pubs for tourists, the south of the river plays an important role with its municipal, education and economic functions. The north of the river has been developed since the 1990s with newly built flats for families, a bus terminal and a university. The separation by the river has influenced residents through the development of a unique lifestyle, where they spend the daytime in the south and the nighttime in the north. Visitors and tourists tend to arrive at the terminal or via roads in the north and proceed to spend their time and money in the south.

The three heritage sites are all located in the centre of the historic area, which will be discussed in this chapter, and are economically and socially connected to each other by sharing the same neighbours as well as economic, educational and religious communities. Like other historic cities, the change of the fabric and the meaning of these heritage sites are closely related to the use of the historic environment and the improvement of daily life. In order to use their historic sites and monuments as a sustainable resource for the present and future generations, residents should be informed about what they would gain or lose in the course of decision making when it comes to managing change. However, decisions on changes to the heritage sites and their setting have been made without their participation. The heritage sites have thus ended up as a 'walled' and 'frozen' territory after the removal of all contemporary urban functions, functions which should be integrated with everyday life of the residents, and instead now only serve as a touristic destination for visitors.

Case 1: No right to live on top of an archaeological site

The first case is a typical conflict between the private right to build a new house on the top of archaeological remains and planning refusal for the scheme. A strong restriction against the contemporary housing need for the sake of preserving historical evidence is now threatening the sustainability of the heritage city by removing local residents and depriving them of their direct ownership of historic environment. To interpret, understand, look after and deliver its significance to the next generation of residents by living within the archaeological area is part of their right and responsibility. Once they are removed from the site, the whole process of interpretation and benefitting from its significance is transferred both legally and morally into the hands of the heritage authority. Such an approach clearly displays that the understanding of heritage as a social process is not yet integrated into actual practice in Korea.

In 2015, the Cultural Heritage Administration of Korea (hereafter CHA), which is a body responsible for identifying and preserving heritage, established a public fund for a group of building owners. The fund was for any resident who intends to replace a severely deteriorated house within the historic town centre of the designated historic area of Gongju City with a traditional timber house. It was the first fund for the CHA to exercise their policy shift in managing heritage. Previously, they had restricted the private rights of development in historic environments without any compensation to the owners and residents. The new fund intends to subside the budget of owners as well as to guide and encourage residents and owners to participate in place-making. Suffering from a governmental policy to restrict rebuilding or extension of homes for the last 50 years, building owners and residents have welcomed the fund. They have started to hire architects to replace old houses with new ones of traditional design as guided by local and central governments. The owner of a house in Banjukdong 197-4 was one of the residents who submitted a rebuilding plan. He then received an official letter from Gongju City Council informing him that his plan was accepted by the planning committee of the local council. At the same time, funds for the new construction were granted by the council.

Because the area was a central part of an ancient capital city, the site and its surrounding area have been marked out on an archaeological map[1]. Both the Cultural Heritage Protection Act (Article 12, hereafter CHPA) and the Act on Protection and Inspection of Buried Cultural Heritage (Article 14) obligate owners to excavate their land if it is marked on the map. Therefore, the owner of Banjukdong 197-4 completed the compulsory excavation using his own expenses. He had understood that an excavation could reveal archaeological objects or building remains on his land. He also strongly believed that he would be able to subsequently use the funds to build a house on the top of the archaeological remains, similar to many other sites within the area who have also been granted development permits.

However, the 2018 excavation revealed a pile of roof tiles of different periods ranging from the ancient period during the reign of Baekjae Kingdom and into the 18th century. The excavation team did not find any remains of the building other than roof tiles and broken sculptures. Because all movable objects from the excavation were transferred to storage for such archaeological finds, there were no remains on site when the archaeological excavation was completed. However, the CHA then issued a preservation order to disallow any type of development.[2] The owner's intention to continue to live in a historic town where he was brought up and on a piece of land that inherited from his parents was shattered. The fund for rebuilding was also withdrawn. The owner is due to leave his land for a new place.

The present law does not obligate the CHA to purchase such sites. He could bring a lawsuit to claim against the heritage authority for compensation for the expenses of excavation and demolition of existing buildings, but he cannot claim for funds to find a new place to live in. Therefore, the owner had to find a new place at his own expense. Many owners in the area have thus given up their plans for rebuilding and continue to live as they have. If such a case repeats, the whole area would remain as a district with a monumental amount of deteriorated buildings or even empty houses in due course.

The CHA had refused the planning application based on new information from one of the excavated roof tiles which contained two Chinese characters, 'daetong(大通)'. The word has dual meanings: (1) the name of the ruined royal temple, and (2) the name of the specific year from AD 527 to 529 during the Northern Wei Period (AD386~534) of ancient China.[3] Archaeologists have interpreted that the two Chinese characters in the roof tile refer to the former without providing any supporting evidence. Daetongsa temple, which was mentioned in historical documents, has been known as the most powerful temple of royal patronage. However, the location had been a mystery. Therefore, when the roof tile was excavated, many archaeologists and historians excluded the possibility that the characters can be the Chinese year. They jumped to a quick conclusion that the roof tile is a fascinating find to indicate the location of the temple that they had desperately been looking for. Despite there being no remains due to all the objects being removed from the site, the historical value of the site has hugely increased by their assumption of it being the location of the ruined royal temple of the ancient period. The National Committee on Planning Permission on Archaeological site, of which members are mostly archaeologists and historians, has used their power to protect the site which ultimately results in the prohibition of any type of development (CHA, 2018). They have even recommended to the CHA to set out a long-term plan to remove the neighbouring buildings for the further excavation in order to confirm the boundaries of the ruined temple.

The owner of the site was invited several times during the excavation. Information on new objects and findings has been shared with him. However, the historical information on the roof tiles did not persuade him enough to understand the significance of the site in the same way as the archaeologists have. In particular, he could not understand why the site, which now has no remains, has such a significant meaning. Archaeologists have placed a great value on the site as a place of a ruined temple and its related historical meanings. However, for the owner, it is just a site where roof tiles of academic informational value have been excavated and removed for study. In both the process of identifying and sustaining values, the heritage authority has fully depended on the voices of experts. They have forced residents to accept that this value should be exchanged with their values of living on the same site over generations of habitation. The CHA and experts have ignored the other historical layers of the land being used as residential homes over centuries till the present. They decided to preserve what cannot be visualised, nor comprehended for the owner. In addition, the historical value that they have claimed will not be apprehended by visitors if the significance of the site is not shared with and agreed to by the visitors. The site will be only meaningful in the excavation report and in history books.

In order for us to practice heritage as a process, the material remains in a living city should be carefully interpreted, remembered and protected for both present and future generations in a way that they could become part of the contemporary life of a local community. To preserve the site without a building and use can indeed leave it as a site of historical evidence of the past. However, it is no longer a heritage place which is alive within the present living city. New information of the past will be recorded but it will hardly be a heritage that can be used as a sustainable resource for residents and wider community. Understanding the place as a process is to strike a balance between the continuity of life of residents and the preservation of historical evidence and meanings. Heritage processes should pursue not only the protection of heritage values and sustainability of tangible and intangible elements, but also the continuity of the process of protecting them. However, in this case, the heritage authority has prioritised the sustainability of an archaeological site of the past while giving up the sustainability of contemporary living. It has also chosen not to pursue a process of sustaining the heritage of the city through contemporary usage by allowing the local community to use and interpreting the site. Since the archaeological site was reburied as ordered by the government, the site has remained as a piece of empty land without any use. The site will be remembered only as a place where important ancient roof tiles were excavated. Only a small information board was erected, but the historical site itself remains overshadowed and enclosed by surrounding residential houses. The neighbours who had decided to live and sustain their lives in the historic environment without rebuilding their homes are now considering leaving the area. The municipal government now worries that the area could be abandoned and slowly turn into a slum.

Case 2: No right to know about the change in heritage values

The second case addresses a problem regarding a decision in favour of a new bridge, located 1.5km away to the north of the first case. It has been made without communication with the residents to have them understand and be part of managing the resultant change, although it can negatively affect both the significance of the heritage site and its setting. Planning permission for a new bridge next to a nationally and internationally designated ancient fortress shows that changes to a site's heritage significance and meaning can be still decided through a top-down governmental process. Without knowing how heritage values can be affected by the new development and what aspects of heritage that they would then possibly lose in exchange for their convenience, the residents will remain as a passive community group in the heritage process.

The historic landscape of the old town of Gongju City has been dominated by the Gongsanseong Fortress, which was a military base of the Baekjae Kingdom. Protecting the capital on the south side of a natural moat, the Geumgang River, the fortress has played an important role not only as an ancient monument of defence, but also as a catalyst for different functions such as markets, hotels, residences, streets, open spaces and transport terminal as part of the historical development of the surrounding area.

At present, three main bridges connect the old and new towns. The first is Gongju Big Bridge, which allows entry to the east side of the old town. The second is Baekjae Big Bridge, which leads to Buyeo City to the west, another former capital city of Baekjae Kingdom. The third is Geumganggyo Bridge, which lands in front of the fortress and leads to the centre of old town. A new bridge has been planned to be constructed next to the Geumganggyo Bridge, to replace its use and function.

Historically, there was no bridge during the period of Baekjae Kingdom. There were two traditional ways of approaching to the south: one was by boat and the other was by a stepping-stone bridge built in area of shallow water level. During the Japanese Colonial Period, Geumganggyo Bridge was built at the edge of the rampart of the fortress to carry people and agricultural products by train from and to surrounding cities. Since it was converted into a bridge for cars and pedestrians, traffic has increased due to the growth of the new town in the 1990s. With only one lane for one-way traffic from the north to the south, it has suffered from traffic congestion in the mornings and evenings. In addition, the two other bridges do not connect directly to the historic centre for visitors and to the public and business offices that serve the residents. Local politicians and campaigners from the residential community have raised the need for a new bridge over the last three years.

With the strong lobbying of a parliamentary member from local community in 2016, the city council and Ministry of Land and Transportation have succeeded in receiving a budget for a new bridge from the Ministry of Economy and Finance. In order to execute the plan, they have submitted a planning application to the CHA.

The location is next to the boundary of Gongsanseong Fortress. The Fortress is both a nationally designated Historic Site as well as being a UNESCO World Heritage Site – 'Baekjae Historic Area'. Based on the CHPA, all private and public developments within a 500-meter zone need to execute a domestic process of impact assessment on the affected Historic Site. Internationally, any development, which can negatively affect to the Outstanding Universal Value (hereafter OUV), should be carefully examined by conducting a heritage impact assessment (hereafter HIA). Considering that UNESCO has adopted in recent years the use of HIA as a necessary tool to assess the possible change of significance of heritage sites by proposed developments, the CHA has, for the first time, requested that the city council proceeds with an UNESCO-guided HIA.

The German institute, Michael Kloos Planning and Heritage Consulting, was commissioned by the Gongju City Council to carry out the HIA, and this was completed in May of 2020. The intermediate report has summarised that the proposed new bridge would have a minor impact on the site's OUV. More importantly, they have strongly advised that the social impact should also be assessed, which includes carrying out a series of public inquiries by involving the local community and residents (Michael Kloos et al., 2020). They have argued that residents and communities need to clearly understand how the OUV can change. In addition, they have requested the city council to submit a new traffic plan for the proposed route which will result in more traffic penetrating the historic area from the landing of the proposed new bridge.

However, without carrying out a public inquiry and developing a traffic plan, the CHA has finally issued planning permission in favour of the local politician and the city council, with a condition that the city council should notify the UNESCO's World Heritage Centre of the planned new bridge.

This decision was made fully based on the advice of the National Committee for Historic Site who were influenced by the result of HIA. The CHA has followed a regulated process as stated in the law, but they have not attempted to reach any consensus on the possible change of meanings and significance of the heritage site with the community who own, use and who will have to deliver the heritage place to the next generation. During the two years of the planning process, there was not a single public inquiry. Residents and other members of the local communities have heard about the construction plan and process from newspapers or TV. However, they have not been informed or been given any information on the possible change of the historic environment and the impact to their present and future lives. They and their children will have to accept the change and impact without knowing how the heritage value of their city would be affected, how traffic can be improved, how views from and towards the fortress would be changed, etc. It is not possible to confirm that they could have sacrificed convenience for the sustainability of historic landscape of the fortress. It could be possible that the residents and communities would willingly commute to the historic centre via a longer route by using other bridges if they knew and had full information regarding the negative impact on the heritage site. However, there was no opportunity where they could participate in weighing the impact on the value as well as comparing between what they will lose and get in terms of economic effect, traffic change, visitors experience and other results caused by the construction of a new bridge. This case demonstrates that the ownership of a heritage site is not in the hands of residents and communities but in that of heritage institutes and experts.

Case 3: No right to be involved in reviving heritage values

The third case displays the problem of the lack of community involvement in deciding on how a ruined site can be revived as a useful tool for sustaining contemporary lives. Located 0.4km away to the east of the first case, it demonstrates that the heritage authority has ignored the needs of the local community and residents who wish to use a heritage site as a resource to improve their quality of life as well as the local economy.

Until it started to fall into ruin during the Japanese Colonial Period and was subsequently replaced in 1910 by a hospital building, the provincial government complex in Gongju City had occupied a vast area of the historic centre. The hospital was originally built in traditional timber structure in 1910, and it was replaced by a much bigger sized concrete building in the 1970s for the public health service of Gongju City.

Although the site has been changed from provincial governmental offices into a public hospital, the surrounding area has remained as the political and economic centre of the city. While the site was used as a hospital, the surrounding area developed as a place for pharmacies and shops for patients and caretakers. The land price has risen not only due to commercial benefit and potential but also due to the pleasant historic and nostalgic environment within the historic centre. This area was also designated in 2011 as a 'Conservation and Promotion Area' by the Special Act on the Preservation and Promotion of Ancient City.

Government-funded repairs of historic buildings which were reopened for commercial use within the area have changed and established the character of the area as a historic town. Such a change has attracted people to visit and enjoy the area, and accordingly, it has helped to increase the price of land. The land price officially posted in the website by the Ministry of Land and Transportation in the area displays that there has been more than 30% rise in price between 2007 and 2017.

When the hospital was moved to the outskirts of the city in 2017, the council began to discuss with CHA on the new use of the site. Owners of shops have paid attention to what plans could be considered by the heritage authorities. Because the area has been suffering from the decrease in business income caused by the relocation of hospital, they have paid full attention to the way that the former hospital site can be redeveloped. Their priority is to overcome the rapid decrease of the business functions of the area. Some residents and business owners have strongly proposed to the city council to renovate the hospital building and re-use it for a different function. Reflecting the local voice, the city council has submitted a planning application to reuse it for cultural and community activities such as a youth career centre, a place running training programmes for local communities, library, meeting rooms and an exhibition hall of historical personalities from the community. However, CHA has refused to accept the proposal, with its decision based on the 'Management Plan on Designated Area of Ancient Capital City of Gongju'. Drafted by the same city council and approved by the CHA in 2015, this plan has stated that the former hospital building should be demolished for the reconstruction of the long disappeared provincial office complex.

The city council has established in 2018 a civic committee for the public participation and gathering public opinion on the future use of the site. There were three rounds of public inquiries which concluded on a set of recommendations to the city council:

1   Demolition of the former hospital building,
2   Speedy excavation collaborated with CHA, and
3   Setting out a plan for site management aiming at rehabilitating the historic centre.

After the planning refusal and public inquiry, the city council demolished the hospital building in 2018. The excavation in 2019 has not revealed any objects or building remains that could provide enough information to restore disappeared provincial office buildings of the Joseon Period (1392–1910). The council then requested the local research institute to draft a plan of reconstruction and rehabilitation of the site as well as the surrounding areas.

The final report on the reconstruction and rehabilitation plan has set out two main schemes:

1   Reconstruction of the main buildings of the former provincial office, and
2   Construction of new functional buildings and facilities to attract visitors and contribute to the sustainable development of the surrounding area.

It has advised that several timber buildings of original administrative function such as the governor's office also be restored (Chungnam Institute of History and Culture, 2019). The design of buildings for reconstruction will be based on historical photos, drawings and documents. In addition, it has recommended the building of an underground parking space at the site for the public. It also proposes to construct a tourist information centre and exhibition area to contribute to the business activities of the area as a part of the latter scheme.

The plan has mainly focused on the recovery and the use of historic and architectural values of the site. It is obvious that once those buildings are reconstructed, the cost of site management will need to be found. The city council does not have any estimation regarding how many visitors would then come nor how the site will be contributing to the economy of the city.

Repositioning residents in a value-based process

The residents in the three cases have been marginalised at each step of the value-based process. The first case shows that experts and academics are still the authority in identifying heritage values. Because residents and community are excluded in value identification, they are again ignored in the processes of maintaining values as the second case has showed, as well as reviving use values as the third case has demonstrated. Despite the existence of their voice, they have not been heard nor their opinions examined and considered.

Authorised heritage discourse claims that heritage is now a useful resource in education, tourism, place-making, economic growth and development. The discourse strongly highlights the critical importance of community involvement. However, many cases have demonstrated that the conventional way of top-down decision-making still dominates the heritage process by privileging the opinions of experts and a limited group of people. If authorised heritage discourse wishes to strongly argue for community involvement, it should then set out the process of identifying communities, setting out a platform to share 'heritage' values from different perspectives and more specifically analysing what values can be sustained or compromised as a result of different decisions. Thereafter, such a communication process with the identified sectors of the community should be properly carried out in practice. For us to achieve sustainability of heritage sites and sustainable heritage processes, the present discourse and practice should be re-examined.

The present discourse should tackle several common issues such as defining who are communities and their roles, and setting out principles of community involvement in each step of value-based decision making. It should also make it obligatory to have a feedback process before and after negotiating the needs between the heritage authorities and communities. The way of implementing such a process in practice can be different based on different cultural context. It can be effected by enacting legal obligations or disseminating guidance notes or just creating a practical toolkit.

Firstly, the definition of community should be revisited and clarified. It is important to identify who is an initiator and an actor in the sustainable management and in the repeated process of managing the change in heritage places. Community involvement can only be possible when we know who the community is, when they know what to expect, as well as how they can participate.

Therefore, we need to identify who are the beneficiaries and their different understanding of values when it comes to identifying values of each heritage site. All perspectives should be considered, recorded and opened to the public so that it can be revisited in the decision-making process of managing the site. In Korea, heritage is still understood as a field for experts. Therefore, the present legal process of value assessment should be revised to include the necessity of having meetings to share information on sites with the identified communities and conducting hearings to listen to different perspectives in valuing the site before officially drafting its value description.

Secondly, there should be more discussions amongst residents as well as between residents and the heritage authorities in deciding the way to best sustain identified values. In this step, we need to analyse who and how different communities can be benefitted or suffered as a result of different decisions. In addition, we need to examine whose values can be compromised and which value can be sustained. Rather than focusing on the agreements between stakeholders, it is more important for them to understand and participate in heritage as a process. It could be more important for the parties to share and exchange information. It would be meaningful for them to work out how to compromise between conflicting needs in the process of identifying values and then deciding on the way to sustain the agreed values. Process-focused management systems should be developed and practiced. The process should be clearly guided in legal statements and implemented by central and local authorities. The sustainability of heritage sites and the sustainable management of heritage in historic towns can be achieved through implementing a pain-staking process of negotiating the different needs and aims of heritage authorities and communities. Heritage authorities understand a historic town as a site with historical evidence from the past, whereas the town's community and residents understand it as a place to reside. The way how and what the former and the latter understand the same place is totally different. In Korea, heritage is still understood as obstacle to comfortable life or as a subject with limited use for the fields of education or tourism. Therefore, authorities need to start from reducing the community's fear in thinking that heritage and their lives are not able to coexist in a historic environment. This can be done through different tools such as organising small talks, educational programmes or prudent decision-making involving local communities. Historical remains are not valued because of its material substance itself. Rather, it is truly valued when communities understand its value for enriching their everyday lives.

Notes

1 The archaeological map has been drawn and updated by the CHA for the last 50 years. The map is open to the public so that anyone can find out which site is included in the map.
2 Under the Korean law for archaeological sites, 'Act on Protection and Inspection of Buried Cultural Heritage', there are three decision options that can be applied to an important archaeological site: (1) to refuse any construction or change to the surface and underground; (2) to allow limited development without any work on underground; and (3) to cut the whole piece of land with archaeological remains and move it to another site.
3 Names of Chinese year, which is named after the Chinese emperor's name, have been used in Korea from the ancient period till the commencement of the Japanese Colonial Period.

References

Australia ICOMOS (2013). *The Burra Charter: The Australia ICOMOS Charter for Places of Cultural Significance*. Australia: Australia ICOMOS.

CHA (2018). *The 5th meeting minute of National Committee on Planning Permission on Archaeological site*. Daejeon: Cultural Heritage Administration (Unpublished document).

Chungnam Institute of History and Culture (2019). *Gongjumok: Final report on restoration and redevelopment plan*. Gongju: Shinchang Printing.

ICOMOS (1987). *Charter for the conservation of historic towns and urban areas (Washington Charter)*. Paris: ICOMOS.

ICOMOS (2011). *The Valletta Principles for the safeguarding and management of historic cities, towns and urban areas*. Paris: ICOMOS.

Michael Kloos Planning and Heritage Consultancy & Chungnam Institute of History and Culture (2020). *Heritage impact assessment: 2nd heritage impact assessment for construction of the planned 'Second Geumganggyo' Bridge*. Gongju: Design Midam.

Neal, S., & Walters, S. (2008). Rural be/longing and rural social organisations: Conviviality and community-making in the English countryside. *Sociology, 42*(2), 279–297.

Orbaşli, A. (2008). *Architectural conservation*. Oxford: Blackwell Publishing.

Smith, L. (2006). *Use of heritage*. London and New York: Routledge.

Waterton, E., & Smith, L. (2010). The recognition and misrecognition of community heritage. *International Journal of Heritage Studies, 16*(1–2), 4–15.

10 Heritage conservation as a social process

Assessing social impacts of participatory cultural heritage conservation

Elia Quijano Quiñones and Kalliopi Fouseki

Introduction

The recognition of culture and heritage as an enabler of Sustainable Development Goals (UNESCO, 2018; UCLG, 2015) has encouraged heritage researchers and practitioners to look more deeply into the specific ways that heritage contributes to this process. Although heritage as a resource for the economy has been widely established in development agendas, new understanding for development, such as the SDGs, has facilitated current studies to advocate for a holistic approach in cultural heritage, highlighting its key role in sustainable development (van Oers, 2015; Guzmán et al., 2018; Gallou & Fouseki, 2019). In relation to this, participatory methodologies are a staple in the effort of merging conservation and development processes by enhancing the quality of heritage as 'social glue', while aiming for a type of conservation that is anchored in the connection and active involvement between people and heritage (Sully et al., 2014; Ferreira, 2018; Vandesande et al., 2018). Nevertheless, exploring long-term benefits beyond theoretical inferences is challenging, and there are ongoing discussions around the improvement of methodologies to investigate the impact of participatory heritage conservation, in order to enhance its role in sustainable development processes (Landorf, 2011; Nocca, 2017; Katrakaziz et al., 2018; Gallou & Fouseki, 2019).

In this chapter, we contribute to this discussion by arguing that the study of the impact of community participation in heritage conservation should be framed around the continuation and evolution of sustainable practices within local contexts. Therefore, we approach heritage conservation as a social process that requires in-depth look into peoples' feelings, attitudes, and behaviours (Landorf, 2011; Wells, 2012; Crouch, 2015; Jones, 2017; Katrakaziz et al., 2018; Nocca, 2017; Gallou & Fouseki, 2019; Fouseki et al., 2020). To do so, we draw on symbolic interactionism (Dennis & Smith, 2015) and social practices (Shove et al., 2012). We address two main questions: (a) What aspects should be included when researching sustainable practices in heritage conservation? (b) How do participatory methodologies in heritage conservation affect communities' social life? To answer these questions, we present and discuss the results of the analysis of 40 in-depth interviews conducted in the Spring of 2018 as part of the first fieldwork season in three municipalities, Chichimila, Tixcacalcupul, and Sacalum located in Yucatán, México. From our results, we propose an evaluation framework of the social impacts of heritage conservation as a social process, derived from the implementation of practices for the sustainable conservation of local heritage, when these are part of participatory methodologies. Finally, we discuss our findings in relation to issues of social sustainability and the SDGs for 2030.

DOI: 10.4324/9781003038955-13

Participatory methodologies: merging sustainable heritage and social sustainability

Sustainable heritage takes from the sustainable development aim the ideal of a responsible use of our resources, to ensure present and future generation needs (Brundtland report, 1987). Likewise, sustainability in heritage aspires to balance access and use of heritage with conservation. In this way, it considers heritage as a non-renewable resource (Loulanski, 2006; Boccardi, 2015). Nonetheless, current approaches to heritage management and heritage studies acknowledge that heritage is not just a 'thing' (Smith, 2006) but rather a social and cultural construct. As such, a value-based approach to heritage management cannot exclude issues of connectivity, continuity, and change, and it rather supports an ideal of sustainability in heritage that cannot be understood only in terms of retaining information about the original fabric or aesthetic features of an object or a site (Loulanski, 2006; Ashworth, 2008; Harrison, 2013; Guzmán et al., 2017).

This has led towards a preference for participatory methodologies as the best approach to capture heritage values and as having additional benefits such as enforcing collaboration between different stakeholders (e.g. local communities, heritage professional, local government, etc.). Consequently, the ultimate aims and outcomes of heritage conservation are more than ever akin to those of social sustainability and development. In this regard, social sustainability is commonly understood as the ability of a society to sustain itself, and it is analysed in terms of coexistence, quality of life, and attachment to a social group. It shares with heritage the relationship between communities and its territory, and the inclusion of sociocultural values within development processes (Landorf, 2011; Forero, 2015, p. 177; Van der Auwera et al., 2015). Specifically, participatory methodologies in heritage conservation enhance the use of cultural heritage in development processes. As Vandesande et al. (2018) argue, the strongest link between conservation and sustainable development is in sharing the task of heritage maintenance with community members.

Thus, the goal of participatory methodologies in conservation, on top of the identification of heritage values, is to produce changes in local attitudes and behaviours to provide ideal conditions for the sustainability of heritage (Wells, 2012; Sully et al., 2014; Della Torre, 2015; Neal, 2015; Wijesuriya et al., 2017). This is achieved through strategies that set the best conditions for interactions between stakeholders at different levels, by targeting actions at understanding use dynamics in heritage, opening channels for communication, and sharing knowledge, skills, and responsibilities (González, 2016; Magar, 2016; Guerrero & Soria, 2018). Thus, it can be argued that participatory methodologies in heritage conservation aim at the creation of communities of practice, which are defined by Wenger and Synder (in Shove et al., 2012) as groups of people informally bound together by shared expertise and passion for joint enterprise. However, previous studies indicate the challenges and the struggle for a long-term commitment of practices implemented through participatory methodologies, which is often due to existing social issues within the groups (Chirikure et al., 2010; Valle, 2014). This urges us to discuss success in heritage conservation as a social process, in terms of the evolution of heritage conservation practices, where the relationship between the sustainability of heritage and development is more symmetrical and associative, mutually affecting each other.

Exploring heritage conservation as a social process

It is essential to conceptualise heritage conservation as a social process delivered through participatory methodologies in order to encapsulate the scope of sustainable practices in heritage conservation. To do so, it is worth presenting at this point the key attributes of participatory methodologies. Firstly, participation is a spatial practice that is transformative since it provides opportunities for the change of social meanings, emotions, experiences, and ways of acting through social interactions (Shove et al., 2012; Handberg et al., 2014). Secondly, participation and heritage conservation, as a social practice, serve a practical or social function (Shove et al., 2012). Therefore, participation in heritage conservation is not a habit, but a routinised behaviour enduring between specific moments of enactments and filled out by multiple single actions. Moreover, participatory conservation, as a social practice, requires three components to emerge and persist: materials, competences, and meanings. If connections between elements of these components are broken or changed, practices can disappear or shift (Shove et al., 2012). In this regard, Fouseki et al. (2020, p.18) have argued that 'heritage conservation is a social and cultural dynamic practice, and its continuation, or disruption, depends on the interconnections between diverse factors such as senses, materials, competencies, space/environments, resources, time, and meanings'.

For this research, we consider that heritage conservation as a social process (the continuation of heritage conservation practices) and its effects in community's social life depend on the subsequent social interactions between people around heritage. To reiterate this point, we draw on symbolic interactionism theories that stress: (1) the social act, not the individual, is the unit of analysis; and (2) new meanings are not solely the product of an individual's mind, but rather the ongoing results of their understanding based on social interactions (de Nooy, 2009; Handberg et al., 2014; Denis & Smith, 2015). Thus, in order to explore this and to answer our research questions, we used three case studies from Yucatán, México, as described below.

Case Studies: the municipalities of Tixcacalcupul, Chichimila, and Sacalum in Yucatán, México

Tixcacalcupul, Chichimila, and Sacalum are located in the state of Yucatán, México, and were chosen as case studies because of their previous involvement in the 2012 heritage conservation project: *'Proyecto de atención a comunidades y participación social: la conservación de objetos de culto del estado de Yucatán'* (Jaspersen, 2012a). The project was carried out by the Yucatecan centre of the National Institute of Anthropology and History, the Secretary of Culture of the state of Yucatán, and the Temporary Employment Program. It involved the restoration of historic religious heritage consisting of polychrome wood sculptures of Catholic saints (circa 17th century) and the mural paintings of a former Franciscan convent (circa 17th–18th centuries). In addition to this, activities common to participatory methodologies, such as sharing decision-making, capacity-building, and sharing preventive conservation knowledge, were applied with the aim of identifying values and uses, and fostering long-term preventive conservation practice (Jaspersen, 2012b, 2014). Although the project delivered immediate results, it did not have continuity and was a one-off experience.

The population of Chichimila (7,952 people), Tixcacalcupul (6,665 people), and Sacalum (4,589 people) are in its majority indigenous Maya (México Government.

INEGI, 2010). Socialisation and organisation around living religious heritage is key to local unity since it is rooted in the Catholic's origins of each municipality. However, a distinctive characteristic is the prevalence of a syncretic behaviour in their traditions, a mixture of Maya heritage and the Catholic religion, which can be traced back to the resilience of the Maya people during the Spanish occupation (Quezada, 2011). Despite the rich heritage and intensive governmental efforts to provide education for all young people, the municipalities face major socio-economic challenges such as marginalisation and migration of the younger generation due to unemployment. Marginalisation is defined by poverty levels, access to health, life expectancy, level of education, and housing. These socio-economic problems are also affecting the community-led organisation of their local festivities (México Government, Fernández, 1995; Poole et al., 2007; Guzmán, 2010; SEDESOL, 2017). Because of the current challenges that the area is facing, we had the interest to explore if there is a continuation of sustainable practices for heritage conservation in these municipalities and if it is possible to identify positive social impacts on community life as a consequence of their participation in the conservation project.

Methodology

To analyse heritage conservation as a social process, that is, to explore the continuation of sustainable practices in conservation and their social impacts on community' life, a qualitative approach through in-depth interviews with members from a case study seemed most appropriate (Villasenor & Magar, 2012; Wells, 2012). For this research, we used the three municipalities aforementioned as cases studies. The interviews were transcribed through ExpressScribet and thematically analysed on the Nvivol 12 software. The interview design was developed using a topic guide to explore areas of interest selected from a literature review (Vanclay, 2002). The interviews aimed to collect the following information: (1) opinions about social environment; (2) attitudes and behaviour regarding cultural heritage; (3) experiences during the conservation project; and (4) perceptions of the benefits of heritage conservation projects. A delivered order from general knowledge to specific situations was applied in an effort to guide the interviewee to describe situations of interest for the research (Vanclay, 2002; McCurdy et al., 2005; Bryman, 2008, p. 239). Key informants, identified from their previous involvement in heritage conservation projects, were interviewed first. Following that, the snow-bowling technique was used by asking each interviewee to name another informant, depending on their involvement or knowledge about previous heritage conservation projects.

Data collection

Forty interviews were conducted with duration between 20 and 60 minutes each. Some interviews included more than one individual at a time, given a total of 49 people interviewed. The majority of the samples were female between the age of 41 and 50 years old. This is not surprising as women around this age group tend to be more involved in religious activities associated with the church, including conservation issues. However, the sample also reflects male members with roles in the organisation of church affairs and community life, as well as other members that were interested in taking part in the research. This allowed the sample to reach a saturation regarding the information

142 Elia Quijano Quiñones and Kalliopi Fouseki

provided by the participants, with recurrent topics of issues, benefits, and current preventive conservation practices.

Results: an evaluation framework for heritage conservation as a social process

The thematic analysis produced 23 social sustainability indicators of eight variables regarding people's perceptions, attitudes, and behaviours around local heritage that allowed us to identify the continuation of heritage conservation practices and its current impact on community life. Variables and indicators were categorised by themes to propose a conceptual framework for the assessment of impacts that heritage conservation has as a social process in four social dimensions: Awareness-Raising, Capacity-Building, Sustaining Civic Relationships, and Reinforcement of Partnerships (Figure 10.1). Additionally, the information of the identified indicators was compared with existing research and measuring methodologies for similarities (Crowther et al. 2008; UNESCO, 2017; Rojas, 2018), to provide consistency in the use of terminology but without compromising the data-driven approach.

Awareness-Raising

The theme came up from comparisons between the current understanding of local heritage as a result of the conservation project, and previous knowledge or the lack of it including also changes in attitudes towards local heritage. This awareness was mentioned as the reasons for renewed interest and pride for their heritage and raised concerns for

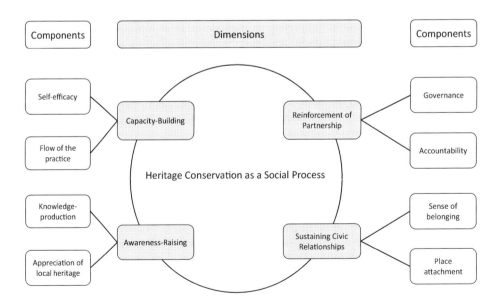

Figure 10.1 Conceptual framework for assessing social impacts of heritage conservation.

other communal areas in their towns. This theme is divided into the following variables and indicators:

- Knowledge production:

  a Historic value: historic information commonly followed with admiration for local heritage's qualities as an antiquity that has uniqueness, and it is worthy of protection.

  > *Chichimila 8: "while they [conservators] were working, they were telling us why ... this is not [original] because it is [made of] plaster. Look, this is a paint that is dissolving thus, it is not dated for the century. What we consider as original? When they are dated for the XIV, XV, XVI century, and so on."*

  b Loss of heritage knowledge: as a direct consequence of heritage's deterioration through the years, either by natural processes, historic events, or changes made by local restorations.

  > *Tixcacalcupul 13: "I did not have that information... I imagined that it was made, like any other [figure], of cement... that is what I thought. But made of wood. I did not know that."*

- Appreciation for local heritage:

  a Aesthetic value: feelings of pride for their heritage as a result of its improved aesthetic after the conservation work. This was in opposition to feelings of shame or disrespectful behaviour mentioned as common while their heritage looked in bad conditions.

  > *Sacalum 6: " Yes, I like to see that... every person that comes from different places said that the Church is beautiful. We like to hear that very much ... On the contrary, imagine if people from other places come and murmur how the church looks neglected, well, it is like you feel bad about it. It is my town, you think".*

  b Curiosity for local heritage: a new interest in knowing more about the history of their local heritage and keeping original features, opposed to the common practice of local restorations carried out by local artisans with a preference for a newer look and the use of inadequate materials.

  > *Sacalum 1: "And you think to yourself, how it came to be? I mean, that old painting was there. For how many years was that painting covered?"*

Capacity-Building

This theme emerged from memories and perceptions about heritage care, changes in general behaviour to comply with the preventive conservation, and the transmission of this knowledge. Interviewees expressed pride and amazement over learning a new skill which allows them to contribute towards what they considered a common good. Also, former female participants of the project spoke positively about the experience of men and women working together for the same purpose, and the opportunity for them, as women, to learn a skill that they perceived is for males. This theme is divided into the following variables and indicators:

- Flow of the practice:

    a   Acceptance of the practice: an agreement or resistance to change a behaviour to comply with preventive conservation measures.

    > *Tixcacalcupul 11: "They are still taking care of it but just a little bit, not like at the beginning. They [local people] would put it down [polychromate wooden sculpture] and used gloves, yes. But now, I saw how they put it down and they did not use gloves."*

    b   Following examples: locals' behaviour around heritage is influenced by figures of authority when these are seen complying with preventive care. This includes parents directing children's behaviours around local heritage.

    > *Chichimila 9: " Sometimes the priest is the first to do something and people copy that behaviour…" … "I have told her [daughter] why we should not, you can look at it but don't touch it…they know [my children] … when they come here with me, it is not like they can touch whatever they want. So, we teach them that. You can look at them, you can venerate them, but you cannot touch them because they are not a toy to be grabbed."*

    c   Responsibility for heritage care: for the case study, the direct responsibility is mostly on their priest and his assistants. General responsibility, nonetheless, was collaborating with money or time, following recommendations, and keeping the area clean.

    > *Chichimila 3: "I think because here it is a parish, so it is the priest and he organizes his people, but in fact, I think we are all responsible."*

- Self-efficacy:

    a   Gender equality: mostly expressed by women that, during the project, learned a new skill in their opinion, not normally associated with their gender. Also, they highlighted that women and men were working together in harmony and equal importance.

    > *Sacalum 6: "I was very proud of doing that type of job. Truth, it gives you joy to enter and see people that would go inside [the church] to see how you are working as a woman. Yeah, we were doing a man's job, but we did not care. We socialized with lots of people, men and young guys that were not the same age than us".*

    b   Preventive conservation knowledge: comments about special treatments, handling procedures, and knowledge about previously unknown materials.

    > *Sacalum 5: "Yes, that it should not get wet…it [mural paintings] had the original paint, but some parts were incomplete…so watercolours were applied. Yes, if it gets wet it will erase again."*

    c   Work and transferable skills: comments around income generation for locals through jobs for the maintenance of local heritage or the use of working experience during the conservation project to try other jobs.

Sacalum 6: "Yes, we were working there and to me was very nice because I learned so much and experienced so many things that I did not know. It was very nice."

Reinforcement of partnership

It came up as a theme through questions around issues of responsibilities over local heritage, petition-making, organisation, and perceptions about communities' role in the decision-making process (e.g. feelings of inclusion). These revealed a stratified web of heritage stakeholders, where major decisions go through a series of vertical interactions that can compromise bottom-up initiatives and the feasibility of partnership in heritage conservation.

This theme is divided into the following variables and indicators:

- Accountability:

  a   Approachability: positive or negatives experiences of interactions with conservators, affecting locals' levels of confidence to approach or to get involved in the conservation process.

  Sacalum 6: "...Because different young people [conservators] came...we socialize with them. They were very kind...Yes, we were not sent people that would look at you angry if you asked something, or that would reply: if you don't know, why are you working here. No, on the contrary, they said to us: do you have any questions? Whatever you feel you can't do, we are very happy for you to ask...Yes, we had a lot of support from them."

  b   Expertise: this was expressed as a level of confidence in the quality of the conservators' work as experts.

  Tixcacalcupul 4: "...here we did not know if the paint used [by local artisans to put on top of the polychromate wooden sculptures] was good or not. Well, the people who came [the conservators] they know. They have more expertise than us, in that way it can be preserved."

  c   Money: a concern for the expensive cost of the project in comparison with local restorations.

  Chichimila 5: "... and in fact, it did not cost 5 pesos [conservation's cost] they were told [general people]. And they were surprised to know how much a restoration cost. It is not that is elevated, it is not that. I told them it was not the case of using overpriced paint."

  d   Services and law prohibition: the necessity of public meetings, around the time of the project, to corroborate that conservators were doing their job. A perception of locals as being subjected to restrictions of use because of their heritage's status as part of the nation.

  Chichimila 5: "The INAH is the one who intervenes so it can be modified. It is something about national heritage, is not, that right? The INAH has to participate because you cannot modify the saints [the polychromate wooden sculptures]".

  e   Time: a concern for the duration of the project, either too long or too short, to achieve the expected results.

> Tixcacalcupul 1: *"Well, they worked well. It was well done but it took a long time. It took a long time for them to finish."*

- Governance:

  a Involvement in heritage affairs: former participants manifested feeling more confident about asking questions if given the opportunity. However, there were no future intentions for proposing or participating in decision-making process.

  > Sacalum 7: *"They did not say anything. I thought people will complain because the presbytery was very beautiful…and they removed it all…people from old times were shier…but now, a lot of people are more awake. If anyone tries to do that now, I am telling you…"*

  b Factors affecting involvement in the decision-making processes: a combination of individual circumstances, bureaucracy, and structures of authority. Factors identified: low income, apathy, lack of continuity, level of education, no time, fear, and problems between actors.

  > Chichimila 9 *"Yes, we did not have any choice more than to sit and wait. I feel in that part… I used to say: Who invented the INAH and considered this as national heritage? It is heritage, but it belongs to my town.*

  c Principal actors: it relates to who was involved during the organisation of the project and who would interviewees approach first if they wanted to raise a concern about their local heritage.

  > Sacalum 9: *"The church is related to the religious part. The priest here is the first authority in that respect, then it goes to the Archbishop. It is like that."*

  d Routes of interactions: the most common connections between actors when decisions regarding heritage are taken.

  > Chichimila 7: *"…Because in the meetings with the priest, he keeps us [local people] informed".*

Sustaining Civic Relationships

It came up as a theme when interviewees shared their perceptions about their personal role during the project and feelings they think have changed as a result of this experience. This revealed that their perceptions about their town and their social group were affected since personal connections were made through social interactions during the project. This experience was enhanced by socialising around a common good, learning as a group, and exercising communal effort. This theme is divided into the following variables and indicators:

- Place attachment:

  a New connections with the place: feelings of emotion, well-being, and personal connection that interviewees experienced through the conservation work.

> Sacalum 8: "...now when we go there [inside the church to look at the mural paintings] ...I always say to my daughter that it feels so nice, like an emotion here [in my heart], yes."

  b Willingness to allocate resources: considering local heritage as worthy of investment in the way of money or time.

> Sacalum 1: "We were paid twice, but I said, even without payment. If I could go and do it, I would; because it is for the benefit of the town, of us."

- Sense of belonging:

  a Connection between generations: the conservation project was seen as helping younger generations to enjoy local heritage as a common asset, to see and learn the same as older generations, while providing new opportunities for social interaction between generations.

> Sacalum 6: "Well, it is for my children. They used to go, sometimes during the school's lunch break, to see what we were doing. They liked to go inside [the church] and see what type of job we were doing."

  b Encouragement to be part of social life: a relationship between an appealing aesthetic of local heritage and engagement in social life. During the project, friendships and social relationships were created.

> Sacalum 12: "...it was a good environment [working in the conservation project] between us and we socialized a lot...I learned a little bit more about Sacalum, I am not from here and yes, it is very interesting their history and what they have here. I did not know anyone before, but I built good friendships [during the project]."

Discussion

This case study analysis shows that heritage conservation is a social process, and as a social process, it encourages sustainable practices in conservation through fostering social interactions between those involved in the process. Applying Shove's et al. approach to social practices (2012), we could argue that the sustainability of participatory heritage conservation depends on the continuous relationship between the materials (i.e. heritage objects and buildings in this case), the meanings (i.e. the values and feelings attached to the process and the heritage) and the competencies (skills) required to conserve heritage and engage communities.

The analysis demonstrated that local participants developed strong social connections with other members and professionals, as well as a strong sense of community and collective identity. However, the complexity of social contexts made people's commitment to sustainable practices easily affected by intra-group and external interactions. The indicators proposed in the conceptual framework should be viewed as interlinked with each other rather than as individual standalone factors. The first two dimensions of the framework relate to personal interaction with heritage (knowledge, feelings, self-efficacy), which immediately become collective since personal performance depends on interactions or feedback given to them by others. For example, some interviewees struggled to keep up with sustainable practices when these are perceived as beyond individual capacities, or when there is a strong disagreement to follow the practices from other members. A consequence of this was situations where the safety of the heritage objects was compromised, to some extent, to avoid problems. Thus, the potential issues of putting the responsibility for the practice into the hands of specific members of the local community (e.g. local leaders) should also be considered before the implementation of such an approach.

On Sustaining Civic Relationships, participatory methodologies showed an effect in peoples' interactions as a group based on connections and collective effort, in this case around local heritage. It was implicit that through the subsequent deterioration of their local heritage, changes in meanings and connection occurred, affecting social behaviour as well. An aesthetic of abandonment was related to negative feelings about their living environment such as sadness and shame, generating anti-social behaviour (e.g. public drinking and urinating in the area, lack of involvement). The relationship between signs of neglect in living environments and anti-social behaviour has been explored under the broken-window theory (see Ling et al., 2019). For the present research, a lack of investment in heritage, and other communal areas, was a source for frustration and a symbol of the underdevelopment. To some extent, this dissatisfaction was previously assessed in the form of local restorations which, in spite of the good intentions, compromised the integrity and long-term conservation of their heritage.

This case study touched meanings of local heritage in its qualities as a common good and as a means to obtain something else. In this case, we provide an Aristotelian (1996) definition of the word *good*, to indicate the use of local heritage as a means for a suitable habitat. Heritage is added to their living environment in two ways: as a source of unity and as public enjoyment of the space through its aesthetics. Thus, the preservation of heritage contributed to unity and tranquillity for locals, which were major concerns for the interviewees. Despite the levels of poverty and marginalisation, locals considered these towns as being very rich in providing its habitants with a place where they can live peacefully with a strong sense of community. Nonetheless, this quality of their environment is currently jeopardised by the increasing loss of social life (e.g. loss of costume and traditions) as a result of economic problems and migration. In this regard, younger generations were considered the most affected. Therefore, local heritage was perceived as necessary for the continuation of the bond between generations and the integration of newcomers. According to the information collected, these necessities were assessed during the heritage conservation project not only by the preservation of local heritage but also by mobilising its qualities as a 'social glue' through exercising common effort and socialisation.

We consider from our results that participatory methodologies in heritage conservation have a direct contribution as a resource to ameliorate social disintegration and other socio-economic problems. This has been previously pointed out by studies on the landscape-based approach (Veldpaus et al., 2013) and cultural resilience (Holtorf, 2018). Therefore, within the current debate of the role of cultural heritage within the SDGs 2030, our results align with the widely explored goal 11 – that is the goal of making cities and human settlements inclusive, safe, resilient, and sustainable. However, it would be useful at this point to draw attention on other areas of comparison between the framework provided here and the SDGs. More specifically, we propose that heritage can have a direct contribution to goals 3 (good health and well-being), 4 (quality education), 5 (gender equality), 8 (decent work and economic growth) (see also Xiao et al., 2018), and goal 16 (peace, justice, and strong institutions). Similarly, we consider a strong contribution to goal 1: ending poverty in all its forms everywhere. We built this point by looking at a definition of poverty – as proposed by Amartya Sen (2001, pp. 85, 89) – in terms of capabilities deprivation. Capabilities are the substantive freedoms that individuals enjoy leading the kind of life that they have reasons to value. A deprivation of basic capabilities results in exclusion from taking part in the life of a community.

Nonetheless, when discussing social sustainability in relation to participatory conservation, we would like to stress that collaboration is key area for creating communities of practice – merging sustainability and conservation. Our proposed framework mostly deals with collaboration under the dimension of Reinforcement of Partnership, which evidenced a complex web of interactions, expectations, and accountability. From the case study analysis, it is possible to argue that professionals are also participants in heritage conservation as a social process, as much as local people are, with the potential for affecting local approachability during the process. Similarly, we can argue that expectations of empowerment through participatory methodologies should approach a definition of power as the ability to achieve a purpose. This can only happen through partnership. Similarly, the continuation and the impact of sustainable practices depend on partnership to remain effective.

Conclusion and future research

Participatory heritage conservation can have social impacts that are possible to frame around the continuation of sustainable practices in conservation. The findings of this qualitative study in Yucatán, México, accord with interpretations of cultural heritage as a social process linked with groups' identity through performances of remembering (Smith, 2006). The study showed that the deterioration and loss of local heritage led to loss of knowledge and meaning amongst local communities in Yucatán around local heritage, limiting opportunities for experiencing a sense of collective identity. This was later reversed as a result of their participation in heritage conservation projects. However, sustainable practices for heritage conservation are also a social process in itself. The components of the framework proposed by this study behave as a compound where their impact not only can be identified and compared, but also that these areas can be approached as the preconditions for the adoption of sustainable practices. Exploring these areas *a priori* can be of benefit for the purpose of identifying potential problems.

Finally, by applying the evaluation framework proposed here in a qualitative methodology, it should be possible to generate data on the evolution of heritage conservation practices at local level. This could help to monitor closely the bridge between heritage conservation and the SDGs. Moreover, this type of research is relevant to the sustainability of rural areas. Currently, the world is experiencing major migration movements from people in rural areas to people in major cities affecting the former in a wide range of social issues. More research on the role of heritage conservation in alleviating this phenomenon could bring about more innovative ways of merging heritage conservation with rural development.

Acknowledgments

This chapter is part of the Ph.D. research entitled *Analysing the impact of participatory heritage conservation on social development: the case of contemporary Maya communities in Yucatán, México*, which is funded by the National Council for Science and Technology

(CONACyT, México), the Secretary for Innovation, Research and Education (SIIES, México), and the Science and Engineering in Arts, Heritage, and Archaeology (SEAHA, UK). Also, we are grateful to Dr. Claudia A. García Solís, from the National Institute of Anthropology and History (INAH), Yucatán-Centre, for her comments and observations on this research. Finally, to the people from Chichimila, Tixcacalcupul, and Sacalum that kindly accepted to be part of this research and shared their experiences and stories, thank you.

References

Aristotle (1996). *The Nicomachean Ethics*. Hertfordshire: Wordsworth Editions.

Ashworth, G. (2008). In search of the place-identity dividend: Using heritage landscapes to create place identity. In J. Eyles and A. Williams (Eds.), *Sense of place, health and quality of life* (pp. 96–185). USA: Ashgate. https://doi.org/10.4324/9781315243474.

Boccardi, G. (2015). From mitigation to adaptation: A new heritage paradigm for the Anthropocene. In M. Albert, F. Bandarin, & A. Pereira Roders (Eds.), *Perceptions of sustainability in heritage studies. Vol.4, Series heritage studies* (pp. 87–97). Berlin: Walter De Gruyter. https://doi.org/10.1515/9783110415278-008.

Bryman, A. (2008). *Social research methods*. New York: Oxford University Press.

Chirikure, S., Manyanga, M., Ndoro, W., Pwiti, G., & Rijhs, F. (2010). Unfulfilled promises? Heritage management and community participation at some of Africa's cultural heritage sites. *International Journal of Heritage Studies*, 16(1–2), 30–44. https://doi.org/10.1080/13527250903441739.

Crouch, D. (2015). Affect, heritage and feeling. In B. Graham & P. Howard (Eds.), *The Ashgate research companion to heritage and identity* (pp. 177–187). London: Palgrave Macmillan. https://doi.org/10.1057/9781137293565_11.

Crowther, J., Lyn, T., & Edwards, V. (2008). *Building connections, getting involved: Measuring social capital outcomes of community learning and development*. Scotland: The Scottish Government. Retrieved from: https://core.ac.uk/download/pdf/4158969.pdf (Accessed on 08 August 2020).

Dennis, A., & Smith, G. (2015). Interactionism, symbolic. In J. Wright (Ed.), *International Encyclopaedia of the social & behavioural sciences* (2nd ed., pp. 352–356). Elsevier: London. https://doi.org/10.1016/B978-0-08-097086-8.32079-7.

Della Torre, S. (2015). Shaping tools for built heritage conservation: From architectural design to program and management. In K. Van Valen & A. Vandesande (Eds.), *Community involvement in heritage: Reflections on cultural heritage theories and practices* (pp. 93–102). Antwerpen: Garant.

de Nooy, W. (2009). Formalizing symbolic interactionism. *Methodological Innovations Online*, 4(1), 39–51. https://doi.org/10.1177/205979910900400105.

Fernández Repeto, F. (1995). Celebrar a los Santos: Sistemas de fiestas en el noroccidente de Yucatán. *Alteridades*, 5(9), 51–61. Retrieved from: https://alteridades.izt.uam.mx/index.php/Alte/article/view/577 (Accessed on 08 August 2020).

Ferreira, T. C. (2018). Bridging planned conservation and community empowerment: Portuguese case studies. *Journal of Cultural Heritage Management and Sustainable Development*, 8(2), 179–193. https://doi.org/10.1108/JCHMSD-05-2017-0029.

Forero, J. (2015). The social function of cultural heritage: Conservation practices on the basis of sociocultural sustainability. In M. Albert, F. Bandarin, & A. Pereira Roders (Eds.), *Perceptions of sustainability in heritage studies, Vol.4, Series heritage studies* (pp. 173–186). Berlin: Walter De Gruyter. https://doi.org/10.1515/9783110415278.

Fouseki, K., Newton, D., Murillo Camacho, S., Nandi, S., & KouKou, T. (2020). Energy efficiency, thermal comfort, and heritage conservation in residential historic buildings as dynamic and systemic socio-cultural practices. *Atmosphere*, 11(6), 604. https://doi.org/10.3390/atmos11060604.

Gallou, E., & Fouseki, K. (2019). Applying social impact assessment (SIA) principles in assessing contribution of cultural heritage to social sustainability in rural landscapes. *Journal of Cultural Heritage Management and Sustainable Development*, *9*(3), 352–375. https://doi.org/10.1108/jchmsd-05-2018-0037.

González Gutiérrez, M. A. (2016). La educación como vía hacia la sostenibilidad del patrimonio cultural. *Conservación y Restauración*, Diciembre (10), 21–24. Retrieved from: https://wwww.revistas.inah.gob.mx (Accessed on 08 August 2020).

Guerrero Baca, L., & Soria Lopez, F. (2018). Traditional architecture and sustainable conservation. *Journal of Cultural Heritage Management and Sustainable Development*, *8*(2), 194–206. https://doi.org/10.1108/jchmsd-06-2017-0036.

Guzmán, P. C., Pereira Roders, A., & Colenbrander, B. J. F. (2017). Measuring links between cultural heritage management and sustainable urban development: An overview of global monitoring tools. *Cities*, *60*(Part A), 192–201. https://doi.org/10.1016/j.cities.2016.09.005.

Guzmán, P. C., Pereira Roders, A., & Colenbrander, B. J. F. (2018). Impacts of common urban development factors on cultural conservation in world heritage cities: An indicators-based analysis. *Sustainability*, *10*(3), 853. https://doi.org/10.3390/su10030853.

Guzmán Medina, V. (2010). Youth, poverty, and exclusion: Health problems of young Mayans in Yucatan. *Social Medicine*, *5*(2), 100–105. Retrieved from: https://www.socialmedicine.info/index.php/socialmedicine/article/view/458 (Accessed on 08 August 2020).

Handberg, C., Thorne, S., Midtgaard, J., Nielsen, C. V., & Lomborg, K. (2014). Revisiting symbolic interactionism as a theoretical framework beyond the grounded theory tradition. *Qualitative Health Research*, *25*(8), 1023–1032. https://doi.org/10.1177/1049732314554231.

Harrison, R. (2013). Forgetting to remember, remembering to forget: Late modern heritage practices, sustainability and the 'crisis' of accumulation of the past. *International Journal of Heritage Studies*, *19*(6), 579–595. https://doi.org/10.1080/13527258.2012.678371.

Holtorf, C. (2018). Embracing change: How cultural resilience is increased through cultural heritage. *World Archaeology*, *50*(4), 639–650. https://doi.org/10.1080/00438243.2018.1510340.

Jaspersen García, G. (2012a). *Conservación y restauración de escultura yucateca en culto: materia, símbolo y función*. Internal Report, Instituto Nacional de Antropología e Historia, Centro Yucatán. Unpublished.

Jaspersen García, G. (2012b). Social participation as a strategy for preventive conservation: The case of current Mayan communities in Yucatán, México. Paper presented at the *Reducing risk to heritage. International meeting*, Amersfoort: The Netherlands ICCROM.

Jaspersen García, G. (2014). Social participation as a way for sustainable projects in conservation of worshiping objects: The case of current Mayan communities in Yucatán, México. Paper presented at the *AIC 42nd annual meeting conscientious conservation: Sustainable choices in collection care*. San Francisco, CA.

Jones, S. (2017). Wrestling with the social value of heritage: Problems, dilemmas and opportunities. *Journal of Community Archaeology & Heritage*, *4*(1), 21–37. https://doi.org/10.1080/20518196.2016.1193996.

Katrakaziz, T., Heritage, A., Dillon, C., Juvan, P., & Golfomitsou, S. (2018). Enhancing research impact in heritage conservation. *Studies in Conservation*, *63*(8), 450–465. https://doi.org/10.1080/00393630.2018.1491719.

Landorf, C. (2011). Evaluating social sustainability in historic urban environments. *International Journal of Heritage Studies*, *17*(5), 463–477. https://doi.org/10.1080/13527258.2011.563788.

Ling R., Jihong, Z., & Ni, H. (2019). Broken windows theory and citizen engagement in crime prevention. *Justice Quarterly*, *36*(1), 1–30. https://doi.org/10.1080/07418825.2017.1374434.

Loulanski, T. (2006). Cultural heritage in socio-economic development: Local and global perspectives. *Environments Journal*, *34*(2), 51–69. Retrieved from: http://citeseerx.ist.psu.edu/viewdoc/download?doi=10.1.1.471.1108&rep=rep1&type=pdf (Accessed on 02 August 2020).

Magar, V. (2016). Sostenibilidad y conservación del patrimonio cultural. *Conservación y Restauración*, Deciembre (10), 8–14. Retrieved from: https://wwww.revistas.inah.gob.mx (Accessed on 08 August 2020).

McCurdy, D., Spradley, J., & Sahndy, D. (2005). *The cultural experience: Ethnography in complex society*. 2nd ed. USA: Waveland Press.

México Government. National Institute of Statistics and Geography (INEGI) (2010). *Censo de Población y vivienda*. Retrieved from: https://en.www.inegi.org.mx (Accessed on 08 August 2020).

México Government. Secretary of Social Development (SEDESOL) (2017). *Informe anual sobre la situacion de Pobreza y rezago social*. Retrieved from: https://www.gob.mx/bienestar/documentos/informe-anual-sobre-la-situacion-de-pobreza-y-rezago-social (Accessed on 08 August 2020).

Neal, C. (2015). Heritage and participation. In E. Waterton & S. Watson (Eds.), *The Palgrave handbook of contemporary heritage research* (pp. 346–365). Hard cover, London: Palgrave Macmillan. https://doi.org/10.1057/9781137293565_22.

Nocca, F. (2017). The role of cultural heritage in sustainable development: Multidimensional indicators as decision-making tool. *Sustainability*, 9(1882), 1–28. https://doi.org/10.3390/su9101882.

Poole, N., Gauthier, R., & Mizrahi, A. (2007). Rural poverty in Mexico: Assets and livelihood among the Mayas of Yucatan. *International Journal of Agricultural Sustainability*, 5(4), 315–330. https://doi.org/10.1080/14735903.2007.9684831.

Quezada, S. (2011). *La colonización de los Mayas peninsulares*. Biblioteca básica de Yucatán. Yucatán: Gobierno del Estado de Yucatán.

Rojas, E. (2018). Sustainable conservation of urban heritage: The contribution of governance-focus studies. In A. Labrador & N. Silberman (Eds.), *The Oxford handbook for public heritage theory and practice* (pp. 55–63). New York: Oxford University Press.

Sen, A. (2001). *Development as freedom*. Oxford: Oxford University Press.

Shove, E., Pantzar, M., & Watson, M. (2012). *The dynamics of social practice: Everyday life and how it changes*. India: SAGE.

Smith, L. (2006). *Uses of heritage*. London: Routledge.

Sully, D., Raymond, R., & Hoete, A. (2014). Locating Hinemihi's people. *Journal of Material Culture*, 19(2), 209–229. https://doi.org/10.1177/1359183513514316.

UNESCO (2017). Measuring culture's contribution in agenda 2030 Sustainable Development Goals. *Workshop on measuring culture in the SDGs*. Retrieved from: https://fr.unesco.org (Accessed on 08 August 2020).

UNESCO (2018). *Culture for the 2030 agenda, United Nations Educational, Scientific, and Cultural Organization, Paris, France*. Retrieved from: https://en.unesco.org (Accessed on 08 August 2020).

United Cities and Local Governments (2015). *Agenda 21 declaration of the inclusion of culture in the sustainable development goals*. Retrieved from: http://www.agenda21culture.net (Accessed on 08 August 2020).

Valle García, S. (2014). Impacto social en los programas de conservación. In María Perevochtchikova (Coord.), *Pago por servicios ambientales en México: Un acercamiento para su estudio* (pp. 197–218). México: El Colegío de México. Retrieved from: https://www.researchgate.net/publication/304501709_Impacto_social_en_los_programas_de_conservacion (Accessed on 08 August 2020).

Vanclay, F. (2002). Conceptualising social impacts. *Environmental Impact Assessment Review*, 22(3), 183–211. https://doi.org/10.1016/S0195-9255(01)00105-6.

Van der Auwera, S., Vandesanse, A., & Koen Van B. (2015). A recent history of heritage community involvement. In K. Van Valen & A. Vandesande (Eds.), *Community involvement in heritage: Reflections on cultural heritage theories and practices* (pp. 7–14). Antwerpen: Garant.

Vandesande, A., van Balen, K., Della Torre, S., & Cardoso, F. (2018). Guest editorial. *Journal of Cultural Heritage Management and Sustainable Development*, 8(2), 78–81. https://doi.org/10.1108/JCHMSD-05-2018-076.

van Oers, R. (2015). Cultural heritage management and sustainability. In M. Albert, F. Bandarin, & A. Pereira Roders (Eds.), *Perceptions of sustainability in heritage studies, Vol.4, Series heritage studies* (pp. 189–202). Berlin/Boston: De Gruyer, Inc. https://doi.org/10.1515/9783110415278-016.

Veldpaus, L., Pereira Roders, A. R., & Colenbrander, B. J. F. (2013). Urban heritage: Putting the past into the future. *The Historic Environment: Policy & Practice*, *4*(1), 3–18. https://doi.org/10.1179/1756750513Z.00000000022.

Villaseñor, I., & Magar, V. (2012) Assessing the performance of conservation activities. In S. Mendes Zancheti & K. Similä (Eds.), *Measuring heritage conservation performance. 6th international seminar on urban conservation* (pp. 1–14). Brazil: CECI&ICCROM. Retrieved from: https://www.iccrom.org/sites/default/files/ICCROM_19_Measuring-HeritagePerformance00_en.pdf (Accessed on 08 August 2020).

Wells, J. (2012). Using sequential mixed social science methods to define and measure heritage conservation performance. In S. Mendes Zancheti & K. Similä (Eds.), *Measuring heritage conservation performance. 6th international seminar on urban conservation* (pp.165–173). Brazil: CECI&ICCROM. Retrieved from: https://www.iccrom.org/sites/default/files/ICCROM_19_Measuring-Heritage-Performance00_en.pdf (Accessed on 08 August 2020).

Wijesuriya, G., Thompson, J., & Court, S. (2017). People-centred approaches. Engaging communities and developing capacity for managing heritage. In G. Chitty (Ed.), *Heritage, conservation and communities: Engagement, participation and capacity building* (pp. 34–50). London: Routledge.

World Commission on Environment and Development (1987). *Report of the world commission on environment and development: Our common future*. Retrieved from: https://sustainabledevelopment.un.org (Accessed on 08 August 2020).

Xiao, W., Mills, J., Gaudi, G., Rodríguez- Gonzálvez, P., Gonizzi Barsanati, S., & González-Aguilera, D. (2018). Geoinformatics for the conservation and promotion of cultural heritage in support of the UN Sustainable Development Goals. *ISPRS Journal of Photogrammetry and Remote Sensing*, *142*, 389–406. https://doi.org/10.1016/j.isprsjprs.2018.01.001.

11 Sustainable heritage through a sustainable community

Chihei Suzuki

Introduction and context

Those of us who are involved in protecting cultural heritage have traditionally been enthusiastic about its sustainability. Archaeological sites or historical buildings with high value have been purchased by governments, and important objects have been stored in museums; that is, efforts have been made to preserve the physical presence of cultural heritage. In extreme terms, even if no one lives there or uses that location or object, as long as the location or object itself remains, then protection of cultural heritage is deemed to have been achieved. Of course, now and in the future, there will continue to be types of cultural heritage for which such preservation methods are suitable. However, in recent years, the notion of cultural heritage has expanded and it is no longer limited to only 'old' and 'monumental' things. Cultural elements directly connected with people's daily life, such as cultural landscapes, are also considered a part of heritage and become objects of preservation today. For these new genres of heritage, called 'living heritage' by some scholars (Miura, 2005; Poulios, 2010, 2014; Wijesuriya, 2015), the sustainability of the heritage and of life itself is closely related.

The Rice Terraces of the Philippine Cordilleras, which were formed over 2,000 years ago, are one example. The beauty of the rice terraces is considered an outstanding example of a cultural landscape that demonstrates the harmony between humans and nature (Nozawa et al., 2008). The terraces were inscribed on the UNESCO World Heritage List in 1995 and represent a heritage whose value is secured by passing the techniques and knowledge of rice terrace creation, cultivation, and repair from one generation to the next. They are a typical example of the case where a community's survival is directly linked to conservation of a heritage. Another example is the Fujian Tulou, earthen-walled circular apartment buildings built in the mountainous area of south-eastern China that were inscribed on the World Heritage List in 2008 (Huang, 2020). As a building type, it was highly regarded from an architectural perspective because of its size and sophisticated construction technology. It was also appreciated for its Outstanding Universal Value as its 'elaborate compartmentalised interiors, some with highly decorated surfaces, met both their communities' physical and spiritual needs and reflect in an extraordinary way the development of a sophisticated society in a remote and potentially hostile environment' (UNESCO, 2008). The fact that people live within it carries an essential value in itself.

Such significant cultural landscapes where nature and human life are in harmony are often found in rural areas. In such areas, the vital issues are increasing income and improving living standards, particularly in developed countries in recent years; dealing

DOI: 10.4324/9781003038955-14

with population decline, which is a pressing issue mainly in rural areas; and sustaining the area itself. Thus, in many cases, the worst-case scenario is that people who are suffering from a shortage of human resources to maintain the community due to the declining population are not in a position to be able to preserve cultural heritage. However, doesn't cultural heritage have the power to solve these local issues? Is it unthinkable to consider changing the idea that maintaining and improving life should get a higher priority than conserving heritage, and that instead conserving cultural heritage can be a plan to rejuvenate and maintain villages?

In this chapter, as the role of the community as the bearer of heritage becomes increasingly more important (Hodges & Watson, 2010), we will discuss the case where local residents' everyday life activities for the sustainability of the community lead to the conservation of heritage.

Diversifying the value of heritage sites has been a global trend in recent decades. In addition to historical 'elite' architecture related to world religions and archaeological sites of huge civilisations, cultural landscapes, cultures of indigenous people, and even local beliefs and cultures that are more closely related to daily life have become regarded as heritage and become targets for conservation (Atkinson, 2008; Schofield, 2014). This increases the importance of local communities as inhabitants who manage heritage sites. If the culture of daily life itself is considered a form of heritage, conserving this heritage and the sustainability of local people's life are then two sides of the same coin (see UNESCO, 2015 Policy on the integration of a sustainable development perspective into the processes of the World Heritage Convention). In other words, conserving the heritage is indispensable for continued life in that area. However, in Japan, particularly in its rural areas, maintaining villages themselves has become difficult due to the declining population. Local revitalisation and sustainable management of local communities centred on, or triggered by, heritage are expected to occur.

In this chapter, it will be introduced the practice of preserving heritage while maintaining local communities by making use of a World Heritage site located in the western part of Japan as an example. In that area, there has been a shift of a managerial paradigm from participatory governance (i.e. where the private sectors participate in the management led by the public sector) towards active management by local residents (i.e. where management is run wholly by local residents' initiative).

Theory and methodology

Hidden Christian Sites in the Nagasaki Region, which were inscribed on the World Heritage List in 2018, are used as an example in this chapter (Japan, 2017). Located in Nagasaki and Kumamoto prefectures in the north-western part of Kyushu Island of the Japanese Archipelago, Hidden Christian Sites in the Nagasaki Region are a serial property comprising 12 components: ten villages, one castle remains, and one cathedral dating from between the 17th and 19th centuries. They reflect the era during which Christianity was banned, as well as when the Christian communities were revitalised after the ban's official lifting in 1873. Hidden Christians survived as communities that formed small villages situated along the seacoast or on remote islands to which Hidden Christians migrated during the ban on Christianity. The Hidden Christians gave rise to a distinctive religious tradition that was seemingly vernacular, yet maintained the essence of Christianity, and they survived, continuing their faith over the ensuing two centuries.

In order to reveal how Hidden Christians lived and built relationships with surrounding settlements and what communities they formed to maintain and inherit the beliefs that led to acquiring a unique cultural tradition, it is required to comprehensively evaluate where they lived, the fields they cultivated, the graveyard where they buried their departed, the traces of their leader's residence, and the surrounding villages with which they maintain a mutual aid relationship. In addition, their descendants have been living in the entire area where the Hidden Christians lived. In Hidden Christian Sites in the Nagasaki Region, the very place where people live, in particular, itself expresses the central value of World Heritage sites, and aiming for heritage sustainability is almost synonymous with local sustainability. Consequently, to achieve local sustainability, it was necessary to strengthen the depopulated and aging communities. However, there is no singular path to strengthening a community; the method varies depending on the characteristics of the location. Also, unfortunately, no absolute method has yet been established for strengthening a community through carrying out a certain activity. Rather, what is closest to our realisation is that, by making multiple attempts from various aspects through repeated trial and error, one effort succeeded while another did not, or that the community as a whole was simply strengthened by stimulating its various aspects. In the next section, we will consider examples of strengthening a community's power through the respective aspects of education, economy, migrant acquisition, and quality of life improvements, which are linked to local sustainability. It should be noted that all cases were carried out under the banner of World Heritage conservation.

I have been investigating this area for more than ten years beginning in 2007 and have been involved in its evaluation and conservation. In that time period, I have been practicing community development and capacity building together with local residents, experts, and government officials. I observed in detail the situation in which people who initially lose confidence in the area where they live and feel that 'it is nothing but a mundane Japanese rural village of agriculture/fishing' gradually recognise that the area has irreplaceable value. They become aware of what they need to do to be able to pass it, as a World Heritage site, on to the next generation. In that case, various efforts were made to conserve the area, which is cultural heritage, resulting in a desire to solve the various problems facing the village. This can be considered an example of contributing to the sustainability of a village/community by promoting the sustainability of its cultural heritage.

Analysis

Education

When aiming for sustainability of a 'heritage', it is first necessary to clarify the heritage's value, how to preserve it, and the challenges that will be faced during its preservation. Education about heritage is necessary to share these ideas not only amongst experts, but also widely within the community. The closer the heritage is to a characteristic of people's daily life culture, the more indispensable are people's initiative efforts for preservation, and inevitably, communities are strengthened through education. Recognising that the place where one lives (where the heritage is located) is highly valuable increases a person's attachment to and pride in the area. The act of preserving it in the community increases unity, and heritage education can be expected to help motivate improving the area where the heritage is located.

Sakitsu Village in Amakusa (Kumamoto Prefecture) is a fishing village, where the fishing deity Ebisu is likened to Zeus, while Maria is found in the pattern of abalone shells. As a village that has inherited its faith with familiar things rooted in the life and livelihood distinct to fishing villages, it is one of the components of the World Heritage. Sakitsu Village is a small settlement with a population of about 600; there was also an elementary school on site until February 2012.

Since 2007, Kumamoto University's civil engineering laboratory in the same prefecture (although it is about 120km away and separated by two islands) has frequently entered the village of Sakitsu (Tanaka, 2019). While they were involved in community development through their study of the area's characteristics, they also pushed forward with student field research. Sakitsu's unique land use is being analysed from a natural, historical, and social perspective, the issues in managing the village are being extracted, and a higher education method that makes practical policy proposals rooted in the local milieu is being developed. In 2012, the vacant teacher's house was rented by the university; called 'Tomitsu Lab: Tomitsu Laboratory' (Tomitsu is the name of the area), it is used as a research base.

Since 2008, the laboratory has frequently organised and held workshops for local elementary school students. Of course, the students learn about World Heritage sites and regional characteristics in classroom lectures, but the series of processes carried out as part of their curriculum also involved going out to the site, discovering the area's attractions and issues, and making proposals for the future. Specifically, they walked around the area in small groups to discover attractive things unique to Sakitsu, as well as those that might obstruct the region's uniqueness and those that need to be improved. In addition, based on their discoveries, they envisioned Sakitsu's past, present, and ten years into the future, and reported them to the class. Through these workshops, the children were able to realise the charms of the area where they live, understand its issues, and acquire an attitude of thinking independently about the area's future image.

In addition, parents were accompanied to the workshops the children participated in. People of working age are often less interested in community activities, but children's activities or encouraging them to work together with their children demonstrates an educational effect even for the active generation, that is, the generation of local leaders who are expected to engage for a long time. Moreover, many elderly people who have lived in the area for numerous years also participated in the workshops with the children and working people. Rediscovering the village's charm, which had become commonplace and usually went unnoticed, conveying the area's history to the younger generation, and having the opportunity to think about Sakitsu's future together can also be seen as a form of social and lifetime education. In 2009, the non-profit organisation (NPO) 'Sainotsu' was formed with the elderly residents at the centre and has led to autonomous activities, such as guiding visitors around the area.

Economic – creating accommodations utilising old Japanese-style houses

Employment is essential for people to continue their lives in a community. Social decline, one of the causes of depopulation, is often due to the lack of attractive and stable employment in the local area. In that respect, can preserving cultural heritage contribute to creating local jobs? The tourism industry comes to mind as an example. In terms of tourism at World Heritage sites, the issue of over-tourism or an excessive number of tourists has recently been called out and tends to be regarded as a problem. Rather than tourism for mass consumption, an alternative is cultural tourism (Pedersen, 2002), where visitors can appreciate the heritage with detailed explanations and get to experience various cultural and natural resources in the area of the heritage site. Would not such tourism greatly contribute to the region's sustainability? Also, cultural tourism would be a 'stay' form of tourism: the longer a visitor stays, the greater the positive impact on the local economy.

The Hidden Christians moved to what are now Remains of Villages on Nozaki Island (Nagasaki Prefecture). Nozaki Island was a sacred area for Shintoism; by pretending to be shrine followers, the Hidden Christians concealed and passed down their own faith. This island has become a World Heritage component that manifests how the Hidden Christians maintained a community. Nozaki Island itself has been an uninhabited island since 1971, but within the island are the remains of the residence of the Shinto priests, the remains of a village occupied during the ban on Christianity, and the former Nokubi Church built after the ban was lifted. The people of the main island of Ojika, which is located about 2km west, manage Nozaki Island.

Ojika Islands' population has also decreased by about three quarters in the last half century, which is a remarkable depopulation. However, it is an island that has prospered in trade and fishing since ancient times, and many traditional wooden structures that are more than 130–140 years old remain. Eight old Japanese-style houses that had been vacant were renovated in 2007, with one of the largest buildings reconstructed as an old folk house restaurant that can also be used by the islanders. The other seven were used for accommodation facilities mainly for visitors to rent. All the old Japanese-style houses in question were donated to the town by their respective owners, who, during the negotiation, supported the purpose for which they would be used by Ojika Islands. Furthermore, although repairs were necessary due to deterioration over time, specialists and engineers, including local carpenters and plasterers, carried out the renovations to improve comfort using modern methods while maintaining traditional taste.

The restaurant and accommodations have been in operation since 2010. While a typical inn on Ojika Island costs about 5,000 yen (50 USD) per night, accommodations that have been renovated from old private houses are priced at 21,000–72,000 yen. In addition, because meals are not provided at these accommodations, guests need to go into town to buy fresh ingredients from the islands and cook them at the inn. They can also eat at the restaurants in the city, including the old folk house restaurant, which have become mechanisms for bringing money to the town. Cleaning the inn, maintaining the garden, repairs, reservation management, and so on are all handled by the islanders, which also creates employment for the townspeople. As a result, as of 2005, tourism income, including product sales and accommodations, was in the tens of millions of yen. Tourism income has increased by about 500 million yen in recent years, and whereas there used to be three full-time employees, there are now about 50 full-time employees.

Economic – adding value to local products

Creating new industries is not the only way to strengthen a region's economic base; promoting the traditional agriculture, forestry, and fisheries industry can also be considered. However, it is not realistic to enhance so much production in areas where depopulation is growing. Instead, increasing value should be considered, and cultural heritage can contribute to this. For example, an appeal that a product is not simply an ordinary product but a special one that was produced in such a charming location with great time and effort or highlighting that the product's purchase contributes to preserving such a valuable heritage makes it possible to add value to products. Moreover, this is a good fit with the cultural tourism mentioned above. Visitors who have stayed in the area and fully experienced its value, including its heritage, can potentially support this appeal. The products may be a little more expensive than regular market products, but such products manufactured in the actual region can be fully expected to sell.

Villages on Hisaka Island (Nagasaki Prefecture) demonstrate a successful use of a feudal lord's migration policy that the Hidden Christians settle in underdeveloped areas and maintain and inherit faith. The villages are a component of the World Heritage site that provides an example of the kind of place people chose as a migration destination where they could maintain a community of faith. Hisaka Island is a small, horseshoe-shaped island and has completely different interior and exterior terrain across the mountains that form the island. On the gentle slope facing the inner bay, a settlement was formed that cultivates the rice terraces as a way of living. A small settlement was also formed on the open seaside where steep slopes are predominant, and cultivation in the terraced fields or fishing has been carried out. A camellia japonica forest, which has high wind resistance, has been formed on the open seaside of the island due to the extremely strong seasonal wind; camellia oil has been produced and used as cooking oil, hairstyling oil, and cosmetic oil since ancient times.

Before the Hidden Christians' migration, the indigenous people were Buddhist fishermen who set up small settlements mainly along Hisaka Island's coast. The Hidden Christians, who had the technology to cultivate farmland, settled mainly in the inland area of the island and reclaimed rice fields and farmlands to avoid unnecessary conflicts with the indigenous fishermen. However, during the fishery's busy season, they worked together with the fishermen as short-term employees, with a goal of maintaining their community while building mutual aid relationships with the fishing settlement. Since the relationship between the Hidden Christians settlement and the Buddhist settlement is also a unique cultural tradition built on the inheritance of faith, Hisaka Island's entire area falls within the boundary of a World Heritage component.

Currently, Hisaka Island has about 300 inhabitants, and depopulation is also evident. There is a shortage of successors in fishery and agricultural processing, and there is a tendency to have a surplus of resources with respect to production and consumption. Hence, the 'Hisaka Island Farm' was established on the island in 2014, and efforts are being made to develop, brand, and sell special products using the closed elementary school building. The island residents have developed products produced on the island, such as the rice harvested from the terraces, camellia oil, and bottled salted sea urchin; they are traded at a higher price than usual and have been well received. For example, the average transaction price of rice nationwide is about 15,000 yen (150 USD) per 60kg, but rice harvested from the rice terraces of Hisaka Island are sold for double that price at about 30,000 yen; moreover, it is said that reservations are sold out every year. These local products are doubtlessly high quality, but they are labelled at the time of sale as having been produced at a World Heritage island. It is thought that the fact that the products are intensively managed by a few people in an attractive location contributes to their high added value. In addition, these products are currently available for purchase only on the island, by pre-order, or at occasional product exhibitions in urban areas, and no aggressive expansion of sales channels, such as conducting sales activities, has been planned. This is based on the stance that the Hisaka Island Farm itself consists of volunteers from the island, and each person carries out activities within a reasonable range. This also contributes to the sustainability of their activities.

Migrant acquisition

Stable community members are necessary to maintain and strengthen the community. Nevertheless, depopulation is growing in Japan's rural areas; this can be attributed to the social decline caused by population outflow and a decline in the natural population increase due to lower birth rates. Hence, migrants are enthusiastically being recruited in various parts of Japan; however, World Heritage sites have an advantage in this environment. In fact, some migrants have stated that they came to the region not only to visit the World Heritage site, but because of their desire to live in the area. Since they come with a strong desire to live in the area, they have a highly positive image of it and great motivation to make it a comfortable place to live. Hence, the migration of people who passionately embark on new initiatives can be expected to have a positive impact on the existing residents.

When the Hidden Christians chose to relocate and were selecting a destination that was more distant from other people, they chose villages on Kashiragashima Island (Nagasaki Prefecture), where people were treated for illnesses. It is a component part of the World Heritage site that demonstrates how people tried to maintain their religious community by avoiding detection. Kashiragashima Island and the two main islands (Nakadori Island and Wakamatsu Island) became connected by land when a bridge was built at the end of the 20th century, and about 17,000 people now inhabit this region (the Northern Goto Islands).

With a population of about 57,000 in the 1950s, this region has also become depopulated. Hence, in recent years, the Northern Goto Islands have been focusing on attracting migrants. They recruit all over Japan by setting up support measures for child-rearing, settling, and employment by establishing facilities where immigrants can experience a short-term stay. At the same time, they promote the area's rich cultural resources, including the World Heritage site, scenic natural environment, and fresh and high-quality ingredients. As a result, in the years from 2011 to 2020, 554 people or more than 3% of the total population settled in the Northern Goto Islands. In particular, they had 135 migrants in 2018, and the year the Hidden Christian Sites in the Nagasaki region was inscribed on the World Heritage List, followed by 136 in 2019 and 88 by October 2020. It is clear that more migrants will come to the islands than before. Migrants are beginning to take root in the area, opening restaurants, teaching marine sports, and pursuing agriculture and fishing.

Looking at the age composition of migrants, the largest proportion (23.9%) are in their 20s, followed by those in their 30s at 21.7%; thus, about half the migrants are of a young and active generation. The next largest group, at 19.8%, includes those under 20 years old, who are likely accompanied by families that are in their 20s and 30s. In addition, 11.8% are in their 40s, and 10.5% are in their 50s. Thus, most migrants are members of the younger generation, and as they put down roots in the Northern Goto Islands, they can be expected to produce next generation, who will bear the community responsibilities. Furthermore, these migrants are expected to bring cultures, customs, and skills that did not previously exist on the islands, which will increase the area's charm. This will stimulate the older island inhabitants to bring about other innovations.

Of course, migrants alone are not enough to curb the social decline resulting from the rapidly progressing depopulation. However, we can see a positive cycle in the Northern Goto Islands, as those who migrated because they were attracted to the islands will increase the area's charm by engaging in various activities never before done on the islands.

Quality of life improvement

Article 5 of the World Heritage Convention (the Convention concerning the Protection of the World Cultural and Natural Heritage) states that 'each State Party to this Convention shall endeavour … to adopt a general policy which aims to give the cultural and natural heritage a function in the life of the community and to integrate the protection of that heritage into comprehensive planning programmes'. If the community's quality of life can be improved by conserving heritage, it is in line with the convention's purpose. Moreover, if the people find conserving heritage is worth living for, their attachment to the area and their civic pride will increase, and the area will become even more attractive. People will not leave such an attractive area, but rather, they may even want to go out of their way to come and live in the area. They will not engage in tourism activities that will place a burden on such an attractive area; rather, they may want to support the area even more by spending money. In regions that have increased their charm to such an extent, community power is strengthened and the settlement's sustainability is ensured, leading to the sustainability of the cultural heritage itself.

The settlement of Kasuga Village and sacred places in Hirado (Nagasaki Prefecture) demonstrate the Hidden Christians' practice of their faith through veneration of an object. During the Christianity ban, the Hidden Christians in Kasuga Village venerated the island where Christians were executed in the early days of the ban as a martyrium. In addition to making it a site for holding rituals to draw holy water, they also practiced their faith through their veneration of a sacred mountain that was also an object of Buddhism and mountain worship before Christianity was introduced. Even after the ban was lifted, people did not return to Catholicism and continued to practice their beliefs. They have been called *Kakure Kirishitan* (literally means Hidden Christian as well but is distinguished from *Sempuku Kirishitan*, Hidden Christian during the ban) whose religious beliefs and customs were transformed significantly during the 20th century.

In Kasuga Village, large-scale rice terraces that have existed since the Christianity ban are still being cultivated. The village includes 22 households that are inhabited, but there are some vacant houses that are no longer inhabited. In 2018, one of those vacant houses was renovated, and a base facility called 'Katarina' was opened to exhibit and explain Kasuga Village's value. The renovation used traditional construction methods to improve the usability of the interior. Panels are placed in the former main building to set up materials display shelves and establish the value of Kasuga Village's history and its World Heritage status while at the same time functioning for product sales. In the adjacent old retirement building, the Kasuga Village's residents are stationed as storytellers, where they freely talk about the village's old lifestyle and entertain visitors with tea and pickles. At the initial opening of 'Katarina' in 2018, there were five storytellers, three of whom were over 90 years old. Presently, four people have retired, including three of the original over 90s storytellers, and another four have joined; hence, there are still five people who are active. The natural hospitality that conveys the village's customs and traditions while serving handmade pickles and tea has been passed down. Although they do not necessarily speak about the value of World Heritage sites, they welcome visitors with a carefree smile, saying, 'welcome, welcome', and each word from their casual conversation impresses the visitors. Also, in a village that was thought to be empty, words such as 'these pickles are delicious' and 'Kasuga is a nice place' encourage them. Initially, they spoke of their worries about the hospitality activity at the information centre, but now, they say that they enjoy interacting with the visitors (Ueno, 2018).

When you realise the charms of the area in which you live as a result of heritage preservation, and, as a member of the community, you feel promoting these charms is worthwhile, even as you grow older, your feelings of affirmation that it is good to live in the area will surely grow even stronger. Furthermore, if both you and other recognise the area as attractive, it will greatly motivate carrying out activities to further refine the area's charms. This positive cycle leads to ease of living, comfort, and improved quality of life in the area.

Conclusions

In this chapter, we examined a framework in which various activities aimed at sustaining cultural heritage contribute to an area's sustainability by strengthening communities, using the World Heritage site, Hidden Christian Sites in the Nagasaki Region, as an example. We were able to observe that the residents are more involved in independent regional management rather than participatory preservation activities. This is precisely the case, since their lives depend on the area's sustainability, and if that is realised through heritage conservation, the residents will, of course, make an effort as it is 'about one's own self'. As the category of cultural heritage expands from being something 'nostalgic' and 'monumental' to the culture of everyday life, heritage stakeholders expand from a limited range of individuals such as owners, experts, and government officials to include local residents and visitors. In that case, the local residents who are stakeholders are responsible for heritage conservation and are also in a position to enjoy the benefits brought about by heritage conservation. If conserving heritage leads to improving the environment of one's daily life, it is only right that the local residents should independently develop heritage conservation activities.

The logic developed in this chapter can be organised in a correlation chart (Figure 11.1). First, heritage conservation efforts manifest a region's value. Enhancing the region through education promotes heritage conservation efforts, bringing the interaction between heritage conservation and education into play. Through education, people will be able to reaffirm their region's charms, and through repeated heritage conservation activities, these charms are then converted into pride and fondness for their region, which is irreplaceable. This pride and fondness will arouse further inquisitiveness and could even motivate heritage conservation. Publicising, promoting, and disseminating information about an area's attractiveness can lead to acquiring and increasing supporters and visitors, as well as adding high value to local products. An attractive area is also linked to easy and comfortable living; thus, some visitors and supporters can be expected to move to the area, making the existing residents want to stay. All of these can lead to improving the area's quality of life, and in these series of processes, the community's functions and ties will be strengthened. As a result, sustainable community

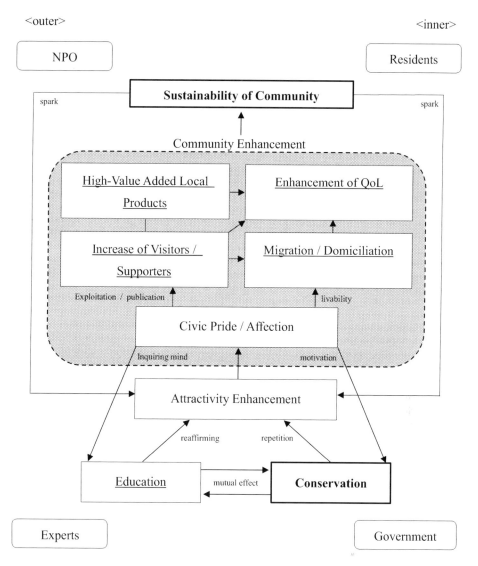

Figure 11.1 Correlation between conservation and sustainability.
Note: Cases are underlined which are dealt with in this chapter.

management becomes possible. This also indicates a cycle in which sustainable regional vitality leads to increased attractiveness and encourages further heritage preservation.

This series of activities is realised by various internal and external actors. In many cases, and even in the case of Hidden Christian Sites in the Nagasaki Region, the trigger was a group of administrative and academic experts who pioneered promoting the World Heritage nomination. At first, it was difficult to encourage the residents to participate, but their initiative was gradually nurtured. Today, active and autonomous activities are being developed in the form of communities or NPOs established by community members.

If this method of developing various activities based on conserving cultural heritage to achieve sustainable community management is to be called the 'World Heritage Approach', is this possible because a community is relatively small and in a rural area whose main industries are agricultural, forestry, and fisheries? One can argue that the answer is negative, as the same approach can also apply to urban areas. Even in an urban area, for instance, due to the shortage of understanding and cooperation by the community around the heritage, the local government may have some difficulty to inject public resources into conservation. Indeed, it is worth generating community activities based on urban heritage conservation because preventing a development project that ignores the value of local heritage will probably be far more expensive than letting the local citizens understand the same value by their direct and active heritage engagement. Moreover, strengthening the community cohesion through local heritage engagement will contribute to enhance the disaster resilience of the city (National Institute for Cultural Heritage, 2015).

Perhaps as the role of the community as the bearer of heritage becomes increasingly more important, the sustainability of the heritage and the local community as the two sides of the same coin may be applicable not only to World Heritage sites, but also to cultural heritage sites in general. The 'World Heritage Approach' that promotes community management through heritage conservation is not limited to the special case of World Heritage sites. If the process of 'making a heritage' (Pearce, 2000) through identifying value in a certain asset and targeting it for conservation can be done with the consensus or initiative not only of experts and government officials but of the community as well, one can expect that heritage management will be promoted in the form of community initiatives in the long term. Indeed, at the Historic Villages of Shirakawa-go and Gokayama which have been inscribed on the World Heritage List since 1995, the activities to preserve the distinguished wooden buildings were begun as an initiative of the local community as the key means to keep the village alive since 1970s. Since then, it is reported that the first generation who has furiously worked on maintenance and conservation of the village has gradually transited to the second generation who has started new efforts, having a more relaxed attitude and a broader perspective of the practice and outcomes of conservation (KURODA, 2016).

References

Atkinson, D. (2008). The heritage of mundane places. In B. Graham & P. Howard (Eds.), *The Ashgate research companion to heritage and identity* (pp. 381–395). Abingdon: Routledge.

Hodges, A., & Watson, S. (2000). Community-based heritage management: A case study and agenda for research, *International Journal of Heritage Studies*, 6(3), 231–243. https://doi.org/10.1080/13527250050148214.

Huang, H. (2020). *Fujian's Tulou: A treasure of Chinese traditional civilian residence*. Singapore: Routledge.

Japan (2017). *Hidden Christian sites in the Nagasaki region*. Retrieved from: http://whc.unesco.org/en/list/1495/documents/ (Accessed on 12 January 2021).

Kuroda, N. 黒田乃生. (2016). Shirakawa-go Gassho-dukuri Shuraku no Hozon 白川郷合掌造り集落の保存, *Sekai-isangaku Kenkyu* 世界遺産学研究, 3, 8–13. https://doi.org/10.15068/00145864.

Miura, K. (2005). Conservation of a 'Living Heritage Site' A contradiction in terms? A case study of Angkor World Heritage site. *Conservation and Management of Archaeological Sites*, 7(1), 3–18. https://doi.org/10.1179/135050305793137602.

National Institutes for Cultural Heritage (2015). *Cultural heritage and disaster resilient communities, the report of the expert meeting in the framework of the Third UN World conference on disaster risk reduction*. Retrieved from: https://ch-drm.nich.go.jp/others/h27culturalheritageanddisaster-resilientcommunities/ (Accessed on 9th January 2021).

Nozawa, C., Malingan, M., Plantilla, A., & Ong, J. (2008). Evolving culture, evolving landscapes: The Philippine rice terraces. In T. Amend, J. Brown, A. Kothari, A. Phillips, & S. Stolton (Eds.), *Protected landscapes and agrobiodiversity values* (pp. 71–93). Heidelberg: Kasparek Verlag.

Pearce, S. (2000). The making of cultural heritage. In E. Avrami, R. Mason & M. de la Torre (Eds.), *Values and heritage conservation: Research report* (pp. 59–64). Los Angeles: The Getty Conservation.

Pedersen, A. (2002). *Managing tourism at World Heritage sites: A practical manual for World Heritage site managers*. Paris: UNESCO World Heritage Centre.

Poulios, I. (2010). Moving beyond a values-based approach to heritage conservation. *Conservation and Management of Archaeological Sites*, 12(2), 170–185. https://doi.org/10.1179/175355210X12792909186539.

Poulios, I. (2014). *Past in the present: A living heritage approach: Meteora, Greece*. London: Ubiquity Press. https://doi.org/10.5334/bak.

Schofield, J. (2014). Heritage expertise and the everyday: Citizens and authority in the twenty-first century. In J. Schofield (Eds.), *Who needs experts? Counter-mapping cultural heritage* (pp. 1–11). London: Routledge.

Tanaka, N. (2019). Design of regional study method for transmission of local identity based on the cultural landscape. *Society of Global Business*. Retrieved from: https://s-gb.net/site/wp-content/uploads/2020/08/11tanaka2019.pdf (Accessed on 6 January 2021).

Ueno, K. 植野健治. (2018). 002・003 Hirado no Seichi to Shuraku (Kasuga Shuraku to Yasumandake)・(Nakaenoshima) 002・003 平戸の聖地と集落（春日集落と安満岳）・(中江ノ島). *Gekkan Bunkazai* 月刊文化財, 662, 23–25.

UNESCO (2008). Fujian Tulou, outstanding universal value. Retrieved from: http://whc.unesco.org/en/list/1113 (Accessed on 25 March 2021).

Wijesuriya, G. (2015). Living heritage: A summary. *ICCROM*. Retrieved from: https://www.iccrom.org/wp-content/uploads/PCA_Annexe-1.pdf (Accessed on 6 January 2021).

//
Part III
Dissonant and "Pacific" heritage

12 Developing international cultural relations through the negotiation of cultural property disputes

A sustainability perspective

Maria Shehade

Introduction

A critical issue faced by many countries worldwide is the request for the return of cultural artefacts to their countries of origin. The last decades saw an increase in such claims made by states, concerning war plunder, artefacts of former colonies, or artefacts of illicit origin, acquired by museums through the market.

Such claims are pursued either through litigation or through alternative processes, with negotiation gaining more ground in recent years. Such negotiation initiatives, resulting in restitution agreements, could offer many advantages to the involved parties apart from resolving the dispute in question. One such potential can be the development and establishment of long-term international cultural relations (ICR) between the involved parties, which can constitute the basis for developing close ties in many other areas of interest.

The involvement of governments in such claims instantly points towards the entanglement of issues of cultural patrimony and legitimacy with issues of cultural diplomacy and ICR. These cases also closely relate to the international museum activity, which may also impact diplomacy practices and soft power. However, despite these connections, very few studies have explored the role of such negotiations and restitution agreements in the development of ICR or their social, economic, diplomatic, and other implications.

This chapter aims to explore these interconnections and the potential of negotiated restitution agreements for establishing ICR, as a first attempt to map all potential implications. Through the example of the Getty–Italy restitution agreement signed in 2007 and its aftermath, the chapter highlights the potential that such resolution processes may have not only in the field of ICR but also in other fields of sustainable development such as the economic development of the involved parties, the physical sustainability and preservation of the contested heritage, the development of tourism initiatives, and the involvement and active participation of all interested stakeholders.

The chapter also discusses the preconditions for the achievement of such benefits and current research agendas on the interconnection of museums and negotiation initiatives with cultural diplomacy, ICR, and soft power, highlighting how such processes can be used as a driver for both social and economic development. The ultimate aim is to shed some light on an understudied aspect of these agreements, which is critical for future research in this area.

DOI: 10.4324/9781003038955-16

Cultural diplomacy, cultural relations, and soft power

Cultural diplomacy attracted much attention in recent years and especially after the 2001 (9/11) terrorist attacks in the US, which created a more intense interest in the role of cultural diplomacy in international relations "as a strategic platform for engaging with other nations and for wielding soft power on the international stage" (Cai, 2013, p. 127). Despite this renewed interest, cultural diplomacy practices are not a recent phenomenon, with examples of such practices being recorded throughout history. The exchange of gifts between kings as a symbol of friendship is recorded as one of the earliest signs of cultural diplomacy practices (Arndt, 2005). Despite this long history, cultural diplomacy has not received much scholarly attention in the wider field of international relations, and a confusion still prevails regarding its objectives and practices.

Cultural diplomacy can be defined as "the deployment of a state's culture in support of its foreign policy goals or diplomacy" (Mark, 2010, p. 64) or as "the exchange of ideas, information, art and other aspects of culture among nations and their people to foster mutual understanding" (Cummings, in Cai, 2013). However, providing a solid definition proves rather difficult, since there is no single definition and no general agreement or clarity regarding what this practice entails or what its objectives are (Mark, 2010; Ang et al., 2015). Cultural diplomacy has been associated with several goals: the cultivation of mutual understanding, the development of long-term relationships between different countries, the combating of stereotypes and the formation of national branding. The term has "broadened considerably over the years", and it now "applies to pretty much any practice that is related to purposeful cultural cooperation between nations" (Ang et al., 2015, p. 366).

The term is also used in conjunction with cultural relations. While cultural diplomacy is designed and exercised by the government and formal diplomats, cultural relations signify cross-cultural exchanges that "grow organically, without government intervention" (Arndt, 2005, p. xviii). Nevertheless, as will be shown, cultural relations can support national goals through the establishment of relationships and mutual understanding.

Cultural diplomacy is usually seen as a subset or particular form of public diplomacy,[1] which is a more "citizen-oriented form of diplomacy", targeting foreign audiences instead of governments (Ang et al., 2015, p. 368). In this context, cultural diplomacy has been acknowledged as an important tool to achieve the goals of public diplomacy (Pamment, 2014). Very often, it is also conceptualised through the notion of soft power, introduced by Joseph Nye. The concept refers to "the ability to get what you want through attraction rather than coercion or payment" (Nye, 2004, p. x) and the ability to persuade others instead of forcing them through military or economic power. As Nye stresses, soft power provides a vehicle through which a nation can shape the preferences of other nations and affect them to obtain preferred outcomes through the power of positive attraction. According to this line of thought, culture, arts, and literature can contribute to a country's attractiveness and therefore can be used to create positive images of the country internationally.

Cultural diplomacy is considered a channel through which soft power can be exercised. Since building an image to which other nations can relate or respect is an important element of soft power, it has been stressed that cultural diplomacy practices can facilitate this endeavour. For example, heritage has become an important component of American soft power, through the US government's support for heritage projects around the world (Luke & Kersel, 2013).

Nevertheless, the notion of cultural diplomacy is still rather vague, with little consensus on how it relates to other notions such as soft power, ICR or even public diplomacy. As emphasised by Nisbett, "it is now accepted that there is an abundance of terms and that these are used vaguely, loosely, and interchangeably" (2016, n.p). This may partly explain the lack of sufficient critical analyses regarding cultural diplomacy's role and how it operates within the wider realm of international relations. Despite the increased interest in public diplomacy, cultural diplomacy was given little attention by both international relations literature and heritage literature, with critical analyses from the perspectives of cultural studies or policy being characterised as "almost non-existent" (Ang et al., 2015, p. 366). This low scholarly attention means that several aspects of cultural diplomacy remain understudied (Mark, 2010, p. 64), including the examination of the connection between cultural diplomacy, restitution negotiations, and ICR development, which is the focus of this chapter.

Museums in the cultural diplomacy realm

Although diplomacy practices are mostly associated with governmental entities, in recent years, there is a rise of non-state actors who increasingly intervene in diplomatic practices. By non-state actors, we refer to "non-sovereign entities that exercise significant economic, political, or social power and influence on the national or international levels" (La Porte, 2012, p. 4). Such actors may include NGOs, social movements, local communities, and museums.

The interconnection of museums with cultural diplomacy became evident during the 18th and 19th centuries, through the establishment of national museums. Newly formed nation states used national museums as a vehicle to strengthen national identities and pride, or a sense of belonging. Apart from preserving cultural heritage and educating the public, museums were seen as "national expressions of identity" (Macdonald, 2003, p. 3). As evident, their contribution in constructing national identities and strengthening patriotism assigned to museums a "sizable political element" (Hoogwaerts, 2016, p. 315).

Thus, throughout their development, museums were interconnected with issues of politics and diplomacy, since through their exhibitions and narrative formation they aimed to promote political discourses of identity, but also to build bridges with other nations, create international partnerships, and even develop cultural tourism. Inevitably, because of these functions, museums are characterised as "heavily political, often involved with or implicated in international relations" (Sylvester, 2009, p. 3).

Nowadays, museums are expanding this role and attempt to communicate national discourses and political messages abroad through activities such as international loans, travelling exhibitions, international programmes, and museum franchises. Such activities allow art exchanges but also the development of platforms for exchanging ideas and developing inter-cultural exchanges. In this context, many scholars focused on the potential of museums as vehicles of soft power (Cai, 2013; Lord and Blankenberg, 2015; Hoogwaerts, 2016). It is often emphasised that museums can contribute to the creation of attractive images for a country and to strong cultural relations through exchanges, such as travelling exhibitions and international loans, that allow the public to engage with aspects of another nation's culture.

Museums are constantly gaining ground as vehicles of cultural diplomacy through activities discussed by scholars, especially from a museum studies perspective: mobility of exhibits and inter-museum loans (Hoogwaerts, 2016), blockbuster international exhibitions (Davidson & Pérez-Castellanos, 2019), exchange of scientific staff, educational programmes, provision of consulting services abroad (Rocco, 2013), and museum franchising, as in the case of the new Louvre in Abu Dhabi (Ajana, 2015; Grincheva, 2020).

However, such activities were also accompanied by intense debate regarding the use of culture as a tool of government propaganda or the development of a modern cultural colonialism through which powerful western museums capitalise on power relations built during colonialism, posing legitimacy questions on the museums' diplomatic endeavours (Reeves, 2007).

Nevertheless, all these activities reflect the different pillars on which the international and political dimension of museums is based: material culture, branding, commercialisation, or propaganda. However, what is usually neglected in current literature is the role of cultural patrimony and restitution as an added pillar in this equation.

Despite the broadness of these museum activities, "academic literature on the phenomenon remained thin", mostly focusing on the activities' operational or economic aspect (Cai, 2013, p. 132). Much scholarly attention was given in museum franchises and international exhibitions, which are considered as "part of the transnational work of museums which is implicated in systems of ICR and politics" (Davidson & Pérez-Castellanos, 2019, p. 15). However, other aspects, equally important to the international museum activity, are usually neglected. The negotiations for the restitution of cultural assets are closely connected to issues of diplomacy and have a strong political aspect that cannot be overlooked (Shehade & Fouseki, 2016). However, very few studies have explored their role in the development of ICR or the social, economic, diplomatic, and other implications of such agreements.

This chapter aims to fill this gap and explore these interconnections. The following section explores the restitution agreement between Italy and the Getty Museum and mainly its aftermath. Although Italy signed several similar agreements with many museums in the US, this particular agreement constitutes one of the broadest ones, including many different aspects of collaboration between the parties, offering a unique opportunity to analyse many different aspects of such collaborations and their effect on the development of ICR.

Negotiations for the restitution of cultural assets as a tool for developing ICR

Recent years have seen an important increase in the use of alternative dispute resolution (ADR) methods for the settlement of cultural property disputes, especially due to the high cost and uncertainty of litigation outcomes. Thus, there seems to be a growing trend towards different resolution methods such as negotiation, mediation or arbitration, with negotiation being the means more frequently used for the settlement of such disputes (Shehade et al., 2016). These alternative methods are also promoted by international organisations,[2] reflected in ADR clauses included in international conventions such as the 1970 UNESCO Convention on the Means of Prohibiting and Preventing the Illicit Import, Export and Transfer of Ownership of Cultural Property, and the 1995 UNIDROIT Convention on Stolen or Illegally Exported Cultural Objects.

This interest in ADR and especially negotiation is not surprising since ADR offers certain advantages such as less expensive and time-efficient procedures, rendering them more accessible to financially weaker parties. They also offer the possibility of considering other issues apart from legal ones such as ethical, historical, moral, political, and religious issues, which is not possible in litigation, which is based on strict legal doctrine. The most strongly emphasised advantage, though, relates to the ability for creative solutions and chances for mutual gain and to the preservation of long-term relationships that allow future cooperation between the involved parties.

When negotiations are successful, they usually lead to the signing of an agreement that includes the details of the resolution achieved regarding the contested artefacts, and also other provisions for loans, exhibitions, collaboration on different levels and other exchanges. Thus, such agreements create avenues for the development of ICR which can in turn facilitate the development of cultural diplomacy endeavours through the established exchanges.

Central to the establishment of these new relationships is the issue of respect. Apart from the need to preserve cultural heritage within the context it was created, many restitution claims are founded on issues of national identity and self-identification. The symbolism behind such artefacts has strong connections to issues of collective identities or social justice, with iconic heritage objects becoming part of the collective identity and memory and eventually of nationhood. Therefore, the removal of such objects is seen as disrupting social justice and collective identity. In such cases, a restitution agreement is not only seen as a reparation of justice but also as an expression of respect, for both the national cultural heritage and the laws of the country.

Thus, such agreements can be the starting point for the development of strong ICR, which in turn can pave the way for other collaborations, exchanges, and cultural diplomacy activities, creating opportunities for both parties involved on different areas of sustainable development: economic development and tourism initiatives, physical sustainability and preservation of the contested heritage, and active participation of all stakeholders (local communities, museums, governments, local authorities). This is clearly evident in the dispute between Italy and the Getty Museum, which provides an example of how museums can "play a more active role in building good relations with countries and communities" and how they can turn repatriation initiatives into a "positive opportunity for public relations and international cooperation" (Green, 2017, p. 6).

The Getty–Italy agreement

The case concerned 52 objects, with a Cult statue of a Goddess being one of the most prominent pieces. The contested artefacts were of illicit origin and came from either clandestine excavations or donations of private collectors whose collections comprised many illicitly acquired artefacts.[3] The two parties started to negotiate in 2002, and after many years of crisis, deadlocks, and turmoil, they eventually reached an agreement in 2007 (for a detailed analysis of the conducted negotiations, see Shehade, 2017).

According to the agreement, Italy received 40 claimed objects,[4] including the Goddess statue, one of its major claims. All legal actions against the museum were dropped, and Italy agreed to offer loans comparable to the objects returned. Italy received the museum's agreement to consult the Italian Ministry of Culture in advance of future purchases, loans, or donations of Italian artefacts. This was a very important step in Italy's attempt to combat the acquisition of unprovenanced artefacts by major US museums, and together with the Getty's adoption of a much stricter acquisitions policy, it was a factor that paved the way for the establishment of a collaboration between the two sides. The parties also agreed to establish a broad cultural collaboration including loans of significant artefacts, joint exhibitions, research and conservation projects, and excavation permits with temporary export licenses for excavated material for research and restoration. The collaboration programme was established through several agreements, complimentary to the initial agreement.

One of the first collaborations was the exhibition "Bernini and the Birth of Baroque Portrait Sculpture", presented at the Getty in 2008 with Italian loans, characterised as the "most complete exhibit of Bernini busts ever displayed outside Italy".[5] This was only the first of a series of exhibitions and collaborations following the 2007 agreement. In March 2009, the Getty and the National Archaeological Museum of Florence signed a long-term cultural collaboration (as a result of the initial agreement), which brought several of the Florence museum's greatest treasures to Los Angeles for the first time. Three exhibitions, with loans from the Florence museum, were held at the Getty with the most prominent one being an exhibition on the Etruscan bronze "The Chimaera of Arezzo". The agreement provided for "not only great exhibits but also the restoration of archaeological finds belonging to Italian heritage",[6] as in the case of a Roman bronze putto treated by the Getty conservation specialists (Lyons, 2014). This collaboration also allowed for the active involvement of many academics, curators, conservators, and other specialists who formed active networks of scientific exchanges through conferences and symposia on Etruscan or Hellenistic art (ibid).

In April 2009, the Getty announced another collaboration with the Naples National Archaeological Museum, which would lend to the Getty two life-sized bronze figures from Pompeii, for two years. The collaboration resulted in a research programme that was closely connected to the physical sustainability of the loaned artefacts, which underwent extensive stability treatment and tests to identify previous restorations before being exhibited (Lyons, 2014).

In 2010, the Getty announced a long-term collaboration with the Sicilian Ministry of Culture and Sicilian Identity. The agreement provided for a "number of collaborative efforts, including object conservation, earthquake protection of collections, exhibitions, scholarly research, and conferences".[7] Among the projects announced was a 2013 exhibition on Sicily, with loans from Sicilian museums and other international institutions. The agreement also included provisions for collaboration on seismic mitigation and a collaborative conservation project with the Agrigento Archaeological Museum. The Getty would receive loans of objects which would be displayed at the museum after going through conservation and after having new custom-made earthquake isolation bases, which would accompany them back to Sicily.

In 2012, the Getty announced another extension of the 2007 agreement, through a bilateral agreement with Rome's Capitoline Museums to create a framework for the conservation and restoration of artworks, future exhibitions, and long-term loans.[8] The parties' good collaboration also led to the return of new artefacts to Italy. In 2013, the Getty announced that a terracotta head of Greek god Hades was being voluntarily sent back to Sicily, after the museum worked with Sicilian officials to determine its provenance.[9] In 2017, the museum announced the voluntary return of a marble statuette of god Zeus, following information provided by Italian officials.[10]

Negotiated agreements in the realm of ICR

Undisputedly, such restitution agreements can solve long-standing disputes on contested artefacts, ameliorate the parties' relationship, and change the involved museums' practices regarding the acquisition of unprovenanced antiquities. However, as evident, these are not the only benefits that can stem from such agreements, which have direct implications on the development of ICR.

The exhibitions organised with Italian loans not only benefited the Getty but Italy as well, since "the works on loan become the new "cultural ambassadors" for the source nation" (Falkoff, 2007, p. 286). These exhibitions promoted Italy to the Getty's large audiences, acting as a tool of nation branding and promoting the Italian heritage to the American public. This may have many effects on multiple levels, from creating an avenue of cultural diplomacy through these exhibitions, to tourism and economic sustainability of Italian regions connected to the loaned or repatriated artefacts, and to the artefacts' physical sustainability and the creation of collaboration networks including different stakeholders.

Many of the artefacts returned to Italy through the initial restitution agreement would return to Aidone, an economically deprived area of 5,000 residents. Following several restitution agreements, signed between Italy and American museums, Italians stressed the repatriations' importance to the area's economic growth and the repatriated artefacts' value in developing sustainable cultural tourism. This was also stressed by Aidone's mayor, noting that the repatriated artefacts could "trigger an unprecedented economic development in Aidone",[11] an expectation also shared by the local community who hoped to increase the visitation to the Aidone Archaeological Museum, which is where the Goddess statue would be displayed. Indeed, the ability of treasure masterpieces to attract visitors from around the world and the subsequent generation of increased economic activity for the hosting museum and the local economy has been stressed very often in relevant literature (Lord & Blankenberg, 2015). The advantages gained regarding the artefacts' physical sustainability and the collaboration networks formed with different stakeholders for advancing research and knowledge were also very important for Italy, that benefited from Getty's conservation expertise.

These exchanges also benefited the Getty, allowing for the creation of important exhibitions and the development of collaborations on research and excavation. Such "blockbuster" exhibitions attract large visitor numbers, with the consequent economic activity, and the collaboration activities developed open avenues of soft power practices, apart from benefiting Getty for its research endeavours.

Collaborations stemming from restitution agreements may also reinforce governmental policies for the establishment of collaboration initiatives with a direct impact on other cultural diplomacy practices. For example, in the US, the Cultural Property Implementation Act (CPIA) includes a mechanism for the establishment of bilateral agreements with other countries for the protection of endangered artefacts. In this context, the US has signed multiple Memoranda of Understanding (MoUs) with many Latin American countries as well as Italy, Greece, Cambodia, and China.[12]

The mentality of these MoUs revolves around both imposing import restrictions for particular types of endangered artefacts and also establishing scientific, cultural, and educational collaborations through the exchange of artefacts, exhibitions, and expertise. Thus, the restitution agreements signed by US museums (whether private or not) can also contribute to the advancement of such policies and to the overall diplomatic moves of the US in the international arena, acting as vehicles for achieving the MoUs' goals. The repatriation of contested artefacts "alongside the formation of cultural heritage partnerships has become an increasingly important tool in the international diplomatic toolkit" (Green, 2017, p. 9). This is especially true for the US, since many of the signed MoUs with countries such as Italy, Greece, or Cyprus, illustrate "strategic geographic partners for the US in the eastern Mediterranean" (Luke & Kersel, 2013, p. 69). This is an example of how ICR developed through agreements between museums and foreign states can also contribute to the cultural diplomacy practices of the state in which the museums are located.

Therefore, restitution negotiations are not just a matter of patrimony, legal jurisdiction, or protection, but may also constitute a key for "efficient and productive cultural diplomacy that builds long-term relationships through archaeological programmes, restoration projects, and museum exhibitions" (Luke, 2012, p. 188). This has a two-fold implication. On the one hand, museums involved in such negotiations should realise the full extent of their potential and effect on issues relating to ICR and cultural diplomacy. Although such negotiations have been thoroughly discussed with regard to the legal or other arguments or the implications of such claims and acquisitions, their effect on issues of ICR or diplomacy has not been fully realised by some museums involved. On the other hand, countries involved in such claims should not only focus on the sheer number of objects claimed, but should see such agreements as a diplomatic opportunity to establish a collaboration with museums that would ensure the development of better and more transparent acquisition practices, and develop alternative diplomatic means with a direct effect on tourism, economy, physical suitability, and heritage protection.

Preconditions for developing ICR through restitution agreements

What does it take for negotiated restitution agreements to have the aforementioned effects on cultural diplomacy and ICR? The Getty–Italy case is also illustrative of the preconditions for the smooth development of ICR through such agreements.

Although the agreement of the Getty with Sicily produced many collaborations and exhibitions, it was not always tranquil. The problems occurred in 2013, because of the exhibition "Sicily: Art and Invention Between Greece and Rome", presented at the Getty. The exhibition consisted of 150 objects, many of which were on loan from Sicilian museums, including a gold phiale and the Mozia Charioteer as major attractions. The exhibition was also scheduled to go to the Cleveland Museum of Art, with the two museums sharing the exhibition costs.

However, in June 2013, Sicilian officials stated that the exhibition's two major attractions should not travel to Cleveland Museum because their absence was depriving Sicily of tourists. This was part of a travel ban on 23 of the island's most important artworks, stating that they should not travel abroad except under extraordinary circumstances and that if they were approved for loan, substantial fees would be imposed to the borrowing institution.[13]

The travel ban reflected a policy shift initiated by the growing concerns that Sicily's most important treasures are frequently on loan abroad, depriving the region of valuable tourism. As expected, the decree created many problems to the collaboration of Sicily with the Getty and other American museums with which Italy signed restitution and collaboration agreements.[14]

In July 2013, the Cleveland exhibition was cancelled, meaning that the Getty would absorb the entire financial responsibility for the exhibition. However, in August 2013, an agreement was reached between Sicily and the Cleveland, after the museum offered to lend 9 Caravaggio paintings for a future exhibition in Palermo (Lyons, 2014), and so the exhibition went on as originally planned.

Sicily's regional government has an autonomous status within the Italian system and is free to make its own cultural policies, with the new policy contrasting with Italy's central lending policies and agreements signed with American museums. Although the collaborations controlled by the central government progressed well and materialised without surprises, despite the succession of ministers, Sicily failed to illustrate the same stability and the agreement's implementation was problematic. As noted by a Getty curator:

> Five cultural commissioners held office, each with different agendas and attitudes about how international exchanges should be used to promote Sicily's artistic heritage. For U.S. museums, where exhibitions take years of advance planning, dealing with last-minute bureaucratic delays and changes of direction is not easy.
> (Lyons, 2014, p. 259)

Evidently, a country's cultural policy regarding such collaborations is extremely important in achieving the development of strong ICR. The instability of cultural policies or the existence of complicated bureaucratic systems impedes the implementation of such agreements and may discourage any future collaborations. Thus, for restitution agreements to contribute to the establishment of strong ICR, the relevant policies of the claiming states should be stable between election cycles to allow for the development of long-term exchanges. This policy stability and the simplification of bureaucratic systems will also facilitate the communication between state ministries and private museums, which can be challenging due to the changes of ministry officials.

Despite these instabilities, Italy is an illustrative example of a country that adopted a restitution model that allows for the creation of ICR through the established agreements. The "Italian model" is based not only on the claims for the restitution of artefacts, but also promotes a wider collaboration with the involved museums through exhibitions, loans, conservation, and other exchanges. This model has a dual benefit, apart from the artefacts' restitution. First, the country has benefits on soft power, tourism development, nation branding, and heritage protection. More importantly, the negotiated agreements constitute the basis for the development of a new, more sincere, relationship with these museums, aiming to the establishment of transparency in acquisition practices, with a positive effect on future acquisitions or claims.

For such a model to be successful, the country's cultural policies should allow for such exchanges, such as international loans. If the procedure for temporary loans is difficult or the country is not willing to support such activities, then it is difficult to achieve the establishment of ICR. In the case of restitution agreements, some nations object to "loans diplomacy" as a resolution method, considering that this may encourage museums to continue their malpractices in acquisitions (Falkoff, 2007). However, such policies hinder the establishment of long-term ICR, and Italy is a good example of the benefits that can be derived from an alternative model.

The aforementioned stability should also characterise the involved museums. As already explained, the development of ICR through restitution agreements closely relates to the issue of respect for a country's cultural patrimony and the connotations this entails. The establishment of further collaborations through restitution agreements requires the devotion of the involved museums towards a new approach regarding their acquisitions and policies. In the case of the Getty–Italy agreement, the museum agreed to consult with the Italian Ministry in advance of future purchases and adopted a stricter acquisition policy in an attempt to limit the acquisition of illicit antiquities. Such initiatives signal the respect towards a country's cultural heritage, a stable intent of the museum to adhere to international norms for protecting cultural property and a sensitivity towards restitution claims with a possible policy change regarding the retention of artefacts. All these can foster a positive ground for the cultivation of ICR through restitution agreements. In this context, a museum's denial to even negotiate such claims may be characterised as an opportunity lost in many different areas, but most importantly for the development of mutually beneficial, multi-level relations.

Conclusion

Restitution claims have usually been dealt with suspicion and defensiveness on the part of museums receiving such claims. However, in recent years, there are many examples of museums that have turned such claims into opportunities for positive engagements with the claiming countries and an opportunity for the development of international cooperation. In these cases, negotiated restitution agreements constituted a vehicle for establishing renewed partnerships and reinforcing governmental policies and cultural diplomacy practices.

As evident from the Getty–Italy case, such agreements could offer many advantages to the involved parties apart from resolving the dispute. The establishment of long-term ICR between the involved parties can constitute the basis for developing close ties with impact on many other fields of sustainable development such as the economic development of the involved parties, the physical sustainability and preservation of the contested heritage, the development of tourism initiatives, and the creation of active networks of different stakeholders.

For this to be achieved, the claiming countries should develop stable cultural policies and systems that facilitate such collaborations and exchanges. Museums on the receiving end of such claims should demonstrate a devotion towards a new approach regarding their acquisitions and policies and grasp the full opportunities that lie ahead of them through such changes. In essence, both parties should realise their potential in the ICR arena and that restitution cases not only affect issues of legitimacy and heritage protection but may also touch upon issues of cultural diplomacy and ICR. Therefore, such negotiations "command strategic, diplomatic attention", and parties involved in such endeavours should "exercise great diplomacy in negotiating access to and/or restitution of cultural heritage" (Luke, 2012, pp. 180, 188). Since many restitution claims are treated with hostility by many museums, involved parties should also have in mind the wider implications of such negotiations and agreements in the sphere of ICR and that failed negotiations, or a denial to negotiate, may lead to "the cessation of dialogue and cultural connections" (ibid) or constitute an opportunity lost for establishing such relations.

Restitution negotiations are by no means simple or easily solvable. Apart from involving complex legal issues, they are strongly affected by many parameters and complex dynamics, which may impact the process positively or negatively (Shehade, 2017). Nevertheless, this chapter aimed to highlight a connection of negotiated restitution agreements, with the development of ICR or diplomatic practices, which has not been emphasised so far to its full extent. As Sylvester stresses, "international relations and museum practices blend into each other" (2009, p. 4); however, cultural heritage literature has not fully grasped this relation. This also applies to restitution agreements and their effect on diplomatic practices, soft power, and ICR.

Thus, more research is needed to understand the full effect of such agreements on ICR, and this chapter constitutes a first attempt to map all potential implications. The attempted mapping of these potential implications and of the interconnections between these agreements and soft power or ICR can also provide valuable information in future attempts to evaluate or measure such activities and implications as well as in models attempting to map and evaluate soft power activities of museums or countries. Apart from the interconnection of museums with issues of branding, tourism, politics, or economic impact, future research should also further investigate the interconnection of negotiated restitution agreements with other goals of sustainable development.

Like blockbuster exhibitions, collaborative programmes, or franchising, the restitution of cultural assets can constitute a powerful tool of cultural diplomacy and another potential avenue of soft power. Therefore, the resolution of cultural property disputes can constitute a driver for the development of ICR between involved parties, which in turn can have a major impact on several sustainable development goals and constitute a driver for both social and economic development. Thus, it is imperative for involved parties to understand this potential and measure their actions and strategies when negotiating such cases, envisioning this not as a threat of loss but as an opportunity of new avenues of cultural engagement.

Notes

1 For public diplomacy, see Melissen (2005).
2 The UNESCO ICPRCP Intergovernmental Committee also promotes ADR methods and recently added mediation and conciliation to its mandate.
3 For a detailed account of the objects' acquisition, see Felch and Frammolino (2011).
4 The Italian demands were later reduced to 46 objects.
5 See Blume (2008).
6 The Getty Museum and the Museo Archeologico Nazionale di Firenze Announce Long-Term Cultural Collaboration, Press Release, 23/3/2009.
7 The Getty Museum and Sicilian Ministry of Culture Announce Major Long-Term Cultural Collaboration, Press Release, 17/2/2010.
8 The Getty and the Capitoline Superintendency of Roma sign agreement creating framework for cultural exchange, Press Release, 9/8/2012.
9 The Getty museum announces the return of a head of Hades from about 400–300 BC, Press Release, 10/1/2013.
10 The Getty Museum and Italian officials announce agreement to return first-century BC sculpture to Italy, Press Release, 14/6/2017.
11 See Povoledo (2010).
12 For the implications of US MoUs, see Luke and Kersel (2013).
13 Decree No.1771.
14 The list also included the Morgantina silver collection returned by the Metropolitan museum to Italy, which would go on display at the Metropolitan again in 2014.

References

Ajana, B. (2015). Branding, legitimation and the power of museums: The case of the Louvre. *Museum and Society*, *13*(3), 322–341.

Ang, I., Isar, Y. R., & Mar, P. (2015). Cultural diplomacy: Beyond the national interest? *International Journal of Cultural Policy*, *21*(4), 365–381.

Arndt, R. T. (2005). *The first resort of kings: American cultural diplomacy in the twentieth century*. Virginia: Potomac Books.

Blume, H. (2008). Italy lending Bernini art to the Getty, *Los Angeles Times*, 2 February.

Cai, Y. (2013). The art of museum diplomacy: The Singapore–France cultural collaboration in perspective. *International Journal of Politics, Culture, and Society, 26*(2), 127–144.

Davidson, L., & Pérez-Castellanos, L. (2019). *Cosmopolitan Ambassadors: International exhibitions, cultural diplomacy and the polycentral museum*. Delaware: Vernon Press.

Falkoff, S. (2007). Mutually beneficial repatriation agreements: Returning cultural patrimony, perpetuating the illicit antiquities market. *Journal of Law and Policy, 16*(1), 265–304.

Felch, J., & Frammolino, R. (2011). *Chasing Aphrodite: The hunt for looted antiquities at the world's richest museum*. New York: Houghton Mifflin Harcourt.

Green, J. (2017). General Treatment: Museums as intermediaries in repatriation. *Journal of Eastern Mediterranean Archaeology and Heritage Studies, 5*(1), 6–18.

Grincheva, N. (2020). Global diplomacy of Louvre Abu Dhabi. *Museum Management and Curatorship, 35*(1), 89–105.

Hoogwaerts, L. (2016). Museums, exchanges, and their contribution to Joseph Nye's concept of 'soft power'. *Museum and Society, 14*(2), 313–322.

La Porte, T. (2012). The legitimacy and effectiveness of non-state actors and the public diplomacy concept. Retrieved from: http://files.isanet.org/ConferenceArchive/58816b94a39845d9a5b618ae52e7c80c.pdf (Accessed on 07 May 2021).

Lord, G., & Blankenberg, N. (2015). *Cities, museums and soft power*. Chicago, IL: University of Chicago Press.

Luke, C. (2012). US policy, cultural heritage, and US borders. *International Journal of Cultural Property, 19*(2), 175–196.

Luke, C., & Kersel, M. (2013). *US cultural diplomacy and archaeology: Soft power, hard heritage*. London: Routledge.

Lyons, C. (2014). Thinking about antiquities: Museums and internationalism. *International Journal of Cultural Property, 21*(3), 251–265.

Macdonald, S. (2003). Museums, national, post-national and transcultural identities. *Museum and Society, 1*(1), 1–16.

Mark, S. L. (2010). Rethinking cultural diplomacy: The cultural diplomacy of New Zealand, the Canadian Federation and Quebec. *Political Science, 62*(1), 62–83.

Melissen, J. (Ed.) (2005). *The new public diplomacy: Soft power in international relations*. Basingstoke: Palgrave Macmillan.

Nisbett, M. (2016). Who holds the power in soft power? Retrieved from: https://theartsjournal.net/2016/03/13/nisbett/#references (Accessed on 07 May 2021).

Nye, J. (2004). *Soft Power, The means to success in world politics*. New York: Public Affairs.

Pamment, J. (2014). Articulating influence: Toward a research agenda for interpreting the evaluation of soft power, public diplomacy and nation brands. *Public Relations Review, 40*(1), 50–59.

Povoledo, E. (2010). A trove of ancient silver said to be stolen returns to its home in Sicily, *New York Times*, 5 December.

Reeves, J. (2007). *Culture and international relations, narratives, natives and tourists*. London: Routledge.

Rocco, F. (2013, December 21). Temples of delight. *The Economist*.

Shehade, M. (2017). *Negotiating cultural property disputes: Bridging the gap between theory and practice, a way forward*. Unpublished doctoral Thesis. London: University College London.

Shehade, M., & Fouseki, K. (2016). The politics of culture and the culture of politics: Examining the role of politics and diplomacy in cultural property disputes. *International Journal of Cultural Property, 23*(4), 357–383.

Shehade, M., Fouseki, K., & Tubb, K. (2016). Editorial: Alternative dispute resolution in cultural property disputes: merging theory and practice. *International Journal of Cultural Property, 23*(4), 343–355.

Sylvester, C. (2009). *Art/Museums: International relations where we least expect it*. London: Paradigm Publishers.

13 Is world heritage politically sustainable?

Kristen Barrett-Casey, Lorika Hisari, and Kalliopi Fouseki

Introduction

Economic, environmental, social and cultural sustainability have become increasingly prevalent in the management discourses of almost every field in the last half of the 20th century. Proponents hope 'to meet the needs of today without compromising the needs of tomorrow' (Brundtland, 1987, Art. 27). The way we have come to understand sustainability has largely been through the Brundtland Report and the 16 Sustainable Development Goals (SDGs) formulated by the United Nations (UN). It is through such international organisations that we hope to see 'sustainability' reached across the world internationally.

However, one of the fundamental and enduring problems of sustainability is no one has yet reached a consensus on what it actually means. Similarly, there are no indicators or metrics which could be pointed to and declare sustainability has been achieved. If these issues are not resolved, the very idea of sustainability itself risks becoming meaningless.

Finding metrics to be used internationally is difficult. In the international context, what is sustainable for one country might not be for another; as such it is unfair and impossible to hold each country up to meeting the same goals. This is especially true for developing countries or countries more prone to conflict. As such this chapter takes the opportunity to use a post-conflict case study to introduce what we hope will be a more nuanced approach to understanding the sustainability of decision-making at heritage sites in post-conflict environments.

We start from the fundamental observation that our understanding of sustainability to date has, at worst, failed, or at the very least is lacking, because we have not taken account of and examined the social and political dynamics which influence and shape the process of decision-making. In the context of the use of heritage as a tool in post-conflict reconciliation, this new political awareness around how conservation decisions are made, including by whom, should encourage academics and heritage practitioners alike to reflect on the current status of heritage management and how this relates to heritage on the international stage, namely the idea of 'World Heritage', a status highly sought after for the perceived economic advantages it brings. We call this 'political sustainability'.

This chapter seeks to analyse both the conservation decisions and criteria for World Heritage status from the point of view of Kosovo and asks if the pursuit of this status is the best thing not only for sustainable management of heritage sites there but also for achieving genuine and long-lasting reconciliation among its multi-ethnic communities.

Globalisation, world heritage and political sustainability

In a global context, sustainability is difficult to achieve homogeneously. Different societies have access to different resources, exist in different social structures and, most importantly, have different conceptions of heritage. Indicators, such as those proposed by the UN (https://undocs.org/A/RES/71/313) only go some way to achieving a measurable idea of sustainability globally. While scholars and organisations have taken turns to create their own frameworks and indicators of sustainability, that we still seek to impose these systems globally is holding many societies up to a level of standardisation which is not appropriate to their specific conditions and needs. In thinking about sustainable heritage, therefore, we should redress the level of importance placed on UNESCO and instead allow sustainability to work from the bottom up. This is not to suggest that international cooperation or standardisation has no place in the heritage sector. Rather, a more meaningful idea of sustainability can be derived from practices which meet present and future needs across societies and cultures. Nederveen Pieterse (2015) believes we are seeing the start of a 'global melange culture' due to the process of hybridisation rather than homogenisation through globalisation. However, we should not be overly optimistic about this theory (Nederveen Pieterse, 2019). Steger reminds us that 'globalisation is an uneven process, meaning that people living in various parts of the world are affected very differently by this gigantic transformation of social structure and cultural zones' (2014, p. 11).

As Harrison rightly observes, 'one cannot properly form new memories and attach value to them without also selecting some things to forget' (Harrison, 2012, p. 579). It is from an 'increasingly broad definition of heritage... [that] we risk being overwhelmed by memory, and in the process making all heritages ineffective and worthless' (ibid.). To combat this, Harrison suggests paying increased attention to management, 'decisions to delist, or cease to conserve particular forms of heritage once their significance to contemporary and future societies can no longer be demonstrated' (ibid.). However, the act of management in itself is a political process, and as such, there is an increasing need to see heritage more broadly in the same light. The management decisions over what to conserve, what to restore and what to forget are in themselves the product of social and political dynamics, perpetually shaping and reshaping conceptions and ideologies of heritage, which find expression in our changing values. Perhaps the best example of recognising contemporary international values is entry to UNESCO's World Heritage List (WHL).

When entry on to the WHL often comes down to meeting certain criteria, what has happened in the last half of the 20th century is that a hierarchy of heritage has been made, where only those capable of meeting very strict criteria are eligible to be considered of world significance. As a result, criticism has been rightly made about how attainable World Heritage status can be for in conflict or post-conflict countries whose heritage sites cannot hope to meet these conditions, no matter how important their heritage is. Further, the nomination process in itself is so costly and time-intensive that such fractured and fragile societies simply do not have the resources to dedicate to the nomination. Both the processes of creating the WHL and the current model of sustainability should lead us to question how our current beliefs, models and practices are impacting the international community and the notion of heritage in general.

When it comes to sustainable heritage, we must be clear about *what* we are trying to sustain. There has been much criticism from scholars about what happens to World Heritage sites once they make the list. The criticism is that there is a homogeneity to these sites, which contain similar styles, visitor centres etc. meaning that a lot of the individuality of these sites has been removed to become more attractive to tourists (Hakala et al., 2011).

In the post-conflict context, in which heritage is at risk of destruction (either deliberately or collaterally), how a site is reconstructed can leave it vulnerable to political manipulation, either to legitimise a new power or to emphasise certain political ideologies. For UNESCO-led reconstructions, or sites which have the aim of becoming World Heritage, there have been criticisms of reconstructions bearing no relation to the historical and political realities of that society. Mostar Bridge, for example, is still celebrated by UNESCO as a symbol of peace and reconciliation in Bosnia; the reality, however, is far from this (Walasek, 2015).

Several important factors for the heritage sector must be considered, and this is why political sustainability is ultimately a necessary inclusion to the model. Firstly, if we work on the contemporary understanding that heritage is a contemporary and perpetual process of meaning-making in the present and if the local community (who ultimately have most interaction and value placed on a site) are neglected for the sake of the idea of *international* heritage, there is a risk they will stop placing meaning and value on the site. By logic, this would mean that, eventually, this site would stop being considered heritage; the choice of how we structure our conservation and administration ultimately decides the future of a heritage site.

Secondly, the narratives and physical remains/reconstructions will also have an impact on how we consider a site to be heritage. Whilst heritage has always been used for political purposes, there is a risk that oversimplification or an emphasis on political ideologies far removed from historical and political realities will also eventually mean people stop valuing a site as part of their heritage.

If these two factors are not considered, we risk losing our heritage to tourism; a site will stop being seen as heritage and become just another place to visit, void of meaning.

The aim of political sustainability is, therefore, not to force artificial change in how heritage sites are organised. Rather, if we understand the social and political dynamics which govern conservation and decision-making processes, we can make further progress in understanding the potential sustainability of how a site is run, as well as making predictions about the difficulties a site might face in the future. On a larger scale, it should lead us to question how we conceive of heritage internationally. With many different cultural understandings of heritage, how long can it be that a western notion of heritage dominates the international stage? Ultimately, perhaps we should question whether UNESCO is a sustainable organisation after all.

Kosovo and the post-war political status

Since 1999, when the war ended in Kosovo, its administration and the political status changed throughout the transitional period. On 10 June 1999, the UN Security Council (SC) adopted a Resolution 1244[1] (1999) that provided a framework for the resolution of the conflict in Kosovo by authorising the deployment of an international civilian and military presence that would provide an international transitional administration and security presence to oversee the return of refugees and the withdrawal of military forces from Kosovo. The resolution also states that the international civilian presence will facilitate a political process to determine the future status of Kosovo (United Nations Mission in Kosovo, 1999).

Based on Resolution 1244, the United Nations Interim Administration Mission deployed in Kosovo (UNMIK)[2] to ensure conditions for a peaceful and normal life for all inhabitants of Kosovo and advance regional stability in the Western Balkans (United Nations Mission in Kosovo, 1999a, 1990b). During the UNMIK, the provisional institutions of self-government in Kosovo were established (UNMIK Regulation No. 2001/9 and No. 2001/19). In 2005, the UN Secretary-General appointed Martti Ahtisaari, a former Finnish President, as his Special Envoy for the future status process for Kosovo that should have culminated in a political settlement that determines the future status of Kosovo (Security Council Report, 2007). The momentum to resolve Kosovo's final status was critical. For more than a year, (2006–2007) Martti Ahtisaari and his team had intensive negotiations with the leadership of Serbia and Kosovo (known as Vienna negotiations) which have been stated to have exhausted the negotiations' potential to produce any mutually agreeable outcome on Kosovo's status with opposing views also related to cultural heritage (Security Council Report, 2007).

However, eight years of Kosovo's status quo under the Security Council resolution 1244 and the UN mission were seen as a factor of political and economic instability which could affect its democratic development and inter-ethnic reconciliation. This could then lead to 'further stagnation, polarizing its communities and resulting in social and political unrest' (Security Council Report, 2007, p. 2). Due to the diametrically opposing stances and the lack of mutual agreement, Martti Ahtisaari recommended "independence as the only viable option" for the future status of Kosovo (Security Council Report, 2007, p. 4). The reintegration into Serbia was not a viable option and the continued international administration, not sustainable (Security Council Report, 2007, p. 3).

In 2008, Kosovo declared independence on the basis of a Comprehensive Proposal for the Kosovo Status Settlement (CSP) known as the Ahtisaari Plan. The independence was recommended to be supervised by the international community for an initial period of time. The International Civilian Office (ICO) established by 25 states who had recognised the Republic of Kosovo (hereafter Kosovo) as an independent state (among them USA, Germany, UK, France and Italy) was established under a mandate to ensure the full implementation of the CSP. The CSP was the basis for the constitution and the legislative framework of a new state and ICO was there to support the Government of Kosovo to implement the latter. ICO ended its mission in December 2012.

Cultural heritage in conflict and the inscription on the World Heritage List

'Culture stands on the front-line of conflict – it should be at the front-line of peace building' – stated Irina Bokova, a former Director-General of UNESCO to UN Security Council (UNESCO, 2015).

The conflict had severe consequences for Kosovo's rich and diverse architectural heritage. During the 1998–1999 armed conflict (war), hundreds of heritage sites with special significance to the Albanian majority population, including at least 207 mosques, more than 500 kullas – traditional stone mansions, historic bazaars and well-preserved urban centres in Kosovo were deliberately targeted and destroyed (Herscher & Riedlmayer, 2000; Nikander & Zirl, 2016). After the war, more than 70 less well-known churches and monasteries in rural areas abandoned by the fleeing Serb minority population became the target of revenge attacks by returning Albanians – most were built in the 20th century, some were listed monuments of the medieval period; the most famous medieval Serbian Orthodox sites were guarded by the international peacekeeping forces (Herscher & Riedlmayer, 2000). Later on, during the 2004 March riots in Kosovo as a consequence of an inter-ethnic violence, approximately 35 churches, chapels and monasteries were destroyed or damaged in only two days (ICOMOS, 2005, p. 161).

The deliberately targeted heritage sites during the war and conflict are more and more expected to play a significant role in recovery, inter-community trust-building and reconciliation as a prerequisite for sustainable peace and development. However, 'reconciliation between conflicting parties is understood as a multifaceted and inter-generational process that the international community can support by acknowledging past wrongs and by providing post-conflict justice through property restitution and/or compensation' (Von Carlowitz, 2005 p. 549). The specific claims e.g., for financial compensations among others can be 'theorised as interruptions' that are about 'the way past atrocities are being felt in the present through global structural inequalities and discrimination' (Joy, 2020, p. 1).

Under strong pressure from the international community (European Union and the United Nations), in June 2004, the Kosovo parliament decided to allocate over three million Euros to repair the damaged Orthodox heritage sites (ICOMOS, 2005; Council of Europe, 2011), a process that was supported by the Reconstruction Implementation Commission, an EU-funded mechanism established in 2004 (Council of Europe, 2011). The UNMIK invited UNESCO, the European Commission and the Council of Europe to conduct the technical damage assessments of the sites that resulted in three technical assessment reports as the basis for the reconstruction work (Council of Europe, 2011).

The 1998–1999 war damages to cultural heritage sites of Kosovo Albanian majority had 'little awareness of or concern' with reluctance by the international community to acknowledge this damage as it was more perceived as a humanitarian support to a population dealt with as 'generic refugees' rather than 'people with distinct and valuable cultural heritage' (Herscher & Riedlmayer, 2000, p. 113). 'Damaging, vandalism, and looting of Kosovo's cultural heritage property during1998–1999 by the Serb forces' is seen to have been completely forgotten (Emancipimi Civil Ma Ndryshe, 2013).

In these peculiar circumstances of a fragile post-war society and despite the post-war political status quo of Kosovo, in 2004 and 2006, four Serbian Orthodox Church (SOC) sites in Kosovo were inscribed into the UNESCO WHL as 'Medieval Monuments in Kosovo (Serbia)' with Serbia as a State Party – the Deçan/Dečani Monastery, the Patriarchate of Peja/Peć Monastery, the Church of the Virgin of Levisha/Ljeviša and the Graçanica/Gračanica Monastery (UNESCO, 2006).

The sites had fulfilled three out of ten selection criteria for the inscription to the WHL:

ii to exhibit an important interchange of human values, over a span of time or within a cultural area of the world, on developments in architecture or technology, monumental arts, town-planning or landscape design;
iii to bear a unique or at least exceptional testimony to a cultural tradition or to a civilisation which is living or which has disappeared; and
iv to be an outstanding example of a type of building, architectural or technological ensemble or landscape which illustrates (a) significant stage(s) in human history;

In 2006, these heritage sites were inscribed into the list of World Heritage in danger 'due to difficulties in its management and conservation stemming from the region's political instability' (Decision 30 COM 8B.54 cited in the State of conservation report by the State Party, UNESCO, 2019, p. 3). The report further states that 'the Committee requested that the State Party (Serbia) work with UNESCO programmes, with the UNMIK and with the Provisional Institutions of Self-Government in Priština in caring for the site' (State of conservation report by the State Party, UNESCO, 2019, p. 3).

World heritage sites, political sustainability, conservation and inter-community dynamics

Over the years during the post-war period in Kosovo, the reality on the ground changed, including institutional transformations and the state-building efforts. The declaration of independence positioned Kosovo in front of the new obligations deriving from the Ahtisaari Plan (CSP) compromising document that were enshrined into its constitution and the legislative framework (Hisari & Fouseki, 2020). One of the important topics that was a subject of Vienna negotiations was the conservation of the SOC sites in Kosovo that made the Annex V of the CSP as a specific chapter on religious and cultural heritage – granting rights, privileges and immunities for the SOC sites including the conservation mechanisms by encouraging communication and cooperation between the SOC in Kosovo and the Kosovo institutions (CSP). As a result, three specific laws were adopted, Law on Special Protective Zones (Law Nr. 03/L-039, 2008, Annex – Maps for Special Protective Zones, 2008); Law on Historic Centre of Prizren (Law Nr. 04/L-066, 2012); and Law on the Village of Velika Hoča/Hoçë e Madhe (Law Nr. 04/L-062, 2012) and 44 protective zones were established around heritage sites mainly orthodox churches and monasteries including a protective zone in an urban area – the Historic Centre of Prizren and a rural area – the village of Velika Hoča/Hoçë e Madhe (Official Gazette of the Republic of Kosovo, 2008 and 2012). These protective zones include the four World Heritage sites in Kosovo inscribed in 2004 and 2006.

This new political reality on the ground that developed over a time span raises the question of how politically sustainable the WH status is of the four WH sites in Kosovo? How does this status contribute to their sustainable conservation and management? Does UNESCO take into consideration the conflict and socio-political circumstances while inscribing properties as a World Heritage site and how might this affect the inter-community dynamics in a post-war recovery process?

Meanwhile, Kosovo is trying to become a UNESCO member. At its first attempt in 2015, it was harshly opposed by Serbia, including the SOC. There was an open public engagement to prevent this with a justification that SOC sites in Kosovo are unprotected and threatened. Kosovo failed to join falling three votes short of the required two-thirds majority among member states. This is an ongoing discussion. Caroline Jaeger-Klein states that one aspect of Kosovo's membership to UNESCO is the further management of the World Heritage Medieval Monuments in Kosovo inscribed by Serbia. In addition, Kosovo must also be well prepared professionally and do the required work about the systematic inventory of its heritage sites as well as the work on the tentative list for UNESCO (Jaeger-Klein, 2016). The author, however, urges Kosovo to UNESCO membership which is also not seen as 'a matter of if and or, but when' (Pavlović, 2017).

The restoration of the Patriarchate of Peja/Peć Monastery – the red facade

There is a fundamental question – how much did the status of World Heritage contribute to the protection of the monument's 'outstanding universal value', 'authenticity' and 'integrity', considering the socio-political circumstances described above and the ongoing discourse. [Ma7] [U8] Furthermore, how did that affect the inter-community dynamics in a post-war recovery process?

The Patriarchate of Peja/Peć Monastery, one of the four inscribed Medieval Monuments in Kosovo also on the List of World Heritage in Danger, is an example to demonstrate that. When during the restoration works in 2008 the façade of the monument was painted in red it triggered a strong public debate. Minister Beqiri of Culture at the Kosovo Government in 2009 had stated at that time that the SOC had started 'to interfere in the renovation of churches and monasteries without a consent from a Kosovo government', by giving an example of painting in red of one part of the Patriarchate (Deutsche Welle in Albanian, 2009).

The physical transformation of the centuries-old cultural monument added to an already ongoing public discourse on the historical narrative of the SOC sites in Kosovo that these monuments were built on the foundations of existing temples. Prof. Dr. Jahja Drançolli (2019) had stated that this intervention is a result of 'insidious politics' to present only one version of history by covering the original architectural and sculptural elements of a cultural monument to make it look like the Serbian monasteries in Serbia; moreover, this is a UNESCO listed monument, therefore, the attention needs to be at presenting what is scientific and not the Serbian politics (cited in Ndërtimi.info, 2019b). Minister Gashi of Environment and Spatial Planning in 2014, in his role of a chairman of the Implementation and Monitoring Council at that time, derived from the Law on Special Protective Zones (elaborated above), had addressed this issue to UNESCO, asking 'for the investigation of this illegal act' (cited in Telegrafi.com, 2014).

Interestingly, the painting of the façade induced wide reactions in Serbia as well. A number of experts dealing with cultural monuments had warned that permanent damage has been inflicted to the monument (Blic.rs, 2008). As reported by RTV B92 (Serbian national news station, television and radio broadcaster), the façade of the Patriarchate is plastered and painted in red, despite being prohibited by the Serbian Institute for the Protection of Cultural Monuments and the Serbian Ministry of Culture (cited in Blic.rs, 2008). Milan Glišić, Professor at the Faculty of Architecture in Belgrade in 2008, had found this to have 'shaken the foundations of a collective memory', calling for the Serbian Institute for the Protection of Cultural Monuments 'to have an attitude on the matter'. The author had seen this to be politically influenced and stated that if this intervention was to prevent the damage of frescos from moisture, then there are contemporary technologies in order to preserve the centuries-authentic look of a building instead of applying a plastered and painted façade (Glišić, 2008).

This example raises the issue about who makes the decision for the restoration of World Heritage sites and on the basis of what this is done. This is particularly critical to consider in post-war fragile contexts as they potentially impact the recovery and inter-community trust building which has shown to be a continued international effort. Its consequences to building peace as 'a permanent ideal and aspiration, as well as a right and a duty, and a foundation for sustainable development' (UNESCO, 2013), are inevitable.

Discussion

Across the literature, there is the predominant presumption that World Heritage status is beneficial for communities, with many citing its contribution to 'sustainability'; a few question whether this is the right decision for the site or the local community. UNESCO is 'convinced that no development can be sustainable without a strong culture component'[3] and that 'culture is both an enabler and a driver of the economic, social and environmental dimensions of sustainable development'.[4] This notion of sustainable development becomes a particularly acute and important issue in the aftermath of conflict, where heritage can risk prolonging the violence (Viejo-Rose, 2013). However, UNESCO has been criticised for its heritage management, from how it behaves on the ground, to its process for inscribing properties to the WHL (Dumper & Larkin, 2012; Brumann, 2015; Brumann & Berliner, 2016).

Scholarship can be roughly separated between those who attempt to make sense of reconstructions in the context of nomination for World Heritage (Khalaf, 2017, 2018, 2019, 2020), and those who seek to understand the agency and role of UNESCO (Keough, 2011; Dumper & Larkin, 2012; Brumann, 2015, 2016). The former are largely uncritical of the predominance of UNESCO, providing no real way of understanding how their theories would work in practice. While the latter are critical of UNESCO, they give no solutions or alternatives to the criticisms they make. Both strands of scholarship need to be used to develop a more holistic, nuanced understanding of the place of reconstructions within international heritage, the place of heritage in reconciliation, and to question the extent to which the idea of international heritage is sustainable politically. While history and heritage have always been used for political purposes (Viejo-Rose, 2013), fundamental questions remain unanswered about the extent to which international organisations with their own political agendas should become involved in fragile societies recovering from conflict.

The main issue which arises is that UNESCO is often politically out-of-step with the local communities they operate in and, to some extent, uses reconstruction projects as an opportunity to impose its own worldview on fragile societies (Kostadinova, 2011; Viejo-Rose, 2013; Walasek, 2015; Isakhan & Meskell, 2019). There is a clear gap between rhetoric and reality on the objectives of reconstructions, leading to more general questions about how we conceive of heritage. Demands for sustainability can be unachievable for vulnerable societies upheaved by conflict, which should lead us to question how sustainable such demands are. Reconceptualising both 'heritage' and 'sustainability' could help us to understand specifically *what* we are trying to sustain, what we are trying to preserve and for what purpose.

One suggestion from this analysis of political dynamics is the development of a theory of political sustainability, where current 'people-centred heritage' is replaced with control and leadership coming from communities themselves (Di Giovine, 2015). This, however, can only come from a reconceptualisation of heritage which does not necessarily put such great demands of sustainability onto it. Indeed, it must be questioned how sustainable such demands are. If we choose rather to sustain historical truth and knowledge, this review will demonstrate how this reconceptualisation can remove some of the tensions in the criteria for World Heritage found in reconstructions.

UNESCO is clear that it does not engage with the inherent political issues of the area. The Director-General, Koichiro Matsuura, said UNESCO 'doesn't want to deal with political issues – we are duty-bound to preserve... authenticity' (Haaretz, 2008)[5]. This is the biggest issue missed by most researchers; UNESCO fails to deal with the underlying political issues that lead to conflict. Therefore, attempts by UNESCO at reconciliation can be nothing but superficial.

Despite UNESCO's intention to 'stay away from political issues', this is almost inevitable. Heritage is a political as well as a social process after all. Indeed, as Meskell notes,

> In theory, UNESCO constitutes the arena where archaeology reaches worldwide attention, and yet archaeologists themselves are largely invisible in the political processes, governance and public profile of the organization... The recognition and value that inscription bestows is remarkably still desired by almost all the nations of the world, regardless of political or religious affiliations, economic status, or historical trajectory.
>
> (Meskell, 2013)

UNESCO has been described as a 'market-place' where bureaucracy masks the fact that 'genuine dialogue is ruled out' (Hoggart, 1978 p. 99). Research is increasingly highlighting its overt politicisation, where World Heritage status is gifted in diplomatic gestures, 'driven by economic and political imperatives' (Meskell, 2015, p. 2). This has led some to doubt the credibility of the World Heritage system at all, claiming that it is merely 'aiming at reinforcing specific systems of power and resulting in social conflict and commercial exploitation of heritage-labelled resources' (Harrison, 2010). However, critics such as Brumann (2015), Meskell (2015) does not question why we continue to view heritage within the World Heritage system. These arguments need to be taken to their logical conclusion to ask if it is a sustainable organisation or enabling conflict to continue. One alternative to explore is an alternative model of 'people-centred heritage' (Schilderman & Lyons, 2011; Maeer, 2017), where local communities take the leadership role and UNESCO acts in a purely advisory capacity. This could reveal more organic processes of heritage in line with contemporary understandings of heritage-making. As Meskell warns:

> the desire for World Heritage and hence the success of the 1972 Convention could ultimately lead to its undoing, since the economic revenues, the national prestige and international bargaining potential that such recognition bestows has eclipsed the very substance and conservation of sites.
>
> (Meskell, 2015, p. 4)

Criticisms made of the commoditisation of heritage should also be seen within the context of UNESCO's worldview (Viejo-Rose, 2013). Further, UNESCO's Cultural Development Indicators reveal much about how it views the purpose of heritage.

If UNESCO is dedicated to sustainable development and local participation, its indicators for proving this should be reflective of these theoretical principles rather than the market value heritage possesses. This would build on the criticism already made of UNESCO's principles of sustainability not being translated into their policies, nor any indication of how they would achieve these principles in reality (Barthel-Bouchier, 2016).

The effects of World Heritage status are 'barely manageable, especially in… historic centres, whose complexity calls for multifaceted solutions and plans… whenever planning and management systems are not fully mature and developed, conservation processes in historic centres have taken controversial forms, usually bearing unsustainable effects.' (Soccali and Cinà, 2020, p. 2). Importantly, 'public participation' has been seen in a neoliberal context, where the poor 'participate by taking personal responsibility while companies and corporations are freed from responsibility for poverty' (Ibid., p. 3). Indeed, there have been many criticisms about UNESCO's white-washing of the living conditions many residents of World heritage sites face (Brouillette, 2015; Rudolff & Buckley, 2016). Questions must be asked about UNESCO's commitment to communities, when even within the nomination process for World Heritage, it is the State party which submits nominations from its own country; this risks the heritage of certain groups being neglected or even harmed because heritage has been internationalised by State Parties (Joy, 2016; Rico, 2016). In the context of reconstruction, it is paramount to remember new political forces who could seek to legitimise their power through re-defining the identity and history of their nation. Without awareness of this process, the violence and conflict could be continued (Viejo-Rose, 2013).

UNESCO risks conflict continuing or re-emerging in societies which are already vulnerable. They refuse to accommodate complex political contexts and instead work to impose their own worldview, making heritage fit to this idea rather than allowing new meanings to emerge or uncomfortable historical realities to remain. Although they have stated their commitment to involving 'local people' in the process of reconstruction, it is by no means clear in what capacity this means, and how much of a lead they are given.

A new theory of political sustainability should be explored in order to build upon current knowledge of how heritage relates to politics, and also what could emerge from local people having real ownership over their heritage. Current understandings of sustainability not only fail to accommodate this, but also lack clear definitions, ultimately rendering it a meaningless concept. For reconstructed heritage in a vulnerable political environment, this exposes it to manipulation and distortion by national and international organisations who would use it in the name of "sustainable development" at the expense of historical truth. The development of a theory of political sustainability can go some way to helping to understand the relationship more between contemporary development and the protection of historical truth. This balance is found best in local leadership. In being able to decide whether and how to reconstruct destroyed heritage, much could be learned about the long process of reconciliation, how heritage is used in a post-conflict environment by people who experienced the conflict, and how important it is to have people within a society who have a long-term commitment to seeing lasting peace being created.

Conclusion

There is no simple answer to our initial question of whether World Heritage is politically sustainable; neither is there a simple answer to what political sustainability actually is or how to achieve it. This concept has been developed in this paper to draw attention to the structural inequalities that exist on the international stage regarding how the heritage of different countries is seen and experienced. The current situation for most scholarship and researchers focuses on how World Heritage status can be achieved by societies and sites who have recently undergone trauma or conflict. No one, to the best of our knowledge, has asked if World Heritage status is the best thing for these sites, or if it is the most 'sustainable' option for both the conservation of the site and for the use of heritage as a tool of reconciliation. It is our opinion that, at least for the case of Kosovo, World Heritage status and the influence of UNESCO have not been helpful for achieving sustainable post-conflict reconciliation.

In post-war and conflict areas, heritage as a complex multidimensional topic has an additional layer of sensitivity as various initiatives, projects and/or activities might undermine the recovery and peacebuilding efforts and unintentionally or not contribute to conflict escalation rather than resolution. This is especially when processes are highly politically driven and when there is a lack of political sustainability. The end of the war does not necessarily mean that the (political) conflict is not present as it might continue appearing in different forms and expressions, and as Viejo-Rose (2013) suggests, cultural heritage can be used as an instrument of 'continuing violence on a symbolic and ideological level' (Viejo-Rose, 2013, p. 144).

The case of Kosovo demonstrates that any efforts at international level such as inscription on the WHL, justified with the need to protect the heritage site without considering historical and political realities on the ground, might potentially affect the sustainable conservation and further deepen inter-community segregation. The change of the political status of Kosovo over time, its state-building and the efforts to join UNESCO also question the political sustainability of the four World Heritage Medieval Monuments.

Consequently, this also raises the need to review and eventually update the threats for which the property was initially inscribed on the List of World Heritage in Danger in 2006 on how well they correspond to the reality on the ground, which are: lack of legal status of the property; lack of legislative protection of buffer zones; lack of implementation of the Management Plan and of active management; difficulties to monitor the property due to political instability, post-conflict situation (visits under the Kosovo Stabilisation Force /United Nations Interim Administration Mission in Kosovo (KFOR /UNMIK) escort and lack of guards and security); and unsatisfactory state of conservation and maintenance of the property.

Based on the above, we propose that further research is urgently needed on how post-conflict political dynamics can be accommodated in the reconciliation process of heritage. We need to know how communities relate to UNESCO, the dialogue between these groups, the freedom given to local communities and how they relate to the finished reconstructions. As 'Communities' is one of the Strategic Objectives of the World Heritage Convention, understanding their role in the process of making heritage is important to an overall understanding of sustainability. One way would be through unstructured interviews with local people which would allow the interviewees to speak about what is important to them.

Notes

1 https://peacemaker.un.org/kosovo-resolution1244
2 https://unmik.unmissions.org/united-nations-resolution-1244
3 https://en.unesco.org/themes/protecting-our-heritage-and-fostering-creativity
4 https://en.unesco.org/themes/culture-sustainable-development
5 UNESCO chief: We are trying to mediate over Mugrabi Gate – Haaretz Com – Haaretz. com

References

Barthel-Bouchier, D. (2016). *Cultural heritage and the challenge of sustainability.* Oxon: Routledge.
Blic.rs (2008). *Pećka patrijaršija obojena u crveno, naneta trajna šteta* (Trans. L. Hisari). Retrieved from: https://www.blic.rs/vesti/drustvo/pecka-patrijarsija-obojena-u-crveno-naneta-trajna-steta/900c8hl (Accessed on 7 December 2020).
Brouillette, S. (2015). *UNESCO and the world-literary system in crisis.* Retrieved from: http://amodern.net/article/unesco-brouillette (Accessed on 03 May 2021).
Brumann, C. (2015). Community as myth and reality in the UNESCO World Heritage convention. In N. Adell, R. F. Bendix, C. Bortolotto, & M. Tauschek (Eds.), *Between imagined communities and communities of practice-participation, territory and the making of heritage* (pp. 273–289). Göttingen: Universitätsverlag Göttingen.
Brumann, C., & Berliner, D. (Eds.) (2016). *World heritage on the ground: Ethnographic perspectives* (Vol. 28). New York: Berghahn Books.
Brundtland, G. H. (1987) Our common future: Report of the world commission on environment and development. Geneva, UN-Document A/42/427. Retrieved from: http://www.un-documents.net/ocf-ov.htm (Accessed on 03 May 2021).
Council of Europe. (2011). *Reconstruction implementation commission for Serbian orthodox religious sites in Kosovo.* Retrieved from: https://issuu.com/councilofeurope/docs/ric_for_serbian_orthodox_religious_sites_in_kosovo (Accessed on 5 December 2020).
Deutsche Welle in Albanian. (2009). *Ministri Beqiri: Trazirat e marsit 2004- përjashtim që nuk duhet të përsëritet* (Trans. L. Hisari). Retrieved from: https://www.dw.com/sq/ministri-beqiri-trazirat-e-marsit-2004-p%C3%ABrjashtim-q%C3%AB-nuk-duhet-t%C3%AB-p%C3%ABrs%C3%ABritet/a-4106245 (Accessed on 7 December 2020).
Di Giovine, M. A. (2015). UNESCO's World Heritage program: Challenges and ethics of community participation. In N. Adell, R. F. Bendix, C. Bortolotto, & M. Tauschek (Eds.), *Between imagined communities and communities of practice-participation, territory and the making of heritage* (pp. 83–108). Göttingen: Universitätsverlag Göttingen.
Dumper, M., & Larkin, C. (2012). The politics of heritage and the limitations of international agency in contested cities: A study of the role of UNESCO in Jerusalem's Old City. *Review of International Studies*, 38(1), 25–52.
Emancipimi Civil Ma Ndryshe. (2013). *An analysis of numerous and continuous faults in cultural heritage.* Series analysis "what went wrong? Emancipimi Civil Ma Ndryshe: Prishtina, Republic of Kosovo. Retrieved from: https://www.ecmandryshe.org/repository/docs/Erroneous_ENG.pdf (Accessed on 10 December 2020).
Glišić, M. (2008). *Uticaj boje na kolektivno pamćenje. In Politika–online media* (Trans. L. Hisari). Retrieved from: http://www.politika.rs/sr/clanak/66020/Uticaj-boje-na-kolektivno-pamcenje (Accessed on 7 December 2020).
Official Gazette of the Republic of Kosovo. (2008). *Law No. 03/L-039 on Special Protective Zones (2008). Annex – Maps for Special Protective Zones (2008).* Retrieved from: https://gzk.rks-gov.net/ActDetail.aspx?ActID=2529 (Accessed on 7 December 2020).
Official Gazette of the Republic of Kosovo. (2012). *Law No. 04/L-066 on Historic Centre of Prizren (2012).* Retrieved from: https://gzk.rks-gov.net/ActDetail.aspx?ActID=2836 (Accessed on 7 December 2020).
Official Gazette of the Republic of Kosovo. (2012). *Law No. 04/L-062 on the Village of Hoçë e Madhe/Velika Hoča (2012).* Retrieved from: https://gzk.rks-gov.net/ActDetail.aspx?ActID=2835 (Accessed on 7 December 2020).
Hakala, U., Lätti, S., & Sandberg, B. (2011). Operationalising brand heritage and cultural heritage. *Journal of Product & Brand Management*, 20(6), 447–456.
Harrison, R. (2010). *Understanding the politics of heritage.* Manchester: Manchester University Press.
Herscher, A., & Riedlmayer, A. (2000). Monument and crime: The destruction of historic architecture in Kosovo. *Grey Room*, 1, 108–22.

Hisari, L., & Fouseki, K. (2020). Post-war cultural heritage preservation in Kosovo: Rethinking the implementation of Ahtisaari Plan Annex V. *Heritage*, *3*(1), 98–115. https://doi.org/10.3390/heritage3010006.

Hoggart, R. (1978). *An idea and its servants: UNESCO from within*. New Brunswick, NJ: Transaction Publishers.

ICOMOS. (2005). *Heritage at risk 2004/2005. Kosovo*. Retrieved from: https://www.icomos.org/risk/2004/kosovo2004.pdf (Accessed on 8 December 2020).

International Civilian Office (ICO). (2012). *State building and exit: The international civilian office and Kosovo's supervised independence 2008–2012*. Prishtina, Republic of Kosovo: ICO.

Isakhan, B., & Meskell, L. (2019). UNESCO's project to 'Revive the Spirit of Mosul': Iraqi and Syrian opinion on heritage reconstruction after the Islamic State. *International Journal of Heritage Studies*, *25*(11), 1189–1204.

Jaeger-Klein, C. (2016). UNESCO World Heritage and Kosovo towards a tentative list for Kosovo. *International Journal of Business and Technology*, *5*(1), 1–10. Retrieved from: https://knowledgecenter.ubt-uni.net/cgi/viewcontent.cgi?article=1090&context=ijbte (Accessed on 7 December 2020).

Joy, C. L. (2016). *The politics of heritage management in Mali: From UNESCO to Djenné*. New York: Routledge.

Joy, C. (2020). Heritage justice. In K. Kristiansen, F. Nyamnjoh, M. Rowlands, S. L. Wang, A. Swenson, & O. Wetterberg (Eds.), *Elements in critical heritage studies* (p. 1). Cambridge: Cambridge University Press.

Keough, E. B. (2011). Heritage in peril: A critique of UNESCO's world heritage program. *Washington University Global Studies Law Review*, *10*, 593–615.

Khalaf, R. W. (2017). A viewpoint on the reconstruction of destroyed UNESCO Cultural World Heritage Sites. *International Journal of Heritage Studies*, *23*(3), 261–274.

Khalaf, R. W. (2018). A proposal to apply the historic urban landscape approach to reconstruction in the World Heritage context. *The Historic Environment: Policy & Practice*, *9*(1), 39–52.

Khalaf, R. W. (2019). World Heritage policy on reconstruction: From exceptional case to conservation treatment. *International Journal of Cultural Policy*, *25*(7), 871–885.

Khalaf, R. W. (2020). Cultural heritage reconstruction after armed conflict: Continuity, change, and sustainability. *The Historic Environment: Policy & Practice*, *11*(1), 4–20.

Klein, C. J. (2016). UNESCO World Heritage and Kosovo towards a tentative list for Kosovo. *International Journal of Business and Technology*, *5*(1), 1–10.

Kostadinova, T. (2011). Cultural diplomacy in war-affected societies: International and local policies in the post-conflict (re-) construction of religious heritage in former Yugoslavia. *Academy for Cultural Diplomacy, Berlin*, 10–17. Retrieved from: https://www.culturaldiplomacy.org/academy/index.php?participants-papers-the-icd-academy-for-cultural-diplomacy-june-session (Accessed on 6 May 2021).

Maeer, G. (2017). A people-centred approach to heritage: The experience of the heritage lottery fund 1994–2014. *Journal of Community Archaeology & Heritage*, *4*(1), 38–52.

Meskell, L. (2013). UNESCO's World Heritage Convention at 40: Challenging the economic and political order of international heritage conservation. *Current Anthropology*, *54*(4), 483–494.

Meskell, L. ed. (2015). *Global heritage: A reader*. New Jersey: John Wiley & Sons.

Ndërtimi.info. (2019a). *"Lonely Planet" ia prezanton botës Patrikanën e Pejës si të rrezikuar* (Trans. L. Hisari). Retrieved from: https://ndertimi.info/arkiva/lonely-planet-ia-prezanton-botes-patrikanen-e-pejes-si-te-rrezikuar/ (Accessed on 7 December 2020).

Ndërtimi.info. (2019b). *Patrikana e Pejës mes versioneve kontradiktore të historisë dhe fasadës që "duhet dënuar"* (Trans. L. Hisari). Retrieved from: https://ndertimi.info/arkiva/patrikana-e-pejes-mes-versioneve-kontradiktore-te-historise-dhe-fasades-qe-duhet-denuar/ (Accessed on 7 December 2020).

Nikander, P., & Zirl, V. (2016). *Building blocks for peace: Cultural heritage protection in Kosovo*. Retrieved from: https://www.osce.org/magazine/271966 (Accessed on 8 December 2020).

Pavlović, A. (2017). Letër nga Beogradi: Të jesh apo të mos jesh në UNESCO. In *Koha Ditore—Local printed and online media* (Trans. L. Hisari). Prishtina: Republic of Kosovo.

Pieterse, J. N. (2015). *Globalization and Culture: Global Mélange.* III. (ed.). Lanham, MD: Rowman & Littlefield.

Pieterse, J. N. (2019). *Globalization and culture: Global mélange.* Lanham, MD: Rowman & Littlefield.

Resolution adopted by the General Assembly on 6 July 2017. Work of the Statistical Commission pertaining to the 2030 Agenda for Sustainable Development. Retrieved from: https://undocs.org/A/RES/71/313 (Accessed on 3 May 2021).

Rudolff, B., & Buckley, K. (2016). World heritage: Alternative futures. In M. Craith, W. Logan, & U. Kockel (Eds.), *A companion to heritage studies* (pp. 522–540). Chichester: Wiley Blackwell.

Schilderman, T., & Lyons, M. (2011). Resilient dwellings or resilient people? Towards people-centred reconstruction. *Environmental Hazards, 10*(3–4), 218–231.

Soccali, G., & Cinà, G. (2020). Heritage policies in the neoliberal arena: Spaces of exclusion and gentrification in urban world heritage sites. *The Historic Environment: Policy & Practice, 11*(2–3), 282–306.

Steger, M. B. (2017). *Globalization: A very short introduction* (Vol. 86). Oxford: Oxford University Press.

Telegrafi.com (2014). *Gashi ankohet në UNESCO: Patrikanës së Pejës, gjatë renovimit i dhanë pamje ruso-sllave* (Trans. L. Hisari). Retrieved from: https://telegrafi.com/gashi-ankohet-ne-unesco-patrikanes-se-pejes-gjate-renovimit-i-dhane-pamje-ruso-sllave/ (Accessed on 7 December 2020).

UNESCO (2006). *Extension of properties inscribed on the World Heritage list (Medieval Monuments in Kosovo). Decision: 30 COM 8B.53.* Retrieved from: https://whc.unesco.org/en/decisions/1013 (Accessed on 10 December 2020).

UNESCO (2013). High-level discussion 2 – background note. *Peace and reconciliation: How culture makes the difference.* Retrieved from: http://www.unesco.org/fileadmin/MULTIMEDIA/HQ/CLT/images/PeaceReconciliationENG.pdf (Accessed on 1 October 2017).

UNESCO (2015). *Meeting of the UN Security Council (2015).* Retrieved from: https://whc.unesco.org/en/news/1269 (Accessed on 25 November 2020).

UNESCO. (2019). *Report on the state of conservation of medieval monuments in Kosovo (Serbia) inscribed in the world heritage list in danger.* Republic of Serbia: Institute for the Protection of Cultural Monuments. Retrieved from: https://whc.unesco.org/en/soc/3880 (Accessed on 10 December 2020). p. 3.

United Nations Mission in Kosovo. (1999a). *United Nations Resolution 1244. S/RES/1244 (1999).* Retrieved from: https://unmik.unmissions.org/united-nations-resolution-1244 (Accessed on 25 November 2020).

United Nations Mission in Kosovo. (1999b). *Mandate of the United Nations interim administration mission in Kosovo (UNMIK).* Retrieved from: https://unmik.unmissions.org/mandate (Accessed on 25 November 2020).

United Nations Security Council. (2007a). *Distr.: General. Report of the special envoy of the secretary-general on Kosovo's future status. S/2007/168.* New York: United Nations Security Council. Retrieved from: https://www.securitycouncilreport.org/atf/cf/%7B65BFCF9B-6D27-4E9C-8CD3-CF6E4F-F96FF9%7D/Kosovo%20S2007%20168.pdf (Accessed on 28 November 2020).

United Nations Security Council. (2007b). Distr.: General. In *Comprehensive proposal for the Kosovo status settlement. S/2007/168/Add 1.* New York: United Nations Security Council. Retrieved from: https://www.securitycouncilreport.org/atf/cf/%7B65BFCF9B-6D27-4E9C-8CD3-CF6E4F-F96FF9%7D/Kosovo%20S2007%20168.Add1.pdf (Accessed on 5 December 2020).

UNMIK Regulations. (2001a). *Regulation No. 2001/9. On a constitutional framework for provisional self-government in Kosovo.* Retrieved from: https://unmik.unmissions.org/sites/default/files/regulations/02english/E2001regs/RE2001_09.pdf (Accessed on 25 November 2020).

UNMIK Regulations. (2001b). *Regulation No. 2001/19. On the executive branch of the provisional institutions of self-government in Kosovo.* Retrieved from: https://unmik.unmissions.org/sites/

default/files/regulations/02english/E2001regs/RE2001_19.pdf (Accessed on 25 November 2020).

Viejo-Rose, D. (2013). Reconstructing heritage in the aftermath of civil war: Re-visioning the nation and the implications of international involvement. *Journal of Intervention and Statebuilding*, 7(2), 125–148.

Von Carlowitz, L. (2005). Resolution of property disputes in Bosnia and Kosovo: The contribution to peacebuilding. *International Peacekeeping*, 12(4), 547–561.

Walasek, H. (2015). Domains of restoration: Actors and agendas in post-conflict Bosnia and Herzegovina. In H. Walasek (Ed.), *Bosnia and the destruction of cultural heritage* (pp. 205–258). New York: Routledge.

14 Political ruptures and the cultural heritage of Iraq

René Teijgeler and Mehiyar Kathem

Introduction

From the legacies of colonial wars and Anglo-American Empires of the 20th Century to the impact of domestic social and political forces to more recently the repercussions of external interventions, how politics shapes state structures can be informative of the condition of heritage landscapes. As politics affects all aspects of culture, studying a country through the prism of the state and its institutions is indispensable to understanding the sustainability of heritage and cultural continuity. Exploring the relationship between politics and cultural sustainability is particularly pressing, given the scale of heritage destruction witnessed in the Middle East. In this chapter, we analyse the case of Iraq by examining key junctures and ruptures in the context of the intertwined relationship between politics and heritage. An overview of major turning points in Iraq's recent history is offered to better understand the ways in which heritage and politics have evolved since the founding of the Iraqi state.

Violent conflicts and changing political structures in Iraq and its neighbours are having a profound impact on cultural heritage, with key points of historical reference in the process of being transformed and in some cases completely erased. Heritage landscapes, including tangible and intangible heritage, have been in rapid flux. Deliberate cultural heritage destruction, including looting, neglect, cultural appropriation and other forms of erasure, is transforming heritage landscapes in the Middle East (Kathem et al., 2020). In the context of Iraq's unfolding cultural and human catastrophe, this chapter calls for a research agenda to integrate more firmly the long-term implications of politics and contestation to examine heritage sustainability.

The growing interest in the politics of heritage is an outcome of a need to better understand the role of culture in everyday life and the ways in which we relate to the world around us. This important turn in academia has underlined the significance of exploring how heritage is used in different political contexts (Smith, 2006). Whilst there has been an increased recognition of politics in heritage research and practice, this has generally not been one about politics as contestation, competition for resources and legitimacy (e.g. DeSilvey & Harrison, 2020). The bulk of studies, especially in reference to the literature on heritage destruction in the Middle East, have generally ignored locally grounded histories and power structures (Kathem, 2020; Kathem & Kareem Ali, 2020). The surge of what could be called the post-Islamic State literature on cultural heritage destruction has instead focused on the criminal nature of non-state actors and deliberate acts of erasure, a trend that disconnects domestic political action from state-building processes and the consequences thereof (e.g. Matthews et al., 2019).

DOI: 10.4324/9781003038955-18

The relevance of state-building has been relegated to insignificance in understanding heritage, overshadowed by studies that explore the role of cultural destruction in relation to the specific moment in question and particular acts of spectacular violence (e.g. Francioni & Lenzerini, 2003). In other cases, a more critical approach to politics is explored which generally isolates one site or theme from an analysis of the political economy of heritage.

The dearth of Middle East-related studies looking at the relationship between state institutions and cultural continuity in crisis contexts has far-reaching consequences. It is reflected in international responses in the fields of culture and the design of major donor-related heritage programmes. In much of these US-European projects designed to 'protect' and 'safeguard' heritage, there is remarkably little if any analysis of the role of politics as contestation. Similarly, notions of sustainability are merely viewed as add-ons to the particulars of funded projects. Sustainability in culture is commonly a project category that only needs to be addressed to secure funding. Consequently, the role of state institutions in supporting cultural infrastructures and the foundations for ensuring cultural continuity is replaced with the idea that external interventions can ensure the protection and safeguarding of heritage. Grounding heritage research in existing and newly emerging structures of politics can help provide a context to the growing body of studies that have dismissed the state altogether and opted to focus on 'ideational projectionism' of what heritage is or how it should look or what it should serve.

Potential tenets of politics as contestation-related heritage approaches should underline the ways in which state resources are distributed, by whom and for what purposes. A political economy of heritage should underline the changing roles of cultural infrastructures as an essential component to fully comprehend heritage continuities. Specifically, how heritage is instrumentalised by the competing state and non-state entities in often fragmented political contexts is indispensable to the study of sustainability. A focus on political elites and their use of heritage, especially as a resource to govern and expand control, is an essential element in these debates (Cesari, 2010, 2015, 2017). Instrumentalisation, often for self-aggrandisement, national unity, or even for national subdivision, affects heritage landscapes and the cultural infrastructures underpinning them in manifold ways. In this context, this paper offers a critical insight into the uses of heritage by political actors, including the Occupation of Iraq, as well as the ways in which changing state structures affect a country's heritage landscape.

The early Iraqi state and cultural heritage

The history of Iraq since the 1920s has been characterised by a continuous struggle of nation-building and self-determination, which was closely intertwined with the negotiation of the past (Baram, 1983, 1994). Social and political groups looked to the country's history to determine their future. Different phases of this negotiation from pan-Arabism, Arab nationalism to Iraqi nationalism, led to a considerable debate over the distinctive characteristic of Iraqi national identity. Those attempts pursued cultural rather than ethnic linkages to unify Iraq's diverse cultural groups under the nation-state. In particular, the archaeological record was used to establish the relationship between specific people and civilisations, and how modern Iraqis were linearly related to these particular histories, their corresponding artefacts and stories (Baram, 1983).

Under British rule, the Mandate period from 1921 to 1932, Iraq's archaeology was considered a largely foreign affair. For the British and other foreign archaeological missions, Iraq antiquities were important in the search for 'progress of civilisation' in order to justify their 'civilising imperial mission' (Bernhardsson, 2005). Colonial presence in Iraq was thus less about the country's history and more about Europe's own attempts to craft its own identity through selected notions of Judeo-Christian heritage and the appropriation of this region's civilisation (Liverani, 2016; Kathem & Kareem Ali, 2020). In this spirit of colonial extraction of Ancient Iraq's heritage, the Iraq Museum, established in 1922, was considered a repository of antiquities to be repatriated to the United Kingdom and the first Antiquities Law (1924) gave the British the opportunity to do so.

Iraq won its independence in 1932 and pursued the establishment of a nation-state (Batatu, 2004). It soon became evident that archaeology could serve as a symbol of unity and a way to ensure that the country's diverse cultural groups were tied to the nation-state through its shared histories. Iraq's attempts to gain sovereignty over its politics was reflected in its attempts to ensure that artefacts were not expropriated by the colonial powers. It wanted to create a sense of belonging to the state and cultural continuity through its heritage was pursued, though colonial plunder was a major thorn as it represented a counterforce to these attempts. More than before, political groups assertively sought to pursue a national identity premised on its past. Politicians wanted to craft stories that could accommodate their master narrative that would form the basis for a new political and cultural community (Kathem & Kareem Ali, 2020). Soon, pan-Arabists and Arab nationalists gained a growing influence on Iraq's politics. The Antiquities Law of 1936 reflected the decline of British influence on Iraq's archaeology and resulted in fewer foreign missions to the country, especially now that the partage, the system for the distribution of artefacts, was less attractive for archaeological expeditions.

With a growing post-World War I political assertiveness, tied to de-colonialisation movements in the region and beyond, Iraq took control over its archaeology and by World War II had full control. In this spirit of greater control over its own cultural affairs, the state prioritised the establishment of new heritage institutions like archaeology departments at Iraq's universities. Similarly, in this post-war period, intellectual and cultural production across the country flourished. Growing societal fissures, including inequalities between those close to the government and Iraq's still largely impoverished population, created the potential for a major break up of the post-mandate political arrangement (Gerke, 1991). When the Hashemite Kingdom, installed by the British in 1932, was finally overthrown in 1958, the newly created republic in its place looked again to Mesopotamian archaeology to foster national identity to bind all Iraqis to a common history that would go beyond ethnicity and religion (Davis, 2005). Again, Iraqi archaeologists played a crucial role in these constructions, which prized its Babylonian, Assyrian, Akkadian and Sumerian civilisations.

Ba'ath rule in Iraq

The rule of the Ba'ath Party in Iraq (1963–2003) was characterised by the promotion of historical narratives that championed notions of a glorious Mesopotamian past. As before, it stressed that Iraq was one nation unified in a shared Mesopotamian-inspired culture. The establishment of a common denominator, a national identity through cultural channels to reach the wider society, was strengthened by prioritising archaeological research. This led to public awareness campaigns and extensive excavations, including in Hatra, Ashur, Nineveh and Babylon. Ancient folklore and paraphernalia based on symbols and designs from ancient Iraq were widely promoted (Baram, 1983; Isakhan, 2011). Provinces were renamed after Iraq's ancient past, and regional museums were supported to educate the public about the country's rich cultural past. The ideological message of all these campaigns was to present a contiguous record of the cultural continuity of Iraqi history.

The Arab-Islamic civilisation was considered the last manifestation of the new-ancient Iraqi people. This 'paradigmatic nationalism' made it possible to lean on sometimes vague and changeable cultural patterns (Bernhardsson, 2005). A new cultural paradigm incorporated poetry and literature of the Islamic period, and statues to commemorate Iraq's Islamic past were erected. The government had strict control over Iraq's built heritage and numerous guards protected nearly all the sites, museums and monuments. Amendments to the Antiquities Law finally forbade the trade in antiquities, which finally put a stop to the looting of Iraq's cultural assets.

During Saddam Hussein's rule (1979–2003), the history of Iraq before Islam received an even stronger interest than his predecessors'. Saddam Hussein's power rested in part on the appropriation of heritage and archaeology which was instrumentalised within a broader system of governance. A key state initiative was the 'The Project for the Re-Writing of History'. Saddam Hussein exploited state-sponsored and sanctioned nationalism to buttress his personality cult (Isakhan, 2011). His personality was projected through leaders such as Salahadeen, liberator of Jerusalem, and King Nebuchadnezzar II, the powerful king of Babylon. Key sites were restructured, including famously Babylon from 1986, where Saddam had his name inscribed into its structures stating 'To King Nebuchadnezzar in the reign of Saddam Hussein'. Saddam Hussein also honoured the achievements of the Abbasid Caliph al-Mansur, who founded Baghdad as an Islamic city. Clearly, these key figures were referred to underscore a collective 'Iraqi' identity and legitimise state power. To foster national identity, all kinds of symbols and monuments were produced. Streets were decorated with numerous memorials, statues, flags, portraits, mosaics, plaques and pictures, all of great symbolic value. The cultural paradigm was also conveniently employed to justify the Iraq–Iran war (1980–1988). Even before the war ended, several monuments were already erected to celebrate Iraq's victory over Iran. It was a conscious and visible effort to show the public the country's glorious past, which was an integral component of Iraq's politics that augmented the pursuit of legitimacy and governance structures.

The tragic events in the Iran–Iraq War in the 1980s and the beginning of the Gulf War in 1991 had many repercussions for Iraq's heritage. As a result of costly wars and UN sanctions following the Gulf War, there was simply not enough money to pay salaries or otherwise fund the heritage sector. At the beginning of the Gulf War, the most valuable artefacts of the Iraq Museum were stored in the vaults of the Central Bank. The war itself caused the destruction of some key historical and cultural sites (Farhan et al., 2020). Next to the ensuing neglect and natural deterioration, the unprotected sites became prey to unprecedented looting. This turned out to be the worst in the No-Fly Zones in the North and the South of the country. Regional museums were looted and some burnt, and over 5,000 artefacts flooded into the international illicit antiquities market. These major political events clearly disrupted the cultural continuity of Iraq's heritage and undermined its cultural infrastructure whose impact would linger for decades.

The desperate economic situation forced people to find other means of sustenance. The plethora of archaeological sites and the prospect of looting offered a way out. As the years passed and the economic sanctions felt increasingly harsh, looting turned into a massive enterprise. Large smuggling networks developed as the market of Iraqi artefacts in the acquiring countries was on the rise and much was smuggled through the Gulf, Jordan and Turkey (Bahrani, 2003; Rothfield, 2008; Stone & Bajjaly, 2008). Sites were bulldozed to facilitate the digging and later also to prepare the land for agriculture. Out of the 600 technicians employed by the state institution, only 50 remained at the beginning of the Iraq War in 2003, and hundreds of well-trained Iraqi archaeologists had left the country (Stone & Bajjaly, 2008; Stone, 2009). The Iraq Museum was closed in 1991 and had lost most of its professional staff. In short, by the end of the millennium, the entire heritage protection system had become severely degraded. To address this unfolding cultural catastrophe, the government attempted from 1998 a new strategy of clear-and-hold. Staff were sent to severely threatened sites to stop the looting and recruit local armed guards. Foreign missions were also trickling back when by the year 2000 the international sanctions regime imposed on Iraq began to crumble. The new Antiquities Law Number 55 (2002) was in part designed to put an end to the massive looting. The penalties for violating the law were very high. Illegal excavation could result in imprisonment for a period of up to 15 years and trafficking in antiquities for a period not to exceed ten years, which had, however, collapsed after the Iraq War of 2003.

Planning the Iraq war

The US-led invasion in Iraq in 2003 did not come as a surprise. Already ten years before, the Bush Senior Administration contracted the neo-conservative *Rendon* group through the Central Intelligence Agency to influence global opinion on Iraq and to start with the unification of the US-European-based Iraq opposition (Hartenian, 2021). In the following years, other influential neoconservative think tanks, such as 'Project for the New American Century and Committee for Peace and Security in the Gulf' as well as hawkish forces urged to remove Saddam Hussein and wage war against Iraq. Influenced by these neoconservative ideas, the Bush government was convinced that the war would be a relatively easy undertaking: that the Iraqi troops would turn against Saddam Hussein and that the Iraqi people would welcome their liberators (Allawi, 2008). In preparation for a post-invasion of Iraq, the 'Future of Iraq Project' was prepared in October 2001 by the State Department. It was one of the most comprehensive US government planning efforts and involved over 200 Iraqi exiles and tens of international experts on Iraq. It organised 16 working groups to design strategies and policies on numerous essential topics. Yet, the project never convened a proposed working group on protecting Iraq's cultural heritage. In January 2003, the project was taken out of the hands of the State Department and handed over to the Pentagon, which immediately discarded it. Whatever post-war planning had been envisaged was dismissed by the occupation administration.

In January 2003, the Office of Reconstruction and Humanitarian Assistance (ORHA) was hastily set up. While the immediate responsibility for administering postwar Iraq fell under the Commander of the US Central Command in the Middle East (CENTCOM), the detailed planning and implementation, the 'expeditionary' office, was left to ORHA. They faced an uphill battle, suffering from lack of time, uncertain funding and incessant interdepartmental strife. The early staff members were military personnel and had little interest in the cultural affairs of the country. ORHA's main task was to integrate all the existing reconstruction plans and coordinate interagency planning at a practical level. The Pentagon expected to be out of Iraq in 90 days, so no plans to safeguard antiquities needed to be prepared.

The invasion and early occupation of Iraq

The adverse effects of the US Occupation of Iraq had far-reaching consequences on the country's cultural infrastructure. During the early occupation, a wave of unprecedented pillage and destruction overran the country. All kinds of buildings were indiscriminately ransacked and set on fire while the occupying forces did not intervene, and in some cases encouraged the mobs and criminal gangs. Unit commanders in fact received an order not to use force to prevent looting. After protests from Iraqi citizens, this was, to some extent, modified later. The culture and heritage sector did not escape the raging crowds. When about a year later the damage could be surveyed, the situation was far more serious than anyone could have been imagined. Striking is the irreparable damage sustained to the cultural infrastructure. The Ministry of Culture was wrecked, burned and forced to move to the House of Fashion. The infrastructure in Iraqi higher-education institutions, including many libraries, has been either looted or severely damaged. The Iraq National Library and Archives (INLA) as many other libraries was partly burned, robbed of practically all fixtures and the card catalogue spread over the floor. Other catalogues were lost in a fire and stolen computers were offered for sale on the sidewalk not far from the institutions. Churches, mosques, palaces and Baghdad's old historical and administrative centres did not escape the bombs or plunder. The archaeological sites similarly suffered as they were left without any protection, especially those in isolated areas. Guards fled, stood by helplessly or joined the looters. Like in 1991, the archaeological sites in the South of Iraq were affected the most.

This cultural attack caused a worldwide outcry, the more as the US and the UK were warned well in advance (Stone & Bajjaly, 2008). Especially, the plunder of the famous Iraq Museum ended in screaming headlines. Numerous foreign archaeologists and archaeological institutions contacted their authorities to share their concerns over the safety of Iraq's heritage. Some went directly to the war planners at the Pentagon, the State Department or the Defense Department. They provided lists of 4,000 crucial sites, museums and monuments throughout Iraq, including their map coordinates (Gibson, 2009). The result of all those warnings and public statements was that the Pentagon drew up a 'no strike list' of 24 sites. When the Coalition forces went into battle, practically all of Iraq's heritage was just about abandoned to their fate.

Like their colleagues abroad, Iraqi heritage staff were afraid of the effects of the foreseeable invasion on Iraq's cultural history. Remembering the 1991 Gulf War, many cultural institutions took appropriate action. Before and even during the attacks, libraries, archives and museums evacuated at least parts of their collection to safer storage locations. The Iraq National Library and Archives, for example, managed to move collections to the Iraqi National Board of Tourism and with the help of many volunteers to a mosque in Sadr City (Johnson, 2005). The Iraq Museum transferred items from various archaeological sites and provincial museums to the capital. Others evacuated their collections to bunkers, basement storerooms or religious buildings. Entrance doors were welded down, and when possible, armed guards were protecting their heritage institutions, often with the help of local citizens. Notwithstanding the foresight and determination of many, massive vandalism and pillage could not be prevented.

Lacking any post-war plan in general and specifically none for the rehabilitation and reconstruction of cultural heritage, the successive US military occupational authorities started to reshape Iraq according to their own interests (Baker et al., 2010). At the bottom of their priority list was the care for Iraq's heritage. During the occupation from April 2003 till the first elected Government in May 2006, the reconstruction of Iraq was successively in the hands of the Office of Reconstruction and Humanitarian Assistance (ORHA), the Coalition Provisional Authority (CPA) and the Iraq Reconstruction and Management Office (IRMO). The CPA was the most powerful US transitional administration office as they also had the responsibility to establish a democratic government of Iraq. The main setbacks the reconstruction efforts met were the lack of pre-war reconstruction and stability planning, insecurity, corruption and the debilitating struggle between the State and Defense Departments. Culture and heritage were clearly nowhere to be found on the main reconstruction budgets. Practically, all the funds were allocated for establishing a new democratic Iraq, improving Iraq's infrastructure, employment and economic recovery. The little money that was available for culture and heritage came from small funds. Most of it went into the rehabilitation of the Ministry of Culture, the reconstruction of the Iraq Museum and some small museums, universities and school libraries.

When in April 2003 ORHA took office in Baghdad, still understaffed, they quickly discovered that conditions on the ground were totally different from what the neoconservatives anticipated. One pre-planned provision was the establishment of the Ministerial Advisory Teams to ensure the Iraqi ministries continued to function. The first cultural advisor discovered he was in charge of restoring the Iraq Museum, and a liaison officer between the Iraqi department of antiquities and ORHA was appointed. There was little they could do. ORHA was phased out in May 2003, less than three weeks after arrival and replaced with the CPA.

The United Nations designated the CPA as the lawful government of Iraq. It replaced the Iraq State without Iraqi participation and vested itself with executive, legislative and judicial authority over the Iraqi government (Dobbins, 2009). The Iraqi Army and other parts of the security structure were disbanded. According to military commanders, this created a pool of disaffected and unemployed people that the insurgency could draw on. The CPA ruled the country by decree, about 100 in total. New commissions and advisors were appointed that effectively took away virtually all of the powers once held by the ministries. The Ministry of Culture in the Iraq Governing Council (IGC) had several successive (deputy) senior advisors, but their hands were tied as they were hardly funded. Some small heritage projects were implemented, but they were more courageous attempts to try and procure part of the enormous funds the CPA had at their disposal. The last CPA advisor though convinced the chief executive officer to clear Babylon from the coalition forces. Yet, it was not his call as the US commander-in-chief was the only person to give the order to clear Babylon. On June 28, 2004, the CPA formally transferred the limited sovereignty of Iraqi territory to the Iraqi Interim Government. The policies of the CPA, disliked by most of the Iraqis, did not lead to rapidly executed and effective reconstruction programs.

Establishing a new political order

The US-led invasion and subsequent Occupation of Iraq was a key turning point in the recent history of the country. It marked a major transformation of Iraq's politics as Saddam Hussein's Baathist Party, which ruled the country for over three decades, was deposed and replaced by Iraqi Diaspora opposition figures (Dodge, 2012). The establishment of Iraqi political institutions under the occupying US administration started with the creation of the IGC. It was established in July 2003 by the CPA and was granted limited powers. It largely existed as an advisory body, and their decisions were subject to the veto of the CPA, which was heavily influenced by dozens of US and Iraqi appointees. Newly appointed ministers working with the IGC were US-handpicked from various Iraqi political parties, representing the new elite. Half were Iraqi exiles, friendly to the occupation, and represented various sectarian interests. The formation of the council and the appointment of the ministers laid the basis for a sectarian dimension of Iraqi politics and deepened sectarian rivalries. Within the area of responsibility of the Ministry of Culture came the Bureau of Tourism, the Antiquities Board, the Convention Center and various other facilities. Several advisors at the CPA advised the ministry on different issues. Their prime concern was rehabilitating the ministry itself and secondly the Iraq Museum. All the ministries were in a very poor state. The bureaucratic infrastructure was almost non-existent. They had no computers, no internet, few phones, and there was no working national phone system in place. The banks were shuttered and there was no way to transfer funds between branches. There were no internal management information systems. No coherent sets of accounts. The ministries did not have their own budgets; thus all procurements and income were to be transferred to the Ministry of Finance. Another familiar problem was that civil servants were reluctant to take responsibility or initiative, a legacy from the deposed totalitarian government and the break-down in authority that the United States pursued in 2003.

With the end of the CPA and the start of IRMO in June 2004, the IGC was replaced with the Iraqi Interim Government (IIG). The Coalition transferred limited sovereignty to the caretaker government, whose first act was to begin the trial of Saddam Hussein. The government began the process of moving towards elections, though the insurgency and the lack of cohesion within the government itself led to repeated delays. The new Iraqi government had little control over its armed forces, lacked the ability to alter or change laws and was unable to make major decisions within specific ministries without tacit US approval. Members of the IIG were again chosen by the CPA with the assistance of a UN special envoy. Though supposedly composed of technocrats, it contained familiar personalities, selected according to sectarian identity and relative closeness to the interests of the US in Iraq. A system of political apportionment, known in Iraq as Muhasa'sa or segmentation, was developed by the US occupation and allies to organise politics and the distribution of power and state resources (Visser, 2007; J. Ismael & Ismael, 2010). Muhasa'sa became an integral convention for the allocation of state assets, including ministries, to competing political groups. Rather than a system designed to support and strengthen national cohesion and state institutions, Muhasa'sa was underpinned by ethnic and religious quotas. This rule applied to the US-established temporary government bodies, such as the IGC and IIG, which were not granted any significant power or resources. In effect, the operation of Iraq's politics relied on an exclusive political pact between newly emerging political elites (Dodge, 2012) which the United States used as a tool to control Iraq's politics.

The consequent fracture of Iraq's centralised state institutions, now controlled by competing political parties, bode poorly for Iraq's cultural infrastructure (Kathem, 2020). Iraq's heritage-related state institutions were similarly, under Muhasa'sa, allocated to different political parties which resulted in major institutional disarray, lack of co-ordination and weak support to rebuilding Iraq's cultural infrastructure. New and old state institutions were consequently altered and now based not on centralisation of power and authority, but based on ethnic and religious backgrounds. What the US Occupation did was to introduce sectarian quotas in Iraq's political system, which previous governments since the founding of the country had sought to prevent (Visser, 2007; T. Y. Ismael & Fuller, 2009; J. Ismael & Ismael, 2010; T. Y. Ismael, 2015). Indeed, rather than attempting to centralise and build credible institutions, Muhasa'sa further fractured power. This was clearly seen in Iraq's first post-invasion elections, in 2005, which legitimised ethno-sectarian political groups in the semi-autonomous region of the Kurdistan Region of Iraq and religiously oriented groups in the rest of the country and organised Iraq's politics based on institutionalised fracture.

The increasing formalisation of Iraq's post-2003 politics meant the deepening of a new political order based on the fragmentation of power (Ismael, 2015). The ratification of the Iraqi Constitution in 2005 allowed for the formal creation of religious endowments in Iraq, and subsequent laws were promulgated through the Iraqi Parliament to organise their work. The Ministry of Religious Endowments and Religious Affairs was disbanded, and its resources were distributed between three new endowments created for the purposes of Muhasa'sa, which were the Shia, Sunni and Minorities Endowments (Hasan, 2019). Much of these institutions, however, were established in violation of existing Iraqi laws, and no real effort was done to harmonise Iraqi legal frameworks regarding such things as archaeology, urban heritage and ownership (Kathem et al., 2020).

In response to the US Occupation and the attempt to establish a new political order, an insurgency, which had been slowly gathering pace from 2003 was within one year creating an untenable situation. Major Iraqi cities, including Fallujah in April and May 2004, and again in November, had been controlled by Al Qaeda and other insurgent groups. Subsequent military operations led by the US military were a key turning point. Levels of violence in the country were rapidly increasing and the United States was now facing a full-scale insurgency. The battles of Fallujah transformed the mindset of US officials about the enormity of the task of continuing the occupation. The CPA was disbanded in June 2004 and a largely ceremonial act of handing power, labelled as a transfer of sovereignty, was carried out (Allawi, 2008). What the US Occupation had realised was that, within one year of the CPA's disbanding of the Iraqi Army and security forces and consequent rising levels of violence, it was increasingly in a quagmire of its own making. In a context of rising levels of violence, the United States aggressively pursued an exit plan but could not easily implement it without first forging a semblance of a political system, having national elections and ensuring that Iraq's security forces were rebuilt. A major consequence of Iraq's insecurity from 2003, especially now in situations of major crises, the protection of archaeological sites was viewed to be of lesser importance. During periods of insecurity, the growth of looting and illicit trafficking of cultural goods was widely noted to be an outcome of Iraq's generalised security conditions.

The elections and concomitant allocation of state resources and institutions on the basis of political quotas did little to quell Iraq's insurgency (Damluji, 2010). Levels of violence further spiked after the bombing of Al Askari Shrine in 2006 and 2007, a revered Shi'i mosque in the city of Samarra. The subsequent months after the bombing saw tens of mosques attacked, sectarian killings and displacement, resulting in a major humanitarian crisis that focused international attention on addressing another emergency. The bombing of the mosque was a clear indication of the importance of heritage to Iraq's post-2003 politics. Indeed, the attack should be viewed as an attempt to undermine Iraq's politics as it was based now on a sectarian system that increasingly folded cultural and religious sites into the operation of formal politics.

In a situation where state institutions were rendered ineffective, a broken political system based on ethnic and religious quotas and a raging, ongoing insurgency and widespread sectarian killings, the US Occupation responded to the worsening situation in the country with the New Way Forward, known as the 'US Surge'. The massive increase in the US troop level from January 2007 was coupled with a nation-wide cash-incentives programme. A new structure tied to US patronage was created, which supported the so-called 'Sunni Awakening Councils'. In practice, this translated into support to local Sunni leaders as well as other communities across the country to join the fight against the insurgency. Support through the Provincial Reconstruction Teams (PRTs), which were civil-military units to manage this new programme, was temporarily successful in turning the tide against the insurgency. By 2008, levels of violence had been reduced significantly, and the United States, with its eyes on exiting the country from its costly intervention in Iraq, formally ended the occupation in December 2011. The departure of the United States from the country did not mean, however, that Iraq's problems were resolved or the legacies of the US Occupation would end.

As part of the US exit plan, the United States invested significant political and financial resources in then Prime Minister, Nouri al Maliki. Al Maliki, representing al Dawa Party, discontinued the Sunni Awakening Councils, which was accused of harbouring anti-Government interests and criminals (Dodge, 2005). Growing discontent about low levels of employment and government services created an increasingly fragile political environment. National elections in April 2014 did little to address Iraq's problems. In June 2014, the Islamic State of Iraq and the Levant, an insurgent group that had previously worked closely with al Qaeda, had taken root in major provinces of Anbar, Kirkuk, Salahadeen, Diyala and Nineveh. In a context of weak state institutions, endemic corruption and an increasingly restive population, the Islamic State took over key provinces and cities, including Mosul. Like the United States did in 2003, it too attempted to transform structures of power in Iraq. It declared itself to be a caliphate. In pursuit of its objectives, the Islamic State undertook widespread cultural cleansing, wiping out what it saw as cultural groups, namely the Yezidis, Christians and Shia. It destroyed hundreds of cultural sites, including churches, mosques, temples, shrines and other major sites. The ensuing destructive war against the Islamic State saw the US military return to Iraq and the creation of a new paramilitary force – the Popular Mobilisation Units. The widespread destruction by the US military and foreign allies, including France, Germany, the UK and Australia, amounted to a huge loss of Iraq's cultural infrastructure. Most glaringly, Mosul's ancient and historic city centre – known as the Old City – suffered severe damage, and no building was left untouched by the effects of the war that were largely destroyed by aerial bombardments (Matthews et al., 2019).

By the time of the liberation of Mosul in 2017, Iraq's political and cultural landscape had once again been altered by the effects of war. State institutions had little, if any, financial support and were, therefore, not pursued by political parties, unlike front-line Ministries. This was the case for the State Board of Antiquities and Heritage (SBAH), which continued to be under-funded and neglected by Iraq's sectarian political groups. It was granted negligent resources to conduct any major reconstruction projects in Mosul and other damaged cities in the country. Without funding and support, SBAH was treated as a peripheral institution, which was in line with the trend in which it was viewed by post-2003 Iraq politics. As has been the case before, SBAH's core financial needs went unaddressed by successive governments, which was also the case for the Ministry of Culture more broadly.

Iraq's rapidly changing politics of winners and losers facilitated increased competition over its cultural resources. In particular, the endowments fought fiercely between each other and with existing state institutions to extract material and cultural resources, including mosques, shrines and monuments, as well as their adjacent estates and lands. The Shia Endowment, managed and controlled by Shi'i clerical families in Najaf and Karbala, expanded its presence in the country and was now both a religious institution as well as a commercial one. It had billions in commercial investments from land it had largely extracted from the state. As the Shia Endowment expanded its investment holdings in hospitals, schools, universities, commercial property and agricultural projects, it increasingly became seen by the general public as a commercial enterprise that used religion as a platform to realise its financial objectives.

The position of the endowments and its relative strength was commensurate with the nature of power in the country, meaning that the Shia Endowment's rapidly growing resources and powers in this period were largely due to the support it received from militia groups and Shia political parties in power (Hasan, 2019). The religious endowments also pursued the restructuring and appropriation of major historical and religious sites. They contracted Iraqi and companies from neighbouring countries to provide conservation services on some sites, which were generally of poor quality. In many cases, major historic and religious sites were restructured in line with the construction of narrowly defined religious and political agendas, which sought to undergird their power base and legitimise their massive expansion in state and society. The Shia Endowment, for example, extracted in 2010 the Shrine of Prophet Ezekiel and its adjacent buildings, known as al Kifl, from SBAH, which it then restructured and converted into an expanded Shia mosque (Kathem et al., 2020). The existing synagogue in the province of Babylon was erased and many buildings, including the Ottoman-era guest houses, were torn down. This overwhelming level of cultural destruction was yet another devastation visited on Iraq's heritage landscape that was a direct outcome of post-2003 politics (Abid, 2016; Farhan et al., 2020).

After the regaining of Mosul by the Iraqi Government, the Shia Endowment pursued a policy in Mosul to capture and control more than 20 mosques, which it said were Shia in identity (Hasan, 2019). It was widely known in Mosul that the Shia Endowment's actions were primarily designed to expropriate large tracts of land and real estate tied to these cultural and religious sites. In addition, it also claimed ownership of the Shrine of Prophet Jonah in the city, which was also contested by the Sunni Endowment and SBAH. These examples attest to Iraq's rapidly changing politics where its newly emerging post-2003 political and religious actors were in the process of expanding and consolidating their positions in the country through the field of cultural heritage.

The internationalisation of Iraq's cultural heritage

The monumental damage visited on Iraq's cultural heritage, especially from 2003, has to date not been adequately addressed. Given its responsibilities as an occupying force, the United States did little to deliver any strategic heritage projects. When compared to other sectors that it prioritised, from women's affairs, human rights, civil society strengthening and governance, heritage and culture was by far the weakest component of its programming in the country. One notable project that the United States spent funds on came about after the significant damage visited on Babylon by the US military. The State Department's Bureau of Educational and Cultural Affairs (ECA) Cultural Heritage Centre and the US Embassy in Iraq responded to media outcries by offering in 2008 a $12.9 million grant. The grant established the Iraq Cultural Heritage Project whose objectives were to offer training in conservation, infrastructural support to the Iraq Museum and assistance to international organisations to work on Babylon's preservation and documentation. A component of this grant was offered to open a conservation institute and training centre. As a result, the Iraqi Institute for the Conservation of Antiquities and Heritage (IICAH) was established in Erbil in 2010 on the basis that the security situation in Baghdad was not conducive to the security requirements of international trainers and partner institutions (Johnson et al., 2016). The institute has largely operated as a training centre and whilst it was primarily designed to serve SBAH, its main focus has been supporting archaeologists within the Kurdistan Region of Iraq.

Whilst IICAH was marketed as an institution serving Iraq, the arrangement for joint management of the institute had broken down and it primarily served the Kurdistan Region of Iraq. The institute's trainers and teaching team is composed of conservation experts from the United States, and indeed, it is largely managed by the US government institution, Smithsonian Institute. In addition to offering conservation training, IICAH is used as a platform for international expeditions and programs and has been funded by the US Department of State, and other US and European organisations. Relative to the size of the damage the United States inflicted in Iraq, support to cultural heritage was nominal and such projects only furthered US interests in the country rather than addressing Iraq's essential and urgent needs. The establishment of IICAH did little to address Iraq's ongoing cultural catastrophe and increased served interests in the Kurdistan Region of Iraq and its US benefactors, namely the Smithsonian Institute.

Whilst the number of international archaeological expeditions grew rapidly in times of relative peace, particularly in the Kurdistan Region of Iraq, there was a remarkable absence of support to Iraq's own national cultural institutions, namely SBAH. By 2020, there was still no national conservation institute that could have supported cultural rebuilding efforts. SBAH continued to be underfunded and it had no international lawyers to support its restitution support. Inventory lists of artefacts were not of the quality acceptable to international institutions. The absence of support to SBAH was contrasted to the increasing number of internationally funded projects to rehabilitate damaged archaeological and historic buildings after the defeat of the Islamic State. The International Alliance for the Protection of Heritage in Conflict Areas (ALIPH), a private foundation financed by French and UAE government funding, was established in the wake of the Islamic State's devastation in Iraq. It partnered mainly with US and European cultural organisations to undertake conservation and cultural rebuilding in Northern Iraq, focusing on Nineveh, and in particular, a number of churches, mosques and other cultural sites were in the process of being restored and rehabilitated. Much of those efforts were led and managed by foreign service-providers who had little interest in supporting national capacities or building sustainable and trust-based partnerships.

Other funding streams were also set up by governments, including the British Council managed Cultural Protection Fund, which offered £30 in support of the Middle East's crisis-affected heritage. One of its flagship projects was the British Museum's Iraq Emergency Heritage Scheme, a £3.2m project that aimed to train Iraqi archaeologists (British Museum, 2017). A major criticism of the scheme was that, whilst the objective was to support Iraq's own emergency-related cultural needs, the project was merely a cover for further excavation work and knowledge extraction as it pursued two major archaeological expeditions in the country to better understand the histories and origins of its Iraqi artefacts that it has in its collections. Given the desperate situation that archaeologists and heritage professionals found themselves in Iraq, such projects were often accepted by SBAH, a nominal partner who was merely required for administrative purposes.

In 2018, UNESCO promoted the 'Revive the Spirit of Mosul', an initiative that was designed to raise awareness of the plurality of the city and the urgent need to rebuild it, particularly its historic sites. Through the initiative, UNESCO secured $50m from the United Arab Emirates to rebuild the iconic Nuri Mosque and minaret that had been destroyed in the war against the Islamic State (Isakhan & Meskell, 2019). A partnership agreement was signed with UNESCO to rebuild the site. In practice, this huge project, whilst officially crafted as a partnership with SBAH and the Sunni Endowment, meant that UNESCO was directly involved in managing service-delivery activities to rebuild destroyed heritage rather than in devising and supporting national institutions and their efforts regarding cultural recovery. The European Union also funded heritage-related projects in the country, often working closely with UNESCO's own plans (Kathem et al., 2020). Taken together, the recent history of foreign-funded projects in Iraq meant bypassing SBAH and central state institutions altogether. Such projects illuminated that heritage institutions in Iraq were ignored or merely needed to rubber-stamp and approve projects. International organisations, including UNESCO, did little in this situation to assess Iraq's own heritage needs, and indeed by 2021, there had been no major assessments or discussions with SBAH about its own priorities.

The future of Iraq's cultural heritage

Prior to 2003, heritage in Iraq was viewed as a platform for national unity, and state institutions, though weakened by multiple wars and sanctions in the 1990s, reflected this political orientation. The championing of symbols of unity, namely around Iraq's pre-Islamic cultural heritage, was designed to invoke a sense of national identity tied to the Iraqi state and the history of the country. Under the US Occupation of Iraq and the concomitant transformation of the state, heritage was transformed. Heritage was now used as a platform and space for the promotion of cultural differentiation, the country's national heritage splintered into fragments competed for by increasingly assertive ethno-nationalist and religiously inclined political groups (Kathem, 2020). In this context, how political actors shape and are themselves beholden to the evolvement of political systems and structures of power is an indispensable purview with which to analyse heritage.

The defeat of the Islamic State came at a heavy human cost to Iraq and devastated large parts of its cultural infrastructure. The damage to Iraq's cultural sites in particular in places such as Mosul, Tikrit and Fallujah have been unprecedented. The absence of sufficient reconstruction funds and support to such institutions as SBAH to rebuild Iraq's cultural icons and landmarks added to the frustration of an ongoing neglect by political parties in control of Iraq's resources (Kathem, 2020). The lingering legacies of cultural destruction on the fabric of Iraqi society in key affected provinces continued as communities were left to rebuild their houses and the areas in which they lived without much government support. Additionally, externally funded projects in the field of heritage and cultural rehabilitation had become heavily politicised and were generally seen as part of US-European cultural expansionism in Iraq. It is important to note that Iran, Turkey and Saudi Arabia also pursued activity in this field, and cultural heritage had become heavily contested, albeit this time within a context of competition for international and regional influence in the country.

In the absence of sufficient government services and growing unemployment, a national protest movement came into prominence in October 2019. Led by secularly leaning youth, and with no known political party organisation, the Iraq Protest Movement demanded radical changes to the country's Muhasa'sa state system. Its main base of activity in Baghdad and Nassiryah was visited by hundreds of thousands of protesters, many of whom camped in these city's main squares. A large number of protestors attested to the fragility of Iraq's post-2003 political order, which had now, after the defeat of the Islamic State, shifted power to institutions outside the control of the central state. In response to human rights violations and more broadly the failure of post-2003 politics, paintings, murals and photographs of pre-Islamic cultures were widely used and shared across the country as a message of national unity. The use of pre-Islamic imagery by Iraqi protestors showed that large swaths of the population were attempting to reclaim symbols of the nation from the dominance of sectarian and ethno-sectarian narratives that had been promoted since 2003.

Conclusion

Politics and cultural heritage evolve in dynamic and closely intertwined ways. What happens to a country's politics affects cultural continuity and cultural heritage in manifold ways that have in the face of devastation visited on Iraq and the wider region not been thoroughly studied. This paper highlighted key periods of change in Iraq's recent history to better understand cultural heritage. It explored cultural heritage degradation and destruction as a relationship to politics and instability. Notions of the state and state-building more broadly have not been understood with a view to safeguarding and protection of cultural heritage, which bodes poorly for our collective thinking and planning for the future. Indeed, considering the unprecedented heritage destruction that this paper has highlighted, especially since the 1990s and again under the US Occupation and afterwards, a focus on local politics and systems of power was clearly an essential lens for developing more effective responses to cultural collapse.

Future research trends in cultural heritage in fragile and war-affected contexts will require making sure that studies are relevant to the everyday realities of those countries. The absence of case studies and detailed analyses of examples that may shed light on those dynamics have, however, not been something that researchers have focused on and much research is based on ideational orientations or international legal frameworks. If this is to change, conversations between different disciplines will be required, particularly for heritage studies to engage with such schools as critical peacebuilding and state-building theories. Negotiating new research that is focused on the long-term impact of instability and conflict will inevitably require the incorporation of politics as contestation, political economy issues, and systems of power as a foundational basis for understanding cultural heritage and its continuity and protection.

References

Abid, S. (2016). *An examination of heritage protection and conservation practices in the pilgrimage city of Najaf* (Doctoral dissertation). Retrieved from: White Rose eTheses Online (etheses.whiterose.ac.uk/17278/). (Accessed on 03 May 2021).

Allawi, A. A. (2008). *The occupation of Iraq: Winning the war, losing the peace*. New Haven, CT: Yale University Press.

Bahrani, Z. (2003). Iraq's cultural heritage: Monuments, history, and loss. *Art Journal*, *62*(4), 11–17. https://doi.org/10.2307/3558482.

Baker, R. W., Ismael, S. T., & Ismael, T. Y. (Eds.) (2010). *Cultural cleansing in Iraq: Why museums were looted, libraries burned and academics murdered*. London: Pluto Press.

Baram, A. (1983). Mesopotamian identity in Ba'thi Iraq. *Middle Eastern Studies*, *19*(4), 426–455.

Baram, A. (1994). A case of imported identity: The modernizing secular ruling elites of Iraq and the concept of Mesopotamian-inspired territorial nationalism, 1922–1992. *Poetics Today*, *15*(2), 279–319. https://doi.org/10.2307/1773167.

Batatu, H. (2004). *The old social classes and the revolutionary movements of Iraq: A study of Iraq's old landed and commercial classes and of its communists, Ba'thists and free officers* (New edition). London: Saqi Books.

Bernhardsson, M. T. (2005). *Reclaiming a plundered past: Archaeology and nation building in modern Iraq*. Austin: University of Texas Press.

British Museum. (2017, November 24). *The Iraq emergency heritage management training scheme: An update* [Blog post]. Retrieved March 5, 2021 from https://blog.britishmuseum.org/the-iraq-emergency-heritage-management-training-scheme-an-update/ (Accessed on 03 May 2021).

Cesari, C. D. (2010). Creative heritage: Palestinian heritage NGOs and defiant arts of government. *American Anthropologist*, *112*(4), 625–637. https://doi.org/10.1111/j.1548-1433.2010.01280.x.

Cesari, C. D. (2015). Post-colonial ruins: Archaeologies of political violence and IS. *Anthropology Today*, *31*(6), 22–26. https://doi.org/10.1111/1467-8322.12214.

Cesari, C. D. (2017). Heritage between resistance and government in Palestine. *International Journal of Middle East Studies*, *49*(4), 747–751. https://doi.org/10.1017/S0020743817000721.

Damluji, M. (2010). 'Securing Democracy in Iraq': Sectarian politics and segregation in Baghdad, 2003–2007. *Traditional Dwellings and Settlements Review*, *21*(2), 71–87.

Davis, E. (2005). *Memories of state: Politics, history, and collective identity in modern Iraq*. Berkeley: University of California Press.

DeSilvey, C., & Harrison, R. (2020). Anticipating loss: Rethinking endangerment in heritage futures. *International Journal of Heritage Studies*, *26*(1), 1–7. https://doi.org/10.1080/13527258.2019.1644530.

Dobbins, J. (2009). Occupying Iraq: A short history of the CPA. *Survival*, *51*(3), 131–162. https://doi.org/10.1080/00396330903011537.

Dodge, T. (2005). Chapter three: Political mobilisation in the new Iraq. *The Adelphi Papers*, *45*(372), 43–56. https://doi.org/10.1080/05679320500115002.

Dodge, T. (2012). Chapter Five: The politics of Iraq: The exclusive elite bargain and the rise of a new authoritarianism. *Adelphi Series*, *52*(434–435), 147–180.

Farhan, S., Akef, V., & Nasar, Z. (2020). The transformation of the inherited historical urban and architectural characteristics of Al-Najaf's Old City and possible preservation insights. *Frontiers of Architectural Research*, *9*(4), 820–836. https://doi.org/10.1016/j.foar.2020.07.005.

Francioni, F., & Lenzerini, F. (2003). The destruction of the Buddhas of Bamiyan and International Law. *European Journal of International Law*, *14*(4), 619–651. https://doi.org/10.1093/ejil/14.4.619.

Gerke, G. (1991). The Iraq development board and British policy, 1945–50. *Middle Eastern Studies*, *27*(2), 231–255.

Gibson, M. (2009). Culture as afterthought: US Planning and non-planning in the invasion of Iraq. *Conservation and Management of Archaeological Sites*, *11*(3–4), 333–339.

Hartenian, L. (2021). *George W Bush administration propaganda for an invasion of Iraq: The absence of evidence*. London: Routledge.

Hasan, H. (2019, March 29). *Religious authority and the politics of Islamic endowments in Iraq*. Retrieved from: https://carnegie-mec.org/2019/03/29/religious-authority-and-politics-of-islamic-endowments-in-iraq-pub-78726 (Accessed on 03 May 2021).

Isakhan, B. (2011). Targeting the symbolic dimension of Baathist Iraq: Cultural destruction, historical memory, and national identity. *Middle East Journal of Culture and Communication*, *4*(3), 257–281. https://doi.org/10.1163/187398611X590200.

Isakhan, B., & Meskell, L. (2019). UNESCO's project to 'Revive the Spirit of Mosul': Iraqi and Syrian opinion on heritage reconstruction after the Islamic State. *International Journal of Heritage Studies*, *25*(11), 1–16. https://doi.org/10.1080/13527258.2019.1578988.

Ismael, J., & Ismael, T. (2010). The sectarian state in Iraq and the new political class. *International Journal of Contemporary Iraqi Studies*, *4*(3), 339–356.

Ismael, T. Y. (2015). The unravelling of the uncivil state: Iraq and the imposition of sectarian governance. *International Journal of Contemporary Iraqi Studies*, *9*(2), 121–137. https://doi.org/10.1386/ijcis.9.2.121_1.

Ismael, T. Y., & Fuller, M. (2009). The disintegration of Iraq: The manufacturing and politicization of sectarianism. *International Journal of Contemporary Iraqi Studies*, *2*(3), 443–473. https://doi.org/10.1386/ijcis.2.3.443_1.

Johnson, I. M. (2005). The impact on libraries and archives in Iraq of war and looting in 2003—A preliminary assessment of the damage and subsequent reconstruction efforts. *The International Information & Library Review*, *37*(3), 209–271.

Johnson, J., Khorsheed, A., & Lione, B. (2016). The Iraqi Institute: Education for archaeological research and conservation. In K. Kopanias & J. MacGinnis (Eds.), *The Archaeology of the Kurdistan region of Iraq and adjacent regions* (pp. 135–138). Oxford: Archaeopress.

Kathem, M. (2020). Cultural (dis)continuity, political trajectories and the state in post-2003 Iraq. *International Journal of Heritage Studies*, *26*(2), 163–177. https://doi.org/10.1080/13527258.2019.1620836.

Kathem, M., Antonelli, G. F., & Selter, E. (2020). *The role of the European Union in the protection and enhancement of cultural heritage in conflict and post-conflict contexts in the Middle East region*. Retrieved from: https://www.culturexchange.eu/system/files/2020-11/Report%2BCultural%2BHeritage%2B1Oct2020.pdf (Accessed on 03 May 2021).

Kathem, M., & Kareem Ali, D. (2020). Decolonising Babylon. *International Journal of Heritage Studies*, 1–15. https://doi.org/10.1080/13527258.2020.1858140.

Liverani, M. (2016). *Imagining Babylon: The modern story of an ancient city*. Boston, MA: De Gruyter.

Matthews, R., Rasheed, Q. H., Fernández, M. P., Fobbe, S., Nováček, K., Mohammed-Amin, R., Mühl, S., & Richardson, A. (2019). Heritage and cultural healing: Iraq in a post-Daesh era. *International Journal of Heritage Studies*, *26*(2), 1–22. https://doi.org/10.1080/13527258.2019.1608585.

Rothfield, L. (Ed.) (2008). *Antiquities under siege: Cultural heritage protection after the Iraq war*. Lanham, MD: AltaMira Press.

Smith, Laurajane. (2006). *Uses of heritage*. London: Routledge.

Stone, P. (2009). Protecting cultural heritage in times of conflict: Lessons from Iraq. *Archaeologies, 5*(1), 32–38. https://doi.org/10.1007/s11759-008-9093-5.

Stone, P. G., & Bajjaly, J. F. (2008). *The destruction of cultural heritage in Iraq*. Suffolk: Boydell & Brewer Ltd.

Visser, R. (2007). The Western imposition of sectarianism on Iraqi politics. *The Arab Studies Journal, 15/16*(2/1), 83–99.

15 Najaf, Iraq

Developing a sustainable approach to threatened heritage

Caroline A. Sandes and Ali N. Attiyah

Introduction

This paper looks at the issues of sustainable urban conservation in the pilgrimage city of Najaf, Iraq (Figure 15.1) – a historic city that sees millions of pilgrims every year but has also suffered all the tragedies and complexities of conflict and other aspects of Iraq's troubled history.

The city of Najaf is located in central Iraq, 160kms south of Baghdad, and 10km west of the Euphrates. Its old town, the historical centre of Najaf, was built around the Imam Ali Shrine and is one of the few Iraqi cities founded in the medieval period that still has this urban texture. It has been considered one of the finest examples of a complete small concentric Islamic town (Al Kubaisy, 2009, pp. 45–47). Just to the north of Najaf is Kufa, Najaf's older sister city, and just south of Kufa is the site of the ancient capital of Hira. Hira, al Manathira, was the capital of the Arab Lakhmids that, between the third and 6th centuries CE was not only a centre for Nestorian Christianity and Arab literature but also played a major role in controlling and settling the tribes of eastern

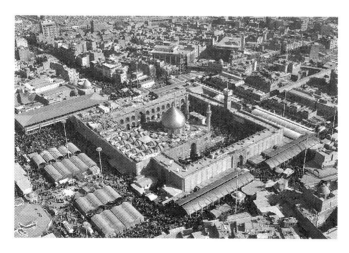

Figure 15.1 Imam Ali Shrine, Najaf, September 2016.
Source: Retrieved from: https://upload.wikimedia. org/wikipedia/commons/9/90/Najaf_City.jpg (Tasnim News Agency, CC BY 4.0 <https://creativecommons.org/licenses/by/4.0>, via Wikimedia Commons).

Arabia (Tabbaa & Mervin, 2014, p. 23). Najaf was an important centre of Christianity, and the evidence of over 30 churches and monasteries have been recorded in the area of Najaf, Kufa and Hira, including remains of one of the earliest churches in Iraq excavated during the expansion of Najaf airport in 2007 (Al Janabi, 2012). Christianity remained active even after Hira was taken over by Muslim armies in 633 AD. Within a few years, Kufa was built as a garrison town and developed from that; Najaf is only recorded as a settlement once it is identified as the burial place of Imam Ali, and it became an officially recognised shrine in 786 AD (Tabbaa & Mervin, 2014, p. 30).

The Shrine of Imam Ali was built on the location believed to be the burial site of 'Ali ibn Abi Talib, the first Shi'i Imam, assassinated in a mosque of Kufa in 661 AD (Nakash, 1994, p. 13). It is consequently of primary importance to Shi'a Muslims, and the third most-visited holy Islamic city by them after Makkah and Madinah in Saudi Arabia. Every year millions of pilgrims visit Najaf from all over the world, particularly from Iran, with an estimated 75,000 per day. On certain days, especially around Ashura – the day that commemorates the battle of Karbala in 680 AD and the martyrdom of Husayn ibn Ali ibn Abi Talib (grandson of the Prophet Mohammed) – up to 3 million pilgrims can descend on the city, though they are mostly concentrated over a total of seven days. Najaf is the most important of four holy cities in Iraq, the others are Karbala, Kazimayn and Samarra. It is to Karbala that the Arbaeen pilgrimage (Arbaeen marks 40 days after the Day of Ashura) makes its way, with up to 27 million pilgrims visiting Karbala, the majority of which walk the 78km from Najaf (Al Arbaeen Encyclopedia, 2020).

Next to Najaf is the Wadi al-Salam, believed to be the world's largest cemetery, which is considered the holiest and sought-out place of burial for the world's Shi'is (Nakash, 1994, p. 18), and is on Iraq's World Heritage Tentative List. In addition, Najaf is, historically and remains so, a major seat of Islamic learning and scholarship (Heern, 2017), and it also retains a number of important crafts and trades connected to these aspects, such as bookbinding and the making of specific clothing.

The development, structure and architecture of Najaf are that of a traditional Islamic city within an arid climate. It was walled with a central mosque, or in this case shrine, that dominated the skyline alongside a grand souq and winding pedestrian routes that connected the centre with the periphery and with clear divisions between public spaces and the private spaces of residential quarters (Al-Kubaisy, 2009, pp. 17, 21). The harsh desert climate determined the layout of residential areas – low-rise, high density, courtyard houses, shaded alleyways and specific features such as wind-towers that caught the breeze and channelled it to the rooms below. This kind of organic and irregular form interconnected with numerous routes and alleyways of Najaf is particularly characteristic of Arab and Islamic cities and provides contrast with the shrine and its golden dome (Abid, 2016, p. 89). The old town contains a great souq, medieval mosques and religious schools, and a number of other buildings of historic or architectural importance (Al-Kubaisy, 2009, p. 46).

Najaf, now with a population of about 1.47 million, has, however, suffered much as have all other historic towns and cities from the dramatic growth in the last hundred years or so and from insensitive redevelopment, particularly since the 1950s. In 1954, four major roads were cut through to the shrine and the city wall was demolished and replaced by a ring road (Al-Kubaisy, 2009, p. 46). There have been a number of development plans put forward since the 1950s and are discussed in detail by Abid (2016), Al-Mosawi (2017, p. 224ff) and Falah (2018, p. 18ff), though most have been considered unsuitable or were not implemented for a variety of reasons, not least Iraq's recent conflicts and unsettled situation. Redevelopment is, however, due to be completed soon on a section of the old town leading to the west in front of the shrine (completely demolished in the 1980s–1990s as part of the repression of Shi'a scholars during Saddam Hussein's regime) as it was acquired by the shrine authorities (Imam Ali Network, 2020). In addition, the consultancy Dewan Architects and Engineers' Najaf Urban Renewal Plan has also been recently approved (Dewan, 2020).

Urban cultural heritage conservation and Islamic holy cities

Sustainable urban redevelopment and sustainable urban heritage conservation are closely linked, and it has been understood for some years now that urban cultural heritage has a major role to play in sustainable cities and in the counterbalancing of some of the issues of rapid urbanisation. A key message from the UN's *Post-2015 Report* is that planning and development that integrates culture increases economic opportunities and enhances the quality of life in urbanising areas (UNESCO, UNPF & UNDP, 2015, p. 48). Much thought has gone into the issues of sustainability, heritage and the urban environment in recent years as all of these ideas have changed in the face of the climate crisis, globalisation and the intrinsic threat of neoliberalism, alongside growing urban populations. As the challenges of managing the urban environment have become more complex, it needs to take into account ever-growing and diversifying populations, sustainability and energy use, changes to work patterns and mobility, and the importance of social inclusion in all aspects (Bandarin, 2015, pp. 2–3). Consequently, the ideas behind urban conservation have also had to change as it has become increasingly difficult to protect just the physical elements of a historic city when it is part of a living and constantly changing system. A greater subjectivity to heritage and its conservation and a division between conservation as a science and conservation as an approach have developed (Orbaşli, 2017, p. 161). This has led to a plurality of values that goes beyond, sometimes contradicts, those espoused by heritage professionals and has required accommodation. This has been further emphasised by the growing understanding of the links between sustainability and cultural heritage, and its place and function in peoples every day, as opposed to something only considered as part of, for example, leisure time. There is a growing appreciation within urban conservation for the more emotional and symbolic aspects attributed to the built environment such as the crafts and trades, the local stories, rituals, practices – what has been called the 'everyday heritage' or the 'heritage of home' (Auclair & Fairclough, 2015, p. 4). As a consequence, there is a need for greater flexibility within heritage practice and theory to accommodate these additional attributes, including increased participation that combine with a movement away from the primacy of the physical and a more sustainable idea of heritage providing a framework for everyday life that is able to change and adapt as needed (Auclair & Fairclough, 2015, p. 209).

In relation to sustainable urban conservation in the Middle East, these changes in approach are important as it moves international conservation thinking closer to what may be considered a more Islamic approach where the intangible is of equal importance, where there is no automatic intrinsic value placed on the material, and where there is a wider 'worldview' approach (Mahdy, 2019a, p. 128). One of the principal considerations when it comes to urban cultural heritage conservation and sustainability in the Middle East, in general, is that of the impact of imported western thought and theories not only in relation to urban conservation but also in relation to urban modernisation and planning in Arabic-Islamic historic cities, particularly from the 1950s onwards (Saliba, 2015; Mahdy, 2019b). This is not, of course, a new issue, nor one confined to the Middle East. Changes in recent decades have had to be made to conservation approaches to make them less western-centric and these are reflected in such documents as the Nara Document on Authenticity (ICOMOS, 1994) and the Australia Charter on Places of Cultural Significance (the Burra Charter (ICOMOS, 1981/2013)) but, in general, neither urban planning nor conservation theories and practices have taken into account the specific character and nuances of historic towns and cities of the Middle East nor of the Islamic worldview that is an intrinsic part of their heritage (Mahdy, 2019b). Arguably, such non-specific theories are, therefore, a plausible threat to the sustainability of such urban heritage in their own right because they undermine or do not value some of the aspects that would in fact aid conservation in this part of the world. There is a demonstrable lack of appropriateness of aspects of western conservation theory to Islamic cities such as Najaf, from quite practical aspects, for example, it is not permitted to reuse historic religious buildings for non-religious purposes (Abid, 2016, p. 325), right up to such broader philosophical concepts such as what constitutes integrity. Sites in the Islamic world may meet World Heritage List conditions of integrity but may not be considered so according to Islamic thought because in Islamic worldview integrity cannot be divorced from intangible, including moral, values –attributes which may be usefully argued for in the protection of sites as with the clashes over the removal of historic material in Ancient Thebes, Egypt (Mahdy, 2019a, p. 131). Here, UNESCO's and ICOMOS's concern was with the loss of historic evidence and the portrayal of a single period of the site due to a project carried out by the Egyptian government, but little was said about the local people displaced and who lost their livelihoods due to the project – this should have been factored into the complaints against the project because intangible values and human well-being are in fact considered superior to tangible values when it comes to conservation and heritage management (Mahdy, 2019a, p. 131).

Historically, western theories and practices have also dominated urban development in the Middle East. From the 1950s, modernist urban planning ideas came to dominate. These caused significant damage to western historic cities, leading to the development of the urban conservation movement there, and likewise affected Middle Eastern cities. They did not take into account wider Islamic philosophies nor Islamic artistic traditions and expression, nor the historic developments, traditional structures, climatic responses or the many other aspects that are fundamental to an Islamic urban settlement (Bianca, 2000, pp. 11–12). From the 1950s, redevelopment plans for Baghdad, for example, by western master planners such as Doxiadis Associates, planned for the demolition of the historic areas of the city (Al-Mosawi, 2017, p. 181).

This combination of imported urban planning and conservation theories and practices and their lack of sympathy to a non-western context such as the Middle East has been compounded by a number of truly global problems that affect cities worldwide. Globalisation and neoliberalism, as already mentioned, combined with the rapaciousness so often evident in corporate development, are by far the greatest threats to the survival and sustainability of urban heritage. In this process of modern urban development, urban heritage has been steadily repositioned into a commodity or consumable most obviously for the tourist industry (Orbaşli, 2017, p. 158) or carefully manipulated in the process of aggressive gentrification (Meskell, 2019), threatening to 'disney-fy' or hollow it out in terms of value and relevance. As a resident commented about Dubrovnik, Croatia, which has seen up to a million tourists in its historic city every year,

I would like it to remain what a city basically is – a living city, inhabited by different kinds of people. This is the starting point for everything. Therefore, life should be the most important symbol of cultural identity. Since there is no life, the only thing that remains is the souvenirs.

(quoted in Mišetić & Ursić, 2015, p. 80)

It is the conservation in Islamic holy cities that is most relevant here. As mentioned, after Makkah and Madinah in Saudi Arabia, Najaf is the third most important pilgrim destination for Shi'a Muslims. All three receive millions of visitors at specific times of the year, doubling if not tripling the usual population of these ancient cities; so finding ways to accommodate the pilgrims is an unavoidable consideration. Makkah in particular has been subjected to massive redevelopment over the last decade or so, both to provide for pilgrims and to harness the huge commercial returns of this. It should be noted that there has always been this tension in holy cities between spirituality and the free provision for pilgrims and the trade and speculation of pilgrim-based enterprises (Saliba, 2015, p. 183). How to balance up these considerations with maintaining the tangible and intangible cultural heritage is a complex problem. The necessity to expand all the provisions for worshippers, both spiritual and logistical, has meant the demolition of historic areas to create large open spaces around the shrines, as seen not only in Makkah and Madinah but also in Karbala, Iraq, and of course Najaf. These large plaza-types spaces are not historically a feature of Islamic cities but a feature of western cities (Saliba, 2015, p. 182) and although they provide much-needed space, what they also do is increase the commercial and highly profitable zone around the shrine, but this does not, in fact, help local tradespeople but instead opens it to neoliberal economic practices and globalised businesses such as chain stores (Saliba, 2015, p. 185). Makkah, in particular, has come to be dominated by its extensive redevelopment, as epitomised by the colossal clock tower. While a paper on the sustainability assessment indicators for Makkah demonstrated that of physical aspects, the retaining of archaeological and cultural heritage sites was the second most important sustainability indicator (along with access to the city and improving connectivity between Haj-related sites) (Anisurrahman & Alshuwaikhat, 2019), it has been suggested that as much as 95% of all of Makkah's historical sites have been demolished, ostensibly for religious, specifically Wahhabism doctrinal, reasons of the fear of polytheism and idolatry, but also in the interests of commercialism (Al Jazeera, 2005; Atassi, 2015). It has been argued that the redevelopment of Makkah has been used then to serve these twin purposes of the Wahhabi proselytization mission and of maximising the financial returns of the Haj, to the benefit of, in particular, Saudi financial elites, and with little consideration given to any other concerns (ADHRB, 2015, p. 1). To that end, the destruction has meant not only the destruction of secular and Islamic heritage sites but also the destruction of thousands of homes and traditional small businesses to be replaced by huge luxury hotels, extensive shopping areas and other such facilities for pilgrims, and that this has cost Makkah its history and its spirituality (Batrawy, 2014). Research suggests that the one aspect not considered in the development plans was the spirituality of the place, although it is possible for economic growth and spiritual processes to co-exist as long as the needs of the worshippers are facilitated and respected (Serafi & Fouseki, 2017, pp. 119, 121). It is also argued that while conservation specialists and those concerned with heritage may deplore the destruction caused by the redevelopment of Makkah, the majority of pilgrims are happy with the developments that facilitate their stay and worship at Makkah and are not worried that it came at the cost of the physical heritage (Mahdy, 2019a, p. 130). But these arguments may be qualified. A small survey at Makkah suggested that both local residents and international pilgrims were specifically interested in the religious material culture relating to the spirituality of the place more so than any non-religious heritage, but whereas international visitors were more favourable towards the redevelopments, the local residents were more negative because their sense of attachment is deeper and for the wider city and its heritage, not just that of the central religious core and its surroundings, and also because they were more aware of the economic interests driving the development (Serafi & Fouseki, 2017, pp. 125, 128). It is evident, therefore, that although places such as Makkah are places of great international importance and do beg the question of whether they 'belong' to a wider community than just the local residents (Saliba, 2015, p. 194), it is vital that, as Dr Sami Angawi, instrumental in recording Makkah's heritage, argues, local residents' needs must be considered alongside those of pilgrims (in Serafi & Fouseki, 2017, p. 128). In addressing local residents' needs, therefore, a much broader range of intangible cultural heritage beyond spirituality, such as collective memories, practices and traditions, that are often also closely associated with the built heritage but are rarely considered in urban planning, should also be addressed (Jigyasu, 2015).

Najaf: combining the tangible and the intangible heritage

> It is necessary to keep the heritage character of those cities and places. Yes, it is necessary also to make an extension to the holy shrines to provide the required services to the pilgrims and to do other intellectual, cultural and service activities. But such an extension should not affect the heritage character of cities.
> (Ayatollah Sistani, statement during Friday Prayers 24th October 2010)

Iraq, as is well known, has suffered intensely, including wars, sanctions and occupation, over the last forty or so years, and continues to do so. Historic cities that have suffered such extensive conflict and trauma are particularly vulnerable – their cultural heritage in its broadest sense is particularly at risk in the face of inappropriate or aggressive redevelopment, and at demolition or neglect for political reasons. The redevelopment of central Beirut post-1990 and the problems inherent in that are well-discussed (for example Sandes, 2017, p. 15) and it is likely that the old town of Mosul is similarly threatened as, for example, there has been deliberate demolition of old buildings and the purchase of damaged properties by developers in order to accumulate the space to build commercial high-rise buildings (Ibrahim & Al-Rubaie, 2019; Meskell, 2019, p. 997).

In theory, Iraq's cultural heritage is well protected by law and under the remit of the State Board of Antiquities and Heritage (SBAH), but it is the enforcement of the law that is problematic due to lack of resources, division of responsibilities between different departments and conflicts of interest (Abid, 2016, 47, p. 265). There is also another significant complicating factor when it comes to the management of Iraq's cultural heritage, and one that is a major problem for Iraq as a whole, which is the political system put in place by the US headed Coalition Provisional Authority in 2003. This system is based on consociationalism – a form of democracy where autonomy is given to various ethnic, religious or other groups to manage their own affairs alongside being granted powers within the national government (Younis, 2011, p. 1). Known in Iraq as the *Muhasasa* system, where power has been divided between the main religious endowments and ethnic groups, such as the Kurds, it has caused nothing but sectarian division and constant competition for resources that should be shared. It has undermined the pre-2003 national approach to cultural heritage (Bernhardsson, 2005, pp. 5ff) and has had a severely detrimental effect on Iraq's cultural heritage as each group vies for control and people are awarded positions based on their affiliations rather than their abilities (Abid, 2016, p. 100; Kathem, 2020). As was evident from the work done for a development plan for old Basra, for example, coordination with the religious endowments is essential if their many historic buildings are to be rehabilitated (LADP, 2018, p. 39), but the greatly increased power given to individual religious endowments under the *Muhasasa* system complicates these issues and gives them relative impunity. There is, for example and particularly amongst the dominant Shi'a endowment, a tendency to completely re-build historic religious institutions in a 'pristine' modern and uniform style (Tabbaa & Mervin, 2014, p. 52).

As mentioned, millions of pilgrims visit the Holy Shrine of Imam Ali in Najaf each year, so this has raised the need to both extend the Holy Shrine to facilitate these extra worshippers and also to provide more hotels and other services for them. There have been a number of development plans put forward for Najaf to accommodate this need to expand. In addition, there has been piecemeal development by local property owners that has seen the insensitive insertion of modern buildings and the simultaneous demolition or neglect of historic properties. This is compounded by the needs of growing families and changing lifestyles for people to move out of the old town, leaving many old houses to decline, hampered by Iraq's complicated inheritance and property ownership laws. It has also caused an increase in car traffic as people need to return to the old city to work, and that of pilgrims, causing additional problems such as pollution and the provision of parking and new roads that damage the cohesion and relationships in the old city and its relationship with the shrine (Abid, 2016 p. 91). The value of Najaf's heritage and these threats to it, and what may be done to alleviate them, were discussed at two conferences in Najaf in 2012 and 2013 and resulted in the *Najaf Charter* (IEEF, 2013).

In addition to its physical fabric, Najaf has a wealth of intangible heritage. The presence of Wadi al-Salam and the shrine and its associated mosques, madrasas and libraries, as well as centuries of attendance by religious pilgrims and scholars, have given rise to specialised crafts such as bookbinding, the making of specific religious clothing such as the highly valued Najaf abaya, jewellery such as the Najaf ring and the sweets that the city is also famous for. These are in addition to the numerous other crafts to be found in the Souq Kabir (Grand Bazaar) and old town that range from perfume-making to traditional bread-making that fulfil the requirements of local residents.

Despite Ayatollah Sistani's instructions quoted above, and the aims of the Najaf Charter (IEEF, 2013) that recognise the importance of protecting Najaf, and indeed the urban heritage of Iraq generally, as part of protecting the identity of the city, the development plans for the old city of Najaf to date make few concessions to either the built or intangible cultural heritage. It can be stated that the urban identity of the city is threatened by the heights of buildings constructed in the last decade. Such high-rise buildings now surround the golden dome of the shrine from all directions undermining its domination of the city's skyline. Before 2003, the urban planning regulations limited the heights of buildings in the old city of Najaf to the height of the shrine's fence, which is about 11 metres. However, due to increasing needs and the pressure from over-commercialisation from one side and the poor control of authorities from the other side, the buildings started to get higher and higher. Buildings with more than three stories in height threaten the Outstanding Universal Value of Wadi al-Salam Cemetery, which was added to the World Heritage Tentative List in 2011. The cultural value of the cemetery was based on its closeness to the shrine and the visual axis extends between the golden dome and its graves. The visual axis can be considered as the main attribute to Criterion III of UNESCO World Heritage Convention of 1972, where the cemetery is the outstanding testimony to the cultural traditions of Muslim people for 1,400 years.

It is not impossible to balance tourism and the needs of religious sites of historical and architectural importance, nor is the way that Makkah has been developed the only way forward. Levi and Kocher (2009, p. 1), using sites in California and Thailand, showed how authentic experiences of sacred places and the spiritual culture of a region can promote sustainable tourism and benefit both the tourists and the community. Elrasoul (2016) in her work in Egypt recommended developing a new applied approach of conservation that takes the interaction between citizens and the city heritage zones into consideration. She concluded that the commercialization concept should be used to motivate, rather than prevent, and to encourage inhabitants to maintain heritage. In addition, the application of sustainable urban development ideas and practices in relation to Iraq's historic cities has also been examined (Al-Mosawi, 2017) and seen in work in, for example, Sulaimaniyah, Kurdistan-Iraq (Amin & Adu-Ampong, 2016) and in Baghdad (Hussein & Abaas, 2013). In a local area development plan drawn up for Old Basra, the sustainable development goals are addressed directly (LADP, 2018, p. 1). Al-Mosawi (2017) has also developed an assessment approach to urban regeneration project proposals to be conducted in Iraqi historic cities. In relation to Najaf, Abid (2016) addressed questions about the significance of Najaf's heritage and city development policies that affect heritage conservation. One of the other main problems he identified, and one that affects heritage conservation generally in Iraq, is the limited understanding of conservation management, the few clear policy implementations, and the overlap of authorities' mandates in decision-making which undermines attempts to prioritise cultural heritage and sustainability. Furthermore, historic cities such as Najaf that attract a huge number of pilgrims are prone to suffer the same problems that World Heritage cities have with large numbers of tourists and the difficulty of balancing the requirements of the latter with those of residents and with development needs of the city concerned (Pendlebury et al., 2009).

As mentioned previously, another problem is the factoring of intangible cultural heritage into the conservation and development of historic cities. Tangible and intangible cultural heritage are interrelated. The intangible heritage should be regarded as the larger framework within which tangible heritage takes shape and significance. In addition, there is a connected dynamism and evolution between tangible and intangible – the practice of the latter helps to ensure the continuing relevance of the former (Jigyasu, 2015, p. 134). Bouchenaki (2003) considered several examples from properties inscribed on the World Heritage List to show that both types of heritage are mixed and carry meanings and the embedded memory of humanity. One of those examples is Jemaa el-Fna Square, Marrakech, where many elements of intangible cultural heritage are practised in the main open space beside the old mosque. Such popular activities give spirit to the place and define the heritage in a living way. The other example considered by Bouchenaki (2003) is the Jongmyo Shrine in the Republic of Korea, where the ancestral rites and ritual music that characterize the spirit of place are still practised.

Ito (2003) stated that intangible culture provides the background of tangible cultural heritage. Using the example of the knowledge and techniques used in Japanese wooden vernacular architecture, he mentioned that there are two types of relationships between tangible and intangible cultures. The first is that the intangible cultural heritage is materialized by the tangible cultural heritage. He used an example of sacred buildings, where such buildings represent the relationship between religion and architecture, but this relationship is not a clear one-to-one correspondence in most cases. The other type of relationship is that the intangible cultural heritage plays a vital role in the establishment of tangible cultural heritage. The research briefly discusses the idea of the correlation between spirit and matter or meaning/value and material from the fact that the strength of urban output is the product of its moral dimension. In a similar vein, Wells (2007) examined the history of conservation charters and the Western bias towards the material that has subsequently been challenged by the Burra Charter and Nara Document on Authenticity, and makes the case for interpretation via a cultural lens that takes in the complexities and interrelationship of the material and the spiritual. The historic centre of Najaf with its society and culture represents a treasure of knowledge and culture, a centre of science and human development – Najaf has been a centre of learning for at least a thousand years (Heern, 2017, p. 11). Hence, there is a need to keep all these values through the revitalization of the historic part of the city but with the threat of insensitive and destructive development, there is a strong need to clearly identify and emphasise the many heritage attributes of Najaf, particularly if its historic old town is to continue as a living city functioning for its residents and not just for visiting pilgrims.

As mentioned previously, western conservation philosophies and theories have considered the physical heritage as intrinsic; there has been little documentation of intangible heritage in cities (Jigyasu, 2015, p. 139). In Najaf, a small survey was carried out as part of a development plan for the Holy Shrine area by the consultant Memar Haram in 2015, of residents of the historic centre to help understand levels of value and importance attached to the heritage was done. The main results showed that the majority of residents wish to stay in the old part of the city despite their suffering from many problems such as decaying housing stock, the difficulties of modernising or enlarging properties to accommodate growing families, the driving up of costs due to the pilgrim industry and other such problems. The key finding was that one of the main reasons that they wish to stay is because the old city is part of their memories (Attiyah et al., 2019).

Following on from this survey, a pilot survey was conducted to explore the potential elements of intangible cultural heritage that are relevant to the sense of place (Attiyah et al., 2019). This formally identified the following intangible elements, practices and practitioners:

- religious students study in a traditional way in specific places
- religious students maintain the traditional dress code
- pilgrims visit the holy shrine
- pilgrims buy handicrafts from shops near the shrine
- craftsmen use traditional ways to bind books in shops near religious schools
- craftsmen make traditional forms and shoes for clerics and religious students
- craftsmen make rings of Dur Al Najaf and inscribe letters and images on them
- craftsmen make traditional sweets

The eight elements of intangible cultural heritage are related to three communities: the pilgrims, religious students and workers/ craftsmen. Moreover, it is seen clearly that elements are interrelated and practised in the same place or nearby. For example, religious students are dressed in the traditional form and shoes manufactured and sold at shops close to their schools. Most of those students buy their books from shops nearby and many books are old so need rebinding many times. Another example is the traditions of the pilgrims when they visit the Holy Shrine and the Cemetery. The pilgrims usually buy handicrafts and traditional sweets after completing their rituals. Combined workshops and shops are used to manufacture and sell such things as traditional clothing, jewellery and sweets near the shrine.

The pilot study (Attiyah et al., 2019) demonstrated that not only is place important for the practice of traditional customs, but also there are interrelations between these practices, places and the people who carry them out or require the services they provide. To further emphasise and promote these connections and to encourage greater interest in Najaf's historic urban fabric, it has also been suggested that a community museum be established (Attiyah et al., 2019).

Conclusion

In conclusion, Najaf's unique heritage is under threat not only due to problems emanating from Iraq's recent history and ongoing issues, but also from redevelopment of the shrine and the need to provide for growing pilgrim numbers, often at the expense of the needs of the residents. These numbers have not been much affected by the ongoing (at time of writing) Covid-19 pandemic. About 12 million attended Arbaeen, the only difference being that pilgrims from outside Iraq were not permitted to attend. Unlike Makkah where the number of pilgrims can be limited, this was not possible at Najaf and despite official attempts, the WHO requirements for holding such gatherings during the pandemic were not followed (Calabrese, 2020; Davison & Al-Deen, 2020). Consequently, the demands of mass pilgrimage on Najaf remains very high and is unlikely to change.

In order to protect Najaf's cultural heritage and ensure its appreciation and relevance to both residents and visitors, it can be stated that a holistic view of heritage and exploring the interrelation between tangible and intangible forms should be taken. In particular, the residents' needs and interests, and the heritage they value, which provides livelihoods, should be a priority. This will also enhance the revitalization of historic cities like Najaf as communities belonging to the culture of such cities will strive to protect their heritage for two main reasons. The first being the place and its characteristics are part of the communities' cultural identity. The second reason is that their livelihood is traditional crafts and activities that they inherited from their ancestors. Sustainable development of cultural heritage will mean in such cases how to develop those traditional activities and promote the products.

Acknowledgements

Many thanks to Dr Hossam Mahdy, Dr Yuri Stoyanov and Guillaume Dreyfuss for their comments on the paper.

References

Abid, S. K. (2016). *An Examination of heritage protection and conservation practices in the pilgrimage city of Najaf.* Unpublished PhD thesis. Sheffield: University of Sheffield.

ADHRB (2015). *Mapping the Saudi State: Chapter 7: The destruction of religious and cultural sites.* Retrieved from: https://www.adhrb.org/2015/10/mapping-the-saudi-state-chapter-7-the-destruction-of-religious-and-cultural-sites/ (Accessed on 31 December 2020).

Al Jazeera (2005). *Makka's historic sites under threat.* Retrieved from: https://www.aljazeera.com/archive/2005/07/2008410115548684950.html (Accessed on 4 June 2020).

Al-Janabi, H. H. (2012, May 7). *Al-Najaf city of churches and monasteries for Christians 1400 years ago.* Retrieved from: https://ishtartv.com/en/viewarticle, 36006.html (Accessed on 13 January 2021).

Al-Kubaisy, F. (2009). *Najaf: The architectural and urban heritage of Iraq's holiest city.* Retrieved from: www.booksurge.com (Accessed on 03 May 2021).

Al-Mosawi, A. Q. (2017). *Towards sustainable urban design strategies for historic city centres in Iraq: Development of an assessment approach for urban regeneration projects.* unpublished DEng thesis. Dortmund: TU Dortmund University.

Amin, H. M. T., & Adu-Ampong, E. A. (2016). Challenges to urban cultural heritage conservation and management in the historic centre of Sulaimaniyah, Kurdistan – Iraq. *Journal of Cultural Heritage Management and Sustainable Development, 6*(3), 1–26. http://dx.doi.org/10.1108/JCHMSD-03-2016-0019.

Anisurrahman, M., & Alshuwaikhat, H. M. (2019). Determining sustainability assessment indicators for the Holy City of Makkah, Saudi Arabia. *Arabian Journal for Science and Engineering, 44,* 5165–5178. https://doi.org/10.1007/s13369-019-03772-3.

Arbaeen Encyclopedia (2020). *Why walking to Karbala in the day of Arbaeen?* Retrieved from: https://www.c-karbala.com/english/Al-Arbaeen-Encyclopedia/4567 (Accessed on 4 January 2021).

Atassi, B. (2015). The man who dreams of old Mecca. *Al Jazeera.* Retrieved from: https://interactive.aljazeera.com/aje/2015/man_who_dreams_of_old_mecca/; (Accessed on 4 June 2020).

Attiyah, A. N., Jabbar, R., & Bnayan, A. (2019). Holistic view of cultural heritage in historic centre of Najaf city. Unpublished paper presented at the Second International Conference for Research and Protection on the Archaeological Heritage of Iraq 28–29 April 2019. Retrieved from: http://2icrpahi.uokufa.edu.iq/.

Auclair E., & Fairclough, G. (2015). *Theory and practice in heritage and sustainability: Between past and future.* New York and London: Routledge.

Bandarin, F. (2015). Introduction: Urban conservation and the end of planning. In F. Bandarin & R. van Oers (Eds.), *Reconnecting the city: The historic urban landscape approach and the future of urban heritage* (pp. 1–16). Chichester, West Sussex: John Wiley & Sons, Ltd.

Batrawy, A. (2014, October 1). 'Mecca-hattan': Islamic's holiest city is being overhauled, and people are furious. *Business Insider.* Retrieved from: https://www.businessinsider.com/mecca-is-being-overhauled-and-people-are-furious-2014-10?r=US&IR=T (Accessed on 31 December 2020).

Bernhardsson, M. T. (2005). *Reclaiming a plundered past: Archaeology and nation building in modern Iraq.* Austin: University of Texas Press.

Bianca, S. (2000). *Urban form in the Arab World: Past and present.* London: Thames and Hudson.

Bouchenaki, M. (2003). The interdependency of the tangible and intangible cultural heritage. *ICOMOS 14th General Assembly and Scientific Symposium, Place – memory – meaning: Preserving intangible values in monuments and sites.* Retrieved from: http://openarchive.icomos.org/id/eprint/468/ (Accessed on 03 May 2021).

Calabrese, J. (2020, December 8). Iraq's fragile state in the time of Covid-19. *Middle East Institute.* Retrieved from: https://www.mei.edu/publications/iraqs-fragile-state-time-covid-19 (Accessed on 4 January 2021).

Davison, J., & Al-Deen, A. D. (2020, October 20). Shi'ite pilgrimage fans fears of spreading Covid-19 in Iraq. *Reuters.* Retrieved from: https://uk.reuters.com/article/us-health-coronavirus-iraq-arbaeen-idUKKBN26T237 (Accessed on 4 January 2021).

Dewan Architects+Engineers (2020). *Najaf Urban Renewal.* Retrieved from: https://www.dewan-architects.com/work/najaf-urban-renewal (Accessed on 06 July 2020).

Elrasoul, R. (2016). City architectural heritage revival: The need of a new applied approach, *Procedia - Social and Behavioural Sciences, 225,* 216–225. https://doi.org/10.1016/j.sbspro.2016.06.021.

Falah, S. M. (2018). *The Shrine that consumed its town: The role of religion and politics in reshaping the Iraqi city of Najaf.* Unpublished PhD thesis. Cincinnati: University of Cincinnati.

Heern, Z. M. (2017). One thousand years of Islamic education in Najaf: Myth and history of the Shiʻi Ḥawza. *Iranian Studies, 50*(3), 415–438. http://dx.doi.org/10.1080/00210862.2017.1285486.

Hussein, W. A. S., & Abaas, Z. R. (2013). Effectiveness of sustainable urban projects in the city of Baghdad: A comparative study of sustainable development projects in Adhamiya and Kadhimiya District. *Iraqi Journal of Architecture and Planning, 12*(1), 44–61.

Ibrahim, A., & Al-Rubaie, A. (2019, July 14). Modern revamp threatens to raze riverfront in Mosul's Old City. *Al Jazeera.* Retrieved from: https://www.aljazeera.com/indepth/features/modern-revamp-threatens-raze-riverfront-mosul-city-190713100636619.html (Accessed on 4 June 2020).

ICOMOS (1994). *Nara Document on Authenticity* Japan. Retrieved from: https://www.icomos.org/en/charters-and-other-doctrinal-texts (Accessed on 08 July 2020).

ICOMOS (2013). *The Australia ICOMOS Charter for the Conservation of Places of Cultural Significance - (The Burra Charter)* (Australia ICOMOS) - 1981, updated in 2013. Retrieved from: https://www.icomos.org/en/charters-and-other-doctrinal-texts (Accessed on 08 July 2020).

IEEF (2013). *Najaf charter for the conservation, restoration and rehabilitation of historic cities, urban areas and historic – heritage monuments.* Retrieved from: http://www.pchiic.ieefoundation.org/ (Accessed on 08 July 2020).

Imam Ali Network (2020). Fatima Courtyard Project. Retrieved from: https://www.imamali.net/?id=2464; (Accessed on 06 July 2020).

Ito, N. (2003). Intangible cultural heritage involved in tangible cultural heritage. *ICOMOS 14th General Assembly and Scientific Symposium, Place – Memory – Meaning: Preserving Intangible Values in Monuments and Sites.* Retrieved from: http://openarchive.icomos.org/id/eprint/484/ (Accessed on 05 May 2021).

Jigyasu, R. (2015). The intangible dimension of urban heritage. In F. Bandarin & R. van Oers (Eds.), *Reconnecting the city: The historic urban landscape approach and the future of urban heritage* (pp. 129–159). Chichester: John Wiley & Sons, Ltd.

Kathem, M. (2020). Cultural (dis)continuity, political trajectories and the state in post – 2003 Iraq, *International Journal of Heritage Studies, 26*(2), 163–177. https://doi.org/10.1080/13527258.2019.1620836.

Levi, D., & Kocher, S. (2009). Understanding tourism at heritage religious sites. *Focus, 6,* 1. https://doi.org/10.15368/focus.2009v6n1.2.

Local Area Development Programme (LADP) (2018). *Old Basra conservation and development plan.* Brussels: UN-Habitat.

Mahdy, H. (2019a). Is conservation of cultural heritage halal? Perspectives on heritage values rooted in Arabic-Islamic traditions. In E. Avrami, S. McDonald, R. Mason & D. Myers (Eds.), *Values in heritage management: Emerging approaches and research directions* (pp. 127–141). Los Angeles: The Getty Conservation Institute.

Mahdy, H. (2019b). Arab values: Towards regional guidelines for ICOMOS doctrinal documents in Arab countries. *Conference Paper: ICOMOS -CIAV&ISCEAH 2019 Joint Annual Meeting & International Conference on Vernacular & Earthen Architecture towards Local Development,* Pingyao, China. Retrieved from: https://independent.academia.edu/HossamMahdy (Accessed on 05 May 2021).

Meskell, L. (2019). Heritage, gentrification, participation: Remaking urban landscapes in the name of culture and historic preservation. *International Journal of Heritage Studies, 25* (9), 996–998. https://doi.org/10.1080/13527258.2018.1542334.

Mišetić, A., & Ursić, S. (2015). Remembering cities: The role of memory in the culturally sustainable development of Dubrovnik. In E. Auclair & G. Fairclough (Eds.), *Theory and practice in heritage and sustainability: Between past and future* (pp. 69–83). New York & London: Routledge.

Nakash, Y. (1994). *The Shi'is of Iraq.* Princeton, NJ & Chichester: Princeton University Press.

Orbaşli, A. (2017). Conservation theory in the twenty-first century: Slow evolution or a paradigm shift? *Journal of Architectural Conservation*, *23*(3), 157–170. http://dx.doi.org/10.1080/13556207.2017.1368187.

Pendlebury, J., Short, M., & While, A. (2009). Urban World Heritage cities and the problem of authenticity. *Cities*, *26*(6), 349–358. https://doi.org/10.1016/j.cities.2009.09.003.

Saliba, R. (2015). Sites of worship: From Mecca to Karbala, the reconciliation of pilgrimage, speculation and infrastructure. In R. Saliba (Ed.), *Urban design in the Arab World: Re-conceptualizing boundaries* (pp. 177–197) London: Routledge.

Sandes, C. A. (2017). Remembering Beirut: Lessons for archaeology and (post-) conflict urban redevelopment in Aleppo. *Ex-Novo Journal of Archaeology*, *2*, 5–32.

Serafi, S. A., & Fouseki, K. (2017). Heritage conservation and sustainable development in sacred places: Towards a new approach. In M.-T. Albert, F. Bandarin & A. Pereira Roders (Eds.), *Going beyond, perceptions of sustainability in heritage studies No. 2* (pp. 115–132). Cham, Switzerland: Springer.

Tabbaa, Y., & Mervin, S. (2014). *Najaf the gate of wisdom: History, heritage and significance of the Holy City of the Shi'a*. Paris & Iraq: UNESCO.

UNESCO, United Nations Population Fund & United Nations Development Programme (2015). *Post-2015 dialogues on culture and development*. Paris: UN.

Wells, J. C. (2007). The plurality of truth in culture, context, and heritage: A (mostly) post-structuralist analysis of urban conservation charters. *City & Time*, *3*(2), 1–14. Retrieved from: http://www.ct.ceci-br.org.

Younis, N. (2011). Set up to fail: Consocational political structures in post-war Iraq, 2003–2010. *Contemporary Arab Affairs*, *4*(1), 1–18. http://dx.doi.org/10.1080/17550912.2011.543780.

16 Sustaining cultural heritage in post-conflict Syria

The case of Aleppo

Hiba Alkhalaf

Introduction

The unprecedented rate of destruction and the continuous targeting of cultural heritage in cities in armed conflict has been one of the challenges facing many countries in the Middle East and North Africa (MENA) region since 2011. In Syria, the largest cities, like Homs and Aleppo, were severely hit, with more than 50% of destruction taking place within (UNESCO, 2018a). After a decade of the conflict, the fate of the country's destroyed cities and its cultural heritage has become the subject of national and international discussion. Cultural heritage includes tangible elements such as monuments, World Heritage sites (WHS), listed buildings, designated areas, archaeological sites and urban heritage as well as intangible elements like traditional crafts, social and economic fabric and the local community. Exploring the sustainable post-conflict heritage reconstruction requires both the theoretical and practice-based dimensions of the topic to be addressed. This chapter, therefore, is structured at three different levels; (1) Macro-level: to clarify the context of heritage during the conflict; (2) Meso level: to understand the relevant practical experience; and (3) Micro-level: to assess how these different dimensions operate in reality. The Macro-level includes national and international literature, which informs the design of this chapter's theoretical framework. The Meso level involves a review of the evolving national practices in post-conflict reconstruction. The Micro-Level will explore the case of Aleppo, focusing on the themes of heritage, values, narratives and stakeholders and local community role in post-conflict reconstruction.

The notion of cultural violence is relevant to heritage studies, as reconstruction after conflict can itself be violent through the advancement of specific narratives by dominant powers (Viejo-Rose, 2011b). Cultural heritage includes tangible and intangible elements, which, when taken together, are a symbiosis of the physical structure, cultural practices, oral traditions, traditional crafts, techniques and knowledge including dance, stories, crafts, designs and even digital heritage. Incorporating or excluding these narratives should illustrate the future identity of a post-conflict society (Viejo-Rose, 2013). Unpacking these narratives created by dominant powers, therefore, calls for culturally sensitive tools to engage the local community in the process of reshaping the post-conflict society. Doing so contributes to achieving the UN's sustainable development goals (SDG), especially number 11 – making cities inclusive, safe, resilient and sustainable. This goal includes the aim to protect and safeguard the world's cultural and natural heritage – not only the tangible elements.

The international community has developed several tools and frameworks since the 1970s in response to disasters and armed conflict in historic cities. A disconnect between place-based and people-centred strategies in city reconstruction and recovery efforts

DOI: 10.4324/9781003038955-20

can still be observed. However, culture has been given little consideration in these processes (UNESCO, 2018a). Recently, UNESCO proposed a framework, Culture in City Reconstruction and Recovery, CURE 2018, that mainstreams culture into post-crisis city reconstruction and recovery, integrating people-centred and place-based policies. Tools such as CURE can help define the relationships between place, people and process, to bind time and place and keep the past alive in the future (Alkhalaf, 2018).

However, 'place' can take different hard and soft physical representations. For example, Punter (1991) and Canter (1977) shaped the most debated place elements through physical structure, activity and meaning. These elements are interlinked and developed throughout the time to create a unique 'sense of place'. In literature, the term 'sense of place' is often used to refer to slightly different conceptualisations of place (Shamai & Ilatov, 2005). It can be related to the 'character' or 'local distinctiveness' of a specific place. It can also emphasise how people experience, use and understand place (Graham et al., 2009). Thus, the term 'sense of place' connects people's experience of a place in the context of cultural heritage. However, historic places are part of the ever-changing environment and might experience significant changes following a natural disaster or armed conflict. Sustaining our heritage requires unpacking the complex relationship between people's experience and place; that requires both place-based and people-centred approaches. Using both approaches facilitates the assessment of tangible and intangible values of a given place as part of the overall management and rebuilding of urban heritage areas.

Based on the previous argument, post-conflict reconstruction efforts need to be programmed to serve as mechanisms for restoring/recreating trust between citizens and the state, by fostering empowerment and inclusion of citizens in decision making and through implementation and monitoring of the reconstruction program at national and local levels (World Bank, 2018, p. 174). The challenge here is to develop instruments that consider the recovery of the post-conflict society, reconstructing the city and cultural heritage as interlinked. Each is dependent on the other. These instruments would also align any reconstruction plans with international principles such as Build Back Better (BBB) and Sustainable Development Goals (SDGs) (Khalaf, 2020). There is, thus, a need to articulate a balanced and comprehensive approach, which would guide both national authorities and international development actors in addressing this challenge. This approach would promote heritage reconstruction to sustain the cultural significance of historic urban landscapes and draw on the potential of cultural and natural heritage to be a catalyst for rapid social and economic recovery (World Bank, 2018, p. 177). Therefore, cultural heritage reconstruction should be an integral part of the city's overall urban revitalisation, development, and planning framework.

This chapter argues that using the concept *sense of place* for historic areas 'as-a-tool' can help define cultural significance and overcome confessional divides and psychological barriers around identity in post-conflict cities. This tool provides an opportunity to negotiate and shape the city's future based on its collective values, centred around physical structures, activities and meanings. Doing so helps build shared memories and places' values that are a starting point for dialogue between people in post-conflict cities. Mapping out the tangible and intangible values, including the designation ones for WHS, can be the first step for peacebuilding and recovery in post-conflict cities. This mapping helps to overcome contested places' conflicts based on moderating the shared values and attributes related to the area from various users' and stakeholders' perspectives. Different methods and means of mapping these values could be used, such as community consultation and stakeholders' workshops and the internet and social media. Using various methods would further our understanding of the diversity of heritage places and attributes within the city and inform decision-making processes in heritage planning and management on both experts' and users' heritage understanding (Ginzarly et al., 2019).

The context of the study

The ongoing conflict in Syria since 2011 has destroyed tangible and intangible assets and left deep marks on the country's social fabric, culture, everyday practices and collective memories. The challenging life conditions have resulted in many internally displaced people and refugees that have changed the country's demographic fabric – half of the pre-war population is now displaced (Overton & Dathan, 2019). With the recent advancement of Syria's political situation and the conflict after a decade, more debates are taking place about the refugees' return and post-conflict reconstruction by the international community – especially countries hosting Syrian refugees.

Reconstruction has been traditionally regarded as a functional process, prioritising urgent needs, including emergency housing for victims of bombings and displaced populations as well as infrastructural recovery (Barakat, 2005). Much of the current attention in Syria focuses on rebuilding historic landmarks and parts of historic cities, as is the case in the reconstruction of the Old City of Aleppo. These choices are driven by political and economic considerations in reconstruction, while cultural and social damage is often neglected. Unlike the scenario in Mosul–Iraq that had a similar devastating impact caused by the conflict. In 2018, in cooperation with the Government of Iraq, UNESCO launched the international initiative 'Revive the Spirit of Mosul' to remedy its current state of destruction. This initiative was admitted internationally and nationally as not only about reconstructing heritage sites but also about empowering the population as agents of change involved in rebuilding their city through culture and education. Heritage reconstruction in conflict areas can have a negative impact on the rebuilding of society in some cases, as it may become an example of war 'through other means'.[1] Therefore, reconstruction's cultural and societal effects need to be understood and guided in Syria by local, national and international stakeholders.

To this end, terms such as post-conflict reconstruction, recovery, rehabilitation, building back better and nation-building are used interchangeably, and differentiating the intention and meaning is very difficult (Barakat, 2005, p.572). In this chapter, the definition of 'reconstruction' is adapted from UNISDR[2] as 'the medium- and long-term rebuilding and sustainable restoration of resilient infrastructure, services, housing, facilities and livelihoods required for the full functioning of a community or a society affected by a disaster' (Saya et al., 2017). The recent CURE framework by UNESCO also adopted this term. 'Reconstruction' is, therefore, associated with *recovery*, which involves the restoring or improving livelihoods and health and economic, physical, social, cultural and environmental assets, systems and activities of the disaster-affected community. Both definitions align with sustainable development principles and 'build back better' to avoid or reduce future disaster risk (UNESCO, 2018a, p. 12).

There are several challenges to post-conflict reconstruction related to the designated heritage and damage and value assessment. For instance, the term "reconstruction" itself, in the World Heritage context, is understood as a technical process for the restitution of destroyed or severely damaged physical assets and infrastructure, following armed conflict or a disaster (UNESCO, 2018b, p. 3). Reconstruction in the World Heritage system is vague because international heritage doctrine rests on unclear terminology (Khalaf, 2017, p. 264). Academics focus on the loss of heritage and its values that are solely defined professionally. In contrast, in the political context, the collateral damage affecting cultural heritage is not yet a central factor in evaluating armed conflict or social unrest, let alone in the prosecution of war crimes (Higueras, 2013, p. 94). These events had an impact on the meaning of heritage and identity. Therefore, looking at reconstruction through the lens of heritage reinforces the notion that heritage is not just the management and conservation of resources, but rather the active 'construction and negotiation of meaning through remembering' (Smith, 2006). However, after conflict, heritage reconstruction is often thought to follow a system of values in the built environment – driven by donors, international aid or local initiatives. The ability to manifest such values is at the same time an expression of power since it takes resources (capital, knowledge and skills, human labour, materials) to rebuild the environment (Viejo-Rose, 2013).

It is crucial to consider that the destruction of cities in Syria, which are a cultural construct, not only represents the destruction of the urban fabric but also marks an end to the set of socio-economic and cultural relations that had sustained and structured the character of the city. The collective memory of the people played a significant role in shaping the city's character (Rossi, 1982). New memories of the Syrian people are being formed in relation to violence, destruction and displacement, especially for a new generation born and raised during the conflict, in displacement or refugee camps. This generation has had no chance to create a positive attachment to their cities and heritage by generating good memories away from destruction and violence. Many studies had explored the impact of armed conflict and displacement on cultural heritage, mostly intangible heritage (Chatelard, 2017). Scholars can only guess the consequences the destruction of irreplaceable cultural treasures will have on the Syrian cultural identity for the young generation (Gibbons, 2017, p. 60). Simultaneously, however, the cycle of loss and destruction has created new memories and practices that emerged during the conflict and added new layers to their cultural heritage, both tangible and intangible.

This chapter focuses on the phases between destruction and reconstruction of the historic place, exploring these new layers of cultural heritage created during the conflict. The principal case study is the Ancient City of Aleppo in Syria, a World Heritage Site where post-conflict reconstruction of historic place as part of the peacebuilding process can be explored around the theme of 'dignity' within its socio-cultural context.

Aleppo (*Halab*)

Until the beginning of the Syrian conflict in 2011, Aleppo was Syria's economic capital with an estimated 3 million inhabitants (UN-Habitat, 2014). The Ancient City of Aleppo is one of the six cultural UNESCO World Heritage Sites (WHS) in Syria; it spans a total land area of 364 hectares (Figure 16.1). It was inscribed as a WHS in 1983 but was added to the list of WHS in Danger since 2013. During the last decade, from July 2012, the historic quarters of the city were the areas where the Syrian Regime and opposition forces clashed, in what was known as the Battle of Aleppo 2012–2016, causing substantial destruction to buildings and infrastructure. As a result, Aleppo was divided into two parts: an eastern one under rebel's control, and a western under the Regime's control, until the end of 2016 (UNESCO, 2017b).

The Syrian government took control of the city by the end of the year 2016. Among Aleppo's worst-hit areas was the UNESCO World Heritage-listed old city with its 12th-century Umayyad Mosque and the Al-Madina Souq that was affected by a fire in September 2012. It was reported that between 700 and 1,000 shops were destroyed by the fire caused by shooting and shelling. The Citadel of Aleppo was damaged during this battle. The damage was assessed using satellite images and shows that more than 10% of the historic buildings of Aleppo have been destroyed during the five years of fighting with high levels of damage to the housing and infrastructure that surrounds the historic buildings (UNESCO & UNITAR, 2018). By 2017, the city has at least 15

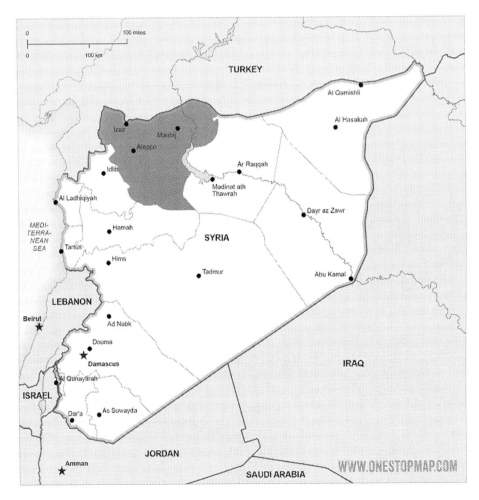

Figure 16.1 A map of Syria showing the location of the city of Aleppo in the North.
Source: https://www.onestopmap.com/syria/syria-21/.

million tons of rubble created by the armed conflict (Overton & Dathan, 2019). Despite that, the international community predicted Aleppo's reconstruction to occur at a rapid pace to meet the urgent needs of the inhabitants (UNESCO & UNITAR, 2018). Now, the Aga Khan Trust and Syrian authority is leading a pilot reconstruction project in the WHS, and working on a five-year masterplan, an advisory document for the Syrian authorities, which champions conserving cultural heritage as a catalyst for Aleppo's social and economic recovery (McGivern, 2018, July 10). These efforts are aligned with the UNESCO's *Actions for Recovery of the Ancient City of Aleppo,* which emerged from the UNESCO Office's meeting in Beirut in March 2017 immediately after the end of the battle in Aleppo (UNESCO, 2017a).

However, the mixed social and cultural community of Aleppo is one of the main features of the city, and its identity and should be integral to reconstruction efforts. Due to Aleppo's strategic location in the north and its rich history, the city was cosmopolitan, multi-cultural, multi-ethnic and multi-religious, with a blend of Arabs, Kurds, Turks, Turkmen, Persians, Circassians, Jews, Armenians and Europeans. Many were refugees, of various Christian and Muslim denominations, fleeing persecution in their homelands. The main challenge here is that the near demise of centuries of multi-cultural co-existence in Aleppo is a worrying sign for the future of mixed communities in Syria (Nallu, 2017). On the other hand, there is a pressing need for 'reconstruction' to accommodate the return of refugees and displaced people. The unplanned return might put more pressure on the city's urban heritage; however, as 'rushed reconstruction often transgresses the delicate process of social healing' (Higueras, 2013, p. 91).

The complexities of the relationship between post-conflict proposals for reconstruction, heritage and identity should be recognised in Aleppo. However, these proposals show how little we understand about its nature and needs. For example, the 2004 masterplan of Aleppo was adopted by the city council in 2018 without taking into account the changes caused by the conflict to local conditions, needs and capacities.[3] Besides, there is an absence of a comprehensive, integrated management plan or a master plan, which has led to different priorities between the official authorities (governorate, antiquities, Awqaf and local community), despite periodic meetings and communication between them (DGAM, 2020). Looking back at the lessons learnt in Iraq, Barakat (2007) argues that reconstruction in Iraq should not be done as reactivation of development, as pre-conflict proposals may address the root causes of conflict to establish an environment of peaceful interaction and a shared vision of the future.

The current practice in Syria raises concerns related to understanding the complex cultural and social change during post-conflict reconstruction, as reconstruction should include efforts towards physical rebuilding and the rebuilding of society. Based on the *UN Post-conflict Recovery,* Aleppo's efforts should be providing security for all citizens and reforming legal and political institutions to meet the new needs (World Bank, 2018). These efforts also include revitalising economic and social structures, assisting refugees' return, peacebuilding and facilitating community participation. This can be achieved by following a local and participatory reconstruction planning process, which is community-owned, driven and consulted on. The process will help rebuild vital trust between the community and authorities and connect divided communities and neighbourhoods again. The challenge remains, how applicable is this argument to the real-world situation in Aleppo? How to assess the cultural and social change of the city's cultural heritage?

New layers of cultural heritage during the conflict

The values that people assign to places are not static; they change over time, including during conflict and political instability. Almost half of Syria's population is either internally displaced or refugees in various countries across the world (Overton & Dathan, 2019). This leads to a debate about the reconstruction of collective memory and identity loss, especially for the new generation. Bevan (2016) has highlighted the long-term impact of conflict and displacement on society's memory and identity "to lose all that is familiar —the destruction of one's environment— can mean a disorienting exile from the memories they have invoked. It is the threat of a loss to one's collective identity and the secure continuity of those identities" (p. 24). It is equally important to understand how heritage has been mobilised in wartime discourse to sow division, which is at the crux of the destruction (Viejo-Rose, 2011b, p. 54). For instance, the Umayyad Mosque, historic souk, and adjoining districts were transformed into the front line during the war, featuring trenches lined with sandbags, fortified tunnels, sniper emplacements and near-daily shelling. During the conflict, these new uses have transformed heritage, added new meanings and even engendered new heritage.

Recent studies that connected heritage to the conflict have argued that conflicts can generate new cultural heritage (Sørensen & Viejo-Rose, 2015). As the cycle of destruction, loss and rebuilding is part of our current heritage (Harrison, 2013), this constant revision and change of heritage may be both sources and outcomes of social conflict (Ashworth & Graham, 2012). This raises significant questions in Aleppo about what its new heritage is and how to avoid turning it into a source of social conflict.

During the Battle of Aleppo 2012–2016, the city was divided within the World Heritage Site's boundary; local neighbourhoods and communities were isolated and experienced various traumatic events. The memory of these traumatic events such as a siege or crossing the checkpoints of different fighting groups around Aleppo becomes part of the city's fabric, its inhabitants, its memorials, and its ruins. There are two crucial points to consider here: *first*, people's perception, attachment to a place and eventually their sense of place that was and is still in a continuous cycle of change; *second*, defining the emerging new sense of place and identity in the post-conflict society. The process of redefining these attributes and values offers the city's people a form of participatory democracy and a negotiation and reconciliation tool.

On the other hand, place attachment has no boundaries as it involves the emotional bond between people and their environment (Lewicka, 2008, p. 214). This bond had been influenced during the siege by increasing social media utilisation by the local community. That was aimed to document and provide photos of the destruction of local heritage and neighbourhoods. These outlets allowed internet users to participate in the construction of local heritage narratives, sometimes even challenging dominant and mainstream discourses (Foth et al., 2008; Giaccardi, 2012; Van der Hoeven, 2019). It was noted that community-driven initiatives during conflict raised awareness of heritage experiences of groups such as minorities and subcultures that are not or insufficiently represented in authorised heritage institutions (Caswell & Mallick, 2014).

There were various initiatives undertaken by the local (remaining) community as responses to destruction. These included the creation of archives using Facebook to map out collective memories related to specific places, events or objects with a timeline covering the conflict and even prior to the conflict. There are other projects such as *News on Live Maps*,[4] *Narrative of Syrian Heritage*, *Syrian Heritage Archive*, which includes

Memories from Aleppo,[5] and *Aleppo Project*.[6] These initiatives provided not only damage recording, but also the required material to map out the new layers of values related to historic places during the conflict, whether it was destroyed or still standing, from the perspective of Syrian people inside and outside the country. Halbwachs (1992) has argued that communities determine the memory of their members, and individuals recall not only what they learn from others, but also what others have told them, which means it must be important and worthy of being told. Thus, it settles in one's memory in reference to others and in the context of social frames – in these cases in Aleppo, the digital framework created by social media. Conversely, things that cannot find a place within the current social framing become forgotten (Assmann, 2008).

Understanding these new layers of cultural heritage concerning the local community requires unpacking two aspects, the attachment and the meaning. Researchers "ought to examine not just how much the place means…but what does it mean?" (Stedman, 2003, p. 826). This is relevant to post-conflict society, where a new layer(s) of meaning has been engendered through the conflict. The *Meaning Theory* by Williams (2014) proposes that each place has various layers of meaning from the surface, inherent, socio-cultural to the deepest levels of meaning, such as identity. Exploring the stories from *Memories from Aleppo*[5] project through the *Meaning Theory* shows that assigned meaning to historic places comes in various types. Still, it does not identify the particular meanings of any given place.

Public understanding of place meanings is intimately tied to the community-based contexts in which the meanings have been developed. However, memories of destruction and traumatic events are no longer contained within the boundaries of the physical sphere in which they occur (Viejo, 2011, p. 59). Given the scale of destruction in Syria, this memory is disseminated via the wave of immigration, displaced Syrian people or through people who work in Syria and travel worldwide. This movement can result in the creation of diasporic 'communities of memory', following Viejo's (2011) argument. Therefore, narratives of war that shape memories and identity can be viewed differently by the various communities, inside or outside the country. For Syrian people inside the country, their memories of war affect their everyday life and the way they imagine the post-conflict state – their hopes and aspirations. In contrast, the perception of the people outside the country might be different. They view their memories of place through a "Kaleidoscopic lens of nostalgia, homesickness and sense of loss" (Viejo-Rose, 2011a, p. 59). These different levels of meaning and memories can divide societies in post-conflict times, especially when the imagined society created does not match the place's reality that has changed during the conflict.

Cities that were destroyed during wartime might tackle their heritage differently in post-conflict reconstruction, either as an integrated part of the city and its society or as separate historic buildings and landmarks. For example, Beirut's historic town's post-conflict reconstruction, after the civil war (1975–1990), known as Solidere, has maintained around 265 ancient structures based on their historical values and potential economic revenue. Lebanese experts and critics criticised the process. The prominent public intellectual Elias Khoury has described it as:

> The huge machine that is reconstructing and regenerating the city is already wiping out the memory of old Beirut, relentlessly tossing the rubble of the old city into the sea. The city's centre, today, is an empty space, a placeless space, a hole in the memory. How are we to preserve the memory of this place in the face of such frightening architectural amnesia?
>
> (Khoury, 1995, p. 139)

This is the main challenge in Syria: how to avoid creating empty and placeless space and a hole in Syrian's memory? Other challenges are related to the struggle of finding a balance between in-city historic areas rehabilitation against rebuilding homes and neighbourhood.

Rebuilding landmarks was one of the main approaches in Dresden (Germany), for example, focusing on the Frauenkirche cathedral, which occurred 50 years after its destruction in World War II. It was rebuilt entirely in its original form after emerging as an anti-war symbol and one of the main sites of war memory and commemoration in divided Germany. Warsaw in Poland witnessed a different scenario after its destruction in World War II. The local community joined forces to start the rebuilding process immediately after the war, focusing on living neighbourhoods rather than landmarks, yet recreating the historical details. In Aleppo, however, the situation and nature of the conflict are complicated. The needs for recovery and reconstruction in Aleppo vary from infrastructure to humanitarian needs and rebuilding the local community and identity. Any post-conflict reconstruction proposal requires the unpacking of many variables, including demographic change resulting from internally displaced people and refugees, traumatic events in significant heritage areas and the new layers of heritage.

Heritage values and local community in Aleppo

Urban heritage within our cities plays a significant role in enhancing place identity, memory and a sense of belonging. A deep human need exists for associations with significant places (Relph, 1976, p. 147). Suppose we choose to ignore that need and allow the forces of placelessness to continue unchallenged in Aleppo. The future will only hold an environment where places do not matter, such as in the examples of some post-conflict cities like Solidere in Beirut as argued by Ilias Khoury.

The term *sense of place* is relevant to the debate about post-conflict reconstruction and the 'placelessness' of the built environment. In general, *sense of place* is often understood as the human interaction with a place or the overall traditional character, national identity or existing socio-cultural atmosphere of a place (Conzen, 1966; Jivé n & Larkham, 2003; Ashworth & Graham, 2012). However, people are not thinking about how they feel about their local place daily. Some disrupting events could make *sense of place* more relevant to people, whether it is a natural or human disaster such as armed conflict or even the current global pandemic of COVID-19. We need to learn more about how experiences beyond disruption or disaster can precipitate increased awareness of our surroundings (Manzo, 2003, p. 53).

Based on the previous review, it is vital to conceptualise what is meant by a historic 'place' in the post-conflict Aleppo, to understand its 'sense of place'. Each place within the city is a territory of significance, distinguished from adjacent and larger or smaller areas by its name, by its particular environmental qualities, by the stories and shared memories connected to it or by the intensity of the meanings people give to it or derive from it (Relph, 1976, p. 7). However, recent debates related to the historic urban landscape approach have broadened the focus to include the conservation of historic urban areas in their settings by considering their distinctive identity, authenticity, character and sense (Roders & Bandarin, 2019). Accordingly, the term sense of place that is socially and culturally constructed can widen the limited perspective of understanding the historic place as a 'territory' within the city. This term can be analytically described and practised through consultation and negotiation with various users inside and outside the country. Therefore, exploring Aleppo's sense of place requires defining the change of perception to the historic place during the conflict.

Approaching an investigation of historic place and post-conflict society through the notion of the sense of place allows for a view of heritage that integrates landscape and culture, the past and the present, the movable and immovable and tangible and intangible elements in any analysis (Waterton, 2005; Smith, 2006). The process of defining a *sense of a place,* therefore, offers both a people-centred and place-centred approach for reconstructing post-conflict society and heritage by unpacking the layers of tangible and intangible values, local culture and the understanding of cultural diversity.

However, Aleppo and its historic centre are not homogeneous. Diverse social and ethnic groups can hold different attachments and memories related to the same place and, consequently, the *sense of place* might be contested. This may create another dimension of the conflict in the urban level. Whether the WHS of Aleppo or the residential neighbourhood, some areas can become a centre of another form of conflict over deeply held place-based values and beliefs, variously held by people who perceive themselves as having an equally strong attachment to the place. This relationship's dynamics have changed during the conflict: of the East and West of Aleppo, the political division, and many internal displacement and refugees. In this case, local people may hold their spiritual value (meaning and memory) above their economic values (use/activity). These concerns become serious threats to the integrity and authenticity of the World Heritage Site when ignored by development planners and international funders. Consequently, the international community might be involved in rebuilding heritage through symbols and narratives that might not necessarily be aligned with those that local communities build in parallel.

Furthermore, the country had seen an increased number of newly established civil society groups and activists in the heritage domain since 2011. These groups have been actively involved in efforts to advocate for the protection of cultural heritage and documentation of heritage destruction using social media, such as the Aleppo National Archive, Turath Halab, Aladyat of Aleppo, etc. In contrast, other groups took actions to protect significant elements like dismantling the 15th Century Minbar in the Great Umayyad Mosque and building a wall to protect the Meharb by the 'Syrian Association for Preserving Archaeology and Heritage' in the rebel-held area (Miznazi, 2015). Civil society groups' practices include diverse programming and activities that utilise culture and heritage as an instrument and resource to provide relief programs that meet the cultural, social and psychological needs and aspirations of the affected communities (Al-jawabra, 2018). These actions have created a collective sense of shared heritage amongst the conflict society.

On the other hand, cultural heritage in Aleppo, and Syria in general, reflects societal divisions with communities living differently (urban vs suburbs, modern vs traditional, temporary vs permanent). During the conflict, several stakeholders have emerged as the key facilitators and influencers in the negotiations related to Syria's future, including the reconstruction phase such as the new National Steering Committee that emerged in 2017, piloting the reconstruction plans in Aleppo.[7] The Ground Zero project in Aleppo was Al-Saqatiyah Souk, funded and supervised by the Geneva-based Aga Khan Trust for Culture, which had promoted the restoration of historic sites as a catalyst for social and economic development and brought together various groups of stakeholders. That

236 Hiba Alkhalaf

is by reviving the traditional Souks and its economy that would encourage the return of the refugees. This is necessary to align recovery and reconstruction efforts in a co-ordinated and effective way. Heritage reconstruction in the case of Aleppo is a critical element in achieving global stability, security and eradication of poverty.

Mapping out the emerged network of stakeholders in addition to the old networks, as well as the civil societies, and their possible roles is needed for an inclusive recovery phase (see Table 16.1). This is a crucial step to identify the local partners in a sustainable heritage reconstruction, to categorise them into groups using the power/interest tool. The Stakeholder Mapping tool categorises stakeholders into four groups according to their levels of both interest and power. Each group has various roles: A. collaborators, B. to be consulted, C. to be involved and D. to be informed. The main aim here is to plan for involving and creating consensus amongst all stakeholders in the Old City of Aleppo. This tool shows that the civil societies groups (in Group C) have a high level of interest but a low power level. By empowering this group, they can be collaborators in the reconstruction process instead of merely being a group to be involved or consulted, hence sharing their individual and collective memories of their heritage to compete and negotiate for their city reconstruction. The main challenge is related to heritage governance and the capacities and skills of primary stakeholders in Syria that have been significantly weakened during the conflict. It is evident that most of the experienced staff in governmental bodies (i.e. the municipality and Old City Directorate), academic institutions and private sector experts, have either left the country or moved to other jobs during the conflict.

To this end, the reconciliation of competing narratives associated with Aleppo's post-conflict life presents one of the most significant challenges to the decision-making process today. These challenges are mainly related to either incorporating or excluding the narratives associated with its various local groups in a way that can illustrate its future identity. These narratives offer the opportunity to construct a political, cultural and ethical framework to understand and plan reconstruction proposals in the WHS and the city. This framework is essential for architects and heritage professionals working on landmarks and for those who are assessing the stability, authenticity and integrity of historic buildings, rehabilitating the souks and demolishing/rebuilding the damaged

Table 16.1 To the left, An attempt to map the main stakeholders at the city of Aleppo, including the local, national, private, professional and international stakeholders as well as the local civil society. To the right, mapping stakeholders using the interest/power tool to put them into groups that each play different roles within the reconstruction process, each group requires different engagement techniques (adapted from Johnson & Scholes, 1999).

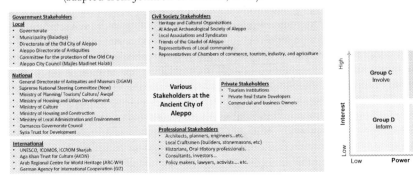

structures. Therefore, any new intervention should challenge communities' segregation and allow individuals and local stakeholders to reflect their collective memory, new layers of cultural heritage, and nurture a renewed sense of common identity.

Conclusion

The destruction of cultural heritage during the armed conflict in Aleppo calls for new and effective participatory approaches to address the main challenges related to a historic place, people, and process to achieve a sustainable rebuilding of the city, its heritage and the society. Several competing priorities exist and identifying the starting point for the process is challenging for both Syrian and international communities, taking into account the process has been begun by the local Syrian government and Aga Khan Trust in the Old City of Aleppo (Al-Saqatiyah Souk). Avoiding urban conflict requires a negotiation between various actors to reach a consensus on a particular set of imperatives for reconstruction and the narratives that give meaning to cultural heritage, i.e., inclusion, reconciliation, local community engagement and ownership. Exclusion of any relevant stakeholders is likely the most severe threat to a successful reconstruction process and should thus be replaced with inclusion and openness. Based on the previous review, this chapter identifies future directions for the field of sustainable post-conflict heritage reconstruction studies related to place, people, and process:

- **Place**: The context of the city's social network became holding places for stories, memories and remembrances. Based on that, the generic top-down assessment of values in post-conflict historic places should be avoided and alternatively be fully contextualised concerning local heritage discourses and the dynamics of heritage governance in Syria. Therefore, there is a need to develop culturally sensitive tools to assess the new layers of cultural heritage and the associated values of the existing historic environment, including the WHS designation values. Further in-depth studies are needed to understand heritage attributes and values, particularly in conflict and displacement situations in addition to exploring communities' role in reconstructing their heritage and mobilising it as a tool for resilience and reconciliation.
- **People**: Sustaining the historic environment's character in post-conflict Aleppo seeks to build up a sustainable local community with a good quality of life. However, the reconstruction process requires consensus amongst the local and international communities. Thus, further work is needed to map out the main stakeholders and civil society that requires an effective participatory engagement approach. These approaches should be aligned with the local legislative framework in Syria to support the planning and implementation of such participatory, inclusive approaches, and their contribution to achieving the Sustainable Development Goals.
- **Process**: The World heritage status of the Old City of Aleppo is an opportunity. Any reconstruction project will be implemented with the World Heritage Committee and Charters' oversight, in line with the Syrian national law. In that context, further work is necessary to explore the implementation of international tools within the World Heritage management plan of Aleppo. This requires assessing the extent to which the existing policy and regulatory frameworks in Syria can accommodate the international participatory and community engagement tools. This assessment will eventually lead to developing new national policies and proposals.

Notes

1 The project CRIC 'Identity and conflict. Cultural heritage and the re-construction of identities after conflict', Retrieved from: https://cordis.europa.eu/project/id/217411/reporting (Accessed on 04 March 2020).
2 United Nations International Strategy for Disaster Reduction (UNISDR).
3 Damas Press, 2018, Retrieved from: http://www.dampress.net/?page=show_det&category_id=6&id=85711, Arabic (Accessed on 15 February 2020).
4 Live Maps, Retrieved from: https://syria.liveuamap.com/en/time/01.09.2018 (Accessed on 17 June 2020).
5 Syrian Heritage Archive, Retrieved from: https://syrian-heritage.org/memories-from-aleppo/ (Accessed on 04 April 2020).
6 The Aleppo Project, Retrieved from: https://www.thealeppoproject.com/ (Accessed on 03 April 2020).
7 The National Higher Steering Committee for the Restoration of the Old City of Aleppo was formed in 2017. It includes the ministries of culture, public works and housing, tourism, experts in the field of archaeology and restoration, aga khan development network, the Syria trust for development and representatives of the local community, Retrieved from: http://dgam.gov.sy/index.php?p=314&id=2317 (Accessed on 15 February 2020).

References

Aljawabra, A. (2018). *Heritage, conflict and reconstructions: From reconstructing monuments to reconstructing societies*. Paris: ICOMOS University Forum.
Alkhalaf, H. (2018). *Sense of place: Towards integrated conservation and sustainable urban development practice*. Unpublished PhD Thesis. Edinburgh: University of Edinburgh.
Ashworth, G. J., & Graham, B. (2012). Heritage and the reconceptualization of the postwar European city. In D. Stone (Ed.), *The Oxford handbook of Postwar European History* (pp. 582–600). Oxford: Oxford University Press.
Assmann, A. (2008). Transformations between history and memory. *Social Research: An International Quarterly, 75*(1), 49–72.
Barakat, S. (2005). *After the conflict: Reconstruction and development in the aftermath of war* (Vol. 1). London: IB Tauris.
Barakat, S. (2007). *Reconstructing post-Saddam Iraq*. New York & London: Routledge.
Bevan, R. (2016). *The destruction of memory: Architecture at war - second expanded edition*. London: Reaktion Books.
Canter, D. (1977). *The psychology of place*. New York: St Martin'S Press.
Caswell, M., & Mallick, S. (2014). Collecting the easily missed stories: Digital participatory microhistory and the South Asian American digital archive. *Archives and Manuscripts, 42*(1), 73–86.
Chatelard, G. (2017). *Survey report: Intangible cultural heritage of displaced Syrians*. Retrieved from: UNESCO: https://ich.unesco.org/doc/src/38275-EN.pdf (Accessed on 03 May 2021).
Conzen, M. (1966). Historical townscapes in Britain: A problem in applied geography. In J. W. House, (Ed.), *Northern geographical essays in honour of GHJ Daysh* (pp. 56–78), Newcastle upon Tyne: Oriel Press.
DGAM. (2020). *State party report- The state of conservation of the Syrian cultural heritage sites*. Retrieved from: UNESCO https://whc.unesco.org/en/soc/3866/&mode=doc (Accessed on 03 May 2021).
Foth, M., Klaebe, H. G., & Hearn, G. N. (2008). The role of new media and digital narratives in urban planning and community development. *Body, Space & Technology, 7*(2), 1–18.
Giaccardi, E. (2012). *Heritage and social media: Understanding heritage in a participatory culture*. New York and London: Routledge.
Gibbons, E. (2017). The Syrian crisis and cultural memory. *Global Societies Journal, 5*(0), 59–69.
Ginzarly, M., Roders, A. P., & Teller, J. (2019). Mapping historic urban landscape values through social media. *Journal of Cultural Heritage, 36*, 1–11.

Graham, H., Mason, R., & Newman, A. (2009). *Historic environment, sense of place, and social capital. Report commissioned by English Heritage*, Newcastle University, Retrieved from: https://historicengland.org.uk/content/heritage-counts/pub/sense_of_place_lit_review_web1-pdf/ (Accessed 03 May 2021).

Halbwachs, M. (1992). *On collective memory*. Chicago, IL: University of Chicago Press.

Harrison, R. (2013). *Heritage: Critical approaches*. New York and London: Routledge.

Higueras, A. (2013). Aid and reconstruction of heritage in the context of post-conflict societies. *Archaeologies, 9*(1), 91–105.

Jivé´n, G., & Larkham, P. J. (2003). Sense of place, authenticity and character: A commentary. *Journal of Urban Design, 8*(1), 67–81.

Johnson, G., & Scholes, K. (1999). *Exploring corporate strategy*. London: Prentice Hall.

Khalaf, R. W. (2017). A viewpoint on the reconstruction of destroyed UNESCO Cultural World Heritage Sites. *International Journal of Heritage Studies, 23*(3), 261–274.

Khalaf, R. W. (2020). Cultural heritage reconstruction after armed conflict: Continuity, change, and sustainability. *The Historic Environment: Policy & Practice, 11*(1), 4–20.

Khoury, E. (1995). The memory of the city. *Grand Street, 54*, 137–142.

Lewicka, M. (2008). Place attachment, place identity, and place memory: Restoring the forgotten city past. *Journal of Environmental Psychology, 28*(3), 209–231.

Manzo, L. C. (2003). Beyond house and haven: Toward a revisioning of emotional relationships with places. *Journal of Environmental Psychology, 23*(1), 47–61.

McGivern, H. (2018, July 10). Aleppo's reconstruction 'deserves international resources'. *The Art Newspaper*. 10. Retrieved from: https://www.theartnewspaper.com/news/global-resources-are-needed-to-rebuild-aleppo (Accessed on 03 May 2021).

Miznazi, D. (2015). Post-conflict reconstruction of Old Aleppo. *News in Conservation*. Retrieved from: https://www.iiconservation.org/node/5495 (Accessed on 03 May 2021).

Nallu, P. (2017). *Analysis: Politics of class and identity dividing Aleppo – and Syria*. Retrieved from: https://www.newsdeeply.com/refugees/articles/2017/03/17/analysis-politics-of-class-and-identity-dividing-aleppo-and-syria-2 (Accessed on 03 May 2021).

Overton, I., & Dathan, J. (2019). *Syria in 2020: The deadly legacy of explosive violence and its impact on infrastructure and health*. Retrieved from: https://reliefweb.int/report/syrian-arab-republic/syria-2020-deadly-legacy-explosive-violence-and-its-impact (Accessed on 03 May 2021).

Punter, J. (1991). Participation in the design of urban space. *Landscape Design, 200*(1), 24–27.

Relph, E. (1976). *Place and placelessness*. London: Pion.

Roders, A. P., & Bandarin, F. (2019). *Reshaping urban conservation: The historic urban landscape approach in action* (Vol. 2). Singapore: Springer.

Rossi, A. (1982). *Architettura della città*. Cambridge, MA: MIT Press.

Saya, S., Hasan, T. M., Mimura, S., Okada, T., Roth, M., Kohler, S., … Missal, R. (2017). *Build back better: In recovery, rehabilitation and reconstruction*. Switzerland: UNISDR. Retrieved from: https://www.unisdr.org/files/53213_bbb.pdf (Accessed on 03 May 2021).

Shamai, S., & Ilatov, Z. (2005). Measuring sense of place: Methodological aspects. *Tijdschrift voor economische en sociale geografie, 96*(5), 467–476.

Smith, L. (2006). *Uses of heritage*. New York and London: Routledge.

Sørensen, M. L. S., & Viejo-Rose, D. (2015). *War and cultural heritage*. Cambridge: Cambridge University Press.

Stedman, R. C. (2003). Sense of place and forest science: Toward a program of quantitative research. *Forest Science, 49*(6), 822–829.

UNESCO. (2017a). *Table of actions for the recovery of the Ancient City of Aleppo*. Retrieved from: UNESCO: https://whc.unesco.org/en/news/1639 (Accessed on 03 May 2021).

UNESCO. (2017b). *UNESCO reports on extensive damage in first emergency assessment mission to Aleppo*. Retrieved from: https://en.unesco.org/news/unesco-reports-extensive-damage-first-emergency-assessment-mission-aleppo (Accessed on 03 May 2021).

UNESCO. (2018a). *Position paper: Culture in city reconstruction and recovery*. Retrieved from: http://documents1.worldbank.org/curated/en/708271541534427317/pdf/131856-WP-REVISED-II-PUBLIC.pdf (Accessed on 03 May 2021).

UNESCO. (2018b). *Warsaw recommendation on recovery and reconstruction of cultural heritage*. Retrieved from: https://whc.unesco.org/en/news/1826 (Accessed on 03 May 2021).

UNESCO, & UNITAR. (2018). *Five years of conflict- The state of cultural heritage in the ancient city of Aleppo*. Retrieved from: https://unitar.org/sites/default/files/media/publication/doc/UNESCO_UNITAR-5-years-conflict.pdf (Accessed on 03 May 2021).

UN-Habitat. (2014). *Aleppo city profile*. UN-Habitat Retrieved from: https://unhabitat.org/node/142348 (Accessed on 03 May 2021).

Van der Hoeven, A. (2019). Historic urban landscapes on social media: The contributions of online narrative practices to urban heritage conservation. *City, Culture and Society, 17*, 61–68.

Viejo-Rose, D. (2011a). Destruction and reconstruction of heritage: Impacts on memory and identity. *Heritage, Memory and Identity, 4*, 53–69.

Viejo-Rose, D. (2011b). *Reconstructing Spain: Cultural heritage and memory after civil war*. Brighton: Sussex Academic Press.

Viejo-Rose, D. (2013). Reconstructing heritage in the aftermath of civil war: Re-visioning the nation and the implications of international involvement. *Journal of Intervention and Statebuilding, 7*(2), 125–148.

Waterton, E. (2005). Whose sense of place? Reconciling archaeological perspectives with community values: Cultural landscapes in England. *International Journal of Heritage Studies, 11*(4), 309–325.

Williams, D. R. (2014). Making sense of 'place': Reflections on pluralism and positionality in place research. *Landscape and Urban Planning, 131*, 74–82.

World Bank. (2018). *Promoting resilience through post-crisis recovery*. Washington, DC: The World Bank Retrieved from: https://reliefweb.int/sites/reliefweb.int/files/resources/WRC3%20Proceedings%20and%20Knowledge%20Report%20FINAL.pdf (Accessed on 03 May 2021).

17 Heritage and peacebuilding

Challenges, possibilities and sustainable practices

Feras Hammami, David C. Harvey, Daniel Laven and Diana Walters

Introduction

The dominance of conflict within narratives of heritage sometimes seems overwhelming. On the one hand, the field of 'conflict heritage' is a significant strand of heritage studies in its own right. This is a field that includes a great deal of what is often termed 'difficult' and 'dark' heritage, as well as work within the realms of commemoration, remembrance and memorialisation, as past wars or battles are marked, accorded value or perhaps become contested. On the other hand, the field of heritage studies more broadly is one in which conflict and contestation are often assumed to be central, both within the popular articulation of heritage and within the critical analysis of heritage.[1] Acting as a powerful resource, therefore, heritage is deployed to support exclusive narratives of self and other, often with divisive ends. And, while heritage scholars critique and deconstruct these deployments, a sense of dissonance and contestation sustains the notion that conflict constitutes an essential element in heritage processes (e.g. see Tunbridge & Ashworth, 1996; Smith & Campbell, 2011). In many ways, this depiction of heritage lies at the heart of agonistic critique; born of an anxiety that in trying to avoid the antagonistic experiences of conflict through the invocations of 'cosmopolitan' heritage, a dangerous gap provides space for extreme nationalism to ferment (Bull et al., 2019, p. 614).[2] As a result, a notion of 'conflict' is seen as essential, and any supposedly 'non-contested' heritage can only ever be the bland cosmopolitan feel-good fare.

Heritage and conflict are charismatic twins, who support each other in public through a noisy conversation that tends to dominate, forever attracting attention, meeting expectations, making promises of reform, negotiating diplomatic accords and road maps towards 'reconciliation'. Love them or hate them, heritage and conflict seem to speak as one. Whether a commentator is critical or supportive, the language of 'heritage-and-conflict' has become normalised, expected and even needed – a repertoire, a narrative and a set of material that has gained legitimacy through an unchallenged consensus of universal logic. While many who aspire to wield power are enthusiastic followers of these siblings, even those that are critical of them often end up using the same language and operating on the same terrain in their assumptions that the two cannot be separated.

Narratives of heritage and peace seem to struggle for space. At least where 'peace' is talked about, it is implicitly or explicitly the sort of 'negative peace' – a position of 'not fighting' – that is invoked.[3] This chapter asks, therefore, whether it is possible to 'do heritage' in a manner that both avoids the bland invocation to say nothing that might offend, while also maintaining a sense of criticality that does not assume an essence of struggle and conflict at the heart of everything. In other words, is it possible to make space for a peaceful – or *pacific* – heritage, which can maintain criticality and a purposeful ambition of social justice, neither accepting perpetual conflict nor promising an unattainable fix of a road map to reconciliation? This would not be to deny the existence of conflict and the deliberate use of heritage to create it, but rather, to do *something else* – perhaps by starting *somewhere else*.

In this chapter, we explore some of the challenges, possibilities and sustainable practices within work connecting heritage studies and peacebuilding. After this introduction, we reflect more fully on the challenges that this ambition implies; just why and how assumptions of essential conflict seem to be so central within so many studies of something called 'heritage'. Instead of decrying or seeking to invalidate much excellent critical work within the field of heritage studies, we embark on an exercise of reflection, with an intention to 'reset' things. Thus, rather than starting from *here*, perhaps it is possible to start from *somewhere else*; an ambition to think what might be possible with a wilful sense of idealism. In terms of these possibilities, first, we have taken strength from some current theoretical trends in other disciplines, especially peace studies, radical theology, resistance studies and cultural geography, thinking through some implications for how a recent rethinking of critical geopolitics might inform a reconsideration of something called 'critical heritage studies'. Second, we feel that an honest and authentic approach would be to offer a few examples, with a strong dose of humility; characterised by being bottom up and heralding from very different contexts; well-known and seemingly obscure. Rather than seeking to offer a 'blistering critique' of what exists, therefore, our approach will merely suggest a few key elements of what we might call sustainable practice for a way forward.

Challenges: Just how did we get here?

Peacebuilding practices have a problem with heritage, since invocations of heritage seem to lie at the heart of so much conflict around the world. Heritage sites have become targets and battlefields, while heritage narratives often sow discord, underscore difference and are used as an excuse for violence and aggression.[4] *Peace efforts* are often pragmatically and technically replaced with *conflict resolutions*; target-driven, using top-down 'road maps' and diplomatic solutions focused on who has control, access and share. The nuances of *dissonance* and contested meanings get lost as heritage *difference* is deployed euphemistically, as an unreflexive backdrop to violence rather than a source of learning. Thus, an imagined – if not invented – mosaic of exclusive identities becomes naturalised and enhanced, or even violently enforced, through the careful conservation of certain tangible and intangible cultural heritage narratives, even where that entails the destruction of others (see, for instance, the implied critique in Krauss, 2015).

On the other hand, heritage studies seem to have a problem with peacebuilding, as the majority of effort – both practical and intellectual – is connected squarely with the marking of conflict. As Gregory Ashworth (2013) argued in his keynote to the Peace Conference in Wageningen, 'peace is boring and war is interesting'. Indeed, within funding regimes where significant anniversaries of particular conflicts are used instrumentally, critical reflection always seems tied to elements of conflict and dissonance. Thus, it becomes difficult to open out a space for heritage to be connected to active peacebuilding. Even where these activities are 'critical of conflict', such as with much recent academic and communal effort associated with the centenary of World War I, the ensuing heritage effort continues to place conflict at the centre, both in terms of critical activity and increasingly in terms of reliance on funding streams.

At present, much of the easy alignment of heritage and conflict tends to reduce heritage to the (literal) *stuff* of battlefields, materials, sites and memorials, together with the singular narratives of national destiny and their associated choreographed rituals. These narratives might be disputed or even critiqued, but that is the point – where a suggestion that to reach a position of *peace* necessarily entails that everyone signs up to a 'road map' towards a 'target'. In a world in which heritage is always about stuff and things (often promoted and circulated through authorised heritage discourse), then it is inevitable that something called 'peace' also becomes a *thing*; a target that is represented by a perceived lack of conflict, rather than a *process* informed by justice and ideals of social equity.

It is within this difficult terrain that ideas of agonism have understandably flourished in recent years (see Mouffe, 2013; Bull & Hansen, 2016). After all, very few would agree with an *antagonistic* pathway (despite the prevalence of such antagonism being implicit in many heritage practices), and for agonists, the cosmopolitan attitude just seems unrealistic at best. Agonism, therefore, seems to be the only game left in town, taken for granted as a good model of governance that can be employed in different contexts and scales. Bregazzi and Jackson (2018, p. 72), however, critique the agonistic assumption that 'moments of violence [are] always present in our assumptions of peace'. As Shields and Soeters (2017, p. 323) exclaim with some exasperation; 'peace research is just the study of war' (see also van den Dungen, 2017). Lamenting how peace research has primarily become an examination of war – 'an overwhelming focus on violence, war, social agonism and conflict' – Bregazzi and Jackson (2018, p. 73) pose the provocative question: 'Why do we always need to start from 'here' – if we start by assuming that sociality always involves struggle then agonism seems to be the only way forward, but why can't we start from somewhere else?'

An imperative to 'start somewhere else' suggests that we need to take the *processual understanding* of both peace and heritage more seriously, as emergent properties that are actually happening on the ground, rather than as stable blueprints directed from above supposedly to achieve measurable goals and targets. For Bregazzi and Jackson (2018, pp. 86–87), peace is much more than an absence of fighting and does not have a binary relationship with war.[5] Crucially, therefore, peace is a non-national entity; always distributed and relational, it is an ongoing project, to be found in the everyday rather than portrayed as a 'target'. Indeed, mirroring established ontological apprehensions of heritage, for Bregazzi and Jackson (2018), peace is a verb – something that is *done* rather than a fixed outcome to be achieved.[6] With this tantalising challenge in mind, is it possible for heritage studies to *start somewhere else*?

Possibilities – of starting somewhere else

> We are better at researching war than peace. For our discipline to play a serious role in addressing the problems wracking twenty-first century humanity, it is imperative that this imbalance is addressed.
>
> (Megoran, 2011, p. 188)

Mirroring the work of Nick Megoran (2010, 2011) and others, who call for a switch from talking about 'critical geopolitics' towards what they call 'pacific geopolitics', perhaps we can invoke an idea of 'pacific heritage' and emphasise why this is *different* from 'critical heritage'.[7] Thus, we are not *against* something called critical heritage, but would rather try and do something a bit different. In many ways, we feel that this ambition is implicit within the work of Viejo-Rose and Sørensen (2015, pp. 289–292) who call for a 'disarming of heritage practice'. Viejo-Rose and Sørensen's body of work (2015) tends to underline how heritage practices run alongside conflict, with both victory and victimhood being used and abused and where UNESCO and other 'landmark' declarations and 'road maps' tend to fix the relationship between heritage and armed conflict as well as what constitutes heritage.[8]

> Only by recognition of the multifocal dimensions of heritage can it be a constructive resource of post-conflict recovery… Rather than closing down options by fatalistically suggesting an inevitable cyclical dimension to conflict, fully acknowledging the inter-relation between cultural heritage and violence can suggest methods for mitigating the potential time bomb effect.
> (Viejo-Rose & Sørensen, 2015, p. 292)

Rather than seeing heritage as a static, authentic object, which manifests the past, but as a *process*, in continual dialogue, operating in the present and with an eye for the future, then possibilities are opened up. Heritage is always about the future – and it can 'do stuff'.[9]

On a macro-scale, perhaps we can also take encouragement from the work of Tim Winter (2015). Using the phrase 'heritage diplomacy', Winter (2015, p. 997) reflects on the implications (and potential) of the 'international mobilisation of heritage'. Noting that 'to date, much of the analysis regarding the politics of heritage has focussed on contestation, dissonance and conflict', Winter (2015, p. 998) argues that such heritage diplomacy can also perhaps enable us 'to read critically what is going on in the space of collaboration, excavate a broader political history of the political and economic, national and international that have received less attention that they deserve. It also helps us move beyond the commonly used frameworks of colonial and postcolonial'.[10] There is a 'hopeful' sense of ambition within Winter's (2015) analysis, which we would strongly endorse, and while he mostly focuses on a macro-scale of UNESCO potential, there is an implicit call for us to look not towards the large scale inter-governmental stuff, but towards more local/bottom-up practices. Indeed, we would argue that the whole notion of seeing heritage as a *process* points towards bringing a sense of agency back towards people. This endorsement of looking towards more bottom-up, local and everyday responses is an authentic invocation towards people making their own history and for 'peace' being a realistic ongoing practice rather than a 'target'.[11] Crucially, however, vested within the notion of starting from *somewhere else*, we are keen not to ascribe to categorised notions of there being inherently 'good' and 'bad' heritage, which we feel, sometimes characterises some aspects of the 'critical heritage' agenda. Perhaps paradoxically, therefore, while we would instinctively seek to support expressions of bottom-up practical heritage work, we should not overlook the pragmatic necessity of top-down activity or even sustain any form of local–global binary. Indeed, imaginative effort and ongoing commitment, institutionalised through 'top-down' activity, would seem to be crucial to streamline with other efforts on different levels of influence and in many of the most intractable conflicts around the world.

Reflecting more broadly on these provocative and stimulating challenges, therefore, rather than rely on a critical exploration of '*heritage-and-conflict*', we feel that an exploration of heritage and peacebuilding *processes* is required. Being critical of the many instances of where and how heritage is invoked and otherwise connected to conflict might reduce the possibilities of providing the space for an exploration of heritage and peacebuilding. As well as being emotionally engaged in what is happening in time and space, authentically situated within worlds of 'practice' and operating at the everyday level, we also feel that such locally sensitive and derived efforts have a greater chance to become sustainable. Indeed, we feel that a focus at the local level can bring together academic theory, professional practice and diplomacy in a manner that is simply not feasible in the grand pronouncements and road maps. Within the remains of this chapter, we further explore our calls to start from somewhere else through three short vignettes. We have chosen these vignettes from our personal and professional experiences to provide contexts for our arguments, explain the affective environment of our analysis and offer case-specific 'hopeful exemplars', that are broad enough to draw bigger conclusions and contribute to a *pacific heritage*.

Walking the High Road to Peace: Reflections from the Vie della Pace

Introduction

The relationship between tourism and peace has received substantial attention in the scientific literature. On one hand, it has been argued for many years that tourism can promote and maintain peace in a variety of settings (e.g. D'Amore, 1988; Wohlmuther & Wintersteiner, 2014). Other studies, however, have questioned the role of tourism in reducing conflicts (e.g. Cho, 2007). Despite this academic debate, policy-makers continue to position tourism as the world's 'peace industry' as evidenced by United Nations Resolution A/RES/70/193. The resolution highlights tourism's potential to foster

> better understanding among peoples everywhere, leading to a greater awareness of the rich heritage of various civilizations and bringing about a better appreciation of the inherent values of different cultures, thereby contributing to the strengthening of peace in the world
>
> (UN General Assembly, 2015, p. 3)

Understanding tourism's potential contribution to peacebuilding is important in relation to contemporary heritage management. Much of the cultural heritage sector relies on tourism to generate revenues for conservation and stewardship efforts (Markham et al., 2016). Thus, for better or for worse, tourism has become an important socio-economic phenomenon that underpins many of the world's heritage experiences.

This vignette discusses the 'hopeful' experience of the Carnic High Route, which stretches along the Austrian–Italian border area.[12] This vignette is 'hopeful' in the sense that it offers an example of what an approach to sustainable, heritage-oriented peacebuilding might look like in a region that witnessed extensive fighting over the last 200 years including the Napoleonic Wars, World War I and the Cold War (Walleczek-Fritz & Fritz, 2015). In this way, this vignette evokes notions of pacific heritage through the practice of outdoor recreation, which now connects a region that has previously been bitterly divided.

Setting the scene

The Carnic High Route is part of a network of 'peace trails' initiated by the Austrian 'Dolomitenfreunde' association (Friends of the Austrian Alps) in the 1970s.[13] Managed as one of Europe's premier outdoor recreation areas and long-distance hiking trails, the Carnic High Route also offers access to several museums, sites and memorials including cemeteries, trenches and bunkers from World War I (Figure 17.1). In their edited volume on tourism and peace, Wohlmuther, Wintersteiner and Wagner (2014) note that these peace trails were established to make the cruel parts of history accessible to visitors, thereby transforming former sites of war and conflict into settings in which these violent pasts can be reconsidered. Wohlmuther et al. (2014) also emphasise that the importance of these peace trails stems from their intention to foster 'interaction and mutual dialogue between the (formerly antagonised and sometimes still conflicting) neighbours' (p. 327). Such interaction and mutual dialogue would seem to be an important part of what pacific heritage is all about; the ability to talk about, listen to and experience difficult pasts in ways that promote learning and understanding instead of othering and violence.

Walking the high road to peace

How might a long-distance hiking trail encourage interaction and mutual dialogue in the wake of two centuries of violent, armed conflict? How might this hiking trail reflect the notion of pacific heritage? First, it may be helpful to understand the essence of hiking, or long-distance walking. Philosophers and adventurers have repeatedly praised the virtues of long-distance walking as a form of reflexive practice, mediation and even pilgrimage. Robert and Martha Manning (2013) express these virtues on the back cover of their popular guidebook, *Walking Distance*:

Figure 17.1 A World War I military cemetery overlooking the popular 'Obstanserseehütte' mountain station along the Obstansersee Lake.

Source: Photo by Daniel Laven, July 2019.

> Walking is simple, but it can also be profound. In a complex and frantic world, long-distance walking helps focus our lives. It is an elegantly sustainable form of outdoor recreation that deepens understanding of the world's cultures and landscapes, stimulates thinking, and improves health.

Indeed, the Carnic High Route's dramatic landscapes, high elevation and challenging terrain, and long distances are highly immersive experiences. These trails offer visitors the opportunity to slowly digest the juxtaposition of the extreme beauty of these landscapes with their extremely violent heritage of armed conflict and human suffering.

Hiking the Carnic High Route is also an intimate experience, even for a solo hiker. Hikers are not permitted to camp along the trail and therefore must overnight in the mountain stations. These mountain stations are communal affairs in which guests sleep in the same room, eat and drink together, and share in the 'everydayness' of the experience. This notion of intimacy has received growing attention in the tourism literature, particularly in understanding the role that tourism might play in peacebuilding. In their work on tourism in the Palestinian–Israeli context, for example, Shepherd and Laven (2020, 2021) describe how hostels provide opportunities for an intimate form of hospitality in ways that more conventional hotels do not. This notion of intimacy is important because it can serve to break down the 'othering' inherent in violent, cross-cultural conflicts while allowing for the reframing of relationships and perspectives (Ibid). In this sense, such intimacy may help 'set the table' for the production of a more bottom-up, pacific heritage through meaningful *in situ* interactions between people and place.

The Carnic High Route also serves to connect communities that were once enemies. For example, between July and October of 1916, the Austrian town of Sillian was reportedly struck by approximately 1,000 artillery shells fired by Italian forces positioned in neighbouring communities (Walleczek-Fritz & Fritz, 2015). Today, visitors and local residents can easily walk along the trail and move across the border, which is completely open. As a result, tourism has now become one of the region's primary economic sectors, which is entirely dependent on the freedom of movement and the freedom to cooperate brought about by peaceful relations. Such cooperative activities include a range of private sector service networks (accommodation, food, guide services, etc.) along with high degrees of coordination between various governmental and not-for-profit entities in terms of trail management, transportation, search and rescue, and joint marketing activities. Not surprisingly, these elements mirror the broader scientific literature and practitioner experience with transboundary protected areas for peace and cooperation, or peace parks (e.g. Sandwith et al., 2001; Ali, 2007). Viewed from this perspective, the Carnic High Route strongly echoes Richmond and Mac Ginty's (2019) notion of 'everyday peace infrastructures', which are characterised by 'on-the-ground activities and networks found in conflict and post-conflict zones whereby individuals and small groups make and maintain peace and engage in the tolerance and conciliation required for society to function' (p. 615). In this sense, the Carnic High Route feels like a promising practice of pacific heritage. These networks and the open mobilities that they support are an explicit nod towards disarming heritage for visitors and local residents alike.

Reflections on a hopeful future

Walking the Carnic High Route was a hopeful experience. I often do my best thinking when I slow things down, and, not surprisingly, I often found myself reflecting on the concepts that frame this chapter while hiking the trail. I am accustomed to doing fieldwork and observing the ways in which ideas interact with landscapes is almost second nature to me. I felt hopeful that the Carnic High Route appeared to confirm in practice many of the scientific principles I often teach; that I was observing a good example of the ways that heritage and tourism might contribute to peacebuilding in a sustainable manner. Beyond my professional persona, I also found myself moved by the entire experience. I grew up in a family deeply scarred by European war, and I realised that, like me, everybody hiking on this trail starts from somewhere else and each participant is given the opportunity to undertake a physical and reflective journey on their own terms. While standing over the cemetery (Figure 17.1), it struck me how unimaginable this scene would be to the soldiers honoured there. The idea that this former war zone, which had witnessed so much bloodshed, now offers such warm, reflective and intimate hospitality was satisfying and deeply emotional. And then it struck me how unimaginable it was for me to be observing this landscape and enjoying the intimate hospitality given my own family's connection to the conflicts of this region. It was an important reminder of the role that heritage can play in realising a more hopeful future, despite it seeming unimaginable at times.

Heritage practice in Kenya and elsewhere – a journey of the self

As a white British academic practitioner, writing about Kenya is often uncomfortable. I have written before about my experiences of working alongside colleagues from the Kenyan Community Peace Museums and Heritage Foundation (CPMHF) and somehow adopted a degree of academic distance, but I find I can't do that anymore. Writing this in summer 2020, in the face of a global pandemic, heightened racial tension, recession and incessant erosion of democracy, I suppose my questions about the 'value of heritage' may now be seen somewhat indulgent. Certainly, I write from a position of privilege and relative safety.

In preparing for this short piece, I contacted Sultan Somjee, ethnographer and founder of the CPMHF and someone with whom I have corresponded for several years. Although we have never met in person, Somjee always gently challenges me, and I have learnt much from him and from many others who work in the same quiet way. Ever generous, Somjee once again reflected back on one of the central issues for heritage practitioners and commentators who seek to work in so-called developing countries. In answer to my question '*How is peacebuilding enhanced through heritage-based approaches?*' He replied '*Heritage-based approaches' to peacebuilding in societies that do not follow western academic traditions, thoughts or religious beliefs must be viewed in the context of Utu*'.[14] In other words, don't start from here. He expanded on this by saying, '*your questioning is from the western academic perspective. If I try to fit into your box, it would not only be dishonest but also pretentious*'.

The domination of selective western thinking and literature relating to heritage is a barrier and also results in a loss. The lens through which heritage-based approaches to issues of peace, conflict resolution, dialogue and social justice are commonly viewed overshadows the depth and wealth of traditions that start from 'another place'. This reinforces the inbuilt bias towards broadly 'western' epistemologies and beliefs that tend to take a reductionist view of approaches openly rooted in faith and spiritual traditions. To an extent, this exposes some of the difficulties that many scholars and practitioners have, or even create when dealing, with the concepts of peace and peacebuilding. The (supposed) absence of 'science' and the lack of exegesis mean that many spiritual- or faith-based texts and practices are skewed and manipulated to the denigration of many of the basic tenets of global belief systems. This creates a paradox that is seemingly intractable. On the one hand, 'religion' is cited as the cause of much conflict, yet by the same token, that spirituality is sidelined as a constituent of identity and a possible source of healing and connection.

Timothy Gachanga and Munuve Mutisya, CPMHF activists, have reflected on the lack of materials for heritage-based interfaith dialogue. In researching this activity, a core approach within CPMHF, they state that 'we gained the impression that very little research has been done in this field' (Gachanga & Mutisya, 2015, p. 277). This was further layered by colonial attitudes that perpetuated a dominant notion of 'heritage' that maintains power today (Coombes et al., 2014). Referring to the central importance of objects as messengers and transmitters of dialogue and healing, Gachanga and Mutisya say that 'despite the role traditional artefacts play in preserving religio-cultural values and narratives, they continue to suffer from a lack of acceptance and inadequate understanding of their role and essence' (Gachanga & Mutisya, 2015, p. 278). This, plus a predominance of western cultural and theological frameworks, has 'hindered an open conversation about the role of peace museums in interfaith dialogue, in promoting community cohesion, and preserving religio-cultural values and narratives' (Gachanga & Mutisya, 2015, p. 279). Olupona (1991), as cited in Gachanga & Mutisya (2015), argues that the two monotheistic traditions to which many Africans have converted over the centuries – Islam and Christianity – have developed a hostile attitude to African traditions (Gachanga & Mutisya, 2015, p. 281). The reductionism in western museological practices views much of the material culture as either 'religious or ritualistic' (Gachanga & Mutisya, 2015, p. 282). This obscures layers of meaning within objects that are often votive or have serious pragmatic purposes in active peacebuilding. Similar healing properties are often within the content of dances, songs, stories, poetry and other intangible expressions of active transformative heritage practices (Figure 17.2).

Figure 17.2 'Peace Park, Nairobi'.
Source: Photo by Diana Walters, March 2013.

Participating in active heritage-based peacebuilding in Kenya was something that put up a mirror to my own beliefs and actions and I began to search for connections. I found that the same reductionism is also reflected within the marginalisation of alternative traditions of theology that share characteristics many with non-western approaches to spirituality. The history of the early Christian church in the UK, for example, can be viewed as a hidden narrative of oppression. It was at the Synod of Whitby in 664 CE that a meeting took place to explore a clash between two missions, the Celtic and the Roman. The Roman mission won, and the Celtic tradition was reduced to the margins by patriarchy and the establishment. Yet the spirituality typical of the Celtic way survives, summarised by J. Philip Newell as 'a spirituality characterised by a listening within all things for the life of God' (2008, p. 7). This experience is mirrored in other 'branches' of theology/philosophy that might be regarded as predominantly western. Richard Rohr, leading scholar and teacher in the Franciscan tradition, advocates Bonaventure's 'univocity of all being' – that everything is connected and that the particular (known as 'thisness') is in fact the route to the universal.[15] In this sense, each stone, leaf, person and animal is interconnected, and by extension, to harm one is to harm all. In this alternative orthodoxy, spaces of freedom are created within existing frameworks but without identifying with them. By definition, these small quiet spaces have the potential to transform, connect and thereby heal.

Which brings me to further difficulty. This is also a genuine dilemma that can leave me swaying in the wind of argument and counterargument; another explanation for my difficulty with the academic certainty and confidence of heritage commentators who write with such clarity and authority. Much is written about emotion and affect in heritage interpretation and understanding, and in many ways, I really welcome this as an expression of the richness and wealth of diversity and difference as a place of encounter, and the recognition of our multiple intelligences (much of loved of museum and heritage educators).[16] Yet I am wary of it. This was heightened for me on a visit to the National Centre for Civil and Human Rights in Atlanta, USA, and the experience of one particular exhibit (Walters, 2017, p. 48). The reconstruction of the Greensboro Woolworth's lunch counter 'sit-in' in 1960 – a peaceful protest by excluded black Americans and their allies – is a powerful, even iconic, image. The 'experience' for the visitor is to sit on a stool at the bar, and while facing pictures from the sit-in (which was violently attacked) to undergo abuse simulated by various technological devices and comprising being verbally and even physically attacked for about 90 seconds.[17] I found this whole experience profoundly disturbing. Maybe this was the intention, but my upset was not based on my own embodied event, but on observing the impact on those around me. It divided people. There were those who thought the whole thing a really good laugh, often these were children. There were others for whom it was clearly deeply traumatic. I don't know what the intention of the experience was, but I suspect that somewhere in there would be the word 'empathy'. Whereas before I had connected to my fellow visitors, chatted in the queue and shared stories, afterwards we actively avoided each other – embarrassed by our reactions, afraid of offending, unsure what or how to 'be' with one another.

I really struggle with how empathy is curated by such experiences. At the heart of that is a suspicion that true empathy is not really possible. How can I possibly understand how it is to be a Kikuyu victim of British colonial violence? How can I really feel the pain and trauma of generations of black Americans that still experience the violence born of systemic racism and segregation? I don't think I can. Psychologist Paul Bloom expresses it thus:

> Empathy is a spotlight focusing on certain people in the here and now. This makes us care more about them, but it leaves us insensitive to the long-term consequences of our acts and blind to the suffering of those we do not or cannot emphasize with.
> (Bloom, 2016, p. 9)

In other words, Bloom is suggesting that empathetic identification with one experience might actually deepen antagonism to those who are regarded as the perpetrators – however justified that might be. Such simplification reduces the nuances of human experience and moves towards the more binary dualistic positions that so hinder potential processes of healing.

In peacebuilding terms, the recognition that there is a spectrum of responses is obvious. But again, this requires us to start from a specific place *vis-a-vis* conflict. So, what might help? Bloom argues for reason and compassion. Liberation theologians, such as Boff and Boff, would argue for compassion, as a position of 'faith confronting injustice' (1986, p. 3). I would add embracing doubt, social justice and human rights, activism, and an understanding that dialogue is more about listening than speaking. If I start from here – that alternative space of co-resistance – rather than from where I may be more comfortable – then maybe I can contribute to a form of pacific heritage or *liberation heritology* from where peaceful change may occur. As Welsh critic and novelist Raymond Williams (1989, p. 118) wrote, '[t]o be truly radical is to make hope possible, rather than despair convincing'.

Justice weaving in Hebron

This vignette is about an emerging new form of co-resistance to the Israeli settlement project in the city of Hebron, located in the southern West Bank, and the potential of this form to navigate new sources of hope to an already fragmented and divided city. Following the Six Day War of 1967 in which Israel captured and occupied the West Bank and Gaza, a few Jewish settlers claimed their right to return to the Jewish history of Hebron.[18] None of the Israeli Jews that resettled in Hebron following the Six Day War were actual descendants of Hebron's Jews or had deeds to any of their properties. Rather, these settlers used theology and archaeology to legitimate their claims (Wolfe, 2006; Weizman, 2007; Hammami, 2019). They also commemorated past experiences of Hebron's Jews to justify the violence enacted on Palestinians, while creating a coherent geography of settlement that reframed Palestinian existence in Hebron as a historically distant and temporary event. The settlement project has fragmented the city, destroyed the local community, produced new traditions and deformed Hebron's cosmopolitan identity, which, as in Palestine more broadly, was rooted in a complex and interconnected religious heritage (Masalha, 2018). Local Palestinian people are subjected to invasive interrogations every day, while Israeli settlers move freely within and between the settlements (Hammami, 2019). This dehumanising practice, as Wolfe (2006) explains, is guided by a settler colonial policy that seeks territorial control, embodiment of indigeneity and the social killing of natives rather than peaceful accommodation. Today, Hebron's Old Town is known as a 'ghost, dangerous or captured city' (Weizman,

Figure 17.3 Al-Shuhada Street, centrally located in the Old Town of Hebron. The Israeli military welded shut the street-facing doors of all the homes and shops, and forcibly displaced most of its residents and shopkeepers.
Source: Photo by Feras Hammami, March 2019.

2004; Hammami, 2019). Figure 17.3 shows the violent closure that has been enforced on the Al-Shuhada Street, which until 1994 was the busiest street of the city of Hebron.

On-site observations, together with interviews with inhabitants, business owners and activists in 2018, uncovered a prevailing *sumud* (steadfastness), with emerging forms of transnational solidarity against the Israeli occupation. *Sumud* refers both to the maintenance of Palestinians on their land, and a more dynamic ideology of resistance that seeks to undermine the Israeli occupation and protect Palestinian collective memory. The escalation of violence in the name of Judaism prompted many Diaspora Jews, Israel-based and international activists to join the Palestinian *sumud*, and form co-resistance.[19]

Mobilised by the slogan '[T]he occupation is not our Judaism', several left-wing leaning Jewish activists' organisations were established in different countries, including Israel. Inspired by the Jewish legacies of the exodus and other expulsions, they seek 'peace' for both Palestinians and Israelis and aspire towards the reparation of the world. These activists participate in the documentation of abuses, provision of legal and social support, organisation of anti-occupation campaigns, the reconstruction of demolished homes, and the exposure of Israeli violence through public and political debates on social media (Waxman, 2017). Although their activism is small scale, emerging from below and largely uncoordinated, these efforts have made settler violence known to the public, protected many Palestinians from deliberate violent acts, and have given a new voice to the Palestinian Nakba.[20]

In this vignette, I will specifically link the Jewish legacies of struggle for liberation with contemporary activism in Palestine to make sense of co-resistance as a new form of activism against Nakba, and provide insights on the potential impact of heritage-led spaces of activism in societal change. Starting from somewhere else – humble, from below, unnoticed, heritage-led and small-scale – rather than top-down diplomatic affairs, seems more sensitive and responsive to the settler colonial project in Hebron, while helping to inform what we call here, *pacific heritage*.

The emergence of social non-movement?

In 1997, the Palestinian Liberation Organisation and the State of Israel signed the so-called Hebron Protocol which resulted in the division of Hebron into two areas (H1 and H2). While Areas-H1 are inhabited by Palestinians and controlled by the Palestinian authorities, Areas-H2 are inhabited by 35,000 Palestinians and 850 Israeli settlers and controlled by the Israeli military. The inclusion of the old town in Areas-H2 and the escalation of violence gave rise to a landscape of activism, much of which was inspired by notions of heritage and religion. Palestinian activists and heritage experts propagated the need to protect the living heritage of Hebron. They established the Hebron Rehabilitation Committee in Areas-H2 in order to restore buildings and the local community (Alazaa, 2016). Many families moved into the old town for *sumud*, while others were attracted by economic incentives or subsidised basic services. At the same time, forty descendants of Hebron's Jews signed a petition, demanding the Israeli government evacuate the illegal settlements founded on their families' properties (Amro & Bunte, 2017, p. 26). Some suggested the use of properties for cultural purposes, commemorating Jewish life and prompting Jewish–Arab understanding (Sipress, 1997), while others decided to donate the properties to the Palestinian authorities.

The misuse of Judaism in the settlement project impelled many Jewish activists from the *All That's Left* collective and similar groups to mobilise protests near religious sites, and during religious holidays.[21] For example, during the weekend of Chayei Sarah of 2013, a few activists erected a tent on Al-Shuhada Street 'to resemble Abraham's Tent, which according to traditional Jewish exegesis was open on all four sides so that anybody passing by would feel welcomed' (Omer-Man, 2013). The Israeli police seized the tent and signs and arrested the activists. Jewish activists would also try to help Muslim Palestinians access the Al-Haram Al-Ibrahimi for Friday prayers. Most of these activities were often practised by Palestinians and Jews separately – though sometimes they would communicate for on-streets protests.

Other co-resistance practices occurred in 2016, when activists from several Jewish Diaspora groups organised a campaign, following the slogan '[T]he Occupation is not our Judaism'. About eighty-five Jewish activists and many Palestinian activists marched through Hebron's streets, protesting the expulsion of locals, and street closures. They also restored one of the buildings, which had been abandoned due to continuous military closures and harassment, transforming it into the first cinema in the old town. In 2018, two Hebron-based resistance groups, Youth Against Settlement and Hebron Freedom Fund, alongside several Jewish activists groups co-organised a Passover Freedom Seder (Grace, 2018).[22] Organising the Freedom Seder in Hebron delivered a highly politicised message against the settler colonial policy in Hebron. While the first group of settlers arrived to Hebron to celebrate the Passover Seder and never left, the activists at Hebron's Freedom Seder refused to leave before liberating Hebron from settlers. After the event, the activists joined several local, small-scale activities that aimed at the restoration of buildings, establishing new public uses in abandoned buildings, helping pupils reach their schools, providing food to families living under military curfew, and raising awareness through educational programmes and debates. Several others debated economic, social and legal means to support local businesses, and community-based organisations, including the women's associations.

These growing local and transnational practices of *sumud* and co-resistance are not particularly visible in the everyday life of Hebron. They are overshadowed by the number and noise of top-down diplomatic peace efforts. I began to notice and track these practices only when I consistently asked my interviewees about prospects for a hopeful future. Though invisible and unseen, the impact of uncoordinated activism in Hebron seems metaphorically comparable to Asef Bayat's 'non-movement' (Bayat, 2013). For Bayat, non-movement represents the 'collective actions of noncollective actors. It grows when a large number of ordinary people and groups whose fragmented but similar activities trigger much social change, although these practices are rarely guided by an ideology or recognizable leaderships and organizations' (Bayat, 2013, p. 15). In Hebron, *sumud* and co-resistance practices are diverse. Some have emerged out of immediate need, while others followed particular political and social agendas. They are fluid, changeable, in a continuous state of becoming. A virtual map of these practices uncovers an emerging social non-movement whose impact is likely to grow through Bayat's (2013) 'quiet encroachment'. Virtually, the affective environments of these non-collective practices enable them to connect in a space outside of the official and diplomatic spaces of change and peace-making. It is crucial to recognise how heritage can contribute to the potential of these non-movements for a gradual and meaningful change. Within the larger political context of the region, it is hoped that these non-movements constitute a purposeful space that is locally situated and aimed at contributing to a new political stage of dismantling the regimes of injustice in the Palestinian–Israeli context instead of trying to fit a language of 'peace' within a system of inherent injustice.

Drawing strings together and trying to find and identify sustainable practices

Rather than pretending to have even some of the answers, which a standard 'conclusion' section ought to try and do, this process of drawing strings together is necessarily hesitant and partial. We are sure that many readers may feel that this is a bit of a cop out; an avoidance of our duty as academics and practitioners; at best displaying a weakness through extreme vagueness and at worst, perhaps even conveying a sense of false humility in order to underline our 'woke' credentials – the humble-brag as a badge of honour. Alternatively, however, we would like to make space for something a bit more powerful – an ambition to imagine the 'unimaginable' – and to do so in an unapologetic manner. Rather than humble comment, therefore, our declaration that we should seek to start *somewhere else* is a powerful proclamation towards making an intervention – to try and change a situation 'as it is' into something that it 'ought to be'…

In its invocation towards transformational change, which reflects social justice, we feel that this stated ambition goes beyond a practice of 'critical commentary'. Taking strength from the activist turn of recent years, and in parallel to the shift from 'Critical Geopolitics' towards 'Pacific Geopolitics', our ambition towards 'Pacific Heritage' provides a challenge, and a prompt towards action. In trying to outline the coordinates of this ambition, we would like to emphasise four elements through which sustainable, and realistically transformational practices can be identified.

First, is the underlining of the significance of the *processual turn* within heritage and peace studies. While the world of 'heritage practice' has perhaps always implicitly or explicitly recognised the importance of understanding heritage and peace in terms of relations, processes and activities – as *verbs* – we feel that further collaborative conversation is required in order to realise more fully what the implication of this parallel recognition of process entails and 'allows'. Rather than dealing with peace as a target that we ought to achieve, the empirical analysis of the Hebron case study showed that such ready solutions often produce more violence. Peace itself is a utopian political stage that is not necessarily what is needed in real time and place. What the activists wanted to do in Hebron is to expand on their solidarity activism, and bridge connections across time and space. They are interested in a safe future and present, as well as not returning to a dark past. Likewise, a processual understanding of both heritage and peace lies at the heart of the experience of peacebuilding practice in Kenya, and the connected critical self-reflection that ensued. Similarly, the inherent 'doing and re-doing' of the heritage tourism experiences on the Vie della Pace is another example of how peace might be practised in landscapes scarred by the shadows of war.

Secondly, and a connected point, we would underline the importance of connecting 'practice' and 'scholarship'. These are always false distinctions, but their imagined separation has sometimes had unhelpful consequences that have limited more creative conversations – perhaps most clearly articulated in the reflexive process displayed in the Kenya vignette. A common investment in a processual understanding of heritage and peace, together with the inherent openness towards activism and dialogue that this implies, provides an opportunity to make the most of this space of learning that hinges upon the affective and emotional capacity of heritage.

Thirdly, as a further rejection of dualistic ways of seeing things, we would like to make space for non-bottom-up actions, including those that range between the top-down and bottom-up actions, policy and viewpoints, as having a necessary (and inevitable) role. It has become a truism that 'bottom-up', unofficial and locally embedded efforts and mindsets are somehow always *better* than the top-down, official and 'authorised'. On the one hand, recent literature has suggested we need a more nuanced attitude to the local and bottom up (see, for instance, Harvey, 2015). On the other hand, by paying more attention to how power works, and in whose interests, we can try to identify and cherish the 'good' over the 'bad' rather than just assume that all bottom up is 'good' (and *vice versa*). Perhaps paralleling Tony Judt's (2010) suggestion that the institution of the nation state needs to be *re-thought* rather than *thrown out*, perhaps we can *re-think* how top-down effort and mindsets can be used to make things better. After all, as the recent global experience of coronavirus has taught us, one cannot solve a pandemic with only bottom-up activism, however authentically embedded, committed and ambitious such activity is.

Lastly, while we are not setting out any Grand Conclusions – or a 'road map' – towards a target of peace, our position is not 'meek'. In making these statements and setting out these ambitions, we are not hiding in the proverbial 'safe space' of the academic ivory tower, but neither are we assuming that violence is inherent nor conceding that everything must focus on the achievement of a negative peace. We feel that this is quite a powerful thing to say and think (with). Implicitly therefore, our task is not one of 'building consensus', but about making transformational dialogue possible and imaginable. This is messy and can have breakages on the way.

In terms of characterising our approach, we see this drawing together of practice and academic effort, conceptual and activist ambition as an exercise of intellectual weaving. Drawing on Jane Addams's feminist peace activist work in the early 20th century, this approach has been termed *peace weaving* (Bregazzi & Jackson, 2018). As such, it is an open-ended practice, perhaps with some knots and frayed ends, but always with potential and intrinsically textured, colourful and collaborative.

Notes

1. See, for example, Logan and Reeves (2009), Gegner and Ziino (2012), Wilson (2013), Windy and Baillie (2013), Roberts and Stone (2014), Sumartojo and Wellings (2014), Stig Sørensen and Viejo-Rose (2015), Viejo-Rose and Stig Sørensen (2015), Wallis and Harvey (2018), and Stig Sørensen et al. (2019).
2. See also Martin (2013), Mouffe (2013), Askins and Mason (2015), and Bull and Hansen (2016).
3. See Bregazzi and Jackson (2018) for definitions of peace and assumptions of *negative peace*.
4. See, for instance, Meskell (2002), Harrowell (2016), Brosché et al. (2017), Kalman (2017), and Isakhan and Zarandona (2018). This work is strongly characterised by an ambition towards criticality, but can sometimes perhaps be a bit caught up with 'heritage-as-physical-site' and 'heritage-as-icon-of-essential-identity'.
5. See also Walters et al. (2017), and Shields and Soeters (2017), who draw on Jane Addams's pioneering work in the early 20th century.
6. For a similar view on heritage 'as process or verb', see Harvey (2001, p. 327). Within peace studies, see, for instance, Shields and Soeters (2017, p. 329), who refer to *positive* peace as always focusing on relationships, networks and social transformation.
7. See Megoran (2010, 2011), McConnell et al. (2014), McGeachan (2014), Brickell (2015), Courtheyn (2018), Horrowell (2018), Osborne (2018), and McDowell and Crooke (2019).
8. While 'landmark' declarations over the UNESCO World Heritage Site bridge at Mostar, or the Al-Haram Al-Ibrahimi mosque and caves are perhaps best known, the same process of *fixing* authorised heritage meanings, relationships and (targeted) outcomes is much broader. In all cases, a supposedly 'natural' connection between heritage and conflict narratives is assumed – for instance, see the US National Park's legal designations of national battlefields, military installations and conflict memorial sites (see van den Dungen, 2017).
9. See also the implications of Giblin (2014, p. 515), who notes that 'heritage should not be assessed regarding if it has healed but analysed regarding how the process of healing is being undertaken'.
10. See also Vos (2015), who reflects upon reconciliatory function of heritage at a local level within a European framework.
11. For work supporting a local scale and bottom-up viewpoint of practice, see for instance, McGeachan (2014), Brickell (2015) and McDowell and Crooke (2019).
12. This trail is also referred to as the Carnic High Trail, Carnic High Mountain Trail, Way of Peace, Peace Trail and the Vie della Pace.
13. See https://www.dolomitenfreunde.at/start-en.
14. This is further developed in Walters (2010).
15. Bonaventure (1221–1247), OFM, born Giovanni di Fidanza, was an Italian medieval Franciscan, scholastic theologian and philosopher. The seventh Minister General of the Order of Friars minor, he was also Cardinal Bishop of Albano. See Rohr (2015, 2018).
16. For work on heritage, affect and emotion, see, for instance, Tolia-Kelly et al. (2017), and Waterton (2014).
17. The stool you are sitting on is shaken unexpectedly.
18. This history connected to the Jews who lived in Hebron's Old Town between 1541 and the massacre of 1929, which led to the expulsion of the Jewish community.
19. This form of activism emerged after the collapse of the Oslo Accords of 1993 and the eruption of Al-Aqsa Intifada (second uprising 2000–2005). It replaces the traditional calls for coexistence which was propagated for during the 1990s by a range of dialogue groups, civil society partnerships, and other diplomacy and peace-making programs.

20 The Nakba began during the 1948 Palestine war and ended with the establishment of the State of Israel as a Jewish national homeland. Though Nakba is often described as a one-single event that between December 1947 and January 1949 resulted in the expulsion of more than 700,000 Palestinian Arabs from their homes and lands (Zochrot, 2014), it is conceived here as 'an ongoing event of erasure, occupation and dispossession'. To talk about Nakba is 'to engage in a complex act of translation that includes moving back and forth between narratives, memories, languages, times and geographies' (Eqeiq, 2012). Its events and realities are still unfolding in Hebron and other places of ethnic cleansing (Pappé, 2006).
21 See https://www.facebook.com/AllThatsLeftCollective/ (accessed on 5th October 2020).
22 The Jewish activist groups included All That's Left, the Centre for Jewish Nonviolence, Jewish Voice for Peace, J Street, Open Hillel, IfNotNow and Students for Justice in Palestine. The Passover is a biblical tradition, which the book of Exodus discusses as the liberation of the Israelites from slavery in ancient Egypt. It inspired the organisation of the first politically motivated Freedom Seder event which was modelled in 1969 in Washington D.C. to mourn the murder of Martin Luther King Jr (Freedom Seder Hebron, 2018).

References

Alazaa, N. (2016). *Forced population transfer: The case of the old city of Hebron*. Bethlehem: BADIL Resource Centre for Palestinian Residency & Refugee Rights.
Ali, S. (Ed.) (2007). *Peace parks: Conservation and conflict resolution*. Boston, MA: MIT Press.
Amro, I., & Bunte, T. (2017). *Hebron, ghost town*. Germany and Palestine: Edition of 3000.
Ashworth, G. (2013). Keynote address: The relationship between ethnic conflict and heritage tourism. *The Peace Conference: Post conflict, cultural heritage and regional development*, 9th–11th October 2013, Wageningen.
Askins, K., & Mason, K. (2015). Us and us: Agonism, non-violence and the relational spaces of civic activism, *ACME: International Journal for Critical Geographies*, 14(2), 422–430.
Bayat, A. (2013). *Life as politics: How ordinary people change the Middle East*. 2nd ed. Palo Alto, CA: Stanford University Press.
Bloom, P. (2016). *Against empathy. The case for rational compassion*. London: The Bodley Head.
Boff, L., & Boff, C. (1986). *Introducing liberation theology*. Maryknoll, NY: Orbis Books.
Bregazzi, H., & Jackson, M. (2018). Agonism, critical political geography, and the new geographies of peace. *Progress in Human Geography*, 42(1), 72–91. https://doi.org/10.1177/0309132516666687.
Brickell, K. (2015). Towards intimate geographies of peace? Local reconciliation of domestic violence in Cambodia. *Transactions of the Institute of British Geographers*, 40(3), 321–333. https://doi.org/10.1111/tran.12086.
Brosché, J., Legnér, M., Kreutz, J., & Ijla, A. (2017). Heritage under attack: Motives for targeting cultural property during armed conflict. *International Journal of Heritage Studies*, 23(3), 248–260. http://dx.doi.org/10.1080/13527258.2016.1261918.
Bull, A. C., & Hansen, H. L. (2016). On agonistic memory. *Memory Studies*, 9(4), 390–404. https://doi.org/10.1177/1750698015615935.
Bull, A. C., Hansen, H. L., Kansteiner, W., & Parish, N. (2019). War museums as agonistic spaces: Possibilities, opportunities and constraint. *International Journal of Heritage Studies*, 25(6), 611–625. https://doi.org/10.1080/13527258.2018.1530288.
Cho, M. (2007). A re-examination of tourism and peace: The case of the Mt. Gumgang tourism development on the Korean Peninsula. *Tourism Management*, 28(2), 556–569. https://doi.org/10.1177/1750698015615935.
Coombes, A., Hughes, L., & Karega-Munene (2014). *Managing heritage, making peace. History, identity and memory in contemporary Kenya*. New York: Tauris.
Courtheyn, C. (2018). Peace geographies: Expanding from modern-liberal peace to radical trans-relational peace. *Progress in Human Geography*, 42(5), 741–758. https://doi.org/10.1177/0309132517727605.
D'Amore, L. J. (1988). Tourism – A vital force for peace. *Tourism Management*, 9(2), 151–154. https://doi.org/10.1016/0261-5177(88)90025-8.

Eqeiq, A. (2012). Epilogue. *Omrim yeshna eretz – Hekayat balad (Once upon a Land)/A Tour Guide*. Sedek. Zochrot: Tel-Aviv, 2012, 500–509. "Al-Nakbah bithalathutumat kalema wa akthar".

FreedomSederHebron (2018). A freedom Seder in Hebron. Retrieved from: https://allthatleft-site.files.wordpress.com/2018/03/a-freedom-seder-in-hebron.pdf (Accessed on 26 January 2019).

Gachanga, T., & Mutisya, M. (2015). Interfaith dialogue at peace museums in Kenya. *Journal of Peace Education*, 12(3), 277–284. http://dx.doi.org/10.1080/17400201.2015.1103395.

Gegner, M., & Ziino, B. (Eds.) (2012). *The heritage of war*. London: Routledge.

Giblin, J. (2014). Post-conflict heritage: Symbolic healing and cultural renewal. *International Journal of Heritage Studies*, 20(5), 500–518. http://dx.doi.org/10.1080/13527258.2013.772912.

Grace, H. (2018). An innovative freedom Seder in Hebron. *The Palestine-Israel Journal*. Retrieved from: http://www.pij.org/details.php?blog=1&id=431 (Accessed on 20 January 2019).

Hammami, F. (2019). Heritage necropolitics and the capture of Hebron: The logic of closure, fear, humiliation, and elimination. In M. Ristic & S. Frank (Eds.), *Urban heritage in divided cities: Contested pasts 1st edition* (pp. 15–34). Oxon & New York: Routledge.

Harrowell, E. (2016). Looking for the future in the rubble of Palmyra: Destruction, reconstruction and identity. *Geoforum*, 69(1), 81–83. https://doi.org/10.1016/j.geoforum.2015.12.002.

Harrowell, E. (2018). Towards a spatialised understanding of reconciliation. *Area* 50(2), 240–247. http://dx.doi.org/10.1111/area.12365.

Harvey, D. C. (2001). Heritage pasts and heritage presents: Temporality, meaning and the scope of heritage studies. *International Journal of Heritage Studies*, 7(4), 319–338. http://dx.doi.org/10.1080/13581650120105534.

Harvey, D. C. (2015). Heritage and scale: Settings, boundaries and relations, *International Journal of Heritage Studies*, 21(6), 577–593. http://dx.doi.org/10.1080/13527258.2014.955812.

Isakhan, B., & Zarandona, J. A. G. (2018). Layers of religious and political iconoclasm under the Islamic state: Symbolic sectarianism and pre-monotheistic iconoclasm. *International Journal of Heritage Studies*, 24(1), 1–16. http://dx.doi.org/10.1080/13527258.2017.1325769.

Judt, T. (2010). *Ill fares the land: A treatise on our present discontents*. London: Penguin.

Kalman, H. (2017). Destruction, mitigation, and reconciliation in cultural heritage. *International Journal of Heritage Studies*, 23(6), 538–555. http://dx.doi.org/10.1080/13527258.2017.1289475.

Krauss, W. (2015). Heritage and climate change: A fatal affair. In D. C. Harvey & J. Perry (Eds.), *The future of heritage as climates change: Loss, adaptation and creativity* (pp. 43–61). London: Routledge.

Logan, W., & Reeves, K. (2009). *Places of pain and shame: Dealing with 'difficult heritage'*. London: Routledge.

Manning, R. E., & Manning, M. S. (2013). *Walking distance: Extraordinary hikes for ordinary people*. Portland: Oregon State University Press.

Markham, A., Osipova, E., Lafrenz Samuels, K., & Caldas, A. (2016). World heritage and tourism in a changing climate. United Nations Environment Programme, Nairobi, Kenya and United Nations Educational, Scientific and Cultural Organization, Paris, France.

Martin, J. (2013). Introduction: Democracy and conflict in the work of Chantal Mouffe. In J. Martin (Ed.), *Chantal Mouffe: Hegemony, radical democracy and the political* (pp. 1–11), London: Routledge.

Masalha, N. (2018). *Palestine: A four thousand year history*. London: Zed Books.

McConnell, F., Megoran, N., & Williams, P. (Eds.) (2014). *Geographies of peace*. London: IB Tauris.

McDowell, S., & Crooke, E. (2019). Creating liminal spaces of collective possibility in divided societies: Building and burning the Temple. *Cultural geographies*, 26(3), 1–17. https://doi.org/10.1177/1474474018817791.

McGeachan, C. (2014). Historical geography I: What remains?. *Progress in Human Geography*, 38(6), 824–837. https://doi.org/10.1177/0309132514546449.

Megoran, N. (2010). Towards a geography of peace; Pacific geopolitics and evangelical Christian crusade apologies. *Transactions of the Institute of British Geographers, 35*(3), 382–398. https://doi.org/10.1111/j.1475-5661.2010.00387.x.

Megoran, N. (2011). War *and* peace? An agenda for peace research and practice in geography. *Political Geography, 30*(4), 178–189. https://doi.org/10.1016/j.polgeo.2010.12.003.

Meskell, L. (2002). Negative heritage and past mastering in archaeology. *Anthropological Quarterly, 75*(3), 557–574. https://www.jstor.org/stable/3318204.

Mouffe, C. (2013). *Agonistics: Thinking the world politically.* London: Verso.

Newell, J. P. (2008). *Christ of the Celts: The healing of creation.* San Francisco: Jossey-Bass.

Olupona, J. K. (1991). *African traditional religions in contemporary society.* New York: Paragon House Publishers.

Omer-Man, S. M. (2013). Activists arrested in Hebron: 'Segregation isn't our Judaism'. Retrieved from: https://www.972mag.com/activists-arrested-in-hebron-segregation-isnt-our-judaism/ (Accessed on 21 May 2020).

Osborne, B. (2018). Reflecting on the Great War 1914–2019: How has it been defined, how has it been commemorated, how should it be remembered? In J. Wallis & D. C. Harvey (Eds.), *Commemorative spaces of the First World War: Historical geographies at the centenary* (pp. 209–224). London: Routledge.

Pappé, I (2006). *The ethnic cleansing of Palestine.* London and New York: Oneworld.

Richmond, O. P., & MacGinty, R. (2019). Mobilities and peace. *Globalizations, 16*(5), 606–624. http://dx.doi.org/10.1080/14747731.2018.1557586.

Roberts, C., & Stone, P. R. (2014). Dark tourism and dark heritage: Emergent themes, issues and consequences. In I. Convery, G. Corsane, & P. Davis (Eds.), *Displaced heritage: Dealing with disaster and suffering* (pp. 9–18). Suffolk: Boydell & Brewer.

Rohr, R. (2015). *The univocity of being* (Center for Action and Contemplation). Retrieved from: https://cac.org/the-univocity-of-being-2015-05-27/ (Accessed on 15 April 2021).

Rohr, R. (2018). *The Franciscan vision* (Center for Action and Contemplation). Retrieved from: https://cac.org/the-franciscan-vision-2018-11-26/ (Accessed on 15 April 2021).

Sandwith, T., Shine, C., Hamilton, L., & Sheppard, D. (2001). *Transboundary protected areas for peace and co-operation.* Gland, Switzerland and Cambridge: IUCN.

Shepherd, J., & Laven, D. (2021). Hostels in hostile territory: The role of tourism spaces in transformative dialogue. In J. T. Da Silva, Z. Breda, & F. Carbone (Eds.), *Role and impact of tourism in peacebuilding and conflict transformation* (pp. 195–217). Hershey: IGI Global.

Shepherd, J., & Laven, D. (2020). Providing counter-narratives: The positive role of hostels in the Israeli-Palestinian context. *Tourism Geographies, 22*(4–5), 848–871. http://dx.doi.org/10.1080/14616688.2019.1669215.

Shields, P. M., & Soeters, J. (2017). Peaceweaving: Jane Addams, positive peace, and public administration. *American Review of Public Administration, 47*(3), 323–339. https://doi.org/10.1177/0275074015589629.

Sipress, A. (1997). Hebron descendants decry actions of current settlers: They are kin of the Jews ousted in 1929. Retrieved from: http://www.angelfire.com/il/FourMothers/Yona.html. (Accessed on 25 October 2019).

Smith, L., Shackel, P. A., & Campbell, G. (Eds.) (2011). *Heritage, labour and the working classes.* London: Routledge.

Stig Sørensen, M. L., & Viejo-Rose, D. (Eds.) (2015). *War and cultural heritage: Biographies of place.* Cambridge: Cambridge University Press.

Stig Sørensen, M. L., Viejo-Rose, D., & Filippucci, P. (Eds.) (2019). *Memorials in the aftermath of armed conflict: From history to heritage.* Basingstoke: Palgrave.

Sumartojo, S., & Wellings, B. (Eds.) (2014). *Nation, memory and Great War commemoration: Mobilising the past in Europe, Australia and New Zealand.* Bern, Switzerland: Peter Lang Publishers.

Tolia-Kelly, D., Waterton, E., & Watson, S. (Eds.) (2017). *Heritage, affect and emotion.* London: Routledge.

Tunbridge, J. E., & Ashworth, G. J. (1996). *Dissonant heritage: The management of the past as a resource in conflict.* London: Wiley.

UN General Assembly. (2015). International year of sustainable tourism for development, 2017: Resolution/adopted by the General Assembly (A/RES/70/193). Retrieved from: https://digitallibrary.un.org/record/821074 (Accessed on 5 May 2021).

van den Dungen, P. (2017). The heritage of peace: The importance of peace museums for the development of a culture of peace. In D. Walters, D. Laven, & P. Davis (Eds.), *Heritage and peacebuilding* (pp. 7–16). Woodbridge: Boydell & Brewer.

Viejo-Rose, D., & Stig Sørensen, M. L. (2015). Cultural heritage and armed conflict: New questions for an old relationship. In E. Waterton, & S. Watson (Eds.), *The Palgrave handbook of contemporary heritage research* (pp. 281–296). Basingstoke: Palgrave.

Vos, C. (2015). Debating the reconciliatory uses of heritage. European post-monumentalism versus regional national monumentalism. *International Journal of Heritage Studies, 21*(7), 716–733. http://dx.doi.org/10.1080/13527258.2014.1001424.

Walleczek-Fritz, J., & Fritz, P. (2015). *Ascend, apprehend, Aspire: The First World War along Carnic Ridge between Sexten and Plöckenpass.* Kartitsch: Sextten/Sesto.

Wallis, J., & Harvey, D. C. (Eds.) (2018). *Commemorative spaces of the First World War: Historical geographies at the centenary.* London: Routledge.

Walters, D. (2010). African peace traditions. In N. Young (Ed.), *Oxford international encyclopaedia of peace.* Oxford: Oxford University Press. Retrieved from: https://www.oxfordreference.com/view/10.1093/acref/9780195334685.001.0001/acref-9780195334685-e-008?rskey=XsLZkn&result=8 (Accessed on 15 April 2021).

Walters, D. (2017). Can museums build peace? The role of museums in peacebuilding and internationalism. In D. Walters, D. Laven, & P. Davis (Eds.), *Heritage and peacebuilding* (pp. 39–52). Woodbridge: Boydell & Brewer.

Walters, D., Laven, D., & Davis, P. (Eds.) (2017). *Heritage and peacebuilding.* Woodbridge: Boydell & Brewer.

Waterton, E. (2014). More-than-representational heritage? The past and the politics of affect. *Geography Compass, 8*(11), 823–33. https://doi.org/10.1111/gec3.12182.

Waxman, D. (2017). Young American Jews and Israel: Beyond birthright and BDS. *Israel Studies, 22*(3), 177–199. https://doi.org/10.2979/israelstudies.22.3.08.

Weizman, E. (2004). The politics of verticality: The West Bank as an architectural construction. *Mute,* 1/27. Retrieved from: https://www.metamute.org/editorial/articles/politics-verticality (Accessed on 5 February 2020).

Weizman, E. (2007). *Hollow land: Israel's architecture of occupation.* New York: Verso Press.

Williams, R. (1989). *Resources of hope: Culture, democracy, socialism.* London: Verso Books.

Wilson, R. (2013). *Cultural heritage of the Great War in Britain.* Farnham: Ashgate.

Windy P., & Baillie, B. (2013). *Locating urban conflicts: Ethnicity, nationalism and the everyday.* Aldershot, Hampshire: Palgrave Macmillan.

Winter, T. (2015). Heritage diplomacy. *International Journal of Heritage Studies, 21*(10): 997–1015. http://dx.doi.org/10.1080/13527258.2015.1041412.

Wohlmuther, C., Wintersteiner, W., & Wagner, C. (2014). Connected by the trails that used to divide us: Peace trails in the Alps-Adriatic region. In C. Wohlmuther, & W. Wintersteiner (Eds.), *International handbook on tourism and peace* (pp. 321–334). Klagengurt/Celovec: Centre for Peace Research and Peace Education, Klagenfurt University in cooperation with the World Tourism Organization (UNWTO).

Wolfe, P. (2006). Settler colonialism and the elimination of the native. *Journal of Genocide Research, 8,* 387–409. https://doi.org/10.1080/14623520601056240.

Zochrot (2014). al-Nakba. Retrieved from: https://zochrot.org/en/contentAccordion/nakba (Accessed on 27 September 2020).

Part IV
Environment, heritage and society

18 Sustainable heritage and climate change

Alessandra Bonazza

Introduction

Despite the considerable European resources committed to research and policy development in the last years aiming at strengthening cultural heritage protection against the impacts of climate change, further steps are still undoubtedly needed in order to ensure a sustainable management and safeguarding of heritage sites at risk. Slow and extreme climate changes impose continuous and new challenges to the effective conservation of cultural heritage, particularly in climate hot spot areas (e.g. Mediterranean Basin, Alpine region), where the effect of multiple pressures is amplified.

Research into adaptation strategies, user-driven solutions and tools is therefore urgently required, based on sound scientific studies, capitalization of achieved knowledge, transferring and dissemination of results and coordinated actions among the different actors involved in the decision-making process for protection and management of cultural and natural heritage (public authorities, scientific community, private sector, rescue bodies).

As highlighted by the European Agenda for Culture, since 2007, cultural heritage has been a priority for European cooperation on cultural policy. Cultural heritage is also considered a strategic resource for a sustainable Europe, as stated in the Council Conclusions of May 2014. Being of value to society from a cultural, environmental, social and economic point of view, the sustainable management of heritage emerges as a strategic century need in the 21st century. In 2018, in the framework of the European Year of Cultural Heritage, cultural heritage has finally been recognized as an incentive for strengthening the resilience of society in facing the impact of catastrophic events and its protection has been highlighted as having a key role in support of socio-economic development and sustainable tourism.

At global policy level, the Sendai Framework for Disaster Risk Reduction 2015–2030 represented a significant turning point, by including among its key priorities the protection of cultural heritage and by inviting national authorities to cooperate in increasing an awareness of cultural heritage impacts in the context of exposure to hazards. The framework was followed by the Action Plan on the Sendai Framework, published in 2016 by the European Commission and covering a five-year period, providing for a more systematic disaster-risk-informed approach in EU policy-making. In particular, in support of the implementation of Sendai Priority 4 (Enhancing disaster preparedness for effective response and to 'Build Back Better' in recovery, rehabilitation and reconstruction), the development of good practices on the integration of cultural heritage in the national disaster risk reduction strategies to be developed by EU Member States is envisaged among implementation measures (Key Area 4; Bonazza et al., 2018).

This chapter illustrates relevant EU-funded research initiatives concerning the impact of climate change on cultural heritage, highlights key points to consider for a proper methodology of analysis and discusses significant results obtained by scientific studies conducted up to now for stone materials employed in cultural heritage. Finally, the still existing gaps and consequently future directions for research are highlighted.

Climate change impacts on cultural heritage: the European research framework

For the first time, research on the assessment of climate change impacts on cultural heritage has been the primary focus in the EU FP6 Project 'Global Climate Change Impact on Built Heritage and Cultural Landscapes' (Noah's Ark, 2004–2007; Sabbioni et al., 2010). The methodology applied encompassed the identification of climate parameters in determining the damage of selected heritage materials and the development of future scenarios of slow deterioration phenomena, such as surface recession and thermal stress on marbles, biological accumulation on architectural surfaces, metals corrosion, based on specific damage functions and models (Bonazza et al., 2009a, 2009b; Gomez-Bolea et al., 2012; Ciantelli et al., 2018; Sesana et al., 2020). One of the major outputs of this project comprises a Vulnerability Atlas and Guidelines document, offering adaptation and mitigation strategies for cultural heritage management in the face of climate change. The Vulnerability Atlas included European maps produced by applying the Global and Regional Hadley climate models (grid resolution of 295 × 278 km and 50 × 50 km, respectively) under the A2 scenarios (IPCC SRES Emission Scenarios used in TAR and FAR). The Noah's Ark project provided a unique know-how on future projections of damage induced mainly by slow climate changes on heritage materials outdoors exposed and introduced for the first time the concept of heritage climatology (Sabbioni et al., 2010).

The scientific approach developed within Noah's Ark constituted the base for research enhancement that was carried out by the FP7 Climate for Culture Project (2009–2014) (Hujibregts et al., 2012; Kramer et al., 2013). In this project, hazard and damage projections were forecasted to assess the impact of the slow ongoing climate change rather than extreme events effects on outdoor and indoor cultural heritage sites. The developed research methodology coupling climatology, building engineering and conservation science expertise allowed the creation of vulnerability maps of historic building envelopes and preserved internal artworks. In addition, projections for sea level increasing up to the year 2100 were calculated using scenario simulation with a global climate model and data from the regionally coupled atmosphere–ocean model (Regional model REMO run on the horizontal grid of 12.5 km, EUR-11; Jacob et al., 2012; Leissner et al., 2015).

Recently, H2020-funded projects (HERACLES, STORM) aimed at enhancing the resilience of cultural heritage towards climate change effects and natural hazards. In particular, the H2020 Project 'Heritage Resilience Against CLimate Events on Site' (HERACLES, 2016–2019, http://www.heracles-project.eu) had as a main objective 'to design, validate and promote responsive systems/solutions for effective resilience of cultural heritage against climate change effects, considering as a mandatory premise a holistic, multidisciplinary approach through the involvement of different expertise'. This was operationally pursued with the development of a system exploiting an ICT platform able 'to collect and integrate multisource information in order to effectively provide complete and updated situational awareness and support decision for innovative measurements improving cultural heritage resilience, including new solutions for maintenance and conservation' (Padeletti et al., 2018). The H2020 Project 'Safeguarding cultural heritage through Technical and Organisational Resources Management' (STORM, 2016–2019, http://www.storm-project.eu/) proposed 'a set of novel predictive models and improved non-invasive and non-destructive methods of surveying and diagnosis, respectively, for effective prediction of environmental changes and for revealing threats and conditions that could damage materials and structures of cultural heritage'. Moreover, the STORM project determined 'how different vulnerable materials, structures and buildings are affected by different extreme weather events together with risks associated with climatic conditions or natural hazards, offering improved, effective adaptation and mitigation strategies, systems and technologies to different materials and structures'. An important result of STORM was a cooperation platform for collaboratively collecting and enhancing knowledge, processes and methodologies on the sustainable and effective safeguarding and management of European cultural heritage. H2020 HERACLES and STORM projects started finally to pay attention on the impacts of extreme climate events.

The newly concluded Interreg Central Europe Project 'Risk assessment and sustainable protection of Cultural Heritage in changing environment' (ProteCHt2save, 2017–2020, https://www.interreg-central.eu/Content.Node/ProteCHt2save.html) aimed at strengthening the resilience of cultural heritage (specifically monumental complexes with related indoor collections) to extreme climate changes through the development of feasible and tailored solutions for supporting regional and local authorities with preparedness measures and evacuation plans in case of emergencies. In this project, 12 different combinations of six forcing global models, driving five regional models, have been taken into account for the elaboration of Web GIS tool for risk mapping of cultural heritage in Europe and in the Mediterranean Basin, exposed to extreme events linked to climate change, particularly heavy rains, flood and fire due to drought periods (Sardella et al., 2020). The results from ProteCHt2save are currently under capitalization and further development within the newly funded Interreg Central Europe Project 'STRENgthening resilience of Cultural Heritage at risk in a changing environment through proactive transnational cooperation, STRENCH' (2020–2022, https://www.interreg-central.eu/Content.Node/STRENCH.html), which also foresees the application of satellite data from the services of the EU Copernicus Programme for the safeguarding of cultural heritage at risk in extreme changing environment.

Climate change impacts on cultural heritage: methodology

Research focused on the assessment of climate change impacts on cultural heritage requires careful consideration of the following key points for setting up a sound and proper methodology of analysis:

1 Identification of environmental (climate and pollution) parameters with priority in causing deterioration in relation to the heritage under threat, by considering the building material, environmental context, exposure, cultural and socio-economic value, is a key starting point. Spatial and temporal resolutions should be defined for

each parameter, whether acting individually or in synergy. The selection should be driven by a heritage climatology concept, where significant parameters are selected in relation to their potential impact on heritage assets and not following the conventional meteorology, which is not always the most useful in making estimates of potential damage.

2 Evaluations on possible combinations of different parameters and on the consequent damage caused on cultural heritage as effect of their synergic action accumulated over time should be foreseen.

3 Identification of critical elements determining the vulnerability of cultural heritage assets in different domains (chemical-physical, cultural, economic and social) should be necessary. Concerning chemical-physical domain for example, organic and inorganic materials are susceptible to different deterioration processes caused by diverse combinations of environmental parameters. The vulnerability rate of cultural heritage categories should be assessed.

4 A clear distinction from the beginning of the analysis should be done between slow climate changes (such cyclic T/RH effects and environmental fatigue, atmospheric pollution changes or slow sea-level rise), for which long-term strategies can be planned and vulnerability and resilience measures have already been developed at research level, and extreme climate events (heavy rain, flooding, fire due to drought periods), for which emergency plans are necessary.

5 The spatial and temporal resolution of the climate and pollution parameters extractable from climate models entails the spatial and temporal resolution of the damage output: future scenarios of deterioration phenomena occurring at local-building scale local with high spatial variability are unlikely developable.

6 Formulation of risk expressions determining the frequency of climate events likely to cause deterioration and development of functions for damage quantification on specific materials is necessary.

Dealing with changing climate as driving factor for damage of built heritage

Research carried out up to now on climate change impacts on built heritage provides evidence of the driving role of climate parameters in determining forthcoming decay processes (Brimblecombe et al., 2006, Brimblecombe and Grossi, 2009). In the future, regulatory pressures are likely keeping the main acidic air pollutants at low levels in European cities, and preventive conservation strategies, including the pedestrianization of areas in which monuments are located, will possibly be more frequently adopted, bringing about a shift in the balance of mechanisms causing damage on buildings. In a recognized situation of changing climate, an understanding of what is expected in the future in terms of decay of heritage stones assumes fundamental importance, especially when considering effective preventive strategies for the conservation and protection of monuments and historic buildings.

European-based maps have been developed representing future trends of the different damage processes on building materials (Grossi et al., 2007; Bonazza et al., 2009a, 2009b), and a number of papers have been published providing more detailed results with intra-regional resolution on specific European areas (Grossi et al., 2008a, 2008b; Brimblecombe and Grossi, 2009; Smith et al., 2010; Grossi et al., 2011). These research works have a key common line indicating that climate will probably dominate in the future over pollution in causing the overall damage on monuments and historic buildings implying a change from the well-known dominating blackening of architectural surfaces in urban areas.

The following sections summarize the major results obtained by the available projections of climate change impact on heritage stone materials: it has to be highlighted that several uncertainties still exist, mainly due to the difficulty of finding suitable models for an accurate and exhaustive quantification of the caused damage. This is true even for the more widely studied and modelled deterioration processes, as in the case of surface recession of carbonate stone induced by chemical dissolution (Bonazza et al., 2009a). The methods and research procedures applied for obtaining the results here summed up and further discussed are detailed in the respective cited references in each section.

Surface recession of carbonate stone

Results from Bonazza et al. (2009a), obtained through a modified Lipfert function (Lipfert, 1989), foresee an increase in karst recession over the 21st century in Europe for carbonate stones of low and medium porosity (<25%). This is the result of a rise in atmospheric carbon dioxide concentrations. The extent of the increase will be most marked in high precipitation areas (e.g. mountains of Central Europe, Scandinavia and Scotland), where carbonate buildings are expected to undergo the highest risk of damage due to 'clean' rainfall. The projected increase could exceed 6 μm $year^{-1}$ in regions where recession was around 20 μm $year^{-1}$ in 1961–1990 (30% increase). The effects of classic acidic air pollutants (SO_2 and HNO_3) and acid rain are forecast to be quite weak, especially in rural areas, but also in urban areas, where more stringent pollution regulation policies have reduced such effects. However, the projection maps presented in Bonazza et al. (2009a) do not consider conditions at local levels in urban areas, where pollutant concentrations could be significant. To this end, research (Grossi et al., 2008a) in Oviedo, Paris and Prague where pollution concentrations are traditionally high predicts a decrease of surface recession ranging from 1 to 10 μm, depending on the damage function adopted for the future up to 2,100, in comparison with 1980 (Lipfert, ICP and MULTI-ASSESS functions) (Lipfert, 1989; Kucera et al., 2007). The minimum values are obtained in the case of use of Lipfert function, which is demonstrated to underestimate the pollution effect. It is therefore arguable that the reduction in surface recession of architectural surfaces in cities is principally linked to lower pollutant concentrations.

Salt crystallization

The crystallization pressure of soluble salts is one of the most important decay factors affecting the durability of porous building materials, such as sandstone, mortar and brick. Salt weathering is mainly driven by a phase change. The damage arises during the crystallization–dissolution cycles, which occur under precise thermohygrometric conditions. Non-hydrated salts, such as sodium chloride, crystallize at a fixed humidity virtually independent of temperature, whereas phase transitions in hydrated salts, such as sodium sulphate, are sensitive to both relative humidity and temperature (Benavente et al., 2008).

Projections at European scale, using HadCM3 and HadRM3 climate models, have been developed plotting the number of cycles per year of relative humidity around 75.5%, which drives the crystallization and dissolution of sodium chloride, as a quantitative indicator of salt crystallization events: the data obtained indicate that such events are likely to increase all over Europe (Sabbioni et al., 2010).

In dealing with salt weathering, Brimblecombe (2010) drew attention to the usefulness of the Köppen–Geiger climate classification when considering some aspects of heritage climatology. The Köppen–Geiger scheme describes climates in terms of codes, with the first letter describing the broad groups of climate, further types mainly depending on seasonality. It was possible to associate the traditional different climate types with potential salt weathering and estimate future salt damage due to sodium chlorides and sulphates at a local scale on the basis of climate change models (Grossi et al., 2011). Temperate fully humid climates (e.g. Cfb) seem to offer the highest potential for salt damage due to possible higher number of transitions in summer (change of phase, e.g. thenardite–mirabilite). Climates with dry summers (e.g. Csa or BS) tend to show a lesser frequency of transitions in summer. A fully humid climate (mainly Cfb) is found across a broad band from North Spain, North France, England across the Low Countries and into Central Europe (Brimblecombe, 2010). Previous work on climate projections (Benavente et al., 2008; Grossi et al., 2008b) suggested a potential increase in this area in the yearly number of transitions over the 21st century. However, a more detailed analysis (Grossi et al., 2011) indicated the importance of seasonal changes, potentially driven by changes in climate types, that are likely to occur in the future: for example, Western European areas with fully humid climates (Cfb) may change to a more Mediterranean or drier climate (Csa) and consequently show different seasonal patterns for salt transitions.

Cryoclastism

A general reduction in frost damage on porous stone is predicted, with the exception of Northern Europe and mountain areas. Grossi et al. (2007) were pioneers in addressing the future evolution of freezing processes on porous stones induced by temperature variations. Adopting the notion that frost damage is also driven by phase transitions, which occur at a fixed temperature of 0°C, and that damage is produced by freezing and thawing cycles that take place below and above this temperature, they evidenced that frost damage on the stone buildings of temperate Europe will decrease significantly in the future (i.e. decrease in the number of transitions). However, higher temperatures in Arctic Europe may drive the loss of permafrost, consequently increasing the risk for the buried archaeological and paleoecological remains (Brimblecombe et al., 2006).

Bio-deterioration

The available projections of future biological colonization on European cultural heritage proposed by Gomez-Bolea et al. (2012) indicated that, in general, Northern Europe will witness an increase in biomass accumulation because of the associated effects of increasing temperatures and higher precipitation. The projections in question are valid for siliceous stones and provide a general overview in line with the more spatially detailed work presented by Smith et al. (2008), who foresee an increase in building greening in Ireland due to more intense biological colonization. It should be emphasized that the building materials considered by the latter authors are siliceous ones (quartz sandstones), in widespread use in Ireland as a raw building material.

Analysis performed in the north UK deserves particular attention as it represents a joint effort of researchers in understanding the impact of climate change on cultural heritage at regional level. Through their experimental work and downscaled modelling,

McCabe et al. (2010) found evidence that increasing winter wetness in the north-western UK may not only already be leading to greater algal growth on buildings, but may also be changing the stone decay patterns considered up to now. They stress the potential implications of the seasonal cyclicity of precipitation events in Ireland, and western Atlantic Europe in general. Smith et al. (2010) propose a new model of surface deterioration due to longer wetted winter surfaces, leading to more biological colonization. It implies an increase in de-cohesion caused by swelling/shrinkage in the interior part of sandstone due to the alternation of winter saturation and summer drying.

Thermoclastism

With regard to thermoclastism, Bonazza et al. (2009b) mapped the scenarios of thermal stress on Carrara marble for the 21st century in Europe. The results evidenced how the Mediterranean Basin will generally continue to experience the highest level of risk. In addition, Central Europe and the South of England will be increasingly affected, while a decreasing risk is predicted for Northern Europe. The authors clearly pointed out the limits of validity of their work, emphasizing the need for a higher degree of spatial resolution at the building level, to obtain a more precise picture of architectural features and their exposure to sunlight. A higher temporal resolution is also required, in order to represent shading by clouds and allow the quantification of thermohygrometric shocks due to sudden rainfall on sun-heated surfaces. In addition, it should be underlined that the work in question concerns a specific stone, Carrara marble, which has a weak calcite texture and polycrystalline isotropic thermal expansion behaviour.

Concluding remarks

Despite the international evidence of the increasing impacts of climate change on cultural heritage, cultural heritage is still not exhaustively considered as affected by climate change that therefore merits protection. There are only few, sporadic examples of national plans for adaptation and mitigation including specific measures for its safeguarding based on damage quantification.

It is therefore imperative that future research should be dedicated on the development of damage functions for quantitative evaluation of the impacts of climate change on specific heritage materials and of multi-risk scenarios for complex systems such as historic centres, archaeological sites, underwater remains, terraced landscapes. Major attention should be directed towards the proposition of preparedness strategies in response to extreme climate events linked to climate change (heavy rain, flooding, drought periods), by mapping risk-prone areas at high-resolution spatial scale. New investigations should build upon the existing knowledge and needs of stakeholders involved in the protection and management of cultural heritage at risk due to climate change.

References

Benavente, D., Brimblecombe, P., & Grossi, C. M. (2008). Salt weathering and climate change. In M. P. Colombini, & L. Tasso (Eds.), *New trends in analytical, environmental and cultural heritage chemistry* (pp. 277–286). Kerala, India: Transworld Research Network.

Bonazza, A., Messina, P., Sabbioni, C., Grossi, C. M., & Brimblecombe, P. (2009a). Mapping the impact of climate change on surface recession of carbonate buildings in Europe. *Science of the Total Environment, 407*(6), 2039–2050. https://doi.org/10.1016/j.scitotenv.2008.10.067.

Bonazza, A., Sabbioni, C., Messina, P., Guaraldi, C., & De Nuntiis, P. (2009b). Climate change impact: Mapping thermal stress on Carrara marble in Europe. *Science of the Total Environment, 407*(15), 4506–4512. https://doi.org/10.1016/j.scitotenv.2009.04.008.

Bonazza, A., Maxwell, I., Drdácký, M., Vintzileou, E., Hanus, C., Ciantelli, C., De Nuntiis, P., Oikonomopoulou, E., Nikolopoulou, V., Pospíšil, S., Sabbioni, C., & Strasser, P. (2018). *Safeguarding cultural heritage from natural and man-made disasters: A comparative analysis of risk management in the EU.* Brussels: European Union. ISBN 978-92-79-73945-3. https://doi.org/10.2766/224310.

Brimblecombe, P., Grossi, C. M., & Harris, I. (2006). Long term trends in dampness in England. *Weather, 61*(10), 278–81.

Brimblecombe, P., & Grossi, C. M. (2009). Millenium-long damage to building materials in London. *Science of the Total Environment, 407*(4), 1354–1361. https://doi.org/10.1016/j.scitotenv.2008.09.037.

Brimblecombe, P. (2010). Heritage climatology. In R. A. Lefreve, & C. Sabbioni (Eds.), *Climate change and cultural heritage* (pp. 54–57). Bari: Edipuglia.

Ciantelli, C., Palazzi, E., Von Hardenberg, J., Vaccaro, C., Tittarelli, F., & Bonazza, A. (2018). How can climate change affect the UNESCO cultural heritage sites in Panama? *Geosciences, 8*(8), 296–314. https://doi.org/10.3390/geosciences8080296.

Gomez-Bolea, A., Llop, E., Arino, X., Saiz-Jimenez, C., Bonazza, A., Messina, P., & Sabbioni, C. (2012). Mapping the impact of climate change on biomass accumulation on stone. *Journal of Cultural Heritage, 13*(3), 254–258. https://doi.org/10.1016/j.culher.2011.10.003.

Grossi, C. M., Brimblecombe, P., & Harris, I., (2007). Predicting long-term freeze–thaw risks on Europe built heritage and archaeological sites in a changing climate. *Science of the Total Environment, 377*(2–3), 273–81. https://doi.org/10.1016/j.scitotenv.2007.02.014.

Grossi, C. M., Bonazza, A., Brimblecombe, P., Harris, I., & Sabbioni, C. (2008a). Predicting twenty-first century recession of architectural limestone in European cities. *Environmental Geology, 56*(3–4), 455–461. https://doi.org/10.1007/s00254-008-1442-6.

Grossi, C. M., Brimblecombe, P., Menendez, B. Benavente, D., & Harris, I. (2008b). Long term change in salt weathering of stone monuments in North-West France. *Proceedings of the 11th Int. Congress on Deterioration and Conservation of Stone* (pp. 121–128). Torun, Poland: Wydawn. Nauk. Univ. Mikołaja Kopernika.

Grossi, C. M., Brimblecombe, P., Menendez, B., Benavente, D., Harris, I., & Deque, M. (2011). Climatology of salt damage on stone buildings. *Science of the Total Environment, 409*(13), 2577–2585. https://doi.org/10.1016/j.scitotenv.2011.03.029.

Hujibregts, Z., Kramer, R. P., Martens, M. H. J., van Schijndel, A. W. M., & Schellen, H. L. (2012). A proposed method to assess the damage risk of future climate change to museum objects in historic buildings. *Building and Environment, 55*, 43–56. https://doi.org/10.1016/j.buildenv.2012.01.008.

Jacob, D., Elizalde, A., Haensler, A., Hagemann, S., Kumar, P., Podzun, R., Rechid, D., Remedio, A. R., Saeed, F., Sieck, K., Teichmann, C., & Wilhelm, C. (2012). Assessing the transferability of the regional climate model REMO to different coordinated regional climate downscaling experiment (CORDEX) regions. *Atmosphere, 3*(4), 181–199. https://doi.org/10.3390/atmos3010181.

Kramer, R., van Schijndel, J., & Schellen, H. (2013). Inverse modeling of simplified hygrothermal building models to predict and characterize indoor climates. *Building and Environment, 68*, 87–99. https://doi.org/10.1016/j.buildenv.2013.06.001.

Kucera, V., Tidblad, J., Kreislova. K., Knotkova. D., Faller. M., Reiss, D., Snethlage, R., Yates, T., Henriksen, J., Schreiner, M., Melcher, M., Ferm, M., Lefevre, R. A., & Kobus, J. (2007). UN/ECE ICP materials dose-response functions for the multi-pollutant situation. *Water Air Soil Pollution: Focus, 7*(1–3), 249–258.

Leissner, J., Kilian, R., Kotova, L., Jacob, D., Mikolajewicz, U., Broström, T., Ashley-Smith, J., Schellen, H.., Martens, M., van Schijndel, J., Antretter, F., Winkler, M., Bertolin, C.,

Camuffo, D., Simeunovic, G., & Vyhlídal, T. (2015). Climate for Culture: Assessing the impact of climate change on the future indoor climate in historic buildings using simulations. *Heritage Science, 3*(38), 38–52. https://doi.org/10.1186/s40494-015-0067-9.

Lipfert, F. W. (1989). Atmospheric damage to calcareous stones: Comparison and reconciliation of recent experimental findings. *Atmospheric Environment, 23*(2), 415–429. https://doi.org/10.1016/0004-6981(89)90587-8.

McCabe, S., Smith, B. J., McAlister, J. J., Viles, H. A., Curran, J. M., & Crawford, T. (2010). Climate change and wet winters: Testing the diffusion of soluble salts in building stone under saturated conditions. *Proceedings of the XIX CBGA congress* (pp. 399–405). Thessaloniki, Greece: Charis Ltd. ISBN 978-960-9502-01-6.

Padeletti, G., HERACLES Consortium Staff. (2018). Heritage resilience against climate events on site - HERACLES project: Mission and vision. In A. Moropoulou, M. Korres, A. Georgopoulos, C. Spyrakos, & C. Mouzakis (Eds.), *Transdisciplinary multispectral modeling and cooperation for the preservation of cultural heritage. TMM_CH 2018* (pp. 360–375). Athens, October 10–13, 2018. Communications in Computer and Information Science, Switzerland: Springer International Publisher. ISBN 978-3-030-12956-9. https://doi.org/10.1007/978-3-030-12957-6.

ProteCHt2save Interreg Central Europe Project. Retrieved from: https://www.interreg-central.eu/Content.Node/ProteCHt2save.html (Accessed on 06 May 2021).

Sabbioni, C., Brimblecombe, P., & Cassar, M. (2010). *Atlas of climate change impact on European cultural heritage.* London: Anthem Press. ISBN 978927909800-0.

Sardella, A., Palazzi, E., von Hardenberg, J., Del Grande, C., De Nuntiis, P., Sabbioni, C., & Bonazza, A. (2020). Risk mapping for the sustainable protection of cultural heritage in extreme changing environments. Special Issue Assessing the Impact of Climate Change on Urban Cultural Heritage. *Atmosphere, 11*(7), 700. https://doi.org/10.3390/atmos11070700.

Sesana, E., Gagno, A. S., Bonazza, A., & Hughes, J. J. (2020). An integrated approach for assessing the vulnerability of world heritage sites to climate change impacts. *Journal of Cultural Heritage, 41,* 221–224. https://doi.org/10.1016/j.culher.2019.06.013.

Smith, B. J., Gòmez-Heras, M., McCabe, S. (2008). Understanding the decay of stone-built cultural heritage. *Progress in Physical Geography 32*(4), 439–461. https://doi.org/10.1177/0309133308098119.

Smith, B. J., McCabe, S., McAllister, D., Adamson, C., Viles, H. A., & Curran, J. M. (2010). A commentary on climate change, stone decay dynamics and the "greening" of natural stone buildings: New perspectives on "deep wetting". *Environmental Earth Sciences, 63,* 1691–1700. https://doi.org/10.1007/s12665-010-0766-1.

STORM H2020 Project. Retrieved from: http://www.storm-project.eu/ (Accessed on 03 May 2021).

STRENCH Interreg Central Project. Retrieved from: https://www.interreg-central.eu/Content.Node/STRENCH.html (Accessed on 03 May 2021).

19 Environmental design strategies for heritage

Rosa Schiano-Phan and Talia Quesada Campaña

The heritage's environmental problem

All over the world, there is a greater understanding of the need to preserve historic buildings and sites. Now more than ever, it is not only expected for heritage sites to have an important role in the cultural and social life of cities, but also to be instrumental in their economic life. This trend has led to the urban regeneration of historic sites and the restoration of historic buildings, being usually linked to tourism. However, some conservationists have lamented that due to this trend, historic authenticity has been corrupted by consumerism (Glendinning, 2013). Furthermore, this already complex scenario is compounded by governments demanding that historic built environments play an active role in the reduction of greenhouse gas emissions and making cities more resilient to climate change. This poses several challenges, which the heritage sector has often struggled to meet.

Heritage conservation is a difficult job that encounters several obstacles for its fulfilment. For Earl (2003), the action of conservation is unnatural because even when the aim is to protect our social memory, it stops the normal course of deterioration. In addition, conservation in the built environment is more than simply maintaining; subjective factors like value and character are crucial. Also, if a historic site must endure time, it has to be useful; thus, contemporary users should be able to comfortably fulfil their needs in them. Due to the above factors, for many conservationists, the hardest task is to balance the clash between conservation demands and the current built environment's legislations (Forsyth, 2007).

For the people who undertake works of restoration on historic buildings, most of the difficulties come from the technical field. Most of the ageing buildings have become redundant with time; therefore, their layout and load-bearing capacity do not match contemporary requirements. In some cases, there is no certainty on the composition of the structural elements. Often, there are problems finding the original materials used for the structure or choosing new materials really compatible with the old ones. Moreover, even with traditional materials, the quality of the works cannot be ensured due to the workforce lacking the knowledge to work with them (Douglas, 2006; Bullen & Love, 2011). This challenge is amplified if the aim of the intervention on the historic building includes environmental and energy performance objectives. Generally, there is not enough information or monitored data on how historic buildings perform before and after restoration works. In addition, it has been shown that sometimes environmental and energy performance data generated by computer modelling and simulation do not accurately reflect how heritage environments really behave; older buildings are

DOI: 10.4324/9781003038955-24

usually thermally heavy and perform differently to new ones on which computational modelling algorithms are based (May & Rye, 2012).

Due to all the above difficulties, materials and components that have been working successfully for years can be seen to fail in many retrofit schemes of historic buildings (Feilden, 1994). For instance, the extensive and inappropriate use of insulation can cause a disruption in the moisture balance of spaces and fabric. Historic masonry was designed and built to be breathable; therefore, arresting the natural-occurring migration of moisture within the fabric can generate condensation, mould growth and even structural damage. These difficulties concern equally walls, roofs and floors – all these are components of a system that work in harmony. Rigorous airtightness is then not advisable in all historic buildings, as a right balance is needed in terms of how much to insulate (Godwin, 2011; Angbota, 2014). Another common strategy in retrofit projects is changing the windows to multiple glazing. This is particularly controversial in the restoration of valuable heritage buildings because the replacement of windows and glazing can damage the existing fabric, and it can produce unwanted double reflection and reduce the amount of daylight. Also, their durability does not go beyond some years and they must be periodically replaced. Therefore, some experts state that the amount of energy they save does not overcome the cost (Angbota, 2014). However, according to some authors, the greatest danger to the image and value of a historic building is the introduction of renewables (Godwin, 2011; Angbota, 2014). Solar panels and other means of producing energy can be very visible if they are located on the main elevation. Nevertheless, a careful intervention and the use of the latest generation technology (e.g. thin film PVs, semi-transparent, etc.) can easily be of no consequence.

Given the various difficulties and problems, it is understandable that many decision makers object to the reuse and retrofit of historic buildings. The perception that knocking down and building a new structure is cheaper and easier still exists in the building industry globally. However, this attitude clearly fails to identify the huge loss in embodied energy and carbon as well as in historic memory and value that demolition and replacement entail. Also, although conservation requirements are seen as real obstacles to the introduction of innovative materials and technologies in the environmental field (Bullen & Love, 2011), the positive attributes that many heritage buildings and sites display, in terms of climate responsiveness and passive design strategies to achieve comfort, are highly underrated and can have a considerable effect on operational energy and carbon emission savings. Therefore, the task of upgrading a historic building in an environmentally responsible way is of paramount importance, because, as Forsyth reminds us (2007), without a purpose a building is condemned to die. Furthermore, if the concept of sustainability is analysed in its three main fields – social, environmental and economical – most historic buildings and sites were conceived to work sustainably. Not only people in pre-industrial buildings used energy from renewable sources like biomass and animal force but also they usually had bioclimatic solutions to condition the interior spaces. Therefore, for some authors, there is no intrinsic clash between preservation of heritage structures and the bases of sustainability (Godwin, 2011). The challenge is to find the right balance between new and old.

Even if the restoration of a historic building is done in the most balanced way, its real impact and success depend on the urban context. The infrastructure around the site must encourage the use of sustainable means of transportation and the correct waste management (Yung & Chan, 2012). Furthermore, regeneration schemes are usually applied to historic districts, which can become gentrified, having a negative impact on the social makeup creating environmental issues of air quality and noise pollution due to the use of cars. On the other hand, sometimes, interventions in old cities apply more 'environmentally friendly' strategies without taking into consideration the character of the area and creating discomfort. For instance, introducing large areas of greenery in compact city districts can change their traditional image and character. There are old urbanisations with very few green areas that, due to their characteristics, have much less harmful environmental impact than others with a greater vegetation volume (Duany, 2010). Sometimes, to decrease the vehicular traffic and to protect the historic environment from pollution, streets in heritage centres become pedestrian. Particularly in Old Havana, this over pedestrianisation trend has become one of the main threads to accessibility (Plan Maestro, 2016). Balance again is needed to achieve a successful sustainable site.

State of the art of environmental design in heritage sites

In the field of heritage conservation, governing bodies and their latest guidance documents, such as the Burra Charter (Australia ICOMOS, 2013), have established that the first and most important step before any intervention is the assessment of the site's cultural and historic significance. Only after all valuable elements are registered and all information is obtained, it is possible to identify what compatible strategies are applicable to achieve sustainability on site (Fouseki & Cassar, 2014).

The main criteria to guide sustainability efforts in historic buildings nowadays start with understanding the importance of the historic elements of the building and its immediate environment. The first step is then to search all possible information in archives, carrying out surveys and even archaeological research. Once the compilation of information is done, then the resulting intervention must be kept to the minimum scale, trying to damage the fabric as little as possible and ensuring that any element of significance is preserved. If something must be removed, records should be made of the action taken. To increase environmental performance, options should be assessed considering traditional materials and techniques as well as their compatibility with new additions. If new equipment or technology needs to be introduced, they have to be hidden from view and with the least disruption of the character (Godwin, 2011).

All actions must be guided by the rules of minimal intervention and reversibility; however, it is argued that preservation standards are too strict in some areas and they should be relaxed for the benefit of the building and the users (Forsyth, 2007). For structures of the highest grade of significance, for example, the conservation regulations are so stringent that the compliance to the law can only be achieved by adopting an approach of 'no intervention' (Fouseki & Cassar, 2014).

Green Building Council (GBC) Historic Building is an instrument born from the Italian experience on historic restoration and the Leadership in Energy and Environmental Design (LEED) international rating protocol. It is the only rating system conceived for pre-industrial structures of great value. The main input of GBC Italy to this new system is the integration of the credit category 'historic value'. This section has the aim of preserving the heritage value of the place and gives significance to the qualities of the historical element. Credits are given for carrying out structural, material and energy surveys. High ratings are considered for the reversibility of the intervention, the existence of a maintenance plan and the presence of a restoration specialist in the project. Moreover, compatibility is credited in three different items: end-use compatibility which evaluates the new use of the building and its benefits to the community, chemical and physical compatibility of materials which focusses on how well the used restoration materials match the existing mortars and structural compatibility that considers the respect to the original structure (Cocco, 2018).

The rest of the credit categories in the GBC Historic Building scheme are very similar to LEED rating system. They encompass items related to the context, water management, energy savings, embodied energy with the use of materials and resources, indoor environmental quality, innovation in design and regional priority (Green Building Council Italy, n.d.). This initiative acknowledges the challenge and potential of making a heritage building sustainable. However, besides adding an evaluation area for the principles that the conservation standards already demand, such as reversibility, authenticity and compatibility, it maintains the usual fields of interventions of any other environmentally conscious rating scheme.

It is understandable then that the interventions done to retrofit an ordinary building are usually also applied in historic ones. These actions include increasing the thermal transmittance (U-value) of the fabric by insulating and increasing the layers on windows with the aim to make the structure as airtight as possible. The introduction of technological equipment and appliances is another common intervention. HVAC systems, including energy-efficient boilers, can be introduced or replaced, and LED lighting is usually put in place. Furthermore, renewable production of energy can be implemented.

In a project of restoration in Old Havana, which is thought to be the first international example of the Italian rating system, the environmental strategies were the same as the ones applied in Italy despite being the building in a tropical climate and in a different context (Cordeiro, 2018). The project introduced a rainwater harvesting system to reduce the use of potable water. To condition the interior spaces, a green labelled air conditioning system controlled by automation was implemented. In addition, double glazing was introduced in windows, and solar panels are predicted to offset 50% of the energy consumed.

A tool that has had a great impact on how historic building restoration is done nowadays is historic building information modelling (HBIM). Developments in the technology for laser scans and photogrammetry have enabled the integration of survey information into the design process. The result is a much more complete project that can be shared with all stakeholders of the building (Khodeir et al., 2016). Despite this, the nature of historic buildings poses a problem for HBIM; in ageing structures, wall sections are not uniform, materials do not appear uniformly and elements are not repetitive. Also, for some practitioners, it is still an insurmountable challenge to work with BIM in heritage preservation. This is an area of future research that can be of much significance in the future development of the heritage sector.

In general, it is argued that in heritage regeneration schemes environmental issues are not in the highest priority (Fouseki & Nicolau, 2018). Furthermore, when a sustainable approach is embraced, the focus is usually placed on technical issues like energy savings and pollution control (Tweed & Sutherland, 2007). This approach has not always been successful; therefore, different methods must be explored.

Innovative strategies for heritage sites

In the building industry, there is a need for alternative approaches to sustainable conservation. If the point of view is changed towards a bioclimatic approach, it can be seen that historic places have great potential. For instance, instead of applying changes to an old building to fulfil the requirements for the efficiency of HVAC systems like airtightness, it would be good to employ equipment and strategies that match the behaviour of the ageing fabric. Taking as reference this way of thinking, several strategies can be proposed to achieve a sustainable restoration in heritage sites.

Simple changes: great impact

There are many strategies and devices to improve daylight access in an interior space and subsequently produce energy savings. However, in historic districts where facades cannot be changed and the inclusion of light wells or solar ducts is not advisable, a change on the urban realm like the increasing of the reflectance of the facades can cause a positive impact on illuminance levels indoors.

Smart zoning

Different functions always concur in a building, each of them with diverse environmental needs. To zone the areas, taking advantage of the climate and the specific requirements can be beneficial from the environmental point of view. For instance, the areas like archives and deposits that do not need high values of illuminance can be concentrated in the centre of the building and those areas where people need lighting can be placed towards the perimeter (Cassar, 2011). This will maximise the use of daylight as much as possible.

Smart zoning also concerns thermal performance. In historic buildings where usually there are ageing artefacts to preserve, it is advisable to differentiate the ones that need a more stringently controlled environment for its preservation from others that can withstand more flexible conditions. The fragile artefacts can be located in smaller spaces where conditioning is less energy intensive or even in enclosed cases, bookstores or other repositories leaving the rest of the space to thermal fluctuate more freely.

Passive and low-energy space conditioning

Interior environments in heritage buildings can be challenging to control due to the composition of their fabric. However, these materials which could be wood, porous masonry and others are buffer composites that can reduce the impact of changes in temperature and moisture (Feilden, 1994). This is especially useful in warm climates with large daily swings; also, heavy thermal mass can provide good opportunities to exploit the difference between outside and inside conditions. For instance, to have lower indoor temperatures and maximise the benefits of thermal mass, the building must be protected from solar radiation during the day and be ventilated during the night to cool down the structure. Subsequently, the structural mass will work as a heat sink during daytime (Givoni, 1994).

The application of passive strategies does not preclude the use of small appliances, such as ceiling fans, which can improve the natural performance of a building (Givoni, 1994). For example, a solution for a very common and complex problem like condensation and moisture build-up in old buildings can be the use of a fan to give a continuous and soft air movement (Feilden, 1994).

Buildings do not need to be fully air-conditioned all the time they are in use. Hybrid and mixed mode systems can be implemented to save energy and to take advantage of changing climate conditions and the features of the structure. The principle of conditioning only when needed and where needed can be complex to implement but advances in automation can help successful implementation (Baker, 2009). Even for conditioning of archives and museums, some researches have implemented less energy-intensive systems, based on scientific evidence which suggests that, for artefact environmental control in warm and humid climates, relative humidity (RH) is the parameter to control. The system is composed of supply and exhaust fans, as well as convective heaters, all connected to a controlled system (Maekawa & García, 2006).

Original features can be instrumental

Traditional ways of achieving comfort should not be discarded without a thorough analysis of how to include them in the project. Internal wooden shutters, for example, can have a major impact on energy conservation in heated spaces; its inclusion not only would have a comfort benefit but also could be of heritage significance (Angbota, 2014). The same happens in the case where spaces were originally fitted out with heavy curtains or where windows have secondary glazing. On the other hand, features like chimneys, cellars and attics can be important for thermal buffering and for the ventilation strategies of indoor spaces.

Methodologies for the environmental performance evaluation of heritage sites

Conservation projects inherently rely on substantial surveying and archival searches to understand what can and cannot be done as an acceptable intervention on a heritage site or building. Traditionally, there is a huge emphasis placed on historical stratification and compatibility of contemporary materials and construction techniques, with a concern towards the requirements of durability and legibility of any new intervention. Performance is a crucial requirement too, but it is mostly interpreted as thermo-hygrometric performance to assure the integrity of the fabric rather than the environmental design performance leading to reduced energy demand for the comfort and preservation of occupants and objects in the building. Similarly, at a larger scale and site level, the concerns are hardly with the preservation of the microclimatic conditions but more with the aesthetic and historical legibility as well as the integration of the new with the old.

Climate analysis

Although climate analysis has played a role in the understanding of typical precipitation levels and the way they could affect moisture content in the fabric and on the envelope, it is important now more than ever to extend the applicability of climate analysis to predictions for climate change scenarios and for climatic factors which were not commonly and routinely considered in the past. The understanding of varying levels of global, direct and diffuse solar radiation, cloud cover, dry and wet bulb temperatures, wind direction and speed as well as ground temperature, for predicted high, medium and low levels of incidence of climate change anomalies, can provide a very useful insight in the performance prediction of heritage buildings and in the forecast of likely issues and planning of adequate maintenance plans. Moreover, the analysis of the typical climate for a given location performed by studying the historical weather data obtained from the closest meteorological station combined with statistical predictions for the future can provide specific directives on the most appropriate climatic responsive and passive strategies adoptable in a building or an entire site. This will allow the quantification of the risks associated with climate change both for the conservation of the building and the site as physical entities but also for the adaptability of their functions to expected future conditions and occupants/inhabitant's requirements.

For example, the analysis of the expected precipitation and frequency of rainfall for the 2050 and 2100 climate change scenarios obtained from weather databases software or national and international repositories (e.g. Meteonorm, [Meteotest, 2020]; World Meteorological Organisation, [Oke, 2006]) can easily provide evidence of the increased intensity of precipitation in shorter events with higher quantities of rain, highlighting the risk of flooding. The analysis of the typical yearly profile of hourly or monthly average dry bulb temperature (DBT) (°C) for a given location can highlight the potential for natural ventilation which combined with the same resolution analysis on relative (RH%) and absolute humidity (AH g/kg) levels and wind velocity (m/s) can identify both risks for the ingress of unwanted humid air or the potential for using air movement to remove said excess moisture. Solar radiation data come in different formats. It is usually found as hourly global and diffuse radiation on the horizontal plane (Wh/m^2), from which the direct radiation is derived, and as incident radiation on a vertical plane for each cardinal orientation (kWh/m^2). The latter is particularly useful to identify the vulnerability of certain building orientations to solar exposure at certain times of the year and of the day or conversely to identify the potential for passive solar heating as potentially originally intended for some heritage buildings.

Site analysis

The analysis of an entire heritage site or the site surrounding a heritage building is a particularly useful exercise to understand the performance of the various components together and in the context of their evolution in time. The analysis of a site can be performed with the objective of capturing a specific environmental performance aspect or to gain a general understanding of the environmental performance of the site with respect to the main microclimatic parameters of thermal, visual and acoustic performance as well as characterisation of air movement and air quality. The collection of data related to the above parameters can be time consuming and might require a level of resolution that is not always possible or desirable to obtain through automatic remote logging but often requires a manual so-called hyperlocal monitoring which can capture the variables at the (hyper) local human scale through point-in-time measurements. Mapping of these measured data is equally crucial and, although there are not many textbooks which document the process of environmental fieldwork, it needs to be rigorous and methodical, availing itself of the methods of surveying and grid mapping and simultaneous data collection at different points on the site. Site analysis can also be performed through 3D modelling of the site topography and building massing followed by computer simulation of a range of microclimatic site characteristics pertaining to, for example, solar geometry and overshadowing, outdoor illuminance, radiation analysis and wind environment. The latter is usually performed using computational fluid dynamic (CFD) simulation, which can give information on both the intensity and direction of the wind flow within the site and their interaction with the buildings, enabling the identification of specific areas of turbulence or downdraught potentially affecting comfort and overall performance. The same type of 3D modelling and dynamic simulation can now be performed for daylighting and solar geometry and radiation analysis (Jankovic, 2017). The advantage of computer simulation is the possibility to investigate the performance of the site during various seasonal periods, identify positive attributes or problem areas and predict the performance of potential strategic solutions and alternative scenarios.

An example of microclimatic site analysis for heritage sites is the fieldwork study conducted in Pakistan in the walled city of Lahore (Sohail, 2020), where the author identified the issue of poor indoor illuminance in the 12th century residential buildings of the old city. There, a combination of high-density urban tissue and a series of historical and more recent stratifications and juxtapositions of building elements, supra-structures and unplanned extensions has obstructed the already very narrow network of streets and alleyways, further reducing the available sky illuminance reaching the buildings. A thorough analysis combining outdoor illuminance (lux) readings at ground level in relation to the urban canyons' proportions (expressed in height to width ratios) can reveal the influence of urban morphology on the availability of natural light not only outdoor but also inside the buildings where most of the problems of under-illumination are experienced (leading to occupants' visual comfort, health and energy demand problems). The study found that alterations in the original height of the surrounding buildings and increased obstructions due to protruding elements from balconies and their unplanned enclosures had further obstructed the proportion of visible sky and reduced the illuminance levels accordingly. Moreover, the lack of maintenance of the exterior render of the fabric had impacted the light reflectance of the building facades, further reducing the amount of illuminance reaching the street and being reflected and diffuse inside the building. The combination of site analysis and computer modelling allowed the exploration of these hypotheses and the quantification of problems and possible solutions. This example also showed the effectiveness of using environmental design methods to quantify strategies and interventions which are small, cumulative and effective for the improvement of the environmental performance of heritage buildings without disrupting the character of the site or other protected building elements such as the size or the typology of the existing fenestration and the appearance of the facade.

Thermal performance analysis

When moving from the larger scale of site and microclimate to the conservation of the individual building and its components, the analysis of its thermal performance can be particularly useful:

a   to pinpoint problem areas of direct intervention for restoration and
b   to identify the potential of the building in achieving comfortable and acceptable conditions for occupants and artefacts with minimum or no reliance on energy-intensive systems.

There are two main methods for the analysis of a building's thermal performance and these are:

i   collection of on-site performance data, commonly known as fieldwork and
ii   computational analysis of the building's current or predicted thermal behaviour.

For the collection of on-site data characterising the thermal performance of the building (Parsons, 2002), there are several performance indicators which can be useful to identify specific problems or potential useful strategies for the building. For example, the indoor air temperature or DBT (°C) would be an easy reference indicator of the thermal performance; however, for determining the effect of the building indoor thermal environment on occupants' comfort as well as for quantifying the level of thermal inertia of the building's fabric, the mean radiant temperature (MRT, (°C), measured with a globe thermometer) can be a very useful performance indicator, especially when in the presence of heavyweight buildings with a large thermal mass, so common in heritage buildings. When analysing specific problems such as condensation, thermal bridging and the specific causes of heat losses or gains through the building elements, surface temperatures (°C) are particularly useful especially if they are collected and logged over a period of time to represent their dynamic performance or if they are recorded for specific but representative moments in time and represented visually through thermal imaging. Thermal imaging can be very useful especially when locating thermal bridging and condensation problems. The correlation between temperature and other collected data on indoor humidity (relative and absolute), wet bulb temperature (WBT) (°C) and dew point temperature can be useful to diagnose potential issues and identify restoration strategies.

The modelling of building geometry and thermal properties of its elements is another crucial step of the thermal performance analysis (Jankovic, 2017). Modelling is usually followed by the simulation of current conditions or predicted scenarios based on assumptions on the level of occupancy, schedules of use and operation and ventilation regimes. Variations to the thermal properties of the building's opaque and transparent elements in thermal capacity, transmittance and solar transmission can be dynamically simulated with specialist software using algorithms characterising the thermal behaviour of the building in relation to hourly weather data. These software can output the thermal performance of the building in a variety of indicators: hourly temperatures for a typical year, average annual energy demand (kWh/m2), peak heating and cooling energy loads (W/m^2).

The analysis of the building thermal performance using dynamic thermal modelling and simulation can become particularly important when assessing the impact of climate change scenarios on heritage buildings and to identify the most suitable strategies to protect the building from extreme weather events and climatic modification. However, this analysis can also identify and quantify the inherent thermal properties of many of these buildings, which were originally operated without the addition of mechanical systems and used natural, climatically responsive and architecturally integrated strategies for the climate control and conditioning of the indoor space. The quantification and appreciation of these properties is one of the tasks of the environmental design discipline which can contribute to the implementation of more effective conservation strategies and at the same time borrow from the well-performing examples of heritage buildings to use lessons from the past to inform future design.

The visual environment

The study of the visual environment of a heritage site or a building can be undertaken for several converging purposes. Similar to the thermal environment, there is a direct relationship between the provision of visual comfort, adequate levels of daylight satisfying functional requirements and the energy savings which can be accrued from the minimisation of artificial lighting use. However, when it comes to visual environment in heritage buildings, there is an additional aspect that needs to be considered relating to the aesthetics of the buildings and the visual effects that were intended for the enjoyment of artefacts and architectural elements of high significance.

On-site measurements of the so-called indoor luminous environment is usually done by collecting and recording levels of illuminance (lux) and mapping them spatially on the building's plans. Illuminance is the amount of light that falls onto a surface and it is usually measured at 0.8 m from the floor. Indoor illuminance is strictly related to the outdoor sky illuminance, which depends on sky conditions, season and latitude. The proportion of the sky illuminance (under uniform overcast conditions) entering a building through its openings in a point in time is expressed by the daylight factor (DF) and is calculated from the ratio between indoor and outdoor illuminance and expressed in percentage. The DF calculation is static and can be obtained from measured or simulated data. As an alternative to the point in time static calculations of the DF commonly used for decades, in the last 20 years, the assessment method called climate-based daylight modelling (CBDM) (Reinhart et al., 2006) is being widely used to evaluate the dynamic daylight performance of buildings, generating annual and hourly predictions over a spatial grid. Similar to dynamic thermal simulations, the dynamic daylighting simulations use weather data for realistic sun and sky conditions that generate output ranges similar to those encountered in real-life analyses. The main performance indicators for the climate-based dynamic modelling are the daylight autonomy (DA) and the useful daylight illuminance (UDI). These two indicators have similarities. DA simulations show how often in a year a minimum illuminance threshold at the reference plane (0.8 m) is achieved without the need for artificial light, whilst the UDI simulations also refer to an upper illuminance threshold, showing how often a certain band is achieved.

The implications of such methods of analysis are primarily to identify and maximise the availability of daylight in an accurate and weather-dependent manner and can be more accurate for arriving at the lighting energy calculations and the study of control strategies for blinds and shading devices. However, in heritage buildings, it might be useful to consider also other performance indicators to characterise the luminous environment,

Figure 19.1 Comparison between conventional photo, HDR photo and false colour luminance plot of historic courtyard in Lahore, Pakistan.
Source: Photo by Sana Aleem, 2018.

which can better characterise the qualitative aspects of a visual environment. Luminance (cd/m^2) measures the amount of light emitted or bounced back from objects and surfaces and it is linked to the properties of reflectance of such surfaces. Luminance mapping from specialist equipment (Pierson et al., 2020) and now also from mobile phone applications linked to the devices' digital cameras (Miri, 2015) is a method able to derive false colour imaging of luminance levels from point-in-time high dynamic range (HDR) photography of real-life objects and settings. The HDR-based false colour mapping and the spatial contrast method (Rockcastle & Andersen, 2013) offer a compositionally dependent method for quantifying local variations in brightness within architectural space, which are perceived in direct relation to their local surroundings (Figure 19.1). The identification of spatial–visual settings that can be perceived as interesting and aesthetically pleasing as well as characterising modelling of objects or building elements can be a useful tool to quantify and evaluate sensory value in heritage sites.

Case studies

The following case studies exemplify environmental evaluation projects of heritage sites and buildings where the brief was: a) to analyse their environmental attributes at microclimatic and building scale respectively; b) to quantify the benefits of their inherently climatically responsive designs; and c) to propose interventions for the improvement of their performance in terms of environmental comfort (for both inhabitants and artefacts) and energy demand reduction, while being sympathetic to heritage value and sensitive to conservation principles.

Golden Lane Estate, London, UK

Golden Lane Estate is a housing complex situated in the city of London (Figure 19.2). Built in the 1950s, it is one of the best examples of modernist architecture in London and has been listed as grade II in the National Heritage List for England (NHLE). Designed by the architecture studio Chamberlin, Powell and Bon (CP&B), it was conceived in the post-war period, with the intention of offering affordable and comfortable housing, along with all services to make an almost self-sufficient development. The main design concept, which is embodied in the master planning of the site, is that of a 'city within the city', creating an introverted- and inward-looking complex of housing blocks facing towards inner courtyards offering greenery, privacy and leisure opportunities.

Several environmental studies have been conducted on this heritage site, exploring both the spatial delight of the outdoor and indoor spaces conferred by the designed environmental attributes (Schiano-Phan et al., 2018) and the effect that the introverted courtyard design can have on the microclimate and air quality in the context of the contemporary urban context (Turci et al., 2019). The overall methodology entailed an understanding of the climate and microclimate and the impact of the design on human comfort. Photographs, sketches and personal experiences were used in the qualitative analysis; on the other hand, a quantitative approach was followed with the help of on-site measurements and software simulations. In addition, historical information and original drawings were consulted to understand the original intentions of the designers.

In a climate in which the average daily temperatures in the warm season are only between 14 and 19°C and a predominant overcast sky, outdoor comfort depends mainly on solar access and wind shielding. These issues were considered in the architects' concept

Figure 19.2 Golden Lane Estate.
Source: Photo by former student Sana Aleem, 2018.

for the internal courtyards by creating different microclimates. The width of the courts is a minimum of two times the height of the buildings to reduce overshadowing and at the same time providing shelter from wind. The performed studies of solar access and overshadowing on site validated the attention paid by the architects to the blocks' orientation, massing and siting to achieve minimum shade in places like playgrounds and allow more overshadowing in tennis courts. The analysis also reveals that most south-facing facades receive considerable sun rays due to the appropriate spacing of the blocks, which helped with passive heating.

The chosen colour palette for the buildings also reflects the intention to maximise daylight reflectance and therefore improve the luminous environment both outdoor and indoors, which is the case especially with Great Arthur House, the highest block inside the estate with a bright yellow reflective curtain wall.

Wind and comfort in outdoor spaces were also studied on the estate. From site measurements, it was found that in the internal courtyards the air speed did not exceed 2 m/s despite higher value at the entrance of the complex. This is caused by the sheltering effect of the blocks' arrangement, which was also further confirmed by CFD simulation. In addition, the effect of wind sheltering on comfort conditions in the courts was assessed using comfort simulation analysis (Schiano-Phan et al., 2018). Input data were based on measurements taken and the results revealed no thermal stress and a neutral grade on the comfort scale in all the studied areas.

The study on the effect of the courtyard configuration of the Golden Lane Estate complex on air quality was performed with a combination of on-site measurements and dynamic computational analysis. The on-site measurements at hyperlocal scale were able to document the extent of the problem and the way that the housing blocks protect the inner courtyards from the ingress of pollutants from the main busy road. These results were verified with the use of the computer modelling (using microclimatic software) which simulated the current conditions for the whole site and confirmed the sheltering effect of the building mass.

Looking at indoor performance, a comparative study was conducted between two similar flats located in different blocks, but with the same orientation, materials and layout. The study focusses on two maisonettes located in two different blocks orientated north–south. One was retrofitted and located in Basterfield House and the other in a condition very close to the original, located in Bayer House. Both maisonettes have a similar layout, lower floors containing living room, dining room and kitchen, while upper floors have a bathroom, two bedrooms and a shared balcony. The apartments have thick solid brick walls towards north with a balcony on the upper floor and towards south there are single-glazed large windows. The general environmental idea of the architects was to use the shallow plan to maximise the use of natural cross ventilation during summer. While the southern-glazed façade would allow for passive solar heating during sunny days in winter and in summer, the protruding balcony would block high sun rays.

The retrofit measures proposed in the Basterfield House apartment were secondary glazing and internal insulation in most of the walls of the maisonette. These interventions were considered the most appropriate and sensitive to the grade II listed status of the site. A comparative study undertaken using dynamic thermal modelling and simulation between Bayer House and Basterfield House apartments revealed that the retrofit actions contribute to a 26% reduction in annual heating energy demand. In terms of comfort, it was determined that the values of temperatures in Bayer House apartment were directly related to the outdoor conditions; this situation was the main indication of the poor performing characteristics of the building's envelope.

The interventions on the Basterfield House apartment are simple and straightforward; however, they have a great impact on the environmental performance of the home. There are no major negative consequences for the image for the grade II listed building; nevertheless, it can be argued that because the block is a modernist example of the architecture, these interventions can be more easily introduced. Moreover, this study also shows that the microclimate and site analysis can be useful to quantify the additional environmental values of an historic site and the way that this can adapt through time in fulfilling the evolving needs of society.

Atarés Castle, Havana, Cuba

The case of Atarés Castle in Havana is an example of sustainable restoration of an old fortress into a museum. The project had three main concerns: to restore the values and character of the structure, to achieve human comfort for the visitors and to offer a good environment for the preservation of the artefacts to exhibit. The main intention was to fulfil all these requirements with the application of passive strategies.

The first step was to analyse the climate of the city. Havana has a hot and humid climate; the behaviour of temperature and humidity is similar throughout the year. However, if temperature and RH are analysed together, it can be observed that during the night, when air temperature reaches its lower values, RH can be higher than 90%, arriving to saturation with high frequency. On the other hand, at daytime, RH reaches minimum values at 1 pm between 30 and 50% when temperatures reach their peak.

Atarés Castle was built in 1767, as part of the new defensive system of Havana (Figure 19.3). The building is an irregular hexagon, surrounded by a ditch, with 4-metre wide walls whose interior spaces are vaults that enclose an interior courtyard. The main construction material is limestone. Along with Old Havana and the rest of its fortifications,

Figure 19.3 Atares Castle, Havana, Cuba.
Source: Photo by Talia Quesada, 2020.

the structure is part of the World Heritage and has grade I of protection (the highest). Therefore, alterations to the image and composition of the structure are strictly forbidden, and all new equipment must be minimal and camouflaged (Quesada, 2015).

The artefacts to exhibit in the interior spaces of the castle were a set of reproductions of artefacts designed by Leonardo da Vinci. Almost all pieces are made up of wood, textile and leather. Despite not being antique, they required a stable environment for their preservation. Based on different studies on the conservation of objects in warm and humid climates, it was decided that the requirement for the pieces was to keep RH under 70%. For human comfort, it was set at a comfort band with a lower temperature limit of 22°C and 30°C for the upper one (Quesada, 2015).

Due to the impracticality of performing on-site measurements, dynamic thermal simulations using specialist software helped to understand the performative characteristics of the building's envelope and interior spaces. The extraordinary character of the edifice caused the decision to have not one base case but several ones to test the influence of each one of the elements of the interior spaces: thermal mass, openings on the facades, openings on the roof and 24-hour ventilation (Quesada, 2015).

In general, it was concluded that when there is no natural ventilation the values for resultant temperature and RH were stable throughout the day. These simulations showed that the heavy mass of the castle can give RH stability but with high temperature values. On the other hand, ventilation made RH to fluctuate but brought lower temperatures (Quesada, 2015).

Daylight was also tested, using the radiance software to check illuminance levels in the vaults. The benchmark was set in 200 lux to prevent light damage in the artefacts. It was concluded that the openings on the roof needed to be covered to reduce the amount of daylight passing to the spaces.

Taking into consideration all previous conclusions, sets of simulations were done testing natural ventilation applied in different moments of the day. These strategies were tested in two different scenarios, the first one with windows and openings in the roof with the current size but covered with louvres to reduce daylight and sunlight. The second scenario was with windows and holes reduced to the original dimensions from when the castle was built.

The case that showed better values for the conservation of the artefacts as well as human comfort was the one with natural ventilation applied during the operation times (9am–5pm); however, temperatures were higher than 30°C on some occasions. Overall, the scenario in which windows and openings were kept at their current size had better performances than the other with windows reduced to the original size (Quesada & Schiano, 2017).

A further passive strategy was tested to see if RH control could be more effective and temperatures could be lowered. The idea was to apply dehumidification during night-time with silica-coated louvres encased in glass panels introduced to the openings on the roof. This time, the results were better in terms of RH and temperature (Quesada & Schiano, 2017).

The evidence-based approach applied in this study was useful to decide whether it could be beneficial to do major changes in the image of the building such as the reproduction of the original openings and windows. It was concluded that both daytime ventilation and silica-coated louvre dehumidifiers with night-time ventilation strategies can be applied in the spaces where the artefacts are located to keep RH below 70%. Nevertheless, the chosen strategy was to maintain the building free-running, opening windows and doors only during operation times and limiting the amount of light that enters into the vaults while allowing air to pass, this was implemented in the building in 2019.

Conclusion & future directions

The current discourse on environmental and energy sustainability in heritage buildings is often focussed on the burden that legislation and directives for the abatement of greenhouse gas emissions creates on the conservation sector. This is due to the achievement of stringent performance standards and the conflicting demands and clashes perceived between the application of energy-efficient measures, renewable technologies and the historic values of authenticity which are often deemed compromised. However, a shift in approach which favours the use of environmental design and bioclimatic strategies in restoration and conservation projects can turn the table and propose environmental design criteria as a paradigm of re-discovery and preservation of the rich environmental heritage that most historic buildings offer. Not only can a passive design and bioclimatic approach help towards the achievement of energy and environmental performance standards, but it can contribute to:

i   Highlighting the climatic responsive and inherently positive attributes of heritage sites which can be rediscovered to suit the adaptive inhabitation of historic buildings and contribute to GHG emission reduction and climate change mitigation;
ii The quantification and de-codification of aesthetic and sensory delight in heritage buildings;
iii Promoting strategies that can avoid expensive and visually and spatially disruptive mechanical systems and can be more compatible with the original building reducing not only operational energy but also the embodied carbon from the avoidance of mechanical equipment.

With the conceptualisation of historic urban landscapes and the shift to a landscape-based approach to the management of historic sites (UNESCO, 2005), heritage buildings are no longer considered in isolation but as part of the urban context they inhabit. This has a direct correspondence in the way that environmental design now concerns itself not only with the energy and environmental performance of individual buildings but of entire urban environments and their microclimates (Schiano-Phan, 2015). Therefore, it is imperative that more research is conducted at the scale of a whole building site, and studies of environmental parameters expanded to include the aspects of outdoor comfort and health as well as the impact of the urban microclimatic environment on conservation strategies. This will enable a better understanding of the vulnerabilities of heritage sites to socioeconomic pressures and climate change and also of strategies to improve resilience.

References

Angbota, H. (2014). Anticipating the unintended consequences of the decarbonisation of the historic built environment in the UK. *The Historic environment: Policy & Practice*, 5(2), 101–115.

Australia ICOMOS. (2013). *The Burra Charter: The Australia ICOMOS charter for places of cultural significance 2013*. Retrieved from: https://australia.icomos.org/wp-content/uploads/The-Burra-Charter-2013-Adopted-31.10.2013.pdf (Accessed on 7 May 2021).

Baker, N. (2009). *The Handbook of sustainable refurbishment. Non-domestic buildings.* London: Earthscan.

Bullen, P., & Love, P. (2011). Adaptive reuse of heritage buildings. *Structural Survey*, 29(5), 411–421.

Cassar, M. (2011). Energy reduction and the conservation of cultural heritage: A review of past, present and forthcoming initiatives. *International Preservation News*, 55(55), 6–9.

Cocco, C. (2018). La restauración sostenible con GBC Historic Building. *Quaderni di Assorestauro-Stories. Escuela en Obra*, 3, 8–11.

Cordeiro, R. (2018). Restauración y sostenibilidad de un edificio histórico en La Habana. *Quaderni di Assorestauro-Stories. Escuela en Obra*, 3, 8–11.

Douglas, J. (2006). *Building adaptation*. 2nd ed. Oxford: Butterworth-Heinemann, Elsevier.

Duany, A. (2010). *Charlas en el Capitolio de La Habana sobre el Nuevo Urbanismo*. La Habana: Ediciones Unión.

Earl, J. (2003). *Building conservation philosophy*. 3rd ed. London: Donhead.

Feilden, B. (1994). *Conservation of historic buildings*. Oxford: Architectural Press. Butterworth-Heinemann.

Forsyth, M. (2007). *Understanding historic building conservation*. Oxford: Blackwell Publishing.

Fouseki, K., & Cassar, M. (2014). Energy efficiency in heritage buildings—Future challenges and research need. *The Historic Environment: Policy & Practice*, 5(2), 95–100. https://doi.org/10.1179/1756750514Z.00000000058.

Fouseki, K., & Nicolau, M. (2018). Urban heritage dynamics in 'heritage-led regeneration': Towards a sustainable lifestyles approach. *The Historic Environment: Policy & Practice*, 9(3), 229–248. https://doi.org/10.1080/17567505.2018.1539554.

Givoni, B. (1994). *Passive and low energy cooling of buildings*. New York: Van Nostrand Reinhold.

Glendinning, M. (2013). *The conservation movement: A history of architectural preservation. antiquity to modernity*. London: Routledge.

Godwin, P. (2011). Building conservation and sustainability in the United Kingdom. The 2nd International Building Control Conference 2011. *Procedia Engineering*, 206(20), 12–21.

Green Building Council Italy. (n.d.). *Historic building: For a sustainable restoration and refurbishment of historic buildings*. Retrieved from: https://www.gbcitalia.org/documents/20182/157656/GB-C+HB_ENG_03.pdf/3e076a9f-c840-4817-9ce5-38b4cf0f9544 (Accessed on 7 May 2021).

Jankovic, L. (2017). *Designing zero carbon buildings using dynamic simulation methods.* London: Routledge.

Khodeir, M., Aly, D., & Tarek, S. (2016). Integrating HBIM (Heritage Building Information Modelling) tools in the application of sustainable retrofitting of heritage buildings in Egypt. *Procedia Environmental Sciences, 34,* 258–270.

Maekawa, S., & Garcia, M. (2006). Low-cost climate control system for museum storage facility in Tenerife island. *Proceedings of the 23rd conference on passive and low energy architecture, PLEA 2006.* Geneva.

May, N., & Rye, C. (2012, September 20). *Responsible retrofit of traditional buildings.* Retrieved from: http://www.sdfoundation.org.uk/downloads/RESPONSIBLE-RETROFIT_FINAL_20_SEPT_2012.pdf (Accessed on 3 May 2021).

Meteotest. (2020). *Meteonorm Software S.* Retrieved from: http://www.meteonorm.com/ (Accessed on 3 May 2021).

Miri, M. (2015). Luminance-based HDR photos to help lighting designers in their everyday work. *Proceedings of the 'Professional Lighting Design Convention (PLDC)' conference 2015.* Rome. Retrieved from: http://aftabsoft.net/presentations/pldc2015.pdf (Accessed on 3 May 2021).

Oke, T.R. (2006). Initial guidance to obtain representative meteorological observations at urban sites. *World meteorological organization, instruments and observing methods, IOM Report No. 81, WMO/TD-No. 1250.* Retrieved from: http://www.wmo.int/web/www/IMOP/publications-IOM- series.html (Accessed on 3 May 2021).

Parsons, K. (2002). *Human thermal environments: The effects of hot, moderate, and cold environments on human health, comfort, and performance.* 2nd ed. Florida: CRC Press.

Pierson, C., Cauwerts, C., Bodart, M., & Wienold, J. (2020). Tutorial: Luminance maps for daylighting studies from high dynamic range photography. *LEUKOS, 17*(2), 140–169. https://doi.org/10.1080/15502724.2019.1684319.

Plan Maestro. (2016). *Plan Especial de Desarrollo Integral 2030. La Habana Vieja, Centro Histórico.* La Habana: Ediciones Boloña.

Quesada, T. (2015). *Conservation of "Santo Domingo of Atares Castle": Sustainable conversion and retrofit of heritage building in Havana.* (Unpublished master's thesis). University of Westminster, London.

Quesada, T., & Schiano-Phan, R. (2017). Conversion of Atares Castle: A bioclimatic evidence-based approach. *Proceedings of the 33rd conference on passive and low energy architecture, PLEA 2017.* Edinburgh.

Reinhart, C. F., Mardaljevic, J., & Rogers, Z. (2006). Dynamic daylight performance metrics for sustainable building design. *LEUKOS, 3*(1), 7–31. https://doi.org/10.1582/LEUKOS.2006.03.01.001.

Rockcastle, S., & Andersen, M. (2013). *Annual dynamics of daylight variability and contrast: A simulation-based approach to quantifying visual effects in architecture.* London: Springer.

Schiano-Phan, R., Lau, B., Pourel, D., & Khan-Phatan, S. (2018). Spatial delight and environmental performance of modernist architecture in London – Golden Lane Estate. *Future Cities and Environment, 4*(1), 1–24. https://doi.org/10.5334/fce.47.

Schiano-Phan, R., Weber, F., & Santamouris, M. (2015). The mitigative potential of urban environments and their microclimates. *Buildings, 5,* 783–801. https://doi.org/10.3390/buildings5030783.

Sohail, L. (2020). *Mitigating daylight issues in organic towns through environmental strategies: A case study of the Walled City, Lahore.* Unpublished master's thesis. London: University of Westminster.

Turci, G., Marchetti, M., Borna, M., & Schiano-Phan, R. (2019). L'impatto della morfologia urbana e della vegetazione sulla qualità dell'aria. *L'Ufficio Tecnico, 18*(10), 9–14.

Tweed, C., & Sutherland, M., (2007). Built cultural heritage and sustainable urban development. *Landscape and Urban Planning, 83*(1), 62–69.

UNESCO (2005). *Vienna memorandum on 'World Heritage and Contemporary Architecture – Managing the Historic Urban Landscape' and decision 29 COM 5D*. Retrieved from: https://whc.unesco.org/en/documents/5965 (Accessed on 7 May 2021).

Yung, E., & Chan, E. (2012). Implementation challenges to the adaptive reuse of heritage buildings: Towards the goals of sustainable, low carbon cities. *Habitat International*, *102*(36), 353–361. https://doi.org/10.1016/j.habitatint.2011.11.001.

20 Energy efficiency in historic buildings

Gustaf Leijonhufvud, Petra Eriksson and Tor Broström

Introduction

Improving energy efficiency in a historic building puts the balance between energy conservation and building conservation to the point. A shallow understanding of the problem and a lack of systematic procedures lead too often not only to frustration but also to poor decisions. This challenge ranges from decisions on single components in a building to designing national policies for large building stocks. A critical mass of research about this issue has emerged in recent years (e.g. de Bouw et al., 2016; Broström et al., 2018). The need for interdisciplinarity has been emphasised, as well as the importance of involving a wider set of values and the perspectives of the users of buildings (Fouseki & Cassar, 2014; Fouseki et al., 2020).

On a global scale, there is an urgent need to reduce greenhouse gas emissions associated with buildings (IPCC, 2014). Indeed, buildings accounted for 28% of greenhouse gas emissions in 2018 with an increase of 45% since 1990 (IEA, 2019). To this end, transformation of the building stock is considered one of the most important and cost-efficient measures in mitigation scenarios (IPCC, 2018). Constructing new, energy-efficient buildings will be far from sufficient, and there is an urgent need to reduce the greenhouse gas emissions from existing buildings.

Historic buildings represent a significant share of the existing building stock in terms of greenhouse gas emissions (Webb, 2017; Herrera-Avellanosa et al., 2019). Historic buildings are also important cultural and material resources. In this chapter, historic buildings are defined as buildings with heritage value, no matter their age. Heritage values, in turn, refer to aspects of aesthetic, historic, scientific, cultural, social or spiritual nature that individuals or society assign to a building (e.g. European Committee for Standardization, 2017). The challenge discussed in this paper is how to reduce energy demand and greenhouse gas emissions without unacceptable effects on heritage values. The problem is not limited to listed buildings. The potential for energy saving as well as the risks to heritage values are generally greater in the general stock of existing buildings where there is no or little formal protection (Herrera-Avellanosa et al., 2019).

The transition to a sustainable energy system requires not only the introduction of renewable energy sources but also increased efforts to reduce energy and power demand in buildings. A picture of a historic district and its individual buildings (Figure 20.1) illustrates the potential conflicts between the goals of a renewable energy system and societal goals for preserving heritage values. Some immediate questions are:

Figure 20.1 Visby is a world heritage town located on the island of Gotland in the Baltic Sea. Gotland will be a Swedish pilot region for a 100% renewable energy system.
Source: Photo by Tor Broström, August 20, 2008.

- To what extent would we be willing to introduce renewable energy sources? Can we accept solar panels on the roofs or wind generators in the neighbourhood?
- What would be the effect on the historic building stock if near-zero energy standards are implemented? Vice versa, what would it cost in terms of money and greenhouse gas emissions by prioritising building conservation before energy conservation?

The aim of this chapter is not to present universal answers to these questions but to bring some clarity to how they can be investigated. Accordingly, the objective of this chapter is two-fold: firstly, the chapter intends to introduce the reader to the different ways in which the above-described challenge has been conceptualised, understood and studied by scholars from different disciplines and to place it in the wider context of sustainable renovation. Secondly, the chapter presents a systematic and interdisciplinary decision-making approach to find a balance between energy conservation and building conservation in the historic building stock. Finally, areas of future research are discussed.

The case studies in this chapter are from Sweden where heating, rather than cooling, is the main challenge. However, both the principal and practical approaches can be applied to places with hot climates (e.g. Raslan, 2019).

Background

Preserving historic buildings and making the building stock more energy efficient are endeavours that have been recognised as both contradicting and aligned (Yarrow, 2018). On the one hand, retaining historic features and aesthetic qualities of buildings came to be perceived as an obstacle to modernisation and energy conservation of the building stock in the wake of the 1970s' oil crisis. Energy efficiency was, correspondingly, perceived as a threat to the historic environment, which sparked actions to protect historic buildings from change. A dispute between energy conservation and heritage conservation was created, and this opposition is enduring in both discourse and practice today (Legnér & Leijonhufvud, 2019). Preservation of historic buildings and energy conservation is, on the other hand, rendered as complementary in the sustainable heritage discourse (Yarrow, 2018). In this latter narrative, historic buildings exhibit inherent sustainable features, such as being naturally adapted to the local climate (Tepper, 2011). Along the same line, it is argued that the historic environment is a source from which we can learn from the past about managing climate change impacts (Heathcote et al., 2017). Also, the notion of 'embodied carbon' of traditional buildings is a recurring argument to legitimise the preservation of historic buildings in more technical debates about climate change mitigation (Historic England, 2019).

The issue of energy efficiency in historic buildings is relevant for various, related, policy spheres, where the most obvious ones are climate change, energy efficiency, housing and heritage preservation. The built environment is, and has for a long time been, perceived as a sector with a large untapped potential for energy efficiency and reduction of greenhouse gases. Studies of the potential for energy efficiency of existing buildings tend to show that cost-efficient energy efficiency measures are not implemented, in spite of generous governmental programmes to stimulate the market since the 1970s (Shove, 1998). The rationales for such programmes have shifted over time from an initial emphasis on economy and energy security to mitigation of greenhouse gases (Economidou et al., 2020). Improving living standards by refurbishing existing buildings is another policy objective often integrated into energy efficiency programmes. More recently, policy programmes promoting 'sustainable renovation' have emerged, where environmental, social and financial aspects are integrated (Thuvander et al., 2012).

Policies for energy efficiency can be divided into regulatory, financial and supportive policies (Tanaka, 2011). These policies exist on a continuum from hard (building codes and taxes) to soft (energy performance certificates [EPC] and voluntary standards). Policies have, in general, shifted from those promoting certain technological solutions to softer ones aimed to rectify market failures (Legnér & Leijonhufvud, 2019).

Energy efficiency in historic buildings is, as discussed above, not a distinct and separate policy domain, and historic buildings have always been treated as a relatively small and difficult-to-treat subset in relation to the whole stock of buildings. A common way of dealing with such buildings in policymaking has therefore been to exempt officially designated buildings from general demands on energy efficiency (Webb, 2017). The substantial drawbacks of such policies have been highlighted in recent years. There is no distinct separation between buildings with and without heritage values, and having heritage values does not imply a lack of energy efficiency potential per se (Herrera-Avellanosa et al., 2019). Exempting officially designated buildings is, therefore, problematic in two ways. Firstly, there is often a possibility to identify energy efficiency measures that are acceptable from a preservation point of view also in these buildings. Such measures will remain unidentified if energy efficiency is neglected in maintenance and renovation. An untapped potential implies unnecessarily high running costs and might eventually lead to the abandonment of the building. Secondly, there are energy efficiency measures that should be avoided, or be better adapted, in non-designated buildings because of their impact on heritage values. In this case, it is important that heritage values also in non-designated buildings are considered early on in the planning process. Standard energy-saving measures can be problematic to use in both cases; nevertheless, the energy efficiency can often be improved considerably if an appropriate package of solutions for the specific building is identified. Soft policy in the form of voluntary decision guidelines has recently been advocated as an alternative to policy based on exemption, but the consequences are not well known (Herrera-Avellanosa et al., 2019).

Leaving the arena of policymaking at a building stock level, the next sections will look into how energy efficiency in historic buildings has been inquired on the building level – as a question about how to balance energy conservation and building conservation in specific buildings. Using a broad brush, we can distinguish two different, but not mutually exclusive, approaches to how this question has been inquired by scholars.

Firstly, scholars have been interested in how energy conservation and building conservation are negotiated in specific projects, including the unfolding of decisions in the messy real world where actions are largely social phenomena. Research of this kind has been relatively scarce but has provided indispensable clues about the links between research, policy and practice. In this approach, the issue has been largely framed as a complex problem embedded in social practices and without easily drawn boundaries. The aim for such inquiry would be to gain further insights into how materials, meanings, norms and actions interact and use such insights to shape policy.

Secondly, the bulk of research in this field has been targeted towards understanding the technical aspects of making traditional buildings more energy-efficient, i.e. identifying relevant energy efficiency measures and figuring out how to integrate them in existing buildings. In parallel to these efforts, researchers have aimed to identify decision-making frameworks and decision support tools to facilitate the planning of energy retrofits. Typically for this approach is that the issue is conceptualised as a multi-criteria decision problem, where the aim of the research is to provide tools for optimisation and for improving the data on which decisions are based.

Energy efficiency in historic buildings as social practice

Questions of how we should go about saving energy are embedded in social practice (Shove et al., 2012; Fouseki et al., 2020). It has been argued that energy efficiency policymaking should be based on a more refined understanding of how decisions on energy efficiency interventions actually are made and how they relate to everyday practices (Gram-Hanssen, 2014). In this context, it is critical to explore how different actors relate to the heritage values of buildings and to understand how such relations can hinder or support the emergence of more sustainable practices (Leijonhufvud & Henning, 2014; Sunikka-Blank & Galvin, 2016; Yarrow, 2016; Fouseki et al., 2020).

The fundamental dilemma about energy efficiency in historic buildings is about finding a balance between a number of entities of one kind (energy use, emissions, thermal comfort, costs, etc.) and an entity that is of another kind: heritage values. While the former kind is more or less objectively quantifiable and transferable, the latter kind is subjective, situational and problematic to quantify. Furthermore, these incommensurable entities are typically identified and assessed by practitioners working within different traditions and with different ways of seeing. The ambiguous and subjective nature of heritage values and the complex relationship between valuation and conservation practice make it evident that heritage values cannot be understood as simple and static attributes of an historic building. This is one important explanation as to why decision-making in practice is diverging from the cost–benefit calculations presupposed in much theoretical work about energy efficiency in buildings.

There is often a need for different professional groups with different competencies to interact when decisions are made in the planning of energy retrofit in historic buildings. Problems and solutions are framed in multiple ways by different actors. The diverse viewpoints of professionals (e.g. architects, building surveyors, heritage experts, engineers) shape, and are shaped by, the situated assessments made in relation to the challenges of specific buildings. While potential conflicts are perhaps obvious between heritage professionals and engineers who represent two different epistemic cultures, they are also present within the cultural heritage field, where for example architects and conservators tend to differ in subtle but critical ways.

Homeowners' perceptions of heritage values are often important for decisions about energy renovation (Fouseki et al., 2020). Understandings of heritage values frame assessments of whether and how modifications are considered acceptable (Yarrow, 2016). It is evident that heritage values, irrespective if referred to as 'significance', 'character' or just 'old', play not only a fundamental but also a complex and ambiguous role when making existing buildings more energy-efficient. In general, policymakers seem to have largely shunned this potential conflict by means of exclusion or ignorance, as discussed in the previous section.

The balancing act between energy efficiency and heritage preservation is often thought of and discussed in terms of a standalone, one-shot decision, where decision-makers commit to energy retrofit interventions for an individual building. To bear in mind is that such decisions are always situated in a wider context. Heritage buildings might be particularly difficult to reach with general energy efficiency policy. Legislative context tends to be downplayed in cases where heritage values are decisive, as the decision-making often is based on professional judgement and informal negotiations about case-specific circumstances rather than from rules inscribed in industry standards (Yarrow, 2016).

Built heritage is clearly at the crossroads of different visions for sustainable futures, viewed as either an obstacle or a solution, depending on who you ask and in which context. Taken together, there are multiple and at times contradictory narratives, viewpoints and attitudes circulating about energy efficiency in historic buildings, and these intersect in the negotiations about how to proceed in specific projects.

Energy efficiency in historic buildings as an optimisation problem

Energy efficiency in historic buildings can be treated as a multi-criteria decision problem where costs (in a general sense) and benefits (of various kinds) are balanced in order to come up with a solution that satisfies, to a reasonable extent, all requirements at the same time (Eriksson et al., 2014). Typically, the main variables would be investment cost and running cost, energy use and greenhouse gas emissions, impact on heritage values and comfort. Each one of these variables should be considered in a life cycle perspective where investments in terms of for example money or embedded energy are weighed against cost or energy savings over a long time period (Berg & Fuglseth, 2018; Wise et al., 2019).

A classic engineering optimisation problem is to find the thickness of added insulation that gives the lowest cost over time. The cost of insulation increases as a linear function of insulation thickness whereas the heat loss, and thus energy cost, decreases exponentially. Solving a second-degree equation will give the optimal thickness with respect to life cycle cost (LCC). A simpler trial and error approach will show that the first centimetres of insulation are very cost effective but the energy saved by going from, for example, 50 cm to 100 cm is so small that it cannot motivate the added cost. Somewhere in between, one can find a favourable balance between the cost and the benefit.

There is a need to find a way to consider many variables at the same time. The opposite is looking at one variable at a time. This approach will lead to suboptimal solutions. If all variables could be described as mathematical functions dependent on each other, it would be a manageable engineering problem. The catch is that all variables cannot easily be quantified, particularly heritage values (Fouseki & Bobrova, 2018). To further complicate things, the optimisation process must deal with packages of measures rather than individual ones.

Thus, the practical approach is to iteratively consider different packages of measures and make an overall assessment based on weighting the different variables in relation to each other. For a historic building, heritage values would, for example, be given more weight (Eriksson et al., 2014).

Energy efficiency in single buildings

The fundamental objective of most buildings is to provide a comfortable indoor climate, separated from the outdoor climate. Energy efficiency is to achieve the desired indoor climate with a minimal cost, energy use and emission of greenhouse gases. In order to maintain a required temperature in a building, heat losses must be balanced with heat supply. We need to understand the building as a technical system of interacting components and look at the potential for improvements both on the demand and the supply side (Abel & Elmroth, 2007; Fouseki & Bobrova, 2018; Fouseki et al., 2020).

Energy efficiency means either to reduce energy use in relation to the benefit or, conversely, to get an increased benefit in relation to the energy use. Improving energy efficiency in buildings generally includes a limited number of general measures that are applicable in all kinds of climates:

- Reducing heat loss through the building envelope
- Reducing infiltration and controlling ventilation, including heat recovery
- Improving the efficiency of heating and/or cooling appliances
- Introducing renewable energy
- Changed behaviour and use of the building

The most frequently discussed measures for improving energy efficiency in historic buildings refer to the building envelope, i.e. walls, roofs, floors, windows and doors. Even a few centimetres of insulation will have a great effect on the energy use of a traditional wall in the Swedish climate (Johansson et al., 2014). External insulation will inevitably have a visual impact (Sunikka-Blank & Galvin, 2016; Leijonhufvud et al., 2018; Legnér et al., 2020). Internal insulation may change the hygrothermal conditions in the wall towards higher risk levels (Blumberga et al., 2016). Insulation of attics and floors can in many cases provide substantial energy savings with a low impact on heritage values. There is a multitude of technical solutions to improve or replace windows

(Kinney & Ellsworth, 2011; English Heritage, 2012). Many of them offer substantial savings, but the visual impact must be considered.

Improving air tightness, particularly in windows and doors, can be an easy and cost-effective measure. Keep in mind that in many historic buildings, infiltration is an essential part of ventilation. Controlling the ventilation by fans, and even installing heat recovery, will both ensure a good indoor climate and lower energy demand. The physical and visual impact of the technical installations must be carefully considered.

Supply-side measures are often forgotten in the discussion about improving the energy efficiency of historic buildings. Replacing direct electric heating with heat pumps will, typically, reduce energy use by a factor of three to five. Introducing renewable energy sources, such as biofuels or solar energy, will not necessarily reduce energy use, but greenhouse gas emissions will be reduced or even eliminated. Again, the physical and visual impact must be considered.

Furthermore, we must also consider the building as part of a larger energy system. The electric grid or a district heating system can provide renewable energy that is produced in a less sensitive location (Milić et al., 2020). The role of buildings to maintain grid stability will be more important with a larger share of renewable energy sources. A practical example would be to use the thermal inertia of the building to maintain thermal comfort and turn on or off the heating for a short period depending on power surplus or power shortage (Lawrence et al., 2016).

Last, but not least, we must consider how the users of a building influence the energy efficiency (Berg et al., 2017). Numerous studies have shown that stocks of technically identical buildings have a large variation in energy use due to differences in how they are used. Building owners insulating their homes might afford to improve the thermal comfort after retrofitting, leading to less savings than calculated (rebound effect), and measured energy use in old buildings might be lower than calculated use (prebound effect). Such effects are important to consider but can be difficult to estimate (Sunikka-Blank & Galvin, 2012). Increased awareness of how behaviour is related to energy use, for example through the setting of thermostats, offers a potential energy saving without negative side effects.

A systematic approach to balance energy efficiency with heritage preservation

From a techno-economic perspective, the aim is to find the combination of measures that give the lowest cost and environmental impact over the life cycle of the building. For a historic building, the techno-economic aspects must be balanced with considerations of how the measures will impact the heritage values of the building.

The European standard *EN 16883 Conservation of cultural heritage - Guidelines for improving the energy performance of historic buildings* presents a working procedure for selecting measures to improve the energy performance in historic buildings (European Committee for Standardization, 2017). By following the procedure, the user assesses the impact of energy efficiency measures on multiple objectives, including the preservation of character-defining elements of the building. The European guidelines comprise the following steps:

1. Building survey and assessment
2. Specifying objectives for energy performance as well as for preservation

3 First assessment of risks and benefits based on a gross list of measures for energy efficiency resulting in a shortlist of measures to be analysed further
4 Life cycle techno-economic optimisation to find the best combination of measures selected in step 3
5 Risk assessment with respect to heritage values and building physics
6 Analysis of the consequences in relation to the targets
7 Iterative adjustments of measures and/or targets

This procedure should not be seen as a mechanical tool for decision-making. Rather, it provides a framework for a systematic and transparent discussion with stakeholders and a multidisciplinary team of experts. The method can be used to show the consequences of different targets in terms of energy savings and impact on heritage values as part of an iterative procedure, aiming to find a solution that balances benefits in relation to risks and costs.

Case study

This section exemplifies the aforementioned approach and discussion in the case of one family, wooden building from the late 19th century in the Swedish capital of Stockholm. It is not formally protected which means that there is a wide room for negotiation on what type of measures that can be accepted. The Swedish Planning and Building Act mandates, without specifications, that renovations in all buildings, no exceptions, should be carried out carefully with respect to the heritage values.

In this case study, the energy saving targets used were 20% and 50%, corresponding to different targets used in the EU and Sweden (Broström et al., 2014).

Based on a gross list of the most common energy efficiency measures in Sweden, a first assessment was made of risks and benefits in relation to the particular building. This step was mainly meant to exclude inappropriate measures and to define a range of acceptable measures. The assessment was carried out by a multidisciplinary group of experts. Based on the existing knowledge and experience, each measure was assessed with respect to energy savings, economic return, impact on heritage values, durability, moisture risk and effect on the indoor environment.

In the next step, a more detailed assessment was made on a shortlist of measures. The basic tool for selecting measures was a program for LCC optimisation (Liu et al., 2018). Using a database with costs and technical specifications for different measures, the program defines a combination of measures that would give the lowest LCC.

The LCC optimisation resulted in the combination of measures that would achieve the given targets at the lowest LCC. The 20% target could be reached with measures that had little or nil effect on heritage values such as weather stripping, heat pumps and attic insulation. In this case, LCC was reduced by 15% (Broström et al., 2014).

In order to reach the national 50% target, exterior wall insulation and window replacement were necessary, in addition to less intrusive measures. In this case, LCC was reduced by 23%. If the exterior insulation and window replacement are excluded in the LCC optimisation, the energy saving would be reduced to 29% and LCC would be reduced by 16% (Broström et al., 2014).

The results illustrate how the method can be used to assess the consequences of policies and plans related to energy efficiency. For this particular building, it could be shown that the European 20% target can be reached without problems. But the more ambitious national target (50%) would necessitate measures, such as external insulation and window retrofits, that would change the visual and material character of the building significantly.

The above method allows for an interaction between the quantitative assessment of the techno-economic optimisation and the qualitative assessment of the impact on heritage values. Through a multidisciplinary dialogue, stakeholders and experts can arrive at a solution that balances energy conservation and building conservation for a given building. This approach can be used to assess the consequences of national targets for energy savings and let stakeholders and experts, through further iterations, decide on an appropriate level of energy saving in relation to the impact on heritage values.

Energy efficiency in building stocks

The previous section showed a method to find an appropriate package of measures to improve energy efficiency in a single building. This section will consider the building stock as a system in order to identify segments of the building stock where substantial energy savings can be realised with an acceptable impact on heritage values, and vice versa, exclude valuable parts of the stock with a small potential. This will facilitate more precise planning and policies, both at a regional and a national level.

Correlating energy use and age on a national level

A study using data from the Swedish database of EPC correlated energy use and energy performance with building age (Donarelli & Broström, Forthcoming). This rather simple analysis gave a number of results that were not obvious or intuitive.

Buildings in Sweden built before 1960 typically have an average energy performance that is twice the national target for new buildings. As 'near-zero energy' is becoming an explicit or implicit standard, incremental measures are no longer sufficient to reach that target. We have to critically consider whether these targets are applicable to historic buildings.

All age groups of buildings built before 1960 have more or less the same energy performance, independent of age. More modern buildings, of course, are much more efficient. This means one cannot, a priori, assume that the oldest buildings have worse performance.

Buildings built before 1850 account for a very small part of the energy use for buildings, 0.9% in single-family houses and 0.2% in multifamily buildings. These buildings would generally have significant heritage values and should not be prioritised when it comes to achieving national or regional targets for energy saving. The energy use in buildings built before 1945 stands for around 20% of the energy use in the Swedish building stock (Donarelli & Broström, Forthcoming). This is part of the building stock with the main challenge in balancing energy savings with heritage values. Well-planned measures can achieve significant energy savings on a national level. Policymakers should be aware that buildings constructed in 1945–1974 are by far the largest age segment in terms of energy use where significant energy savings can be achieved with an acceptable impact on heritage values.

The above results are specific to Sweden and should not be extrapolated to other countries but the simple method can be generally used. For instance, a study of the building stock in Basel, Switzerland, showed that buildings built before 1921 had a better energy performance than more modern buildings (Aksoezen et al., 2015). Weak relationships between building age and energy performance were also detected in a study in the United States (Webb et al., 2018).

Correlating heritage values with energy use on a regional level

In the above investigation, building age was correlated with energy use and energy performance. Age may or may not reflect heritage values. A more precise analysis requires an investigation on how energy use is correlated with heritage values. Since the 1920s, the City Museum of Stockholm has classified historic buildings using a three-step scale. All classified multifamily buildings (more than 15,000) in Stockholm were analysed (Eriksson & Johansson, 2021). These buildings stand for 75% of the energy used for all multifamily buildings in Stockholm. Only 3% of the energy is used in buildings with the highest heritage classification. These buildings should therefore not be a priority when it comes to achieving national or regional goals. Buildings in the second and third heritage classification use as much as 72% of the total energy and must be addressed in order to reach general targets.

This case study shows that a relatively simple analysis of energy use in relation to age and heritage classification can be used to establish a priority among the different segments of large building stocks for the purpose of planning and policymaking.

Combining top-down with bottom-up analysis on a local level

This section shows how the bottom-up single building approach can be combined with a top-down building stock analysis for a historic district. The historic district of Visby, with 1,235 heritage buildings, was used as a case study (Eriksson et al., 2019).

During the first step, the building stock was categorised based on construction type and spatial features (Broström et al., 2017). This resulted in 12 categories representing 88% of the building stock. For each category, an archetype building was defined with typical properties in terms of heated area, U-value, etc. For each archetype building, the LCC optimisation (as described above) was used to find the package of solutions that would result in the lowest LCC. In the next step, preservation objectives, based on the degree of change that the buildings can withstand before the heritage values are completely changed or lost, were transformed into two levels of restrictions:

- Restricted – LCC optimal but excluding window replacement and facade insulation
- Balanced – LCC optimal but excluding window replacement, limited insulation thickness

For each building category, the cost of imposing restrictions on energy measures in relation to the techno-economic optimal can be predicted, in terms of energy, greenhouse gas emissions and LCC. Thus, an informed decision on what restriction levels are appropriate can be facilitated. For example, the restrictions had a major impact on energy use but the difference in LCC was quite small.

Furthermore, this type of analysis facilitates differentiated goals and strategies for each building category. The results showed that, within the restrictions, stone buildings had a larger potential for energy saving than wooden buildings. The results for each building archetype can be extrapolated to predict the effects for the whole category and in turn for the district as a whole.

This section on building stock analysis has shown a way to systematically balance energy saving and preservation in historic building stocks by combining a detailed building analysis with an overview of the whole building stock. The method can be used to support local planning and policymaking.

Discussion

As researchers in this field, we stand between a critical and a normative/operative perspective on improving energy efficiency in historic buildings. On a principal level, we cannot and should not try to bridge this gap once and for all. However, when we bring it down to a practical level, the dilemma becomes less complex, although not necessarily less contested.

Decisions about how to save energy are situated in complex and multilayered interactions of sociotechnical systems and embedded in social practices. If real cases are studied, commitment to a certain course of action is often related more to habits, conventions and norms than to deliberate and explicit decision-making. There can be multiple factors influencing building owners to adopt energy efficiency measures, such as replacing old windows with energy-efficient ones, and these factors often do not align well with accounts where decision-makers make calculations of pros and cons as a basis for their commitments. This is even more salient in buildings with heritage values, as these values by nature are subjective and resonate with affective capabilities. Approaches to preserving heritage are parts of social practices and discourses that are essentially different from those involved in the technical understanding of buildings.

For the heritage practitioner working with practical problems, there is a need to channel the often somewhat abstract results of heritage value assessments into practically applicable procedures. We have in this chapter shown that the practice of improving energy efficiency in historic buildings does not have to be very complicated. By planning and assessing energy efficiency measures in systematic ways, it is possible to manage the many competing objectives involved, both in individual buildings and in building stocks. Heritage values are treated as given inputs in the instrumental decision framework we have advocated. It is up to the involved stakeholders to perform a heritage value assessment in the individual case. This point of view implicitly takes for granted that heritage values, to at least some extent, are pre-existing, stable and possible to uncover through a deliberate analysis, often performed by a heritage expert. We have, however, argued that such assessments are both situated and limited in scope.

We suggest that a fundamental challenge in both research and practice is to merge a critical perspective on cultural heritage with operative procedures that will be broadly accepted by professional actors as well as users and the public. Much research has been made from the outset of the instrumental, and in effect, normative perspective. There is a lot of useful information based on such research efforts and we have provided some examples based on case studies. A problem is that existing procedures for systematic planning are not widely used (Herrera-Avellanosa et al., 2019). Research is needed to understand and reduce the gap between what is advocated in instrumental decision frameworks on how this issue should be managed and the ways that energy efficiency in historic buildings is played out in practice.

A building stock perspective is needed for planning and policymaking at national, regional and local levels. The types of analyses of building stocks described in this chapter are necessary but they require access to correct and aggregate data on large building stocks. Relevant data is in many cases non-existent or placed in different non-compatible databases. The only way the type of methods described here can be commonly used is by providing aggregated and updated data on the building stocks.

Energy efficiency in historic buildings is a subset of energy efficiency in buildings which in turn is a subset of building renovation. In developing and establishing energy efficiency in historic buildings as a field of inquiry, it has been deemed important to focus on the aspects that set it apart from the general context of energy renovations. However, the main societal challenge is not improving energy efficiency in medieval monumental buildings but rather the much larger stock of more modern buildings. Both researchers and professionals in this relatively narrow field have to consider a strategy of mainstreaming heritage aspects in overarching processes to have a broader impact on both greenhouse gas emissions and the preservation of historic buildings.

References

Abel, E., & Elmroth, A. (2007). *Buildings and energy-a systematic approach.* Stockholm: Formas.

Aksoezen, M., Daniel, M., Hassler, U., & Kohler, N. (2015). Building age as an indicator for energy consumption. *Energy and Buildings, 87,* 74–86. https://doi.org/10.1016/j.enbuild.2014.10.074.

Berg, F., Flyen, A.-C., Godbolt, Å. L., & Broström, T. (2017). User-driven energy efficiency in historic buildings: A review. *Journal of Cultural Heritage, 28,* 188–195. https://doi.org/10.1016/j.culher.2017.05.009.

Berg, F., & Fuglseth, M. (2018). Life cycle assessment and historic buildings: Energy-efficiency refurbishment versus new construction in Norway. *Journal of Architectural Conservation, 24*(2), 152–167. https://doi.org/10.1080/13556207.2018.1493664.

Blumberga, A., Kass, K., Kamendere, E., Žogla, G., Kamenders, A., Blumberga, D., ... Hansen, E. J. de P. (2016). *State of the art on historic building insulation materials and retrofit strategies.* Retrieved from: http://ribuild-web.sbi.aau.dk/sites/default/files/RIBuild_D1.2_1.0.pdf (Accessed on 03 May 2021).

Broström, T., Donarelli, A., & Berg, F. (2017). For the categorisation of historic buildings to determine energy saving. *AGATHÓN International Journal of Architecture, Art and Design, 1,* 135–142. https://doi.org/10.19229/2464-9309/1212017.

Broström, T., Eriksson, P., Liu, L., Rohdin, P., Ståhl, F., & Moshfegh, B. (2014). A method to assess the potential for and consequences of energy retrofits in Swedish historic buildings. *The Historic Environment: Policy & Practice, 5*(2), 150–166. https://doi.org/10.1179/1756750514Z.00000000055.

Broström, T., Nilsen, L., & Carlsten, S. (2018). *Postprints to the 3rd international conference on energy efficiency in historic buildings (EEHB2018), Visby, Sweden, September 26th to 27th, 2018.* Retrieved from: http://urn.kb.se/resolve?urn=urn:nbn:se:uu:diva-378215 (Accessed on 03 May 2021).

de Bouw, M., Dubois, S., Dekeyser, L., & Vanhellemont, Y. (2016). *Postprints to the 2nd international conference on energy efficiency and comfort in historic buildings (EECHB2016), Brussels, Belgium, October 19th to 21th, 2016.* Retrieved from: https://www.eechb.eu/wp-content/uploads/2016/12/Proceedings_EECHB.pdf (Accessed on 03 May 2021).

Donarelli, A., & Broström, T. (Forthcoming). Does age matter? How building age influences energy performance in the Swedish residential building stock. *Accepted to the 4th Conference on Energy Efficiency in Historic Buildings (EEHB 2022), Benediktbeuern, Germany, May 4th to 5th, 2022.*

Economidou, M., Todeschi, V., Bertoldi, P., D'Agostino, D., Zangheri, P., & Castellazzi, L. (2020). Review of 50 years of EU energy efficiency policies for buildings. *Energy and Buildings*, *225*, 110322. https://doi.org/10.1016/j.enbuild.2020.110322.

English Heritage. (2012). Energy efficiency and historic buildings: Secondary glazing for windows. Retrieved from: https://historicengland.org.uk/images-books/publications/eehb-secondary-glazing-windows/ (Accessed on 03 May 2021).

Eriksson, P., Hermann, C., Hrabovszky-Horváth, S., & Rodwell, D. (2014). EFFESUS methodology for assessing the impacts of energy-related retrofit measures on heritage significance. *The Historic Environment: Policy & Practice*, *5*(2), 133–149. https://doi.org/10.1179/1756750514Z.00000000054.

Eriksson, P., & Johansson, T. (2021). Towards differentiated energy renovation strategies for heritage-designated multifamily building stocks. *Heritage 4*, 4318–4334. https://doi.org/10.3390/heritage4040238.

Eriksson, P., Milić, V., & Brostrom, T. (2019). Balancing preservation and energy efficiency in building stocks. *International Journal of Building Pathology and Adaptation*, *38*(2), 356–373. https://doi.org/10.1108/IJBPA-02-2019-0025.

European Committee for Standardization. (2017). *EN 16883:2017 Conservation of cultural heritage – guidelines for improving the energy performance of historic buildings*. Brussels: CEN.

Fouseki, K., & Bobrova, Y. (2018). Understanding the change of heritage values over time and its impact on energy efficiency decision-making at residential historic buildings through system dynamics. In: *Postprints to the 3rd international conference on energy efficiency in historic buildings (EEHB2018), Visby, Sweden, September 26th to 27th, 2018*. Retrieved from: http://urn.kb.se/resolve?urn=urn:nbn:se:uu:diva-378215 (Accessed on 03 May 2021).

Fouseki, K., & Cassar, M. (2014). Energy efficiency in heritage buildings — future challenges and research needs. *The Historic Environment: Policy & Practice*, *5*(2), 95–100. https://doi.org/10.1179/1756750514Z.00000000058.

Fouseki, K., Newton, D., Murillo Camacho, K. S., Nandi, S., & Koukou, T. (2020). Energy efficiency, thermal comfort, and heritage conservation in residential historic buildings as dynamic and systemic socio-cultural practices. *Atmosphere*, *11*(6), 604. https://doi.org/10.3390/atmos11060604.

Gram-Hanssen, K. (2014). Retrofitting owner-occupied housing: Remember the people. *Building Research & Information*, *42*(4), 393–397. https://doi.org/10.1080/09613218.2014.911572.

Heathcote, J., Fluck, H., & Wiggins, M. (2017). Predicting and adapting to climate change: Challenges for the historic environment. *The Historic Environment: Policy & Practice*, *8*(2), 89–100. https://doi.org/10.1080/17567505.2017.1317071.

Herrera-Avellanosa, D., Haas, F., Leijonhufvud, G., Brostrom, T., Buda, A., Pracchi, V., … Troi, A. (2019). Deep renovation of historic buildings. *International Journal of Building Pathology and Adaptation*, *38*(4). https://doi.org/10.1108/IJBPA-12-2018-0102.

Historic England. (2019). Heritage counts 2019- there's no place like old homes: Re-use and recycle to reduce carbon. Retrieved from: https://historicengland.org.uk/research/heritage-counts/2019-carbon-in-built-environment/carbon-in-built-historic-environment/ (Accessed on 03 May 2021).

IEA. (2019). Tracking buildings 2019, IEA, Paris. Retrieved from: https://www.iea.org/reports/tracking-buildings-2019.

IPCC. (2014). Climate change 2014: Mitigation of climate change. contribution of working Group III to the fifth assessment report of the intergovernmental panel on climate change [Edenhofer, O., R. Pichs-Madruga, Y. Sokona, E. Farahani, S. Kadner, K. Seyboth, A. Adler, I. Baum, S. Brunner, P. Eickemeier, B. Kriemann, J. Savolainen, S. Schlömer, C. von Stechow, T. Zwickel and J.C. Minx (Eds.)]. Cambridge: Cambridge University Press.

IPCC. (2018). Global warming of 1.5°C. An IPCC special report on the impacts of global warming of 1.5°C above pre-industrial levels andrelated global greenhouse gas emission pathways, in the context of strengthening the global response to the threat of climate change, sustainable development, and efforts to eradicate poverty [Masson-Delmotte, V., P. Zhai,

H.-O. Pörtner, D. Roberts, J. Skea, P.R. Shukla, A. Pirani, W. Moufouma-Okia, C. Péan, R. Pidcock, S. Connors, J.B.R. Matthews, Y. Chen, X. Zhou, M.I. Gomis, E. Lonnoy, T. Maycock, M. Tignor, and T. Waterfield (Eds.)]. Cambridge: Cambridge University Press.

Johansson, P., Hagentoft, C.-E., & Sasic Kalagasidis, A. (2014). Retrofitting of a listed brick and wood building using vacuum insulation panels on the exterior of the facade: Measurements and simulations. *Energy and Buildings*, 73, 92–104. https://doi.org/10.1016/j.enbuild.2014.01.019.

Kinney, L., & Ellsworth, A. (2011). The Effects of energy efficient treatments on historic windows. Retrieved from: https://www.energystar.gov/sites/default/files/Attachment%20to%20Lynn%20Bingham.pdf (Accessed on 03 May 2021).

Lawrence, T. M., Boudreau, M.-C., Helsen, L., Henze, G., Mohammadpour, J., Noonan, D., ... Watson, R. T. (2016). Ten questions concerning integrating smart buildings into the smart grid. *Building and Environment*, 108, 273–283. https://doi.org/10.1016/j.buildenv.2016.08.022.

Legnér, M., & Leijonhufvud, G. (2019). A legacy of energy saving: The discussion on heritage values in the first programme on energy efficiency in buildings in Sweden, c. 1974–1984. *The Historic Environment: Policy & Practice*, 10(1), 40–57. https://doi.org/10.1080/17567505.2018.1531646.

Legnér, M., Leijonhufvud, G., & Tunefalk, M. (2020). Energy policy and conservation planning in Sweden: A longitudinal evaluation. *International Journal of Building Pathology and Adaptation*. 38(4), 555–572. https://doi.org/10.1108/IJBPA-11-2019-0096.

Leijonhufvud, G., & Henning, A. (2014). Rethinking indoor climate control in historic buildings: The importance of negotiated priorities and discursive hegemony at a Swedish museum. *Energy Research & Social Science*, 4(0), 117–123. https://doi.org/10.1016/j.erss.2014.10.005.

Leijonhufvud, G., Tunefalk, M., & Legnér, M. (2018). What's behind the façade? A long-term assessment of the Swedish energy efficiency programme 1977–1984 and its impact on built heritage. In *Postprints to the 3rd international conference on energy efficiency in historic buildings (EEHB2018), Visby, Sweden, September 26th to 27th, 2018*. Retrieved from: http://urn.kb.se/resolve?urn=urn:nbn:se:uu:diva-378215 (Accessed on 03 May 2021).

Liu, L., Rohdin, P., & Moshfegh, B. (2018). Investigating cost-optimal refurbishment strategies for the medieval district of Visby in Sweden. *Energy and Buildings*, 158, 750–760. https://doi.org/10.1016/j.enbuild.2017.10.002.

Milić, V., Amiri, S., & Moshfegh, B. (2020). A systematic approach to predict the economic and environmental effects of the cost-optimal energy renovation of a historic building district on the district heating system. *Energies*, 13(1), 276. https://doi.org/10.3390/en13010276.

Milić, V., Ekelöw, K., & Moshfegh, B. (2018). On the performance of LCC optimization software OPERA-MILP by comparison with building energy simulation software IDA ICE. *Building and Environment*, 128, 305–319. https://doi.org/10.1016/j.buildenv.2017.11.012.

Raslan, E. (2019). Energy retrofit and conservation issues in hot arid climates: The case of 19th and 20th centuries heritage residential buildings in Cairo, Egypt. (Unpublished PhD Thesis). Milan: Politecnico di Milano.

Shove, E. (1998). Gaps, barriers and conceptual chasms: Theories of technology transfer and energy in buildings. *Energy Policy*, 26(15), 1105–1112. https://doi.org/10.1016/S0301-4215(98)00065-2.

Shove, E., Pantzar, M., & Watson, M. (2012). *The dynamics of social practice: Everyday life and how it changes*. Los Angeles: SAGE.

Sunikka-Blank, M., & Galvin, R. (2012). Introducing the prebound effect: The gap between performance and actual energy consumption. *Building Research & Information*, 40(3), 260–273. https://doi.org/10.1080/09613218.2012.690952.

Sunikka-Blank, M., & Galvin, R. (2016). Irrational homeowners? How aesthetics and heritage values influence thermal retrofit decisions in the United Kingdom. *Energy Research & Social Science*, 11, 97–108. https://doi.org/10.1016/j.erss.2015.09.004.

Tanaka, K. (2011). Review of policies and measures for energy efficiency in industry sector. *Energy Policy*, 39(10), 6532–6550. https://doi.org/10.1016/j.enpol.2011.07.058.

Tepper, A. T. (2011). *The secretary of the interior's standards for rehabilitation & illustrated guidelines on sustainability for rehabilitating historic buildings.* Washington DC, USA: Government Printing Office.

Thuvander, L., Femenías, P., Mjörnell, K., & Meiling, P. (2012). Unveiling the process of sustainable renovation. *Sustainability, 4*(6), 1188–1213. https://doi.org/10.3390/su4061188.

Webb, A. (2017). Energy retrofits in historic and traditional buildings: A review of problems and methods. *Renewable and Sustainable Energy Reviews, 77,* 748–759. https://doi.org/10.1016/j.rser.2017.01.145.

Webb, A., Beckett, L., & Burton, M. D. (2018). Examining the energy performance of older and historic buildings using municipal benchmarking data. In *Postprints to the 3rd international conference on energy efficiency in historic buildings (EEHB2018), Visby, Sweden, September 26th to 27th, 2018.* Retrieved from: http://urn.kb.se/resolve?urn=urn:nbn:se:uu:diva-378215 (Accessed on 03 May 2021).

Wise, F., Moncaster, A., Jones, D., & Dewberry, E. (2019). Considering embodied energy and carbon in heritage buildings–a review. *IOP Conference Series: Earth and Environmental Science, 329,* 012002. Bristol: IOP Publishing.

Yarrow, T. (2016). Negotiating heritage and energy conservation: An ethnography of domestic renovation. *The Historic Environment: Policy & Practice, 7*(4), 340–351. https://doi.org/10.1080/17567505.2016.1253149.

Yarrow, T. (2018). How conservation matters: Ethnographic explorations of historic building renovation. *Journal of Material Culture, 24*(1), 3–21. 1359183518769111. https://doi.org/10.1177/1359183518769111.

21 Balancing heritage values, thermal comfort and energy efficiency in world heritage sites

The case of Mexico City

Krisangella Sofía Murillo Camacho, Kalliopi Fouseki and Hector Altamirano

Introduction

Although listed buildings are exempted from minimum energy efficiency requirements, energy-related interventions may have an irreversible impact on the heritage value of both listed and unlisted historic buildings. A universal solution for energy-saving potential and building conservation does not exist due to the uniqueness of each historic building (Cassar, 2011). The European Union (EU) has made gradual progress towards achieving savings in its stock of heritage buildings by encouraging building sectors to reduce CO_2 emissions and combat climate change (Berg et al., 2017). Various energy-efficient retrofit programmes have been developed in Europe with solutions that aim to balance the need for heritage conservation and energy efficiency (Commission of the European Communities, 2006; Berg, 2017; Lidelöw et al., 2019). Historic England also addresses the needs of the heritage sector in this way by publishing energy efficiency guidelines for heritage buildings. These advocate a 'whole house retrofit approach' that takes into account a building's materials and performance, alongside its values. However, a pre-established set of values may easily neglect the significance of individual occupants' own attitudes and behaviours (Fouseki et al., 2020).

Renovations for energy efficiency may not only help reduce energy consumption but also provide a means of preserving a building for future generations (Webb, 2017). However, there are still technical and economic barriers that lower the energy-saving potential of these renovations (Berg et al., 2017). This chapter aims to illustrate some of these barriers as they are experienced and perceived by residents in the historical centre of Mexico City. By doing so, it augments past research endeavours with a non-European perspective. Due to the unique nature of each heritage building, a holistic solution is required to make historic buildings, or buildings with a cultural significance, more energy-efficient and sustainable.

There are two main challenges to this end. First, there is a difficulty in developing and integrating less invasive energy efficiency technologies into the historic fabric of a building. Second, understanding users' attitudes towards the value they assigned to heritage buildings and the performance for thermal comfort of the buildings (an understanding that is important for drafting relevant policies) is an additional complexity. This study showed that occupants of heritage buildings in Mexico City attached attracted high cultural values to their buildings. In most cases, they would mitigate thermal discomfort in winter through passive actions such as using heavier curtains and putting on more clothing. Due to the original features, aesthetics and architectural value of their buildings, along with the high cost of interventions, the occupants resisted any upgrades. Nevertheless, they would consider making changes for thermal comfort as long as they had the resources; the changes were minimal and the government would approve the work. There are hitherto no studies in Mexico – much less guidelines – concerning energy efficiency in heritage-protected areas for which the views of the occupant are taken into account.

In Mexico, the regulatory frameworks governing heritage and energy work separately. Apart from the federal government's programme to certify sustainable buildings, Mexico has no comprehensive, nationwide certification programme or evaluation system for historic buildings that is based on an institutional framework and environmental standards. The Mexican government only provides building performance standards for modern buildings, which limits their occupants' ability to understand and execute refurbishing programmes easily. New models must be developed to address the lack of public policy governing the implementation of energy-efficient measures in historic buildings that are being retrofitted. To this end, a greater understanding of heritage values and energy efficiency could inform how the energy infrastructure of such buildings could be adapted while preserving their significance to their occupants.

This chapter seeks to complement the limited knowledge we have on how occupants in world heritage sites choose to adopt measures for energy efficiency and thermal comfort. It does so by focusing on the social and cultural values that occupants of listed and non-listed buildings in Mexico City attach to their residences and the values they prioritise for energy efficiency interventions. This chapter aspires to serve as a model for future research into the integration of occupants of the historic centres of Latin America into the decision-making processes behind sustainable programmes for heritage buildings and energy efficiency.

This study is founded on the premise that social meaning, spatial structures, heritage values, sustainability preservation and energy efficiency are interconnected. When reinforced over time, these factors may drive or prohibit changes in energy efficiency. Moreover, tension will eventually arise between the limitations placed on a listed building (that impede change) and the lack of interventions applied (that affect the building directly). The present study, applied in the historic centre of Mexico City, assumes that values associated with the tangible characteristics of a building (e.g. architectural, historic and aesthetic) and sentimental, symbolic values (e.g. family attachment) increase the overall value of a residence over time and determine which building characteristics residents are willing to change, compromise or maintain as they strive to improve their building's energy performance.

The chapter utilises the method of system dynamics for mapping the interconnections listed above. This approach enables the capture of the complex causal structure of a system (in this case, the system is the building) in a formal model that can be simulated and validated against real-world observations (Forrester, 1987). It is grounded in the theory of non-linear dynamics and feedback control developed in the fields of mathematics, physics and engineering. As shown in Fouseki et al. (2020), understanding the dynamic interrelationship between values, users and energy makes the management plans for heritage areas more sustainable.

Context

There has been a growing number of EU-funded research programmes surrounding climate change, energy efficiency and heritage in recent years. Energy Efficiency for

EU Historic Districts' Sustainability (EFFESUS), Climate for Culture and Efficient Energy for EU Cultural Heritage (3ENCULT) are among the few existing examples of projects that aim to address the impact of a changing climate on heritage while proposing energy-efficient retrofit solutions that respect the heritage values of historic buildings. In the United Kingdom, the technical guidance from Historic England (2018) seeks to improve historic buildings' energy efficiency by taking a whole-building approach. In the EU, standards and technical documents for improvements in energy performance are provided by the Comité Européen de Normalisation (CEN, 2017).

The programmes mentioned above rely on research that aims to integrate new technologies into historic buildings. A historic building should be understood as being part of a social–cultural organisation (i.e. the values and communities that inhabit or use it) and an architectural system with materials and physical components (that evolve over time and are subject to users' behaviours and approach to preservation). In addition, most legislation does not entail an in-depth discussion of users' values.

The case of the historic centre (*Centro Histórico*) of Mexico City – one of the largest and most populated capitals in the world – shows the need for good design policies that include sustainability since the site has been declared a World Heritage Site and gone through many management plans without much progress. The site has a rich architectural and archaeological wealth that includes vestiges of the five Aztec temples located so far. It is also home to the most significant cathedral in the Americas. Different styles of the Viceroyalty of New Spain period can be observed in the same space. The site offers significant opportunities for reducing energy consumption and greenhouse gas emissions through the retrofitting of existing buildings. Thus, integration of energy efficiency heritage values and user-assisted decision-making processes is needed to develop energy efficiency and management plans.

This chapter aims to address these gaps by looking specifically at residents' behaviours towards energy efficiency and heritage conservation. This, it will be shown, is merely a process of negotiating the need to improve thermal comfort while reducing energy bills and conserving the heritage of their residences. To explore this negotiation process, it is important to look at what is meant by the conservation of heritage values, thermal comfort and energy behaviour for energy efficiency in heritage buildings and in a world heritage site.

Conservation of heritage values

One way to think about values is to consider them as a perceptual mechanism identifiable to the object attributed to them across different periods of time. The concept of values is dynamic and appears and disappears according to an intellectual, cultural, historical and psychological framework that varies according to people, groups and time. The fact is that the symbolic charge acquires different connotations in each historical moment, producing at the end a sequence in time of distinct interpretative figures (Ballart et al., 1996, p. 215).

In the literature, values have different characteristics assigned to them. For instance, heritage values are seen as being plural, in recognition of the fact that heritage is considered significant for a range of different reasons, and a comprehensive, universally applicable value typology is an impossibility (Fredheim & Khalaf, 2016). However, values are different depending on the context in which they are viewed.

For protection and management, the universal characterisation of cultural property began in 1931 with the Athens Charter (Le Corbusier & José Luis Sert, 1942). The cultural attributes transcended international standards and acquired particular relevance in the 20th century. As a result, several international organisations were created to safeguard heritage with the objective of its management by the EU and the United Nations Educational, Scientific and Cultural Organization (UNESCO, 2014).

In this sense, the definition of values is seen as the social attribution of the qualities of things to their relationship with the cultural good, the site, the context and its evolution over time (Riegl, 1996; Martínez Yáñez, 2006; UNESCO, 2014). Values also refer to a mental construction that one person or a group of persons has about the importance of humanity's cultural manifestations and, in general terms, they determine the meaning that values give to the sense of life for communities (Palma Peña, 2017).

Avrami et al. (2000) explain that each act of conservation is shaped by how an object or place is valued, its social contexts, the available resources and local priorities. Thus, if the object is preserved because it has value, then its conservation is related to meanings and values attached to it (Pye, 2001; Appelbaum, 2007). Conservation becomes a value-based activity that can be understood as an expression of values that change with social constructions (Pearson & Sullivan, 1995; Avrami et al., 2000; Richmond & Bracker, 2009). For instance, the meanings given to values and objects by society influence the conservation process and, simultaneously, with the new research values such as cultural diversity, identity, memory and authenticity.

Thermal comfort and energy behaviour

There is a presumption that old buildings do not offer good thermal comfort, though this is subjective to an individual. Thermal comfort perceptions can be described as a condition of mind that expresses satisfaction with the thermal environment and is assessed by a subjective evaluation (Martínez-Molina et al., 2016). A good indoor climate is important for the occupant not only because it makes a space more comfortable but also because thermal comfort is part of the decision-making process towards energy consumption and retrofit solutions for energy efficiency. It is important to balance energy efficiency, conservation procedures and the thermal comfort of occupants because they inhabit the buildings.

One approach to getting greater energy efficiency within buildings is found in the literature as 'energy behaviour'. Defined as 'actions taken by owners in the use of energy in their homes', these need little or no financial investment and require only daily habit awareness. Typical ways to achieve such energy savings include consuming less hot water, adjusting the inside temperature and controlling natural ventilation. Yohanis (2012), Lopes et al. (2012) and Berg et al. (2017) show that energy behaviour can reduce energy consumption, while also having a less physical impact on the building. They focus on conventional aspects of user behaviour that can be integrated into a procedural approach for energy rehabilitation in historic buildings.

However, users' behaviour is not uniform, which creates several research challenges, especially in the context of heritage buildings whereby the physical characteristics and cultural context vary (Stephenson et al., 2010). The way residents use a building is more important than the type of changes they make for energy efficiency (Fouseki & Cassar, 2014). Research on user behaviour shows that non-intrusive measures could produce more significant energy savings than physical improvements alone (Ben & Steemers, 2014).

Heritage buildings and energy efficiency

Heritage value, and how this is conserved when technologically advanced solutions are implemented, has always been a problematic area of discussion within the sphere of building conservation. A universal solution is not possible for the various problems that affect historic buildings (Cassar, 2011). The EU has made gradual progress towards this by promoting the use of different strategies in all productive sectors to achieve the goals set for reducing CO_2 gas and combating climate change (Berg et al., 2017). These practices could also be implemented in other parts of the world.

The literature on energy efficiency in heritage buildings has led to the development of decision-making frameworks to assess energy consumption and performance (Lopes et al., 2012; Ma et al., 2012; Berg et al., 2017). Webb (2017) reviewed the criteria for different methods of analysis and decision-making processes used to evaluate energy modifications in heritage and traditional buildings. Eriksson et al. (2014) have developed such a methodology for the European EFFESUS project that compares heritage significance evaluation, impact definitions and balancing processes. Broström et al. (2014) investigated political energy targets and their effects on built heritage in a Swedish research project entitled 'Potential and Policies for Energy Efficiency in Swedish Historic Buildings'.

From a social point of view, the findings of an ethnographic study by Yarrow (2016, 2019) showed the different perspectives of professionals and owners involved in changes in historic buildings. Yarrow highlighted how the past value was negotiated in different social forms concerning climate change and energy efficiency. This approach showed a series of dynamics that have received little attention, demonstrating conservation as a practice grounded in various material, bodily, emotional, ethical and conceptual forms.

Recent studies (Fouseki & Bobrova, 2018; Koukou & Fouseki, 2018; Fouseki et al., 2020, Murillo et al., 2022) showed how homeowners' priorities for renovation (e.g. their need for thermal comfort and attachments to cultural features of their house) drove or precluded their energy efficiency interventions on traditional listed or unlisted buildings. Fouseki et al. (2020) investigated the parameters that affected inhabitants' decisions over time and reiterated how heritage policies and energy efficiency measures should be more socially conscious.

Users make an important contribution to historic buildings (they maintain, conserve and care for buildings for future generations) and have the capacity to adapt while minimising their energy demand and physical impact on buildings. When international institutions in the EU and UK propose measures and programmes to reduce the operational energy use of an individual heritage building, they are required to integrate occupants in the process. In Mexico, creating research programmes and sustainability guidelines for heritage buildings and their occupants would promote better design practices for energy efficiency and ensure their continuity.

The case of the historic centre of Mexico City

Energy efficiency and thermal comfort in Mexico

In Mexico, energy efficiency and thermal comfort have become a national concern in the residential sector. Research focussing on energy in Mexico has increased since the development of the national energy policy between 2012 and 2018 (*Reforma Energética*).

Measures by the Mexican government to reduce nationwide electricity consumption include the creation of the Trust for Thermal Insulation of Housing and Integral Systematic Saving Programmes. Both of these, led by the Secretary of Energy (Secretaria de Energia, 2014), have achieved an estimated reduction of 3,410.72 GWh and 1,534,824 fewer tonnes of concentrated CO_2 in the atmosphere (Ma et al., 2012; Medrano & Izquierdo, 2017). Despite such progress, alongside the goals that have been set to retrofit existing buildings, much work remains to be done in Mexico within the heritage sector and for its occupants.

Mexico City's historical centre was chosen for an in-depth case study (Figure 21.1). This location provides an ideal opportunity for researchers to examine how owners and tenants of historic houses in a world heritage site negotiate decisions between conserving heritage and improving energy efficiency. The world heritage status imposes certain restrictions upon users regarding what changes they can and cannot make. It is therefore interesting to examine how residents intervene in this specific context. The existence of social housing (intended for people with limited resources) also makes this area compelling to study.

The historic centre of Mexico City (*Centro Histórico de la Ciudad de México*) was declared a World Heritage Site in 1987 by fulfilling at least six of the cultural and four of the natural criteria of the operational guide of the 1972 Convention Concerning the Protection of World Cultural and Natural Heritage by UNESCO. The area demonstrated a significant exchange of human values and the essential development of architecture, art and urbanism, for instance through its unique testimony of a cultural tradition or civilisation that is alive or has disappeared (Unesco, 1972), such as The Aztec Empire. The historic centre of Mexico City is an ensemble of architecture and Spanish cultures that are in danger.

The area comprises 668 blocks that include approximately 1500 buildings catalogued for their artistic and historical value, although the present study only pertains to housing. According to the *Autoridad del Centro Histórico de la Ciudad de México* (The Historic Centre of Mexico City Authority), the government's general objectives between 2018

Figure 21.1 The Zócalo in the historic centre of Mexico City.
Source: Photo by Murillo, March 2021.

and 2024 are to conserve, restore and protect the cultural and historical heritage of the area through urban and economic regeneration, citizen participation, strengthened habitability and cultural identity. The historic centre's management is complex, with both local and federal governments in charge of the area. This management system influences decision-making at all levels in the Historic Centre including the national monuments, conservation sites and nearby areas. This process gradually began the valuations of this polygon as a historical and patrimonial space in the second half of the 20th century.

In 1978, with the discovery of the remains of the Templo Mayor of Tenochtitlán, public interest was reactivated due to the historic centre's importance and to preserve and protect its physical conditions. As a result, in 1980, it was declared a Zone of Historic Monuments called the Historic Center of Mexico City (CDMX, 2011). The management is understood as a process capable of strengthening technical and legal protection mechanisms and fostering social participation with a comprehensive conservation and sustainable use approach, which considers the improvement of the quality of life of society in a World Heritage Site.

Since the historic centre of Mexico City was declared a World Heritage Site in 1987, it has faced social, political, environmental and economic challenges, given its geographical location and the historic transformation of its social context. The creation of management instruments have been needed and, due to political interest, local strategies have changed depending on the government of the time.

The axes that govern the historic centre (Mendiola, 1989) and have provided the structure for the local management programmes since 1987 are the normative, strategic, sectoral and instrumental co-responsibility (policies). The normative axis includes a general management plan and urban development objectives. The management plan has changed with each new government, thereby complicating the continuity of any program. The strategy has implemented short- and long-term goal programmes and brought in penalties and fiscal and financial incentives (Mendiola, 1989). The plan also provides public services and updates on the inventory of existing buildings by identifying those that require renovation and conservation.

Methods and materials

Through a sociotechnical method of system dynamics, social data (related to residents' attitudes towards heritage values and energy efficiency) were collected, analysed and synthesised alongside environmental and building condition data. The environmental data relate to the environmental impact of decisions on energy consumption, users' thermal comfort and the perceived current physical condition of the building itself.

This chapter is based on an analysis of a sample of five buildings that are part of the local regeneration programme, have demonstrably preserved the original typology of the architectural unit and are part of the World Heritage Site (a total of 134 housing buildings comprise the total housing unit). The sample includes one listed monument, one unlisted but protected building apartment and three listed apartment buildings.

The architectural styles of these buildings range from the 16th to 19th century (predominantly baroque and colonial style; Figure 21.2) and have influenced each one's materials (*tepetate, cantera* and brick). Very few buildings are in the modern style. The houses vary in size, but most occupy three-storey buildings.

Figure 21.2 The buildings and their architectural styles.
Source: Photo by Murillo, March 2021.

Participants were recruited for the study based on at least one of the following criteria: they must either have lived in the building since the Declaration of World Heritage in 1987 or their buildings must have belonged to the government-listed building and land use inventory. The study involved questioning the participants about their buildings in the form of semi-structured interviews and photo-elicitation (i.e. interviewers were shown energy efficiency options upon which they were prompted to comment). The interviews were transcribed and translated from Spanish to English, and the subsequent data were analysed through NVivo coding, enabling us to identify cause-and-effect relationships between factors that affected certain interventions (or lack thereof; Table 21.1).

The cause-and-effect relationships were then visualised with Vensim software in the form of a 'causal-loop' diagram following the principles of system dynamics (see also Fouseki & Bobrova, 2018; Fouseki et al., 2020). The semi-structured questionnaire was designed to understand each building's condition, the values attributed to it and energy efficiency and thermal comfort interventions undertaken. Complementary data regarding the buildings' physical condition were collected and combined with environmental data collected through a thermal imaging camera and environmental monitors (tiny tags). Following the interviews with the participants, the monitors were placed the same day in bedrooms and living rooms (spaces where there is less interior comfort for them). These spaces had been identified by the participants as the coldest and most humid.

The monitoring period took place over 28 days in winter from December to January, and permission was obtained from the residents to install the equipment inside their buildings. Relative humidity and temperature were monitored to explore environmental conditions indoors and to contrast these data with the users' perceived thermal

Table 21.1 Interview number and building status

Building number	Interview number	Building status	Style	Number of floors	Year of construction	Level of protection	User
1	MX-U1	Listed	Colonial	1	1885	Protected	Owner family of eight
2	MX-U2	Listed	Barroque	1	1887	Protected	Owner family of two
3	MX-U3	Not Listed	Modern	4	1941	Only exteriors protected	Owner family of three
4	MX-U4	Listed	Barroque	2	1890	Protected	Owner family of four
5	MX-U5	Listed	Colonial	2	1957	Only exteriors protected	Tenant family of two

comfort. After 28 days, the tiny tags were collected and the data were downloaded. This made it possible to compare and contrast the participants' desired and perceived thermal comfort with actual data on interior temperature and relative humidity.

The interview data were coded following the methodology proposed by Fouseki et al. (2020). This coding resulted in 251 core codes with different subcodes (nodes). The codes were then grouped into four major categories, including decision-making process, energy efficiency actions, heritage buildings' conditions and users' values/values over time. During the coding process, cause-and-effect relationships between the nodes were identified and recorded in Excel. Table 21.2 shows the codes grouped into 12 categories.

Table 21.2 Coding groups

Category	Codes	Category	Codes
Time	Time in the property, changes over time	**Materials**	Façades, walls, humidity, porosity
Needs	Thermal comfort in winter and summer, perceived thermal comfort	**Practice**	Qualified interventions
Feelings	Satisfaction, guardianship, family attachment	**Cost**	Rent, cost of changes
Value	Sentimental, aesthetics, historic, symbolic, originality	**Risk**	Earthquakes, humidity, lack of maintenance
Place/space	Urban context, size of the rooms	**Ownership**	Owner, tenant
Actions	Preventive maintenance, corrective maintenance, thermal comfort actions, ventilation, preservation	**Building type**	Listed, not listed

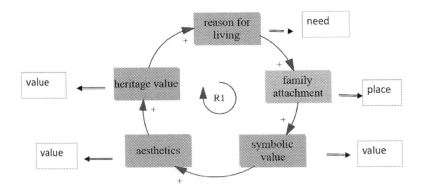

Figure 21.3 Reinforcing causal loop created by the authors.

The relationships were then mapped with Vensim, a commonly used software for system dynamic analyses (Figure 21.3), to illustrate cause-and-effect relationships identified through the analysis. This produced a diagram to show the dynamic system between heritage buildings, values, thermal comfort and energy efficiency (decision-making process). This process was used for all interviews. An aggregate version summarises all the interconnected variables identified during the interviews (Fouseki et al., 2020). A positive (+) or negative (−) symbol is used for reinforcing or balancing relationships. A reinforced relationship illustrates users' values in the heritage building, e.g. the more original the façades, the more aesthetic value for the user.

The type of ownership is reinforced in the system by the reason for living (an owner prefers to stay in the property, instead of selling it). This relationship is reinforced by time. The more time spent in the property, the more responsible the user will feel about any changes that could affect any value (tangible or intangible) in the building. The preservation is reinforced by the original and architectural value that satisfies the user: '*The architecture and the shape of everything—the windows, the doors—I like all of those a lot, so we don't touch them*' (MX-U2). A balancing relationship is shown in thermal comfort. The lower the thermal comfort in winter, the colder the atmosphere. In this way, a heater would be used more often in winter to provide better thermal comfort: '*We have a heater, which we use in the evenings for a while to warm the place up a little. It's not a big deal. If I'm freezing, I will turn it on*' (MX-U1).

Overall heritage values

The analysis illustrates that the original façade reinforces the aesthetic value of the building, which in turn enhances the tranquillity of the surrounding space and the residents' overall satisfaction with living in the building. This encourages more time being spent on the property and strengthens sentimental values attached to the house: '*The façade is pretty, the building is beautiful and well-preserved ... It is in a lovely location on the street. I consider it beautiful*' (MX-U3).

In addition to the beauty of the buildings, the values attributed by their occupants include architectural value, historical value and originality. The exterior characteristics they most appreciate included the façades, orientation and number of floors; for the interiors, these were the structure, distribution, amount of space, materials, natural ventilation and natural light. Quality of life is a priority and valued as part of what the building offers:

> *The first priority is that the structure of the building is in good shape. In that way, any earthquake or events will not affect our safety. The second is that the house does not have any cracks. It has to be in good condition to be safe. The third is that we have to avoid risks inside the apartments. The other two priorities are maintaining the doors and windows. The last one is to have clean water, which keeps us healthy and with quality of life.* (MX-U3).

Users prioritise the preservation of the original façades as providing a valuable opportunity to maintain continuity for future generations:

> *The façades have been preserved impeccably, so they are as they were many, many years ago. No entries or exits have been modified; they are as they were when the family of my children's great-grandmother bought the house.*
>
> (MX-U3)

Heritage values assigned to historic buildings

The value that residents attach to the house varies according to their lifestyles. The historical value is the primary value associated with the house, followed by its sentimental value: '*For the year it was made and for its architecture—its balconies, its corridors—it is very beautiful. I like everything. I like it a lot*' (MX-U2). When mapped on Vensim, family attachment to the property shows a significant relationship: the owner has a reason for living in the property that becomes stronger over time, despite the fact that the physical condition of the original materials requires more maintenance. The historic context also reinforces the owner's relationship with architectural, aesthetic and symbolic values over time. The residents endeavour to keep the building as original as possible because of its architectural and sentimental value, in part due to family ties: '*The greatest meaning to me is sentimental because I have lived here [for many years], and this is where my father died … I met my husband here; my son was born here*' (MX-U2).

How users think about changes to the condition of buildings over time

Based on the occupants' perceptions, dampness deteriorates the materials of all the buildings' original façades and raises the cost of changes due to their aesthetic value and the old structures that need specialised interventions and maintenance. The occupants were aware of the relationship between humidity, deterioration and use and the need to maintain the original façades and materials in good condition: '*It is because the use of buildings has changed over time. For example, the [original materials] need a lot of maintenance*' (MX-U5).

The actions undertaken by residents are also driven by the estimated cost of change. While changes to the windows were made due to deterioration and humidity, the changes were further driven over time by the durability of the new materials. Despite the cost, deterioration and time, the occupants display a willingness to preserve the original materials and have turned to government programmes for interventions for the façades.

'The roofs need waterproof paint and the façade need some painting. The FIDEICOMISO [Historic Centre Trust] gave approval to carry out the interventions on the façade' (MX-04).

The risk of a building collapsing (because of an earthquake or time) is reinforced by the age of its construction. Despite these conditions, the residents feel satisfied with the maintenance carried out and have resisted making substantial changes. Such resistance is also related to the architectural value and high cost of interventions for heritage buildings: Costs 'are high because all the interventions have to be done by specialists' (MX-U2). The residents consider heritage value to be important and are willing to preserve it. They are also aware of the rising economic value of future interventions in the building.

Negotiating thermal comfort, energy efficiency and heritage conservation

Most of Mexico City's urban area (87%) has a temperate subhumid climate. The remaining parts have a dry or semi-dry climate (7%) or a humid temperate climate (6%). Most of the residents' ideal thermal conditions expressed in the interview are between 20°C and 30°C. 'I would say 28–30°C [would be the ideal temperature for me]' (MX-U1). '28°C' (MX-U5). During the month of monitoring, the interior temperature was higher and constant compared with the external temperature. The interior temperature range was 16°C to 19°C and the exterior temperature range was 10°C to 21°C.

In terms of indoor environmental conditions, thermal comfort is important for the residents. Monitoring of relative humidity and temperature showed differences with the users' perception of the building performance. Most perceive their building as being cold in winter but cool and pleasant during the summer. As reported by one respondent:

> During the winter, it is a little cold, but nothing happens; I always wear my coat. In summer, it's a delight. I tell you: it's a delight to go inside from the street and find natural air conditioning, thanks to the height of the ceilings and the walls that keep a delightful temperature. It's very nice.
>
> (MX-U1)

Another resident comments:

> Right now, it is a little cold, but here the temperature is very stable. When it is very warm, I open the windows and it gets cooler. During the winter, I close them, and it keeps the inside temperature warmer because the roof retains the interior heat, so the climate is stable.
>
> (MX-U3)

The residents control the inside temperature by opening and closing the windows or using heating during the night (as a last resort).

The visual condition assessment made in the visit with the users showed the problems with damp expressed by the materials' natural properties on the surface of the walls, ceilings, and façades (Figure 21.4). The interior temperature was also perceived differently depending on the floor the participants occupied, although there was a general consensus that the thermal comfort indoors was cold. A thermal differentiation between the levels is recognised (generally, the second level is more comfortable during winter). The residents recognise sunlight as a relevant factor associated with a feeling of warmth.

Figure 21.4 Works in walls due to damp presence.
Source: Photo by Murillo, December 2018.

However, this element cannot be projected onto the whole of the house and cannot be controlled entirely. Some apartments lacked direct sunlight during the winter due to their orientation and interior layout. The thickness of the walls and the height and dimensions of the windows also influenced temperature. For instance, a north-facing, first-floor façade was perceived as being cold, compared with a façade with the same orientation but on the top floor, which was perceived as being warm.

The environmental data collected and contrasted with the occupants' perceptions confirmed that thermal comfort in winter was affected by floor level, façade orientation, ventilation, room size, natural light and humidity. The façades that do not receive direct sunlight throughout the year might make the building feel cooler in both winter and summer. Ground floor apartments were perceived as being colder than those on higher floors. Ventilation and the size of rooms would increase users' perception of thermal comfort, which is also reinforced by time: the more time residents spend in the property, the more aware they are of the seasonal changes affecting its thermal properties and improved comfort.

Despite the discomfort described above, the residents are not prepared to make interventions to the buildings for three main reasons. The first reason is an attachment to the building and its value (due to its sentimental, architectural and historical value). The second is the listed status of the historic area being an additional barrier to interventions. The third reason relates to preserving the property for future generations. However, in general, if residents feel thermally uncomfortable inside, they will consider making minimal improvements, such as changing the floor and ceiling materials to wood. They also express willingness to make interventions for reducing energy consumption (e.g. solar panels on the roof). However, this has not materialised due to restrictions imposed by the government for heritage site protection and because of the associated costs.

Thermal comfort actions and attitudes in heritage buildings

What impact does prioritising heritage value have on energy efficiency? In this case, most of the residents' interactions with thermal comfort and heritage value are low-impact and low-cost. The actions and attitudes of the occupants towards thermal comfort are somewhat passive, and they prioritise protecting the building's heritage value. There are, however, unconscious activities that effectively give them more comfortable indoor temperatures. If it is too hot, opening windows and doors are the main options, while if it is too cold, closing the windows is the most common strategy. In one case, the heat from the sun could be so strong that a resident would use a fan in a room without natural ventilation.

The occupants employ various low-cost and passive thermal actions during different seasons of the year. While opening the balcony windows seems to suffice during the summer, blocking cold air from entering by stuffing clothes under doors (and closing windows) is often used in winter. The natural light creates a warm atmosphere that reduces the usage of electricity:

> *In the living room, I like it more because we have windows and natural light, and the atmosphere is very warm ... We have windows everywhere, and that allows us always to be illuminated, and always without artificial light.*
>
> (MX-U1)

Natural light and ventilation enable lower electricity consumption, enhance aesthetics and increase satisfaction from living in the building. The rooms' size as well as the height of the walls and ceilings all contribute to natural light (even if they also contribute to colder temperatures in winter).

While the respondents adopt passive measures during the day to improve thermal comfort, the use of heating at night is inevitable for some users: '*We have a heater that we use in the evenings for a while to make it warmer*' (MX-U1). '*I put heaters all over the house. I think that the rooms that do not have a lot of furniture get colder. Anyway, I use the heaters*' (MX-U5). '*The heritage building layout and design allow it to be more natural*' (MX-U2).

Conclusion

Current research on how residents of historic buildings negotiate their need for thermal comfort, lower energy bills and heritage conservation shows that the values they attribute to their building can be critical in that negotiation process. Values are also changeable. For instance, residents may initially prioritise the aesthetics of original windows over their need for thermal comfort. However, this may change over time (Fouseki & Bobrova, 2018). The diversity and dynamic nature of values is not encapsulated by current policies on energy efficiency in historic buildings (see Fouseki et al., 2020). A preset of non-negotiable heritage values is provided to which the energy efficiency policies need to adhere. The non-negotiable nature of heritage values is inevitably more prominent in protected areas. The case of Mexico City, a world heritage area, evidently showed that there is a minimal scope of intervention and change while the need for energy efficiency changes can be rather high for some residents. This issue becomes even more challenging in the context of social housing, which was examined as part of this chapter. The heritage area is occupied by social housing inhabited by people with limited resources, while other buildings that have been conserved and renovated have commercial uses.

The dynamic relationships explored here show that heritage values, preservation, thermal comfort and energy efficiency interplay when residents make decisions about energy and thermal comfort interventions. The values associated with a building's tangible characteristics and meanings unquestionably affect the elements that residents are willing to change or maintain. However, there is a diversity of values, and each case is unique. It is due to this uniqueness that a participatory approach to sustainable design and renovation is needed that assesses the individual's heritage values and needs (financial, social, energy or thermal comfort), alongside the individual characteristics of the building, instead of applying a universal, standardised approach to the property. Current and future guidelines need to integrate the need for a participatory approach to sustainable design in the context of historic buildings.

In the case of Mexico, it was shown that the owners and tenants appreciate similar values to heritage professionals, such as architecture, history and aesthetics, and they value the aesthetics of the façade in particular. This may be explained by the fact that the area has been a World Heritage Site since 1987, so the residents have had time to adopt this value system. Alongside the values attached to the physical attributes of the building, there is also a strong family attachment. By preserving the original materials, residents can sustain the sentimental value associated with personal and family memories, and on a practical level, they can sustain the building for future generations.

This study adds to the knowledge base on this subject by using an example from Latin America, where no previous studies on this topic exist. Current studies on how occupants' heritage values drive or preclude energy efficiency interventions in historic buildings are derived mainly from Europe. Therefore, by examining Mexico, we can gain new insights into the subject matter. The study in Mexico reaffirms previous findings (e.g. Fouseki & Bobrova, 2018; Fouseki et al., 2020) that call for an in-depth understanding of the values attached to a historic building before designing and implementing an energy efficiency strategy. Environmental monitoring shows that the buildings' temperatures are constant but not comfortable enough for users' expectations. The residents used their own means to adapt their needs to space, temperature and humidity (e.g. wearing more clothes, closing windows and blocking the cold air from outside). These actions passively produce thermal comfort and low-cost solutions. The final objective for energy efficiency in heritage buildings, in any case, must be to balance users' different needs and values comprehensively and effectively.

This work developed methods and a strategy to conserve the built environment's historical and cultural value in Mexico City. Its sociotechnical approach comprehensively evaluated and explored the parameters of values, energy efficiency and users alongside building components. The present research will help the energy efficiency decision-making process when aspects of the importance of heritage are integrated within it.

Given the rapid growth of cities leading to greater consumption of resources, the federal and local legislative sustainability framework will need to be continuously updated. As buildings are sociotechnical systems, a sociotechnical methodology must be applied to retrofit existing heritage buildings, to avoid consequences for a historic centre. Practices and tools for historic centres should put heritage buildings and users in Mexico on a sustainable path to continue mitigating climate change, thus serving as a model for other Latin American countries.

Funding

This research has been supported by the *Secretaria de Energía y Consejo Nacional de Ciencia y Tecnología* (SENER-CONACyT Mexico) and the EPSRC Doctoral Training Centre in Science and Engineering in Arts, Heritage and Archaeology (EP/L016036/1).

Acknowledgements

To SENER-CONACyT Mexico and EPSRC Doctoral Training Centre in Science and Engineering in Arts, Heritage and Archaeology for supporting the research fieldwork in Mexico. Special thanks for the time given by the anonymous participants in the study.

References

Appelbaum, B. (2007). *Conservation treatment methodology*. Oxford: Butterworth-Heinemann.

Avrami, E., Mason, R., & Torre, M. de la. (2000). *Values and heritage conservation research report* (pp. 1–11). Los Angeles: The Getty Conservation Institute. Retrieved from: http://hdl.handle.net/10020/gci_pubs/values_heritage_research_report (Accessed on April 2021).

Ballart Hernández, J., Fullola Pericot, J. M., & Petit i Mendizábal, M. D. Ángels. (1996). El valor del Patrimonio Histórico. *Complutum*, (2), 215–224. Retrieved from: https://revistas.ucm.es/index.php/CMPL/article/view/CMPL9696330215A (Accessed on April 2021).

Ben, H., & Steemers, K. (2014). Energy retrofit and occupant behaviour in protected housing: A case study of the Brunswick Centre in London. *Energy and Buildings*, *80*, 120–130. Retrieved from: https://doi.org/10.1016/j.enbuild.2014.05.019 (Accessed on April 2021).

Berg, F., Flyen, A.-C., Godbolt, Å. L., & Broström, T. (2017). User-driven energy efficiency in historic buildings: A review. *Journal of Cultural Heritage*, *28*(1), 188–195. https://doi.org/10.1016/j.culher.2017.05.009.

Broström, T., Eriksson, P., Liu, L., Rohdin, P., Ståhl, F., & Moshfegh, B. (2014). A Method to Assess the Potential for and Consequences of Energy Retrofits in Swedish Historic Buildings. *The Historic Environment: Policy & Practice*, *5*(2), 150–166. https://doi.org/10.1179/1756750514Z.00000000055.

Cassar, M. (2011). Energy reduction and the conservation of cultural heritage: A review of past, present and forthcoming initiatives. *International Preservation News*, (55), 6–9.

CDMX, A. (2011). Plan Integral de Manejo del Centro Histórico de la Ciudad de México. *Gaceta Oficial Del Distrito Federal*, 3–111. https://doi.org/10.1287/deca.1080.0132.

CEN Comité Européen de Normalisation (2017). *Conservation of cultural heritage – Guidelines for improving the energy performance of historic buildings TC 346*, EN 16883:2017. Brussels: CEN Comité Européen de Normalisation.

Commission of the European Communities (2006). Action plan for energy efficiency: Realising the potential, Brussels, *Eur-lex*. Retrieved from: https://eur-lex.europa.eu/LexUriServ/LexUriServ.do?uri=COM:2006:0545:FIN:EN:PDF (Accessed on April 2021).

Eriksson, P., Hermann, C., Hrabovszky-Horváth, S., & Rodwell, D. (2014). EFFESUS methodology for assessing the impacts of energy-related retrofit measures on heritage significance. *The Historic Environment: Policy & Practice*, *5*(2), 132–149. https://doi.org/10.1179/1756750514Z.00000000054.

Forrester, J. Wright (1987). Lessons from system dynamics modelling. *System Dynamics Review*, 3(2), 136–149. https://doi.org/10.1002/sdr.4260030205.

Fouseki, K., & Bobrova, Y. (2018). Understanding the change of heritage values over time and its impact on energy efficiency decision-making at residential historic buildings through system dynamics. In T. Broström & L. Nilsen (Eds.), *Proceedings of the 3rd International Conference on Energy Efficiency in Historic Buildings (EEHB2018), Visby, Sweden, 26–27 September 2018* (pp. 11–21). Sweden: University of Upsalla.

Fouseki, K., & Cassar, M. (2014). Energy efficiency in heritage buildings—future challenges and research needs. *The Historic Environment: Policy & Practice*, 5(2), 95–100.

Fouseki, K., Newton, D., Murillo Camacho, K. S., Nandi, S., & Koukou, T. (2020). Energy efficiency, thermal comfort, and heritage conservation in residential historic buildings as dynamic and systemic socio-cultural practices, *Atmosphere*, 11(6). https://doi.org/10.3390/atmos11060604.

Fredheim, L. H., & Khalaf, M. (2016). The significance of values: Heritage value typologies re-examined. *International Journal of Heritage Studies*, 22(6), 466–481. https://doi.org/10.1080/13527258.2016.1171247.

Historic England (2018). *Energy efficiency and historic buildings: How to improve energy efficiency*. London: Historic England. Retrieved from: https://historicengland.org.uk/images-books/publications/eehb-how-to-improve-energy-efficiency/heag094-how-to-improve-energy-efficiency/(Accessed on 30 January 2021).

Koukou, T., & Fouseki, K. (2018). Heritage values and thermal comfort in Neoclassical residential buildings of Athens, Greece: Tension or co-existence? In T. Broström and L. Nilsen (Eds.), *Proceedings of the 3rd international conference on energy efficiency in historic buildings – EEHB2018*. (pp. 463–471). Sweden: University of Upsalla.

Le Corbusier, & José Luis Sert. (1942). Carta de Atenas. *Ciam*, 1–35. https://doi.org/10.1017/CBO9781107415324.004.

Lidelöw, S., Örn, T., Luciani, A., & Rizzo, A. (2019). Energy-efficiency measures for heritage buildings: A literature review. *Sustainable Cities and Society*, 45: 231–242 https://doi.org/10.1016/j.scs.2018.09.029.

Lopes, M. A. R., Antunes, C. H., & Martins, N. (2012). Energy behaviours as promoters of energy efficiency: A 21st century review. *Renewable and Sustainable Energy Reviews*, 16(6), 4095–4104. https://doi.org/10.1016/j.rser.2012.03.034.

Martínez-Molina, A., Tort-Ausina, I., Cho, S., & Vivancos, J.-L. (2016). Energy efficiency and thermal comfort in historic buildings: A review. *Renewable and Sustainable Energy Reviews*, 61, 70–85. https://doi.org/10.1016/j.rser.2016.03.018 (Accessed on April 2021).

Martínez Yáñez, C. (2006). *El patrimonio cultural: los nuevos valores, tipos, finalidades y formas de organización*. (Unpublished doctoral thesis). Retrieved from: https://digibug.ugr.es/handle/10481/1343 (Accessed on April 2021).

Ma, Z., Cooper, P., Daly, D., & Ledo, L. (2012). Existing building retrofits: Methodology and state-of-the-art. *Energy and Buildings*, 55, 889–902. https://doi.org/10.1016/j.enbuild.2012.08.018.

Medrano-Gómez, L. E., & Izquierdo, A. E. (2017). Social housing retrofit: Improving energy efficiency and thermal comfort for the housing stock recovery in Mexico. *Energy Procedia*, 121, 41–48. https://doi.org/10.1016/j.egypro.2017.08.006.

Mendiola, A. (1989). *Architectural urban preservation of the neighborhood of the Santisima Historic Center of Mexico City* (Unpublished Bachelor's thesis). National Autonomous University of Mexico, Mexico. Retrieved from: https://repositorio.unam.mx/contenidos/254257 (Accessed on April 2021).

Murillo Camacho, K. S., Fouseki, K., & Altamirano Medina, H. (2022). Decision-Making Processes of Residents in Preservation, Thermal Comfort, and Energy Efficiency in Heritage Buildings: A Pilot Study in Mexico City. *Applied Sciences* 12 (3), 1486. https://doi.org/10.3390/APP12031486.

Palma Peña, J. M. (2017). Valores sociales y valores patrimoniales: Elementos para determinar la significación del patrimonio documental. *Biblioteca Universitaria, 16*(1), 34–45 https://doi.org/10.22201/dgb.0187750xp.2013.1.18.

Pearson, M., & Sullivan, S. (1995). *Looking at heritage places: The basics of heritage planning for managers, landowners and administrators.* Carlton: Melbourne University Press.

Pye, E. (2001). *Caring for the past: Issues in conservation for archaeology and museums.* London: James & James.

Riegl, A. (1996 [1903]). The modern cult of monuments: Its essence and its development. In N. Stanley-Price, M. Kirby Talley Jr, & A. Me Melucco Vaccaro (Eds.), *Historical and Philosophical Issues in the Conservation of Cultural Heritage.* (pp. 69–83). Los Angeles, CA: The J. Paul Getty Trust.

Richmond, A., & Bracker, A. (Eds.) (2009). *Conservation: Principles, dilemmas and uncomfortable truths.* London: Butterworth-Heinemann.

Secretaria de Energía. (2014). *Fondos Sectoriales de Energía – Estrategia [Secretary energy funds – strategy 2014–2018].* Retrieved from: https://www.gob.mx/sener/acciones-y-programas/fondos-sectoriales-de-energia (Accessed on April 2021).

Stephenson, J., Barton, B., Carrington, G., Gnoth, D., Lawson, R., & Thorsnes, P. (2010). Energy cultures: A framework for understanding energy behaviours. *Energy Policy, 38*(10), 6120–6129. https://doi.org/10.1016/j.enpol.2010.05.069.

UNESCO (1972, 2014). Resoluciones, recomendaciones, Convención para la Protección del Patrimonio Mundial Cultural y Natural, Organización de las Naciones Unidas para la Educación, la Ciencia y la Cultura. *Actas de la Conferencia General, 17a reunión, París, 17 de octubre-21 de noviembre de 1972, v. 1:* Paris, France Retrieved from: http://unesdoc.unesco.org/images/0011/001140/114044s.pdf#page=139. (Accessed on April 2021).

Webb, A. L. (2017). Energy retrofits in historic and traditional buildings: A review of problems and methods, *Renewable and Sustainable Energy Reviews, 77,* 748–759. https://doi.org/10.1016/j.rser.2017.01.145.

Yarrow, T. (2016). Negotiating Heritage and Energy Conservation: An Ethnography of Domestic Renovation. Historic Environment: Policy and Practice, 7, 340–35. https://doi.org/10.1080/17567505.2016.1253149.

Yohanis, Y. G. (2012). Domestic energy use and householders' energy behaviour', *Energy Policy, 41,* 654–665. https://doi.org/10.1016/j.enpol.2011.11.028.

22 Food heritage as a catalyst for environmental sustainability

Reflections on the cultural value imbued by citizens to food and its role in supporting scientific debate about food security

Georgios Alexopoulos, Suzanne Kapelari and Theano Moussouri

Introduction

In the last two decades, the concept of food heritage, often conceived within the wider notion of intangible cultural heritage (ICH), has gained a lot of currency in the international heritage discourse. The growing pressure for all types of organisations to embed sustainability through their long-term planning and day-to-day operations is affecting also cultural institutions despite the previously dominant notions that tended to separate nature/environment from culture. Overcoming this nature–culture divide, food heritage seems to constitute an interesting interface between environmental and cultural sustainability. Focussing on the role of food heritage in citizen's relationship with food and drawing from research conducted for the BigPicnic project, funded by the European Union's Horizon 2020 Research and Innovation Programme, this paper aims to underline the significant connection and interplay between cultural and environmental sustainability in the context of food heritage. Research conducted for the aforementioned project demonstrated a strong link between food culture and environmental concerns expressed through sustainability as a value.

The aim of this chapter is to critically reflect on the importance of the notion of food heritage to the heritage sustainability discourse and to explore how the cultural and social values that citizens attach to food can contribute to discussions of and scientific debate on food security. More specifically, we investigate how foodways, traditional knowledge and practices, food memories and the social context of eating as well as food well-being underline the connection between cultural and environmental sustainability. We argue that food systems contain both intangible and tangible aspects that go beyond science, technology and politics/governance by touching on human behaviour that reflects cultural values and identities. This relationship underlines the importance of better understanding how heritage and culture can constitute a catalyst for environmental sustainability.

The arguments presented in this chapter are based on qualitative and quantitative evidence and a meta-analysis, which synthesised findings from studies conducted by 15 botanic gardens during the co-creation of exhibitions, science cafés and other activities as well as post-opening. What makes the findings even more pertinent to the discussion of sustainable heritage is the fact that the heritage dimension of food did not feature in the initial aims and objectives of the project. Nevertheless, this dimension emerged strongly through the engagement of the project partners with a range of citizens – an engagement that was guided by the responsible research and innovation (RRI), co-creation and team-based inquiry (TBI) approaches (see below for an explanation of these terms). This paper, therefore, intends to unweave from the aforementioned process the cultural associations and meanings attributed to food by various citizens and their importance in defining eating habits, nutritional choices and attitudes towards food production and consumption that, at the same time, have an impact on environmental issues and concerns. Furthermore, a very important finding of the BigPicnic research project is the recognition of food heritage as one of the interconnected concepts that define food security[1] and the impact that such projects can have for promoting recommendations to policymakers on sustainable heritage. This analysis strongly supports the argument for culture being recognised by cultural heritage policies as the fourth, so-called, pillar of sustainable development.

This chapter starts by contextualising the notion of food heritage and particularly its relationship with and contribution to discourses of sustainability. This section emphasises the importance of the growing literature on food heritage (within the wider heritage discourse) but also underlines the necessity for transdisciplinary approaches for addressing heritage sustainability issues and challenges, both global and local. A short overview of the BigPicnic project is presented, followed by the research methodology that led us to the findings that will be discussed in this chapter. We then focus on evidence that addresses the aspects of food heritage – such as traditional knowledge, food memories and food well-being in the context of diaspora communities – and examine how these support the role of food heritage in motivating, triggering and promoting environmental sustainability through food security.

The notion of food heritage in the context of heritage sustainability discourses

Before proceeding to discuss the specific example of the BigPicnic project, it is essential to summarise the relationship between the notion of food heritage and its central

Table 22.1 Definitions

Intangible Cultural Heritage (ICH): 'the practices, representations, expressions, knowledge, skills – as well as the instruments, objects, artefacts and cultural spaces associated therewith – that communities, groups and, in some cases, individuals recognize as part of their cultural heritage' (UNESCO ICH Convention, 2006, Article 2.1). This ICH is manifested in the following five domains: (a) oral traditions and expressions, including language as a vehicle of ICH; (b) performing arts; (c) social practices, rituals and festive events; (d) knowledge and practices concerning nature and the universe; (e) traditional craftsmanship (ibid, Article 2.2).
Foodways: 'patterns of diet, nutrition, cooking, eating, feasting and fasting' (Fischer, 1989, p. 9), 'the beliefs and behavior surrounding the production, distribution, and consumption of food' (Counihan, 1999, p. 2). According to Engelhardt (2013, pp. 1–2) foodways deals with 'the study of what we eat, how we eat, and what it means' and describes the cultural processes, social interactions and cultural exchanges that 'define food, drink and nutrition'.
Food security: Food security 'exists when all people, at all times, have physical and economic access to sufficient, safe and nutritious food to meet their dietary needs and food preferences for an active and healthy life' (FAO, 2009).
Food sovereignty: 'the right of people to healthy and culturally appropriate food produced through ecologically sound and sustainable methods, and their right to define their own food and agriculture systems' (Nyéléni, 2007, p. 9).

role in the growing discourse on heritage sustainability. Table 22.1 (see Appendix) also outlines the definitions of certain important concepts and terms that are central to the discussion of this topic.

Sustainable development and the concept of intangible cultural heritage

The notion of sustainable development has been deemed as 'the most important guiding principle for the 21st century' (Albert, 2015, p. 11). In the last couple of decades, the idea that heritage could be a driver, a vector or a tool for sustainable development has emerged strongly through ideological shifts and a range of new approaches, practices and policies in both the heritage sector and academia (Clark, 2008; Barthel-Bouchier, 2013; Albert, 2015, 2017; Auclair & Fairclough, 2015; Dessein et al., 2015; Labadi & Logan, 2016; Larsen & Logan, 2018). This was considered a 'profound transformation' (van Oers, 2015, p. 189). To begin with, the discourse of both sustainability and heritage has some common characteristics. Since 1987, sustainability has been widely described as 'meeting the needs of the present without compromising the ability of future generations to meet their own needs' (UN, 1987). Discourses on heritage preservation on the other hand have been underpinned by the notion of intergenerational equity that emphasises the need to manage cultural heritage (that comes from the distant or even recent past) for both the present and the future generations (Throsby, 2002, p. 107; Auclair & Fairclough, 2015, p. 9). Heritage has been deemed to always act as a bridge between culture and the environment (Clark, 2008, p. 94). With the rise of environmental concerns such as climate change, Harvey and Perry (2015, p. 3) have described the merging of the two as the 'heritage–climate change nexus' in which heritage involves a 'present-centred and future-orientated processing of a tangible and intangible sense of the past'. Heritage and sustainability also share common ground when both are perceived as ongoing processes and as being people-centred (Auclair & Fairclough, 2015, p. 9). Nevertheless, some scholars have been sceptical of the link between heritage and sustainable development and have criticised the division between natural and cultural heritage (Bushell, 2015, p. 504). Going even further, considering the inevitability of change (environmental, social, political) and the extent to which both natural and cultural heritage management have not adequately addressed what they specifically define as 'the future', it has been suggested that conventional heritage practices are indeed unsustainable (Harrison et al., 2020, pp. 485–486).

Notwithstanding, a significant development from the initial conceptualisations of sustainability was the recognition of culture as a distinct pillar in addition to the original three 'pillars' of environment, economy and society (Hawkes, 2001; Duxbury & Gillette, 2007; UCLG, 2010). UNESCO had long argued that sustainable development and the blossoming of culture are interdependent (UNESCO, 1998, p. 13) and later directly addressed the importance of introducing cultural heritage in the sustainable development agenda (UNESCO, 2013). As a consequence, within the heritage sector, the sustainability paradigm consists of the four cornerstones of environmental, economic, social and cultural development. Heritage can be viewed both as a vital component of the so-called cultural (fourth) pillar of sustainable development and also a contributor to the other three pillars (Clark, 2008, p. 95; Albert, 2015, p. 11).

Initially, it was thought that the sustainable development goals (SDGs) adopted in 2015 by the United Nations as an international instrument for the implementation of

326 *Georgios Alexopoulos et al.*

its 2030 development agenda (Robert et al., 2005) did not adequately address the role of cultural heritage for sustainability (Petti et al., 2020, pp. 6–7). However, the heritage sector has more recently come to realise that cultural heritage can contribute to achieving most of these goals (Engels, 2017, p. 50), while various heritage organisations have also strived to address the SDGs more effectively through their work (ICOMOS, 2017, p. 7; UCLG, 2018; ICCROM, 2020; McGhie, 2019, 2020). On an international level, UNESCO (as a normative and standard-setting organisation) has pursued in the last decade to operationalise the sustainability concept through culture by directing its policies and practices through the axes of both 'heritage' and 'creativity' (van Oers, 2015, p. 192).[2] Indeed, 'heritage' has also been included as one of the seven interrelated policy dimensions in the Culture for Development Indicators Suite methodological toolkit (UNESCO CDIS, 2020).

As we stress in the next section, the notion of food heritage is inextricably linked with the emergence of the concept of intangible cultural heritage (henceforth ICH; see Table 22.1). Although introduced earlier by UNESCO, the notion of ICH achieved global recognition in 2003 with the adoption of the 'Convention for the Safeguarding of Intangible Cultural Heritage' (henceforth ICH Convention). The ICH Convention acknowledged the importance of ICH as a 'mainspring of cultural diversity and a guarantee of sustainable development' in the face of threats such as globalizing processes and social transformations (UNESCO ICH Convention, 2006). It was widely acknowledged that certain intangible heritage practices and expressions can generate revenues (e.g. through cultural tourism and the market value of craftsmanship) which in turn can secure their viability while the process of safeguarding can protect their cultural character from globalising processes that homogenise and over-commercialize (Erlewein, 2015, p. 75). With the amendment of June 2018, the relevant 'Operational Directives' of the ICH Convention have clearly highlighted the contribution of intangible heritage to the social, economic and environmental pillars of sustainable development as well as peace (see Table 22.3). In addition to the concept of the universal heritage promoted by UNESCO, the common European heritage promoted by the Council of Europe has also recognised ICH as a factor in sustainable development and cultural diversity (Brown, 2018, pp. 108–109). Scholars from natural history museums have even suggested that the term 'intangible natural heritage' would be better placed to address human relationships with nature that are passed down from generation to generation encompassing, among other things, ecological food webs, human food security and health and global climate (Dorfman, 2012, p. 4).

Another aspect that is vital to pinpoint when identifying the parameters that enable heritage to be sustainable is the active role of citizens in decision-making. Supporting bottom-up approaches (as opposed to 'top down') in heritage practices, management and planning is deemed to work better as this model recognises the needs of and gives a voice to the various 'non-expert' stakeholders and is, therefore, more responsible and ethically sound (Smith, 2006, pp. 34–37; Alexopoulos, 2013, p. 70; Schofield, 2014, pp. 5–6; Logan, 2016, pp. 256–257). It is increasingly acknowledged that only through such participatory and inclusive approaches both cultural and natural heritage can be sustained in the future (de Merode et al., 2004, p. 9).

The emergence of food heritage

When reflecting on food, one can underline that the production, elaboration and consumption of food is a process 'common to all human beings' and we all must eat to sustain ourselves (Di Giovine & Brulotte, 2014, p. 1). Belasco (1999, p. 27) has noted that apart from being one of the essentials of life, our biggest industry and most frequently indulged pleasure, food is also a cause of disease and death. Food is a 'universal medium' (Watson & Caldwell, 2005, p. 1) and among the most foundational elements of culture (Timothy & Ron, 2013, p. 275). However, beyond sustenance and nutrition, food is 'packed with social, cultural and symbolic meanings' (Bell & Valentine, 1997, p. 3) with eating being a cultural and social as much as a biological activity (Food & Foodways, 1985, p. I; Watson & Caldwell, 2005, p. 1). Bessière (1998, p. 24) has underlined that because both food and cooking are culturally determined they place the eater in a social universe and a cultural order.

The parameters that can render food as a form of cultural heritage extend, however, beyond its social value. Firstly, food-related traditions and cultural values can be situated within the past-present-future nexus, which is a common way to conceptualise heritage. This happens as there exists a value linked to culturally prescribed ways (e.g. recipes) of making food with people being conscious of the history behind this process and the relevant inherited traditional practices (Carr et al., 2018, p. 145). Already in the 1990s, researchers employed the notion of food or culinary heritage in order to describe the transmission of culinary know-how, the preservation of traditional dishes, the connection to culinary roots and the notion that eating so-called natural or traditional products enables the eater to appropriate and embody the nature, culture and identity of an area (Bessière, 1998, p. 25). Furthermore, it is argued that food constitutes a form of heritage exactly because it is integral to the formation, performance and confirmation of identities: personal, group, regional, national, etc. (Di Giovine & Brulotte, 2014, pp. 1–4). Food is also a strongly mnemonic and sensory device that offers a sense of place and belonging (Carr et al., 2018, p. 145) and offers, as an 'edible chronotope' (a sensory space-time convergence), experiences that bind people together through space and time (Kirshenblatt-Gimblett, 2004, p. xiii). Other scholars have viewed certain types of foods as 'edible souvenirs' exactly because their consumption can elicit precise memories of people, places and events in the past (Di Giovine & Brulotte, 2014, p. 18). The aforementioned qualities have led many scholars to emphasise that food as an element of heritage is fundamentally different from other forms of heritage because it is multisensory (its consumption and preparation call for the use of all five senses: sight, smell, taste, touch and hearing) and it is both tangible and intangible (Ron & Timothy, 2013, p. 235; Counihan, 2014, p. 220; Matta, 2019a, p. 51). Therefore, food has been considered to be culture, materiality and a central element of sociability at the same time (Carr et al., 2018, p. 145). However, food can also be part of a heritage-making process for future generations, and this has been observed in efforts such as the safekeeping of crop seeds in the Svalbard Global Seed Vault – a 'final backup' for securing future food supplies and agrobiodiversity (Harrison, 2017, p. 86).

The systematic academic study of food only emerged from the second half of the 20th century despite the fact that many disciplines were directly or indirectly preoccupied with the study of eating and cooking.[3] About a decade ago, it was noted that although food heritage has been studied in different disciplines, using various approaches and methodologies, it had just emerged as a new field of research (Bessière & Tibère, 2011). This was followed by a growing literature that articulates food as heritage and various national and international attempts to claim food-based heritage which altogether have been described by Demossier (2016, p. 89) as a worldwide 'food heritage fever' and reflect what has been deemed a 'heritage turn' in food studies (Geyzen, 2014). The beginning of the 21st century brought the worldwide phenomenon of food heritagisation and patrimonialisation, and together with the recognition of food as an element of ICH by UNESCO, these developments have both placed food cultures into heritage frameworks (Di Giovine & Brulotte, 2014; Matta, 2016, 2019b; Romagnoli, 2019, pp. 163–165). Although 'alimentary heritage' was not specifically identified by the ICH Convention, food practices were gradually included on the ICH list under the condition that they are part of a process from production to consumption (rather than isolated practices) and that they intersect with the five cultural domains identified by the convention (Csergo, 2018, pp. 450–451; see Table 22.1). So, what is food heritage specifically? A general overview of the food heritage elements inscribed shows that there are examples of food preparation out of various ingredients and using particular practices and tools; food that is prepared as part of specific rituals, religious or cultural celebrations and festivals; traditional practices of food production and consumption that characterise the foodways of specific groups of people, regions or nations.[4]

A very significant aspect of the framework supported by the UNESCO's ICH Convention is the emphasis on people as tradition bearers and practitioners, i.e. what is important to preserve is the skills and the know-how of the people rather than the intangible (or tangible, for that matter) elements surrounding heritage. This has been exemplified in various cases ranging from the Neapolitan 'Pizzaiuoli' (UNESCO ICH, 2020), the hereditary Iemoto Masters of the Japanese Tea Schools (Cang, 2008) and the Mexican traditional female cooks (Matta, 2019b) to the notion of olive growers as 'Mediterranean food ambassadors' (Alonso & Krajsic, 2013). What seems to be important in most of these examples is not the food itself but the associated cultural practices which promote commensality (eating together), communality, neighbourliness and hospitality and are accompanied by various forms of music, dance, poetry, customs, legends, tales, art and craftsmanship. Furthermore, traditional agriculture and its resulting farming landscapes are often considered a form of cultural heritage, as demonstrated by the inclusion of, for example, traditional rice terraces in China and the Philippines in both the UNESCO's World Heritage List and the FAO's 'Globally Important Agricultural Heritage Systems' (GIAHS) with the latter containing 62 designations from around the globe since 2005 (FAO, 2020).

Food heritage and sustainability

Food has always had an important role in global challenges and politics that affect the environmental, economic and social aspects of sustainability. Many concerns have been raised and several national and worldwide interventions have addressed the impact of the food industry, food consumption and production on issues such as the climate crisis, environmental pollution, threats to biodiversity, the generation of food waste, food poverty and deprivation, the obesity pandemic, etc. (Rehber, 2012, pp. 353–354; FAO, 2015). Food has also played a key role in pandemics of foodborne diseases from the so-called 'mad cow disease' in the UK and the avian flu (Watson & Caldwell, 2005, pp. 2–4) to the most recent outbreak of the COVID-19 virus (Smith & Wesselbaum, 2020). Debates about these global challenges have also fuelled food activism movements that, among other issues, oppose genetically modified organisms, promote organic and fairtrade products or advocate for 'slow food' (Mann, 2014, pp. 1–2; Siniscalchi & Counihan, 2014, pp. 3–4). The efforts to achieve sustainable food consumption have gained a new dynamic with the emergence of the so-called green, political and ethical consumer (Boström & Klintman, 2009). We have argued elsewhere (Kapelari et al., 2020, pp. 14–15) that the worldwide discourse on food security has, to a great extent, evolved with little reference to the discourse surrounding the cultural aspects of food and the notion of food heritage. This is despite the fact that food security has been addressed by foodscape studies (Vonthron et al., 2020, Table 22.2) and despite the recognition that food security is directly linked with, for example, intangible cultural and natural heritage (Dorfman, 2012, p. 4; Hosagrahar, 2019, p. 14) or city development planning (Pearson & Pearson, 2016, p. 175).

Research on the phenomenon of food heritagisation has demonstrated various examples of how food and its role in cultural sustainability are intertwined with complex economic, social and political issues. It has been argued that through processes of heritagisation, food is often being rediscovered and reinvented as cultural heritage both as a means of local development and as a bearer of collective territorial identities (Grasseni, 2011). Culinary tourism, which can take various forms such as wine tourism, food routes and trails and agritourism (Timothy, 2016, pp. 14–17), is seen to contribute to cultural sustainability (Reynolds, 1993). At the same time, it has been argued that the so-called 'new tourist' (van Westering, 1999, pp. 78–79) is more culturally orientated and environmentally aware and therefore culinary encounters and gastronomy constitute 'a force in sustaining, developing and promoting heritage' (ibid, p. 81). What's more, comparative regional studies have shown that food, through culinary tourism and food festivals, can help foster movements that contribute to sustainable cultural revitalisation programmes (Di Giovine et al., 2017, p. 214). UNESCO's food heritage nominations also show that turning food into heritage can create a special dynamic. Pfeilstetter (2015, p. 224) has argued about the Mediterranean diet that naming and promoting it as cultural heritage has rendered it a distinctive kind of food (different from ordinary food and other gastronomic practices) both qualitatively – because it is sustainable, ecological and healthy – and culturally – because it is Mediterranean, traditional and authentic. Finally, food has acquired importance in a global arena not only as an element of cultural heritage but also as part of the creative industries as demonstrated by UNESCO's Creative Cities Network (UCNN). Within this wider programme, 'gastronomy' was

Table 22.2 The 15 botanical garden partners of the BigPicnic project

Name	Country
Botanical Garden of the University Vienna	Austria
University Botanic Gardens of Sofia University 'Saint Kliment Ohridski'	Bulgaria
Hortus botanicus Leiden	The Netherlands
University of Warsaw Botanic Garden	Poland
Juan Carlos I Royal Botanic Gardens, University of Alcalá de Henares	Spain
Botanical Garden and Botanical Museum at Freie Universität Berlin	Germany
Natural History Museum of the University of Oslo	Norway
National Museum of Natural History and Science at the University of Lisbon	Portugal
Royal Botanic Garden of Madrid	Spain
Royal Botanic Garden Edinburgh	United Kingdom
Balkan Botanic Garden of Kroussia	Greece
Botanic Garden Meise	Belgium
School Biology Centre Hannover	Germany
Bergamo Botanic Garden	Italy
Tooro Botanical Gardens	Uganda

identified as one of the seven creative fields and, in this case, creativity is recognised as a strategic factor of sustainable development (UNESCO, 2018b, p. 10).

All of the aforementioned examples attest to the societal, economic and political issues surrounding food cultures and food heritage but also to the relevance of the latter towards the concept of sustainability.

Methodology and the case study of the BigPicnic project

This paper will discuss some elements from the research findings of the BigPicnic project (the project was titled 'Big Picnic: Big Questions – engaging the public with Responsible Research and Innovation on food security') that took place from May 2016 to April 2019 and was funded by the European Commission's Horizon 2020 Programme (BigPicnic, 2020). This was a collaboration between a consortium of 19 partners, among which 15 are botanical garden partners (see Table 22.2). The aim of the BigPicnic project was for the partners to co-create with their local communities and their chosen target audiences a series of exhibitions, science cafés and events that would generate dialogue and build a greater understanding of food security issues (Wippoo & Dijk, 2016; BigPicnic Project Consortium, 2019). The project strived not only to give a voice to the public on RRI[5] in food security but also to communicate the findings to policymakers (BigPicnic Recommendations, 2019). During the aforementioned co-creation activities, an extensive set of qualitative studies was carried out by the partners and this was also combined with a large-scale survey with 1,189 respondents (Kapelari et al., 2020, pp. 4–5). In this paper, we will focus mostly on the findings from the qualitative studies.

As mentioned above, the 15 botanical garden partners (henceforth BG partners) organised a wide range of activities and received training and guidance from the University College London for the collection of data – primarily qualitative – following various methods (observations, interviews, focus groups, photographs, concept maps, etc.) and for analysing this following the TBI evaluation framework. TBI is a form of action evaluation, originally developed by the NISE net (Nanoscale Informal Science Education Network), which is employed in complex social interventions in order to assist practitioners and other stakeholders to define and then formatively redefine project effectiveness, and to forge effective action/practice (Rothman, 1998; Pattison et al., 2014). In the case of the BigPicnic project, the application of the TBI evaluation framework aimed to help botanic garden practitioners to evaluate their projects and reflect on their practice (Moussouri et al., 2019). This paper addresses the findings from a sample of 76 TBI reports completed by the BG partners, all of which were aligned with the key food policy priorities identified by Food 2030 and the SDGs.[6] The authors of this paper conducted a meta-analysis of these datasets.

The cultural and social dimensions of food: findings from the BigPicnic project

The BigPicnic project generated very interesting findings on various aspects that relate to food security and sustainability but the data gathered also frequently touched on cultural and social values attributed to food. The co-creation activities undertaken covered a diverse range of topics – ranging from food waste and labelling, urban gardening, crop sustainability to traditional medicinal plants, superfoods, edible insects, etc. What is particularly important to emphasise is that the whole process of allowing citizens of various groups to co-create activities (such as exhibitions) and to engage in dialogue with scientists (through science cafés) was underpinned by a directly bottom-up approach. This approach is in tune with the increasing awareness of how top-down approaches pose a risk to sustainable heritage practices (de Merode et al., 2004, p. 9) mentioned earlier. In terms of the overall significance of food towards achieving sustainability, SDGs 4 (quality education) and 3 (good health & well-being) were by far deemed to be the most significant. It is noteworthy that more than half of the TBI reports and nearly all of the botanical gardens received responses that revealed links with the notions of 'food cultures', 'foodways' and 'food heritage' (as these were described earlier). These responses were categorised in the meta-analysis stage under the broad term 'culture and food'. We will further elaborate here on three important themes that emerged from the data which relate to [1] 'traditional foodways, knowledge and practices', [2] 'food memories, stories and the social context of eating', and [3] 'migration and food: well-being of the diaspora communities'. This will subsequently lead us to a discussion of these findings and particularly the role of food heritage in fostering debates about food security and sustainability.

Traditional foodways, knowledge and practices

We already mentioned above that UNESCO's promotion of traditional food practices has been part of a wider heritigisation process that has not only boosted efforts to protect cultural diversity and the transmission of traditional food-related know-how but has also rendered food heritage a valuable asset for local communities, whole regions or nations. Nevertheless, since conventional heritage practices are increasingly viewed as unsustainable (Harrison et al., 2020, p. 486), alternative perspectives and approaches (to the ones held by heritage professionals) need to be more widely considered. The traditional ways of eating appeared very important among participants to the BigPicnic activities when they engaged in discussions about food security topics. Several respondents mentioned specific examples of foodstuffs that they were familiar with or for which they had cultural attachment. They underlined how important these types of food, plants or dishes were to them either because they grew up eating them or because these were associated with special situations (events, celebrations, rituals) related to their family, region or country.

The partners from the city of Bergamo, in Northern Italy – which coincidentally was awarded City of Gastronomy status in October 2019 (UCNN, 2020) – reported in the context of bread making in particular, that the traditional way of eating was valued along with dialogue and a relationship of trust between consumers and bakers/sellers. The appreciation of the Italian Mediterranean diet – belonging to the broader Mediterranean diet that has been inscribed in UNESCO's representative list of intangible heritage elements (Pfeilstetter, 2015) – was also evident and here the element of traditional eating was also associated with the notion of seasonality and the link between territory and culture. Another participant from Bergamo while participating in an activity about food plants stated that many of the names included in the available checklist were familiar through childhood experiences. Reports from Vienna, Austria, suggested that the knowledge of the older generations about food and nutrition was significant in influencing food choices but this knowledge was considered both an asset and a burden. In the

Botanic Garden Meise (BGM) in Belgium and the Tooro Botanical Gardens in Uganda, members of the African diaspora and the local communities respectively provided details about specific dishes made out of the plants they were seeing and engaging with and the specific occasions or celebrations where this culinary practice would take place:

> *Kahunga can also be prepared and eaten as solid food, for example this is our staple sauce as Bakonzo tribe, we prepare sombe and bundwe (casava flour) as a special meal for the visitors*
> (Comment by an Ugandan village farmer)

In a previous section of this paper, we mentioned how UNESCO conventions have strongly promoted traditional food practices and gastronomic creativity by celebrating and calling for the protection of the cultural diversity that exists in various countries and communities (UNESCO, 2005; UNESCO ICH Convention, 2006). Maintaining cultural diversity in the use of food was deemed important by participants to the BigPicnic activities and this diversity was seen as a parameter that affects how people use and consume food. Comments recorded in both Spain and Greece emphasised that people should relish both their own culinary traditions but also the diversity within individual countries. What is also interesting is that members of the public in Bulgaria expressed their keenness for the preservation of traditional recipes of edible plants and this seems to align with the notion that globalising processes and social transformations can often pose threats to the sustainability of certain traditions.

Finally, an interesting example of a form of traditional knowledge that was considered a useful solution for offering sustainable solutions to Uganda's food security issues emerged from data collected by the Tooro Botanical Gardens. More specifically, the revival of traditional ways of storing crops, like the traditional Enguli granaries, was considered a useful way for tackling contemporary food storage problems and a potential solution to avoid food spoilage in periods of famine. Such an approach echoes the principles advocated by UNESCO's ICH Convention. More specifically, UNESCO's recently published 'Thematic Indicators' have suggested that 'knowledge and practices transmitted from generation to generation' in areas such as agriculture, food systems, traditional medicine, natural resource management, ecosystem services and ecological resource management have the potential to contribute to food security and the SDG 2 (zero hunger; Hosagrahar, 2019, p. 14). With regard to the same SDG, the International National Trusts Organization (INTO) has also advocated for the importance of cultural factors, such as traditional knowledge and practices and has supported, for example, the planting of traditional 'heritage stocks', the promotion of sustainable farming practices along with genetic preservation of heritage livestock species (Canovas & Maurice, n. d., p. 3). In the African context, Webber Ndoro (2004, p. 84) has argued for the importance of traditional and customary heritage systems (using various examples, including Uganda), claiming that a management ethos that arises from the local social environment is the finest system. Such solutions provided by non-expert communities relying on traditional knowledge have often the potential to achieve truly bottom-up approaches. We will return to this point in the discussion section that will follow.

Food memories, stories and the social context of eating

Food has the special quality of triggering memories from the past and for carrying associations and stories that people often automatically relate to eating and to their own culinary traditions. Not surprisingly, in the BigPicnic activities, food was deemed to have strong associations with specific memories and stories that people keep and remember. This echoes the so-called 'proust effect' whereby a person can be taken back to childhood when the taste and smell of food or other senses stimulate the memories (Campen, 2014, p. 2). It also confirms the mnemonic function of food that renders it an 'edible chronotope' (Kirshenblatt-Gimblett, 2004, p. xiii) or 'edible souvenir' (Di Giovine & Brulotte, 2014, p. 18). The role of food in the sensory experience of places – with this sensory heritage element constituting a smellscape – can also trigger memories and emotions as demonstrated by a study of the Spice Market Quarter of Istanbul (Davis & Thys-Şenocak, 2017).

The comments received from the TBI reports pointed to the vital role of childhood memories in defining attitudes towards as well as knowledge about food. From data collected by the BG Partner in Warsaw, Poland, respondents mentioned that food triggered nostalgic thinking about home (for example, grandma's baking) and specific tastes that were now lost. Furthermore, other people tended to believe that food was tastier in the past and some believed that people who had experienced hunger would hold greater respect towards food. In both Italy and Spain, the lack of specific food memories from childhood were considered to explain the lack of knowledge about specific types of plants by some people. Findings from Hannover, Germany, and Greece also acknowledged the senses (e.g. taste/flavour, smell) as an important trigger for food memories as people automatically remember eating things in a specific way at a certain point in time.

I'm back in my home!

(Statement of an immigrant from Kosovo, living in Germany, smelling lemon balm)

When thinking about my childhood I always remember eating tomatoes from my granny's gardens

(Comment recorded by the BG Partner in Sofia, Bulgaria)

As already indicated, food is imbued with social values and meanings, and eating not only constitutes a social activity but places also the eater in a social universe. In the international heritage discourse, UNESCO has identified social practices as one of the five domains of ICH (see Table 22.1), and most of the elements inscribed in its 'Representative List of the Intangible Cultural Heritage of Humanity' demonstrate the importance of communality. This 'social context of eating' (what the food heritage literature has underlined as the commensality of food) was another important aspect that emerged from the BigPicnic project data. Respondents appeared to place value on the social interaction that takes place when people share food and eat together. Such comments stressed that communities appreciate how pleasant and useful it is for people to be connected through occasions that involve making food or eating together. Quite interestingly, at the BGM where the culinary practice of eating insects (as an alternative food) was addressed in some activities, it was observed that social norms often dictate whether or not something is acceptable as a food source. In this case, parents and grandparents had a significant impact on the decision of the children: whether they would try out or not a dish that contained insects. From data collected in Poland, it was noted that people felt an obligation to offer high-quality food when serving food to someone they knew personally. It is no surprise to be reminded that etymologically both the Greek and Latin words for 'companion' are related to the idea of sharing food: 'syntrofos' (σύντροφος) a person with whom you share food and 'com panis' meaning a person whom you share bread with (Travlou, 2020, p. 173).

Migration, food heritage and well-being in the diaspora

Food-centred and gustatory nostalgia is a very common element within diasporic or expatriate populations and this is evident both in processes of identity construction and in their experience of displacement (Holtzman, 2006, pp. 366–367). David Sutton's research on diaspora communities has highlighted how some of its members can have a longing for a lost homeland through food, e.g. the past is stored in the smell of olive oil or the taste of a fresh-cut fig (Sutton, 2005, 2010).

A very interesting element that emerged from the BigPincic data was the relationship between migration and food heritage. This was directly related to the foodways of diaspora communities and how these were affected by the fact that living far away from the country of origin makes access to certain ingredients difficult. This was an aspect that highlighted also the importance of certain plants, food products and dishes for contributing to the formation of cultural identities. These findings came about primarily from activities undertaken by the BGM in Belgium that worked closely with members of the African community living in the country. Indeed, the BGM, in collaboration also with the non-governmental organisation FoodBridge, engaged with the African diaspora of Belgium in various projects. These projects highlighted the potential of food as a tool for building bridges between people and across cultures. Maureen Duru, who was actively involved in the BGM's activities, has studied the daily food habits and culturally prescribed food norms of the Nigerian diaspora communities in Belgium and has stressed the importance for migrants of eating familiar (related to the homeland) food and of recreating home in the diaspora context through food (Duru, 2017, pp. 259–265). Her research has confirmed that food can be used by people in order to both distinguish themselves from others (a marker of identity) but also in order to engage with others within their environment (Duru, 2017, p. 15).

The co-creation activity 'The Face behind the food' aimed to help people undergoing (forced) migration to cope with 'lost' foodways and to feel at ease in their new situation. In this case, the BGM also aimed to contribute to helping these people 'reconnect'. For the purposes of this project, participants visited the botanical garden in order to share their experiences about the African plants. In this process, not only they presented stories and their favourite food memories but also their traditional knowledge about the production, consumption and sociocultural significance of the plants in question offering an alternative narrative/story to the scientific knowledge usually presented at the botanical garden. This activity was also combined with a shared cross-cultural meal while a smaller group of participants came back for a second more in-depth visit to share their knowledge with guides and educators. These participants were also offered a gardening course during which they expressed their desire to learn how they could grow tropical vegetables in Belgium.

From data received by the participants of the aforementioned activity, it was stated that people of the African diaspora miss the feeling of cultivating, harvesting and gathering their own food. Often, they feel they have insufficient access to healthy, clean and reasonably priced food of their desire. In this case, people expressed their will to somehow be able to control the relevant food market. In addition, food was considered a way to 'reconnect' with the home country while concerns were also raised about the agrofood sector and the well-being of the African farmers producing the plants that are then sold to other countries. Overall, these participants enjoyed sharing their food and plant knowledge, and the space of the BGM appeared to become a place where African diaspora people could do gardening but also showcase their crops, traditional knowledge and national or local cuisine.[7]

To come back to some points raised earlier, the memories, stories and traditional knowledge of the members of the African diaspora were very much valued by the BGM staff and managed to add additional layers to the existing scientific knowledge presented in the botanical garden. Furthermore, it offered additional perspectives and dimensions to discussions about food security and sustainability during the science cafés that brought together the experts and members of the public. In recent years, UNESCO's operational directives for the implementation of the ICH Convention (see Table 22.3) have considered traditional knowledge and practices to be useful to both inclusive social sustainability and environmental sustainability. Certain forms of ICH such as 'farming, fishing, hunting, pastoral, food gathering, food preparation and food preservation knowledge and practices, including their related rituals and beliefs' are deemed to contribute to food security and nutrition, agro-biodiversity and resilience to climate change (UNESCO, 2018a, Paragraph 178). On the other hand, the same directives have stressed the importance of 'knowledge and practices concerning nature and the universe' for achieving environmental sustainability (ibid). In any case, ensuring the transmission of these practices is a vital aspect which underlines the importance of the human element: the notion of people as knowledge and cultural bearers (as emphasised earlier in our chapter). What's more, this close relationship between sociocultural elements and wider human perceptions of nature, flora, fauna and food demonstrates one of the central arguments of this paper: that food heritage can serve as a catalyst for environmental sustainability.

We close this subsection with a reference to findings that pointed the close connection between food heritage and well-being of diaspora communities. It has been suggested that heritage can contribute to SDG 3, 'good health and well-being' (Hosagrahar, 2019, p. 14). According to the World Health Organization's Constitution, 'health is a state of complete physical, mental and social well-being and not merely the absence of disease or infirmity' (WHO, 2020). Indeed, existing research on heritage and well-being supports the importance that heritage places have on the social well-being of people (Power & Smyth, 2016) with some heritage organisations encouraging the preservation of meaningful places for people for the sake of the relevant health benefits (Reilly, Nolan & Monckton, 2018; Canovas & Maurice, n. d., p. 5).

Table 22.3 Safeguarding intangible cultural heritage and sustainable development at the national level (UNESCO, 2018a, Chapter VI, Paragraphs 170–197)

VI.1 Inclusive social development	VI.1.1 Food security
	VI.1.2 Health care
	VI.1.3 Quality education
	VI.1.4 Gender equality
	VI.1.5 Access to clean and safe water and sustainable water use
VI.2 Inclusive economic development	VI.2.1 Income generation and sustainable livelihoods
	VI.2.2 Productive employment and decent work
	VI.2.3 Impact of tourism on the safeguarding of intangible cultural heritage and vice versa
VI.3 Environmental sustainability	VI.3.1 Knowledge and practices concerning nature and the universe
	VI.3.2 Environmental impacts in the safeguarding of intangible cultural heritage
	VI.3.3 Community-based resilience to natural disasters and climate change
VI.4 Intangible cultural heritage and peace	VI.4.1 Social cohesion and equity
	VI.4.2 Preventing and resolving disputes
	VI.4.3 Restoring peace and security
	VI.4.4 Achieving lasting peace

The food-related activities organised in the context of the BigPicnic project and the subsequent qualitative studies rendered some interesting insights on the contribution of food heritage and the preservation of specific foodways for the well-being of citizens. The work of the Royal Botanic Garden Edinburgh (RBGE) in Scotland was particularly revealing in this topic. More specifically, food was seen as a medium for communication that enables members of the diaspora to create social contacts with Scottish people and improve their knowledge of the English language and local accent. The RBGE organised a series of co-created exhibitions and activities employing what they termed as digital storytelling. This project aimed to give a voice to people who experience food insecurity and to work with communities to increase access to nutritious food.[8] Within this context, the digital story created by an Iranian immigrant was very revealing about the intersection between food culture, traditional eating habits, health problems and well-being. This participant underlined how his eating habits were defined by the traditional food of his home country but the fact that he was recently diagnosed as a diabetic had a significant impact on his diet. He particularly emphasised how certain types of food characterise the ethnic cuisine of certain groups of people:'...*And I told her I am Iranian – rice is like coffee for you. Everything is rice*' – comment by the Iranian participant in RBGE.

Through the digital storytelling project, this person managed to take back control of his food choices and he actually turned his whole life around. The activity in which he participated made him improve his confidence and to feel he benefited from engaging with Scottish people and that he could integrate to the community. Although the food culture that informed his eating habits initially seemed to be an obstacle for tackling his health problem, the active involvement in RBGE's activities allowed for a meaningful reflection on health, nutrition and identity and for a consideration of various alternative food options.

Discussion

Reflecting on the findings of the BigPicnic research and their relevance to food heritage sustainability, there are two points that we would like to emphasise. First of all, the social and cultural dimensions of food should not be neglected for their contribution to heritage sustainability as their influence is not restricted merely to cultural or social sustainability but clearly extend to environmental sustainability as well. The other point is that participatory/bottom-up approaches to heritage management and practice can further enhance the opportunities to engage with and promote food heritage in a truly sustainable manner.

The link between cultural and environmental sustainability, as discussed above, has already been acknowledged by key players in the heritage sector, such as UNESCO, and has also been supported by an abundance of literature (albeit not always converging) that comes from various disciplines. What the BigPicnic data indicate or, rather, reinforce is that the traditional foodways of communities and the food practices that are passed down from generation to generation are certainly important for identity making and the sense of belonging. Preserving these social and cultural values that render food as heritage is important for the people concerned (the people who value food as heritage). However, the importance of these dimensions that food embraces extends beyond that. The right to access culturally appropriate food and to sustain the social and cultural activities that surround food can ensure that also different value systems, cosmologies and forms of knowledge are protected. The latter can improve our availability to adopt sustainable approaches in heritage, food and environmental policies.

The importance of traditional knowledge came across strongly in the interactions of the botanical gardens and the communities they engaged within the BigPicnic activities, as seen by the examples already mentioned. Undoubtedly, several examples internationally point towards the contribution of traditional agriculture and indigenous/traditional/local knowledge to sustainable and climate-smart food production (Dweba & Mearns, 2011; Magni, 2016; Singh & Singh, 2017, pp. 301–302). However, that is not to say that all traditional ways of food production and consumption as well as land management are inherently ecologically sustainable or constitute a panacea for sustainability (Clarke, 1990, pp. 235–236; Nygren, 1999, p. 268). When discussing food heritage and sustainability, it is best neither to support the idealised notion of the 'ecologically noble savage' – the idea that all indigenous populations have lived in harmony with the environment (Hames, 2007, pp. 178–179) – nor to exacerbate any divisions between 'traditional' as opposed to 'scientific' knowledge or local as opposed to universal knowledge (Nygren, 1999, pp. 267–268; Moore et al., 2003, p. 23). Traditional food knowledge can coexist and play a significant role alongside other forms of knowledge.

In a similar manner, different food memories, stories and the social context of eating are also integral in any effort to preserve food heritage. It has been argued that food is much more than sustenance and nutrition and therefore the mnemonic and sensory processes that are involved in the heritage-making process that surrounds food are vital. Understandings of how citizens can adopt healthier eating habits, avoid food waste and proactively contribute to efforts to tackle the climate crisis are intertwined with how people perceive themselves and their sense of community and belonging through food. In today's globalized conditions, many cities and countries encompass culturally diverse societies. In this context, the findings that relate to food heritage in the context of diaspora communities are important because they indicate that food value sharing can truly become the opportunity for dialogue and knowledge exchange that can further sustainability goals. In the case of BGM in Belgium, members of the diaspora were allowed to engage in activities that demonstrated their know-how, traditional knowledge and food practices, and by using food security as a point of discussion, they managed to enhance cross-cultural dialogue but also bring the views of both the food experts and the non-experts to the table. In the case of the RBGE activities in Scotland, certain citizens managed to even turn their lives around not only improving their well-being but also seeking to promote food security for the greater good.

Moving to the second point that we would like to raise, the BigPicnic project fostered a series of activities that brought together various citizens (the 'non-experts') with botanical gardens, scientists, educators, professionals of the food industry, etc. The principles of co-creation were at the centre of this approach. It was through this approach that members of various communities (local communities, families, school children, members of the diaspora and farmers) were able to reflect on and engage in dialogue about food security, sustainability and eventually their own perceptions of food heritage. This proved that engaging citizens in scientific debates about the aforementioned matters is important as it broadens the understanding of what food heritage sustainability should actually entail. The overall impact of the BigPicnic project can be considered through various outputs: the numbers of people engaged through the project's outreach activities, the potential for organisational change from the part of the participating partners and the end result of these activities which, among other deliverables, produced some policy recommendations (BigPicnic Recommendations, 2019) but also though some long-term impacts. The latter, for example, included the transformation of the lives of some project participants from citizens with concerns about food security to food activists.[9] All of these aspects support the idea that engaging citizens in discussions about food heritage in bottom-up approaches can contribute to wider understandings of food heritage sustainability.

In this paper, we summarised a range of sources that support the contribution of food heritage to sustainability with a significant emphasis on how this was achieved through looking at food as part of ICH. Although the potential for intangible heritage to truly achieve sustainable development has often been questioned (Boswell, 2011), there have been projects, such as the development of local/indigenous oral traditions, performing arts and traditional craftsmanship in India (Bhattacharya, 2015), that have provided livelihood opportunities to local communities. It has been suggested that in order to properly address sustainable development the heritage discourse needs to focus more on the concepts of empowerment and participation (Albert, 2015, pp. 17–18). These two concepts emphasise the responsibility that people must assume individually and collectively for the future of the planet and for the management and use of common resources that are scarce, including natural and cultural resources (ibid). Indeed, the work of political economists, such as Elinor Ostrom (1990), who have promoted the notion of the commons – where the public or the community are made up of responsible citizens – has inspired discussions of the potential of a 'heritage of the commons' (Lekakis, 2020). The acknowledgment of the importance of the notion of food sovereignty in discourses that deal with food and sustainability (Nyéléni, 2007; Gordillo & Jeronimo, 2013) also points towards this direction and so do the discussions of the notion of 'food democracy' (Renting, Schermer & Rossi, 2012).

We have already mentioned how supporting bottom-up approaches in heritage practices, management and planning is key to sustaining both cultural and natural heritage in the future (de Merode et al., 2004, p. 9). The approaches adopted through the Big-Picnic process of co-creation seem to align with opportunities for communities to have a stronger voice in food security – here lies a common point between participatory heritage and the movement for food sovereignty. With the Declaration of Nyéléni (2007, p. 9) recognising culturally appropriate food, produced through ecologically sound and sustainable methods, as a right we are perhaps entering a stage at which food heritage should be considered more strongly beyond the boundaries of heritage studies or any other single discipline for that matter. The notion of food sovereignty supports the view of food as a commons (Vivero-Pol et al., 2019) – a resource shared by communities but also the shared social practices for governing this resource (Ferrando & Vivero-Pol, 2017, p. 51) – and recent conceptualisations of heritage as commons have also considered cooking and culinary practices as both ICH and an act of commoning (Travlou, 2020).

It is probably appropriate to close this discussion by mentioning an aspect of food heritage that would require further research. This is the role of food heritage in promoting gender equality (relating to SDG 5). We have already underlined the significance of traditional food knowledge and practices for both cultural and environmental sustainability. In this context, research conducted by the Cross-Cultural Foundation of Uganda (CCFU, 2017, p. 3) has suggested that prior to the current marginalisation of women and girls that came about from the early 1960s, culturally defined female rights were protected through traditional Acholi cultural norms and principles. According to these cultural norms and practices, women were recognised and respected for their important role in various aspects that among others included their traditional roles in the agricultural sector and their traditional responsibility for food security (ibid, p. 23). This is only one example from one specific country and region but this offers food for thought for further implications that the notion of food heritage may have for sustainability.

Conclusions

This paper has argued that the notion of food heritage can be an important catalyst – if not a mainspring, as UNESCO has advocated – for discussing global and local challenges to sustainability and promoting food security. The findings from the BigPicnic research project and the examples employed in the analysis highlight that viewing food as heritage requires considering food's tangible and intangible dimensions and values along with its multisensory elements. All of these parameters exist along with but extend beyond science and technology and the politics and governance surrounding food. Food and food heritage is directly attached to human behaviour that reflects sociocultural values and identities and therefore understanding and sometimes changing the behaviours of citizens are the key to achieving sustainability. In the context of food security that engages with many global challenges and the notion of sustainability, science and technology offer many solutions and are vital. However, paying closer attention to how cultural values and identities interact with food and food choices can shed light to overlooked aspects and can offer a greater variety of solutions to the global food challenges. By looking at cultural heritage and environmental sustainability together and in an environment that allows various voices and concerned stakeholders to express their opinions, issues of sustainability can be better communicated.

A reflection on the findings of the BigPicnic project demonstrated the significant role that traditional food knowledge and practices hold for various communities when they think about and ponder about the sustainability of their foodways and systems. Furthermore, the food memories and stories attached to cooking and eating and the social context of eating are not only important for the relevant communities but also for preserving additional layers of value that are not necessarily always shared with the experts (be it food or heritage experts). This implies that when discussing food heritage and sustainability it is not enough to target individuals/consumers but the focus should be rather placed on sociocultural groups and wider networks (e.g. families and communities) and their role in shaping individual values, identities and, consequently, behaviour.

Addressing and incorporating the diverse and enormous literature on food cultures, foodways and food heritage into the policies that tackle food security perhaps may be deemed to complicate things by significantly expanding the parameters that need to be considered. Nevertheless, this task is worthy as shown by the emergence of a strong discourse on food heritage and its contribution to sustainability and as demonstrated by the findings of the BigPicnic project discussed in this paper. We would further argue that research on food heritage would benefit from closer engagement with other transdisciplinary approaches and fields of study, such as social ecology[10] and agroecology,[11] with mutual cross-fertilisation of ideas and findings providing additional useful insights.

Acknowledgements

This work was supported by the European Commission's Horizon 2020 Programme ['Science with and for society'. ISSI.1.2014.2015 – 'Pan European public Outreach: exhibitions and science cafés engaging citizens in science'] under grant agreement No 710780. The authors would like to thank all the project partners in the BigPicnic project as well as the various communities and individuals who contributed their ideas and knowledge of food. Georgios Alexopoulos would also like to thank the University of Innsbruck's Area of Science, Geography, Computer Science and Mathematics Education (DiNGIM) for funding his involvement in the writing of this chapter.

Appendices

Notes

1 The four pillars of food security are 'availability', 'access', 'utilization' and 'stability' (WSFS, 2009). The BigPicnic consortium has viewed food security as an umbrella term for three interconnected concepts: 'access', 'safety', 'sovereignty' (BigPincic, 2020).
2 The link with the cultural and creative industries was promoted with the guiding principles of the 'Convention on the Protection and Promotion of the Diversity of Cultural Expressions' which states that cultural diversity is a 'mainspring for sustainable development for communities, peoples and nations' (UNESCO, 2005, Article 2).
3 Literature from the so-called 'Anthropology of Food' (Mintz & Du Bois, 2002; Watson & Klein, 2016, pp. 2–3), 'Geography of Food' (Bell & Valentine, 1997; Freidberg, 2003) and 'Sociology of Food' (Goody, 1982; Mennell et al., 1993; Carolan, 2012) has investigated the cultural and symbolic dimensions of food as well as contemporary social issues in food production and consumption systems. Since the beginning of the 2000s, the notion of 'food culture' gained importance in the social and human sciences and, even though scholars from different disciplines have often emphasised different aspects, it has been acknowledged that global themes emerging in public debates about food require approaches that go beyond clear-cut disciplinary frameworks (Fumey et al., 2016). Interdisciplinary research on food has led to the emergence of 'food' or 'culinary studies' (Julier et al., 2019, pp. 1–2), a thriving scholarly field even if sometimes hard to define precisely (Albala, 2013, p. xv), and the last two decades have seen the development of various relevant academic degrees (Watson & Caldwell, 2005, p. 1; ASFS, 2020). Folklorists and historians have also employed the term 'foodways' (see Table 22.1) in order to address the exploration of socially bounded lifestyle groups whose social network includes the sharing of a food system (Brown & Mussell, 2001, p. 11). Food has also played an important role in spatial conceptualisations of the environment through the notions of 'foodscape' that addresses the nexus of 'food-people-places' (Vonthron et al., 2020, p. 8) and 'smellscape' that investigates the nexus between 'people-odours-environment' (Henshaw, 2013, p. 2).
4 Some of these include the Mediterranean diet, the gastronomic meal of the French, the beer culture in Belgium, traditional Mexican cuisine, the know-how of cultivating mastic on the island of Chios (Greece), and the tradition of kimchi-making (Democratic People's Republic of Korea).
5 According to the European Commission, RRI 'implies that societal actors (researchers, citizens, policymakers, business, third-sector organisations, etc.) work together during the whole research and innovation process in order to better align both the process and its outcomes with the values, needs and expectations of society' (EC, 2020).
6 Food 2030 was launched after the 2015 Milan World Expo and is the European Union's research and innovation policy response to international policy development for food sustainability (Fabbri, 2017, pp. 2–3). The four key Food and Nutrition Security priorities are the following (Food, 2030): 'nutrition for sustainable and healthy diets', 'climate smart and environmentally sustainable food systems', 'circularity and resource efficiency of food systems', and 'Innovation and empowerment of communities'.
7 The following video was produced by the Meise Botanic Garden in July 2018 and provides interesting insights about the experience of working with the African, diaspora. It also includes interviews from a visit to the Tooro Botanical Gardens in Uganda (another partner of the BigPicnic project) and reflects on the shared experiences and ideas about food security and the involvement in the BigPicnic project: https://www.bgci.org/resources/bgci-tools-and-resources/bigpicnic-resources/
8 Some of the very interesting stories created by the RBGE can be found on the following website: https://www.rbge.org.uk/news/big-picnic/big-picnic-stories/
9 Some of the stories created by the RBGE's digital storytelling activities were presented in 2019 to Members of the Scottish Parliament and this coincided with a period of consultation over Scotland's ambition to become a good food nation (RBGE, 2019).
10 Social ecology, following a transdisciplinary approach, has made important contributions to sustainability research by focussing on the interrelations between societies and their natural environment in an overall effort to view the human world and society systematically (Wheeler, 2012, p. 2; Kramm et al., 2017, pp. 2–4; Stokols, 2018, p. 15). For example, the notion of the 'Mediterranean diet' – recognised as intangible cultural heritage by UNESCO since 2013 – and its contribution to the sustainability of local communities has been recently viewed from the perspective of social ecology (Petridis & Huber, 2017).

11 Agroecology encompasses transdisciplinary approaches that 'consciously seek to combine the experiential knowledge of farmers and indigenous people with the latest insights from the science of ecology' (Pimbert, 2015, pp. 287–288).

References

Albala, K. (2013). Introduction. In K. Albala (Ed.), *Routledge international handbook of food studies* (pp. xv–xvi). Abingdon and New York: Routledge.

Albert, M.-T. (2015). Mission and vision of sustainability discourses in heritage studies. In M.-T. Albert (Ed.), *Perceptions of sustainability in heritage studies* (pp. 11–20). Berlin: De Gruyter.

Albert, M.-T. (2017). The potential of culture for sustainable development in heritage studies. In M.-T. Albert, F. Bandarin, & A. Pereira Roders (Eds.), *Going beyond: Perceptions of sustainability in heritage studies no. 2* (pp. 33–43). Cham: Springer.

Alexopoulos, G. (2013). Management of living religious heritage: Who sets the agenda? The case of the monastic community of Mount Athos. *Conservation and Management of Archaeological Sites*, 15(1), 59–75.

Alonso, A. D., & Krajsic, V. (2013). Food heritage down under: Olive growers as Mediterranean 'food ambassadors'. *Journal of Heritage Tourism*, 8(2–3), 158–171.

Auclair, E., & Fairclough, G. (2015). Living between past and future: An introduction to heritage and cultural sustainability. In E. Auclair & G. Fairclough (Eds.), *Theory and practice in heritage and sustainability: Between past and future* (pp. 1–22). Abingdon and New York: Routledge.

ASFS (Association for the Study of Food and Society) (2020). Food studies programs. Retrieved from: http://www.food-culture.org/food-studies-programs/ (Accessed on 16 September 2020).

Barthel-Bouchier, D. (2013). *Cultural heritage and the challenge of sustainability*. Walnut Creek: Left Coast Press.

Belasco, W. (1999). Why food matters. *Culture, Agriculture, Food and Environment*, 21(1), 27–34.

Bell, D., & Valentine, G. (1997). *Consuming geographies: We are where we eat*. London and New York: Routledge.

Bessière, J. (1998). Local development and heritage: Traditional food and cuisine as tourist attractions in rural areas. *Sociologia Ruralis*, 38(1), 21–34.

Bessière, J., & Tibère, L. (2011). Editorial: Patrimoines alimentaires. *Anthropology of Food* [Online], 8. Retrieved from: http://journals.openedition.org/aof/6782 (Accessed on 15 September 2020).

Bhattacharya, A. (2015). Art for life: Intangible cultural heritage and livelihood development in India. In P. Basu & W. Modest (Eds.), *Museums, heritage and international development* (pp. 83–95). Abingdon & New York: Routledge.

BigPicnic (2020). *BigPicnic: Big questions – engaging the public with responsible research and innovation on food security*. Retrieved from: https://www.bgci.org/our-work/projects-and-case-studies/bigpicnic/ (Accessed on 2 December 2020).

BigPicnic Project Consortium (2019). *Exhibition case studies*. BigPicnic Deliverable D3.1 London: BGCI. Retrieved from: https://www.bigpicnic.net/resources/exhibition-case-studies/ (Accessed on 3 July 2020).

BigPicnic Recommendations (2019). BigPicnic recommendations – seven policy briefs. Retrieved from: https://www.bgci.org/resources/bgci-tools-and-resources/bigpicnic-resources/ (Accessed on 4 January 2021).

Boström, M., & Klintman, M. (2009). The green political food consumer: A critical analysis of the research and policies. *Anthropology of Food* [Online], S5: Can consumers save the world? https://doi.org/10.4000/aof.6394.

Boswell, R. (2011). Challenges to sustaining intangible cultural heritage. *Heritage & Society*, 4(1), 119–124.

Brown, A. E. L. (2018). ICH, cultural diversity and sustainable developmentI. In C. Waelde, C. Cummings, M. Pavis, & H. Enright (Eds.), *Research handbook on contemporary intangible cultural heritage: Law and heritage* (pp. 106–138). Cheltenham and Northampton, MA: Edward Elgar Publishing.

Brown, L. K., & Mussell, K. (2001). Introduction. In L. K. Brown & K. Mussell (Eds.), *Ethnic and regional foodways in the United States: The performance of group identity* (pp. 1–16). Knoxville: The University of Tennessee Press.

Bushell, R. (2015). Heritage and sustainable development: Transdisciplinary imaginings of a wicked concept. In E. Waterton & S. Watson (Eds.), *The Palgrave handbook of contemporary heritage research* (pp. 492–506). London: Palgrave Macmillan.

Campen, C. van (2014). *The proust effect: The senses as doorways to lost memories*. Oxford: Oxford University Press.

Cang, V. G. (2008). Preserving intangible heritage in Japan: The role of the Iemoto system. *International Journal of Intangible Heritage*, *3*(2008), 71–81.

Canovas, A., & Maurice, O. (n. d.). Heritage conservation and the sustainable development goals. INTO (International National Trusts Organization). Retrieved from: https://intoorg.org/about-into/knowledge-library/documents-and-publications (Accessed on 12 September 2020).

Carolan, M. (2012). *The sociology of food and agriculture*. London & New York: Routledge.

Carr, G., Sørensen, M. L. S., & Viejo Rose, D. (2018). Food as heritage. In E. Lightfoot, X. Liu, & D. Q. Fuller (Eds.), *Far from the hearth: Essays in honour of Martin K. Jones* (pp. 145–151). Cambridge: McDonald Institute for Archaeological Research.

CCFU (2017). *Women, culture & rights in acholi*. The Cross-Cultural Foundation of Uganda. Retrieved from: https://crossculturalfoundation.or.ug/report-women-culture-rights-acholi-launched-gulu/ (Accessed on 23 August 2020).

Clark, K. (2008). Only connect – cultural heritage and sustainable development. In G. Fairclough, R. Harrison, J. H. Jameson Jnr., & J. Schofield (Eds.), *The heritage reader* (pp. 82–98). London & New York: Routledge.

Clarke, W. C. (1990). Learning from the past: Traditional knowledge and sustainable development. *The Contemporary Pacific*, *2*(2), 233–253.

Counihan, C. M. (1999). *The anthropology of food and body: Gender, meaning, and power*. New York and London: Routledge.

Counihan, C. (2014). Cultural heritage in food activism: Local and global tensions. In R. Brulotte & M. Di Giovine (Eds.), *Edible identities: Exploring food and foodways as cultural heritage* (pp. 219–230). Farnham: Ashgate.

Csergo, J. (2018). Food as a collective heritage brand in the era of globalization. *International Journal of Cultural Property*, *25*(4), 449–468.

Davis, L., & Thys-Şenocak, L. (2017). Heritage and scent: Research and exhibition of Istanbul's changing smellscapes. *International Journal of Heritage Studies*, *23*(8), 723–741.

de Merode, E., Smeets, R., & Westrik, C. (2004). Introduction. In E. de Merode, R. Smeets, & C. Westrik (Eds.), *Linking universal and local values: Managing a sustainable future for world heritage* (pp. 9–15). World Heritage Paper No 13. Paris: UNESCO World Heritage Centre.

Demossier, M. (2016). The politics of heritage in the land of food and wine. In W. Logan, M. N. Craith, & U. Kockel (Eds.), *A companion to heritage studies* (pp. 87–100). Chichester: Wiley Blackwell.

Dessein, J., Soini, K., Fairclough, G., & Horlings, L. (Eds.) (2015). *Culture in, for and as sustainable development: Conclusions from the COST Action IS1007 investigating cultural sustainability*. Jyväskylä: University of Jyväskylä.

Di Giovine, M. A., & Brulotte, R. L. (2014). Introduction: Food and foodways as cultural heritage. In R. Brulotte & M. Di Giovine (Eds.), *Edible identities: Exploring food and foodways as cultural heritage* (pp. 1–27). Farnham: Ashgate.

Di Giovine, M., Mabry, J. B., & Majewski. T. (2017). Moveable feasts: Food as revitalizing cultural heritage. In H. Silverman, E. Waterton, & S. Watson (Eds.), *Heritage in action: Making the past in the present* (pp. 201–216). Cham: Springer International Publishing.

Dorfman, E. (2012). Intangible natural heritage: An introduction. In E. Dorfman (Ed.), *Intangible natural heritage: New perspectives on natural objects* (pp. 1–15). London & New York: Routledge.

Duru, M. (2017). *Diaspora, food and identity: Nigerian migrants in Belgium*. Brussels: P. I. E. Peter Lang.

Duxbury, N., & Gillette, E. (2007). *Culture as a key dimension of sustainability: Exploring concepts, themes, and models.* Vancouver, BC: Creative City Network of Canada – Centre of Expertise on Culture and Communities. Retrieved from: http://cercles.diba.cat/documentsdigitals/pdf/E130054.pdf (Accessed on 2 July 2020).

Dweba T. P., & Mearns, M. A. (2011). Conserving indigenous knowledge as the key to the current and future use of traditional vegetables. *International Journal of Information Management*, 31(6), 564–571.

EC (2020). *Responsible research and innovation* (RRI). Retrieved from: https://ec.europa.eu/programmes/horizon2020/en/h2020-section/responsible-research-innovation (Accessed on 2 August 2020).

Engelhardt, E. (2013). Introduction - redrawing the grocery: Practices and methods for studying southern food. In J. T. Edge, E. Engelhardt, & T. Ownby (Eds.), *The larder: Food studies methods from the American South* (pp. 1–6). Athens, GA & London: University of Georgia Press.

Engels, B. (2017). Natural world heritage and the sustainable development goals. In M.-T. Albert, F. Bandarin, & A. Pereira Roders (Eds.), *Going beyond: Perceptions of sustainability in heritage studies no. 2* (pp. 45–56). Cham: Springer.

Erlewein, S. (2015). Sustainable development and intangible cultural heritage: Integrating culture into development. In M.-T. Albert (Ed.), *Perceptions of sustainability in heritage studies* (pp. 71–84). Berlin: De Gruyter.

Fabbri, K. (Ed.) (2017). *Food 2030: Future-proofing our food systems through research and innovation.* Luxembourg: Publications Office of the European Union. Retrieved from: https://fit-4food2030.eu/food-2030/ (Accessed on 02 September 2020).

FAO (2009). *Declaration of the world summit on food security.* Food and Agriculture Organization (FAO) of the United Nations. Retrieved from: http://www.fao.org/tempref/docrep/fao/Meeting/018/k6050e.pdf (Accessed on 10 June 2020).

FAO (2015). *FAO and the 17 sustainable development goals.* Rome: Food and Agriculture Organization of the United Nations. Retrieved from: http://www.fao.org/3/a-i4997e.pdf (Accessed on 4 September 2020).

FAO (2020). *Globally important agricultural heritage systems* (GIAHS). Retrieved from: http://www.fao.org/giahs/background/en/ (Accessed on 12 September 2020).

Ferrando, T., & Vivero-Pol, J. L. (2017). Commons and "commoning": A "new" old narrative to enrich the food sovereignty and right to food claims. *Right to Food and Nutrition Watch*, 50–56. Retrieved from: https://www.righttofoodandnutrition.org/files/02.rtfanw-2017_eng_17_12_article-5_web_rz.pdf (Accessed on 10 March 2021).

Fischer, D. H. (1989). *Albion's seed: Four British folkways in America.* New York & Oxford: Oxford University Press.

Food (2030). Food 2030- Website of the European Commission. Retrieved from: https://ec.europa.eu/research/bioeconomy/index.cfm?pg=policy&lib=food2030 (Accessed on 2 September 2020).

Food & Foodways (1985). English editorial. *Food and Foodways*, 1(1–2), i–ii.

Freidberg, S. (2003). Not all sweetness and light: New cultural geographies of food. *Social & Cultural Geography*, 4(1), 3–7.

Fumey, G., Jackson, P., & Raffard, P. (2016). Introduction: Food cultures and spaces. *Anthropology of Food* [Online], 11. https://doi.org/10.4000/aof.8048

Geyzen, A. (2014). Food studies and the heritage turn: A conceptual repertoire. *Food and History*, 12(2), 67–96.

Goody, J. (1982). *Cooking, cuisine and class: A study in comparative sociology.* Cambridge: Cambridge University Press.

Gordillo, G., & Jeronimo, O. M. (2013). *Food security and sovereignty: Base document for discussion.* Rome: Food and Agriculture Organisation (FAO) of the United Nations. Retrieved from: www.fao.org/3/a-ax736e.pdf (Accessed on 12 September 2020).

Grasseni, C. (2011). Re-inventing food: Alpine cheese in the age of global heritage. *Anthropology of Food* [Online], 8. https://doi.org/10.4000/aof.6819.

Hames, R. (2007). The ecologically noble savage debate. *Annual Review of Anthropology, 36,* 177–190.

Harrison, R. (2017). Freezing seeds and making futures: Endangerment, hope, security, and time in agrobiodiversity conservation practices. *Culture, Agriculture, Food and Environment, 39*(2), 80–89.

Harrison, R., DeSilvey, C., Holtorf, C., Macdonald, S., Bartolini, N., Breithoff, E., Fredheim, H., Lyons, A., May, S., Morgan, J., & Penrose, S. (2020). *Heritage futures: Comparative approaches to natural and cultural heritage practices.* London: UCL Press.

Harvey, D. C., & Perry, J. (2015). Heritage and climate change: The future is not the past. In D. C. Harvey & J. Perry (Eds.), *The future of heritage as climates change: Loss, adaptation and creativity* (pp. 3–22). London & New York: Routledge.

Hawkes, J. (2001). *The fourth pillar of sustainability: Culture's essential role in public planning.* Victoria, Australia: Common Ground Publishing.

Henshaw, V. (2013). *Urban smellscapes: Understanding and designing city smell environments.* New York: Routledge.

Holtzman, J. D. (2006). Food and memory. *Annual Review of Anthropology, 35,* 361–378.

Hosagrahar, J. (Ed.) (2019). *Culture 2030 indicators: Thematic indicators for culture in the 2030 agenda.* Paris: UNESCO. Retrieved from: https://unesdoc.unesco.org/ark:/48223/pf0000371562 (Accessed on 17 May 2020).

ICCROM (2020). ICCROM's commitment to the sustainable development goals. Retrieved from: https://www.iccrom.org/iccroms-commitment-sustainable-development-goals (Accessed on 2 June 2020).

ICOMOS (2017). ICOMOS action plan: Cultural heritage and localizing the UN Sustainable Development Goals (SDGs). Retrieved from: https://www.icomos.org/en/focus/un-sustainable-development-goals/8778-cultural-heritage-and-sustainable-development (Accessed on 2 June 2020).

Julier, A., Counihan, C., & Van Esterik, P. (2019). Introduction: The continuing salience of food and culture. In C. Counihan, P. Van Esterik, & A. Julier (Eds.), *Food and culture: A reader* (pp. 1–9). Abingdon & New York: Routledge.

Kapelari, S., Alexopoulos, G., Moussouri, T., Sagmeister, K. J., & Stampfer, F. (2020). Food heritage makes a difference: The importance of cultural knowledge for improving education for sustainable food choices. *Sustainability, 12*(4), 1509. https://doi.org/10.3390/su12041509

Kirshenblatt-Gimblett, B. (2004). Foreword. In L. M. Long (Ed.), *Culinary tourism: Material worlds* (pp. xi–xiv). Lexington: University Press of Kentucky.

Kramm, J., Pichler, M., Schaffartzik, A., & Zimmermann, M. (2017). Societal relations to nature in times of crisis—Social ecology's contributions to interdisciplinary sustainability studies. *Sustainability, 9*(7), Article 1042. https://doi.org/10.3390/su9071042.

Labadi, S., & Logan, W. (Eds.) (2016). *Urban heritage, development and sustainability: International frameworks, national and local governance.* Abingdon and New York: Routledge.

Larsen, P., & Logan, W. (Eds.) (2018). *World heritage and sustainable development: New directions in world heritage management.* London: Routledge.

Lekakis, S. (Ed.) (2020). *Cultural heritage in the realm of the commons: Conversations on the case of Greece.* London: Ubiquity Press.

Logan, W. (2016). Whose heritage? Conflicting narratives and top-down and bottom-up approaches to heritage management in Yangon, Myanmar. In S. Labadi & W. Logan (Eds.), *Urban heritage, development and sustainability: International frameworks, national and local governance* (pp. 256–273). Abingdon and New York: Routledge.

Magni, G. (2016). Indigenous knowledge and implications for the sustainable development agenda. Background paper prepared for the 2016 Global Education Monitoring Report '*Education for people and planet: Creating sustainable futures for all*'. Retrieved from: https://unesdoc.unesco.org/ark:/48223/pf0000245623 (Accessed on 12 September 2020).

Mann, A. (2014). *Global activism in food politics: Power shift.* Basingstoke: Palgrave Macmillan.

Matta, R. (2016). Food incursions into global heritage: Peruvian Cuisine's slippery road to UNESCO. *Social Anthropology*, *24*(3), 338–352.

Matta, R. (2019a). Heritage foodways as matrix for cultural resurgence: Evidence from rural Peru. *International Journal of Cultural Property*, *26*(1), 49–74.

Matta, R. (2019b). Mexico's ethnic culinary heritage and *cocineras tradicionales* (traditional female cooks). *Food and Foodways*, *27*(3), 211–231.

McGhie, H. A. (2019). *Museums and the sustainable development goals: A how-to guide for museums, galleries, the cultural sector and their partners*. Curating Tomorrow, UK. Retrieved from: https://curatingtomorrow236646048.files.wordpress.com/2019/12/museums-and-the-sustainable-development-goals-2019.pdf (Accessed on 2 June 2020).

McGhie, H. (2020). The sustainable development goals: Helping transform our world through museums. *ICOM Voices*. Retrieved from: https://icom.museum/en/news/the-sustainable-development-goals-helping-transform-our-world-through-museums/ (Accessed on 2 June 2020).

Mennell, S., Murcott, A., & van Otterloo, A. H. (1993). *The sociology of food: Eating, diet and culture*. London: Sage Publications.

Mintz, S. W., & Du Bois, C. M. (2002). The anthropology of food and eating. *Annual Review of Anthropology*, *31*(1), 99–119.

Moore, D., Kosek, J., & Pandian, A. (2003). Introduction: The cultural politics of race and nature: Terrains of power and practice. In D. Moore, J. Kosek & A. Pandian (Eds.), *Race, nature and the politics of difference* (pp. 1–70). Durham, NC: Duke University Press.

Moussouri, T., Alexopoulos, G., & Francis, D. (2019). *Team-based inquiry practitioners manual*. London: Botanic Gardens Conservation International. Retrieved from: https://www.bgci.org/resources/bgci-tools-and-resources/bigpicnic-resources/ (Accessed on 7 January 2021).

Ndoro, W. (2004). Traditional and customary heritage systems: Nostalgia or reality? The implications of managing heritage sites in Africa. In E. de Merode, R. Smeets, & C. Westrik (Eds.), *Linking universal and local values: Managing a sustainable future for world heritage* (pp. 81–84). World Heritage Paper No 13. Paris: UNESCO World Heritage Centre.

Nyéléni (2007). *Nyéléni, 2007: Forum for food sovereignty, Sélingué, Mali, February 23–27, 2007*. Retrieved from: https://nyeleni.org/DOWNLOADS/Nyelni_EN.pdf (Accessed on 7 September 2020).

Nygren, A. (1999). Local knowledge in the environment-development discourse: From dichotomies to situated knowledges. *Critique of Anthropology*, *19*(3), 267–288.

Ostrom, E. (1990). *Governing the commons: The evolution of institutions for collective action*. Cambridge: Cambridge University Press.

Pattison, S., Cohn, S., & Kollmann, L. (2014). *Team-based inquiry: A practical guide for using evaluation to improve informal education experiences*. Retrieved from: http://nisenet.org (Accessed on 5 May 2020).

Pearson, D., & Pearson, T. (2016). Branding food culture: UNESCO creative cities of gastronomy. *Journal of Food Products Marketing*, *23*(3), 342–355.

Petridis, P., & Huber, J. (2017). A socio-metabolic transition of diets on a Greek Island: Evidence of "Quiet Sustainability". In E. Fraňková, W. Haas, & S. J. Singh (Eds.), *Socio-Metabolic perspectives on the sustainability of local food systems: Insights for science, policy and practice* (pp. 263–289). Cham: Springer International Publishing.

Petti, L., Trillo, C., & Makore, B. N. (2020). Cultural heritage and sustainable development targets: A possible harmonisation? Insights from the European perspective. *Sustainability*, *12*(3), 926. https://doi.org/10.3390/su12030926.

Pfeilstetter, R. (2015). Heritage entrepreneurship. Agency-driven promotion of the Mediterranean diet in Spain. *International Journal of Heritage Studies*, *21*(3), 215–231.

Pimbert, M. (2015). Agroecology as an alternative vision to conventional development and climate-smart agriculture. *Development*, *58*(2–3), 286–298.

Power, A., & Smyth, K. (2016). Heritage, health and place: The legacies of local community-based heritage conservation on social wellbeing. *Health & Place*, *39*, 160–167.

RBGE (2019). *Royal botanic garden Edinburgh: The people's food stories*. Retrieved from: https://www.rbge.org.uk/news/big-picnic/ (Accessed on 2 September 2020).

Rehber, E. (2012). Food for thought: "four Ss with one F": Security, safety, sovereignty, and shareability of food. *British Food Journal, 114*(3), 353–371.

Reilly, S., Nolan, C., & Monckton, L. (2018). *Wellbeing and the historic environment: Threats, issues and opportunities for the historic environment*. Historic England. Retrieved from: https://historicengland.org.uk/images-books/publications/wellbeing-and-the-historic-environment/ (Accessed on 2 September 2020).

Renting, H., Schermer, M., & Rossi, A. (2012). Building food democracy: Exploring civic food networks and newly emerging forms of food citizenship. *International Journal of Sociology of Agriculture and Food, 19*(3), 289–307.

Reynolds, P. C. (1993). Food and tourism: Towards an understanding of sustainable culture. *Journal of Sustainable Tourism, 1*(1), 48–54.

Robert, K. W., Parris, T. M., & Leiserowitz, A. A. (2005). What is sustainable development? Goals, indicators, values, and practice. *Environment: Science and Policy for Sustainable Development, 47*(3), 8–21.

Romagnoli, M. (2019). Gastronomic heritage elements at UNESCO: Problems, reflections on and interpretations of a new heritage category. *International Journal of Intangible Heritage, 14*, 157–171.

Ron, A. S., & Timothy, D. J. (2013). The land of milk and honey: Biblical foods, heritage and Holy Land tourism. *Journal of Heritage Tourism, 8*(2–3), 234–247.

Rothman, J. (1998). Action evaluation and conflict resolution in theory and practice. *Mediation Journal, 15*(1), 119–131.

Schofield, J. (2014). Heritage expertise and the everyday: Citizens and authority in the twenty-first century. In J. Schofield (Ed.), *Who needs experts?: Counter-mapping cultural heritage* (pp. 1–11). Farnham: Ashgate.

Singh, R., & Singh, G. S. (2017). Traditional agriculture: A climate-smart approach for sustainable food production. *Energy, Ecology and Environment, 2*(5), 296–316.

Siniscalchi, V., & Counihan, C. (2014). Ethnography of food activism. In C. Counihan & V. Siniscalchi (Eds.), *Food activism: Agency, democracy and economy* (pp. 3–13). London and New York: Bloomsberg.

Smith, L. (2006). *Uses of heritage*. London and New York: Routledge.

Smith, M. D., & Wesselbaum, D. (2020). COVID-19, Food insecurity, and migration. *The Journal of Nutrition, 150*(11), 2855–2858.

Stokols, D. (2018). *Social ecology in the digital age: Solving complex problems in a globalized world*. London: Academic Press.

Sutton, D. E. (2005). Synesthesia, memory, and the taste of home. In C. Korsmeyer (Ed.), *The Taste culture reader: Experiencing food and drink* (pp. 304–316). Oxford: Berg.

Sutton, D. E. (2010). Food and the senses. *Annual Review of Anthropology, 39*, 209–223.

Throsby, D. (2002). Cultural capital and sustainability concepts in the economics of cultural heritage. In M. de la Torre (Ed.), *Assessing the values of cultural heritage: Research Report* (pp. 101–117). Los Angeles, CA: The Getty Conservation Institute.

Timothy, D. J. (2016). Introduction: Heritage cuisines, foodways and culinary traditions. In D. J. Timothy (Ed.), *Heritage cuisines: Traditions, identities and tourism* (pp. 1–24). Abingdon: Routledge.

Timothy, D. J., & Ron, A. S. (2013). Understanding heritage cuisines and tourism: Identity, image, authenticity, and change. *Journal of Heritage Tourism, 8*(2–3), 99–104.

Travlou, P. (2020). From cooking to commoning: The making of intangible cultural heritage in OneLoveKitchen, Athens. In S. Lekakis (Ed.), *Cultural heritage in the realm of the commons: Conversations on the case of Greece* (pp. 159–182). London: Ubiquity Press.

UCLG (2010). *Culture: Fourth pillar of sustainable development. United cities and local governments*. Retrieved from: http://www.agenda21culture.net/sites/default/files/files/documents/en/zz_culture4pillarsd_eng.pdf (Accessed on 11 June 2020).

UCLG (2018). *Culture in the sustainable development goals: A guide for local action*. United cities and local governments. Retrieved from: https://www.uclg.org/sites/default/files/culture_in_the_sdgs.pdf (Accessed on 11 June 2020).

UCNN (2020). *UNESCO Creative Cities Network (UCNN)*. Retrieved from: https://en.unesco.org/creative-cities/creative-cities-map (Accessed on 10 September 2020).

UN (1987). *Our common future: Report of the world commission on environment and development*. Oxford: Oxford University Press.

UNESCO (1998). *Intergovernmental conference on cultural policies for development: Final Report*. Stockholm, Sweden, 30 March–2 April 1998. Retrieved from: https://unesdoc.unesco.org/ark:/48223/pf0000113935 (Accessed on 18 May 2020).

UNESCO (2005). *The 2005 convention on the protection and promotion of the diversity of cultural expressions*. Paris: UNESCO. Retrieved from: https://en.unesco.org/creativity/convention/texts (Accessed on 17 September 2020).

UNESCO (2013). *Introducing cultural heritage into the sustainable development agenda*. United Nations Educational, Scientific and Cultural Organisation. Retrieved from: http://www.unesco.org/new/fileadmin/MULTIMEDIA/HQ/CLT/images/HeritageENG.pdf (Accessed on 11 June 2020).

UNESCO (2018a). *Operational directives for the implementation of the convention for the safeguarding of the intangible cultural heritage*. Paris: UNESCO. Retrieved from: https://ich.unesco.org/en/directives (Accessed on 12 September 2020).

UNESCO (2018b). *UNESCO creative cities programme for sustainable development*. Paris: UNESCO. Retrieved from: https://unesdoc.unesco.org/ark:/48223/pf0000264238 (Accessed on 10 September 2020).

UNESCO CDIS (2020). UNESCO culture for development indicators (CDIS). Retrieved from: https://en.unesco.org/creativity/activities/cdis (Accessed on 17 September 2020).

UNESCO ICH (2020). Art of Neapolitan 'Pizzaiuolo'. Retrieved from: https://ich.unesco.org/en/RL/art-of-neapolitan-pizzaiuolo-00722 (Accessed on 15 September 2020).

UNESCO ICH Convention (2006). Convention for the safeguarding of the intangible cultural heritage. Retrieved from: https://ich.unesco.org/en/convention (Accessed on 15 September 2020).

van Oers, R. (2015). Cultural heritage management and sustainability. In M.-T. Albert (Ed.), *Perceptions of sustainability in heritage studies* (pp. 189–202). Berlin: De Gruyter.

van Westering, J. (1999). Heritage and gastronomy: The pursuits of the 'new tourist'. *International Journal of Heritage Studies*, 5(2), 75–81.

Vivero-Pol, J. L., Ferrando, T., De Schutter, O., & Mattei, U. (Eds.). (2019). *Routledge handbook of food as a commons*. Abingdon and New York: Routledge.

Vonthron S., Perrin C., & Soulard C.-T. (2020). Foodscape: A scoping review and a research agenda for food security-related studies. *PLoS ONE*, 15(5), e0233218. https://doi.org/10.1371/journal.pone.0233218.

Watson, J. L., & Caldwell, M. L. (2005). Introduction. In J. L. Watson & M. L. Caldwell (Eds.), *The cultural politics of food and eating: A reader* (pp. 1–10). Malden, MA: Blackwell Pub.

Watson, J. L., & Klein, J. A. (2016). Introduction: Anthropology food and modern life. In J. A. Klein & J. L. Watson (Eds.), *The handbook of food and anthropology* (pp. 1–27). London: Bloomsbury Academic.

Wheeler, S. M. (2012). *climate change and social ecology: A new perspective on the climate challenge*. London and New York: Routledge.

WHO (2020). *Constitution of the World Health Organization*. Retrieved from: https://www.who.int/about/who-we-are/constitution (Accessed on 10 September 2020).

Wippoo, M., & van Dijk, D. (2016). *Blueprint of toolkit for co-creation*. BigPicnic Deliverable D2.1 London: BGCI. Retrieved from: https://www.bigpicnic.net/media/documents/BigPicnic_-_D2.1_Blueprint_of_toolkit_for_co-creation.pdf (Accessed on 10 May 2020).

WSFS (2009). *Declaration of the world summit on food security*. WSFS 2009/2. Retrieved from: http://www.fao.org/wsfs/wsfs-list-documents/en/ (Accessed on 2 September 2020).

23 The search for virtue

Sustainability and systemic protection of agricultural heritage

Isabelle Anatole-Gabriel

Historical and intellectual development of sustainability in the United Nations context

The search for sustainability has been a motto of the United Nations system since the publication in 1987 of the Brundtland Report entitled 'Our common future'. It was first applied to the Environment to address threats posed by uncontrolled industrial development on natural resources, it then pervaded most of the United Nations agencies and their respective programmes to give birth to a new paradigm of action: sustainable development. In the 1990s, annual reports from United Nations Development Programme contributed to unfold theoretically and programmatically the sustainable development paradigm within the UN mandate, including UNESCO's specific fields of action, such as Education and Culture.[1]

Although debates on the contribution of culture to development go back to the seventies, it is only during the nineties that a strong internal paradigmatic shift allowed to address culture no more as a mere additional – and not priority – factor contributing to the development once all others, such as health, employment and education, have been fulfilled but as the objective of development. It has been argued that the lack of cultural dimension in the development programmes would explain their failures in the long run. Subsequently, a strong advocacy movement arose to include Culture into the global development agenda in order for development projects to be better aligned with the actual needs of communities. Major UNESCO reports paved the way to the encounter between development, sustainability and culture. The report of the World Commission on Culture and Development, *Our creative Diversity* (1995), followed by the two *World Culture Reports* (1998, 2000), explained the operational relationships between core societal values such as identity and diversity and societies' capacities to accept and integrate changes. Heritage and cultural industries were identified as the main vectors of these values and became UNESCO's main Culture programmes.

On the assumption that heritage is a source of identity and an expression of diversity, heritage's contribution to ensuring the sustainability of cultures became an uncontested assertion. The main objective of sustainable heritage preservation policies was to ensure sufficient funding allowing people to continue enjoying a major vector of identity and diversity. The rationale for its physical conservation subsequently changed, giving away the inherited European Enlightenment concept of cultural heritage as intellectual and aesthetic achievements of the past. Sustainability of heritage meant its physical preservation in the long run.

The 1972 Convention for the protection of cultural and natural heritage (the 1972 Convention, UNESCO, 1972) was created to ensure the physical preservation of heritage with Outstanding Universal Values. It is a unique international legal instrument for the protection of both cultural and natural heritage and the most universally ratified international legal instrument. As a matter of fact, since the creation of UNESCO, heritage has been a field of international commitment and geopolitical consensus. The highly visible and popular Lists of World Heritage assets, whether tangible (the 1972 Convention) or intangible (the 2003 Convention, UNESCO, 2003b) have become useful soft power instruments, compared to the still highly disputed field of creative industries (the 2005 Convention). This consensual quality explains why world heritage is the only cultural sector mentioned in the 2015 United Nations Sustainable Development Goals (UN SDG) adopted for the coming 15 years, until 2030, to respond to long-lasting as well as new global threats (United Nations, 2015c). The 1972 World Heritage Convention with more than 1,000 sites (1,121 in December 2020) in 167 countries appears a powerful and ready-made instrument to address global threats such as the loss of biodiversity, development pressure and climate change. Recent studies have indeed underlined the importance of world heritage properties in vital phenomena such as storing carbon and providing vital ecosystem services or in being laboratories to test resilient management strategies to face sea-level rise, habitat shifts, more frequent extreme weather events, uncontrolled urbanisation and rural emigration (UNESCO, 2007). Contributing to UN SDG has required specific policy-oriented action, and in that aim, the integration of a sustainable development perspective into the processes of the World Heritage Convention was considered an adequate response.

In 2015, the General Assembly of States Parties to the World Heritage Convention adopted a Policy document that took stock of all policies developed to address global challenges in the field of World Heritage (UNESCO, 2015b). The Policy defined three dimensions of sustainable development: (a) environmental sustainability, (b) inclusive social development and inclusive economic development and (c) the fostering of peace and security. The introductory paragraph of the Policy states that:

> with a view to ensuring policy coherence with the UN sustainable development agenda [...] States Parties should ensure an appropriate and equitable balance between conservation, sustainability and development, so that World Heritage Properties can be protected through appropriate activities contributing to the social and economic development and the quality of life of our communities.

Activities to achieve this balance are further detailed for each of the three dimensions[2]

Although introducing new vocabulary into the heritage preservation field, these policy recommendations to operationalise sustainability in World Heritage properties management remain abstract and prescriptive. In practice, actions and policies are aimed, on the one hand, to ensure the physical sustainability of heritage and on the other to favour heritage as a tourism asset. Despite the genuine interest in the subject, the contribution of heritage to sustainability remains unclear and its physical conservation is the main objective of updated international policies.

Much literature has already been written in the last decade on heritage and sustainability, especially on world heritage and sustainability (Barthel-Bouchier, 2013) discussing and presenting the main arguments of the institutional debates as well as its limits and contradictions. As a matter of fact, allocation of funds to cultural activities including heritage preservation is still described as proof of achievement of a 'cultural turn' within the UN and of the key role of Culture in sustainable development (Labadi, 2020).

The intention of this paper is not to discuss further the present place that heritage conservation occupies in the sustainable development equation driving international policies – assuming it will be covered by other chapters – but to build upon empirical studies on a specific category of heritage sites to put forward a new proposal for applied research led by the idea of virtue. Virtue – the word is remote from the heritage field. Referring to a code of behaviour, virtue sounds conservative and slightly reactionary. Referring to processes, it means efficiency and power. In that sense, a virtuous approach might prompt to reorganise innovatively the various components of a complex equation so that it would free an unexpected solution. Indeed, the main result of the literature mentioned above demonstrates clearly that sustainability is a multifactorial process.

The specific category of agricultural heritage addresses the challenges posited by the sustainable development paradigm and at the same time exemplifies this particular virtuous approach. It provides multifarious case studies allowing to forge an approach based on synergies between various international standards of protection matching the common sustainability pillars.

The 1972 Convention and the sustainable protection of agricultural heritage: successes, shortfalls and future developments.

Agricultural landscapes and cultural landscapes

In the framework of the 1972 World Heritage Convention, agricultural heritage belongs to the category of cultural landscapes (UNESCO, 2003a). Although the Convention recognises the way in which people interact with nature,[3] it was not before 1992 that cultural landscapes were formally integrated into the World Heritage List as a new category of sites. While the political rise of indigenous people on the international stage coincides with environmental concerns and the very last phase of decolonisation, mainly in the Pacific region, international experts acknowledged the need to include values granted by indigenous people to specific landscapes which directly contributed to protecting biological diversity by capitalising on diverse cultural knowledge. The cultural landscape's invention was therefore mainly motivated by the need to get away from a monumental Eurocentric approach of heritage and favour a more anthropological understanding of heritage which will ultimately compensate for the growing unbalance of representation on the World Heritage List between Europe and the rest of the world. If various political elements therefore supported the integration of this new category within the mechanism of the World Heritage Convention, this decision was a milestone achievement in many ways. Cultural landscapes embrace not only the recognition of the diversity of interactions between humankind and its natural environment as well as the acceptance of the living heritage of indigenous people and of spiritual relationships to nature, but also it opened ways to ensure maintenance of biological diversity through cultural diversity.

As set out in the *Operational Guidelines* of the 1972 Convention, cultural landscapes are 'illustrative of the evolution of human society and settlement over time, under the influence of the physical constraints and/or opportunities presented by their natural environment and of successive social, economic and cultural forces, both external and internal' (UNESCO, 2015a, p. 83). They 'often reflect specific techniques of sustainable land use' and 'protection of cultural landscapes can contribute to modern techniques of sustainable land use and can maintain or enhance natural values in the landscape' (p. 83). Recognition of traditional forms of sustainable land use which produced cultural landscapes opened a conceptual renewal of the conservation field and practices, and it can then be argued that this major step paved the way towards integrating the idea and notion of sustainability into the World Heritage Convention.

To date, 103 properties on the World Heritage List are cultural landscapes. Africa has the highest percentage (16.1%) of cultural landscapes in all regions although this figure is relative because of the limited number of sites for the region on the List. In Europe, which has 514 registered sites, 52 are cultural landscapes, representing 10.3% of the total number. Asia, for its part, accounts for 9.5% of cultural landscapes. Cultural landscapes appear to be relatively evenly distributed across regions. Among the 103 cultural landscapes inscribed on the World Heritage List, more than 60% have agricultural and/or pastoral components. Agricultural landscapes play a vital role in providing food or cash resources, *i.e.*, socioeconomic revenues, to farmers' communities. Although attention to social and economic forces is clearly mentioned in the definition provided by the 1972 Convention, we question the relevance of the World Heritage Convention to ensure the protection of social and economic forces as a 'factor of production of cultural landscapes'. For these agricultural landscapes, the main challenge of World Heritage recognition is the ability to maintain production and farming activities within a framework led by the notion of outstanding universal value defining elements and attributes that must be kept unchanged.

Asian rice fields and European vineyards: examples of challenges for sustainable agricultural land use

The rice terraces of the Ifugao region, located in the mountain region of northern Philippines, were among the first agricultural landscape inscribed on the World Heritage List in 1995. The justification for the inscription of the Rice Terraces of the Philippine Cordilleras makes redundant use of the sustainability notion. It mentioned that,

> The rice terraces are a dramatic testimony to a community's sustainable and primarily communal system of rice production, based on harvesting water from the forest clad mountain tops and creating stone terraces and ponds, a system that has survived for two millennia.

It further explained that

> The rice terraces are a memorial to the history and labour of more than a thousand generations of small-scale farmers who, working together as a community, have created a landscape based on a delicate and sustainable use of natural resources.

The rice terraces form a sustainable system in the full sense. It is a cooperative system since the farmers help each other during the various stages of agricultural life. Buildings and lands are bequeathed to families and villages and transmitted from generation to generation. The conservation of the terraces is based on this cooperative system specific to the indigenous communities living on the site. If farming systems have shaped remarkable landscapes whose value has been recognised by their World Heritage status, however, their social, economic and environmental sustainability rely on cultural knowledge and social systems which are not physical attributes as defined by the Convention. In order to compensate for this deficiency, the *Operational Guidelines* for the implementation of the 1972 Convention integrated the involvement of local populations in the nomination and the management mechanisms.[4] The main objective of management based on local knowledge is to support change while preserving universal values at a time when anthropic, climatic and development pressures are multiplying in rural areas. Was this enough to ensure sustainability?

Even if the inclusion of communities is foreseen, the lack of attention to other key components, such as the products, might result in major conservation problems. In the case of the rice terraces of the Ifugao region, under the pressure of tourism that targeted the harvesting of the rice fields by the local communities and the desire for the profitability of the international label of the rice fields, the government sponsored the use of new high-yield varieties of rice and the endemic variety, Tinawon, was abandoned step by step over the years after the inscription. In multiplying annual harvests, the new variety brought pests, farmers had to use chemical fertilisers. Ultimately, the combined pressure of numerous harvests and pests partly disrupted the fragile ecosystem of the terraces and traditional agricultural practices. The increasing demand for wood crafts by tourists and the export market caused watershed deforestation, thus diminishing water supply to the rice fields. However, the economic benefits of tourism did not filter down to the farmers who maintain the rice terraces. This cascade of negative effects resulted in the inscription of Rice Terraces on the List of World Heritage in Danger in 2001. In order to keep the World Heritage site status attached to the physical structures of the landscape, the government offered grants to restore the terraces and local farmers' communities contributed less and less to the maintenance of the site causing the loss of the traditional cooperative labour required to maintain the irrigation system and terrace walls: the entire cultural, social, environmental and economic ecosystem at the origin of the landscapes was jostled, partly destroyed, by the lack of adequate protection of one component of the agricultural landscape (SITMo, 2008; Anatole-Gabriel, 2016).

Lessons have been drawn from the exemplary case of the Ifugao Rice Terraces as evidenced by the analysis made of the cultural landscape of the province of Bali, inscribed in 2012, and whose title – the Subak systems as a manifestation of the Tri Hita Karana philosophy – attests to the attention paid to the social system[5]. The heritage statement of the landscape explained that

> None of the component parts is under threat but the terraced landscape is highly vulnerable to a range of social and economic changes, such as changes in agricultural practices and increasing tourism pressures. The management system will need to provide support to sustain the traditional systems and to provide benefits that will allow farmers to stay on the land.[6]

However, beyond the declarative, the capacity of the heritage protection system to regulate and maintain a complex system of interacting phenomena, many of whose components are outside its scope, is a challenge. The case of European vineyards, for which local social and economic actors have triggered their inscription on the World Heritage

List often against the opinion of heritage professional experts, provides an example of complementary analysis.

The European vineyards

Soon after the inclusion of cultural landscapes into the categories of world heritage, European countries proposed renowned vineyards for inscription. Since the mid-1990s, their number on the list has been steadily increasing and has become a competitive designation used to boost international recognition of commercial brandings, such as the Italian Prosecco. The first vineyard designated as World Heritage in 1997 was already Italian. The agricultural landscape of Portovenere, Cinque Terra and the Islands was inscribed

> as a cultural landscape representing the harmonious interaction between people and nature to produce a landscape of exceptional scenic quality that illustrates a traditional way of life that has existed for a thousand years and continues to play an important socio-economic role in the life of the community.

The inscription of the French vineyards of Jurisdiction de Saint Emilion followed in 1999 and acknowledged 'an exceptional landscape devoted entirely to wine growing, with many fine historic monuments in its towns and villages'. Since these first vineyards, many others have been inscribed: Haut-Douro and Pico in Portugal in 2001 and 2004, Tokaj in Hungary in 2002, Lavaux in Switzerland in 2007, Langhe-Roero and Monferrato in the Piedmont region of Italy in 2015; Climats, Terroirs of Burgundy in France in 2016, until the highly contested inscription of Italian Conegliano and Valdobbiadene Prosecco Hills in 2019[7].

The case of the renowned European vineyards is interesting to compare to the rice fields of Southeast Asia as their heritage value is linked to agricultural practices that have been similarly in use for centuries. However, renowned vineyards deploy a system of production of the agricultural resource driven by a scarcity-based market economy maximising economic income. Wine production being not a food resource for the farming communities, the maintenance of vineyard cultural landscapes is entirely conditioned by the economic valuation of the product. However, is the importance of the economic factor recognised in the heritage value?

The reputation of the wine is mentioned in both Portuguese world heritage files, Alto Douro and Pico. For Alto Douro Wine Region, the analysis of the heritage value mentions that

> Wine has been produced by the traditional landowners in the Alto Douro region for some 2000 years. Since 1756, its main product, port wine, has been world famous for its quality. This long tradition of viticulture has produced a cultural landscape of outstanding beauty that reflects its technological, social and economical evolution

while for the landscape of the Pico Island vineyards culture, the heritage value – 'the extraordinary beautiful human-made landscape of small, stone-walled fields is a testimony to generations of small-scale farmers who, in a hostile environment, created a sustainable living and much-valued wine' – is complemented by the legal national wine protection, 'Laws protecting the vineyards and wine production standards on Pico Island were passed in 1980'. A rather general reference to the quality and reputation of the wines is made in the other world heritage files, such as that of Tokaj, which resolutely emphasises the quality of the wines –

> this complex network of vineyards, farms, villages and small towns with its historical labyrinth of wine cellars, illustrates all the facets of the production of the famous Tokaj wines, whose quality and management have been strictly controlled for almost three centuries

but in the spirit of a universally shared reputation. There is no characterisation of the specific qualities of the wine produced for the vineyard of Lavaux, apart from the mention of a 'highly appreciated wine' and no reference to wine is made in the file of the Vineyard Landscape of Piedmont. Despite the acknowledgement of economic forces in the making of the cultural landscapes and the crucial role of the agricultural product in the maintenance of the vineyards, wine is almost never mentioned in the nomination files.

When we study the reputation of wines, *i.e.* their market value, before the recognition of their world heritage value, we can see that the inscription has come to crown a predominant position on the wine market in the world. This is particularly the case for the Hungarian, French and Italian vineyards: Tokay (Hungary), Vineyard Landscape of Piedmont: Langhe-Roero and Monferrato (Italy), Jurisdiction of Saint Emilion (France), the Climats, Terroirs of Burgundy and Champagne Hillsides, Houses and Cellars (France). For other vineyards, the inscription has helped to strengthen the wine's reputation. This is the case for the vineyards of Haut-Douro (Portugal), Lavaux (Switzerland), Wachau (Austria) and Val d'Orcia (Italy). When researching the possible effects of World Heritage listing on the reputation of wines on the wine market, it appeared that the impact of World Heritage listing on the reputation of wines is not very significant in the short and medium term. It can, as the case of Pico seems to indicate, have an impact in the long term. The direct impact of heritage value on the absolute market value of the wine depends mainly on the ability to link the wine reputation, i.e. economic value, to the heritage value of the vineyard landscapes. While in the case of some vineyards, the reputation of the wines promotes and reinforces the physical sustainability of the World Heritage site; in the majority of cases, it can be seen that socioeconomic sustainability is disconnected from the heritage distinction (Anatole-Gabriel, 2019).

It is therefore imperative to reconcile the heritage protection of an agricultural landscape with the socioeconomic sustainability of all its components including the product that it produces. There is therefore a need to further the research agenda on theoretical models based on empirical analysis to propose innovative responses to global threats posed to Heritage, as well as to meet expectations embedded in the concept of sustainable Heritage. The next section of this paper presents the theoretical hypothesis that sustains the development of synergies between the various international standards of protection matching the common sustainability pillars and explains why an enhanced and systemic protection through multi-designation will achieve holistic sustainability of agricultural landscapes.

Multi-designation and systemic protection of agricultural landscapes: the heritage leverage

Going beyond heritage valuation limited to its physical sustainability requests a conceptual shift. The development of Cultural Economics, which emerged in the 2000s, allowed this conceptual shift. Within the cultural field, a specific literature flourished aiming at identifying internal vectors of sustainability, especially economic and social vectors.

Sophisticated models and new notions drawn from Economics, such as notions of cultural capital and positive amenities, have opened the door for a robust reconceptualisation of heritage value linkages with sustainability (Throsby, 2002, 2016).

Cultural heritage as fourth pillar?

Sustainability has been conceptualised as reliant on three pillars: social (people), environmental (planet), and economic (profit). Authors argue that

> while there is some attention for the positive externalities of culture in terms of environmental sustainability, if culture is regarded as the fourth pillar of sustainability, the viability of the sector is an inherent part of sustainability and not a mere external contribution to it.
>
> (de Beukelaer & Spence, 2019, p. 164)

This theoretical assertion has been particularly exemplified by agricultural heritage sites whose specific sustainable land use techniques, mastered by people know-how (social sustainability), take into account the characteristics and limits of the natural resources (environmental sustainability) and can maintain, and even enhance, the associated cultural values (cultural sustainability) while producing subsistence and livelihood for local populations (economic sustainability). In a nutshell, some agricultural landscapes have demonstrated their potential to promote sustainable livelihoods while contributing to biodiversity. Therefore, agricultural landscapes not only bring together the three capitals (environmental, social and economic) that constitute the three pillars of sustainability in Economics theory, but their global recognition as Heritage highlights the fourth – cultural capital.

Shortfalls of heritage tourism as valuation of cultural capital

When further exploring the link between socioeconomic and cultural sustainability, it may be recalled that cultural tourism has been proposed as an answer to the search for the economic sustainability of heritage. However, in addition to the fact that the cultural capital conceptual innovation has only been operationalised by additional income brought by tourism, cultural tourism contribution to heritage sustainability has shown its limits. The pros and cons of cultural heritage tourism – the ambiguous nature of the share of cultural tourism in the sustainability of heritage (providing revenues that eventually would sustain the physical conservation of heritage while at the same time threatening its preservation) – have been discussed extensively (Bartel-Bouchier, 2013). The new motto of *sustainable tourism* taken on board by the World Heritage agenda seems to mainly develop a narrative claiming for sustainable uses of heritage through reinforced planning and diversification of touristic offers, leaving unresolved the attribution of a decent share of the economic benefits to actual heritage stewards or, in the case of agricultural landscapes, to heritage makers.

Ensuring the balance and virtuous complementarity of all capitals identified by economic theory may seem like a major challenge. Despite its attempt to encompass human interactions with nature at large, the 1972 Convention fails to fully recognise and consequently protect specific core features of agricultural landscapes. Such features relating to other capitals might fall under the scope of other conventional instruments. Therefore, cooperation developed with other instruments aimed at protecting the various capitals of agricultural landscapes represents a positive evolution.

Cooperation with other protection systems

The use of multiple designations contributes to protecting the values of different capitals including the intellectual property rights, which secure the market value of the agricultural product. There are at least three main systems complementary to the World Heritage Convention: the UNESCO Biosphere Reserves, the Globally Important Agricultural Heritage Systems (GIAHS) coordinated by the Food and Agriculture Organization of the United Nations (FAO), and the Geographical Indication (GI) system.

In 2002, the FAO established an international partnership to ensure the conservation and progressive management of remarkable agricultural systems, labelled GIAHS. To date, 62 systems have been designated in 22 countries. These systems value local natural resources and their management is based on practices adapted to the local context. This labelling does not distinguish the cultural character of the landscape or the know-how but the technical knowledge of production, maintenance and care of agricultural lands with attention to the livelihoods that communities derive from their work. Typically, these traditional farming systems are selected based on the ingenuity of their management systems, their importance to local food security, their high level of agricultural biodiversity and associated biodiversity, as well as their role as conservation reserves of indigenous knowledge. According to FAO, GIAHS is characterised by the dynamic conservation and sustainable management of outstanding land-use systems, with a focus on balancing conservation, adaptation and economic development, and GIAHS represent not only outstanding natural landscapes but also agricultural practices that generate rural livelihoods while uniquely combining biodiversity, resilient ecosystems, traditions and innovations as cited by Koohafkan and Altieri (2011).

A few local and national authorities have embarked on this process and some World Heritage sites benefit already from the multiplier effects and source of sustainability of multiple labelling. Rice fields were among the first agricultural landscapes to be labelled GIAHS, such as the Rice Terraces of Ifugao in the Philippines, in 2011, six years after their inscription on the World Heritage List. China, but also Japan, use GIAHS labelling in a preferential way for their agricultural landscapes, as opposed to the World Heritage or Intangible Heritage labelling that may be applied for at a later date. The cultural landscape of the Honghe Hani Rice Terraces obtained the GIAHS label in 2010 and the World Heritage status in 2013 while the Jasmine and Tea Culture System of Fuzhou City obtained the GIAHS label in 2014 but is not yet inscribed on the World Heritage List. Europe makes much less use of this label and it is mainly the countries of Mediterranean Europe (Spain, Italy and Portugal) that apply to it, sometimes due to failure to register on the World Heritage List, as was the case of the Añana salt production system in Spain, registered with GIAHS in 2017.

Protecting intellectual property rights with geographical indications

In addition to the protection of the environmental and sociotechnical capitals, intellectual property rights, protecting the origin, the name of the product and the know-how of the production, complete the mechanisms available for agricultural landscapes.

Names of agricultural products, such as the Tea Darjeeling or the Champagne wine, used as powerful commercial labels represent a substantial part of the economic capital drawn from a global market economy. Intellectual property rights are secured through GI or appellation of origin. The appellation of origin and the GI identify an intellectual right rewarding the producers–creators of the reputation of the country of origin. It acknowledges and guarantees the local production of the product and mainstreams financial revenues to their owners. It appeared in France for winemaking in 1919 as Appellation d'Origine Protégée (AOP) and extended to Europe in 1992. The World Trade Organisation created the GI in 1995. GI is nowadays widely adopted by countries to protect their heritage and know-how.

In 2015, the Climats, Terroirs of Burgundy vineyards were inscribed on the World Heritage List (Figure 23.1). The success of the economic model of the Burgundy vineyards, which gather the most expensive wines of the world, is based on the one hand on the invention in 1936 of the concept of protected designation of origin which established the link to the geographical origin[8], and, on the other hand, on a model of direct distribution between the winemakers and the producers, and the market. As underlined above, synergies between intellectual property rights protection and heritage recognition are explicit in all vineyard landscapes listed on the World Heritage List. Each of these winemaking territories has installed, prior to the World Heritage designation, a legal system for the protection of the winemaking know-how and for the protection of the land that led to the designation of the wines themselves. These legal protections ensure property rights to the local winemakers, and consequently, a say in the global market to defend and protect winegrowers and winemakers' sufficient economic resources.

Remote from Europe, the World Heritage Colombian Coffee Cultural Landscapes are also a demonstrative successful case of the systemic approach linking the protection of the various cultural, environmental, social and economic capitals. The Coffee Cultural landscapes cover a vast territory including, in addition to plantations, 411 rural districts and 24,000 agricultural holdings supporting more than 80,000 people. The social

Figure 23.1 The Corton Hill in the World Heritage property Les Climats du vignoble de Bourgogne, Burgundy, France. © Michel Joly.

and cultural organisation of the coffee producers is at the origin of the production of high-quality coffees, using the tools of global marketing of highly differentiated terroir products to gain market share. The specific community model of coffee production, which encompasses not only the know-how of coffee growing but also the market organisation, has become one of the attributes of the World Heritage nomination. This is a unique case where the social capital is fully acknowledged as part of the heritage value. The Colombian Coffee Landscapes have obtained an Appellation d'Origine Controlée for the coffee brand in 2005, a GI in Europe in 2007 and the World Heritage status in 2011. The cultural concept of Coffee Cultural Landscapes combined with the property rights protection of the product ensures a worldwide fame to the coffees originating from this region. In Vietnam, the Complex of Hué Monuments has been a World Heritage Site since 1993, as an outstanding example of eastern feudal capital and a tribute to its history as a political, cultural and religious centre during the Nguyen Dynasty. This dynasty attracted a particular craft, illustrated among other things by the conical hat of Hué adorned with poems. The hat has been protected since 2010 as GI, which terms of reference refer clearly to the heritage value recognition of Hué Imperial City so as to enhance the fame of the hat. Conversely, to this day, no national law or any property rights protect the Tinawon endemic rice variety, which is marketed by a Seattle-based US company, Eight Wonders, which promotes the world heritage status as a market engine.

Attention given to the intangible components in heritage valuation of physical sites might well fall under the scope of the UNESCO Convention for the safeguarding of the intangible cultural heritage. Since its adoption in 2003, a substantial consideration of the intangible aspects of sites registered under the 1972 Convention has made it possible to draw attention to social and economic practices such as traditional craftsmanship skills or, as in the case of the Bali and Ifugao rice terraces, to rituals and festive events closely related to the maintenance of cultural landscapes. This is also the case for the Georgian method of traditional winemaking in kvevris, an oval-shaped clay container that is used to vinify, age and store wine, which was inscribed in 2013 on the Representative List of the Intangible Heritage of Humanity attached to the 2003 Convention. The inscription of Georgia's vineyards on the World Heritage List is still under consideration to complete the protection of this traditional Georgian technique.

Conclusion

A systemic model driven by the economic notion of capital is emerging from these diverse experiences. For the most successful cases, heritage protection intervenes to strengthen the holistic protection of all components of agricultural landscapes, thus leading to optimal sustainability of resources and of their controlled exploitation. The critical idea to conceptualise theoretical relationships between heritage and sustainability would be to develop more systematic synergies between the various international standards of protection matching the sustainability four pillars. Indeed, it is through the combination of designation, able to ensure the conservation and protection of environmental, cultural, social and sociotechnical and economic capitals in a complementary and virtuous way, that a sustainable approach can truly be operationalised. However, the added value of the heritage approach in triggering a virtuous process of multi-designation cannot be underrated. The assets of heritage protection, and in particular of World Heritage, is to be a source of participative governance for all stakeholders in a territory, a factor in attracting investment and global visibility and, above all, a source of pride, identity and local belonging, which constitutes the last pillar of sustainability – the human pillar.

Post scriptum: Most of the agricultural landscape study cases can be found in the publication *The heritage value of terroir economies* (*La valeur patrimoniale des économies de terroir*, Editions Universitaires de Dijon, 2016). The research project developing the theoretical framework presented in this paper will be led by the transdisciplinary research laboratory HERITAGES of the Cergy Paris University (CYU), together with a consortium of leading universities, in the framework of the European research programme Horizon Europe.

Notes

1 Over the last two decades, authors have underlined the role of international organisations in the making of heritage objects (tangible and intangible) especially through the establishment of List associated with normative frameworks aiming at their protection. The neologism of heritisation (*patrimonialisation* in French) was created to account for and describe the process of entrusting specific value and legal status to a constantly growing ensemble that constitutes the Heritage (of Mankind, of Humanity…). See Kirshenblatt-Gimblett (2004), Smith and Akagawa (2009), Heinich (2009), Harrison (2013) *Heritage: Critical Approaches*.

2 For Environmental Sustainability, actions consist of: (a) protecting biological and cultural diversity and ecosystem services and benefits and (b) strengthening resilience to natural hazards and climate change. For Inclusive Social Development, among other objectives: (a) in enhancing quality of life and well-being and (b) respecting, consulting and involving indigenous people and local communities. For Inclusive Economic Development: (a) in ensuring growth, employment, income and livelihoods, (b) promoting economic investment and quality tourism, and (c) strengthening capacity building, innovation and local entrepreneurship.

3 Cultural landscapes are cultural properties and represent *"the combined works of nature and of man"* designated in the Article 1 of the Convention which stipulates

> For the purposes of this Convention, the following shall be considered as "cultural heritage" […] - sites: works of man or the combined works of nature and of man, and areas including archaeological sites which are of Outstanding Universal Value from the historical, aesthetic, ethnological or anthropological points of view.

> See Operational Guidelines for the Implementation of the World Heritage Convention - Guidelines on the inscription of specific types of properties. Annex 3.

4 Paragraph 12 states:

> General criteria for protection and management are equally applicable to cultural landscapes. It is important that due attention be paid to the full range of values represented in the landscape, both cultural and natural. The nominations should be prepared in collaboration with and the full approval of local communities.

5 See Lansing and de Vet (2012), and Lansing et al. (2009).

6 All heritage statements or Statement of Outstanding Value of the world properties mentionned in this paper are accessible for each respective World Heritage properties on the World Heritage Web site at http://whc.unesco.org/en/list/

7 Thirteen European vineyards have been included in the List of World Heritage Cultural Landscapes over a period of 22 years. The request for the inscription of Conegliano and Valdobbiadene Prosecco Hills on the World Heritage List in 2018 was rejected by the international heritage experts but inscribed one year later after an intensive lobbying of the Italian government.

8 A legal translation of the notion of *terroir*, the appellation of origin, aims to protect the quality of a product resulting from its geographical origin and, implicitly, from the cultural know-how associated. See Jacquet (2019).

References

Anatole-Gabriel, I. (Ed.) (2016). *La valeur patrimoniale des économies de terroir*. Dijon: Editions Universitaires de Dijon.

Anatole-Gabriel, I. (2019). World heritage listing: An actor of economic and cultural prescription? *International conference on wine market and cultures of consumption*. Honk Kong Polytechnic University, UNESCO Chair on "Culture and Wine Traditions: Under Publication."

Barthel-Bouchier, D. (2013). *Cultural heritage and the challenge of sustainability*. Walnut Creek, CA: Left Coast Press.

De Beukelaer, C., & Spence, K.-M. (2019). *Global cultural economy*. London: Routledge.

Harrison, R. (2013). *Heritage: Critical approaches*. London: Routledge.

Heinich, N. (2009). La fabrique du patrimoine - De la cathédrale à la petite cuillère. *Ethnologie de la France, 35*. Paris: Editions de la Maison des Sciences de l'Homme.

Jacquet, O. (2019). The Development and promotion of controlled designations of origin (Appellations d'Origine Contrôlées) in Burgundy: The recognition of Terroir Wines (1884–1970). In S. A. Conca Messina, S. Le Bras, P. Tedeschi, & M. Vaquero Piñeiro, (Eds.), *A history of wine in Europe, 19th to 20th centuries – markets, trade and regulation of quality* (pp. 67–91). London: Palgrave Macmillan.

Kirshenblatt-Gimblett, B. (2004). Intangible heritage as metacultural production. *Museum International, 56*(1–2), 52–65.

Koohafkan, P., & Altieri, M. A. (2011). *A methodological framework for the dynamic conservation of agricultural heritage systems*. Roma: FAO Publishing. [Online] Retrieved from: http://www.fao.org/family-farming/detail/fr/c/285324/ (Accessed on 14 May 2021).

Labadi, S. (Ed.) (2020). *The cultural turn in international aid – Impacts and challenges for heritage and the creative industries*. London: Routledge.

Lansing, J. S., Downey, S. S., Murray, P. C., Jansen, M., & Schoenfelder, J. W. (2009). A robust budding model of Balinese water temple networks. *World Archaeology, 4*(1), 110–131.

Lansing, J. S., & de Vet, T. A. (2012). The functional role of Balinese Water Temples: A response to critics. *Human Ecology, 40*(3), 453–467.

SITMo (Save the Ifugao Terraces Mouvement). (2008). *IMPACT: The effects of tourism on culture and the environment in Asia and the Pacific: Sustainable tourism and the preservation of the World Heritage site of the Ifugao Rice Terraces, Philippines*. Bangkok: UNESCO Publishing.

Smith, L., & Akagawa, N. (Eds.) (2009). *Intangible heritage*. London: Routledge.

Throsby, D. (2002). *Economics and culture*. Cambridge: Cambridge University Press.

Throsby, D. (2016). Les économies de terroir comme atouts économiques et culturels: valeur, valorisation et durabilité. In I. Anatole-Gabriel (Ed.), *La valeur patrimoniale des économies de terroir* (pp. 79–87). Dijon: Editions Universitaires de Dijon.

UNDP. (1987). *Report of the world commission on environment and development: Our common future*. New York: UNDP Publishing. [Online] Retrieved from: https://sustainabledevelopment.un-.org/content/documents/5987our-common-future.pdf (Accessed on 03 May 2021).

UNDP. (1990–2020). *Human development reports*. New York: UNDP Publishing. [Online] Retrieved from: http://www.hdr.undp.org/en/global-reports (Accessed on 03 May 2021).

UNESCO (1972). *Convention concerning the protection of the world cultural and natural heritage*. Adopted by the General Conference at its Seventeenth session Paris, 16 November 1972. [Online] Retrieved from: https://whc.unesco.org/en/conventiontext/. (Accessed on 03 May 2021).

UNESCO. (1995). *Our creative diversity. Report of the world commission on culture and development*. Paris: UNESCO Publishing. [Online] Retrieved from: https://unesdoc.unesco.org/ (Accessed on 03 May 2021).

UNESCO. (1998). *World culture report: Creativity and markets*. Paris: UNESCO Publishing. [Online] Retrieved from: https://unesdoc.unesco.org/ark:/48223/pf0000112074?posInSet=1&queryId=3daaf18c-b0e7-4345-bbad-ff7048440dce (Accessed on 03 May 2021).

UNESCO. (2000). *World culture report: Cultural diversity, conflict and pluralism.* Paris: UNESCO Publishing. [Online] Retrieved from: https://unesdoc.unesco.org/ark:/48223/pf0000121058 (Accessed on 03 May 2021).

UNESCO. (2003a). Cultural landscapes: The challenge of conservation. *World Heritage Papers Series.* 7. Paris: UNESCO Publishing. [Online] Retrieved from: http://whc.unesco.org/en/series/7/ (Accessed on 03 May 2021).

UNESCO. (2003b). *Convention for the safeguarding of the intangible cultural heritage.* Adopted by the General Conference at its thirty second session Paris, 17 October 2003. [Online] Retrieved from: https://ich.unesco.org/en/convention. (Accessed on 03 May 2021).

UNESCO. (2005). *Convention on the protection and the promotion of the diversity of cultural expressions.* Adopted by the General Conference at its thirty third session Paris, 20 October 2005. [Online] Retrieved from: https://en.unesco.org/creativity/sites/creativity/files/passeport-convention2005-web2.pdf (Accessed on 03 May 2021).

UNESCO. (2007). *Policy Document on the impacts of climate change on world heritage properties.* Paris: UNESCO Publishing. [Online] Retrieved from: https://whc.unesco.org/en/news/441/ (Accessed on 03 May 2021).

UNESCO. (2015a). *Operational guidelines for the implementation of the world heritage convention.* Paris: UNESCO Publishing. [Online] Retrieved from: http://whc.unesco.org/en/guidelines/ (Accessed on 03 May 2021).

UNESCO. (2015b). *Policy document for the integration of a sustainable development perspective into the processes of the world heritage convention.* [Online] Retrieved from: https://whc.unesco.org/en/sustainabledevelopment/ (Accessed on 03 May 2021).

United Nations (2015c). *Transforming our world: The 2030 agenda for sustainable development.* New York: United Nations. [Online] Retrieved from: https://sdgs.un.org/publications/transforming-our-world-2030-agenda-sustainable-development-17981 and https://sdgs.un.org/goals(Accessed on 03 May 2021).

Part V

Sustainable heritage-led transformation

24 A boundary approach to urban heritage

The case of Egyptian antiquity

Amr El-Husseiny and Kalliopi Fouseki

Introduction

While boundaries and borders are organising features of nation-state territories and manifest in urban environments globally for a variety of reasons and uses, boundaries have become a particular urban feature of Egyptian cities and especially surrounding sites of historic significance and other elements of urban heritage. A boundary's spatial manifestation challenges the relationship between a site and the notion of heritage, not only in its material being but also as a representation of the past (Figure 24.1).

While it can be argued that boundaries and buffer zones complement urban heritage sites by an 'amplification' of representation and meaning (Turner, 2009), such a conception may be challenged when boundaries are observed in different social and spatial contexts. Hence, the nature of boundaries of heritage discourses can be seen

Figure 24.1 The Egyptian public 'dwelling in the boundaries' of urban heritage sites in the context of historic Cairo. The boundary marks the distinction between the early 16th century complex of Sultan Al-Ghouri, a symbolic representation of Egyptian history and identity, and the public everyday life of street vendors selling mainly undergarments to passers-by.

Source: Photo by Elhusseiny, September 18, 2020.

as a question of difference and rupture rather than that of universality and continuity. This echoes with the contested and conflicting nature of heritage as reflected in its 'dissonance'; such a view highlights its conceptual complexity and interdisciplinarity (Graham et al., 2000).

This chapter aims to introduce the idea of boundaries as an analytical tool to understand heritage theories and practices. The main objective is to examine some of the boundaries that define heritage discourses; it is an attempt to explore the theoretical frontiers of heritage practices and what such locations might add to our understanding of the concept. Boundaries emerge between the theoretical purposes of heritage and the contextual sociopolitical conditions in which these practices are deployed. This is directly related to recent attempts to understand the relationship between the purposes of heritage as a practice and the public in the specific conditions of the Global South, the Middle East and, specifically, Egypt (Meskell, 1998; Exell & Rico, 2014; Langer, 2017; Hammond, 2020). A basic introduction to boundary theories from a social sciences' perspective, what boundaries are and how they are deployed, will be discussed in the first part of the chapter to set the groundwork for understanding their manifestations in the field of heritage. The second part deals with some of the established boundaries mentioned within the existing theories of heritage practices, highlighting their extensions and limitations beyond the context of their production. The final part addresses the opportunities of exploring the concept of heritage boundaries beyond the current discourse, focussing on the Global South, specifically Egypt, suggesting different sets of boundaries that may be at work and could be addressed to expand our understanding of the workings of heritage practices within a global context. While sites like Egypt may have been at the heart of the production of institutionalised antiquities and heritage practices in the modern sense throughout the 19th century (Colla, 2007), a closer investigation of its contemporary conditions may contribute to our understanding of how such concepts may have unfolded in Egypt in relation to its 'universal' understanding.

The concept of boundaries in social sciences

The field of boundary studies has expanded throughout the last century from a limited understanding of physical borders and boundaries as lines on the ground-defining nation states from a geographic perspective to a field that has shifted towards a wider understanding of the concept (Lamont & Molnár, 2002; Van Houtum, 2005; Wilson & Donnan, 2012; Kolosov & Scott, 2013). Nevertheless, the notion of boundaries has always played an integral part in social theorising through the works of Durkheim, Marx and Weber (Lamont & Molnár, 2002). The idea itself has been applied in numerous fields across social sciences referring to various concepts that generate new theoretical insights about an entire range of social processes (Lamont & Molnár, 2002). Despite its diverse uses by researchers in different areas of social sciences, there have been few attempts at a theoretical integration of the idea of boundary theories. The review produced by Lamont and Molnar, at the core of boundary studies, is considered one of the attempts to connect the theoretical similarities underlying different locations where the idea is being deployed.

While drawing interrelations between boundaries as they appear in different locations within the discipline, Lamont and Molnar attempt to distinguish between symbolic and social boundaries in their introduction to boundary theories. They define symbolic boundaries as 'conceptual distinctions' formulated by various social actors to differentiate and categorise entities be they objects, people or even time and space (Lamont & Molnár, 2002). On the other hand, social boundaries are considered the 'objectified forms' of these differences and distinctions, where material and non-material differentiations are produced in certain contexts. Both concepts seem to be related since symbolic boundaries are considered an underlying cause within the construction of social boundaries, yet the relation between them is mostly implied rather than made explicit in academic literature (Lamont & Molnár, 2002). For example, to make the differentiation clearer, any assertion in the conceptual space may be perceived as a symbolic boundary in terms of its distinction from other assertions; it becomes a social boundary as soon as such assertion is activated in the formation of a certain group that expresses such abstract idea in a concrete form of social practice. Also, a clearer definition of 'boundaries' may contribute to a better understanding of each level of boundaries and the existing connections between them.

From a sociological perspective, Andrew Abbott attempts to redefine groups by their boundaries suggesting the role of boundaries in the creation of social entities (Abbott, 1995). Accordingly, Abbott defines boundaries as the 'locations of difference' where any two adjacent points at least differ in some respect (Abbott, 1995). In that sense, boundaries are created through any sort of difference, or what he may refer to as a 'proto-boundary', that emerges in any kind of local interaction (Abbott, 1995). Hence, a boundary as an analytical tool is not limited to its contribution in terms of understanding the nature of social entities, but also the process of its emergence and continuous alteration in a 'world of events' and constant change from a processual social perspective (Abbott, 2016). Dealing with interaction and change as fundamental functions of social identity, boundaries defined in this sense are freed from any static conception of the term and deviate from the concepts of social reproduction present in the studies of class for instance. Also, by defining the nature of boundaries as locations of difference, the concept is released from its widespread understanding as a direct consequence of the existence of entities – proto-boundaries can be identified before any entity even exists through a preliminarily observed difference in a social field.

Echoing Abbott's idea of 'things of boundaries', Star and Griesemer propose the analytical tool of 'boundary objects' to discuss the work of Joseph Grinnell in the Museum of Vertebrate Zoology at the University of California Berkley from a social perspective. Boundary objects, which may be abstract or concrete, are defined as 'objects which are both plastic enough to adapt to local needs and the constraints of the several parties employing them, yet robust enough to maintain a common identity across sites' (Star & Griesemer, 1989, p. 393). The main features of such objects are identified as failing to present a strong structure in common use between different social worlds, yet strongly structured within each individual site (Star & Griesemer, 1989). Star and Griesemer study the objects of natural history research museums as boundary objects where professional, amateur, academic and lay worlds intersect, and the objects survive multiple translations between distinct groups. The introduction of this conception of boundary objects resonates strongly with David Thelen's notion on the boundaries' double nature, not necessarily only about division and differentiation but as sites of communication and negotiation for social worlds that fits the suggested definition of boundary objects (Thelen, 1999). This clarification of the nature of boundaries contributes to overcoming one of the main challenges to applying the concept in social enquiries, which lies in its metaphorical representation.

From a critical perspective, the idea of boundaries does have its limitations when it is used as an analytical tool in the humanities. The main problem with boundaries, according to Riesch, is the fact that the precise meaning and nature of what 'boundaries' are supposed to be, gets lost somewhat in the 'intuitiveness of the metaphor' (Riesch, 2010). This apparent limitation may be overcome by attempting to clearly identify the ontological nature and agency of boundary objects, be they material or non-material, and also relying on social identity and social representation theories as a strategy to provide explanations and theorising on the precise nature of boundaries and their functioning components (Riesch, 2010). In his work, Hauke Riesch demonstrates how both social representation and social identity theories may contribute to the definition of the functions of boundary objects and the dynamics of their production (Riesch, 2010).

In summary, the utilisation of boundaries as an analytical tool to understand social processes may be beneficial due to the concept's multiple qualities. First, its double nature as a categorisation and identification method is to understand social entities and their dynamics along with it being a tool to identify communication and relations between different social worlds, as in the concept of boundary objects (Figure 24.1). The second quality would be its capacity to describe change, as a location of difference, rather than defining social entities and practices in a static manner. Its applicability to both the symbolic and social, and the conceptual and material, as well as drawing relations between them contributes to the understanding of the complexity of causal relations in a critical sense rather than a classic simple chain of causality. Finally, the adaptability of the concept across disciplinary boundaries fitting different narratives and modes of inquiry contributes to a multidisciplinary approach to questions allowing the coexistence of multiple viewpoints and drawing on their relations. Such qualities relate to various challenges that are discussed within heritage discourse, especially those related to heritage as a social process in non-Western contexts and their representation, as will be discussed in the next section.

Boundaries in the heritage discourse

The idea of boundaries is not exactly new to the field of heritage studies. Of course, it is widely discussed in terms of heritage site management when dealing with buffer zones and a site's physical borders. In Egypt, these date back to the early practices of the *Comité* in Egypt and are considered the spatial manifestations of underlying conceptual aspects of heritage in the Egyptian context (Figure 24.2).

While urban boundaries are erected around heritage sites in Egypt, their symbolic foundations may be produced and manifested elsewhere within global heritage discourses. On a theoretical level in relation to heritage as a discourse, the idea of boundaries does appear in some of the works on heritage theories. Smith utilises the idea of 'boundaries' in its symbolic sense in her description of the Authoritative Heritage Discourse (AHD) (Smith, 2006, p. 12). According to Smith, boundaries are constructed by the AHD as a necessity for its self-realisation being an established set of values and meanings that constitute how we think about heritage and the past (Smith, 2006). The first set of boundaries is produced where 'heritage' is associated with the past and removed from its present context due to its 'innate qualities', according to Smith. The second instance of their production is when heritage has been established as a profession and a certain expert group is assigned as the 'stewards of the past' (Smith, 2006, p. 29). Heritage professionals are established as a social entity and their authority operates as

Figure 24.2 The mosque of Al-Saleh Tala'i' in Cairo, one of the first interventions of the Comité de Conservation des Monuments de l'Art Arabe in 1919 in the process of establishing modern heritage preservation practices in Egypt. Their approach included clearing all structures adjacent to the mosque's walls, labelling them as 'parasite structures', and introducing the idea of a fence or boundary to surround the mosque and separate it from its urban surroundings. The later layers of practice added two more boundaries surrounding the original one resulting in its current state, surrounded by three layers of boundaries and interspaces.
Source: Photo by Elhusseiny, July 2, 2020.

a form of expert knowledge in the production of the AHD (Smith, 2006). Such processes of production of heritage discourses are accused of limiting the debate concerning meaning and value when it comes to heritage, allowing a limited space for debate on the interpretations and management of heritage sites and objects (Smith, 2006).

The notion of an AHD has become one of the recent cornerstones of the field of critical heritage studies through which Smith's symbolic boundaries of heritage discourse resonate in a wide range of the literature addressing political aspects of representation in heritage practices every time the AHD is referred to in the field (Smith, 2006; see also Harrison, 2013). Smith draws on discussions of discourses as institutionalised methods of thinking in a Foucauldian sense, relying on the communication of expert knowledge, to explore the role of archaeological heritage management in relation to the technologies of governance (Smith, 2006; see also Harrison, 2013). One cannot dismiss the fact that the critical heritage discourse and the boundaries identified by Smith have contributed significantly to the understanding of the political position of heritage practices and some of the major shortcomings of the international practice such as the excessive focus on materiality and a Eurocentric inclination. While acknowledging such contributions, there are certainly limitations to the ideas presented especially when deployed in different global contexts (Harrison, 2013; Wells & Stiefel, 2019).

The first limitation is concerned with the fact that critical heritage studies' focus on the politics of representation and the discursive aspects of heritage has led to shortcomings in producing accounts that adequately theorise the role of material objects (Harrison, 2013). This relates to Smith's first type of boundary of the heritage object being removed from its present (Smith, 2006). Accordingly, critical heritage studies scholars have been focussing less on the affective qualities of material heritage and our fundamental physical relationship with objects as part of 'being in the world' (Harrison, 2013). In return, a perceivable gap is present between heritage practices and critical theories where theories have failed to engage with the realities of heritage objects and the practice that has such objects at the core of its focus (Wells & Stiefel, 2019). This concern points towards the focus on the essential meanings of heritage as the study of a process of social identification with practices and material objects of the past as a basic function of inhabiting and 'being in the world'. Such criticism is a rather general one that applies to all contexts, whether situated in 'the West' or the Global South, where the variations of understanding heritage extend.

The fundamental function of identification – the concept of objects embodying the personality of a group or being bound up with their identity – is an idea that underlies most conceptions of heritage (Gillman, 2010). It is also considered the ethical basis for decisions of heritage conservation as an act of consolidating certain aspects of individual or group identities. The basic idea is rooted as an existential self-conception of being, where a person identifies oneself by continuity over time – a conception that can only be realised, according to Radin, by the relationship with the external environment (Radin, 1982). Hence, certain objects are considered to be closely bound with our personhood and the way we constitute ourselves as continuous 'being in the world' (Radin, 1982). Based on this view of self-conception, the ownership and value of these objects labelled as cultural properties are manifested as a group's right while the absence or destruction of such objects would be considered 'psychologically intolerable' (Moustakas, 1988, p. 1196). Accordingly, this fundamental function is deeply rooted in the Western modern philosophies of understanding basic identification and our relationship with the external world. The extent to which this belief has been established in non-Western contexts, where it may be crowded with the existing views of 'being in the world' and various understandings of the notions of property, ownership and the human relationship with objects, is a boundary that may explain diverse relationships in terms of receiving the notion of cultural heritage and its progression within such diverse and distant locations. In most Western contexts, there is also a continuity with such centrality of material objects that may be considered an implicit component of modern heritage practices.

For instance, much of the theorisation of heritage conservation practices from a global perspective relies on the notion of 'continuity within discontinuity' with the pre-modern past. For example, Jokilehto traces the sacredness of materials dating back to European pagan history and its manifestation in religious practices of Christianity and Judaism as the origins of modern heritage conservation movements (Jokilehto, 1999). Similarly, Wells refers to the Christian past as the origin of all preservation practices whether it is realised by contemporary practitioners or implicitly embedded as the case of most contemporary practices (Wells & Stiefel, 2019). The Christian tradition extends to the cult of relics throughout the European Middle Ages sparking a fascination with physical remains of Christian antiquity ranging from pieces of the True Cross to locks of Jesus' hair (Silberman, 1990). Such continuity of a relationship with the material world has established a certain world view in most Western contexts where the production of heritage involves a form of 'sacralisation' that extends to our modern times (Meyer & de Witte, 2013) and heritage objects play an essential role in self-identification and the sense of 'being in the world'. Accordingly, this specific history has a significant role in shaping the ideas and values of the 'modern' heritage conservation movement can be described as a form of 'continuity within discontinuity' in relation to 'bygone past' traditions (Heller, 1982).

Also, the fact that heritage, as a discipline of the present concerned with 'objects of the past', extends the relationship with such objects beyond the material being by adding a representative function. The representation of heritage objects is also related to a certain relationship with the past or a 'historical consciousness'. According to Heller, historical consciousness develops through several stages, each outlining a certain structure for answering fundamental questions: 'where have we come from, who are we, where are we going?'(Heller, 1982). The global heritage discourse usually engages with two main stages of Heller's six-stage model which are the historical consciousness of a 'particularity reflected in generality' and a 'reflected universality'. The former reflects the manifestation of symmetrical historiographies each with the freedom to choose its own past to answer the mentioned questions. Such historiographies are produced by each nation state or institution, where their present is born from 'times bygone', bearing in mind the plurality and generality of the concept of history in cultures identified with rationality and freedom. The latter is characteristic of the consciousness of only 'History', a universal world history where all human histories are unified under a universalistic concept (Heller, 1982). Both modes are reflected in UNESCO's general guidelines when dealing with heritage by relying on nation states as the main representative agents and the notion of 'outstanding universal value'. Hence, such ideals of historical consciousness that frame a specific relationship with the past form a symbolic boundary where the whole body of global heritage theories has such conceptions at its core while it barely engages, if at all, with other modes of understanding the past expressed in the more 'primitive' stages of Heller's progressive model.

Such modes of consciousness differ significantly from the previous stages addressed by Heller, the last of which is an 'unreflected universality' associated with religious world views. This is a consciousness where the answer to fundamental questions all lie in what Heller refers to as the 'universal myth' that provides 'complete answers that cannot be challenged' (Heller, 1982, p. 13). Heller implies that such a state of consciousness already points towards the 'reflected universality' of a universal world history in a progressive sense. Yet, the reliance on a belief system to answer the fundamental questions of identity alters the role of heritage objects as identifiers and therefore such objects need to be revisited from a contextual local lens rather than the universalist historic lens of global heritage discourses. Such consciousness manifests in the relationship with historic mosques in Egyptian local contexts. Paradoxical practices reflect a continuity in the act of building, as a valued religious act, combined with the destruction of the material historic fabric that is central to the resurrection of a 'bygone past'. Also, the heightened local interest in mosques compared to other registers of the past in the Egyptian context, as in the case of the city of *Esna* in Upper Egypt, reflects the differential relationship with the past that does not conform to practices of choosing a 'bygone past' or an ancient era to be revived (Figure 24.3).

The second limitation is that critical engagement with heritage discourses in non-Western contexts has remained minimal (Exell & Rico, 2013) and almost non-existent from a local perspective in the Middle East and its current state of silence regarding heritage discourses (Rudolff, 2003; Bernbeck, 2010). As a result, such

Figure 24.3 Within the preference of the Temple of Khnum in the old city of Esna in Upper Egypt, the only building in a maintained condition, other than the temple isolated in its pit, is the green dome of Ibrahim Maghazi Mosque, an early 20th-century structure that was recently rebuilt except for its minaret. Not only is the mosque rebuilt by local funding and craftsmanship amidst the decaying area but it is also the only exception to the ban of all construction work in the area surrounding the temple along with the ongoing restoration work of a few houses funded by the United States Agency for International Development. The mosque is only one example of many other mosques being rebuilt and renovated by local families in Esna but with no attention to the structures' material, historic significance or other ancient registers of the past.

Source: Photo by Elhusseiny, September 23, 2020.

contexts are usually marginalised and play a limited role in theorising heritage and its core conceptions. Silence marks the boundaries of discursive and conceptual space the same way physical boundaries mark heritage spaces and monuments in their wider urban context. Bernbeck points out the contradiction where the curation of the social context of heritage is explicitly encouraged in global conventions – on the other hand, such curation is only accepted in case the symbolic order of global heritage discourse is preserved and the foundations of heritage ideology are left unchallenged (Bernbeck, 2010). International institutions have been scrutinised for adopting Eurocentric approaches and silencing some of the non-Western contexts that lack representation and depart from the 'Western' conception of heritage. While such assertions reflect a serious concern regarding the inclusivity of global heritage discourses, the proposed approach questions the processes and boundaries producing silence involving the marginalised non-Western contexts as well as global heritage institutions.

A pertinent example of such silence in the Egyptian context is the case of *Al-Azhar* mosque in Historic Cairo. Even though it is Egypt's most well-known mosque and university from the *Fatimid* period, it has gone through a series of controversial restoration and renovation projects that have clearly contradicted global conservation practice guidelines. The restoration work starting in 1996 attracted a significant amount of criticism and halt to all work in the mosque was requested by UNESCO. The claims raised included the work being best described as 'renovation' rather than 'restoration' and that the authenticity of the historic mosque had been destroyed by the work done. While such concerns may reflect an evident lack of understanding of the position of mosques and religious practices in Muslim communities, the local responses were void of any engagement with the key concepts of authenticity and heritage in relation to conservation practices in Egypt (Rudolff, 2003). More recent restoration works completed in 2018 have not been mentioned in any of the annual World Heritage Conventions, yet

Figure 24.4 Bab Al-Muzayinīn – the current main entrance of Al-Azhar Mosque, one of the most significant Egyptian national symbols dating back to the Fatimid 10th century and the formal religious institution in Egypt. In its recent renovation works, completed in 2018, a significantly higher boundary replaced the existing one, limiting the accessibility for the public and rendering the historic mosque out of reach. Also, the mosque's World Heritage Site status seems ambiguous in local media as Al-Ahram, one of Egypt's leading official newspapers, states that it has lost its status after the 1996 renovation work while other sources claim it is still maintained. In any case, the recent works have not been addressed by the World Heritage Committee and were executed in a state of silence.

Source: Photo by Elhusseiny, July 2, 2020.

have been celebrated in the press locally as a restoration of *Al-Azhar* to its 'original form 1070 years ago' (Abdel Hady, 2018). Such practices have stirred very limited conversation about the conservation of Islamic heritage in a case where silence marks the spatial boundaries established around the mosque as well as the symbolic boundaries of global heritage discourses (Figure 24.4).

The third limitation is associated with the sociopolitical context of the issues addressed within critical heritage studies. This relates to Smith's contention (second type of boundary) that the emergence of heritage as a profession produced a discourse that is considered a technology for power and governance (Smith, 2006). This, and similar accounts addressing the postmodern condition, the production of the past in the present, or the role of expert knowledge as a governance technology are mainly focussed on the European and North American context – the West, where 'modern states' are established within the social conditions specific to their 'modern nations' (Walsh, 1992; Lowenthal, 1998; Smith, 2006). While shedding light on some issues at the core of heritage practices in their context, the applicability of the ideas presented in the Global South, specifically in Egypt and the Middle East, is questionable according to their specific modes of governance and their underlying legitimate perception of authority and power. To better understand the proposed enquiry, an introduction to what constitutes the modern state and how it might vary in the Global South is necessary to further explore the boundaries of such discourse.

What distinguishes the political condition of modern states, which is bound to the Western context addressed in heritage theories, is the basic principle on which the state's organisation and legitimacy rest. Such a principle is described in a Weberian sense as 'rational-legal authority' (Weber, 1947). It simply indicates that modern states rely on the law to provide the element of authority or 'morally justified power'. The law is directed towards publicly acknowledged goals that are based on rational reasoning. Accordingly, the structure relies on the individual's strict division between private and public roles (Clapham, 1985). This means that officials would abide by the lawful roles of their positions based on the powers and restraints decided according to lawful criteria. On the other hand, officials would revert to the status of private individuals outside the office with a separate set of personal ambitions and obligations (Clapham, 1985). This 'ideal type' is nowhere to be found as it requires a dual nature that goes against basic intuitive instincts; yet, the existing modern states would approximate to such ideal rational-legal mode of operation (Clapham, 1985). Based on such a model of operation, the Foucauldian concepts of discourse, power and governmentality apply and, accordingly, the AHD establishes its boundaries and gains its authority over the public by defining the 'publicly accepted' goals of heritage practices (Smith, 2006). This boundary comes to the foreground in the developing contexts and the questions of applicability to other modes of authority and distribution of power.

Clapham suggests that the basic issues with understanding the 'third-world' state, and hence its politics, are its failure to approximate to the ideal rational-legal mode of operation (Clapham, 1985). Clapham proposes a model of *neo-patrimonialism*, a term established by Shmuel Eisenstadt in 1973 (Eisenstadt, 1973), where the relationships of a broadly patrimonial type persist within a structure that belongs to a rational-legal system (Clapham, 1985). In this case, individuals hold power within their official positions of bureaucratic structures, yet utilise such power in a rather private form creating personal vassal-and-lord-like relationships rather than those of subordinate and superior

(Clapham, 1985). Hence, the concept of power in such a context is relatively understood in personalistic logic that is characterised by patronage, clientelism and more recently corruption (Pitcher et al., 2009). Again, the system described is another form of an ideal state that does not exist in such pure form, yet reflects a tendency to flout rational-legal authority in favour of a more personalised mode where the boundary between public and personal is less clear and where such boundary varies significantly from one context to another (Pitcher et al., 2009).

In the case of Egyptian politics, since the 1980s, scholars have tended to describe the Egyptian state as a relatively 'soft' state, with limited capacities of governance, leading to the current state of a weak regime and underdevelopment (Amin, 2011). This assumption was based on the positive relationship between state capacity and autocracy. The following generation of scholars is more likely to favour a more negative relationship between the two concepts, emphasising political continuity and the stagnation of a highly authoritarian order (Dorman, 2007). On the other hand, Springborg highlights the confusion between authoritarian and patrimonial modes of governance, where patrimonial rulers may sustain great power but little authority that has to be compensated by 'enveloping clients in an intricate web of reciprocities' (Springborg, 1979). His views suggest a patrimonial nature of Egyptian politics within its modern state structure since the rule of Nasser and through Sadat till its present condition (Springborg, 1979, 2017). The implications of these political conditions on the dynamics of heritage practices and their produced meanings are central to understanding the development of heritage as a concept within the Egyptian context.

From a heritage perspective in most Western contexts, the idea of the modern nation state is still the dominant focus at the centre of the heritage discourse (Graham et al., 2000) – an idea that relies heavily on the separation of public and private domains or the creation of a 'public sphere'. Lahouari Addi proposes the failure of the creation of such public sphere, in the modern sense, within the Global South's political structures – specifically countries of the Middle East currently in an in-between state that is neither 'traditional' nor 'modern' (Addi, 1997). Accordingly, when analysing the sociopolitical function of heritage discourses in some non-Western contexts, as the case with the Egyptian state, much of the literature on heritage theories intended to explain the production of specific relationships of governance within heritage practices may not match the existing conditions of authority and power situated in various contexts of the Global South. This is not to ignore the influences and exchange between the West and the developing world as an ongoing process of mutual change, but to highlight the boundaries of different power dynamics and governance systems which might explain the different meanings and functions of heritage in Egypt as a non-Western context of the Global South (Figure 24.5).

Discussion: the global south as a space for expanding the boundaries of heritage

Understanding heritage as a 'value-loaded concept' has been a challenge within the relatively recent field of heritage studies (Harvey, 2001; Carman, 2003; Harrison, 2013). While a great deal of research investigates the notion of heritage and its manifestations, much of it has defined the term purposely using vague and malleable definitions or rather focussing on a certain economic or sociopolitical aspect of the concept. Such

Figure 24.5 The Aga Khan Historic Cairo project is celebrated as the most successful community-oriented conservation project by Egyptian practitioners. Despite the massive component of community work involved in the project, the status of urban heritage objects and their spatial manifestations remain unaltered – the Eastern side of the Fatimid wall (the main heritage component of the first phase of the project) remains beyond reach to its neighbouring community of Al-Darb Al-Ahmar, surrounded by boundaries complemented with barbed wire resembling high-security borders. The absence of the idea of a 'public space' in the modern sense and the specific sociopolitical context produces forms of participation and community components in heritage practices that are fundamentally different from the global conceptions of such concepts.

Source: Photo by Elhusseiny, July 2, 2020.

positions are understandable considering the acknowledged ambiguity of the concept. The reason why the question of understanding heritage as a concept is crucial is that it is being studied in contexts where it is claimed to be imported through cultural contact with the West, where the idea has originated and translated across the globe in its modern sense. Thus, any attempt to deal with 'the duality of heritage' (Graham et al., 2000) practices in a contemporary or historic sense requires an establishment of what the concept represents and the shared meanings and purposes it holds as a 'boundary object' where the tension lies clearly within its coexistence as a universal and peculiar concept in diverse contexts. While heritage remains an ambiguous concept in a global sense between different social contexts, it can be strongly comprehended within an individual context and compared to the translated global origins of its idea. For example, on a national level, in order to be 'modern', nations must prove that they have a history that has its share in shaping their modern identity (Mitchell, 2002). In this sense, proclaiming Egyptian heritage, for instance, constitutes part of the manifestation of a 'modern nation' in the global context. Accordingly, the Egyptian society's investments and the institutions it inherited from its colonial past to care for its heritage prove that Egypt deserves to be among other civilised and modern nations. The concept of heritage thus works as a seemingly unifying and verifying notion that connects and communicates between modern states, their societies and their pasts, even if their representations of the concept itself vary greatly from one nation to the other.

The distinction between heritage as a modern representation and its practice as a continuous function of identification and being does not go unchallenged. In the Egyptian context, the representation of heritage is intertwined with reclaiming authority and autonomy from colonial practices, the making of a nationalistic movement and the production of a modern state (Mitchell, 1988; El-Habashi, 2001; Reid, 2002, 2015; Colla, 2007). Furthermore, it cannot be removed from the traditional beliefs and the eschatological nature of identification that is contributing to a fundamental sense of 'being in the world' within the wider Egyptian public. The questions concerning the position of heritage in the contemporary Egyptian context and actual dynamics of production are at the core of the proposed approach seeking to clarify the intersecting boundaries of heritage in a non-Western context and what it may contribute to the understanding of the concept of heritage in a global sense. This approach resonates with the emerging scholarship tackling the relative paucity of research on urbanisation in the Global South and cities of the Middle East, and the academic potential gained from their deeper investigation, especially in relation to understanding and dealing with wider global challenges (de Satgé & Watson, 2018; Wippel & Steiner, 2019).

The review of boundaries expressed within Western heritage discourse and their limitations beyond the context in which they have been produced suggests the presence of certain gaps when addressing the concept of heritage in other contexts. The perceived gap is not the product of a Eurocentric discourse alone, but a noticeable absence of the non-Western contributions, specifically the Middle East and the Arab context, from the debates on the nature of heritage and its key measures such as demarcation and authenticity in the World Heritage Convention for instance (Rudolff, 2003). Such silence is persistent even when conflicts arise concerning the practices of conservation, especially the ones related to Islamic sites (Rudolff, 2003) as if to mark the boundaries of the heritage discourse and mask the absence of key concepts within their local practices where these concepts may have been replaced with distorted versions of heritage ideas (Schröter & Taylor, 2017). The reasons for such silence and the fact that these issues are considered secondary to the heritage discourse are yet to be examined and understood as a key component of the global heritage discourse.

This focus on Egypt is not intended to make the facile argument of cultural relativism that heritage works in the West, but since other cultures are different, they need their own theories of heritage and its practices. On the contrary, the proposed perspective seeks to locate the issues of translation, global expansion, modernisation and colonialism within the understanding of the already established field of heritage studies, rather than outside it. The local populations of Egypt as part of the Global South have one of two choices: either to identify themselves as part of an 'Eternal Egypt' that is 'modern' and 'progressive' due to its ancientness and the active presence of its past in the present times or be identified as people who have no care or respect to their own past condemned with backwardness and regression or even iconoclasm. The research seeks to problematise such a dichotomy, adopting the idea of Egyptian populations 'dwelling in the boundary', a notion that resonates with recent explorations of the nature of urban frontiers, interfaces and boundaries in cities (Breitung, 2011; Vis, 2018; Sema, 2019). This proposition suggests boundary thinking as a means of exploring this in-between space, exploring a third way that capitalises on the remnants of the past to link people with their history and a shared human past at large.

The use of a boundary approach as an analytical tool to investigate urban heritage in the Global South in the context of Egypt provides three main opportunities for expanding the current global discourse on cultural heritage in general and historic urban environments in specific. The first would be the possibility of defining a rather under-represented area in the global literature in the field of heritage studies which is the Middle Eastern context and specifically Egypt, not as a separate entity but as an integral to the global discourse. The second potential to such an approach would also be related to global discourses; boundaries as a concept that is attentive to differences provide a useful analytical tool for an implicit comparative study of the heritage scene in the Global South against the wider global canvas of heritage – a type of study that has been perceived as a gap in the wider field of heritage (Mualam & Barak, 2019). The third advantage of a boundary approach is the versatility of the boundary to be identified between conceptual, social and physical spaces and disciplines which allows a perception of urban heritage objects as collectives. It also illuminates the relationships between such spaces through their various manifestations in social processes, discourse and urban space. Physical boundaries have become a defining feature of Egyptian urban heritage spaces. While boundaries are an embodiment of all sorts of differences that are implicated in the production of heritage space in the Egyptian context, it may also be perceived as an opportunity to define how such context informs and is informed by global heritage discourses.

References

Abbott, A. (1995). Things of boundairies. *Social Research, 62*(4), 857–882.
Abbott, A. (2016). *Processual sociology.* Chicago, IL: University of Chicago Press.
Abdel Hady, S. (2018, March 5). Al-Tarmīm yoʿīd gamiʿ Al-Azhar ʿila souratoh al-ʾoula qabl 1070 ʿaman. *Al-Ahram.* Retrieved from: https://gate.ahram.org.eg/News/1836133.aspx (Accessed on 25 April 2021).
Addi, L. (1997). The failure of third world nationalism. *Journal of Democracy, 8*(4), 110–124.
Amin, G. (2011). *Egypt in the era of Hosni Mubarak.* Cairo: American University in Cairo Press.
Bernbeck, R. (2010). Heritage politics: Learning from Mullah Omar? In R. Boytner, L. S. Dodd, & B. J. Parker (Eds.), *Controlling the past, owning the future: The political uses of archaeology in the Middle East.* (pp. 27–54). Tucson: University of Arizona Press.
Breitung, W. (2011). Borders and the city: Intra-urban boundaries in Guangzhou (China). *Quaestiones Geographicae, 30*(4), 55–61. https://doi.org/10.2478/v10117-011-0038-5.
Carman, J. (2003). *Archaeology and heritage: An introduction.* London and New York: Continuum.
Clapham, C. S. (1985). *Third world politics: An introduction.* Wisconsin: University of Wisconsin Press.
Colla, E. (2007). *Conflicted antiquities: Egyptology, Egyptomania, Egyptian modernity.* Durham, NC: Duke University Press.
de Satgé, R., & Watson, V. (2018). *Urban planning in the global south: Conflicting rationalities in contested urban space.* London: Palgrave Macmillan.
Dorman, W. J. (2007). *The politics of neglect: The Egyptian State in Cairo, 1974–98.* (Doctoral dissertation). Retrieved from: SOAS Research Online https://eprints.soas.ac.uk/155/1/Dorman_Politics_of_Neglect.pdf (Accessed on 25 April 2021).
Eisenstadt, S. N. (1973). *Traditional patrimonialism and modern neopatrimonialism.* London and Beverly Hills, CA: Sage Publications.
El-Habashi, A. (2001). *'Athar' to Monuments: The Intervention of the 'Comité de Conservation des Monuments de l'Art Arabe.* (Unpublished PhD thesis). Retrieved from: University of Pennsylvania. doi: 10.16953/deusbed.74839 (Accessed on 25 April 2021).

Exell, K., & Rico, T. (2013). "There is no heritage in Qatar": Orientalism, colonialism and other problematic histories. *World Archaeology, 45*(4), 670–685. https://doi.org/10.1080/00438243.2013.852069

Exell, K., & Rico, T. (Eds.) (2014). *Cultural heritage in the Arabian Peninsula: Debates, discourses and practices.* Surrey: Ashgate Publishing Limited.

Gillman, D. (2010). *The idea of cultural heritage.* Cambridge: Cambridge University Press.

Graham, B., Ashworth, G. J., & Tunbridge, J. E. (2000). *A geography of heritage: Power, culture and economy.* London: Arnold (A Hodder Arnold Publication).

Hammond, T. (2020). Heritage and the Middle East: Cities, power, and memory. *Geography Compass, 14*(2), p. e12477. https://doi.org/10.1111/gec3.12477.

Harrison, R. (2013). *Heritage: Critical approaches.* London: Routledge.

Harvey, D. C. (2001). Heritage pasts and heritage presents: Temporality, meaning and the scope of heritage studies. *International Journal of Heritage Studies, 7*(4), 319–338. https://doi.org/10.1080/13581650120105534.

Heller, A. (1982). *A theory of history.* London: Routledge.

Jokilehto, J. (1999). *A history of architectural conservation.* Oxford: Butterworth-Heinemann.

Kolosov, V., & Scott, J. (2013). Selected conceptual issues in border studies. *Belgeo, 1.* https://doi.org/10.4000/belgeo.10532.

Lamont, M., & Molnár, V. (2002). The study of boundaries in the social sciences. *Annual Review of Sociology, 28*(1), 167–195. https://doi.org/10.1146/annurev.soc.28.110601.141107.

Langer, C. (2017). The informal colonialism of egyptology: From the french expedition to the security state. In M. Woons & S. Weier (Eds), *Critical epistemologies of global politics* (pp. 182–202). Bristol: E-International Relations Publishing.

Lowenthal, D. (1998). *The heritage crusade and the spoils of history.* Cambridge: Cambridge University Press.

Meskell, L. (1998). *Archaeology under fire: Nationalism, politics and heritage in the Eastern Mediterranean and Middle East.* London: Routledge.

Meyer, B., & de Witte, M. (2013). Heritage and the sacred: Introduction. *Material Religion, 9*(3), 274–280. https://doi.org/10.2752/175183413X13730330868870.

Mitchell, T. (1988). *Colonising Egypt.* Berkeley: University of California Press.

Mitchell, T. (2002). *Rule of experts: Egypt, techno-politics, modernity.* Berkeley: California: University of California Press.

Moustakas, J. (1988). Group rights in cultural property: Justifying strict inalienability. *Cornell Law Review, 74*(6), 1179–1227.

Mualam, N., & Barak, N. (2019). Evaluating comparative research: Mapping and assessing current trends in built heritage studies. *Sustainability, 11*(3), 1–27. https://doi.org/10.3390/su11030677.

Pitcher, A., Moran, M., & Johnston, M. (2009). Rethinking patrimonialism and neopatrimonialism in Africa. *African Studies Review, 52*, 125–156. https://doi.org/10.1353/arw.0.0163.

Radin, M. J. (1982). Property and personhood. *Stanford Law Review, 34*(5), 957–1015. https://doi.org/10.2307/1228541.

Reid, D. M. (2002). *Whose Pharaohs?* Berkeley: University of California Press.

Reid, D. M. (2015). *Contesting antiquity in Egypt : Archaeologies, museums & the struggle for identities from World War I to Nasser.* Cairo : The American University in Cairo Press.

Riesch, H. (2010). Theorizing boundary work as representation and identity. *Journal for the Theory of Social Behaviour, 40*(4), 452–473. https://doi.org/10.1111/j.1468-5914.2010.00441.x.

Rudolff, B. (2003). *Our universal sacred? Authenticity in conservation policies on Islamic religious sites in the context of global conservation strategies.* (Unpublished master's thesis). Cottbus: Brandenburg University of Technology.

Schröter, M., & Taylor, C. (2017). *Exploring silence and absence in discourse: Empirical approaches.* Retrieved from: https://link.springer.com/book/10.1007%2F978-3-319-64580-3 (Accessed on 25 April 2021).

Sema, S. (2019). Interfaces/intersections in architecture and urbanism. *Archnet-IJAR: International Journal of Architectural Research, 14*(1), 1–4. https://doi.org/10.1108/ARCH-03-2020-218.

Silberman, N. A. (1990). *Digging for God and country: Exploration, archeology, and the secret struggle for the Holy Land, 1799–1917.* New York: Doubleday (Anchor Books).

Smith, L. (2006). *Uses of heritage.* New York: Routledge.

Springborg, R. (1979). Patrimonialism and policy making in Egypt: Nasser and Sadat and the tenure policy for reclaimed lands. *Middle Eastern Studies, 15*(1), 49–69.

Springborg, R. (2017). *Egypt.* Cambridge: Wiley.

Star, S. L., & Griesemer, J. R. (1989). Institutional ecology, "translations" and boundary objects: Amateurs and professionals in Berkeley's Museum of Vertebrate Zoology, 1907–39. *Social Studies of Science, 19*(3), 387–420.

Thelen, D. (1999). Rethinking history and the nation-state: Mexico and the United States. *Journal of American History, 86*(2), 439–452. https://doi.org/10.2307/2567038.

Turner, M. (2009). Introduction. In O. Martin & G. Piatti (Eds.), *World heritage and buffer zones - international expert meeting on world heritage and buffer zones Davos, Switzerland 11 –14 March 2008.* (pp. 15–18). Paris: UNESCO - World Heritage Centre.

Van Houtum, H. (2005). The geopolitics of borders and boundaries. *Geopolitics, 10*(4), 672–679. https://doi.org/10.1080/14650040500318522.

Vis, B. N. (2018). *Cities made of boundaries: Mapping social life in urban form.* London: UCL Press.

Walsh, K. (1992). *The representation of the past: Museums and heritage in the post-modern world.* London: Routledge.

Weber, M. (1947). *The theory of social and economic organization.* New York: Free Press.

Wells, J., & Stiefel, B. (2019). Introduction. In J. Wells & B. Stiefel (Eds.), *Human-centered built environment heritage preservation: Theory and evidence-based practice* (pp. 1–30). New York: Routledge.

Wilson, T. M., & Donnan, H. (Eds.) (2012). *A companion to border studies.* Retrieved from: https://onlinelibrary.wiley.com/doi/pdf/10.1002/9781118255223 (Accessed on 25 April 2021).

Wippel, S., & Steiner, C. (2019). Urban development in the Middle East and North Africa: Deconstructing visions, politics and identities. *Middle East - Topics & Arguments, 12*, 5–16. https://doi.org/10.17192/meta.2019.12.8022.

25 Sustaining heritage places

Crossroads between urban imaginaries, heritage use and sustainability

Tu-Chung Liu and Kalliopi Fouseki

Introduction

Imagination is one of the most crucial mental processes of meaning-making and possibility envisioning for human thinking and action. Without imagination, there is no correlation of human experience between the past, the present and the future (Ricoeur, 2007; Zittoun & Gillespie, 2016). The historic urban environment is the meeting point of the past, present and future where imagination is being enacted. As the ever-changing nature of people, place and city contributes to the accumulation, creation and re-creation of human and historic legacies, a wide range of values related to the past would and could be imagined and produced within a historic city.

It is the aim of this chapter to reconsider heritage and heritage places through the lens of imagination. It will do so by investigating the relationship between imagination, heritage-led urban regeneration and sustainability. More specifically, this heritage–imagination–sustainability inquiry will examine how urban heritage is being used to relate the past, the present and the future of a physical place, especially in the context of heritage-led urban regeneration.

The underpinning premise of the chapter is that heritage is both a real and an imagined praxis of value attributed to historical legacies. As such, heritage is a process of simultaneously *seeing* the past and *being* (meaning having a distinct identity). This premise reaffirms the proposition that heritage is created by human imagination (Ashworth, 2013) and that heritage is constructed with extrinsic motivations (Smith, 2006; Harrison, 2010). In other words, any uses of heritage could be regarded as 'a strategy of appropriation' (Macdonald, 1997, p. 174) to demonstrate specific attitudes towards the past and related identities.

Having outlined the link between heritage and imagination, it is now important to interconnect this relationship with the concept of sustainability. Linking heritage with sustainability is not a new endeavour, especially in the realm of conserving and regenerating historic districts for the sustainable development of neighbourhoods, cities or nations. Indeed, the contribution of culture to various dimensions of sustainability – environmental, economic and social – has been highlighted in recent years (e.g. Hawkes, 2001; UCLG, 2004; UNESCO, 2010; Soini & Birkeland, 2014; Dessein et al., 2015). However, the more fundamental concern about what are the essential factors that can enable the advancement of the culture–city–sustainability relationship has received less attention. Sustainability implies holistic and integrated perspectives to confront environmental and societal challenges and to rethink the nature of development (Brocchi, 2008; Thiele, 2013; Borowy, 2014). To this end, the concept of sustainability implies advancing human capabilities to confront future global challenges (Sen, 2010, 2013). Given that the future is defined by cultural facts (Appadurai, 2013), what could enable human cultures to face the constantly changing world is imagination, anticipation and aspiration.

Our imagination process and related meaning-making are affected by historical, social and geographical contexts (Ricoeur, 1976; Mills, 2000; Harvey, 2005). Zittoun and Gillespie (2016) contend that 'our imagination is nourished by representations, discourses, images, and ideas that come through social interaction, through social media (reading, watching and playing) available in our sociocultural environment' (p. 45). Moreover, given the focus on historic urban environment, the chapter will also draw on the concept of urban imaginaries in order to link human imagination with a city. Urban imaginaries, as the medium of imagination for a physical place, is related to how our mental mappings or the interpretive grids, with the intersection of images, myths and stories, affect the act of imagining the spaces and related urban life in which we live (Soja, 2000; Tate, 2012). In other words, urban imaginaries act as catalytic contents in an imagining process for urban affairs; more specifically, the related urban imaginaries here may refer to personal descriptions about the past and/or the present of a city.

Therefore, through connecting the issue of urban imaginaries and imagination with the implementation of related heritage-led urban regeneration initiatives in a heritage place of Dadaocheng, Taipei City, Taiwan, the chapter will demonstrate the link between policy implementation, as one of the important collective–cultural practices, and individual imagination. Consequently, this chapter will conclude that the main concern of urban cultural policy with heritage use for sustainability should be reoriented from producing a collective imagination with a unified image to cultivating urban imaginaries with diverse living stories for multiple approaches of place-identity.

Heritage use and imagination in urban regeneration

Over the past few decades, the role of urban heritage in city-making is becoming increasingly vital due to the cultural turn in urban affairs, from cultural identity, cultural governance, cultural economy, cultural tourism and culture-led urban regeneration to cultural sustainability. This gradual shift reveals that heritage use as cultural strategy has been adopted and promoted in various urban development strategies, especially in the context of policy intervention (Pendlebury & Porfyriou, 2017). In other words, the potential contribution of cultural resources for urban physical regeneration has been emphasised, and this has been integrated into urban planning as well as related governance (Bianchini & Parkinson, 1993; Griffiths, 1993; Zhong, 2016; Morató & Zarlenga, 2018).

With a broader perspective of the built environment, Ashworth (1997, 2011, 2013) seeks to rethink the relationship between heritage use and urban planning. Ashworth utilises three paradigms to explain different ways of viewing and valuing the past in the context of urban development since the 1960s onwards, namely the preservation paradigm, the conservation paradigm and the heritage paradigm.

For Ashworth (2013), the preservation paradigm aims to protect individual artefacts or buildings from harm, as these historical remains possess intrinsic historical values, whilst the conservation paradigm widens the lens on both the holistic ensemble preservation and adaptive reuse. As for the heritage paradigm, Ashworth further proposes the concept of the heritage paradigm to interpret a continuation of paradigm shifts, particularly at the end of 20th century, in which histories, memories or relics are regarded as heritage resources to be selected and packaged to reveal messages in relation to contemporary social, political and economic demands. In other words, as Ashworth (2011) states, 'the use determines and, in that sense, creates the resource rather than use being a subsequent action for something already preserved' (p. 10). Accordingly, heritage is both imagined and fashion-driven, as the past and the future are imagined and created with diverse purposes within the form of heritage in the present; it also means that constantly imagining new pasts to meet changing needs makes heritage resources inexhaustible. Thus, 'the supply of heritage in total is limited only by the limits of the human imagination to create it' (Ashworth, 2011, p. 12). In a sense, through the power of imagination, historical legacies seem to be functional choices and packages to be rendered for the urban development.

Reconsidering the heritage paradigm proposed by Ashworth (2011), the message of heritage use in the contemporary urban context is mainly concerned with the flexible use of historical resources rather than historic preservation of any intrinsically historical values. This is not to say that the contemporary heritage paradigm does not relate to issues of authenticity and reliability of historical remains and related preservation/conservation activities. On the contrary, it seems to be based on a more strategic and mutable perspective on the present/future possibilities of urban development with heritage resources. In other words, the relationship between heritage resources or products and the past of a city has been broadened from maintaining historical remains through historical evidence to representing the past for strategic interpretation, in any possible interpretive ways and forms, to satisfy contemporary needs.

Meanwhile, if we agree that urban regeneration can not only transform the physical environment but also change the social and cultural meanings of places (Dumitrache & Nae, 2013; Ujang & Zakariya, 2015), the question of how a site has been and is being promoted as a historic district with heritage products becomes central to the issue of heritage-led urban regeneration. This is because the prefix 're'- of the word 'regeneration' not only concerns generating the novel possibilities to use and integrate historical legacies into urban development but also reflects or redefines the existing contents and contexts of a historic environment. Likewise, as Reeve and Shipley (2014) note, a long-term heritage investment of urban regeneration, especially through the analysis of a heritage-led regeneration scheme of the Townscape Heritage Initiative, funded by the Heritage Lottery Fund of the UK, has affected attitudes and perceptions of local people regarding the quality of the local environment.

Defining heritage as real-and-imagined praxis and place as a historically contingent process, this section has demonstrated the quality of change and process in relation to both heritage and place. In this regard, if we seek to situate imagination within heritage-led urban regeneration, we could argue that imagination, in this context, relates to the future-based action of envisioning the future of a place through heritage. In light of this, the process of imagination in the context of urban regeneration implies understanding and re-assessing the values attributed to the past as well as the relationships between individuals, historical legacies and places for the development of the future.

Heritage-led urban regeneration initiatives in Dadaocheng, Taipei City

Dadaocheng, located in the current administrative district of Datong District, west of Taipei City, has been one of the most important commercial areas and trading ports in Taipei City since the mid-nineteenth century. With the change of commercial centres of Taipei City in the 1970s and the preservation movements of Dadaocheng in the 1980s, Dadaocheng constitutes the first case for the government to introduce a specific law of urban planning in 2000, the Plan of Dadaocheng Historic Specific-Use Zone, to preserve the historic environment in Taiwan. The aim of this plan is to advance cultural memory and heritage resources in terms of preserving historic streets (Taipei City Government, 2000). The plan instigated activities of policy implementation of heritage-led urban regeneration concerned with four dimensions and promoted by different departments of the Taipei City Government, namely urban regeneration, commercial development, cultural affairs and cultural tourism.

In terms of policy rhetoric, we can further explore the relations between heritage use and place value. Whilst the main task of the Plan of Dadaocheng Historic Special-Use Zone in 2000 is to preserve a historical streetscape, Dadaocheng Old Townscape Redevelopment Plan between 2007 and 2009 seeks to deal with regeneration issues of both the old townscape and traditional industry to make Dadaocheng a living historic district, and then a tourist destination. Accordingly, the former regards a historic district as a valuable resource for urban competitiveness and memory, whilst the latter emphasises how historical legacies help promote local identity and tourism.

After 2010, the focus of related policies seems to have shifted to that of historical legacies as the means of cultural and creative industries. The most obvious case of this trend is the Urban Regeneration Station (URS) project in which different regeneration stations aim to search for creative ways, including art exhibitions, the local cuisine, teaching workshops and cultural souvenirs, reuse of historical buildings and engagement with local communities. With interdisciplinary cooperation and participation, this policy intends to promote Taipei as a creative city (Department of Urban Regeneration, Taipei City Government, 2013). Until now, the policy has produced five stations in Dadaocheng, including four bases at Dihua Street and one at Yanping North Road.

Furthermore, the policy interventions of the Department of Cultural Affairs since 2011 onwards have encompassed a threefold aspect in Dadaocheng. The first and main policy is to run the Dadaocheng Theatre to promote traditional performing arts, especially Taiwanese Opera as well as glove puppetry. This space, refurbished from an old community activities centre and reopened in 2011, aims to represent the past historical glory of Dadaocheng as an important hub for performing arts. Moreover, as Dadaocheng is regarded as one of the potential creative hubs, the Department of Cultural Affairs has been hosting a short-term festival annually since 2013. This is part of Taipei City's cultural and creative policy, the Plan of Taipei Cultural and Creative Clusters, to promote creative districts in terms of exhibitions, workshops and related activities. Finally, this department also provides some subsidies to private historic buildings for refurbishment. Thus, the Department of Cultural Affairs focuses mainly on the strategies of marketing Dadaocheng as a creative cluster and a base for traditional arts. Meanwhile, this department has tried to regard historical legacies as cultural and economic resources and as the material for cultural and creative industries.

As such, the policy initiatives of different departments through the lens of culture and creativity have been promoted over the past two decades. Introducing the official designation of a historic district and the following policy initiatives in relation to Dadaocheng, the chapter intends to provide the contexts regarding the intervention of urban governance whilst opening space for reflection regarding the effects of policy discourse and action, producing some specific urban images and imaginaries, on imagining the place of Dadaocheng and using historical legacies with individual heritage actors.

The relationship between urban imaginaries, heritage use and place vision

In order to explore the heritage–person–place relationship of human imagination, thirty-five semi-structured interviews were conducted in 2016 and 2017. The interviewees are leading forces and figures consciously using historical legacies in and for the district of Dadaocheng, including seven government officers of different departments, three project managers for conducting the government project in Dadaocheng, fourteen local entrepreneurs, seven staff members of local cultural organisations or groups and four local older inhabitants. Accordingly, in addition to the understanding about individual experiences of heritage use in Dadaocheng, the interview data are also concerned with both the individual descriptions of Dadaocheng and the vision of the future development of Dadaocheng.

Through the thematic analysis of the data (Braun & Clarke, 2006; Bryman, 2016), two main thematic areas were identified including (1) public discourses for promoting Dadaocheng between policy practice and individual heritage use; and (2) heritage use with place values concerned with urban imaginaries and with vision of the future of Dadaocheng.

Firstly, in terms of public discourses for promoting Dadaocheng, the government officers tended to describe their heritage policy actions using terms such as 'creative city', 'culture-led urban regeneration', 'ecological museum' or 'sustainable development'. This also means that Dadaocheng has the potential and related resources to fit the contents and visions of specific discourses. For example, one of the government officers stated:

> the idea of regeneration, compared with the term of redevelopment promoted in United States, could inspire us to think the future of Taipei City and Dadaocheng towards sustainable development in different perspectives, such as society, environment and economy.
>
> (Interviewee A1, personal communication, 8 August 2016)

This quote also reveals the significance for the regeneration scheme to adopt a holistic and integrated approach to sustainable development. Moreover, another government officer mentioned 'when it comes to promoting Dadaocheng (with the URS policy), it at least consists of two policy contexts, with the discourses of historic preservation and creative city or cultural and creative industries' (Interviewee A2, personal communication, August 10, 2016). Accordingly, Interviewee A2 advocates for the merging of two policy areas which may in a way be in conflict with each other, and yet this merging is essential in the context of historic areas.

An additional public discourse for the promotion of Dadaocheng draws on its performing arts and traditional products, such as tea, which could potentially form part of an ecological museum. As indicated by Interviewee A7:

> now the Taipei City Government is trying to run a project with the discourse of Dadaocheng as an ecological museum, in which tea, traditional performing arts, and the democratic movements of the 1920s are key elements to be explored and promoted.
>
> (Interviewee A7, personal communication, 22 June 2017)

In addition, the aforementioned public discourses for promoting Dadaocheng seem to provide general ideas penetrating in the narrative content of local people. For instance,

Interviewee C2 sought to use the term of cultural and creative industries to rethink the development of Dadaocheng: 'the development of an old street could not work just for tourists, and historic atmosphere does not become the decoration of cultural and creative industries' (Interviewee C2, personal communication, August 15, 2016). Likewise, Interviewee D6 utilised the idea of sustainable development to envision the development of Dadaocheng by stating that 'we need to reconsider the issue of sustainable development of historic districts over time, especially for maintaining the physical environment and traditional businesses' (Interviewee D6, personal communication, June 24, 2017). Therefore, discourses for a place through the lens of policy-making could influence how we narrate a place in different ways.

Secondly, the thematic analysis of thirty-five interviews resulted in the identification of fourteen urban imaginaries of Dadaocheng. Dadaocheng was imagined as follows: (i) a place of historical stories; (ii) a well-preserved historic district; (iii) living old streets, combining old and new; (iv) policy initiatives of preservation and regeneration; (v) vintage feelings; (vi) mixing Chinese and Western cultures; (vii) a tourist destination; (viii) a place of family memory; (ix) a place of prosperous business circles, once the most important commercial centre; (x) a place associated with the early history of Taipei City; (xi) a place associated with the early history of Taiwan; (xii) a place of good community relationships; (xiii) once the most important base of social movements; and (xiv) a traditional performing arts hub. The most predominant imaginaries in the interviews was the idea of Dadaocheng as a place of historical stories and as a place of prosperous business circles.

Moreover, as for the coherence between urban imaginaries, the action vision of heritage use and the vision of the future of Dadaocheng, the thematic analysis showed that whilst about two-thirds of the research participants tended to link their heritage action visions with the ideal future image of Dadaocheng to varying degrees, others presented their place visions in a more general sense. This difference, for instance, could be further distinguished between participant C2 and D2.

Cultural entrepreneur C2 sought to organise an international arts festival each year and engage new microbusinesses associated with the decade of the 1920s and the historical legacy categories he had identified, such as architecture, tea, fabric, theatre and foods. In this way, he aimed to promote Dadaocheng as an area of cultural entrepreneurship, and this action vision also corresponded to his imagination for the future of Dadaocheng: a district of cultural entrepreneurship. In contrast, Interviewee D2, as a leader of puppet theatre museum, thought that 'it (Dadaocheng) could be sustained in an organic way without too much government intervention' (Interviewee D2, personal communication, June 20, 2017), rather than the emphasis on relating his action vision of historic preservation, particularly for traditional performing arts. However, this indirect correlation between action vision and place vision could be linked to both topics of urban imaginaries and place values. This is because interviewee D2, for example, has imagined Dadaocheng as 'a living representation' of old Taipei, which means that many traditional industries, old shops and religious festivals continue to be maintained in Dadaocheng. So an organic way of urban development with little new policy implementation and/or intervention seems to be a better method to keep this area as a living precinct.

As such, this research suggests that the relationship between urban imaginaries, heritage use and place vision is mutual, especially regarding aspects of heritage practice and place envisioning. This implies that if heritage is being used strategically in order to achieve a higher level of place vision, urban imaginaries seem to play a facilitating role for using historical legacies and understanding the past of a physical environment. In other words, different visions of the future of Dadaocheng, to a large extent, could be regarded as the extension of urban imaginaries.

Discussion

Concentrating on the exploration of a relationship between individual imagination, urban imaginaries and a heritage place, the related discussion of this case study seeks to respond to the aim of this chapter regarding how urban heritage is being used to relate the past, present and future of a physical place, especially in the context of heritage-led urban regeneration. More specifically, the chapter outlines the link between policy implementation and individual imagination. This link, as shown above, may be based on the construction of urban imaginaries and place values.

Linking policy implementation with the promotion of urban imaginaries and place values, urban heritage itself is not only value-driven but also value-generative. This is to say that if sustaining economic/cultural viability through the lens of creativity is the most significant current concern about the development of Dadaocheng for policy-makers or those heritage actors who live and/or work in this area, the long-term policy practices with discursive promotions as well as heritage products may play a role in affecting the behaviours and related meaning-making in this area, especially for using historical legacies. In other words, these discursive or material results of the urban cultural policy have formed the potential elements of semantic circles for our imagination, especially contributing to the construction of urban imaginaries and place values.

In the context of Taiwan, the discourse of the cultural and creative industries not only affects the cultural policies of the central government[1] but also influences the urban governance of the major cities in Taiwan. The use of culture and the concept of the creative city have become new strategies for urban development. To develop the creative economy, the five municipalities of Taiwan have adopted different approaches in accordance with their local advantages. According to Chung's list (2014), their policy rhetoric includes 'Taipei: becoming a design capital', 'New Taipei city: a happy city', 'Greater Taichung: a city for creative living', 'Greater Tainan: cultural capital, creative city' and 'Greater Kaohsiung: an ocean city of creativity'. From the slogans mentioned above, it is easy to recognise the influences of the related concepts. In other words, the culture/creativity-led strategies have become crucial elements of the transformation of cities in Taiwan, at least for local decision and policy-makers as well as creative entrepreneurs.

In addition, the very close association of the economy and culture between constructed urban imaginaries and the discourses of cultural economy as well as creative city is here based on at least three aspects of imagining the past prosperous period of Dadaocheng: cultures of economic activities of Dadaocheng, cultural and artistic activities due to economic support in Dadaocheng and cultural means for economic benefits for Dadaocheng. Meanwhile, these three features for describing the relations between economy and culture in the past of Dadaocheng seem to resonate with contemporary discourses of cultural economy and creative city. In view of this, urban imaginaries associated with the historical glory of Dadaocheng provide reasons and inspirations for any regeneration projects with contemporary discourses of the cultural economy, leading to using historical legacies through the lens of cultural creativity. In other words, from the perspective of regeneration officers, regenerating Dadaocheng implies that Dadaocheng could 'return' to its former glory and beyond, whether cultural and/or economic. Accordingly, urban heritage is being used to connect historical prosperity and cultural identity, and urban heritage is being used to advance cultural and economic vitality for the future.

As such, any heritage-led regeneration activities, promoted by the government or by individual actors, seem to rediscover glorious stories about Dadaocheng, reinterpret the story elements with individual interests and then generate new stories with historical legacies. For example, in a government report reviewing the results of urban regeneration projects in Taipei City between 2006 and 2014, the case of URS44 was introduced with the statement that 'based on historical resources, Dadaocheng Story House (URS44) has sought to explore and collect the past stories of Dadaocheng. By doing this, these materials could further help create new stories of urban regeneration and future development' (Taipei City Urban Regeneration Office, 2014, p. 170). For Dadaocheng, historical legacies are the core medium to channel the ways of imagining the place of Dadaocheng and its possible values. The research participants often made references to historical elements, from historical events, famous organisations, celebrities, architecture, traditional industries and music troupes to the time of the 1920s, that enabled them to imagine the past of Dadaocheng and define specific place values.

Subsequently, a suggestion that urban heritage is being used to create values for the story contexts of both a physical place and individual life histories may imply that historical legacies are not only the medium to connect historical stories but that they are also the catalyst for them to be integrated into the context of everyday life to enrich individual life history. In terms of a heritage place, story-making relates to not only an individual intervention of heritage use in a heritage place but also a collective construction of urban imaginaries for a historic environment. This also means that whilst telling individual stories with a historic place is a kind of dialogue between past-making and time experience (Pavličić, 2014), ongoing narratives could make us play the role of co-author with historical legacies for the future, and then, we would also be informed by those historical resources of the story (Walter, 2014).

Therefore, in terms of the context of urban regeneration of Dadaocheng, urban heritage is being used to link the historical prosperity of Dadaocheng to project future prosperity and to advance cultural and economic vitality, responding to the contemporary discourses in and for a historic district. Furthermore, urban heritage is being used to cultivate a better future for Dadaocheng with more inspiration and more opportunities for creating diverse individual living stories with time and place.

The effect of urban cultural policy on imagining a city compromises the tension between encouraging and constraining the act of imagination. Being a potential element for defining urban imaginaries and place values, urban heritage within the policy context is related to at least two concerns: the uses of heritage in the policy and the uses of heritage for the city. For example, the URS policy with different stations has various approaches to heritage use and community connection, yet diverse proposals are indeed based on cultural creativity or creative economy. This also means that the 'creativity' represented in these historical spaces is related to artistic exhibitions, cafés, cultural workshops and design product shops. Consequently, the cultural or creative turn for the process of city-making may become problematic. More specifically, it raises some critical questions such as whose culture and city? (Zukin, 1995), whose urban renaissance? (Porter & Shaw, 2009) or culture for the city? (Colomb, 2011).

Therefore, the main concern of urban cultural policy with heritage use for sustainability should shift from producing a collective imagination with a unified image or discourse to cultivating urban imaginaries with diverse living stories, since, as Healey (2002, p. 1778) states, 'in any city, there are many cities which could be imagined and articulated as expressions of the city'. This is to say that our imagination with heritage use aims to produce diverse approaches of place-identity in individual ways. In other words, the more places the individuals identify with a city, the more living histories a city contains for reconnecting both time and the city with the self.

Conclusion

This chapter aimed to reconnect the subjects of heritage and heritage place with the concept of imagination. Using the case of Dadaocheng in Taipei, we showed that planning officers and creative entrepreneurs drew on the historic prosperity of an area (past) in order to advance cultural and economic vitality (present) whilst enriching the place with individual life histories and visions about the future. In other words, we demonstrated that urban imaginaries shaped by the linkage between heritage uses and urban future visions denote the ideal and imagined futures of a city that is often associated with the culturally imagined past and present.

The case of Dadaocheng in Taiwan revealed that the long-term policy practices of heritage-led urban regeneration have had some effects on discourse reference and material representation for the construction of urban imaginaries and place values. To this end, the aim of an urban cultural policy could and should create the space of inspiring and promoting more diverse urban imaginaries instead of the implementation of guiding collective identity. More specifically, regarding transformations for sustainability, future policy initiatives are expected to continuously engage more narrative stories between the individual story context of life history and the collective story context of urban development. More research is needed on exploring how urban imaginaries are being shaped and transformed in relation to the heritage of a place and how such imaginaries affect policies and individual visions for the future of a place. It is also important to capture the 'imaginaries' of local communities which may very well differ from those of the planning officers. Finally, we hope that more research will emerge in the future linking anthropological and sociological theories of imagination with heritage studies.

Notes

1 In 1995, The Council of Cultural Affairs (the predecessor of the Ministry of Culture) proposed the concept of 'Culture Industrialisation, Industries Culturalisation', leading to the connection of local features and cultural industries. Subsequently, the discourse of the creative industries has been promoted by the UK Government since 1997, and the Taiwanese Government adopted the term 'cultural and creative industries' to respond to the trend of the creative economy. With the development of the cultural and creative industries, three policy documents were published, namely Challenge 2008: National Development Plan (2002–2007) by The Executive in 2002, White Paper on Cultural Policy by The Council for Cultural Affairs in 2004 and Creative Taiwan: Cultural and Creative Industries Development Action Plan by The Council for Cultural Affairs in 2009. Consequently, in 2010, the Cultural and Creative Industries Development Law was promulgated.

References

Appadurai, A. (2013). *The future as cultural fact: Essays on the global condition*. London: Verso Books.

Ashworth, G. J. (1997). Conservation as preservation or as heritage: Two paradigms and two answers. *Built Environment*, 23(2), 92–102.

Ashworth, G. J. (2011). Preservation, conservation and heritage: Approaches to the past in the present through the built environment. *Asian Anthropology*, 10(1), 1–18. https://doi.org/10.1080/1683478X.2011.10552601.

Ashworth, G. J. (2013). Heritage in planning: Using pasts in shaping futures. In G. Young & D. Stevenson (Eds.), *The Ashgate research companion to planning and culture* (pp. 185–202). Farnham: Ashgate.

Bianchini, F., & Parkinson, M. (1993). *Cultural policy and urban regeneration: The west European experience*. Manchester: Manchester University Press.

Borowy, I. (2014). *Defining sustainable development for our common future: A history of the world commission on environment and development (Brundtland Commission)*. London: Routledge.

Braun, V., & Clarke, V. (2006). Using thematic analysis in psychology. *Qualitative Research in Psychology*, 3(2), 77–101. https://doi.org/10.1191/1478088706qp063oa.

Brocchi, D. (2008). The cultural dimension of sustainability. In S. Kagan & V. Kirchberg (Eds.), *Sustainability: A new frontier for the arts and cultures* (pp. 26–58). Frankfurt am Main: Vas Verlag.

Bryman, A. (2016). *Social research methods*. 5th ed. Oxford: Oxford University Press.

Chung, H.-L. (2014). Developing the creative economy: The network approach of the five municipalities in Taiwan. In H. K. Lee & L. Lin (Eds.), *Cultural policies in East Asia: Dynamics between the state, arts and creative industries* (pp. 193–209). London: Palgrave Macmillan.

Colomb, C. (2011). Culture in the city, culture for the city? The political construction of the trickle-down in cultural regeneration strategies in Roubaix, France. *Town Planning Review*, 82(1), 77–98. https://doi.org/10.3828/tpr.2011.3.

Department of Urban Regeneration, Taipei City Government (2013). *URS: Creative clusters in the city*. Retrieved from: https://www-ws.gov.taipei/001/Upload/public/MMO/URO/20100706_臺北市都市再生前進基地推動計畫.pdf (Accessed on 14 May 2021).

Dessein, J., Soini, K., Fairclough, G., & Horlings, L., (Eds.) (2015). *Culture in, for and as sustainable development. Conclusions from the COST Action IS1007 investigating cultural sustainability*. Finland: University of Jyväskylä. Retrieved from: http://www.culturalsustainability.eu/conclusions.pdf (Accessed on 14 May 2021).

Dumitrache, L., & Nae, M. (2013). Urban regeneration and affective connections to place in Bucharest city center. Analale Universitatii Bucuresti. Geografie. Retrieved from: http://annalsreview.geo.unibuc.ro/2013/Dumitrache.pdf. (Accessed on 14 May 2021).

Griffiths, R. (1993). The politics of cultural policy in urban regeneration strategies. *Policy and Politics*, 21, 39–46.

Harrison, R. (2010). What is heritage? In R. Harrison (Ed.), *Understanding the politics of heritage* (pp. 5–42). Manchester: Manchester University Press.

Harvey, D. (2005). The sociological and geographical imaginations. *International Journal of Politics, Culture, and Society*, 18(3/4), 211–255.

Hawkes, J. (2001). *The fourth pillar of sustainability: Culture's essential role in public planning*. Melbourne: Common Ground.

Healey, P. (2002). On creating the "city" as a collective resource. *Urban Studies*, 39(10), 1777–1792.

Macdonald, S. (1997). A people's story: Heritage, identity and authenticity. In C. Rojek & J. Urry (Eds.), *Touring cultures: Transformations of travel and theory* (pp. 155–175). London: Routledge.

Mills, C. W. (2000). *The sociological imagination* (40th anniversary edition). Oxford: Oxford University Press.

Morató, A. R., & Zarlenga, M. I. (2018). Culture-led urban regeneration policies in the Ibero-American space. *International Journal of Cultural Policy*, 24(5), 628–646. https://doi.org/10.1080/10286632.2018.1514037.

Pavličić, J. (2014). Narrative of cultural heritage: Theory and practice-church of St. Nikola in Pristina. Култура/*Culture*, 8, 121–129.

Pendlebury, J., & Porfyriou, H. (2017). Heritage, urban regeneration and place-making. *Journal of Urban Design*, 22(4), 429–432. https://doi.org/10.1080/13574809.2017.1326712.

Porter, L., & Shaw, K. (Eds.) (2009). *Whose urban renaissance? An international comparison of urban regeneration strategies*. London: Routledge.

Reeve, A., & Shipley, R. (2014). Heritage-based regeneration in an age of austerity: Lessons from the Townscape Heritage Initiative. *Journal of Urban Regeneration & Renewal*, 7(2), 122–135.

Ricoeur, P. (1976). Ideology and utopia as cultural imagination. *Philosophic Exchange*, 7(1), 17–28.

Ricœur, P. (2007). *From text to action*. Evanston: Northwestern University Press.

Sen, A. (2010). Sustainable development and our responsibilities. *Notizie di Politeia*, 26(98), 129–137.

Sen, A. (2013). The ends and means of sustainability. *Journal of Human Development and Capabilities*, 14(1), 6–20. https://doi.org/10.1080/19452829.2012.747492.

Smith, L. (2006). *Uses of heritage*. London & New York: Routledge.

Soini, K., & Birkeland, I. (2014). Exploring the scientific discourse on cultural sustainability. *Geoforum*, 51, 213–223.

Soja, E. W. (2000). *Postmetropolis: Critical studies of cities and regions*. Oxford: Blackwell.

Taipei City Government. (2000). *The plan of Dadaocheng historic special-use zone*. Retrieved from: https://uro.gov.taipei/News_Content.aspx?n=BE199E778DC570E6&sms=BF71818320ED-C104&s=8E949820E1470C8F (Accessed on 14 May 2021).

Taipei City Urban Regeneration Office. (2014). Reviewing the results of urban regeneration projects in Taipei City between 2006 and 2014. Retrieved from: https://www ws.gov.taipei/Download.ashx?u=LzAwMS9VcGxvYWQvcHVibGljL0F0dGFjaG1lbnQvNTIxMjE0NT-E0Njk2LnBkZg%3d%3d&n=NTIxMjE0NTE0Njk2LnBkZg%3d%3d&icon=..pdf. (Accessed on 14 May 2021).

Tate, S. (2012). Everyday life, tinkering, and full participation in the urban cultural imaginary. *Environment, Space, Place*, 4(2), 104–129.

Thiele, L. P. (2013). *Sustainability*. Cambridge: Polity.

Ujang, N., & Zakariya, K. (2015). The notion of place, place meaning and identity in urban regeneration. *Procedia- Social and Behavioral Sciences*, 70, 709–777. https://doi.org/10.1016/j.sbspro.2015.01.073.

United Cities and Local Governments (UCLG) (2004). *Agenda 21 for culture*. Retrieved from: http://www.agenda21culture.net/sites/default/files/files/documents/en/c21_015_en.pdf. (Accessed on 14 May 2021).

UNESCO (2010). *The power of culture for development*. Retrieved from: http://unesdoc.unesco.org/images/0018/001893/189382e.pdf. (Accessed on 14 May 2021).

Walter, N. (2014). From values to narrative: A new foundation for the conservation of historic buildings. *International Journal of Heritage Studies*, 20(6), 634–650. https://doi.org/10.1080/13527258.2013.828649.

Zhong, S. (2016). Artists and Shanghai's culture-led urban regeneration. *Cities*, 56, 165–171. Retrieved from: https://doi.org/10.1016/j.cities.2015.09.002.

Zittoun, T., & Gillespie, A. (2016). *Imagination in human and cultural development*. London: Routledge.

Zukin, S. (1995). *The cultures of cities*. Oxford: Blackwell.

26 FORT ST ANGELO is not a billboard

Image-driven media and the resilience of the project

Erica Giusta and Guillaume Dreyfuss

Introduction

Heritage and image in the age of social media

In 1979, in her seminal essay 'On Photography', Susan Sontag wrote that the camera has been so successful in 'beautifying the world that photographs, rather than the world, have become the standard of the beautiful' (Sontag, 1979, p. 85). Ever since its invention in 1839, photography – the daguerreotype at the time – has maintained close links with the built environment and with architectural heritage in particular. For instance, the French *Commission des Monuments Historiques* created only two years before took a vivid interest in the new invention, which eventually led to the commissioning of the first series of photographs of historical buildings in 1851. Known as the *mission heliographique*, this project was the work of five photographers and resulted in the compilation of over 250 photographs. The relation intensified in 1873 with the request to include photographs of the buildings proposed to be scheduled (Gui, 2016, p. 2), initiating a practice that has been at the core of heritage documentation since then.[1]

However, from the early 20th century, architects were already lamenting the effect of photography and of the illustrated journal on architecture. Adolf Loos, for example, criticised his 'contemporary and rival Austrian architect Josef Hoffmann because in Loos's view his houses appeared to be made for the camera; they were two-dimensional and had lost all tectonic qualities' (Colomina & Wigley, 2017, p. 260). One can also mention Le Corbusier, a serial photographer who in the 1930s shot endlessly his travels but also his work, whether models or on site. In 1936, during his journey from Brazil across the Atlantic, he would take about 600 photographs (Benton, 2014, p. 63). Previously, in 1933, Le Corbusier had taken part in a cruise around the Mediterranean on board the *SS Patris II*. Organised by the *Congres Internationaux d'Architecture Moderne* (CIAM), the cruise was designed to take the passengers around several monuments and sites from the 'Neolithic' and 'Antiquity' times. With on-board Sigfried Giedion, Laszlo Moholy-Nagy, Alvar Aalto and Fernand Léger amongst others, the entire journey was the subject of several photographs and video reportages, documenting both on-board activities and discussions, and sites and heritage during the visits on land (Fernandez, 2020, p. 133). Owing to the rise of mass media and the democratisation of photography with the development of compact and instant cameras, photographs in the late 1970s were almost as ubiquitous as they are today, to the extent that Sontag was already describing photographing as a 'a way of certifying experience' (Sontag, 1979, p. 9). For Sontag, photographing was also a way of refuting it at the same time, by reducing the temporal aspect of any experience into a series of centralised images, leaving everyone in search for the perfect image, the photogenic image. Social media accelerated this phenomenon and its dynamics due to their instant and almost simultaneous real-time production and consumption, which, in the past decade, became swiftly available to almost anybody, globally. We have become so obsessed with reducing reality to sets of focused imagery that we began seeing objects, spaces and ourselves photographically, in constant search of the photogenic; creative professions, and architecture in particular, have been affected by this endless quest for the perfect image very deeply, at the core of their initial production stages too. In the 21st century, the development of visualisations and virtualisation and their insertion in the regular design workflows has also contributed to the advent of changing viewpoints, which in turn may sought to be reproduced when the project reaches completion.

The image-driven processes that prioritise 'the image of something' over 'the object/subject itself' originated at that point in time when cameras became an affordable and very popular object, and a tool for what was widely acknowledged as transparent and truthful documentation of reality. The concept of a faithful image of the existing, a reproduction, was already the driver behind the early uses of photography in documenting the monument, as even a mediocre photographic print of a monument is better than the best of engravings (Wey, 1851, cited in de Mondenard, 1994). Before the birth of the concept of autonomy of photography, but not without some discussions around authorship, photography was tasked with documenting and preserving the stone monuments of the world, paradoxically on glass, coated paper and celluloid film. Ever since then, and uninterruptedly until the advent of social media, the quantity of images produced seems to have grown proportionally to the ever-widening gap between different 'categories' of photographs. If this could be broadly categorised into mainly documents, memories or fine art, the more approachable the photographic process became, the more categories and numbers of photographs increased. This leads inevitably to a certain amount of vacuity of the subject, which in many instances requires the support of narratives to impart meaning and coherence to the visual production.

Image-based social media platforms like Instagram and Pinterest have blurred all boundaries between these informal categories, but also between public and private aspects of photographs (Berger, 2013, p. 52) and between 'sources of emission' and 'points of reception'. According to Roland Barthes, the photograph that appears in the press is in fact one which is 'formulated by a source of emission, a channel of transmission and a point of reception' (1982, p. 194). The sources of emission are the operators that run the magazine, for example the ones which organise the taking of the photograph, the subsequent editing, the addition of commentary and captioning, while the point of reception is the public which consumes the paper. The linearity of this process has been completely disrupted by social media, where anyone is simultaneously a source of emission and a point of reception – or rather, a producer and a consumer, given the highly commercial nature of some online platforms. The end result of this new non-linear process is a sheer amount of haphazard, fast-paced and image-based exchanges, controlled by algorithms whose main function is to capitalise on those exchanges. The consequent commodification of the exchanges and their contents is therefore unavoidable: an image of a heritage building posted on Pinterest, for instance, is ranked and distributed multiple times according to parameters which belong to the realm of marketing and online advertising, and which are alien to heritage values, and to architectural and conservation theories or sustainability policies. This phenomenon defines a new 'value' of images and raises two crucial questions, which are investigated in this chapter: How is the process of production and preservation of heritage affected? And how is its perception, from both professionals and general public, challenged?

The chapter also aims at exploring the digital context in which the above-mentioned cultural shift in the 'value' of the architectural image unfolds. Strategies to shift the focus back on the project, the 'real thing', rather than its image and its communication, are investigated also as a way to understand the broader impact of contemporary image-driven culture on the built environment.

Literature context

While there is an abundance of manuals, data and policies regarding technical aspects of sustainability in relation to heritage, very little can be found regarding other aspects of what should be a holistic approach to the production, regeneration and preservation of heritage. As Cassar notes: 'We have complex issues to grapple with, such as the growth of cultural tourism from the (still comparatively) affluent developed world, the enormity of the threat of climate change, and the extent to which the rapid advance of technology can help us manage natural and anthropogenic effects on cultural heritage' (2009, p. 4). Because of this complexity, it is clear that a truly holistic and sustainable approach should examine the impact of intangible elements, and their relationship with physical elements too, when assessing heritage interventions. Among the intangible elements, new media and their widespread image-based culture play a role which has not yet been academically investigated in relation to heritage and to the need for sustainability as an overarching characteristic of the project.

In parallel, more work has been published about architecture and social media: biennales from Istanbul to Chicago have been organised on the theme, and countless adaptations of media theorist Marshall McLuhan's 'The Medium is the Message' have been discussed. On one hand, positive repercussions of the so-called democratisation of architecture thanks to social media platforms have been praised (Ferrando, 2018), while on the other hand, concerns about the 'vicious circle' of 'Instagram producing copycats and environments being ravaged by selfie-taking' (Lange, 2019) have been expressed.

Accessibility of online contents, mainly images, 'has allowed a growing number of editors to give birth to a new generation of architecture magazines published on social platforms', which considerably opened up the debate in doing so; the resulting image-based curated archives mostly gather photographs and illustrations 'that can come from (sometimes very) distant historical periods. Such an openness of scope, which is a direct expression of the immediate availability that all images have once they are on the internet, is quite relevant for architectural culture' (Ferrando, 2018). As architecture critic D. T. Ferrando argues in his article in *The Architectural Review* (2018), 'these atlases follow a fertile tradition in the history of Western art, which was initiated in the 1920s by Aby Warburg's *Mnemosyne Atlas*. In such tradition, visual montage is used to express a certain idea, by means of the dialectical tensions produced by the association of images gathered in one place'. Warburg's compositions show an interest in creating visual systems that would help us understand our relationship with the past and with the natural world, while laying the foundation of what he described as 'a new theory on the function of human visual memory' (Schwendener, 2020)[2]. He was therefore assigning images a specific 'value', originating in the methodology he adopted in selecting them, combining them and ultimately elevating them to 'universal memories'. Images were in Warburg's practice a rigorous research tool, and their different associations the result of critical interpretation. *Mnemosyne Atlas* can be considered the first attempt at blurring the boundaries between photographs as technical documents and photographs as memories, owing to the new value they acquire through meaningful association.

The Pinterest architect

In this respect, the way in which image-based online platforms (like Pinterest, Instagram and Tumblr) allow users to organise their contents can be envisaged as a digital evolution of the Warburgian board, to the extent that this phenomenon also generated a rekindled interest for the art historian's explorations. There is, however, a big difference in the methodology behind the collection of images online. Suggestions for contents similar or somehow related to the images which we 'post' and 'like' or 'pin' are in fact calculated by algorithms which flatten the complexity of any picture to a series of 'hashtags' and which, even when running a search by specific key words, give priority to either most popular or advertised contents, because of the commercial nature of the platforms profiting from the exchanges. Pinterest, in particular, has never attracted as much attention and media scrutiny as the likes of Instagram and Facebook, but it is certainly not less controversial and influential, especially when it comes to the initial production stages of design and architecture. Pinterest places itself somewhere half-way between social media and search engine, and it organises millions of user-generated images in boards, accurately labelled by themes and tagged with a number of key words. When used at the initial stages of any architectural project, it is the combination of key words that generates the project's visual references, on the basis of algorithms calculations and associations. This process in isolation may appear devoid of any thorough and constructed intellectual or artistic endeavour.

When searching 'cultural heritage', for example, some of the options given for more key words that will refine the research are as diverse as 'intangible', 'Indian' and 'family'; suggestions for the most common associations of key words populate the screen in what becomes a 'mix match' of bidimensional, formal references ranging from travel guide covers to anonymous monuments. From time to time, images with words appear and more sophisticated matches are made by the algorithm. It all depends on the words inserted in the search bar and their consequent combinations, which lay the foundations of a sequence of statistically successful associations of colours, shapes and compositions. What counts is the series of tags, not the concept or the theory underpinning the project, and even less the context of what is shown. The more an image is 'pinned' (saved) on somebody's board, the more often it will be proposed to other people in search of similar combinations of key words while assuring the utmost standardisation of final outcomes in doing so. The suggestions are endless, just like the repetition of similar types of images. The way Pinterest is experienced can vary a lot from architect to architect; however, what is certainly more objective is the fact that its use is widespread at all stages, and among the general public too. At the end of 2019, Pinterest reported having four billion active boards and 335 million monthly active users worldwide, 51 million more than at the end of 2018. Since its launch in the USA in 2010, the platform grew exponentially and is now more popular than Tumblr, one of the first image-based blogging platforms with 327 million monthly users. It would be interesting to evaluate both the immediate and lasting effects of the current COVID-19 pandemic on the usage and spread of these image-based platforms. Several professionals rely heavily on this platform, for a number of reasons, compulsively collecting enormous amounts of arbitrary images determined by calculations of intelligent machines, on the basis of what is statistically more successful from a purely formal point of view. This results in scores of students, architects and professionals in general, currently copying the same image without questioning its qualities, just reproducing it like mere 'builders of pretty pictures'. All the automated calculations at the origin of so many mood boards and presentations are alien to the complexity of the discipline and, their use and abuse, will contribute to make the creative side of the profession irrelevant in the long run. When the process of production of architecture and heritage starts as a purely visual and bidimensional exercise, it can only end up in a slightly different version of that same thing.

The ubiquitous formatting of the social

Historically, the 'value' of images in the architectural field has always evolved as a consequence of technological progress. Over the course of the past five centuries, machines have shaped the built environment, first with the standardisation of images thanks to the invention of printing press with movable types, then with the standardisation of things resulting from it (Carpo, 2001). If we consider Sebastiano Serlio's architectural manual 'Five Books of Architecture' as a precursor of modern illustrated architectural publications and Gutenberg's invention as the first technological revolution to influence the production and communication of architecture, then photography can be looked at as the second transfiguring invention while the rise of new media and digital images as the third one, currently unfolding. If, thanks to Gutenberg's invention, printed images were considered exact reproductions of an original mould and the technological innovation element was standing as guarantee of that fidelity to reality, the process is today inverted. The wider sociocultural context in which digital images are produced, be they photographs, graphics or scanned documents, also influences the way in which exchanges generate and develop. According to media theorist Geert Lovink, we have reached 'the hegemonic era of social media platforms as ideology' (Lovink, 2016, p. 2); the elevation of social media to ideology acknowledges the supremacy of the digital and the 'incomputable impacts of the ubiquitous formatting of the social. Treating social media as ideology means observing how it binds together media, culture and identity into an ever-growing performance of gender, lifestyle, fashion, brands, celebrity, and news – all of this, imbricated with the entrepreneurial values of venture capital and start-up culture' (Lovink, 2016, p. 2).

In a digitally produced image, the technological innovation element stands as a guarantee of manipulation and non-adherence to reality. The inversion in the perception process has not been followed by an equal inversion in the 'utilisation process'. We live in 'a post-literate world ruled by images' (Lovink, 2016), and, as demonstrated by Carpo, the architectural realm in particular has always been one of the most affected disciplines. Inevitably, not only it is important to question *how* images are researched and shared, but also *why,* for what purpose. The 'formatting of the social' operates as a standardising agent able to convert a critically valid observation into an ordinary self-gazing exercise. In this scenario, 'the value of the photograph as a marker of memory is debatable as it does not constitute anymore the trace of something we wish to remember but, on the contrary, it has developed into a curated moment in a parallel existence' (Dreyfuss & Buhagiar, 2018, p. 29).

Photography, heritage and its commodification

Buildings and monuments, besides landscapes and portraits, were the first obvious subjects of 19th-century photographers, firstly due to the long exposure times required and then to the interest of institutions in photography as a reliable recording tool allowing easy reproduction of its outcomes. Photographers in the 19th century were interested in leaving behind a trace of what could disappear (Walunsinski, 1994, p. 11). These were at the time perceived as non-interfering observers offering an impartial vision of reality; the opposite quickly became apparent, in that any photograph is the result of a number of choices (sight, light, exposure, etc.) in a specific context, at a specific point in time, and that it therefore embodies a specific way of seeing (Berger, 1972). Photography, intended as a tool which rather than representing reality produces a new reality thanks to its many interpretative layers (Colomina, 1994), is key in relation to the evolution of the production and preservation of heritage.

The reproduction and widespread distribution of certain architectural imagery, throughout mass media first and social media more recently, offer the possibility of elevating the status of an architectural design to that of an iconic state (Rattenbury et al., 2002, p. 57), thus contributing to the production of heritage on two main levels. On one level, the production of built heritage, through the crystallisation of the actual built fabric in a particular state and in context – here the image – will take the place of the object in the reality of the moment, inferring its heritage status to the building. On another level, the photographer's role in elevating the status of the design can be so successful that the life of the built fabric lives beyond its physical presence and permanently becomes imprinted within our collective memory through iconic imagery: in this case, the image itself may become heritage. The German pavilion for the 1929 International Exposition in Barcelona, designed by Mies Van der Rohe, is the most popular example of the phenomenon: the original structure had been dismantled soon after the event but due to the success of a few iconic photographs that were reproduced countless times through mass media, the pavilion was rebuilt between 1983 and 1986, on the basis of those initial photographs (Blundell Jones, 2012). Today, 90 years later, the majority of the snapshots that visitors to the Barcelona Pavilion take and post online are manifestly influenced by that first set of images, reproducing a specific vision now permanently imprinted within our collective memory. With respect to the mechanical reproduction which makes these processes possible, it is a founding principle of photography. This is aptly summarised in the expression *photographic reproduction* which has influenced the debate around the authenticity of photography and its value as an art form, especially following Benjamin's seminal work (1968). On this characteristic, John Berger argued that when the artwork is multiplied through technical reproduction, so does its meaning, which is then further fragmented through its various modes of distribution (Berger, 1972, p. 19). If the same principle is applied to the reproduction of heritage photographs on algorithm-dominated social media for lucrative purposes, the different layers of critical understanding and potential interpretations will not only be fragmented, but flattened and commodified too, resulting in an eventual commodification of heritage itself, or to the least of its representation.

In an ironic twist, the widespread use of well-known heritage buildings, facades and monuments as advertising spaces for commercial purposes is a clear symptom of how the perception, from both institutions in charge and general public, is being affected. Fort St Angelo in Malta, for instance, a well-known 16th-century bastioned fort that withstood the Great Siege of Malta in 1565 and from which Caravaggio audaciously escaped in 1608, was recently promoted on social media as a 'unique advertising space' (Heritage Malta, 2019); projections promoting different kind of events or urging people to stay indoors during the pandemic, frequently appeared on the bastions overlooking the Grand Harbour, in a manner that is reminding of a common billboard. If, on the one hand, it is more than legit to put heritage structures to use and improve their financial viability, on the other hand, it is important to question the system which makes the brazenly commercial nature of these initiatives seem obvious and almost inevitable. It fits into the current all-encompassing commodification processes that social media are part of and that they encourage.

Even though the role of cultural heritage as a central element in processes of regeneration and sustainable development of cities and regions is increasingly explored the current focus on online presence, image and communication, risks that some interventions in heritage context become, in some cases, reduced to a search for the photogenic, the 'likable'. The ascent of social media further unlocked new levels of the cultural shift in the value of images, now from the image as a document or as a memory to the image as a commercial product and a marketing vehicle.

Rediscovering images as a research tool

Today, taking a photograph has become so common and informal that we are underestimating the value pictures can have, including both in terms of the interpretation of heritage and as a document for posterity. There is an immediate propension to use the phone's camera to record almost anything happening, from events to the next new dish. As a result of the current uninterrupted flow of production and consumption of digital image-based exchanges, the value of photographs as research tools has also been neglected. The prevalent use of camera phones of ever-improving technical abilities could nevertheless instigate new documentation and research project, using citizen science models. There are methods to look at buildings, heritage and their interaction with the photographic medium, which go beyond the immediate consumption and gratification from other people's validation of our own vision. However, these have increasingly included automated processes such as total stations with inbuilt photographic camera, drone photography and 3D scanning, further increasing the divide between scientific documents and photography. In these instances, the photographer is mainly reduced to a role of operator, required to leave out any level of interpretation. This seems in contradiction with the initial intents of the 19th-century photographic missions. Many of the photographs taken are aimed at social media, as an attempt at expressing individuality – or testing its validity – within specific parameters. In most cases, the aestheticisation contingent to achieving social media success reduces heritage to the status of mere prop or backdrop to the performance. Phenomena like the rehabilitation of Brutalist architecture – probably a success story in this instance – brought to the fore by social media groups publishing aesthetically pleasing photographs of concrete buildings, or the cataloguing of abandoned heritage sites on Instagram under the hashtag 'ruin porn' are just two examples of how the aestheticisation of photographs values formal qualities over any other. A photograph which lacks understanding of all the different layers and which does not allow cross-contamination between disciplines will not be able to maintain value beyond the moment of its publishing.

At the same time, it is difficult to deny the prominence of social media and visual culture and to ignore its potential to generate new fields of research in heritage sustainability. It is, for instance, the combination of low-cost flights and global mobility in general, as well as the increase in mobile photography that have pushed people to create (or 'curate') a World Heritage Sites bucket list, increasing visitor pressure on already very popular sites, where most people take the same photograph in front of 'the' heritage attraction. The thousands of images that social media users generate reiterate a standardised and almost neo-conservative perspective on heritage as something static. In truth, heritage is not static at all by definition, being the link between past present and future, in constant evolution.

Heritage management is going through a process of change in both theory and practice, from focusing on isolated built heritage assets towards a landscape-based approach (Veldpaus et al., 2013), and so should heritage visual documentation: intangible notion like social context, participation and economic sustainability, which are being adopted as driving parameters for decision-making processes, should be given priority at all stages, from initial documentation to communication and promotion of the project. Heritage-related images taken for investigative purposes, be they technical or artistic, are crucial in defining interventions, but are also a future heritage in support of heritage that will perhaps inform and influence future choices. Photographs of 1933s CIAM have provided key insight information on a turning point of architectural culture while becoming heritage in their own rights as testimony, for example. Photographs of the *Mission heliographique* are also considered heritage, while having been used to take important restoration decisions regarding some of the most prominent French monuments. Architects have always looked at images and photographs as a source of precious historical information. What photographs will be used in the future for heritage projects? Will today's images be able to transmute into heritage supporting heritage or will they be just masses of data to mine? In a context in which Instagram filters and Pinterest boards are the new visual and primary language and tool for self-expression, consequences on the knowledge currently being created have to be analysed.

New forms of knowledge

Architecture and conservation are forms of knowledge, and as any form of knowledge, they should be developed and implemented through scientifically valid research. There are different methods of judging and recognising valid research. Jeremy Till suggests in 'Architecture research: three myths and one model' (2008) to apply the triple test of originality, relevance and rigour. When applying this triple test to the photographs and visual documents produced for and exchanged through algorithm-dominated platforms, one finds that none of the conditions are met. Algorithms do not provide originality, for what we see when typing any key word is the same series of images that millions of users are looking at out there. Also, as proven by scandals such as the Cambridge Analytica in 2018 (Varela-Rodriguez & Vincente-Marino, 2020), algorithms tend to feed users always and increasingly with the same typology of content. For instance, the more images of a specific restoration approach are 'pinned', the more the Pinterest Smart Feed will show similar images with the same hashtags – hence promoting the above-mentioned pervasive formatting of the social that standardises and flattens complexity to purely formal and superficial elements. This excludes the element of relevance too: a purely formal and superficial element cannot add to the store of knowledge of architecture in a relevant way, unless perhaps as an analytical outlook on a trend. What algorithms have is rigour, the third fundamental element of the test, but it lacks the critical filter essential to research. Perhaps artificial intelligence will eventually manage to emulate a human, informed, critical outlook, but until now criteria such as number of visualisations, number of clicks, leads and search engine optimisation have been the dominating parameters of social media rigour. These criteria do not belong to the sphere of architecture but to the one of marketing only. In order to rediscover the value of images as research tools, it is crucial to prioritise one's own critical filters over what is suggested by the online platform, and to treat the latter as nothing more than a support or dissemination tool. Pinterest, for example, can be a valid support tool during non-initial design stages, just like a Tumblr page or an Instagram profile can be curated spaces for meaningful archiving and debate. The profile Text Catalogue (2020), for instance, has managed to make use of social media, and Instagram in particular, as spaces for the synthesis and questioning of the approach to Maltese heritage and built environment in general. They describe themselves as a group 'Analysing, documenting, critiquing and imagining architectures of varied territories' (Fleri Soler & Darmanin, 2019). In doing so, they challenge the status quo of the perception and preservation of local heritage, they advance new themes of debate that should be central to heritage sustainability like postcolonial issues, and they raise awareness on the importance of reviewing and perhaps widening the current definition of what qualifies as 'heritage' and what the role of architects really is in that respect. Text Catalogue is just an example of how images can be vehicles for the communication and debate of complex research questions and how social media can make their message very direct without flattening it. The problem lies in the way that message is spread (or not) by the algorithm, which will always prioritise contents on the basis of commercial parameters.

If knowledge depends on the tools that are used to create it and disseminate it, it is clear that the nature of image-based knowledge currently being produced is changing drastically and that the lack of critical understanding that social media seem to have contributed to cause might enhance commodification processes that will not be sustainable in the long run because of the ephemeral criteria on which they are based.

Barthes in his essay dedicated to an inquiry into the nature and essence of photography, in which he reviewed how different layers of meaning within the photographic message can be manufactured at different levels of production, listed among others, aestheticism, 'photogenia' and syntax as primary procedures (Barthes, 1982, p. 200). In an effort to mitigate the potential negative impacts of social media dynamics and understand the nature of the images that are shaping a new form of knowledge, aestheticism, 'photogenia' and syntax can be employed as guiding analysing parameters when using and producing images.

Aestheticism involves using visual coding systems borrowed from other art forms – for instance mimicking the visual order of a painting. John Berger also acknowledged the similarities between public images and oil paintings by saying that 'Publicity has in fact understood the tradition of the oil painting more thoroughly than most art historians' (Berger, 1972, p. 135). Social media images descend from this tradition, having adopted the most recent marketing and advertising strategies in combination with digital postproduction tool like filters and effects. The 'likable' is deeply rooted in aestheticism. The consequences of aestheticism on the validity of heritage 'images as producers of knowledge' are of an interpretative nature: formal composition of images and visual codes are a fundamental layer to be read and understood, in order to extrapolate additional information. The contribution of this particular aspect to the validity of an image depends on the understanding and interpretation of aestheticism elements in it.

The concept of 'photogenia' in this context refers to the fact that subjects are deliberately made beautiful in order to send out a more convincing message by manufacturing an image that is more charming in its reception, through the use of certain gear – or, in the case of Instagram photographs, of certain filters and/or hashtags. Common portable devices have become so sophisticated, and postproduction filters and editing software for photographs so readily available, that photogenia has reached an unprecedented level of democratisation. Photogenia can be the most problematic and deceptive characteristic of social media images for knowledge creation purposes, as it shifts the attention on artificial elements that might be antagonistic to the understanding of the building portrayed by the image.

Syntax refers to a number of photographs which may come together and form a sequence offering an alternative reading or interpretation, in a way that is reminding of Aby Warburg's visual experiments and the most recent squares of Instagram profiles. Syntax may become the key to discovering new values in image-based exchanges: perhaps the quantity of available photographs will be so great, and their individual originality so poor, that only by associating them in alternative ways, not based on commercial analytical calculations of algorithms, they will be able to support a new form of knowledge. Syntax is also a great opportunity for experimentation in ways of interpreting and creating knowledge.

Conclusions

All images, when they are the result of a critical exercise and not of a casual act, have the potential to become 'heritage in support of heritage'. Photographing and production of images processes can provide valid research tools when informed by theoretical parameters which test and question the value of the final outcome; aestheticism, photogenia and syntax are just three basic parameters that can help generating a distinction between the mere 'likable' and valuable research tools. Critical understanding of the context needs to be prioritised over superficial, aesthetically pleasing techniques, whose nature is intrinsically commercial.

In the sheer number of images that flood our everyday lives, it is key to identify marketing techniques like knolling, the process of arranging different objects so that they are at 90-degree angles from each other and photographing them from above or focusing on details for the sake of the likable – which differs from sampling, and the casual acts which are (most of the times) unconsciously influenced by those techniques. This typology of images, between the casual and the commercial, might acquire different value, as a mass of data to explore and extrapolate technical information from. The huge quantity of images produced on or in relation to heritage can provide useful insights on movements of people in specific sites, potential overcrowding and perhaps even pollution monitoring. Advanced technologies like artificial intelligence can also support data mining processes of social media image-based exchanges in relation to heritage, thus creating new values which turn negative aspects of the above-mentioned commodification processes into useful tools in support of research.

The current shift in the value of the image towards its commodification, caused by the digital revolution and the rise of new media, needs added theoretical attention in relation to the safeguarding of sustainable heritage. While some positive aspects have been highlighted in relation to opportunities especially in the architectural field, to open up the debate and fuel new and independent criticism, potentially negative impacts seem far more reaching and lasting. An image-obsessed and fast-paced contemporary culture is in fact prioritising instantaneity over durability, refuting a basic principle of sustainability and antagonising the potential value of the visual with its immediacy. To be able to redefine the value of the image, its modes of production and dissemination need to be probed in relation to the multiple contexts in which it is being articulated. The digital revolution is setting new standards that need to be taken into account in updating the framework for sustainable heritage (and architecture in general). Intangible elements that go beyond the technical and the factual or the represented have to be investigated as they evolve in relation to the technological progress which is propelling them. With the relative accessibility to new digital tools, new categories of heritage may arise, and their sustainability may be ensured – not just a hashtag – if based on scientifically valid research and procedures. The parameters of scientifically validated design research will need to be reviewed as image-based knowledge itself will continue evolving just like it historically always did in relation to technological progress. The transdisciplinary approach that this shift requires might also encourage cross-fertilisation between sectors only apparently distant, like sustainable heritage and start-up culture, and reengage with the community. The place of the image within the heritage realm offers the opportunity to build bridges which transcend the traditional taxonomy of representation. Relatedly, the recurring presence of heritage within visual media also cross-pollinates a potentially vigorous research field for sustainable heritage if it is interrogated in all its complexity and not purely as material support. An instance to reconsider the place of time in the image for it potentially embraces the interactions between design and sustainable heritage, addressing the divide between the object and the project.

Notes

1 Findeli (1998, p. 67) states that design views the world as a *project*, whereas the sciences tend to view it as an *object*.
2 This quote, often repeated, finds its origin in a letter from Aby Warburg to Karl Vossler dated 12 October 1929. For context and a full reference to the letter, see Quiviger (2013, pp. 242–243 and fn. 26).

References

Barthes, R. (1982). The photographic message. In S. Sontag (Ed.), *A Roland Barthes reader* (pp. 194–210). London: Farrar, Straus and Giroux, Inc.
Benton, T. (2014). Le Corbusier photographe secret. In M. Bedarida (Dir.), *Le Corbusier: aventures photographiques* (p. 63). Paris: Editions de La Villette.
Benjamin, W. (1968). The work of art in the age of mechanical reproduction. In H. Arendt (Ed.), *Illuminations* (pp. 217–251). New York: Harcourt, Brace & World.
Benjamin, W. (1980). A short history of photography. In A. Trachtenberg (Ed.), *Classic essays on photography* (pp. 199–216). New Haven, CT: Leete's Island Books.
Berger, J. (1972). *Ways of seeing*. London: Penguin Group.
Berger, J. (2013). *Understanding a photograph*. London: Penguin Group.
Blankenbehler, B. (2015, September 4). Post-modernism of Venturi and Philip Johnson: Engaging the image of culture. *Architecture Revived* [Blog post]. Retrieved from: http://architecturerevived.com/post-modernism-of-venturi-and-philip-johnson-engaging-the-image-of-culture/ (Accessed on 03 May 2021).
Blundell Jones, P. (2012). The Photo-dependent, the photogenic and the unphotographable: how our understanding of the modern movement has been conditioned by photography.

In A. Higgott, & T. Wray (Eds.), *Camera constructs: Photography, architecture and the modern city* (pp. 46–60). Farnham: Ashgate.

Buhagiar, K., & Dreyfuss, G. (2018). The syllable of a stammerer. In J. Abela, & E. Buttigieg (Eds.), *Parallel existences: The Notarial Archives. A photographer's inspiration. Alex Attard* (pp. 27–30). Malta: Kite Group.

Carpo, M. (2001). *Architecture in the age of printing.* Cambridge, MA: MIT PRESS.

Cassar, M. (2009). Sustainable heritage: Challenges and strategies for the twenty-first century. *APT Bulletin: Journal of Preservation Technology, 40*(1), 3–11.

Colomina, B. (1994). *Privacy and publicity: Modern architecture as mass media.* Cambridge, MA: MIT PRESS.

Colomina, B., & Wigley, M. (2017). *Are we human? Notes on an archaeology of design.* Zurich: Lars Muller Publishers.

De Mondenard, A. (1994). Des documents d'une apparente objectivité. In Coll. *Photographier l'architecture 1851–1920* (pp. 17–32). Paris: Editions de la Reunion des Musees Nationaux.

Ferrando, D. (2018). Spreading the word: Victor Hugo's Tumblr account. *Architectural Review.* Issue December 2018–January 2019. Retrieved from: https://www.architectural-review.com/essays/spreading-the-word-victor-hugos-tumblr-account (Accessed on 03 May 2021).

Fernandez, H. D. (2020). Le Corbusier. Towards the origins of architecture. In K. Buhagiar, G. Dreyfuss, & J. Bruenslow (Eds.), *The founding myths of architecture* (pp. 132–147). London: ArtificePress Limited.

Findeli, A. (1998). Will design ever become a science? Epistemological and methodological issues in design research, followed by a proposition. In P. Strandman (Ed.), *No guru, no method: Discussion on art and design* (pp. 63–69). Helsinki: UIAH.

Fleri Soler, E., & Darmanin, A. (2019). What is text catalogue? *TextCatalogue* [website post]. Retrieved from: https://textcatalogue.com/manifesto/ (Accessed on 03 May 2021).

Gadamer, H. (1975). *Truth and method.* New York: Seabury Press.

Gui, I. (2016). Photographies et monuments historiques: les fonds photographiques de la médiathèque de l'architecture et du patrimoine. *Livraisons de l'histoire de l'architecture, 31,* pp. 91–101. https://doi.org/10.4000/lha.618.

Heritage Malta (2019, July 2nd). A unique advertising space! #FortStAngelo #GrandHarbour. [Facebook post]. Retrieved from: https://www.facebook.com/HeritageMalta (Accessed on 03 May 2021).

Lange, A. (2019, September 7). Is Instagram ruining architecture? *The New York Times,* Retrieved from: https://www.nytimes.com/2019/09/07/opinion/sunday/instagram-architecture-design.html?smtyp=cur&smid=tw-nytopinion (Accessed on 03 May 2021).

Lovink, G. (2016). On the Social Media Ideology. *E-flux Journal.* Issue 75, September 2016. Retrieved from: https://www.e-flux.com/journal/75/67166/on-the-social-media-ideology/ (Accessed on 03 May 2021).

Mitchell, W. J. T., & Hansen, M. B. N. (2010). Introduction. In W. J. T. Mitchell, & M. B. N. Hansen (Eds.), *Critical terms for media studies* (pp. xiii, xxii). Chicago, IL: The University of Chicago Press.

Parnell, S. (2016). Post-truth architecture. *The Architectural Review,* Issue Dec 2016. Retrieved from: https://www.architectural-review.com/essays/post-truth-architecture.

Quiviger, F. (2013). From Paris to Rome, Hamburg and London. Aspects of the afterlife of Giordano Bruno in the twentieth century. In H. Hufnagel, & A. Eusterschulte (Eds.), *Turning traditions upside down: Rethinking Giordano Bruno's enlightenment* (pp. 232–248). Budapest: CEU Press.

Rattenbury, K., Cooke, C., & Hill, J. (2002). Iconic pictures. In K. Rattenbury (Ed.), *This Is Not architecture: Media constructions* (pp. 57–90). London: Routledge.

Schwendener, M. (2020, May 14). This atlas of art & memory is a wonder of the modern world. *The New York Times,* web article.

Sontag, S. (1979). *On photography.* London: Penguin.

Till, J. (2008). Architectural research: Three myths and one model. *Building Material, 17,* 4–10. Retrieved from: https://jeremytill.s3.amazonaws.com/uploads/post/attachment/34/2007_Three_Myths_and_One_Model.pdf.

Valery, P. (1939/2015). The centenary of photography. In J. Mathews (Trans.), *Collected works of Paul Valery, Volume 11: Occasions* (pp. 158–167). Princeton, NJ: Princeton University Press.

Varela-Rodriguez, M. & Vincente-Marino, M. (2020). Automated image extraction from Instagram for social research: a technical and ethical exploration. In *TEEM'20: Eighth International conference on technological ecosystems for enhancing multiculturality* (pp. 588–592). New York: Association for Computing Machinery.

Veldpaus, L., Pereira Roders, A., & Colenbrander, B. (2013). Urban heritage: Putting the past into the future. In *The historic environment: Policy & practice, Volume 4* (Issue 1), pp. 3–18. https://doi.org/10.1179/1756750513Z.00000000022.

Walunsinski, G. (1994). 'Photographie, photographe, architecture, le point de vue d'un photographe'. In A. de Mondenard, P. Néagu, G. Walunsinski (Eds.), *Photographier l'architecture 1851–1920* (pp. 11–16). Paris: Editions de la Reunion des Musees Nationaux.

27 Integrating urban conservation into urban planning

Arthur Parkinson and Mark Scott

Introduction

Historic urban environments are critical repositories of cultural heritage, not only in their buildings, monuments, public spaces and townscape but also for their intangible features such as creating local identity and sense of place. Historic urban environments have been formed through incremental change over many years in response to shifting urban dynamics to produce a complex and highly differentiated urban fabric in terms of structure, ownership and the historic periods represented. The shared characteristics of historic places make them primary cultural assets, which are increasingly being reframed and recognised for their potential to deliver wider economic, social, cultural and sustainability benefits (UNESCO, 2011). Within this context, conservation of built heritage has become firmly established as a central urban planning goal within the management of contemporary urban space. In this chapter, we aim to chart the evolution of conservation-planning and position its practice within wider urban planning discourses and agendas. We examine the emergence of conservation as a social movement and its early institutionalisation into international charters and legislation and highlight how conservation-planning shifted from the margins to the mainstream of planning practice in recent decades. We critically examine key shifts in conservation-planning practice, notably the influence of the so-called authorised heritage discourse (AHD), the instrumental use of urban heritage for urban regeneration purposes and the emergence of the Historic Urban Landscape (HUL) approach. The final part of the chapter places conservation-planning within the context of neoliberalisation and its impact on shifting conservation practices and the governance of urban heritage, and how 'we make use of' heritage in contemporary urban planning processes.

The emergence of urban conservation

Within an international context, conservation of the built environment first emerged as a social movement in the late 18th century resulting from political, social and economic modernisation in Enlightenment Europe (Glendinning, 2013). The movement gained further traction in 19th-century Europe as a reaction against the destruction of the historic built environment in the name of progress. This was particularly evident in the competing perspectives of Eugène Viollet-le-Duc on the one hand and John Ruskin on the other. Viollet-le-Duc, a French architect, and others outside France propounded the view that historic buildings could be improved through the use of modern techniques and materials. In contrast, Ruskin (who as primarily an art critic, and not an architect) – and later William Morris and the Society for the Protection of Ancient Buildings (SPAB), formed in England by a small group of enthusiasts in 1877 – were vigorous advocates of a more conservative approach, which rejected the validity of applying modern techniques to historic buildings for their improvement. Ruskin, Morris and SPAB viewed buildings as artistic creations and stressed the importance of tradition, craftsmanship and protection of the original fabric of buildings, leading to a technical focus on 'conservation repair' and the preservation of authenticity (Pendlebury, 2008). This approach has dominated professional conservation practice, which tends to focus on traditional craftsmanship, materials, design and artistic merit, and the preservation of these attributes as cultural patronage in accordance with professional knowledge. This discourse sees buildings as the work of a professional architect, 'retaining exemplar properties worth preserving' (Tait & While, 2009, p. 721). Thus, this approach to conservation-planning is focused on the *tangible* dimensions of built heritage such as building materials, façades, structures and building ensembles, which represent cultural significance (Veldpaus et al., 2013).

The early institutionalisation of built heritage protection into legislation in the early 20th century relied heavily on this social movement advocacy (generally dominated by wealthy, middle-class groups). For example, in the Netherlands, early efforts to protect built heritage were led by the Heemschut Heritage Association, established in 1911 by leading figures in Amsterdam civil society, which for decades campaigned against the radical transformation of Dutch urban centres (Nadin et al., 2014). Later, in the 1930s, an urban-orientated preservation lobby began to emerge in Britain, notably the Georgian Society established in 1937 (Phelps et al., 2002). Similarly, in the North American context, the National Trust for Historic Preservation in the United States and Heritage Canada first emerged as heritage advocacy groups (Kalman, 2014).

Influenced primarily by trends and conservation actors in Europe, at least until the middle of the 20th century, a significant number of international charters and conventions sought to establish key principles in relation to built heritage conservation. The earliest of these was the Athens Charter for the Restoration of Historic Monuments, which emerged out of the First International Congress of Architects and Technicians of Historic Monuments, 1931, to provide recommendations on the conservation and restoration of historic sites (Stubbs & Makaš, 2011). However, more significant was the signing of the International Restoration Charter (the Venice Charter) and the establishment of the International Council on Monuments and Sites (ICOMOS) at the Second Congress of Architects and Specialists of Historic Buildings in Venice in 1964. This charter set out principles for architectural conservation and restoration and had a significant influence upon the development of conservation practice in the decades that followed (Ahmad, 2006). Over subsequent decades, a series of international charters and conventions further formalised and advanced built heritage conservation practice, outlined in Box 27.1 below.

In post-war Europe, while there were significant efforts to record and protect historic structures (Larkham, 2003), the need for urban reconstruction following the devastation of war led to an emphasis on large-scale comprehensive redevelopment schemes for central cities. Comprehensive redevelopment of the historic urban fabric during this period was also linked to the need to respond to increasing urbanisation, increasing car ownership (and the desire to accommodate the motor car in urban areas) and investment

BOX 27.1 KEY INTERNATIONAL CHARTERS AND CONVENTIONS IN THE EVOLUTION OF BUILT HERITAGE CONSERVATION

- *Athens Charter for the Restoration of Historic Monuments* (First International Congress of Architects and Technicians of Historic Monuments, 1931) establishing agreed recommendations on the conservation and restoration of historic sites.
- The *International Restoration Charter (the Venice Charter)* and the establishment of the *International Council on Monuments and Sites (ICOMOS)* (Second Congress of Architects and Specialists of Historic Buildings, 1964). This charter set out principles for architectural conservation and restoration and had a significant influence upon the development of conservation practice in the decades that followed.
- The *World Heritage Convention* (UNESCO, 1972) was important in broadening the scope of cultural heritage to include not only monuments and sites but also groups of buildings and urban settings.
- The *Australia ICOMOS Charter for the Conservation of Places of Cultural Significance (Burra Charter)* (ICOMOS, 1979, revised 1981, 1988, 1999) was particularly significant in introducing social value alongside aesthetic, historic, scientific and spiritual heritage value.
- The *Convention for the Protection of the Architectural Heritage of Europe* (the Granada Convention) (Council of Europe, 1985) sought to ensure integration of architectural heritage protection within 'town and country planning' (article 1) and defined architectural heritage according to historical, archaeological, artistic, scientific, social or technical categories of interest.
- The *Charter for the Conservation of Historic Towns and Urban Areas (Washington Charter)* (ICOMOS, 1987) sought to clarify the application of the principles contained in the Venice Charter to historic urban areas.
- The *Convention on the Protection of the Archaeological Heritage of Europe (Valletta Treaty)* (Council of Europe, 1992) is notable in that it is primarily focused upon 'ascribed values' rather than on the material heritage.
- The *Nara Document on Authenticity* (ICOMOS, 1994) recognised the need to acknowledge a greater cultural diversity in the understanding of the concept of authenticity in heritage conservation.
- The *European Landscape Convention (Florence Convention)* (Council of Europe, 2000), concerned with all dimensions of the European landscape (natural, urban, peri-urban) balancing preservation and economic use value.
- The *Convention for the Safeguarding of the Intangible Cultural Heritage* (UNESCO, 2003) which built on the earlier 1989 *Recommendation on the Safeguarding of Traditional Culture and Folklore* (UNESCO, 1989), to establish the scope of intangible cultural heritage and makes recommendations on its protection.
- The *Council of Europe Framework Convention on the Value of Cultural Heritage for Society* (Faro Convention) (Council of Europe, 2005), which is concerned

> with the relationship between cultural heritage and human rights and democracy and, crucially, that 'objects and places are not, in themselves, what is important about cultural heritage. They are important because of the meanings and uses that people attach to them and the values they represent' (Council of Europe, 2020).
> - The *Recommendation on the Historic Urban Landscape* (UNESCO, 2011) defines urban heritage as the product of the layering of diverse values, traditions and experiences by successive and existing cultures.

pressure for residential and commercial developments (Freestone, 2000). In this context, conservation was at the margins of planning policy and practice and urban renewal was often characterised as 'starting from scratch', resulting in urban clearance and insensitivity to traditional urban morphology (Albers, 2006). For example, Larkham (2003) notes that during and immediately after World War II, several hundred reconstruction plans were drawn up for the majority of UK towns and cities, including both those suffering bomb damage and those relatively or completely unscathed. While post-war austerity initially prevented widespread clearance in practice, the 1960s witnessed a surge in interest in urban renewal. However, this period was short lived as a professional concern with conservation emerged alongside a rise in the popular challenge to comprehensive redevelopment and top-down planning (Taylor, 1998).

This contested landscape of the 1970s shifted again in the 1980s with a turn to the instrumentalisation of conservation (Pendlebury, 2008), and a wider recognition of the relationship between conservation and urban regeneration outcomes as heritage was reframed as a driver of economic development. This real estate–heritage nexus is long established in the United States, with Redaelli (2020) noting prominent examples of redeveloping historic buildings in the late 1970s/1980s for retail, dining and entertainment purposes, such as Faneuil Hall Marketplace in Boston and Harbor Place in Baltimore. In the United Kingdom, early emblematic examples include the redevelopment of Covent Garden, London, while the redevelopment of the Liverpool's Royal Albert Dock provided a template for post-industrial cities through revalorising industrial heritage combined with a new cultural economy, including the opening of Tate Liverpool (Fageir et al., 2021). Figure 27.1 illustrates the example of Nyhavn in Copenhagen, where the restoration of quayside merchant houses and warehouses is now one of the city's main tourist attractions. These successes resulted in heritage professionals further promoting the economic potential of heritage conservation, such as English Heritage's influential publication, 'The Heritage Dividend' (English Heritage, 1999). Similarly, the Dutch Belvedere Programme (1999–2009) provides a further example of shifting the framing of urban heritage management as a potential cost to promoting heritage assets as an economic opportunity. This programme established a national incentive programme focused on the reuse or transformation of historic buildings with the potential to boost an area's economy. For Janssen et al. (2017), this approach is part of a longer-term shift in Dutch practice from value assessment (or culture of loss) to value creation (a culture of profit).

Figure 27.1 Nyhavn, Copenhagen – an example of restoration of built heritage for regeneration. However, the area is now described as a tourist enclave and entertainment strip for visitors.
Source: Roberts and Eldridge (2012); Photo by Arthur Parkinson, 2008.

Values and knowledge

Traditionally, built heritage conservation *practice* has been dominated by conservation professionals, with an 'expert' knowledge and status that has allowed them to frame decision-making (Pendlebury, 2002; Smith, 2006). Rooted in a historical, Eurocentric heritage ideal, it tends to focus on expert knowledge and skills, outstanding universal value, a hierarchy of significance privileging architectural, historic, scientific and aesthetic value, and protecting the authenticity of tangible assets. Smith (2015) identifies the dominant actors and approach in heritage conservation over time, from an antiquarian bias dominant in the 18th century, wherein the archaeologist led decision-making, towards a commemorative bias in the 19th century, with the historian taking a leading role, to the emergence of an aesthetic bias in the 20th century, whereby the architect and architectural historian became the key professional groups. This parallels wider heritage debates whereby 'experts' have been able to control decision-making through an AHD – the expert account of the heritage story (Waterton et al., 2006). In this context, Waterton and Smith (2010) argue that dominant political, professional and academic practice in heritage has adopted particular narratives and meanings related to heritage, underpinned by the assumptions of these expert groups. Moreover, these expert narratives of built heritage often reflect dominant elite interests in society, leading to a prioritisation within conservation-planning that reflects elite tastes and values, for example the interests of a small group of enthusiasts in 19th-century Britain (Pendlebury, 2008) or the dominance of middle-class interests in contemporary Britain (Waterton & Smith, 2010).

Graham (2002) therefore argues that built heritage decisions are value judgements, made on the basis of one set of subjective and contestable values. However, subordinate groups in society do not possess the specialist knowledge to meaningfully participate in heritage discourse or, put another way, they have insufficient social and cultural capital to engage substantively or to influence those who define official notions of heritage. In this unequal context, Waterton et al. (2006) contend that heritage experts use their expert knowledge to maintain their position of power through an AHD, which becomes routinised and institutionalised in policy and practice. Though heritage experts may use their expert knowledge to reinforce their own power position deliberately, Waterton et al. further argue that this also happens inadvertently. Even with the best of intentions, the *expert* nature of the AHD – and therefore *exclusionary* nature – prevents parity of engagement between experts and non-experts. Therefore, where alternative values and priorities are not understood, these may not be properly represented in policy and practice. Furthermore, the division between 'expert' and 'non-expert' is problematic because so-called 'non-experts' possess a unique reservoir of experiential knowledge about the places they inhabit. Thus, people 'with strong ties to a place … are experts' (Schofield, 2014, p. 8) as much as the professionals who have traditionally valued heritage in the positivist epistemological tradition of the conservation movement. This suggests the need for more inclusive approaches that emphasise multiple values and the discursive construction of heritage (Wells, 2015).

While much of the literature suggests that heritage decisions are framed by an 'official' discourse that is contested and power-laden, it would be wrong to uncritically assume that the heritage community operates within only *one* authorised discourse, rather than to varying degrees being influenced by, and participating in multiple discourses. Similarly, to assume that the heritage community is entirely and exclusively coincident with one AHD would be a mistake: the discursive territory is complex. Nevertheless, the above discussion has implications for social justice and the democratic credentials of heritage policy and practice within the wider spatial planning system.

As While (2007, p. 646) observes, although there increasingly appears to be a societal consensus surrounding the value of preserving buildings of recognised significance for the wider public good and for future generations, 'questions about what precisely should be protected, why and in what form are much less clear cut'. Within the context of wider place-making strategies, heritage policy and related conservation practices are not only framed by professional discourses, but also intersect with wider societal values (Farmer & Pendlebury, 2013). These tend to be rooted in non-expert experiential knowledge (Fazey et al., 2006), often emphasising associative attributes such as symbolism, collected memories, place-attachment and place-identity, or manifest as nostalgia, and interconnected with intangible forms of heritage (Parkinson et al., 2016a). Moreover, where they evolve in contentious political contexts, conservation and place-making practice can reveal historic or latent social conflicts reflecting collective remembering, cultural politics and identities intertwined with the symbolic representation of the built environment (Neill, 2011). This is particularly heightened in divided societies or in places containing a legacy of past conflicts (Parkinson et al., 2016b). Despite this, and reflecting the dominant discursive terrain, spatial planning has traditionally paid only limited attention to intangible qualities of place (Stephenson, 2010), and social value has similarly often tended to be marginal in heritage practice (Jones, 2017).

Recognising this complexity, and reflecting the wider cultural turn in the humanities and social sciences, the heritage field witnessed a shift as far back as the late 1980s and 1990s towards a greater emphasis upon the plurality of heritage and the social dimensions of its value (Harrison, 2013), though it has been slower to have a substantive impact upon conservation practice. It is only with the turn of the 21st century that what Smith (2015) terms as an ecological bias in contemporary conservation has begun to emerge, suggesting a reorientation away from the heritage artefact towards the relationships between these objects and with people. According to this perspective, the conservation gaze broadens from a focus primarily upon the physical, to engage with the economic, social and cultural context, conceiving of urban settlements as a dynamic process as well as physical objects, where architecture and landscape become interwoven with cultural practice, and where the diversity of layered cultural experiences is emphasised. As such, the historic environment must be understood through this experience, and not simply professional measurement and observation. This does not mean that experts should defer to non-experts, but that they become more self-critical of their expert role, and of their normative commitment to, and knowledge of, socially inclusive heritage practices (Høland & Skrede, 2019). The practicalities of integrating this approach into heritage practice have been the subject of recent study (see, for example, Fouseki et al., 2020; Robinson & McClelland, 2020), yet it remains relatively marginal in heritage practice (Jones, 2017).

This ecological bias or paradigm is central to the HUL approach, based on the influential work of Francesco Bandarin and Ron van Oers. As one of its authors argues:

> Slowly it seems that we are moving away from 19th and early 20th Century concepts derived from the rather static approach to monuments preservation (i.e. the 'do-no-harm' posture) ... towards more dynamic processes in which the safeguarding of cultural significance plays a key role.
>
> (van Oers, 2010, p. 13)

The HUL approach, therefore, shifts emphasis from monumental architecture to the conservation of *urban values* that underpin the life of the city – these values need to be defined and redefined by each generation in order to arrive at sustainable interventions that protect and enhance these values to individuals, communities and wider society (Bandarin & van Oers, (2012). The focus on wider societal values signals a move away from an expert-dominated heritage narrative towards a democratisation of conservation-planning practice, emphasising the importance of exploring, negotiating and debating the meanings of built heritage. Within the HUL approach, the historic urban environment is viewed as a cultural landscape and as a repository of social history and community values (Taylor, 2016), with intangible aspects of heritage considered alongside tangible assets (Ginzarly et al., 2019). This approach also recognises that historic places are not static collections of buildings or heritage assets but are subject to dynamic forces that continue to shape the urban landscape. Therefore, the HUL approach represents a shift to understanding the longer-term evolutionary processes shaping historic environments, recognising that historic places will keep evolving, and where the emphasis is upon '*how* to develop sustainably without damaging ... cultural significance, instead of *if* one should develop at all' (original emphasis) (Veldpaus et al., 2013). Moreover, reflecting shifts within the wider heritage field, the focus shifts from the 'object' (i.e. an individual building or ensemble) to the larger scale and the relationship between the historic environment and the wider city and its geographical setting. Noting that urban conservation had become a specialist field, isolated from urban development and governance, the HUL approach therefore also seeks to integrate heritage conservation into wider processes of urban management, renewal, and economic and social development. Box 27.2 outlines the key steps in the HUL approach as outlined by UNESCO (2011). Despite its broader focus, the HUL is still underpinned by the

> **BOX 27.2 CRITICAL STEPS IN THE HISTORIC URBAN LANDSCAPE APPROACH (UNESCO, 2011)**
>
> The HUL approach
>
> 1. undertakes a full assessment of the city's natural, cultural and human resources;
> 2. uses participatory planning and stakeholder consultation to decide on conservation aims and actions;
> 3. assesses the vulnerability of urban heritage to socioeconomic pressures and impacts of climate change;
> 4. integrates urban heritage values and their vulnerability status into a wider framework of city development;
> 5. prioritises policies and actions for conservation and development including good stewardship;
> 6. establishes the appropriate public–private partnership for local management frameworks; and
> 7. develops mechanisms for the coordination of the various activities between different actors.

traditional object-focus narrative characteristic of the conservation movement. While this does not negate the significant contribution of the HUL to achieving inclusive conservation-planning, there is still some distance to travel (Fouseki, 2019).

The neoliberalisation of conservation-planning

The political–economic ideas of neoliberalism have become deeply entrenched within public administration in most parts of the world (Sager, 2011) with profound sociospatial consequences as increasingly the belief that the market should discipline politics at a variety of spatial scales has become adopted as orthodoxy (Jessop, 2002). The extensive literature on neoliberalism seeks to explore the shifting relations and balance between the state and market, as market ideology has become increasingly dominant within spheres of action traditionally led by government. This shifting balance has been characterised as both 'roll-back' and 'roll-out' neoliberalism (Peck & Tickell, 2002), combining both a commitment to extending markets while also deploying state power in the pursuit of market interests as a form of meta-regulation.

Over the last decade or more, the spatial planning literature has witnessed widespread accounts of the application of neoliberal ideas to understanding both spatial governance processes and development outcomes. Allmendinger (2016, p. 1), in a detailed critique of the neoliberalisation of planning, argues that planning has

> shifted incrementally but perceptively away from an area of public policy that was an arena where [urban development] issues could be determined in the public interest to one that legitimises state-led facilitation of growth and development by superficially involving a wide range of interests and issues

and that planners are themselves directly complicit in these shifts and changes. In this way, planning has increasingly moved towards acting as a facilitator of market-led development aided by a 'delivery state' ethos (Parker et al., 2020).

Extensive studies demonstrate that planning has been both the *subject* of neoliberalism (specifically deregulation and privatisation tendencies) and a *method* of neoliberalism as planning reforms provide an enabling agenda for business-friendly policies. In this regard, planning policy and practice has developed a wide ranging repertoire of neoliberal toolkits including entrepreneurial governance (Scott et al., 2018), fast-track planning legislation (Fox-Rogers et al., 2011), introduction of business-friendly zones and flexible zoning along with market-based incentives (Sager, 2011), emergence of new 'soft spaces' and governance arrangements for planning (Allmendinger & Haughton, 2013), privatisation of urban space (e.g. Biddulph, 2011) and increased use of planning gain within development management (e.g. Fox-Rogers & Murphy, 2015). Conservation-planning is located within this wider planning regime, with practice shaped and challenged by roll-back and roll-out neoliberal practices.

Similar to wider planning processes, conservation-planning has been reshaped through shifts in the governance of urban heritage and the shifting role of the local state in managing the historic urban environment. From leading and managing heritage, the role of the local state has increasingly moved to steering, facilitating and encouraging non-state actors to move centre stage. This includes an increased emphasis on business-friendly arenas or the use of 'soft spaces' to lead heritage initiatives. For example, Veldpaus and Pendlebury (2019) outline the English case and the increasing role of business improvement districts (BIDs) – business-orientated urban partnerships – in managing historic urban centres. They highlight how BIDs will often draw on and exploit (selective) heritage narratives for place-branding while emphasising delivery of regeneration outcomes and flexibility in managing or protecting heritage assets, noting that BIDs are accountable only to levy-paying businesses. Similarly, Scott et al. (2018), drawing on the Irish experience, highlight the role of business-friendly urban development partnerships in developing local growth strategies based on heritage-based place-branding. In this case, their research identifies the use of intangible heritage narratives to sell the city, while calling for more flexible planning decisions on actual tangible heritage assets. These governance arrangements provide examples of planning's new 'soft spaces', as outlined by Allmendinger and Haughton (2013). These are spaces that lie beyond and in parallel to formal statutory planning arenas, characterised by lower levels of political accountability, fewer opportunities for statutory public participation and 'fuzzy' or undefined responsibilities or boundaries.

These governance shifts highlight the dominance of business interests in shaping urban governance, with the rise of public–private partnerships, and an agenda focused on enabling growth rather than a concern with welfare or wealth redistribution. Under these conditions, interventions often focus on projects and spaces more readily experienced by 'outsiders' – tourists, place-mobile capitalists – than by local citizens (Scott et al., 2018). In this context, Ward (2010) identifies the importance of discourses and representations of the city in terms of the marketable urban experience and place-branding. From this perspective, heritage assets and historic places are increasingly mobilised by central and local state actors as a resource in competitive economic development strategies – heritage thus becomes central to how cities make reference to 'authenticity', 'uniqueness' and 'identity' as marketable resources aligned explicitly with place-marketing and branding (Biddulph, 2011). This moves beyond the so-called heritage dividend positioned by heritage agencies as a justification for conservation interventions, towards a wider mobilisation of heritage to re-image the city. This approach presents challenges for conservation practitioners as key urban strategies suggest the need for 'flexible' conservation approaches to facilitate wider development goals, while other social values are downplayed (Scott et al., 2018). As a result, Lagerqvist (2016) argues that long-term strategies and heritage projects that may not lead to measurable revenues in terms of tourism or employment may become side-lined. For more traditional conservation-planning approaches, this raises concerns surrounding wider cultural values or diminishing importance attached to cultural significance in heritage management. Moreover, associating 'heritage' with place-branding also neglects opportunities to foster a meaningful engagement between local citizens and a city's heritage.

The diminished role of the local state has also provided an opportunity for the voluntary sector to emerge as managers and owners of heritage as the state withdraws from its traditional roles. Negussie (2006), for example, charts the increasing role of voluntary bodies in built heritage conservation in Sweden since the 1990s due to a shrinking public sector and overall deregulation. Similarly, Pendlebury et al. (2020) note the growing role in the United Kingdom of long-established voluntary sector actors as local government resources become increasingly scarce, highlighting the example of York Conservation Trust, which now owns substantial property assets in York's historic urban centre. While Pendlebury et al. (2020) suggest that this example demonstrates the potential of the voluntary sector to provide stability to long-term heritage management, it also provides an alternative to a purely growth-based agenda for exploring heritage values. However, this potentially more empowering role is dependent on voluntary sector capacity and resources, which varies significantly across geographical contexts.

These trends intensified further in the wake of the 2008 global financial crisis and the emergence of a politics of austerity following widespread sovereign debt crises. Peck (2012) suggests that austerity 'reforms' represent a new and deepening phase of neoliberalisation with fiscal retrenchment going beyond the immediate management of a global financial crisis, instead forming part of the longer-term neoliberal project to reshape and redefine the state at a national and local level. As noted by Peck (2012), austerity measures were often devolved to municipalities or city administrations, thereby imposing local government spending cuts on municipalities already bearing the physical scars of disinvestment, disuse and decline, and in vacant and abandoned spaces of stalled urban development and public retreat. In a study of three Irish cities faced with these conditions, Scott et al. (2020) illustrate how urban heritage was 'rediscovered' by local authorities as a foundation for local growth strategies, but the actual means of protecting heritage is weakened through 'flexible approaches' and reduced resources and capacity. While heritage was being promoted to stimulate the local economy, central government heritage protection programmes were being drastically cut, while local authorities experienced staff reductions leading to a diminished capacity to act. Veldpaus and Pendlebury (2019) also demonstrate the impacts of austerity on further hollowing out the local state's conservation-planning capabilities and resources, further intensifying a planning roll-back agenda that had gathered pace over decades (see Allmendinger, 2016). In a case study of Newcastle, UK, Veldpaus and Pendlebury describe how a proactive local authority conservation team, involved in preparing conservation area management plans, character statements, modernising historic environment records and developing interpretive programmes, had experienced a halving of their staff in the post-crash period. This led to a narrow focus on core tasks, such as listed building consents. However, notably, the impacts of austerity across different countries are uneven, often depending on different baseline positions pre-crash. For example, in a comparative study of municipalities across the Netherlands, Ireland and England, Pendlebury et al. (2020) note how each local authority experienced austerity measures, but with Dutch municipalities experiencing cutback measures from a much higher base level of resource. In contrast, austerity had been imposed on English local authorities with a deep-seated exposure to ongoing neoliberal reforms over decades, while Ireland's heritage sector has traditionally been under-resourced within the context of a largely pro-development planning system.

Conclusion

This chapter charts the development of conservation-planning over the 20th century through to the present day. We illustrate the emergence of conservation-planning as a social movement towards its institutionalisation in international charters and conventions. Understanding the evolution of practice, as advanced by its proponents, gives an understanding of how professional ideas have evolved, from a focus on individual buildings or monuments of cultural significance to a broader focus on ensembles and the integration of heritage within the wider urban landscape. In more recent years, conservation-planning has moved away from a purely expert-dominated control of heritage debates, with the HUL approach promoted by UNESCO (2011) representing a key example of a more progressive and inclusive approach to exploring the multilayered values that underpin historic places. However, it is also important to consider conservation-planning as a constituent element of wider planning systems to understand its impact on broader urban development processes and outcomes. From a marginal position in the mid-20th century, the conservation of built heritage has increasingly moved centre stage, as 'heritage assets' were revalorised for their potential to underpin property-led regeneration and wider place-branding strategies. Key to the mainstreaming of conservation-planning has, therefore, been the ability of heritage professionals to demonstrate its value in terms of economic potential and thereby a political justification for sustaining heritage protection and financial supports (Pendlebury et al., 2020). Early successes, such as Covent Garden in London, Temple Bar in Dublin, Nyhavn in Copenhagen (Figure 27.1) and Faneuil Hall Marketplace in Boston, provided a global template for the adaptive reuse of heritage assets as the cornerstone of flagship regeneration projects, emphasising the economic use of urban heritage. Through demonstrating economic potential, conservation practice shifted from the margins to the mainstream of planning and urban policy, and in this shift, 'urban conservation has become an inescapable element of the way cities remake themselves in the twenty-first century' (Pendlebury & Strange, 2011, p. 361).

The wider political landscape dominated by neoliberal ideas provides an important backcloth to understanding both the framing of urban heritage and the role of conservation-planning in urban development. This political project presents various challenges for conservation-planning practice through roll-back and roll-out initiatives. Roll-back conservation-planning, heightened over the last decade through austerity measures, has often resulted in significant cutbacks in financial and human resources at the local state level, reducing technical capacity and a refocus on narrow core functions, while national funding for heritage faced similar reductions. However, as Allmendinger (2016, p. 10) argues, 'neoliberal planning is not simply about a crude removal of regulation but a sophisticated capture and roll-out of spatial governance'. In this way, roll-out neoliberalised conservation-planning legitimises pro-market values, often focusing on exploiting heritage in place-branding and marketing to demonstrate place 'authenticity' and 'uniqueness', but with less emphasis on heritage protection and management (Scott et al., 2018). However, the neoliberal influences on conservation-planning are not coherent or uniform across different spatial contexts, but are experimental (Allmendinger, 2016) and overlap spatially dependant inherited institutions and practice norms (Brenner et al., 2010). In this regard, neoliberalised conservation-planning tends to co-exist alongside rather than displace more traditional approaches to heritage management (Janssen et al., 2017). Table 27.1 provides a summary of the key characteristics and policy approaches in the evolution of conservation-planning, which now overlap in practice, rather than representing 'clean breaks'.

Neoliberalisation processes are not stable, however, and are subject to flux and challenge, clearly evident in recent years. For example, the impact of COVID-19 is unclear; however, the pandemic has continued to cause much uncertainty over 2020 and beyond in relation to the future of retail, tourism and other aspects of the urban consumption economy, alongside uncertain future demand for office space in urban centres. These factors present challenges to historic urban cores in terms of shifting consumption patterns, business closures and vacancy within historic districts. Drawing on the lessons following the financial crisis suggests that tangible heritage assets may be exposed to diminished resources, while once again heritage narratives are mobilised as historic urban cores attempt to bounce back from the economic consequences of the pandemic.

Table 27.1 A typology of conservation-planning policy approaches. While each column represents a chronological shift, the different approaches tend to co-exist and overlap, rather than representing 'clean breaks'

Traditional conservation-planning	Conservation and regeneration	HUL approach	Neoliberalisation of conservation-planning
Top-down professional control	Instrumental use of heritage	Integration of the goals of urban conservation into wider economic and social development.	Impacts of roll-back and roll-out neoliberal repertoires
Emphasis upon expert knowledge and skills, outstanding universal value, a hierarchy of significance (privileging architectural, historic, scientific and aesthetic value) and protecting the authenticity of tangible assets	Emphasis on demonstrating the economic potential of built heritage to justify conservation		Diminished central and local state resources
	Non-statutory urban regeneration strategies, property-led regeneration and design-led frameworks	The cultural landscape as a dynamic process that includes built structures (architecture) interwoven with cultural practices.	Increased involvement of non-state actors
			Market-based instruments (e.g. tax incentives)
Statutory protection of structures and areas of heritage significance through the planning system	Real estate–heritage nexus	Cultural heritage value understood through mapping of people's experience and not simply professional measurement.	'Flexible' and 'soft' spaces outside of statutory arenas
	Rise of facadism		A concern with heritage as place-branding
	A concern with profit		
A concern with loss and preservation		A concern with values	

Another issue of prominence in the 2020s has been the 'Black Lives Matter' movement, which in the United Kingdom encompassed debates on the symbolic importance of built heritage and monuments and included the high-profile toppling of a statue of Edward Colston, a slave trader, in the city of Bristol in June 2020. This demonstrates the importance of counterbalancing the emphasis on the economic potential of heritage with the importance of debating and exploring wider heritage narratives, including the decolonialisation of dominant discourses (McClymont, 2022). Advancing the democratisation of conservation-planning and demonstrating its potential social value – as a means of exploring and challenging the dominant modes of framing heritage – and also examining the potential for community-building and creating new discourses about justice *through* heritage are critical for the future of conservation-planning. All of these themes, ranging from the democratisation of heritage to the Black Lives Matter movement, the decolonisation of built heritage debates, and the impact of COVID-19 upon historic town centres, present significant and ongoing challenges that constitute a basis for future research at the interface of urban conservation and urban planning.

References

Ahmad, Y. (2006). The scope and definitions of heritage: From tangible to intangible. *International Journal of Heritage Studies*, *12*(3), 292–300. https://doi.org/10.1080/13527250600604639.

Albers, G. (2006). Urban development, maintenance and conservation: Planning in Germany – values in transition. *Planning Perspectives*, *21*(1), 45–65. https://doi.org/10.1080/02665430500397295.

Allmendinger, P. (2016). *Neoliberal spatial governance*. New York: Routledge.

Allmendinger, P., & Haughton, G. (2013). The evolution and trajectories of English spatial governance: 'Neoliberal' episodes in planning. *Planning Practice & Research*, *28*(1), 6–26. https://doi.org/10.1080/02697459.2012.699223.

Bandarin, F., & Van Oers, R. (2012). *The historic urban landscape*. Oxford: Wiley Blackwell.

Biddulph, M. (2011). Urban design, regeneration and the entrepreneurial city. *Progress in Planning*, *76*(2), 63–103. https://doi.org/10.1016/j.progress.2011.08.001.

Brenner, N., Peck, J., & Theodore, N. (2010). After neoliberalization? *Globalizations*, *7*(3), 327–345. https://doi.org/10.1080/14747731003669669.

Council of Europe (1985). *Convention for the protection of the architectural heritage of Europe (Granada Convention)*. Strasbourg: Council of Europe. Retrieved from: http://conventions.coe.int/treaty/en/Treaties/Html/121.htm (Accessed on 14 April 2011).

Council of Europe (2000). *European landscape convention (Florence convention)*. Strasbourg: Council of Europe.

English Heritage (1999). *The Heritage dividend: Measuring the results of English Heritage regeneration 1994–1999*. London: English Heritage.

Fageir, M., Porter, N., & Borsi, K. (2021). Contested grounds: The regeneration of Liverpool waterfront. *Planning Perspectives*. *36*(3), 535–557. https://doi.org/10.1080/02665433.2020.1804989.

Farmer, G., & Pendlebury, J. (2013). Conserving dirty concrete: The decline and rise of Pasmore's Apollo Pavilion, Peterlee. *Journal of Urban Design*, *18*(2), 263–280. https://doi.org/10.1080/13574809.2013.772884.

Fazey, I., Fazey, J. A., Salisbury, J. G., Lindenmayer, D. B., & Dovers, S. (2006). The nature and role of experiential knowledge for environmental conservation. *Environmental Conservation*, *33*(1), 1–10. https://doi.org/10.1017/S037689290600275X.

Fouseki, K. (2019). Re-conceptualizing 'historic urban landscapes. In V. Bharne, & T. Sandmeier (Eds.), *Routledge companion to global heritage conservation* (pp. 26–38). London: Routledge.

Fouseki, K., Guttormsen, T. S., & Swensen, G. (Eds.) (2020). *Heritage and sustainable urban transformations: Deep cities*. London: Routledge.

Fox-Rogers, L., & Murphy, E. (2015). From brown envelopes to community benefits: The co-option of planning gain agreements under deepening neoliberalism. *Geoforum, 67*, 41–50. https://doi.org/10.1016/j.geoforum.2015.09.015.

Fox-Rogers, L., Murphy, E., & Grist, B. (2011). Legislative change in Ireland: A Marxist political economy critique of planning law. *Town Planning Review, 82*, 639–668. https://doi.org/10.3828/tpr.2011.37.

Freestone, R. (2000). *Urban planning in changing world: The twentieth century experience*. London: Spon Press.

Ginzarly, M., Houbart, C., & Teller, J. (2019). The historic urban landscape approach to urban management: A systematic review. *International Journal of Heritage Studies, 25*(10), 999–1019. https://doi.org/10.1080/13527258.2018.1552615.

Glendinning, M. (2013). *The conservation movement: A history of architectural preservation: Antiquity to modernity*. New York: Routledge.

Graham, B. (2002). Heritage as knowledge: Capital or culture? *Urban Studies, 39*(5–6), 1003–1017. https://doi.org/10.1080%2F00420980220128426.

Harrison, R. (2013). *Heritage: Critical approaches*. New York: Routledge.

Hølleland, H., & Skrede, J. (2019). What's wrong with heritage experts? An interdisciplinary discussion of experts and expertise in heritage studies. *International Journal of Heritage Studies, 25*(8), 825–836. https://doi.org/10.1080/13527258.2018.1552613.

Janssen, J., Luiten, E., Renes, H., & Stegmeijer, E. (2017). Heritage as sector, factor and vector: Conceptualizing the shifting relationship between heritage management and spatial planning. *European Planning Studies, 25*(9), 1654–1672. https://doi.org/10.1080/09654313.2017.1329410.

Jessop, B. (2002). Liberalism, neoliberalism, and urban governance: A state–theoretical perspective. *Antipode, 34*(3), 452–472. https://doi.org/10.1111/1467-8330.00250.

Jones, S. (2017). Wrestling with the social value of heritage: Problems, dilemmas and opportunities. *Journal of Community Archaeology & Heritage, 4*(1), 21–37. https://doi.org/10.1080/20518196.2016.1193996.

Kalman, H. (2014). *Heritage planning: Principles and process*. New York: Routledge.

Lagerqvist, M. (2016). Reverberations of a crisis: The practical and ideological reworkings of Irish State heritage work in economic crisis and austerity. *Heritage & Society, 9*(1), 57–75. https://doi.org/10.1080/2159032X.2016.1246155.

Larkham, P. (2003). The place of urban conservation in the UK reconstruction plans of 1942–1952. *Planning Perspectives, 18*(3), 295–324. https://doi.org/10.1080/02665430307975.

McClymont, K. (2022) The Fall of Statues? Contested Heritage, Public Space and Urban Planning. *Planning Theory & Practice, 22*(5), 767–795. https://doi.org/10.1080/14649357.2021.1995255.

Nadin, V., Van Der Toorn Vrijthoff, W., & Zhou, J. (2014). *A sustainable future for the historic urban core - first briefing paper*. Delft, The Netherlands: TU Delft.

Negussie, E. (2006). Implications of neo-liberalism for built heritage management: Institutional and ownership structures in Ireland and Sweden. *Urban Studies, 43*(10), 1803–1824. https://doi.org/10.1080%2F00420980600838168.

Neill, W. J. (2011). The debasing of myth: The privatization of titanic memory in designing the 'post-conflict' city. *Journal of Urban Design, 16*(1), 67–86. https://doi.org/10.1080/13574809.2011.521014.

Parker, G., Wargent, M., Linovski, O., Schoneboom, A., Gunn, S., Slade, D., Odeleye, N.-D., Maidment, C., Shepherd, E., Doak, J., Elliot, T., Nicholls, V., Street, E., Dobson, M., Platts S., & Tasan-Kok, T. (2020). The future of the planning profession. *Planning Theory & Practice, 21*(3), 453–480. https://doi.org/10.1080/14649357.2020.1776014.

Parkinson, A., Scott, M., & Redmond, D. (2016a). Competing discourses of built heritage: Lay values in Irish conservation planning. *International Journal of Heritage Studies, 22*(3), 261–273. https://doi.org/10.1080/13527258.2015.1121404.

Parkinson, A., Scott, M., & Redmond, D. (2016b). Defining "official" built heritage discourses within the Irish planning framework: Insights from conservation planning as social practice. *European Planning Studies*, 24(2), 277–296. https://doi.org/10.1080/09654313.2015.1077782.

Peck, J. (2012). Austerity urbanism: American cities under extreme economy. *City*, 16(6), 626–655. https://doi.org/10.1080/13604813.2012.734071.

Peck, J., & Tickell, A. (2002). Neoliberalizing space. *Antipode*, 34(3), 380–404. https://doi.org/10.1111/1467-8330.00247.

Pendlebury, J. (2008). *Conservation in the age of consensus*. London: Routledge.

Pendlebury, J. (2002). Conservation and regeneration: Complementary or conflicting processes? The case of Graninger Town, Newcastle-upon-Tyne. *Planning Practice & Research*, 17(2), 145–158. https://doi.org/10.1080/02697450220145913.

Pendlebury, J. (2013). Conservation values, the authorised heritage discourse and the conservation-planning assemblage. *International Journal of Heritage Studies*, 19(7), 709–727. https://doi.org/10.1080/13527258.2012.700282.

Pendlebury, J., & Strange, I. (2011). Centenary paper: Urban conservation and the shaping of the English city. *Town Planning Review*, 82(4), 361–393. https://doi.org/10.3828/tpr.2011.23.

Pendlebury, J., Scott, M., Veldpaus, L., van der Toorn Vrijthoff, W., & Redmond, D. (2020). After the crash: The conservation-planning assemblage in an era of austerity. *European Planning Studies*, 28(4), 672–690. https://doi.org/10.1080/09654313.2019.1629395.

Phelps, A., Ashworth, G. J., & Johansson, B. O. H. (2002). *The construction of built heritage: A north European perspective on policies, practices and outcomes*. London: Ashgate.

Redaelli, E. (2020). Including preservation in planning: Albina and Portland's comprehensive plan. *Journal of the American Planning Association*, 87(2), 266–281. https://doi.org/10.1080/01944363.2020.1822753.

Roberts, M., & Eldridge, A. (2012). *Planning the night-time city*. New York: Routledge.

Robinson, J. S., & McClelland, A. G. (2020). Troubling places: Walking the "troubling remnants" of post-conflict space. *Area*, 52(3), 654–662. https://doi.org/10.1111/area.12616.

Sager, T. (2011). Neo-liberal urban planning policies: A literature survey 1990–2010. *Progress in Planning*, 76(4), 147–199. https://doi.org/10.1016/j.progress.2011.09.001.

Schofield, J. (Ed.) (2014). *Who needs experts?: Counter-mapping cultural heritage*. Ashgate: Farnham, Surrey.

Scott, M., Parkinson, A., Redmond, D., & Waldron, R. (2018). Placing heritage in entrepreneurial urbanism: Planning, conservation and crisis in Ireland. *Planning Practice & Research*. Advance online publication. https://doi.org/10.1080/02697459.2018.1430292.

Scott, M., Parkinson, A., Waldron, R., & Redmond, D. (2020). Planning for historic urban environments under austerity conditions: Insights from post-crash Ireland. *Cities*, 103, 102788. https://doi.org/10.1016/j.cities.2020.102788.

Smith, L. (2006). *Uses of heritage*. London: Routledge.

Smith, J. (2015). Applying a cultural landscape approach to the urban context. In K. Taylor, A. St Clair, & N. Mitchell (Eds.), *Conserving cultural landscapes. Challenges and new directions* (pp. 182–197). New York: Routledge.

Stephenson, J. (2010). People and place. *Planning Theory & Practice*, 11(1), 9–21.

Stubbs, J. H., & Makaš, E. G. (2011). *Architectural conservation in Europe and the Americas*. Hoboken, NJ: John Wiley & Sons.

Tait, M., & While, A. (2009). Ontology and the conservation of built heritage. *Environment and Planning D: Society and Space*, 27(4), 721–737. https://doi.org/10.1068%2Fd11008.

Taylor, K. (2016). The historic urban landscape paradigm and cities as cultural landscapes. Challenging orthodoxy in urban conservation. *Landscape Research*, 41(4), 471–480. https://doi.org/10.1080/01426397.2016.1156066.

Taylor, N. (1998). *Urban planning theory since 1945*. London: Sage.

UNESCO (1989). *Recommendation on the safeguarding of traditional culture and folklore*. Retrieved from: http://portal.unesco.org/en/ev.php-URL_ID=13141&URL_DO=DO_TOPIC&URL_SECTION=201.html (Accessed on 24 October 2020).

UNESCO (2003). *Convention for the safeguarding of the intangible cultural heritage.* Paris: UNESCO. Retrieved from: https://ich.unesco.org/en/convention (Accessed 24th October 2020).

UNESCO (2011). *The historic urban landscape approach.* Paris: UNESCO.

Van Oers (2010). Managing cities and the historic urban landscape initiative – an introduction. In UNESCO (Ed.), *World heritage papers, managing historic cities* (pp. 7–17). Paris: UNESCO.

Veldpaus, L., & Pendlebury, J. (2019). Heritage as a vehicle for development: The case of Bigg market, Newcastle upon Tyne. *Planning Practice & Research.* Advance online publication. https://doi.org/10.1080/02697459.2019.1637168.

Veldpaus, L., Pereira Roders, A. R., & Colenbrander, B. J. (2013). Urban heritage: Putting the past into the future. *The Historic Environment: Policy & Practice,* 4(1), 3–18. https://doi.org/10.1179/1756750513Z.00000000022.

Ward, K. (2010). Entrepreneurial urbanism and business improvement districts in the state of Wisconsin: A cosmopolitan critique. *Annals of the Association of American Geographers,* 100(5), 1177–1196. https://doi.org/10.1080/00045608.2010.520211.

Waterton, E., & Smith, L. (2010). The recognition and misrecognition of community heritage. *International Journal of Heritage Studies,* 16(1), 4–15. https://doi.org/10.1080/13527250903441671

Waterton, E., Smith, L., & Campbell, G. (2006). The utility of discourse analysis to heritage studies: The Burra Charter and social inclusion. *International Journal of Heritage Studies,* 12(4), 339–355. https://doi.org/10.1080/13527250600727000.

Wells, J. C. (2015). In stakeholders we trust: Changing the ontological and epistemological orientation of built heritage assessment through participatory action research. In B. Szmygin (Ed.), *How to assess built heritage? Assumptions, methodologies, examples of heritage assessment systems* (pp. 249–265). Lublin: Lublin University of Technology.

While, A. (2007). The state and the controversial demands of cultural built heritage: Modernism, dirty concrete, and postwar listing in England. *Environment and Planning B: Planning and Design,* 34(4), 645–663. https://doi.org/10.1068%2Fb32085.

28 Foregrounding ethics in conservation in Singapore

Issues, questions, and framework

Kang Shua Yeo and Jeffrey Kok Hui Chan

Introduction: architectural conservation *vis-à-vis* ethics

One major reason why conservation ethics is under-developed in Singapore is that we are often unaware that we are discussing ethics during the practice of conservation. Architectural conservation in Singapore generally references the general principles for the conservation of historic structures of the *Society for the Protection of Ancient Buildings Manifesto* (Morris, 1877), the *Athens Charter* (ICOMOS, 1931), and the *Venice Charter* (ICOMOS, 1964) in the formulation of "3R Principle" of "Maximum Retention, Sensitive Restoration and Careful Repair" by the Urban Redevelopment Authority in the late 1980s (Yeo, 2018, pp. 647–648, 652; URA, 2020a). Subsequently, *The Nara Document on Authenticity* (ICOMOS, 1994) as well as *The Burra Charter* (ICOMOS, 2013) were also referenced by conservation practitioners. The standards of ethics developed by the English conservator, architect, and educator, Bernard Feilden (Feilden, 2003, p. 6), were also introduced by educators who graduated from schools in the United Kingdom:

1 The condition of the building before any intervention and all methods and materials used during treatment must be fully documented.
2 Historic evidence must not be destroyed, falsified, or removed.
3 Any intervention must be the minimum necessary.
4 Any intervention may be governed by unswerving respect for the aesthetic, historical, and physical integrity of the cultural property.
5 All methods and materials used during treatment must be fully documented.

Architectural conservators also referenced ethical discussions from their counterparts in the museum sector, namely those codes and guidelines published by various professional institutions (AICFAIC, 1994; ICOM, 2004; CAC & CAPC, 2009; ICON, 2014). It is these "codes of ethics that are seen as the profession's necessary process of articulating collective ideals… while ethics in conservation practice are applied through a rationalistic scientific approach based on evidence and hard facts" (Malkogeorgou, 2006). According to Maurice Lagueux's heuristic for ethics, these could be considered internal issues about the practice of conservation (Lagueux, 2004). They form the paradigm of conservation practice underpinned by a rationalistic approach in the examination of past evidence and documentation in order to transmit knowledge into the future.

However, most of the time, these internal issues concern professional judgement on appropriate technical treatments or other material-based matters (Choay, 2001). Ethical dilemmas arise as Barbara Appelbaum rightfully pointed out that "conservation training is material-based, and yet ... dilemmas are not primarily material ones. ... Codes of ethics say that conservation treatments should be 'appropriate'. Appropriate to what exactly?" (Appelbaum, 2013, p. xviii). Such dilemmas are also complicated by external issues that are present in the practice of architecture, often in the form of tension resolution between different stakeholders of a project, mediating between the wishes of the client with legitimate building safety codes, design, aesthetics, authenticity, and integrity (Mansfield, 2008, p. 272).

Furthermore, most writings and codes on conservation ethics do not indicate how ethical dilemmas or even conflicts between ideals can be resolved but merely describes what ethical ideals are. Consider an example by Kalman (2014):

1   Conservation ethics as a moral practice: acting responsibly to humanity;
2   Conservation ethics as best practice: acting responsibly to the historic place;
3   Conservation ethics as best professional practice: acting responsibly to clients and colleagues.

Taken individually, each is a consistent tenet. But in conditions of professional practice characterized by "complexity, uncertainty, instability, uniqueness, and value-conflict" (Schön, 1983, p. 39), these tenets can be in conflict. For instance, how should conservation as a moral practice be reconciled with conservation as best practice when the valorization of historical places provide sustained employment and cultural enjoyment but compromises the authenticity of these places? In this way, and like many other writings on conservation ethics, Kalman (2014) has only offered a version of *descriptive ethics* for conservation. These tenets describe what is required in conservation as a moral or best practice; they do not yet prescribe *what ought to be done* in a practical situation. To begin to resolve the conflict between these principles will require further thinking on the *normative ethics* of conservation, where, for example, a professional conservator resolves this conflict by relying on exemplary and normative frameworks of utilitarianism, deontological ethics, or virtue ethics.

Definitions of ethics and morality in architectural conservation

At this point, it is worth providing a working definition of ethics and morality in architectural conservation. Defining these terms is important because both terms are commonly used to refer to the domain of values, and they are also used interchangeably. However, we argue that distinguishing these terms can reveal more meanings than conflating them.

Here, and among many meanings, ethics refers to the relations that we should have with other people (Margalit, 2010, p. 3). Invariably, these relations vary across different cultures. Nevertheless, relations characterized by respect for human dignity and autonomy, obligation to pursue justice, and maintain non-malevolence, constitute the core tenets of ethics in the "West." These relations can be defined either as acceptable, or aspirational. Different societies maintain different standards for acceptable behaviour, and exceeding these standards, for instance, in performing supererogatory acts, then sets the example for aspirational standards. On the other hand, to treat others below acceptable standards is deemed unethical.

Following this, morality refers to how human relations should be in virtue of our nature as human beings (Margalit, 2010, p. 2). To dehumanize another person, or to create conditions that lead to dehumanization, is therefore patently immoral. Morality demands fairness, dignity, autonomy, compassion, non-maleficence—surely among many other foundational values—for other people.

But in non-ideal conditions of everyday life, these moral demands can be in conflict with each other. Relying on an analogy from land-use planning (see Beatley, 1994), planners are agonized by land-use choices that require a trade-off between these fundamental moral values—for example between improving the dignity and autonomy of people by creating new employment opportunities through a massive shopping mall development and advancing fairness and perhaps non-maleficence by creating a green park instead. When moral demands conflict in this way, ethics is required for us to decide on how to satisfy each of these conflicting demands adequately. What counts as adequate, how adequate and why, and for whom it is adequate, then define the ethical relations between us and these moral others. In this way, different individuals, organizations, and societies are likely to adopt different ways of making ethical decisions despite sharing a more or less common basis of human morality.

Types of ethical questions in architectural conservation

Within moral philosophy as a discipline, there are three major kinds of ethics. They are, namely, descriptive ethics, normative ethics, and meta-ethics (Frankena, 1973, pp. 4–5). Each kind of ethics reveals a different aspect of conservation ethics and, in tandem, elicits a different set of questions.

First, descriptive ethics relies on empirical inquiry, rendered through historical or scientific methodologies, to describe or explain the moral or ethical dimensions of architectural conservation (see Frankena, 1973, p. 4). For example, the conservation of St. Pancras Railway Station in London required difficult choices to be made between the preservation of historic fabric and readapting this fabric so that it is fit for contemporary use (Allington-Jones, 2013). As far as these hard choices were concerned, they had been made. A study focusing on the descriptive ethics in the architectural conservation of St. Pancras Railway Station would entail the following questions such as where, and how, were these hard choices made? Are the trade-offs between preservation and adaptive re-use justified, and what were the criteria used to evaluate these trade-offs? To rely on an analogy, the examination of conservation ethics through the frame of descriptive ethics is like doing forensics; it is to study the nature and the process of hard choices that had already been made. The goal of descriptive ethics is to describe or explain the ethical insights and pitfalls of these choices and then draw out instructive lessons for future decision-making.

Second, normative ethics aims to find out what ought to be done in a difficult or problematic situation. In this way, normative ethics anticipate practical action. Although a historical case could still be hypothetically dissected by relying on the lens of normative ethics—for instance, by asking what ought to be done before a challenging situation in the past—this is, however, not the most useful way of applying normative ethics. To be useful, normative ethics provides action-guiding recommendations and the reasons that one should take this course, before a challenging situation in the present. In urban planning where heritage is valorized (see Choay, 2001), conservators often have to consider the hard question of whose interest to privilege, and why. To privilege the interests of tourism and new business owners (e.g. new hotels) may mean compromising the interests of lived-in residents and businesses. Livelihoods and welfare of different groups are at stake; they compete in uneven ways where very often, those being asked to bear the costs are not the ones who would benefit from the conservation effort.

In this situation, it is paramount that conservators engage with normative ethics in order to systematically clarify the difficult issues at hand, and bring to bear the different frameworks such as utilitarianism, deontological ethics, and virtue ethics, on these issues. The conservators may still not discover an unequivocal answer on what to do. But at the very least, they would have performed their professional due diligence in weighing and appraising these issues through different ethical frameworks and then share these findings with their stakeholders. To formulate this differently, without normative ethics, conservators would have to either rely on their intuitions or gut feelings or else succumb to power or the domineering political interests in these issues. Neither is appealing before these high-stakes issues—and this leaves normative ethics as the only practical and accountable option.

Finally, meta-ethics is neither concerned with empirical or scientific inquiries nor does it involve making or justifying any normative judgment (Frankena, 1973, p. 5). Instead, it is concerned with reflecting on the ethical concepts and categories used in conservation ethics. For example, Semes (2009) exemplifies the tension in conservation practices as a tug-of-war between the canonical position of liberal adaption of the historical fabric and the position of treating the historical fabric as an inviolable, quasi-sacred entity. A meta-ethical inquiry in conservation ethics would examine the ethical groundings of these positions and then explain if these positions could be justified as general action-guiding principles. A significant contribution to conservation ethics through the meta-ethics frame might entail the creation of a completely new principle to guide conservation and then proceed to explain why this is ethically superior to classical canons.

Here, we suggest that meta-ethics does play an important role in conservation ethics. But given the nascent state of conservation ethics, descriptive ethics and normative ethics are comparatively more pressing areas of inquiry. While the former provides case studies and reports that enable reflective practice and pedagogy in conservation ethics, the latter offers conceptual tools that, with repeated practice, can clarify morally tricky issues and improve the accountability of decision-making in conservation practices. The remaining sections of this chapter will focus on drawing out descriptive and normative ethical issues and questions through the two case studies.

Ethical dilemmas in conservation practices

An ethical dilemma presupposes a conflict between at least two moral requirements, and no matter what one does, one cannot satisfy both requirements. This results in a moral residue—signs of an unavoidable moral failure (Tessman, 2017). As a highly contested domain, the built environment is beset with many ethical dilemmas. For instance, should one build a mixed-use development on a greenfield site, further supposing that this development will create much-needed employment opportunities, or to conserve the existing greenery into a green park that is open and accessible to all (Chan, 2019)? No matter the decision, some regrettable residue will remain.

But not all ethics involve moral quandaries. There are issues where a particular moral demand is unequivocal and paramount relative to all other competing demands of the moment. The restoration of heritage features that have been eroded by time and decay is one such example. The dilemma is not about whether to restore or not to restore, but rather, it is about how to do it sensitively and appropriately while striking a prudent balance between authenticity, artistry, affordability, and future cost of maintenance. In other words, how not to botch the job (see Davis-Marks, 2020).

And there are other ethical issues where the moral demands are at the outset, unclear, and confounded. For example, the increasing presence of wildlife in cities appears to suggest a win-win outcome for people and fauna, but this also leaves a lingering doubt if the welfare of fauna can be advanced in a thoroughly artificial environment. But which framework—to cite only three, utilitarian ethics, deontological ethics, or virtue ethics—is most appropriate to clarify these issues? Unlike the previous question of "how," the question here is, "*What* is the most appropriate ethical framework that can be applied to further clarify these issues?" These ethical issues do not require immediate action, but, instead, a willingness to step back, reflect, and probe entangled and difficult issues to arrive at a clearer vision of what is morally at stake. After this, action can then ensue. Many other ethical issues do not immediately invoke normative action, and the scope of this chapter does not permit a fuller explication on this front. Nevertheless, ethical dilemmas warrant a rightful focus here because they are most commonly encountered in conservation practices.

Following this and closely related to ethical dilemmas are three orders of questions in conservation practices. First-order questions relate to internal and technical issues of conservation, and they are relatively straightforward. The solution is typically discovered by rational methods. For example, what sort of plaster composition should one use in the restoration and/or repair of historic lime plaster? A rational answer would be to analyse the historic lime plaster for its composition and the properties and composition of the materials for repair using lime, sand, and other aggregates if they are present in the historic sample. Unless these technical and material issues are ethically questionable (e.g. using a dye that relies on environmentally polluting production techniques), first-order questions are usually not ethically significant.

But these first-order questions often set the stage for second-order questions of *appropriateness,* where the ethical dimensions are far more salient. In the choice of *appropriate* materials for conservation, long-term chemical stability is one of the main criteria. For this reason, generic materials whose properties are well documented are preferred over proprietary products, where reliable and detailed information on their compositions are not easily available. Yet, in the world of construction where warranties against materials are demanded by clients, proprietary products with their material warranties have an ironic advantage over generic materials that do not have any warranties. Consequently, the choice is between choosing a proprietary lime plaster product without much documentation and track record over an extended time but comes with product warranty versus using a tried and tested slaked lime. Proprietary lime plaster products generally have undisclosed additives added. It is these undisclosed chemicals that pose a certain "risk" to the long-term chemical stability of the product when applied to the historic building fabric. In other words, choosing the appropriate materials will entail very different stakes and consequences in the future. This choice not only sets a precedent for future professional practices but also has the potential to irreversibly harm the historic fabric.

There are also third-order questions, where choices in conservation practices have large and significant social, political, and moral impacts. For example, how should one conserve the birthplace of Adolf Hitler, if at all (Katz, 2019)? Destroying the building or leaving it to dereliction will be tantamount to the deliberate destruction of Austria's Nazi history, but restoring it as a historical monument appears to endorse the values associated with Nazism, and may unintentionally create a rallying point for neo-Nazis. Consider another example, which is the UNESCO

World Heritage designation of George Town, Penang. While this designation has undoubtedly conserved the physical and tangible heritage, it has, however, triggered the erosion of much of the cultural and intangible heritage of George Town. The position of the boundary line between the core zone and buffer zone of conservation has also impacted existing residents and businesses unevenly, where certain stakeholders benefited disproportionately at the expense of others, especially renters, who were displaced to make space for new hotels, cafes, and other amenities catering to mass tourism (see Lee et al., 2008; Shamsuddin et al., 2012, 2018). In sum, third-order questions have recognizable ethical features and will require moral reasoning to clarify the stakes and consequences of decision-making.

Case Study 1: Traditional architecture heritage—Tan Si Chong Su (陈氏宗祠 or Ancestral Hall of the Tan Clan), Singapore

Tan Si Chong Su, also called Po Chiak Keng (保赤宫), is a Chinese temple located at 15 Magazine Road not far from Chinatown and the Singapore River (Figure 28.1). Constructed between 1876 and 1878 as an ancestral temple for the Tan clan, the building was gazetted as a National Monument on 19 November 1974. Between 1998 and 2001, the temple's custodian, Mr. Tan Khuan Seng, carried out work on the temple without obtaining permission from relevant authorities. Apart from repairs to the timber structures that were damaged as a result of termite activities and moisture, additional architectural ornaments were also added. From Mr. Tan's perspective, the work was a matter of spiritual merit-making. This concept is considered fundamental

Figure 28.1 Tan Si Chong Su's entrance hall before the 1998–2001 works, circa 1980. Source: Courtesy of Kang Shua Yeo.

to Buddhist and Taoist ethics, where merit is accumulated as a result of good deeds, acts, or thoughts. The act of repairing and enhancing religious buildings that house deities is considered a good deed. On the other hand, the then Preservation of Monuments Board (now Preservation of Sites and Monuments, National Heritage Board) and the Urban Redevelopment Authority, however, regarded the works as illegal alterations. Mr. Tan was subsequently prosecuted by the state on "a charge of renovating the temple without obtaining the consent of the Preservation of Monuments Board" and was fined $500 (Anon, 2003, p. H13; Chan, 2003; Yeo & Chong, 2015). This seemingly light penalty led the state to pass amendments to the Preservation of Monuments Act in 2009, which allowed it to levy harsher penalties for such offences (Chan, 2003; Chuang, 2009; Lee, 2009).

The question on whether the temple ought to be restored to the state before the works is left hanging in the air. Taking into account the fact that the temple had to raise considerable funds to carry out the works between 1998 and 2001, and since the building is currently in a relatively good state, is it justifiable to request that the temple restore its building, which they will have to fundraise? If the authorities allow this to pass, will this set a negative precedent and a mockery of the conservation regime? Or is an undertaking by the temple custodians to restore only at their next major planned works an acceptable compromise? How should one proceed?

In this case, it is first important to distinguish descriptive and normative ethics. Insofar as the descriptive ethics for conservation practices is concerned, this case highlights not only an example of acting irresponsibly to a historic place (see Kalman, 2014) but also the divergent moral perspectives of the custodian (or the client) and the regulatory authorities. While the former views the independent restoration of the temple as an integral act of his religious faith, the latter frames restoration as a scientific and technical endeavour, which has to further conform to established regulatory and international best practices. In other words, the former believes in enacting his faith to the deities as a moral good that is paramount above all other goods, which contrasts with the latter's definition of the moral good as the public good of a well-ordered society regulated by legitimate rules, standards, and laws.

Admittedly, these are not always independent spheres of moral good. After all, there is such a political entity as a theocracy. Nevertheless, there is usually also an intractable gulf between morality as religious good and morality as the public good. These are often incommensurable goods. It is prudent to give room to the religious good while creating institutions that can secure the public good—and ensuring that as far as possible, these two moral goods do not come into open conflict. On this, ethics is about what one *ought not to do* as well as typically about what one *ought to do*. A prudent action to take is to prevent being drawn into a hard dilemma between the religious good and the public good, where there is neither politically nor morally satisfactory resolution. Ineluctably, the shift to a harsher penalty suggests that the regulatory authorities understood this possibility and worked to prevent being again drawn into the hard choice of conserving the public good but at the expense of disparaging the religious good. In this case, descriptive ethics suggest scrutiny on what one ought not to do to prevent falling into a dilemmatic situation where there can only be unavoidable moral failures. A conservator who perceives this dangerous possibility ahead of time can then create institutional safeguards (e.g. the harsher levies), which may greatly prevent these hard dilemmas from surfacing in the first place.

On the other hand, the normative ethics of this case concerns the future of this temple and conservation practices in Singapore. What position ought a conservator take in this case? As a demonstration of moral reasoning, the application of ethical frameworks to this case is instructive. Here, utilitarian ethics and deontological ethics are applied to illustrate the process of moral reasoning.

Simply defined, utilitarian ethics connotes the moral good in terms of good outcomes for as many people as possible. Given that the temple had spent a considerable sum to carry out the questionable restoration, and given that it is a functional building still in use by many members who will doubtlessly be financially burdened by a corrective restoration—a project that will also deprive the use of the temple for a protracted period—utilitarian ethics suggests relenting from this corrective restoration. The justification of this suggestion is derived from the benefits and incentives of *not carrying out* a new restoration, which is the continuous usage of the temple, and freedom from being burdened with the costs of this restoration. Conversely, deontological ethics, which relate to the moral good in terms of duties and right actions independent of good outcomes, suggests that the basis of embarking on a corrective restoration is a matter of principle consistent with what is required in proper scientific and legal conservation. There is a principled necessity to rectify the questionable restoration in this temple and to set right a precedent for the future.

What ought a conservator do? Applying ethical frameworks to this case clarifies the moral stakes involved, and different ethical frameworks are likely to draw out different moral stakes. Frankena (1973, p. 3) suggests that in a scenario where there are conflicting duties, one can consider which duties should take precedence. In this illustrative analysis by utilitarian ethics and deontological ethics, which should take precedence? Should one leave the temple as is, for as many members as possible, and for as long as possible, take precedence over the rectification of the temple as a matter of principle, and in this process, nullifying the benefits of sunk costs and regular use? A conservator is correct to think that doing the right thing entails embarking on a proper restoration of the temple. However, this should not come at the cost of conceivable harm in terms of the new and considerable financial burdens imposed on the members, and depriving them of using the temple during the restoration project—especially when the temple is functional and does not suffer from any safety lapses. Unless the members are willing to bear these considerable costs, leaving the temple until the next major planned works are due can be justified as an ethical thing to do in this case.

Case Study 2: Modern architectural heritage—Golden Mile Complex, Singapore

Golden Mile Complex, originally known as Woh Hup Complex, is a strata-titled mixed-use development located between Beach Road and Nicoll Highway (Figure 28.2). It was developed as part of urban renewal plans in the late 1960s. The brutalist architecture was designed by Design Partnership (now DP Architects) and was completed in 1973. Foreign workers, especially those from Thailand, frequented the complex due to the area being a long-distance bus terminal plying the Singapore–Haadyai, Thailand route, since the late 1970s. Gradually, the complex started to have eateries and shops catering to these workers (Gutierrez & Wong, 1985). Controversy over the building started in 2006, when Nominated Member of Parliament Ivan Png described the building as an "eyesore," "vertical slum," and "national disgrace" due to the makeshift enclosures of the balconies, with the then Parliamentary Secretary for National Development Mohammad Maliki Osman agreeing with him in Parliament (Chang, 2006; Leng, 2006; Ng & Huang, 2006). Such scathing views were disputed by some residents and architects (Ng & Huang, 2006). Later in the same year, a collective sale of the building for the redevelopment of the site was first mooted (Fang, 2006; Teo, 2006). Several attempts at collective sale failed as potential developers found it risky, partly because the land's 99-year lease started in 1969 and millions are required to top up

Figure 28.2 The brutalist architecture of Golden Mile Complex.
Source: Courtesy of Darren Soh.

the lease (Lim, 2007; Tan, 2009). It was also difficult to reach the statutory 80% consensus by share value and strata area among owners for the sale to go through (Teo, 2006).

However, as buildings age, and especially when they are wrapped up in redevelopment disputes, their maintenance is not of priority by their respective management committees. The degradation of the living environment due to aging amenities will also contribute dissenting owners to change their minds on redevelopment. This is partially why a decade later in the mid-2010s, several architecturally significant buildings including Golden Mile Complex are again at risk of collective sale and demolition. Advocacy for the preservation of these buildings is led by the Singapore Heritage Society with the issuance of their advocacy paper *Too Young to Die* (SHS, 2018). Unfortunately, one such building, Pearl Bank Apartments, completed in 1976, was successfully sold for redevelopment in February 2018 (Luo, 2018) and was subsequently demolished in 2019. The widespread advocacy as well as the loss of Pearl Bank Apartments prompted the state to explore "options to facilitate conservation" for Golden Mile Complex (Anon, 2018).

However, conservation of the Golden Mile Complex is a far more complex proposition in no small part because of its groundbreaking design (Koolhaas & Mau, 1995, pp. 1069–1073). Designed with the notion of the megastructure, where aspects of architecture and urban design are merged into a single system (Maki, 2008, p. 40), the Golden Mile Complex was not only the first of its kind in the Southeast Asian context but also represented a regionally sensitive and novel approach in combining and concentrating a diversity of urban functions within its brutalist frame. In this way, the Golden Mile Complex has always accepted the disquieting reality of negotiating conflicts between different users, and where constant concession and compromise are prioritized over the certainty of any resolution. Among many negotiations, how should the needs of the residents be reconciled with that of the business owners, and with that from a multitude of the public? In rendering a more humanist modern architecture in the wake of Team X, the architects of the Golden Mile Complex had only provided an open template where life, and the chaos, of the city, is constantly fomented by cross-purposive encounters, conflicts, and serendipities within its frame.

This reading of the Golden Mile Complex is corroborated by the observation of its present predicament. Specifically, there are two salient issues. First, the Golden Mile Complex represents a category of modern heritage—brutalist architecture—that is "difficult" for the public to "like" (Chan, 2020, p. 115; cf. Mould, 2016). In other words, there are controversial disagreements on the value of conserving such modern heritage. Brutalist works of architecture are rarely perceived as aesthetically pleasing. But to appreciate these works on aesthetics ground is missing the point. Especially for the Golden Mile Complex, its architectural value resides in the transient negotiations and formation of urban life within its brutalist frame. But this perspective can be criticized as elitist and informed by a privileged reading of architecture history. For many—especially for the immediate community that relies on the Golden Mile Complex for their livelihoods—the Complex is a home and a community anchor that should be immune from both conservation and redevelopment. But for many others, the Complex is just a decrepit, unsanitary, and arguably inhumane development that should make way for redevelopment. Therefore, in the face of such controversies, who decides what is heritage? And what is a fair process for negotiating these differences? Here, the first task of descriptive ethics aims to map the contours of these disagreements, before proceeding to discover the different kinds of procedural institutions that can be used to address diverging values in disagreements.

Second, the Golden Mile Complex is a strata-titled development. This status is exposed to divergent owners' opinions on retention of status quo, which is favourable to conservation, or conversely for redevelopment, which threatens the Golden Mile Complex. As discussed, this strata-titled status is consistent with the original design intention, where different users could come to own a piece of this development and thereby were able to create the most diverse urban life as possible. But in the context of prospective conservation, this status exposed the Golden Mile Complex to intractable complications. Consider the commercial space owners of similar development, Queensway Shopping Centre in Singapore, that prefer status quo, where commercial footfall is high and shops enjoy brisk businesses. Most shops are privately owned and occupied by the owner. Most are not tenanted out. These owners would have since paid up for their strata-titled property in full. This meant that their overheads can be kept relatively low. This is opposed to owners of residential space in the same strata-titled development that favour redevelopment (Chong, 2019; Lim, 2019; Low, 2019).

Returning to the Golden Mile Complex, these conflicts are nearly symmetrical: shops and businesses are content with the status quo, but where owners of residential space tend to favour redevelopment. Will the onus of resolution then fall on the authorities, or by litigations? And, who is to pay to maintain such heritage? Are there institutional and civic processes that can be created to negotiate such differences, and are there safeguards that can be integrated into these processes to ensure equitable participation and voices? The choice on how to resolve these differences is predicated on conservation as best practice. But it is also an ethical choice. Normative ethics will therefore first attempt to clarify these choice scenarios and weigh their moral significance through different ethical frameworks, before sharing the results of this process with the different stakeholders. In all likelihood, there will still be winners and losers should the Golden Mile Complex be ultimately conserved. Here, normative ethics will next attempt to locate an ethical compromise (Chan & Protzen, 2018) and, if not, then allocate proper restitution to parties that will be made worse-off by the decision for conservation.

However, in 2019, the state issued planning advice that the Golden Mile Complex's "landmark main building will have to be retained" but it may be developed as an integrated project with a new building constructed next to the main building (Leong, 2019; Lim, 2019; Wong, 2019). On 9 October 2020, the Urban Redevelopment Authority proposed that the building be legally conserved with unprecedented clarity upfront on the incentives provided by the state "to building owners or prospective buyer-developers should the conserved building be sold." These include "bonus floor area which will allow the building of an additional 30-storey tower within the existing site, a partial development charge waiver on the additional floor area, a development charge waiver for the enhancement in value of the conserved gross floor area, adjustment of the site boundary to be more regular, the option to top up the lease on the land to 99 years, and the flexibility to adapt the building to a mix of possible uses" (URA, 2020b). Advocacy groups have lauded this proposal and commented that apart from the building's historical and architectural values, the adaptation and reuse of aging buildings are also about being sustainable (Lim, 2020; SHS, 2020). This announcement by the authority signalled the commencement of the statutory process to amend the Master Plan for legal conservation. The proposal was made available for public inspection and comments for "a period … within which objections to and representations concerning the proposed amendment may be made" (Singapore, 1999). Whatever the outcome, this changed the ethics of conserving the Golden Mile Complex considerably in the following two ways.

First, the 2019 planning advisory, as well as the 2020 conservation proposal, introduce a new question of how to integrate a new work of architecture with the Golden Mile Complex. In turn, the ethical onus also shifts to the architects that will undertake the development of this integrated development. While this advisory statement appears consistent with the evolving life of a megastructure, it nevertheless brings about the delicate question of how to add new fabric without subtracting from the landmark Golden Mile Complex. And because of the "deliberately" vague conservation framing of the advisory and conservation proposal—that is, "to conserve the main building with its signature terraced profile atop the podium block as key features to be retained" can be framed in a multitude of ways—the onus will be on the architects of the new development to carefully thread the delicate line between satisfying the requirements of the new development while "retaining" the Golden Mile Complex within one integrated project. After all, "retaining" can span from adopting discrete strategies, for instance, by providing a simple connection between the new development and the Golden Mile Complex, all the way to gutting out the Golden Mile Complex as an act of "facade-ism" (cf. Bloszies, 2012), and then deprioritizing it to the new development. There is no way to predict what prospective architects will do, although it is not likely with proposed conservation that it will be "façade-ism." On this, architects attentive to conservation may wish to reflect on Silman's (2007) ethical question of limit: If anything is possible by technology, ought one do it? Given the immense interpretative room of retaining the landmark Golden Mile Complex main building, what then defines the limit of what an architect *should not do*?

In parallel, the planning advisory and conservation proposal also shift the ethical onus from people to the market. Faced with the prospect of recalibrating real-estate development to reflect market demands, prices, and cost in this prime location, the possibility of a new 30-storey tower adjacent to the Golden Mile Complex will result in a one-third increase over the existing development density. In turn, prices in the existing Golden Mile Complex may have to be commensurably adjusted to reflect the price consistency of this new integrated project. As a result, most of the existing renters, who either count the Golden Mile Complex as their home or their place of livelihood, may in turn be displaced. In defaulting to settlement by market mechanisms this way, important questions on "equitable stakeholdership" in modern heritage, as well as the civic art of continual and dialogic negotiation that produces not only conflict but also surely trust and collaboration, are prematurely eroded. According to Sennett (2018, p. 197), dialogics is an ethical practice of communication; it fosters respect of others, cultivates cooperativeness rather than competitiveness, and is outward rather than inward-turning. In settling for a continuous process of conservation once and for all by market mechanisms, the weightier prospects for an ethic of the city are nullified. In parallel, the original and intangible intention of the Golden Mile Complex as an open megastructure coevolving architecture, city, and life will also be unwittingly lost.

Conclusion: a call to have open discussions on ethics

This chapter begins by highlighting the increasingly important role of ethics in architecture conservation today. Existing research in conservation studies frames ethics as a moral, responsible, and best professional practice. But this approach to ethics fails to provide concrete guidance in complex cases where ethical tenets are in conflict with each other. The choice of dissecting the respective case studies of a traditional and a modern architectural heritage are deliberate attempts to generate a discussion on complex cases and then to demonstrate how ethical issues could be clarified and systematically broached in conservation. Engaging with the frame of ethics in conservation enables one to draw out new issues that are important for accountable decision-making.

In both cases, while the legalistic approach to circumvent ethical dilemmas—for instance, a harsher penalty for non-compliance to conservation rules in the former case and statutory planning advice and conservation proposal in the latter case—is not completely inapposite, it is the open discussion of ethics that is valuable and critical. There will always be questions on whether penalties and planning advice prescribed are adequate. In particular, in the case study on Golden Mile Complex, the ostensibly "light touch" planning advice in 2019 was purportedly formulated under the premise of not unduly interfering with market forces. This is a fallacy as the market will price in the cost of complying with the planning advice. Or the 2020 conservation proposal where the state was more assertive and proposed to gazette the building for conservation with upfront clarity on the incentives accompanying conservation status. Is there a risk that these incentives then sidestep previous ethical dilemmas of the Golden Mile Complex but end up inviting new uncertainties that can compound the future of this building? Complex problems of conservation are unlikely to disappear even with more legal stipulations. Instead, these problems tend to change because of these new constraints. Engaging in ethical reasoning and open discussions on conservation in the public sphere may be the safest warrant in this situation.

Finally, ethics cannot be substituted by either legal stipulations or regulations. Fundamentally, ethics is not about regulations. The objective of ethics is not to impose a particular behaviour or to ensure these behaviours are complied with. By asking ethical questions and engaging in open and sustained discussions about ethics, it helps us look deeper into conservation issues, be less passive and in a better position to broaden our scope, uncover blind spots, and ultimately foster conviction in making the right decision and doing the right thing.

Postscript: Since the writing of this chapter, there have been further developments regarding Golden Mile Complex. The building has since been conserved. However, this does not alter the chapter's arguments.

References

(AICFAIC) The American Institute for Conservation and the Foundation for Advancement in Conservation, (1994). *Code of ethics and guidelines for practice*. Retrieved from: https://www.culturalheritage.org/about-conservation/code-of-ethics (Accessed on 15 April 2021).

Allington-Jones, L. (2013). The Phoenix: The role of conservation ethics in the development of St Pancras Railway Station (London, UK). *Journal of Conservation and Museum Studies, 11*(1), 1–21. https://doi.org/10.5334/jcms.1021205.

Anon (2003). Charged over temple changes. *The Straits Times*, p. H13. Retrieved from: https://eresources.nlb.gov.sg/newspapers/digitised/issue/straitstimes20031114-1# (Accessed on 15 April 2021).

Anon (2018). Golden Mile Complex may be conserved even as en bloc tender is launched. *ChannelNewsAsia*. Retrieved from: https://www.channelnewsasia.com/news/singapore/golden-mile-complex-may-be-conserved-en-bloc-tender-ura-10882580 (Accessed on 15 April 2021).

Appelbaum, B. (2013). *Conservation Treatment Methodology*. Lexington (K. Y.): [s. n.].

Australia ICOMOS (2013). *The Burra Charter — The Australia ICOMOS Charter for Places of Cultural Significance* (Burra Charter). Retrieved from: http://australia.icomos.org/wp-content/uploads/The-Burra-Charter-2013-Adopted-31.10.2013.pdf (Accessed on 15 April 2021).

Beatley, T. (1994). *Ethical land use: Principles of policy and planning*. Baltimore, MD: John Hopkins University Press.

Bloszies, C. (2012). *Old buildings, new designs: Architectural transformations*. New York: Princeton Architectural Press.

Canadian Association for Conservation of Cultural Property and of the Canadian Association of Professional Conservators (CAC & CAPC) (2009). *Code of ethics and guidance for practice*. Retrieved from: https://www.cac-accr.ca/download/code-of-ethics/ (Accessed on 15 April 2021).

Chan, C. (2003). Temple official accused of illegal changes to monument. *The Straits Times*. p. 4. Retrieved from: https://eresources.nlb.gov.sg/newspapers/digitised/issue/straitstimes20031113-1# (Accessed on 15 April 2021).

Chan, J. K. H. (2019). *Urban ethics in the anthropocene*. London: Palgrave.

Chan, Y. K. (2020). The golden mile complex: The idea of little Thailand in Singapore. *Austrian Journal of South-East Asian Studies, 13*(1), 103–121. https://doi.org/10.14764/10.ASEAS-0031.

Chan, J. K. H., & Protzen, J. (2018). Between conflict and consensus: Searching for an ethical compromise in planning. *Planning Theory, 17*(2), 170–189. https://doi.org/10.1177/1473095216684531.

Chang, C. (2006). It's so unlike orderly S'pore. *The New Paper*. p. 5. Retrieved from: https://eresources.nlb.gov.sg/newspapers/digitised/issue/newpaper20060307-1# (Accessed on 15 April 2021).

Choay, F. (2001). *The invention of the historic monument*. Cambridge: Cambridge University Press.

Chong, A. (2019). Some shop owners resistant as Queensway Shopping Centre gears up for collective sale. *ChannelNewsAsia*. Retrieved from: https://www.channelnewsasia.com/news/singapore/queensway-shopping-centre-en-bloc-shop-owners-resistant-11491862 (Accessed on 15 April 2021).

Chuang, P. M. (2009). Heavier penalties to protect monuments from destruction. *The Business Times*, p. 13. Retrieved from: https://eresources.nlb.gov.sg/newspapers/digitised/issue/biztimes20090414-1# (Accessed on 15 April 2021).

Davis-Marks, I. (2020). Botched art restoration in Spain renders smiling statue unrecognizable. *Smithsonian Magazine*. Retrieved from: https://www.smithsonianmag.com/smart-news/amateur-restorer-botches-carving-spain-180976271/ (Accessed on 15 April 2021).

Fang, J. (2006). Golden Mile complex planning collective sale. *The Straits Times.* p. S36. Retrieved from: https://eresources.nlb.gov.sg/newspapers/digitised/issue/straitstimes20061014-1# (Accessed on 15 April 2021).

Feilden, B. (2003). *Conservation of historic buildings.* Amsterdam: Elsevier.

Frankena, W. (1973). *Ethics.* 2nd ed. Upper Saddle River, NJ: Prentice-Hall.

Gutierrez, L., & Wong, K. C. (1985). Little Bangkok at the Golden Mile. *The Straits Times.* p. 3. Retrieved from: https://eresources.nlb.gov.sg/newspapers/Digitised/Article/straitstimes19850106-1.2.9.8 (Accessed on 15 April 2021).

International Council of Museums (ICOM) (2004). *ICOM Code of Ethics for Museums.* Retrieved from: https://icom.museum/wp-content/uploads/2018/07/ICOM-code-En-web.pdf (Accessed on 15 April 2021).

International Council of Monuments and Sites (ICOMOS) (1931). *The Athens Charter for the restoration of historic monuments – 1931 – International Council on Monuments and Sites.* Retrieved from: http://www.icomos.org/en/charters-and-texts/179-articles-en-francais/ressources/charters-and-standards/167-the-athens-charter-for-the-restoration-of-historic-monuments (Accessed on 15 April 2021).

International Council of Monuments and Sites (ICOMOS) (1964). *International Charter for the Conservation and Restoration of Monuments and Sites (The Venice Charter 1964).* Retrieved from: http://www.icomos.org/charters/venice_e.pdf (Accessed on 15 April 2021).

International Council of Monuments and Sites (ICOMOS) (1994). *The Nara Document on Authenticity (1994)* (Nara Document). Retrieved from: http://www.icomos.org/charters/nara-e.pdf (Accessed on 15 April 2021).

The Institute of Conservation (ICON) (2014). *The institute of conservation's professional standards.* Retrieved from: https://www.icon.org.uk/asset/ED1586F0%2D74AB%2D4540%2D9CF8A-C50A35217D6/ (Accessed on 15 April 2021).

Kalman, H. (2014). *Heritage planning.* London: Routledge.

Katz, B. (2019). Hitler's birthplace will be converted into a police station. *Smithsonian Magazine.* Retrieved from: https://www.smithsonianmag.com/smart-news/hitlers-birthplace-will-be-converted-police-station-180973618/ (Accessed on 15 April 2021).

Koolhaas, R., & Mau, B. (1995). *Small, medium, large, extra-large: Office for Metropolitan architecture.* Rotterdam: O10 Publishers.

Lagueux, M. (2004). Ethics versus aesthetics in architecture. *The Philosophical Forum, 35*(2), 117–133. https://doi.org/10.1111/j.0031-806X.2004.00165.x.

Lee, S. H. (2009). Stiffer penalties to protect monuments. *The Straits Times,* p. 30. Retrieved from: https://eresources.nlb.gov.sg/newspapers/digitised/issue/straitstimes20090414-1# (Accessed on 15 April 2021).

Lee, L. M., Lim, Y. M., & Nor'Aini, Y. (2008). Strategies for urban conservation: A case example of George Town, Penang. *Habitat International, 32(3),* 293–304. https://doi.org/10.1016/j.habitatint.2007.09.001.

Leng, A. (2006). 'Eyesore' at Beach Road: No relief in sight. *The Straits Times,* p. H7. Retrieved from: https://eresources.nlb.gov.sg/newspapers/digitised/issue/straitstimes20060307-1# (Accessed on 15 April 2021).

Leong, G. (2019). Developers get more clarity on Golden Mile en bloc site. *The Straits Times.* Retrieved from: https://www.straitstimes.com/business/developers-get-more-clarity-on-golden-mile-en-bloc-site (Accessed on 15 April 2021).

Lim, J. (2007). Golden years. *The Straits Times,* p. 6. Retrieved from: https://eresources.nlb.gov.sg/newspapers/Digitised/Article/straitstimes20071002-1.2.132.4.2 (Accessed on 15 April 2021).

Lim, J. (2019). Golden Mile Complex will 'not be the same even if building is kept'. *TODAY,* p. 6. Retrieved from: https://www.todayonline.com/singapore/golden-mile-complex-will-not-be-same-even-if-building-kept-say-residents-shop-owners-and (Accessed on 15 April 2021).

Lim, J. (2020). URA proposes to conserve Golden Mile Complex, offers incentives for redevelopment. *TODAY*, Retrieved from: https://www.todayonline.com/singapore/developers-interested-redeveloping-while-conserving-parts-golden-mile-complex-get-govt (Accessed on 15 April 2021).

Lim, K. (2019). Queensway Shopping Centre's en-bloc bid falls through; some shop owners rejoice. *TODAY*. Retrieved from: https://www.todayonline.com/singapore/queensway-shopping-centre-en-bloc-sale-falls-through-after-lukewarm-response (Accessed on 15 April 2021).

Low, Y. J. (2019). Queensway Shopping Centre en bloc attempt faces challenges from shop owners. *TODAY*. Retrieved from: https://www.todayonline.com/singapore/queensway-shopping-centre-en-bloc-attempt-faces-challenge-shop-owners (Accessed on 15 April 2021).

Luo, S. (2018). Pearl Bank Apartments in Outram sold en bloc to CapitaLand for S$728m. *The Straits Times*. p. 6. Retrieved from: https://www.straitstimes.com/business/companies-markets/capitaland-acquires-pearl-bank-apartments-for-s728m-q4-profit-falls-38 (Accessed on 15 April 2021).

Maki, F. (2008). *Nurturing dreams: Collected essays on architecture and the city*. Cambridge: MIT Press.

Malkogeorgou, T. (2006). The ethics of conservation practice: A look from within. *Conservation Journal*, Spring, 52. Retrieved from: http://www.vam.ac.uk/content/journals/conservation-journal/issue-52/the-ethics-of-conservation-practice-a-look-from-within/ (Accessed on 15 April 2021).

Mansfield, J. (2008). The ethics of conservation: Some dilemmas in cultural built heritage projects in England. *Engineering, Construction and Architectural Management*, *15*(3), 270–281. https://doi.org/10.1108/09699980810867424. Retrieved from: www.emeraldinsight.com/0969-9988.htm (Accessed on 15 April 2021).

Margalit, A. (2010). *On compromise and rotten compromises*. Princeton, NJ: Princeton University Press.

Morris, W. (1877). *The Manifesto of the Society for the Protection of Ancient Buildings (SPAB)*. Retrieved from: https://www.spab.org.uk/about-us/spab-manifesto/ (Accessed on 15 April 2021).

Mould, O. (2016). Brutalism redux: Relational monumentality and the urban politics of Brutalist architecture. *Antipode*, *49*(3), 701–720. https://doi.org/10.1111/anti.12306.

Ng, D., & Huang, E. (2006). National disgrace? But the view is great. *The New Paper*, pp. 4–5. Retrieved from: https://eresources.nlb.gov.sg/newspapers/digitised/issue/newpaper20060307-1# (Accessed on 15 April 2021).

Schön, D. A. (1983). *The reflective practitioner: How professionals think in action*. New York: Basic Books.

Semes, S. W. (2009). *The future of the past: A conservation ethic for architecture, urbanism, and historic preservation*. New York: W.W. Norton & Company.

Sennett, R. (2018). *Building and dwelling: Ethics for the city*. London: Allen Lane.

Shamsuddin, S., Sulaiman, A. B., & Che Amat, R. (2012). Urban landscape factors that influenced the character of George Town, Penang UNESCO World Heritage Site. *Procedia – Social and Behavioral Sciences*, *50*, 238–253. https://doi.org/10.1016/j.sbspro.2012.08.031.

Shamsuddin, S., Sulaiman, A. B., & Che Amat, R. (2018). Historic urban landscape and the character of George Town, Penang UNESCO World Heritage Site. *Asian Journal of Environment-Behaviour Studies*, *3*(9), 27–35. https://doi.org/10.21834/aje-bs.v3i9.297.

Silman, R. (2007). Is preservation technology neutral? *APT Bulletin*, *38*(4), 3–10.

Singapore Heritage Society (SHS) (2018). *Too young to die: Giving new lease of life to Singapore's modernist icons*. p. A16. Retrieved from: https://www.singaporeheritage.org/wp-content/uploads/2018/08/SHS-Position-Paper-Too-Young-To-Die-Aug-2018.pdf.

Singapore Heritage Society (SHS) (2020). *Proposed conservation of Golden Mile Complex: Joint statement by the Singapore Heritage Society, ICOMOS Singapore, and the DoCoMoMo Singapore Working Group*. Retrieved from: https://www.facebook.com/notes/singapore-heritage-society/proposed-conservation-of-golden-mile-complex/10159343251436614 (Accessed on 15 April 2021).

Singapore, Republic of (1999). *Planning (Master Plan) Rules 1999 (S 245/1999)*. Subsidiary Legislation Supplement. Retrieved from: https://sso.agc.gov.sg/SL-Supp/S245-1999/ (Accessed on 15 April 2021).

Tan, H. Y. (2009). Golden Mile owners try backdoor route to en bloc sale. *The Straits Times*, p. A16. Retrieved from: https://eresources.nlb.gov.sg/newspapers/Digitised/Article/straitstimes20071002-1.2.132.4.2 (Accessed on 15 April 2021).

Teo, J. (2006). Ageing malls in prime spots head for collective sale. *The Straits Times*, p. H20. Retrieved from: https://eresources.nlb.gov.sg/newspapers/Digitised/Issue/straitstimes20061017-1# (Accessed on 15 April 2021).

Tessman, L. (2017). *When doing the right thing is impossible*. New York: Oxford University Press.

Urban Redevelopment Authority (URA) (2020a). *Conservation guidelines*. Retrieved from: https://www.ura.gov.sg/Corporate/Guidelines/Conservation/Conservation-Guidelines (Accessed on 15 April 2021).

Urban Redevelopment Authority (URA) (2020b). *Supporting the conservation and commercial viability of golden mile complex*. Retrieved from: https://www.ura.gov.sg/Corporate/Media-Room/Media-Releases/pr20-28 (Accessed on 15 April 2021).

Wong, D. (2019). Golden Mile complex seeks to sell en bloc again. *The Straits Times*. Retrieved from: https://www.straitstimes.com/business/property/golden-mile-complex-seeks-to-sell-en-bloc-again (Accessed on 15 April 2021).

Yeo, K. S. (2018). Bridging positivist and relativist approaches in recent community-managed architectural conservation projects in Singapore. *Journal of Social Issues in Southeast Asia*, *33*(3), 647–676. https://doi.org/10.1355/sj33-3e.

Yeo, K. S., & Chong, T. (2015). What makes a national monument?. *The Straits Times*, p. A21. Retrieved from: http://www.straitstimes.com/opinion/what-makes-a-national-monument (Accessed on 15 April 2021).

29 Urban values-centred regeneration in the perspective of the circular economy model

An overview of the key issues

Luigi Fusco Girard and Francesca Nocca

Introduction

Although the wealth of a country/region is increasingly represented by the wealth produced in its cities, it is equally true that cities are also great generators of entropy: they are the most important source of pollution, environmental degradation and climate destabilization, which negatively affect health and well-being. They are the places where energy consumption and pollution are highest (www.unhabitat.org), as well as conditions of social fragmentation (Fusco Girard & Nocca, 2019). This situation in which cities are today necessarily requires the identification of new approaches and strategies for the development and regeneration of urban system.

The thesis here proposed is that the new required approaches have to be coherent with the new 'humanistic' paradigm that is emerging, framed in the perspective of an 'ecological economy' to preserve the environmental and eco-systemic quality while producing (and redistributing) wealth.

The 'traditional' economy has contributed to the degradation of many cultural landscapes and specific cultural assets/sites; it has generated damage to the environment, to health and to the social system. In brief, the traditional economy has been a source of high entropy (Georgescu-Roegen, 1971, 1976).

One of the solutions to the challenges of our time lies in the implementation of models inspired by the natural system that offers efficient management, production and consumption models (Pauli, 2014). More precisely, we need a circular economy, which is an economy inspired by nature's organizational cycles/processes.

The principle of the closure of loops, of which Barry Commoner was the first to deal with in *The Closing Circle* in 1971 (Commoner, 1971), is fundamental in the identification of the ecological paradigm, but, to date, it has not been fully implemented.

It is necessary to feed organizational/entrepreneurial processes that are capable of promoting a system of relationships, generating new symbiotic links. In short, the circular economy is a reducer of entropy on different levels as well as a producer of efficiency and better resilience (Fusco Girard et al., 2019).

The circular economy model can represent a key element to move towards the 'new humanism'. The human scale city (i.e. the city of humanization, 'of the human being for human being' (Lazzati, 2000, p. 15) can be interpreted as the city that valorizes the local identity, its memory and its cultural heritage/landscape, often located in its 'places'; valorizes its public spaces, making them as catalysts for relationships/bonds, through the care and management of their inhabitants; enhances the cooperative capacity of citizens; realizes processes of real democratic participation in self-management and

DOI: 10.4324/9781003038955-35

self-government (based on the principle of subsidiarity); invests in the solidarity economy and in the economy of relationships; invests in the ecological economy, through circularization and symbiosis processes; invests in human and social capitals, that is on training/education, and therefore on creativity; and realizes a redistribution of wealth, reducing poverty. So, the 'new humanism' implies more bonds, more relationships and thus more collaborative and cooperative capacities. The latter are at the basis of the circular economy model, characterized by mutual exchanges and relationships, and collaborative and cooperative bonds.

It is common to interpret circular economy as an economy that reduces/eliminates waste and underutilization through the activation of 'closed' loops. Indeed, the essence of circular economy should be sought rather in the ability to activate processes of collaboration/cooperation/symbiosis among different subjects (economic, public, social). For example, it is usual to focus on the different amount of waste in the building sector or in the production of urban waste; or the amount of metals, glass, paper, pipes and wood that can be reused/recycled.

However, this is only one aspect of the circular economy. The most interesting aspect is that the above-mentioned processes of reuse, repair, recycling, etc., involving different actors, are based on cooperation and therefore on the density of interpersonal relationships and on mutual trust which, in turn, generates such reciprocal exchanges. Thus, the circular economy is configured as a 'hybrid' model that combines cooperation and competition, and efficiency and co-responsibility. It brings together different processes that increase performance, being generators of creative energy, new opportunities, original complementarities that produce cooperative relationships in which each subject 'receives' and in turn 'offers' in a circuit of reciprocity, that is benefits for all.

The 'new economy', in order to become actually inclusive and ecologically sustainable (Bruni & Zamagni, 2004; Hart et al., 2010; Porter & Kramer, 2011), should be characterized by

1 conservation/valorization of existing local resources (and also their waste); this means maintenance/restoration/recovery/regeneration of local resource assets;
2 de-carbonization processes to decouple wealth production from negative environmental impacts;
3 new models of business organization that are configured as a 'hybrid' between profit and non-profit, between public, private and social.

Circular economy and the value-centred approach

The challenge of humanization is now associated with the capacity to simultaneously reduce poverty/inequality conditions, to preserve the vitality of natural ecosystems and to guarantee inclusive economic growth and well-being in the future. The traditional values of humanism (freedom, justice, inclusion, prosperity) are therefore enriched in a broader perspective across the space (including marginal/poor people and natural ecosystems) and over time (including future generations).

The transition towards the 'new humanism' and the concept of circular economy (as key model in this transition process) evokes a set of values. This value-centred approach includes all the values of a resource. The circular economy is linked also the concept of the complex social value (CSV) (Fusco Girard, 1986; Fusco Girard, 1987; Nijkamp, 1997) as a combination of use values and 'intrinsic value'.

The CSV, usually applied in cultural heritage studies, integrates the concept of total economic value (TEV) including the 'intrinsic value' component, a non-economic value that is recognized by the community as a 'permanent value' during the long term.

The notion of 'intrinsic value' draws its foundation from the ecological economy and the recognition of the system's autopoietic capacities (Zeleny & Hufford, 1992; Turner, 1993; Maturana & Varela, 2001; Costanza et al., 2014). Indeed, Ruskin and Morris had already introduced this notion (Morris, 1889; Ruskin, 1989), later taken up by Riegel as the value of memory (Riegel, 1903). It was the Burra Charter (ICOMOS, 1979, 2013) that opened the perspective of intrinsic value in the field of cultural heritage conservation.

In ecological economy, the 'intrinsic value' is interpreted as a 'primary value' that exists independently of the presence of human beings, that is, independently of its usefulness for the human-social system. In the urban context, the 'intrinsic value' has to be interpreted in a way that does not coincide with this definition but represents an 'extension' in the context of that dynamic, complex and adaptive system that is the city with its landscape.

It also represents a way of thinking in a circular logic (a rationality linked to virtuous circuits, that is capable of regenerating, redesigning, re-organizing and thus of contributing to the resilience), which goes beyond linear rationality, much more attentive to changing contexts, conditions and circumstances. Therefore, it represents the foundation of each new use value – that is of each 'hybridization' between different use values. It is a value that cannot be renounced to 'manage change'.

The formulation of the 'intrinsic value' was recently taken up in the European Commission's document which considers the dual dimension of culture as a value 'in itself and for itself', distinguishing it from the instrumental one (such as economic and social values; European Commission, 2014).

The 'intrinsic value' is therefore linked to the 'spirit of places' (Norberg-Schulz, 1998) and to knowledge rooted in a site that reflects a centuries-old (if not millenary) history that has given physical–spatial form to the relationship between man and nature, generating a specific identity. It expresses the sense of belonging of a community to a place and vice versa. For example, the 'intrinsic value' of the Matera (Italy) city landscape lies in its circular millenarian processes linked to the water cycle, the material cycle and the culture of recycling/reuse. The system of water conservation is an example of circular process in Matera: taking advantage of the surrounding territory, composed mostly of tufa, the inhabitants in this area have been able to dig, in addition to housing, even the ducts and tanks that would allow the water supply to the houses of the 'Sassi' (Fusco Girard et al., 2019). The cisterns were private, that is inside the houses, or larger to supply entire neighbourhoods or districts. The reuse of water is not the only form of circular economy in the history of the city of Matera. The walls of the city, no longer necessary and already transformed into real quarries of materials for new construction, are an example of recycling economy. Moreover, because of the extreme poverty of the city, families were forced to eliminate any form of waste, recycling and reusing waste. There are many other examples demonstrating the millennial culture of circular economy that characterizes the city of Matera, and which expresses the 'intrinsic value' that has shaped the city for centuries (Fusco Girard et al., 2019).

The 'intrinsic value' survives even when a cultural asset is no longer used. Being the result of a social 'construction', its interpretation and evaluation require the participation of the local community in its various components. Therefore, a neo-positivist approach is not possible. A participatory dialogic communicative process is necessary, and it can also lead to the identification of different coexisting 'intrinsic values', even if characterized by common elements.

The role of 'intrinsic value' is to guide the development of a site, a historic centre, a city. It helps to identify a direction for its use and management (Fusco Girard & Vecco, 2019; Fusco Girard, 2020a). In this sense, by offering a perspective to new strategies for local development and regeneration, the 'intrinsic value' is the foundation on which to articulate any new use value (or combination of several use values) connected to a new project/strategy. In this way, the new project is in continuity with a territorial urban history and offers the 'energy' for a creative synthesis, for processes of hybridization between memory and innovation. The 'intrinsic value' offers a 'threshold' within which the change is admissible.

A creative hybridization between the 'intrinsic value' – that determines the 'spirit of places' itself (Norberg-Schulz, 1998) – and the forms – that express innovation aimed at improving the quality of life – is needed. This hybridization project cannot ignore certain general conditions. Some of them are represented by the achievement of the strategic goals and targets of the UN Agenda 2030 (United Nations, 2015), COP 21 and the New Urban Agenda (United Nations, 2016).

Assuming the unifying approach of the landscape as a 'prism' that converges all the different perspectives from which the city system can be interpreted, this contribution intends to propose some strategies for the regeneration of the city based on the model of the circular economy. In particular, the focus is on the operational strategy of adaptive reuse of cultural heritage, highlighting the relationship between intrinsic value and use values. In this perspective, the evaluation activity plays a fundamental role in identifying the multidimensional productivity of the different alternatives and their impacts, beyond the reductive perspective of the archaeological/museum park.

The landscape perspective for sustainable regeneration of the city: historic urban landscape

The implementation of a local development strategy and the achievement of the above objectives related to the model of circular urban economy require a holistic-systemic and multidimensional approach that covers economy, ecology, society, territory, technology and institutions. However, we could take the unifying approach of the 'landscape' as a prism or as a lens to integrate the different perspectives from which the city system can be interpreted. It offers and obliges a systemic vision of reality and its transformation, including all the complex interdependencies between natural and man-made components and the people who live them.

The landscape represents the reality perceived by each subject, but it also becomes the foundation of a collective vision in which community knowledge with scientific knowledge is integrated. It recognizes the growing role and importance of the local community but, at the same time, it is recognized as an increasing weakness of the 'connective infrastructure' of our society and cities that need to be enhanced and regenerated (Fusco Girard, 2018). The landscape is the result of a hybridization process between the 'work' of nature and the 'work' of man. A disciplinary hybridization is required for its understanding and interpretation: among different scientific disciplines and between these scientific disciplines (expert knowledge) and common knowledge.

The landscape perspective, being focused on subjective perception, is fundamentally anthropocentric and humanistic. This approach based on landscape is therefore consistent with the new paradigm that is emerging today: the humanistic paradigm (see §§ 15, 24, 25 of the New Urban Agenda) (United Nations, 2016) and, at the same time, the ecological paradigm (European Green Deal) (European Commission, 2019). In fact, the landscape is the reality perceived by each subject who lives a specific space, but then becomes a collective vision.

The landscape reflects all the challenges of our time, from social inequalities to climate change and migration flows. The landscape indicates the state of health of a territory or a city. The winning and losing values/interests of a society can be read in the landscape: the culture of a community can be read. The landscape therefore represents a complex and synthetic indicator of the sustainability (or un-sustainability) of an urban/territorial/regional system.

The systemic notion of HUL proposed by UNESCO (UNESCO, 2011) is an example of the interpretation of 'active' landscape conservation that integrates (under certain conditions) with economic development.

HUL emphasizes the focus on the tangible and intangible values of cultural heritage/landscape. Some of these intangible values are represented by the so-called intrinsic value (Fusco Girard & Nijkamp, 1997).

The landscape is a 'function' of circular processes. Terraced landscapes are an ancient example of 'circular' model in using resources (Fusco Girard et al., 2018). They constitute a type of landscape characterized by a particular beauty that is the reflection of circular processes. Organized through circular processes, they produce a landscape of high aesthetic quality and, at the same time, they provide many ecosystem services to local communities by allowing production where nature does not allow it (because of its form; Fusco Girard et al., 2018).

It is recognized that a high-quality landscape contributes to urban productivity because it becomes an 'attractive force' for a wide range of activities. This means that cultural heritage/landscape is increasingly interpreted as a resource for local/regional economic development, being able to stimulate the localization of creative activities, to produce new employment, to contribute to social inclusion and cohesion.

The implementation of the above-mentioned HUL approach proposed by UNESCO to 'manage change' is possible if a creative hybridization between old and new is achieved. This is possible from the perspective of the circular economy (Chertow, 2000; Preston, 2012; Ellen MacArthur Foundation, 2015). As already noted, the circular economy can be interpreted as a form of hybridization between cooperation and competition. Both circular economy and heritage conservation aim to extend the useful life of resources as long as possible: heritage conservation is more effectively achieved through circular economy processes.

The above is determining a new attention to the forms of interdependence between urban regeneration of the cultural landscape and the circular economy. The approach proposed by HUL implies the ability to identify the limits (that is the threshold) within which the change is admissible. The 'management of change' (UNESCO, 2011) has to ensure coherence (continuity) with the past, with identity, with memory – that is, with the 'intrinsic value'. The 'intrinsic value' of the landscape gives physical–spatial shape to the urban structure and configures its specific identity. Today, it is necessary to re-interpret on different levels the systemic perspective suggested by HUL in order to identify a concrete development strategy capable of capturing the 'spirit of places', making it an 'engine' of vitality, regeneration processes and planning that is configured as an effective hybridization between old and new.

The 'CSV' (Fusco Girard, 1987) of cultural heritage/landscape, based on use values and 'intrinsic value', can represent an approach that orients possible solutions (Fusco Girard et al., 2019).

The recommendations on HUL recognize first of all a close relationship between the design/planning of cultural heritage conservation and economic development: heritage conservation and management should be included within the framework of economic development strategies (part VI) and in particular within sustainable development strategies (part VII, paragraphs 10 and 11). The notion of sustainable development recurs several times in the text (paragraph 24/b etc.), as does the economic strategy in which conservation activity is to be incorporated/integrated.

The HUL recommendation suggests a landscape-based approach to conserve and manage cultural heritage (see HUL Preamble). As mentioned above, such a landscape perspective is structurally 'unifying', because multiple and multidimensional approaches converge in it: it integrates the social perspective with the environmental, economic, physical, cultural and legal perspectives.

All the current problems are incorporated into the landscape: from climate change to pollution, health, social marginality, economic wealth production, poverty, etc. The landscape is a prism that allows us to assume a human-centred perspective that is focused on the human dimension (Fusco Girard, 2020b). There is no landscape if there is not its perception by a human being, through its different senses. On the other hand, landscape is the result of a series of choices made by each subject and the community, reflecting the culture of a society. Culture represents the human product, the human creation par excellence. The landscape is therefore configured as a vital/living resource, able to change continuously under the pressure of people's needs, interests and hopes.

Actually, in HUL, the economic model is not only that of sustainability. The HUL recommendations repeatedly suggest that the landscape should not be preserved, but rather carefully managed in its transformations (paragraph 12) in a careful way, that is sustainable, taking into account all the values at stake: cultural, social and economic values (paragraph 5). All the above is necessary to produce development that is not only sustainable, but also 'human and social' (paragraphs 10 and 18), attentive to people's well-being and quality of life (paragraphs 17 and 18).

In the UNESCO document, the different values and attributes to evaluate the landscape in an integrated and comprehensive way emerge (paragraph 10): cultural and environmental values (paragraphs 3,21), intangible and visual values (paragraph 9), perceptive values (paragraph 13), social values (paragraphs 24,6), environmental values (paragraph 19), ancient values and contemporary values (paragraph 22).

Emphasizing the key role of creativity (paragraph 12), HUL calls for the integration of old and new. In a sense, the notion of hybridization is evoked here, albeit implicitly (Fusco Girard, 2020b). It is not a mere juxtaposition, but a reflection of complementarity, generating synergies, circularity; it is not only multifunctionality, but a juxtaposition between heterogeneous programs: between calculating rationality and relational rationality, between cash flow control and creativity and between economic/financial convenience analysis and sense/meaning construction.

The circular economy model is not formally made explicit, but many principles that characterize the circular economy model are nevertheless evoked (Fusco Girard, 2020b). In paragraph 11, the need for a productive and sustainable use of space resources is stressed. Paragraph 19 makes reference to the efficient use of the environmental resources represented by water and energy. Paragraph 2 calls for a strategic vision that goes beyond the short term: it stresses the importance of a long-term vision, which is typical of the circular economy.

Furthermore, paragraph 24/d stresses the need for self-financing – that is self-regeneration of financial resources in order to conserve the regenerated heritage over time. In addition, paragraph 22 stresses the need for 'harmonious' cooperation between different private and public actors. These are two typical characteristics of the circular economy model.

Finally, it should be noted that together with the notion of sustainable development within which the conservation/management of heritage/landscape is placed, the notion of 'human development' is mentioned in paragraph 12 and the notion of human and social development is mentioned in paragraph 18. They are characteristics consistent with the model of the circular economy, careful to avoid waste and underutilization of the different forms of capital, from natural to man-made, to human to social capitals.

In conclusion, the HUL approach finds its realization within the model of circular economy, which is evoked even if indirectly (Fusco Girard, 2020b).

The above has implications in terms of evaluation tools, which are increasingly required to 'support decision-making processes' (paragraph 24/b) both in planning and in 'change management'. In paragraph 10, it is recognized that evaluation has to be integrated and comprehensive.

In part IV, concerning tools, it is underlined (in paragraph 24/b) that evaluation tools are necessary to 'monitor and manage change in order to improve the quality of life and urban space'. It introduces an assessment of cultural (heritage impacts), social and environmental impacts to improve choices, also calling for new financial tools (paragraph 24/d). They are therefore 'integrated' evaluation tools. More precisely, they are 'hybrid' evaluation tools that combine heterogeneous approaches and procedures.

Towards integrated assessment methods in the circular economy perspective

The above has implications in terms of evaluations. It is necessary, in fact, that the evaluation processes are able to 'capture' the multiple values produced by the circular model implementation. This value-centred approach includes all the values of a resource.

In particular, dealing with the circular economy, we refer also to the CSV (Fusco Girard, 1987; Fusco Girard & Nijkamp, 1997), as a combination of use values (assessed considering the impacts of each function) and 'intrinsic value'. The challenge is to assess taking into account both the 'intrinsic value' of the cultural landscape and the 'instrumental values' for local communities (ecosystem services – supply, regulation, maintenance and cultural services (TEEB, 2010)).

Circular economic systems have the capacity to conserve/develop the 'intrinsic value' over time, strengthening the relationships among man, community and natural ecosystems. It is necessary to develop a new assessment framework for the 'circularization' of the landscape system, in order to support decisions regarding the conservation/transformation of heritage, considering the specific attributes of functional reuse.

The attributes of functional reuse in the perspective of the circular economy are the following:

- activation of a symbiotic relationship with the natural ecosystem and the social system: between cultural heritage and its context, based on mutual complementary systemic interdependencies, economic, social and environmental ones;
- autopoietic (that is regenerative) capacity of many activities located in the cultural asset (which depends on its context);
- activation of a network of interaction and formal, informal, direct, indirect and induced relationships between the re-functionalized cultural asset and other (more or less) nearby activities;
- saving in the consumption of natural resources, energy, etc. for the use of cultural heritage, thus minimizing entropy;
- activation of synergic and cooperative processes between activities and subjects involved in a dynamic and adaptive perspective;
- self-financing capacity coming from some functions, localized without depending on external public contributions.

These attributes correspond to specific evaluation criteria and therefore require specific indicators (see CLIC research, www.clicproject.eu).

Circular economic processes reduce negative impacts on the environment and, at the same time, on the differential and Marxian rent, thus avoiding/reducing the use of natural and man-made capitals (Fusco Girard & Nocca, 2018).

In particular, the circular economy has impacts on the real estate market (Fusco Girard & Nocca, 2018):

- the reduction of underutilized land/spaces thanks to the implementation of the principles of the circular economy;
- the reduction of the rent in the city centre due to redistributive phenomena;
- more the building/space functions are flexible, more the value of the property increases as the long-term adaptation costs are reduced;
- the use of renewable resources produces a reduction in management costs. The use of materials that improve comfort inside buildings has a positive impact on the health and well-being of the occupants of the building, thus resulting in real estate added value.

Evaluation methods are necessary to integrate the change in local metabolism with the change in the well-being of the inhabitants, with changes in economic, social and environmental conditions.

Actually, if we move from the microscale to the mesoscale and the macroscale, we can integrate the above with evaluation tools ranging from life cycle assessment (LCA) (Björklund, 2012) to material flow analysis (Hendriks et al., 2000), to the evaluation of ecosystem services (Maes, 2013) and to the more general metabolic impact assessment (Ruas et al., 2017).

Evaluation methods capable of reflecting also the long term, incorporating not only the economic and financial benefits but also the perceived impacts at a subjective level (in addition to incorporating the intrinsic value), are necessary in any functional reuse interpreted in the circular model perspective.

A reference assessment framework should therefore include the circularization of material, energy and water flows, as well as historical–cultural aspects and socioeconomic processes, expressing the adaptation and development of local communities in the ecosystem.

Within this framework, evaluation methods should be 'people-oriented'. They should assess the variation in well-being for different social actors with a medium-/long-term time horizon (Fusco Girard & Nocca, 2018).

The circular model implementation necessarily requires an integrated evaluation tool able to capture the multidimensional impacts that it is able to produce and considering all the subjects/actors involved in its implementation, that is a multicriteria, multigroup, quanti-qualitative and short- and long-term evaluation tool (Zeleny, 2011). Multicriteria and multigroup evaluations are key hybrid tools (Fusco Girard et al., 2014) to manage and compare the positive and negative impacts (Coccossis & Nijkamp, 1995; Lichfield, 2005; Fusco Girard & Nijkamp, 2009) and to balance them for the different stakeholders (public, private, financial, social and civil). This evaluation tool has to be able to overcome the limitations of the current economic approach (Porter & Kramer, 2011), 'capturing' also the relational aspects. It is necessary to develop an evaluation tool which is able to integrate the economic dimension with the environmental, social and cultural ones. This integrated assessment is both quantitative and qualitative, as subjective and perception indicators have also to be considered, including an appropriate set of criteria and indicators that differs passing from the macroscale (metropolitan city) to mesoscale (historic district) and microscale (single sites or building).

New use values for the regeneration of cultural heritage/landscape

The urban regeneration linked only to the tourism sector (the consequent commercial, handicraft, transport sectors, etc.) and the related real estate rent produce benefits that do not tend to be redistributed in the city, but concern only some areas/subjects. Moreover, considering that the 'net' benefits of the tourism economy are much lower than the overall ones, it should be stressed that tourism should also be organized according to the processes of the circular economy (Fusco Girard & Nocca, 2017).

Possible circular economy initiatives can be referred to

- the development of an integrated network of urban greenery including roof gardens (where possible), urban gardens, shared gardens, therapeutic gardens and other forms of greenery inspired by emerging 'nature-based solutions' to connect the different components in a systemic perspective;
- the valorization of short supply chains and zero-kilometre products for the production, promotion and sale of local agri-food products;
- reuse of abandoned spaces for new functions, enhancing existing activities (craft, commercial, recreational, etc.) as drivers and activating 'multiplier' effects not only in the tourism sector;
- valorization of all underutilized spaces, such as public spaces and good common;
- adaptive reuse of cultural heritage/landscape;
- promotion of productive activities linked with: repair, maintenance, reuse, recovery, recycling and regeneration of materials for a new urban metabolism; as well as linked with 'regenerative/multifunctional agriculture' and with the reuse of agricultural/forestry residues to create bio-products.

The 'intrinsic value' represents the element on which the perspective of regeneration should be developed today: it represents the 'vector' that should guide urban and economic development.

This means that the regeneration of cultural heritage is not only a problem of stone conservation, but is linked to other values that reflect its specificity and identity that have to be creatively reinterpreted in the transformation/regeneration projects (Fusco Girard & Nocca, 2019; De Medici et al., 2020).

The choice of new use values that can be deduced from the good practices of adaptive reuse (consistent in particular with the Burra Charter) refers to functions such as museums, research and education activities, public libraries, cultural and community centres, cultural and creative clusters/industries, incubators and ICT clusters. The Burra Charter suggests to enhance the relationship between communities and places through the participation and involvement of the local community.

The search of the more satisfying compromise between 'intrinsic value' and many use values of the landscape can be identified (also through appropriate evaluation methods) as the final goal of an urban circular economy.

The requalification and adaptive reuse of religious cultural heritage

An operational proposal for the regeneration of the city through adaptive reuse in a circular economy perspective refers to the subject of religious cultural heritage.

This heritage, provided with a very high intangible and material value belonging to the whole community (Gerhards, 2019), represents a very frequent category of UNESCO sites of international relevance. It represents a sign of particular beauty that strongly characterizes the urban and extra-urban landscape. If it is left to degrade, it causes damage to one of the most important resources for its attractiveness (that is the landscape resource). In any case, the multiplies maintenance and management costs, making them unsustainable (Fusco Girard & Nocca, 2019).

A special attention deserves to be paid to the issue of religious cultural heritage that is being disposed of. However, as a result of the processes of demographic decrease and aging, of increasing urbanization of urban coastal areas with the abandonment of internal areas, with the absence of traditional religious practice, it becomes increasingly required the identification of new non-liturgical uses for buildings once intended for worship.

However, it is necessary that this functional reuse takes into account the 'intrinsic value' of cultural heritage in a systemic perspective. In other words, it is necessary to analyse and interpret the tangible and intangible values and the characteristics of the impacts determined by the proposals for functional reuse of cultural asset (churches, convents, etc.) in a perspective that is as participatory as possible with the local community, identifying the use values that are most consistent with this intrinsic value.

The evaluation activity becomes fundamental to identify the overall productivity of functional reuse in social, cultural, symbolic and economic terms, considering the impacts resulting from the change determined by alternative projects (Fusco Girard & Nocca, 2019).

The framework of the circular economy model offers the perspective in which to include project proposals based on the notion of sustainability, resilience, inclusion, hybridization, systemic interdependence and intrinsic value. It also offers the criterion on the basis of which to develop alternative combinations of functional/adaptive reuse of a site, a cultural asset and a landscape. The maximization of intrinsic value and possible use values should be done together with the search for maximum adequacy between it and possible alternative use values (Fusco Girard et al., 2019).

This type of adaptive reuse approach involves different values and actors/stakeholders in the conservation process (Lueg, 2011). The different stakeholders are generally all those who recognize a value to the religious property, such as owners, users and community members.

Unlike the regeneration of other types of buildings, adaptive reuse of religious buildings is often controversial because the ideas and values that different social stakeholders attribute to the building itself (and the link between the building and its sacred use) are different. Therefore, the analysis of the values linked to religious property is fundamental and affects the project of reuse. Several case studies show that the success of the reuse project depends not only on the new use attributed to the building, but also on the general approach related to its regeneration (Lueg, 2011; Gerhards, 2019).

The example of adaptive reuse of the Martinikirche in Bielefeld (dating back to 2004/2005) with a new use value represented by catering activities shows that a new use, which can be perceived 'in contradiction to church values' and not suitable for a specific type of building such as a church (restaurant), can instead become successful, at certain conditions (Lueg, 2011).

The success of this good practice is mainly linked to the involvement of all stakeholders and in particular of the church community itself which, having been actively involved during the whole regeneration process, has contributed to its success (Lueg, 2011).

The most important prerequisite that the church considers in a reuse project is that the new use does not contradict the original use as a church and the values it represents. Most of the guidelines (analysed by Lueg (2011)) consider demolition as the last choice (and only in cases where the building has no particular cultural and architectural significance), while they prefer the preservation of religious use. If the latter is not possible, it is possible to propose other uses that are in any case linked to the ecclesial reality ('extended use' as a meeting room for ecclesial groups, for administrative purposes or for concerts) (Lueg, 2011).

However, 'extended use' is not always the solution and therefore it becomes necessary for the owner to leave the property. In this case (whether the church remains the owner or the building is sold to a third party), it would always be preferable that the choice falls on uses that are not in conflict with religious values. From this point of view, cultural use (such as cultural centres, concert halls, museums) would be consistent with the role of the church as a historical supporter of art and culture. Another congruent use is the social one: educational centres, charitable structures, shelters for the homeless, etc. (Lueg, 2011). Even residential use is often accepted by the church if it has a social aspect (e.g. social housing). Commercial use, on the other hand, is considered the least favourable in a reuse process. A good practice of adaptive reuse of a religious building is the St. Elisabeth Church (Aquisgrana, Germany), nowadays known as the 'Digital Church' (Gerhards, 2019), characterized today by co-working spaces for start-ups and medium-sized companies and whose main hall serves as a meeting place for cultural or private events.

The role of the community in the regeneration processes

The community plays a key role in the regeneration processes and the identification of choices. The community's participation is expressed first of all in the process of identification and interpretation of the 'intrinsic values', which 'go beyond' the use values and values independent of use. The 'intrinsic value' of cultural heritage was recently evoked in the European Commission's Communication 'Towards an integrated approach to cultural heritage' (2014) as a value other than economic and social values. Economic value is defined as an instrumental value because it contributes to the creation of value and employment and generates many external effects also in other sectors, from the tourism industry to the cultural/creative industry.

The social value of cultural heritage is interpreted in the light of its ability to contribute (thanks to specific sociocultural impacts) to the regeneration of the sense of belonging, inclusion, active citizenship; to social cohesion, to the regeneration of microcommunities, and more generally to a relational culture, capable of reducing the social degradation of decaying urban areas (especially thanks to processes of real involvement of the local community in the choices of redevelopment and management). While trying to bring the inhabitants of a territory back to their common cultural roots, by valorizing cultural heritage, new 'horizons of meaning', collective memory and sense of identity, aesthetic emotions and sense of psychological well-being and prospects of social justice can be opened up.

Each social group can perceive the 'intrinsic value' of memory according to its particular perspective. But, probably, in these differences, there are common elements that can represent a potential energy to guide the regeneration processes.

The role of the 'intrinsic value' is essentially to help manage the change of the HUL, identifying not only the type of functional reuse consistent for a given cultural asset, with its particular history, but above all to guide local development, in both its material and immaterial components, thus combining the preservation of the roots with a dynamic and creative perspective, in a circular logic.

Alternative solutions of valorization are identified to guarantee – through a participatory process – old and new, memory and future and creativity of past generations and creativity of current generations, through a process of critical discernment by the community. It changes the priorities deriving from purely technical and economic approaches.

In the light of the above, the 'intrinsic value' becomes the foundation of any reuse/regeneration project, the reference point attentive to history/tradition/memory and also to the needs of the local community, capable of an authentic propulsive energy, mediating between memory and innovation.

The different focus groups, Delphi procedures, deliberative arenas and Living Labs are the dialogic–communicative processes through which if and how much the 'intrinsic value' is perceived today by the different stakeholders can be verified.

These participatory processes are critical because they help us compare a resource's values with the costs of conserving them; that is, they help us understand how much a community is willing to pay for conservation to keep a resource's values alive.

Assuming the landscape in a constructivist and non–neopositivist perspective (where the social component plays a fundamental role in identifying the elements of memory, identity and authenticity by integrating expert and civic knowledge; UNESCO, 2011), the CSV helps us to choose in a context characterized by multiple interests, values, objectives and constraints. But, above all, it allows place-making processes and the promotion of their attractiveness by offering, as already highlighted, the direction/orientation for a new development strategy of the city system (and also of the 'change management' of this system), considering together the ancient centre, the historical city and the extra-urban territory. In essence, it interprets those intangible elements that have shaped the physical space and the landscape itself.

The beauty of the landscape as an 'attractive force'

Beauty is the 'red thread' that connects the different perspectives of the landscape, from architecture to urban planning, ecological, economic/social, historical, anthropological, etc. The beauty of the urban/built landscape and of the natural landscape can represent an effective entry point through which to promote a sustainable human development strategy. Beauty is seen in a multidimensional perspective, from economic to social to cultural ones. It 'opens' also to an experience characterized by less conflicting, more collaborative/cooperative behaviours; it opens to 'We' in a society where too often only the 'I' triumphs.

This beauty determines a 'field of gravitational forces', that is where it exists, it generates a sort of 'attractiveness' for people, visitors, tourists, investments, innovative activities, specialized work, etc. Where beauty is present, it is easier to trigger virtuous processes of development. Where beauty is absent, it is difficult to detect these processes. The beauty produces economic values and, at the same time, social, cultural, environmental, aesthetic and civil values (Fusco Girard & Nocca, 2019).

Beauty can become a key factor in the economic development of a site. Through tourism, for example, the aesthetic values of a site are turned into economic values.

The public spaces of the city (squares, places, monuments, landscape, etc.) to be maintained, managed, preserved, enhanced, redeveloped and designed in a perspective of common good are the element through which the overall living conditions of the inhabitants can be improved. In fact, they contribute to improve economic conditions (generating more intense rates of use, which are reflected, for example, in real estate added values), employment, but also social cohesion, because they promote a sense of co-ownership, pride, as well as representing a concrete meeting place between different subjects, where they exchange information, experiences, knowledge, emotions and evaluations.

This 'generating capacity' of beauty is at the basis of the strategy for the 'beautiful city', already experimented in Paris in the last century and then reworked in Barcelona on the occasion of the Olympics. It is realized through the regeneration of places (historical centres, landscape sites, waterfronts, etc.).

Beauty is a notion always evoked but also elusive and perhaps even ambiguous and, in any case, 'fragile'. It communicates a sense of fullness that derives from the perception that certain attributes/criteria are satisfied at the highest level (Zeleny & Hufford, 1992).

It is not only harmony of proportion, order or symmetry (a concept that evokes a static interpretation of beauty). Beauty is born as a reflection not only of geometric proportions, but rather of systemic components. Beauty evokes a dynamic and vital dimension: it is linked to regenerative processes involving different components in a systemic and autopoietic perspective. Maturana and Zeleny recognized that beauty is the result of neg-entropic processes (Zeleny & Hufford, 1992; Maturana & Varela, 2001), being the reflection of an organizational structure where the individual components integrate each other circularly and harmoniously, as in living systems. The result is a 'sense of completeness', of satisfaction, of harmony.

In this perspective, beauty has to be linked to intrinsic, eco-systemic values. Beauty is a reflection of circular self-regenerative/autopoietic capacity. From the beauty follows the capacity to generate (generative capacity) positive external effects that is instrumental values of economic, social, environmental, civil and cultural types, as well as the same systemic resilience over time.

Empirical evidence shows that beauty is a 'generator' of vitality (Fusco Girard et al., 1989). For example, it attracts the localization of activities, stimulates the strengthening of existing ones and discourages their delocalization. In addition, beauty also contributes to building a 'good life' of the community that is it promotes social and civic values. It 'opens' the 'I' to others, to 'We', contributing to overcoming isolation because it produces 'Sense of Us'. Being a multiplier of values, it 'helps' life, contributes to unite what is divided and therefore contributes to the humanization of the lived experience.

Beauty is the perception that each element is interconnected with all the others determining a sense of unity among all the components: between heaven and earth and between people and nature. Beauty can be seen, but also heard and touched. In short, it should be understood not only as a factor of aesthetic fruition nor as a mere embellishment of reality, through scenographic games or scenic combinations (in an ephemeral and decorative sense), but as a result of the variety and plurality of interrelated dynamic elements that constitute the natural/built environment.

Beauty can itself become a factor in a site's economic development and, at the same time, a factor that contributes to social development (because it creates employment, but also interpersonal relationships, reduction of the sense of isolation, sense of community, social capital, etc.). A high-quality environment can increase the perception of well-being/health in a place. On the contrary, where there is no beauty, there is degradation, violence and illegality.

On the one hand, beauty attracts the localization of activities, stimulates the strengthening of existing ones and discourages their relocation, as many concrete experiences show. Stimulating creativity (directly and indirectly) produces economic utility, contrary to what Maynard Keynes claimed in 1936, when he stated that 'beauty is a luxury to be postponed to future/better times because it is not useful'.

At the same time, beauty contributes to social development, in the sense that it also contributes to building the 'good life' of the community, offering a relational (and not conflictual) openness among subjects: stimulating a sense of collaboration, co-evolution, co-ownership, improving the perception of the quality of life, as well as contributing to well-being and health (Geddes, 1915).

Beauty characterizes those areas of urban space defined as 'places'. It is primarily due to the human scale of these sites, that is, the proportions between public spaces (i.e. squares) and private spaces (i.e. residences). This human scale makes social inclusion easier than what happens in the new suburbs characterized by tower-shaped typologies, which are more and more high and anonymous.

Finally, beauty promotes cultural values because it changes the ranking of priorities between values/preferences/objectives in the choices, 'weakening' the priority of the useful over other criteria of choice.

Beauty of the landscape and utility: how to overcome conflict

Beauty is 'threatened' by the economy. There is a conflicting relationship between the beauty of the landscape and its instrumental utility, that is, between aesthetics and economy. We read this conflict, for example, in the speculations that tourism and real estate economy produce, reducing the landscape from a public/common good to a private good, often destroying its aesthetic quality.

This conflictual relationship arises from the dualism between intrinsic values and instrumental values of the cultural landscape/heritage (Fusco Girard & Nocca, 2019). The intrinsic values reflect the ecosystemic harmony and the functioning of the landscape in a systemic logic, guaranteeing autopoietic characteristics. Instrumental values, instead, are the values that are attributed to 'something' because it is useful. Instrumental values are related to the 'I'. The intrinsic values are related to 'We', to reciprocity, to cooperative, synergistic and symbiotic capacity.

However, the current economy considers only the instrumental values, neglecting their foundation which is in the intrinsic values. The (traditional) economic vision establishes priorities that can be completely overturned if, together with instrumental values, intrinsic values are also considered.

The model offered by the economy of nature, the 'circular economy', is here proposed as an economy that safeguards beauty, as a co-evolutionary economy, which does not require the traditional trade-off between efficiency and protection. It limits the production of waste/disposal/dumps, contributing to the quality of the landscape. On the other hand, the continuous maintenance action avoids the degradation of the landscape and, vice versa, a quality landscape 'opens' to collaborative/cooperative relationships between different subjects. In essence, beauty and economy are interdependent.

The circular economy is also the economy of cooperation, solidarity, co-evolution and the long term. It is based, for example, on common goods that require management practices based on the cooperation/collaboration/coordination of different subjects, opening up to a vision of relational values, of use values, independent of use values, of values in and of themselves (that is intrinsic ones) as well as to the co-production of market economic values.

The great challenge of the regeneration of cities/territory is based on cultural heritage/landscape as a great 'beauty repository' (Fusco Girard & Nocca, 2019): on the ability not to lose, but to recover and regenerate this infrastructure as a 'connective infrastructure', capable of regenerating community relations through the renewal of cultural memory and its celebration. Not taking care of and enhancing the beauty of the natural and built landscape means giving up the benefits that beauty itself can generate. But it also means losing the actual identity of our cities and territory, which is expressed by the cultural landscape/heritage. To this end, today, we need to re-learn from nature and its circular processes: the circular economy preserves and enhances the cultural heritage/landscape.

Conclusions and further research

The redevelopment of the quality urban landscape is of interest to all social actors and requires an integrated development strategy. This integrated 'circular regeneration' strategy involves not only public institutions but also enterprises. Many innovative activities, start-up hubs and incubators of innovative activities can find their best location in historical, urban and extra-urban heritage.

The circular economy produces a number of benefits ranging from improving the urban economy (thanks to economies from synergies) to reducing pollutant/climatic impacts, improving employment, reducing the pressure on municipal budgets, etc.

As already pointed out, evoking the circular economy model does not only mean proposing a model of creation, distribution and capture of value that avoids waste/underuse, minimizing waste. This is only one aspect of the circular economy.

The most important aspect is to identify some complementarities between the activities of different subjects, which thus enjoy reciprocal benefits due to the use of by-products that are transformed into resources and/or energy on the basis of specific relationships of systemic interdependence. The above allows a reduction/minimization of waste and an improvement in overall productivity.

The development strategy therefore is based on the ability to identify systemic development projects through symbiotic/systemic processes.

This model, in coherence with the HUL approach, and even before with the Burra Charter, encourages the creation of a 'community of relationships', which is a relevant element not only in determining the quality of life but also for generating new economic value chains. In addition to cooperative values, it stimulates attention to 'intrinsic values', as well as a widening of the time horizon of choices that is extended to the long term. It also represents a formidable opportunity for cultural education, training in the ability to relate, to critical discernment.

The above implies a series of consequences starting from the attention to all urban public spaces and common goods, to be produced, reproduced and valorized, but also an attention to systemic and synergistic relationships between the historical city and the 'modern' city, between cities and territory and between cities and smaller towns/villages.

It was stressed how the circular economy promotes efficiency and co-evolution together. But it is also the economy of the community. It is the economy that is founded (and in turn promotes) on cooperative, collaborative, solidarity values: it is the economy of 'with'. It shows with clear empirical evidence that 'cooperation is economically, socially, ecologically convenient' (Fusco Girard, 2014).

The regeneration of the city/territory certainly requires technical and technological innovations. They are necessary, but they are not enough. At the same time, regeneration on a cultural level is required.

A new 'culture' is needed, characterized above all by a long-term horizon, as well as the recognition of intrinsic and not only instrumental values. We need to think in an open, systemic, relational way, focusing on the relationships of interdependence between private interests and general interests.

In this perspective, the development of a cultural project supporting the regeneration of the link between man and man and between man and nature/ecosystems assumes a central role. In this way, it is possible to avoid that the reference to the beauty of the city/territory becomes only a reference to quality furniture, to superficial cosmetics, to spectacular embellishment, which generates economic benefits, but which does not reduce the negative impact on daily life in the city. It is also necessary to be able to produce external effects on city life, in terms of the ability to live together, to work together. We need to be able to generate an impact on the art of life: on the ability to transform each inhabitant into a citizen, into an artist of citizenship, capable of creatively combining beauty, utility and justice in his choices, that is private interests and general interests. By this strategy of civil aesthetics, public spirit, critical thinking and therefore responsibility can be produced. From the ability to transform aesthetic values into civic values follows the possibility of concretely realizing the paradigm of the new Humanism. It is, therefore, a matter of reproducing 'Sense of Us', that is, the ability to interpret own rights/needs from a relational perspective.

Finally, the value-centred approach necessarily requires integrated evaluation tools in order to consider monetary and non-monetary, quantitative and qualitative dimensions and all values of a resource. Furthermore, in a circular economy perspective, we need an evaluation tool that integrates 'traditional tools' (that were born and are used in the linear economy field, such as the heritage impact assessment (HIA), environmental impact assessment (EIA), social impact assessment (SIA) with tools characterized by a matrix linked to the circular economy model that is capable to capture the multidimensional impacts that this model is able to produce.

The evaluation framework linked to circular economy and thus able to capture the multidimensional impacts represents a fertile and interesting field of research. To date, the few official circular economy assessment frameworks are mostly sectoral, focusing on individual dimensions (economic, social, environmental), and do not include their interrelationships. Indeed, it is necessary to demonstrate the multidimensional benefits of the circular economy and the circular city model in order to convince policy-makers/community/companies that investing in it is convenient. So, the assessment field represents a fertile area for further researches to identify evaluation tools able to capture the multidimensional impacts of the circular economy model implementation, also considering the relationships among the different dimensions involved.

References

Björklund, A. (2012). Life cycle assessment as an analytical tool in strategic environmental assessment. Lessons learned from a case study on municipal energy planning in Sweden. *Environmental Impact Assessment Review*, 32, 82–87.

Bruni, L., & Zamagni, S. (2004). *Economia Civile. Efficienza, equità, felicità pubblica*. Milano, Italy: Il Mulino.

Chertow, M. R. (2000). Industrial symbiosis: Literature and taxonomy. *Annual Review of Energy and the Environment*, 25, 313–337.

Coccossis, H., & Nijkamp, P. (1995). *Sustainable tourism development*. Aldershot: Avebury Press.

Commoner, B. (1971). *The Closing circle: Nature, Man, and Technology*. New York: Knopf Press.

Costanza, R., Cumberland, J. H., Daly, H., Goodland, R., Norgaard, R. B., Kubiszewski, I., & Franco, C. (2014). *An introduction to ecological economics*. Boca Raton, FL: CRC Press.

De Medici, S., De Toro, P., & Nocca, F. (2020). Cultural heritage and sustainable development: Impact assessment of two adaptive reuse projects in Siracusa, Sicily. *Sustainability*, 12(1), 311.

Ellen MacArthur Foundation. (2015). *Growth within: A circular economy vision for a competitive Europe*. Retrieved from: www.ellenmacarthurfoundation.org (Accessed on 03 May 2021).

European Commission. (2014). *Communication from the Commission to the European Parliament, the Council, the European Economic and Social Committee and the Committee of the Regions. Towards an integrated approach to cultural heritage for Europe. Brussels, 22.7.2014*. Brussels: Belgium.

European Commission. (2019). *The European Green Deal. European Commission*. European Commission: Brussels: Belgium. https://doi.org/10.1017/CBO9781107415324.004.

Fusco Girard, L. (1986). The Complex Social Value of the architectural heritage. *ICOMOS Information*, 19–22.

Fusco Girard, L. (1987). *Risorse architettoniche e culturali: valutazioni e strategie di conservazione*. Milano: Franco Angeli.

Fusco Girard, L. (2014). The role of cultural urban landscape towards a new urban economics: New structural assets for increasing economic productivity through hybrid processes. *Housing Policies and Urban Economics*, 1(1), 3–27.

Fusco Girard, L. (2018). Discourse at High-level European Parliament Conference "Cultural heritage in Europe: Linking past and future. Brussels, Belgium.

Fusco Girard, L. (2020a). The circular economy in transforming a died heritage site into a living ecosystem, to be managed as a complex adaptive organism. *Aestimum*, 77, 145–180. https://doi.org/10.13128/aestim-9788.

Fusco Girard, L. (2020b). Towards the implementation of the circular economic model in metropolitan cities: The case of Naples. In S. Suzuki & R. Pattuelli (Eds.), *Broad view of regional sciences- essays in honor of peter Nijkamp* (pp. 303–328). Heidelberg Germany: Springer.

Fusco Girard, L., & Nijkamp, P. (1997). *Le valutazioni per lo sviluppo sostenibile della città e del territorio*. Milano: Franco Angeli.

Fusco Girard, L., & Nijkamp, P. (2009). *Cultural tourism and sustainable local development*. London: Ashgate.

Fusco Girard, L., Nijkamp, P., & Voogd, H. (1989). *Conservazione e sviluppo. La valutazione nella pianificazione fisica*. Milano, Italy: Franco Angeli.

Fusco Girard, L., & Nocca, F. (2019). La rigenerazione del "sistema Matera" nella prospettiva dell'economia circolare. In L. Fusco Girard, C. Trillo, & M. Bosone (Eds.), *Matera, città del sistema ecologico uomo/società/natura: il ruolo della cultura per la rigenerazione del sistema urbano/territoriale* (pp. 69–100). Naples, Italy: Giannini Editore.

Fusco Girard, L., Nocca, F., & Gravagnuolo, A. (2019). Matera: City of nature, city of culture, city of regeneration. Towards a landscape-based and culture-based urban circular economy. *Aestimum*, 74, 5–42.

Fusco Girard, L., & Vecco, M. (2019). Genius loci: The evaluation of places between instrumental and intrinsic values. *BDC - Bollettino Del Centro Calza Bini*, 19(2), 473–495.

Fusco Girard, L., De Rosa, F., & Nocca, F. (2014). Verso il piano strategico di una città storica: Viterbo. *BDC. Bollettino Del Centro Calza Bini*, 14(1), 11–38.

Fusco Girard, L., Gravagnuolo, A., & De Rosa, F. (2018). The multidimensional benefits of terraced landscape regeneration: An economic perspective and beyond. In M. Varotto, L. Bonardi, & P. Tarolli (Eds.), *World terraced landscapes: History, environment, quality of life* (pp. 273–293). Cham, Switzerland: Springer.

Fusco Girard, L., & Nocca, F. (2017). From linear to circular tourism. *Aestimum*, 70, 51–74. https://doi.org/10.13128/Aestimum-21081.

Fusco Girard, L., & Nocca, F. (2018). Circular city model and its implementation: Towards an integrated evaluation tool. *BDC – Bollettino Del Centro Calza Bini*, 18(1), 11–32.

Fusco Girard, L., & Nocca, F. (2019). Moving towards the circular economy/city model: Which tools for operationalizing this model? *Sustainability*, 11(22), 6253. https://doi.org/10.3390/su11226253.

Geddes, P. (1915). *Cities in evolution: An introduction to the town planning movement and to the study of civics*. London: Williams & Norgate.

Georgescu-Roegen, N. (1971). *The entropy law and the economic process*. Cambridge, MA: Harvard University Press.

Georgescu-Roegen, N. (1976). *Energy and economic myths: Institutional and analytical essays*. New York: Pergamon.

Gerhards, A. (2019). Spazi sacri: valori immateriali in favore della comunità intera. Un progetto per chiese dismesse. In *International conference on "Riuso adattivo e gestione integrata del patrimonio culturale religioso dismesso". 7 November 2019*. Naples, Italy.

Hart, K., Laville, J. L., & Cattani, A. D. (2010). *The human economy. A citizen's guide*. Cambridge: Polity Press.

Hendriks, C., Obernosterer, R., Müller, D., Kytzia, S., Baccini, P., & Brunner, P. H. (2000). Material flow analysis: A tool to support environmental policy decision making. Case-studies on the city of Vienna and the Swiss lowlands. *Local Environment*, 5(3), 311–328.

ICOMOS. (1979). *The Burra Charter, The Australia ICOMOS charter for places of cultural significance*. Australia.

ICOMOS. (2013). *The Burra Charter, The Australia ICOMOS charter for places of cultural significance*. Australia.

Lazzati, G. (2000). *La città dell'uomo. Costruire da cristiani la città dell'uomo a misura d'uomo*. Rome, Italy: AVE Editore.

Lichfield, N. (2005). *Community impact evaluation: Principles and practice*. London: Taylor & Francis.

Lueg, R. (2011). *Houses of God… or not?! Approaches to the adaptive reuse of churches in Germany and the United States. Masters Final Project, degree of Master of Historic Preservation*. University of Maryland, College Park, United States.

Maes, J. (2013). *Mapping and assessment of ecosystems and their services. An analytical framework for ecosystem assessments under action 5 of the EU biodiversity strategy to 2020*. Luxembourg.

Maturana, H. R., & Varela, F. J. (2001). *Autopoiesi e cognizione. La realizzazione del vivente*. Padova, Italy: Marsilio Editori.

Morris, W. (1889). *Address to the annual general meeting of SPAB*. Retrieved from: www.spab.org.uk.

Norberg-Schulz, C. (1998). *Genius Loci. Paesaggio, Ambiente, Architettura*. Milano, Italy: Electa.

Pauli, G. (2014). *Blue economy. 10 anni, 100 innovazioni, 100 milioni di posti di lavoro*. Milano, Italy: Edizioni Ambiente.

Porter, M. E., & Kramer, M. R. (2011). Creating shared value. *Harvard Business Review, 89*(nos. 1–2 (January–February 2011)), 62–77.

Preston, F. (2012). A global redesign? Shaping the circular economy. *Energy, Environment and Resource Governance, 2*(2), Retrieved from: https://www.chathamhouse.org/sit. (Accessed on 03 May 2021).

Riegel, A. (1903). The modern cult of monument: Its character and its origin. Trans. Kurt W. Forster and Diane Ghirardo. *Oppositions, 25*, 21–51.

Ruas, P. H. B., Cardoso, A. M. P., & Nobre, C. N. (2017). Persuasive technology in online social networks: A systematic literature review. *International Journal of Web Based Communities*, (13), 4.

Ruskin, J. (1989). *The seven lamps of architecture*. New York: Dover Publications.

TEEB. (2010). *The economics of ecosystems and biodiversity: Mainstreaming the economics of nature: A synthesis of the approach, conclusions and recommendations of TEEB*. Birkirkara, Malta. Retrieved from: www.teebweb.org.

Turner, R. K. (1993). *Sustainable environmental economics and management*. London: Belhaven Press.

UNESCO. (2011). *Recommendation on the historic urban landscape*. Paris: UNESCO World Heritage Centre.

United Nations. (2015). *Transforming our world: The 2030 agenda for sustainable development. Resolution adopted by the General Assembly on 25 September 2015*. New York: United Nations. https://doi.org/10.1007/s13398-014-0173-7.2.

United Nations. (2016). *Draft outcome document of the United Nations conference on housing and sustainable urban development (Habitat III)*. New York. https://doi.org/10.1257/jep.27.4.187.

Zeleny, M. (2011). Multiple Criteria Decision Making (MCDM): From paradigm lost to paradigm regained? *Journal of Multi-Criteria Decision Analysis, 18*, 7–89. https://doi.org/10.1002/mcda.

Zeleny, M., & Hufford, K. D. (1992). The application of autopoiesis in systems analysis: Are autopoietic systems also social systems? *International Journal of General Systems, 21*(2), 145–160.

Part VI
Inter-temporal and inter-spatial, dynamic heritage research methods

30 Using system dynamics in heritage research

Kalliopi Fouseki, Katherine Curran and Josep Grau-Bové

Introduction

This chapter aims to critically discuss the application of the method of system dynamics in heritage research. The ultimate goal is to unveil new research avenues and new insights in the heritage field resulting from this application. The method of system dynamics is underpinned by the theory of systems thinking. This theory contends that social and other phenomena, events and behaviours are driven by various elements which interact with each other forming a complex and dynamic system. The interconnections of the various elements are not linear – but non-linear (Monat & Gannon, 2015, p. 11). More recently, 'critical systems thinking' advocates for critical approaches to systems thinking and research by committing to critical self-reflection on chosen methods and by accepting that all system approaches have a contribution to make (Jackson, 2001).

Accordingly, the method of system dynamics provides a tool to unpack the interconnected elements that shape the dynamic structures of a phenomenon, event or object; here, we consider a 'system' in the broadest sense. One of the key outputs of the use of this method is the development of a model which maps and simulates the change of a system's parameters over time. The basic steps in forming a system dynamics model include (a) problem definition and system conceptualization; (b) model formulation and representation; (c) model evaluation, policy analysis and model use; and (d) policy formulation and evaluation (Sterman, 2000). Each of the steps is explained in greater detail in the following sections.

Before exemplifying the use of system dynamics in three, diverse heritage case studies, it is worth noting the core elements of this method: (a) *complex and non-linear systems*, (b) *feedback structures*, (c) *feedback loops* and (d) *stocks and flows*. As aforementioned, a 'system' consists of multiple elements that are interlinked with each other and which change constantly over time. This interlinkage results in non-linear relationships (meaning that one relationship that dominates the present may disappear in the future; Forrester, 1987, p. 107). The non-linear relationships are also *cause* and *effect* relationships in that each element provides the cause of a subsequent element (Randers, 1980, p. 120). In order to understand the behaviour of a 'system', we need to unfold its underpinning feedback structure. To put it simply, if we use the metaphor of an 'iceberg', what we may observe as behaviour is the top of an 'iceberg' and not what is underpinning this top. By trying to understand how the various elements connect and change over time, we can understand the behaviour of the system. There are three main feedback structures. A positive feedback structure is associated with exponential growth. A negative feedback structure results in goal seeking to bridge the gap between a desired and an actual goal, and a negative feedback structure with delays generates oscillation (Forrester, 1987, p. 133).

DOI: 10.4324/9781003038955-37

The modelling of the cause and effect relationships requires the identification of *stocks and flows* (Forrester, 1987, p. 191). *Stocks* connote anything (tangible or intangible) accumulated over time. What drives the accumulation of stock over time is known as *flow*. The sections below illustrate the use and applicability of system dynamics in three heritage-related case studies. The first case study discusses the development of a system dynamics model that simulates how the cultural values of residents inhabiting listed or non-listed heritage buildings change over time and what the impact of that change is on energy efficiency decisions. The second case study investigates the applicability of system dynamics in preventive conservation. The third case study deals with the system dynamics modelling of degradation of heritage plastic collections. Although the three cases are related to heritage, they deal with different problems and thus use different approaches to the collection, analysis and modelling data.

Case study 30.1 Mapping the dynamic change of heritage value and energy efficiency decision-making of residents of heritage buildings

The aim of this study was to investigate how residents of heritage buildings negotiate their decisions on improving the energy efficiency of their residences while conserving the heritage features they value the most (for detailed papers on the project, see Fouseki & Bobrova, 2018; Fouseki et al., 2020). The study is cross-cultural and draws on 59 in-depth, semi-structured interviews with owners and tenants of heritage, listed and non-listed buildings in Greece, UK and Mexico. The aim of this section is not to present the findings of the study which have been published elsewhere (see Fouseki et al., 2020) but rather to critically discuss the process adopted in applying the method of system dynamics.

PROBLEM STATEMENT

As aforementioned, the first step in developing a system dynamics model is to define the problem and the boundaries of the system to which the problem relates (e.g. Luna-Reyes et al., 2003, p. 275). In this specific case study, the problem was defined as follows: 'how do cultural values, with which original features of an old building are associated, change over time, and how does that change affect decision-making on energy efficiency interventions and the preservation of heritage values?' It is worth mentioning here that, although having set the problem is critical, an 'open eye' needs to be maintained while looking at the data, as this can lead to refining the problem even further.

Methodology

Following the principles of grounded theory according to which the data drive the theory (Glaser & Strauss, 1967), the interview data which totalled to 206,771 words were coded using first an *open coding* process. An open coding process allows the identification of themes and variables related to the key research problem (Corbin & Strauss, 2014). This process usually leads to hundreds of variables which can then be grouped into broader categories through *axial coding*. By doing so, we were able to refine the problem and the boundaries of the 'system' which consist of the building material, the people and the values assigned to the original features of the building. It soon became apparent that, while at the beginning of the research, the 'system' under exploration was the building itself, the real system with which the problem was concerned was the 'practice' of negotiation between energy efficiency and heritage conservation. Having completed the grouping of the variables in broader categories, we continued with the identification of cause and effect relationships among the variables using the analytical process developed by Kim and Andersen (2012; an example is illustrated in Table 30.1).

The first row depicts the 'cause' variable, and the second row, the 'effect' of the cause. Our personal coding/description of the variable is inserted in brackets, while direct quotes extracted from the interview are in italics. Table 30.1, for example, shows that the presence of 'original features' is a cause variable in that it contributes to the visual aesthetics of the house. The more extensive the original features, the stronger the visual aesthetics. This relationship is therefore positive and reinforcing. The 'visual attractiveness' which is listed as an effect can be a cause for another type of effect, such as 'likelihood to move to the house', as shown from the interviews. Therefore, the cause and effect variables are all interconnected, creating multiple loops. Once the various cause and effect variables and their interconnections are mapped, a *causal loop diagram* is designed to visualize those interconnections. A causal loop diagram is a diagram which visualizes the feedback loops that are assumed to have caused behaviour of key variables over time (Randers, 1980, p. 119). Each cause–effect relationship is indicated with a + or − sign depending on whether the relationship is positive and reinforcing (e.g. the more…the more) or balancing (e.g. the more…the less).

RESULTS

The causal loop diagram, in effect, visualizes the *dynamic hypothesis* of the problem that is a hypothesis on how the 'system' under exploration behaves over time. In this particular study, one of the *dynamic hypothesis* is that while the 'original features' are highly valued by the residents in the early years when they move to the

Table 30.1 Example extrapolated from the detailed matrix of cause and effect variables. This formed the basis for developing the causal loop diagram

Cause variable	(Original features, e.g. sash windows)	(Visual attractiveness) *I love the look of it*
Effect variable	(Visual attractiveness) *I love the look of it*	(Satisfaction)
Relationship type	Positive (the more the original features, the higher the visual attractiveness)	Positive

Source: Created by Fouseki & Bobrova (2018).

building, this value declines because the need to reduce energy bills and improve the thermal comfort is higher. The original features are viewed as a contributor to decreased thermal comfort. This was one of the first *dynamic hypotheses* that emerged when we completed the first study in an East London area (Walthamstow; Fouseki & Bobrova, 2018). The question that immediately emerged is to what extent is this *dynamic hypothesis* valid in other contexts. The more and diverse the case studies included in the study, the more the various *dynamic hypotheses*. In relation to this particular hypothesis, we observed that the value attached to 'original features' did not decline over time (as in the case of Walthamstow) in conservation or world heritage areas (Fouseki et al., 2020). Although residents acknowledged thermal comfort challenges associated with preserving original windows, they highlighted alternative means to improve thermal comfort without compromising the preservation of original windows. The type of thermal comfort strategies varied once again among different case studies. Individuals, for instance, who have been living in the buildings for more than 30 years tended to use passive means (such as warmer clothes, blankets and heavy curtains). We thus noted that the negotiation process between heritage conservation, energy efficiency and thermal comfort is a complex and dynamic sociocultural practice of the following interconnected elements: materials (e.g. original features), competencies (e.g. restoration skills), resources (e.g. costs), values, space/environment (e.g. natural light), senses (e.g. thermal comfort) and time (e.g. years living in the house). The connection or disconnection of those elements will depend on (a) the nature of the context (e.g. rural, urban, conservation area); (b) the listing status; (c) age and construction materials of the building; (d) local climate; and (e) ownership status (Fouseki et al., 2020, p. 1).

A system dynamics model simulates the *dynamic hypothesis*. In this particular study, the model did not intend to predict the future but rather to simulate the growth and decline of the key components (e.g. values, material change and thermal comfort) and their interactions over time. By doing so, the model provides a tool that enables dialogue with heritage professionals and heritage or energy policy-makers developing guidance on energy efficiency in historic buildings that corresponds to the particular characteristics of each context. The development of the model requires the development of a stock and flow diagram which illustrates the variables that accumulate over time in boxes (stocks) and the rates that drive this accumulation (rates; Fouseki & Bobrova, 2018). The interrelationship of the variables is then described using simple mathematical equations (Sterman, 2000). Translating the relationships of our study – relationships which were often intangible and abstract – into mathematical equations provoked lengthy debates and conversations. A conventional scale from 0 to 1 was decided upon in order to declare how high or low a particular heritage value or the need for thermal comfort was. Although it is impossible to attribute a numeric value that shows the degree of importance placed on original features, a change in the value (which is what matters) can be shown and simulated by attributing the different scores (Fouseki & Bobrova, 2018).

Case study 30.2 Systems modelling for preventive collection management

Collection managers are faced with many decisions that concern the totality of the collection. Occasionally an artefact, because of its value or condition, may require individual attention. But it can be argued that the fate of the collection as a whole depends on how the collection is stored, accessed and managed. That is, what defines the longevity of a collection are global policies or decisions, rather than the treatment of individual objects. Complexity emerges when we observe the consequences of this type of decision. For example, the durability of a collection increases by improving environmental conditions. Retrofitting or ventilating storage areas has an impact on energy consumption. Energy consumption has an impact on the budget, which may have an impact on the frequency of conservation treatments, which, finally, has an impact on the value of the collection. Systems modelling can help managers tackle this complexity.

Problem statement

Paper collections, such as archives and libraries, provide a perfect case study for this type of modelling. They are one of the few types of collections that have been studied in this manner, in the pioneering research of Cristina Duran-Casablancas (Duran-Casablancas et al., 2021). They are ideal for three reasons. Firstly, we understand the chemical and physical degradation of cellulose well. Existing models describe how paper will degrade over decades or centuries, given storage conditions and frequency of handling (Strlič & Colar, 2015). Secondly, paper collections are relatively homogeneous. In comparison with a modern art gallery, for example, an archive contains fewer materials in fewer combinations. Finally, collection managers have at their disposal a range of whole-collection policy options that have been well studied.

To understand this problem in detail, it is useful to draw a quick sketch of these options. One method to slow down the chemical degradation of a large collection of paper is to improve storage conditions by reducing temperature and humidity. Another common method is mass deacidification: reducing the acidity of the paper, which plays a key role in its degradation (Strlič & Kolar, 2005). Mass deacidification is conducted in large batches, in purpose-built machines. These methods prevent the paper from becoming fragile and brittle and therefore reduce the risk of wear and tear during handling. Obviously, another way to reduce wear and tear is to avoid handling: for example, a collection may be digitized to reduce instances of access. There are, in turn, many different approaches to digitization: digitize popular items, digitize on demand or digitize everything. Each option has different energy and staff demands.

We could develop similar arguments for other collections. For example, the glass panels in stained glass windows or multipanel windows have similar properties. They are relatively homogeneous chemically. They are subject to a risk of mechanical failure and a (much lower) risk of chemical degradation. They are also subject to collection-wide policies: the use of rooms, the frequency of cleaning, the handling of windows, the criteria of replacement of broken glass and the proximity to visitors. The list of examples could go on: bricks on a historic building, bronze statuary in a city and canvases on a painting gallery. In all these cases, managers ask themselves similar questions: What are the long-term consequences of decisions? And, more importantly: What combination of decisions maximizes the durability and minimizes the cost?

Methodology

Constructing a causal loop diagram is a good first step. This can be done in conversation with a diversity of stakeholders, ideally including collection managers with first-hand experience on the consequences of their decisions. In the case of libraries and archives, one can involve individuals in charge of digitization, environmental management and collection treatment. The main advantage of causal loop diagrams is their accessibility: everyone is able to express cause and effect relationships. Some key disadvantages are the lack of hierarchy, where all causes and effects seem equally important, and their boundless nature, as any type of interaction can be added.

In the case of collection modelling, it is unavoidable that causal relationships will be of a wildly diverse nature. Some are eminently physical, for example, the relationship between storage temperatures and reaction rates. But some involve social processes, for example, the relationship between digitization and physical access. It is certainly possible that a percentage of users will always choose the reading room, even if their document of choice is available online.

After these hypotheses are formulated, the modeller needs to choose the scale of the system. This is an art, not a science. A researcher oriented towards the physical sciences may see the social interactions as the limit of their model. The availability of data for validation is also important when setting the limits of the model. In the case of paper collections, it is often possible to obtain data of document requests online and in reading rooms. These data can potentially be used to validate the behavioural aspects of the model. In the case of a collection of glass, patterns of visitor access or occupancy could be recorded. But a researcher could legitimately choose to not be anchored to validation and continue further, exploring dynamic hypotheses that cannot be quantified. There is value in all these approaches.

RESULTS

Collection modelling aims at providing actionable advice. This aim often requires a functioning predictive model. In the case of paper collections, this has been achieved by solving three different mathematical models together: (1) a model that predicts the degradation of paper as a function of storage conditions and access – this model is based on the collection demography model (Strlič et al., 2015), (2) a model that predicts physical access as a function of digital availability and the number of requests and (3) a model that predicts energy expenditure as a function of desired environmental conditions and building geometry. Some examples of the scenarios that can be simulated with part of this model can be found in Duran (2021).

However, system dynamics is more than a set of equations and the methods to solve them. It is also a logical narrative to describe the dynamic behaviour of a system. New vocabulary emerges during the process of creating a mathematically consistent description of the system. Often, this vocabulary is more precise than previous ways of referring to separate aspects of the problem. Another benefit is the clear identification of areas that require further research. For example, the three models listed above have different levels of certainty: (1) is heavily validated with real-world data; (2) is a newly proposed model, validated only against a handful of real-world case studies. Many existing models can do the role of (3), but the question is which ones are applicable to real archives and libraries. In other words, by attempting to connect different processes, we discover which pieces match well and which ones require further work.

While the final objective of collection modelling is to support decision-making, there is a lot to learn along the journey. A system dynamics approach helps define a problem well, identify the current limits of knowledge and assess the uncertainty of predictions.

Case study 30.3 Applying system dynamics to the degradation of plastic museum artefacts

This case aimed to explore the use of system dynamics to understand and develop policies for the conservation of plastic artefacts in museum collections. It is based on work from the ERC Starting Grant funded project COMPLEX: *The Degradation of Complex Modern Polymeric Objects in Heritage Collections: A System Dynamics Approach*.

Problem statement

The development of plastics over approximately the last 150 years has revolutionized society, leading to significant scientific advances, enabling the development of new art forms, such as cinema, but also causing substantial environmental damage. Plastic objects thus form an important part of our material past that need to be protected so that they can be studied and understood. However, plastics are unfortunately some of the most fragile materials in collections, degrading more rapidly and catastrophically than more familiar heritage materials such as metals, ceramics or stone (Shashoua, 2008). The chemical and physical mechanisms of degradation in heritage plastics have been studied for some time; however, there remains a significant gap in understanding. This was articulated clearly by Madden and Learner (2014, p. 8):

> our understanding of plastics stability remains rudimentary. We have a menu of mechanisms that potentially explain degradation, but there is a tendency to default to them and recite them, rather than investigate skeptically what is actually going on.

This defines the problem. While there is a knowledge of the mechanisms that cause degradation, these exist as a 'menu' or a list, with the interactions between them unexplored. This means that this knowledge is difficult to translate into an understanding of *what is actually going on*. Questions remain as to what are the most important factors in determining the vulnerability of a plastic artefact and as to what are the best storage and display strategies. A systems approach is thus very appropriate for exploring this problem, considering a plastic object within its environment as a complex system of interactions. The research aimed to understand the structure of the system and to identify key 'leverage points' in material degradation processes so that practical modifications such as temperature or humidity control can be used to improve collection care.

Methodology

One of the first steps within the research was to construct a causal loop diagram that mapped the relationships between different variables within the system. This was co-created during a workshop with academics and museum professionals, all working in the area of plastics conservation and facilitated by a system dynamics expert. Participants were asked to identify variables they thought were important and ways in which those variables could change over time, and these were then brought together into a causal loop diagram via a group discussion. Standard methods from system dynamics such as Graphs Over Time, Dots and Initiating And Elaborating A Causal Loop Diagram scripts were used (Andersen & Richardson, 1997; Wikibooks, 2020).

Following this, different parts of the model were translated into mathematical models. Given that many of the relevant processes, e.g. diffusion or chemical reactions, can be described by well-defined partial differential equations, modelling was done using Matlab and gPROMS (a chemical engineering software). In addition, significant experimental work was done in order to measure values for key parameters such as diffusion coefficients and to provide data for validating the model. The validation data consisted of measurements of material properties over time from accelerated degradation experiments. At the time of writing this chapter, the project is still ongoing and the experimental and mathematical modelling work is not yet complete.

Results

It is beyond the scope of this chapter to describe all of the results from the project. Here, we provide a brief discussion of the way in which the project developed and an example of the type of results produced and the way in which they have been used.

Project development

The development of the causal loop diagram identified key aspects of the system to be modelled. These included the impact of relative humidity on chemical degradation, the physical process of plasticiser loss from artefacts and the overall 'value' of the artefacts. An important decision made early in the work was to develop separate mathematical models for different processes with the aim of re-integrating them at a later phase of the project. This enabled specific processes to be understood in more detail and made practical sense in terms of project management. However, it does mean that some of the key interrelationships between decay mechanisms were not explored in the initial stages of the project.

Another important decision was taken to focus on the physical, quantifiable aspects of the system. The co-created causal loop diagram included connections between physical variables such as moisture content and more abstract concepts such as the perceived value of the artefact. It was felt that it was not feasible to model such different variable types within the scope of the project. This means that the different aspects of the model are consistent with each other. However, it means that while we are capturing the way in which the physical and chemical properties of the material change over time, we are not relating these changes to the significance of the artefact, e.g. its aesthetic value.

Example results

One part of the system that has been modelled mathematically is known as the 'vinegar syndrome'. This is a well-known degradation process in the historic plastic cellulose acetate, which involves a chemical reaction between the plastic, water and acetic acid (or vinegar), generated from the reaction (Allen et al., 1987). The acid acts as a catalyst for the chemical reaction, meaning that the rate of the reaction increases over time, representing a reinforcing loop within the system. The mathematical model was validated using experimental data from within the pre-existing literature (Bigourdan & Reilly, 1997), and the results were published in Ahmad et al. (2020). Predictions of the lifetimes of cellulose acetate cinematic film at different temperatures and relative humidities were made and compared to predictions from an existing model developed at the Image Permanence Institute (IPI; Reilly, 1993). We predict that even using cold storage for historic film, significant damage will happen within decades rather than the centuries predicted by the IPI modelling. We attribute this striking difference to the inclusion within our model of the reinforcing acid catalysis during the whole period of the degradation. Acid catalysis is also considered within the work of the IPI but is not felt to be significant until a later period of degradation (referred to as the auto-catalytic point). Our conclusions suggest that historic cellulose acetate film is at risk and that strategies such as digitization need to be considered as a matter of urgency.

Conclusion

This chapter demonstrated the applicability and potential of system dynamics in exploring heritage-related problems. It mainly focused on outlining some of the advantages and disadvantages of using the method as well as on the need to exercise critical thinking in the use of the method. In all three cases, the decision to apply system dynamics dictated new ways of looking at heritage change. It forced us to move beyond the linear examination of change to a more complex and dynamic study of change which brought together social, environmental, cultural, physical and economic factors.

Moreover, the value of the modelling process did not lie so much on the final output (the model itself) but more on the process of developing the model which allowed the identification of multiple parameters contributing to the dynamic behaviour of the system in question.

One of the key issues of debate is 'what is the added value' of using system dynamics, especially in the analysis of qualitative data. Through discussions and experimentation with the systemic modelling process, we concluded that the unique characteristic of system dynamics is that it looks at the underpinning structure of a systemic problem and the system itself, and its change over time through a non-linear manner. It allows the exploration of the various elements, not in isolation but in interconnection. This allows new interconnections to be identified and also enables existing relationships to be explored in more depth. Moreover, it affirmed what has been discussed in literature that heritage is a complex and dynamic system. A heritage item, as small as an object or as large as a building, or city is a complex and dynamic systems which require methods such as system dynamics to unveil their complexity.

Furthermore, in all cases, the understanding of the problem in question and the underpinning structure of the system had decision-makers (e.g. residents, conservators, collection managers) and policy-makers in mind as one of the target audiences. Indeed, the key driver in all three projects was not just to better understand a problem but more to understand human behaviour and human action in relation to the problem. By doing so, future strategies and policies can be better informed.

Another issue of debate was what one of the co-authors identified as the paradox of system dynamics. Given that system dynamicists claim to depict complex systems, they do so through a rather reductionist approach. Indeed, as Sterman argues, a system dynamics model 'must address a specific problem and must simplify rather than attempt to mirror an entire system in detail' (Sterman, 2000, p. 89). This 'reductionist' approach can though be useful for unveiling what really drives change. A related point is the need to work with different models within the same system, as discussed in Case Studies 30.2 and 30.3. There is a risk here that the previously discussed benefits of a holistic approach that reflects the realities of a complex and interconnected system will be lost if we proceed to break that system into parts. However, in some cases, this is essential for practical reasons, to build fundamental understanding of differing aspects of the system and to manage a research project. A system dynamics approach thus involves a balance between taking a broad perspective of the system as a whole and a 'zooming-in' process that goes into detail on specific aspects of it.

Finally, the 'modelling' processes in each case provoked interesting discussions. In the first case study, the issue of 'modelling behaviour and perception' provoked debate in that the nature of the data would not allow to predict the change of the system in future. In this case, modelling provided the tool for exploring whether the understanding of the system gained through the modelling process is capable of generating a behavioural pattern of interest. In other words, the process of modelling enhanced the initial *dynamic hypothesis* and offered an additional tool for engaging with policy-makers. The third case study could potentially produce a predictive model. The key parameters such as diffusion coefficients and equilibrium moisture contents can be measured, allowing a specific scenario to be simulated over time. However, as the research developed, it became clear that the value of the research approach was not so much in the ability to predict the outcomes of specific scenarios, which could be difficult to generalize but in our ability to identify the key parameters that have the greatest impact on the system (e.g. *how important is the diffusion coefficient*) and to compare differing scenarios (e.g. *does it make a big difference if I store this object in an open or closed container?*). Tools such as sensitivity analysis are of use here.

To summarize, using a system dynamics approach allows the full dynamic complexity of heritage to be taken into consideration. Significant challenges arise with this approach such as the need to define the limits of the system and the level of detail of interest. However, at its best, system dynamics is a powerful tool for engaging with policy-makers and providing an evidence base for future heritage management strategies.

References

Ahmad, I. R., Cane, D., Townsend, J. H., Triana, C., Mazzei, L., & Curran, K. (2020). Are we overestimating the permanence of cellulose triacetate cinematographic films? A mathematical model for the vinegar syndrome. *Polymer Degradation and Stability, 172,* 109050. https://doi.org/10.1016/j.polymdegradstab.2019.109050.

Allen, N. S., Edge, M., Appleyard, J. H., Jewitt, T. S. (1987). Degradation of historic cellulose triacetate cinematographic film: The vinegar syndrome. *Polymer Degradation and Stability, 19*(4), 379–387.

Andersen, D. F., & Richardson, G. P. (1997). Scripts for group model building. *System Dynamics Review, 13*(2), 107–129.

Bigourdan, J.-L., Reilly, J. M. (1997) *Environment and enclosures in film preservation.* Rochester, NY: Image Permanence Institute, Rochester Institute of Technology.

Corbin, J., & Strauss, A. (2014). *Basics of qualitative research: Techniques and procedures for developing grounded theory.* Thousand Oaks, CA: Sage Publications.

Duran-Casablancas, C., Strlič, M., Beentjes, G., de Bruin, G., van der Burg, J., & Grau-Bové, J. (2021). A comparison of preservation management strategies for paper collections. *Studies in Conservation, 66*(1), 23–31.

Forrester, J. W. (1987). Lessons from system dynamics modelling. *System Dynamics Review, 3*(2), 136–149.

Fouseki, K., & Bobrova, Y. (2018). Understanding the change of heritage values over time and its impact on energy efficiency: Decision-making at residential historic buildings through system dynamics. In *The 3rd international conference on energy efficiency in historic buildings (EEHB2018), Visby, Sweden, September 26th to 27th* (pp. 11–21). Sweden: Uppsala University.

Fouseki, K., Newton, D., Murillo Camacho, K. S., Nandi, S., & Koukou, T. (2020). Energy efficiency, thermal comfort, and heritage conservation in residential historic buildings as dynamic and systemic socio-cultural practices. *Atmosphere, 11*(6), 604. https://doi.org/10.3390/atmos11060604.

Glaser, B. G., & Strauss, A. L. (1967). *Grounded theory: Strategies for qualitative research.* Chicago, IL: Aldine Publishing Company.

Jackson, M. C. (2001). Critical systems thinking and practice. *European Journal of Operational Research, 128*(2), 233–244.

Kim, H., & Andersen, D. F. (2012). Building confidence in causal maps generated from purposive text data: Mapping transcripts of the Federal Reserve. *System Dynamics Review, 28*(4), 311–328.

Luna-Reyes, L. F., & Andersen, D. L. (2003) Collecting and analyzing qualitative data for system dynamics: Methods and models. *System Dynamics Review, 19*(4), 271–296.

Madden, O., & Learner, T. (2014). Preserving plastics: An evolving material, a maturing profession. *Conservation Perspectives: The GCI Newsletter,* 4–9.

Monat, J. P., & Gannon, T. F. (2015). What is systems thinking? A review of selected literature plus recommendations. *American Journal of Systems Science, 4*(1), 11–26. https://doi.org/10.5923/j.ajss.20150401.02.

Randers. J. (Ed.) (1980). *Elements of system dynamics method*. Cambridge, MA: Wright-Allen Press.

Reilly, J. M. (1993). *IPI storage guide for acetate film*. Rochester, NY: Image Permanence Institute, Rochester Institute of Technology.

Shashoua, Y. (2008). Degradation of plastics. In *Conservation of plastics: Materials science, degradation and preservation* Oxford: Butterworth-Heinemann.

Sterman, J. D. (2000) *Business dynamics: Systems thinking and modeling for a complex world*. New York: Irwin/McGraw-Hill.

Strlič, M., & Kolar, J. (Eds.) (2005). *Ageing and stabilisation of paper*. Ljubljana: National and University Library.

Strlič, M., Grossi, C. M., Dillon, C., Bell, N., Fouseki, K., Brimblecombe, P., Menart, E., Ntanos, K., Lindsay, W., Thickett, D., & France, F. (2015). Damage function for historic paper. Part II: Wear and tear. *Heritage Science*, *3*(1), 1–11.

Wikibooks. (2020). *Scriptapedia*. Retrieved from: https://en.wikibooks.org/wiki/Scriptapedia (Accessed on 14 May 2021).

31 Port city resilience

Piloting a socio-spatial method for understanding, comparing and representing linked maritime heritage

Carola Hein, Lucija Ažman Momirski and Yvonne van Mil

Introduction

Port cities are pivotal nodes of history, heritage and culture. For many centuries, they have facilitated global trade at the edge of land and water; numerous maritime heritage practices and objects are embedded in contemporary cities. Port cities have continuously been economically strong cities, often with dense population concentrations that have responded quickly and creatively to diverse external and internal crises (Konvitz, 1978; Broeze, 1989, 1997; Lee & Lee, 1998; Hein, 2011, 2012, 2016, 2019, 2020; Miller, 2012; Bosa, 2014). This maritime heritage speaks to the resilience of port cities through the ages and the capacity of public and private, port and city stakeholders to adapt to diverse environmental, political, economic, social and cultural changes. Maritime practices have always brought together multiple stakeholders; however, the maritime heritage that we currently protect is often disconnected from its environment and from other elements of the historic practices that generated it. A better understanding of this unique heritage and the networks that created them can inform sustainable practices today, both for individual sites and for maritime networks.

The lessons of port city heritage are multiple; they are not evident in single buildings, but are systemic. There is no single form of a port city; each one varies in location, function, size and form depending on local particularities. Each city has found its own approach to integrating shared shipping systems, creating a unique heritage that is interconnected with other sites around the world through its primary function of accommodating shipping. The location of ports in relation to cities varies extensively. Some ancient ports of Rome, such as Ostia, were independent of urban centers. In other cases, such as Carthage, Leptis Magna and Alexandria, the port had a synergistic relationship with its city (Martino et al., 2015). Commercial ports were often connected to urban buildings: piers with *horrea* (warehouses), *tabernae* (workshops), galleries, arcades, markets, fish ponds, tanks, aqueducts and lighthouses, but also through programs such as *hospitia* (places of shelter for travelers). Meanwhile, military ports were separated from the urban tissue by defensive elements such as ramparts. Evolving technology, changing politics and other societal factors have often led to innovation. When the Free and Hanseatic City of Hamburg, for example, joined the German Reich, it received funding to build a mono-functional and duty-free warehouse district, the Speicherstadt, in Hamburg, the largest warehouse district in the world. Its construction was paired with the creation of a mono-functional office area, the Kontorhaus District, and together, they have received UNESCO World Heritage status. Many port structures corresponded to the demands and practices of ports and cities at the time of their construction; however, as part of an industrial and constantly evolving system, they risk becoming obsolete as the technology and functions of port cities change.

The resilience of ports and their neighboring cities is a result of their adaptability and of environmental, economic, social and political sustainability. A shared maritime mindset, or port city culture, plays an important role in such processes, and a better understanding of this heritage can benefit future sustainable development (Hein, Luning, van de Laar 2021 a, b). Port and city stakeholders have long responded collectively to internal and external challenges. As rivers and coastlines changed, as ships became larger, and as containers transformed the port's operations, many historic port areas (e.g. Miletus and Brugge) became obsolete; local stakeholders left behind former port infrastructures to develop new sites, often to incorporate them in new municipal boundaries. At times these sites were abandoned or closed down, in other cases, their function changed to non-maritime use for industrial, logistic and commercial activities, or they began hosting different maritime activities, such as cruise shipping. This change in administration and function has required new approaches to waterfront regeneration and heritage. An architectural masterpiece of the port, the magnificent Corderie Del Tana in the Venetian Arsenal, for example, since 1980 has been reused as a location for the International Art Exhibition and the International Architecture Exhibit. However, the complex of former shipyards and arsenals, which covers almost a sixth of the city, has long been abandoned and is no longer in use. The maintenance and improvement of port heritage has become critical, also because of its typological heterogeneity: Fortress, palaces, customs houses, warehouses, arsenals, docks and piers are some examples of this endangered heritage (Martino et al., 2015). Finding new purposes for these sites is difficult and requires more than individual interventions.

In this chapter, we argue that a sustainable approach to port city heritage requires systemic investigation and intervention involving actors of the port, city and region, acknowledging historical networks may help. Over the last half century, the preservation of port city heritage has largely been left to the individual cities that host these ports. As Sánchez and Daamen (2020) put it, the adoption of the UNESCO Historic Urban Landscape (HUL) approaches has triggered deeper reflection among key urban actors, for example, on the links between Lisbon's city and port, notably on the waterfront. The HUL approach can also serve as an inspiration to explore the maritime links among cities and use these to inspire preservation and new development activities. Since 1981, UNESCO has inscribed 50 marine protected areas (in 37 countries) on the UNESCO World Heritage List—recognized for their unique marine biodiversity, ecosystem, geological processes or incomparable beauty (World Heritage Marine Programme, 2020). Since 1992, significant interactions between human and natural environment have been recognized as cultural landscapes. Marine protected areas are preserved from a natural heritage perspective, but they are also adjacent to cultural sites, such as port cities. An integrated approach to ports and cities, land and water will aid the search for sustainable solutions. Sea-level rise and other climate-related changes will occur simultaneously in these cities.

Protecting World Heritage sites for their Outstanding Universal Value (OUV), UNESCO recognizes buildings, urban areas and landscapes as much as knowledge, wisdom, values, beliefs and the behavior of individuals and communities. These dimensions—physical structures, institutional knowledge and culture—are closely interrelated in port cities. This includes specific knowledge of fishing, of navigation and of the sea itself (e.g., traditional skills of building and sailing Iranian Lenj boats in the Persian Gulf [Urgent Safeguarding List, 2020]). For example, coastal communities have long known the warning signs of tsunamis after earthquakes. The importance of this experiential knowledge was demonstrated by an island community in Indonesia in 2004, when 80,500 people fled the coast for nearby hills. As a result, only seven people in this island community died from the tsunami, while a total of 163,795 people died in the rest of Indonesia's northern province of Aceh (de León et al., 2006). Port cities, in addition to containing localized forms of knowledge, are also sites of cosmopolitanism and integration. They have traditionally welcomed transient populations and migration (Crul et al., 2019). Understanding the culture and values of communities living on and with water systems is an essential for making heritage resilient.

Various research projects, strategies and courses have been developed by authorities and institutions to protect the heritage of port cities. These include the course port heritage: neighborhoods with ports/ports with town (AADIPA-COAC 2020), which deals with three thematic areas (the history and the port, interventions of restoration in the port area and urban management of the port heritage: the neighborhood and the port). The Global Architectural History Teaching Collaborative (GAHTC) hosts a course on port cities between global networks and local transformations (Hein & Mager, 2021). Historic England (2020), a public institution that helps people maintain, enjoy and celebrate England's spectacular historic environments, lists the most important historic places in England, including coastal, marine and maritime heritage. The priority of this section is to "assess the survival, character and importance of England's port and harbor heritage to support effective ways" to ensure its survival as a positive contribution to the future distinctiveness of its ports. Historic England (2020) points out that much of this rich heritage is poorly understood and yet is under considerable pressure from coastal erosion, development and harmful activities. A project for the preservation of the maritime cultural heritage, Remember (2020), involves eight Adriatic ports under the leadership of the Port of Venice. It demonstrates that cultural heritage can serve as a driving force for the sustainable development of port cities through proposals such as one virtual museum of archaeological heritage, the desire of ports to reconnect to their surroundings and for more competitive and balanced development paths.

Currently, many attempts at enhancing the particular character of port city culture and maritime heritage focus on selected, localized interventions. While it is important to examine the reconstruction of a specific waterfronts and urban areas, it is also necessary to analyze these projects in the larger context of sea and land, port and city, and as part of global maritime networks. We need to assess the role that contemporary ports, preserving and building on their shared heritage, can play in the sustainable development of port city regions. Such an approach requires new methodologies that allow us to examine port city development over time and through space from the perspective of shared maritime systems and change.

Mapping as a means for a deeper spatial understanding

Ports, cities and regions are linked by water. They all have to address the water-related challenges of climate change and sea-level rise, but shared approaches to water structures, practices and heritage (tangible and intangible) are only slowly beginning to emerge. Port cities are not just about buildings. They are the expression and testimony of an entire way of life that involves exchange, communication and conviviality, as well as a mixture of material and immaterial heritage traditions centered on the natural element of water. The preservation of maritime (or water) heritage, as addressed in new waterfront developments, lacks the study of larger networks on which that heritage depends. Planning of port and waterfront areas is often carried out by specialists from different disciplines (Hein, 2016), but current heritage practices repeatedly ignore traditional infrastructural and maritime links, indicating a systemic approach to water heritage is missing. Port cities and port areas have a special potential for future-oriented sustainable development, but to achieve this, their heritage must be understood and protected.

We propose a socio-spatial methodology as a new approach to water and port city heritage, with the intention of collecting existing knowledge and developing new knowledge, values, attitudes and preferences. The objective of the methodology is to gain an advanced understanding of historic water- and urban development-related challenges, to create a new awareness of maritime practices, and to help ports and cities to rethink their maritime heritage. This approach can demonstrate the potential for heritage institutions, including UNESCO, to consider port cities and water structures in the context of the networks that made them grow, and to acknowledge the interconnected challenges that heritage sites will face due to climate change, such as sea-level rise. The port city can thus be seen as a networked entity, a port cityscape (Hein, 2019), in which several functions and their spaces—including ships and pipelines, port facilities and warehouses, industrial and logistics structures, main administrative and retail buildings, but also residential and leisure facilities—are brought together to form a port city culture that can facilitate sustainable development for all stakeholders.

To understand how port city regions have developed in response to diverse challenges, and what this means for their sustainable development, it is helpful to use a methodology based on the analysis of long-term historical data. A close analysis of the historic transformation of the built environment (e.g. land use, land ownership, infrastructures), the development of institutional structures (municipal boundaries) and the narrative that accompanies them (as embedded in maps and plans) through historical geospatial mapping can facilitate the identification of "gaps," where spatial, institutional or cultural opportunities and challenges exist and where planning can be useful. We propose the use of maps to help visualize the impact of overlapping political, economic, social and other interests and changes. Historical maps and contemporary mapping offer a language that transmits spatial data and historical changes (Miller, 2003) both to protect heritage and learn from it.

Historical maps are heritage in themselves, but they also convey the historic patterns of settlement and land use that influence the character of the contemporary environment, that shape current institutions and that determine the survival and development of the heritage. Maps are specifically important for understanding hidden designers, such as land use. Historical information on land use is essential for recognizing the effects of anthropogenic changes in land use. Such an understanding, we argue, can provide novel insights into the conditions and complexity of climate change and multiple related transitions (energy, digital, technological). For example, "The earliest approach to integrate heritage preservation into the urban land use planning system can be dated back to 1968 in New York City, which was the first city on the globe, [that] implemented transfer of development rights (TDR) to allow the preservation of heritage properties and opened the path for urban redevelopment at the same time" (Song, 2018, p. 276). In-depth knowledge of the long-standing technical and social conditions in port city regions can lead to better comprehension of resources and systems, better future design and the planning of measures to strengthen those systems. Such a methodology supports a new holistic approach to port city heritage and a foundation to meet the water challenges threatening the livability of cities.

The methodology: big data on heritage for the design of the future

The proposed methodology (developed by a group working on Digital Humanities at Delft University of Technology and notably in the Chair History of Architecture and Urban Planning in an interdisciplinary workshop) aims to problematize our understanding of the past to both protect and develop historic cities in urban deltas based on long-term comparative analysis. Additionally, its intent is to build greater awareness of the interconnection of water and cities through time, underscoring the role of socio-cultural development in shaping the future. To get a better understanding of how port cities developed in, on and along the water, how the water between sea, river and land has changed over time and how key stakeholders have dealt with water-related issues and heritage, we need a methodology.

The data wheel methodology (Figure 31.1) stands for the continuous process in geospatial mapping of collecting, preparing, analyzing, visualizing and sharing data. It also emphasizes the circular quality of the approach, which allows for the process to consistently add on new knowledge and integrate findings from one round of analysis to another. It is in line with the UNESCO HUL approach (and other approaches, e.g., hydrobiography, a term coined by Eric Luiten in 2014) using historically grounded investigation and geospatial mapping as a basis for informed planning and policy-making, education, outreach and training. It consists of five steps:

Step 1—definitions, collection, assessment

Finding—or building—an appropriate and reliable dataset is one of the biggest challenges in research. To establish a dataset that shows the changing relationship of sea and land in port cities, we need to evaluate historical maps and align them with each other. We also need to establish a glossary of appropriate terms and definitions for the identification of water- and port-related (heritage) sites, intangible practices and water-related challenges and crossing this information with other (historical) visual, written or data sources. Existing datasets and historical maps imply not only definitions, but also decisions; they reflect local particularities and historical choices that may already shape answers. Establishing this dataset requires the right humanities and social science-based knowledge and labor to process datasets in a way that will lead to meaningful and reliable results.

Step 2—preparation of the collected data

Preparation of the collected spatial data on water–land spaces in urban deltas is time-consuming and involves scholarly expertise. The data need to be adjusted for inclusion in a GIS-based mapping through georectification of historic maps, geolocalization of non-spatial information, optimization of the database and semantic enrichment of the data. This results in a new organization of historic data related to water and port heritage. Following the datawheel circular methodology, the database is open to further enrichment and for investigation in diverse localities. Collaboration with computer scientists

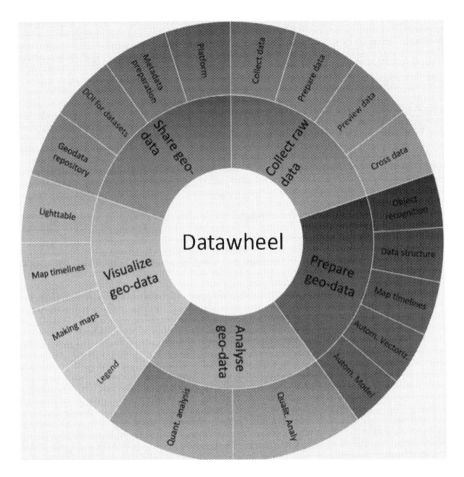

Figure 31.1 The datawheel methodology.
Source: By Carola Hein and a group working on Digital Humanities at Delft University of Technology.

using crowdsourcing and artificial intelligence allows scholars to more comprehensively understand historic cities in light of changing water patterns and port form and function.

Step 3—analysis of the collected and organized data

The geospatial datasets allow for a (big) data analysis using a combination of qualitative and quantitative approaches to understand societal long-term effects of decisions and to connect spatial, social and cultural aspects of water and port impact on natural and built heritage. Social scientists, historians and humanities scholars can use the datasets to draw a more complete picture of short- and long-term effects of water and port developments in urban deltas and port cities. Analysis of sites in relation to water-related challenges using a comparative perspective can allow for an assessment of contemporary proposals in the light of long-term developments and heritage preservation.

Step 4—data visualization

Once collected, organized and analyzed, the findings need to be visualized through geospatial mapping and infographics to show the complex correlation between spatial structures, spatial changes through time and social phenomena. The visualizations serve as a gap-finder (Hein & Van Mil, 2020), helping to connect sites and urban areas to new climate change-based challenges and to existing patterns of resilience. In-depth analysis can reveal the long-term effects of decisions on water systems and allow research findings to be made accessible to large and diverse audiences.

Step 5—sharing, dissemination, pilot studies

The collected data, its analysis and visualization can be presented open source, which allows for the co-creation of interpretations. The datasets and their visualization can then be used to educate academics, professionals and general citizens. An online platform allows stakeholders to explore the visualizations to develop new spatial perspectives for living with water. Summer schools or short-time workshops can bring together academics and practitioners and help connect local partners with global ones. Such a comprehensive investigation of urban deltas and port cities can promote the integration of longitudinal knowledge into design and become a foundation for future turns of the wheel.

Pilot case studies: Rotterdam, Hamburg and London

Using ongoing research into North Sea port city regions by the Chair History of Architecture and Urban Planning at Delft University of Technology, we have started to apply the methodology. We have selected three port city regions that developed in relation to one another around the North Sea, and we use them as pilot studies. The choice of a shared body of water, the North Sea, as the foundation for comparative research allows us to test the methodology for long-term analysis of water and land interaction in urban deltas and port cities as a foundation for developing sustainable heritage approaches. This pilot study concentrates primarily on the relation between sea and land and between port and city at a regional scale (Figure 31.2). It serves as foundation for an analysis of select themes, notably water–land relations, the historical development of infrastructure such as dikes, waterways and land roads, and specific forms of land cover (besides port and built-up area, it identifies mixed-use port areas and reclaimed port areas).

City spaces with no clearly defined forms of land use often turn out to be places with heritage values. Before the industrial revolution, many urban areas had multiple functions and there was no clear distinction between public and private and port areas were not always dedicated with fixed infrastructure. The quay on the water had several functions—such as mooring ships, storing goods and hosting public gatherings. In the late 20th century, the growth of the port in remote areas and the abandonment of traditional inland ports led to the transformation of former port areas into urbanized waterfronts, the reclaimed port area. Historic warehouses and old water and harbor structures were used to maintain and/or strengthen a city's identity and became key elements in redevelopment projects, such as Speicherstadt in Hamburg, the Maritime District and Wilhelminapier in Rotterdam or Canary Wharf in the Docklands of London.

This pilot study in mapping shows the cities of Rotterdam, Hamburg and London at an increasing scale and at selected moments in time, which we identified in relation to major political, technological and social changes. All three cities experienced historical shocks and often the same maritime-related ones. The comparative mapping method shows that the three cities responded to these changes in distinctive ways developing different patterns of resilience in regard to the relation between city and port. The maps allow insight into the potential opportunities of spatial mapping as a tool to recognize specific shared maritime challenges and their impact over time, such as the impact of ship size or of the maritime and port technologies on historical port cities. The maps show the importance of comparatively investigating how each city responded to these challenges in unique and local ways in order to predict the impact of these spaces and institutions on the future. This approach can be refined by datasets that include information on socio-cultural and economic developments and thematic spatial data, such as drinking water supply, drainage, land reclamation and flooding on a citywide scale. The approach can also be refined at a smaller scale to investigate water and heritage issues at a building level.

Preparation of the collected data, analysis of the collected and organized data and data visualization for the three pilot case studies

For the comparative study of the three port city regions, we used historical geospatial mapping (HisGIS), overlaying different data layers, providing a similar level of abstraction, and a uniform legend. To establish the analytical geospatial historical maps, we started with contemporary GIS datasets for the year 2020, such as global and continental GIS datasets covering several nation-states with sufficient spatial resolution to analyze and compare them in a consistent and systematic way. We used EuroGlobalMap (Eurogeographics 2017) for the infrastructure, the Global Administrative Boundaries (2018) for political boundaries and CORINE Land Cover (Copernicus 2016) for the built-up areas. National and regional data can be more detailed and accurate, but has its own definitions and criteria, making it difficult to combine and compare. Critical interpretation of the data is necessary, as is editing the GIS data and providing it with new coding to obtain comparable and uniform maps. Since global and European datasets are often generalized based on satellite imagery, the data are manually made more accurate using OpenStreetMap (2020), to better match the scale of the maps and reference year.

From the obtained dataset for the year 2020, we generated spatial data for the earlier periods, based on national sources and historical maps of different scale and quality. For the year 1900, national topographic maps, such as the maps of the Ordnance Survey, the national mapping agency of Great Britain, are reliable sources that allowed us to identify changes in infrastructure and land use. For earlier periods, historical city maps were among the few available resources. The use of historical maps as a source for the reconstruction of history requires careful handling of those maps. In the 1970s, J.B. Harley (1968) first touched on what he considered to be unreflective use of maps and mapping practices by cartographers and historical cartographers. Other authors continued this argument (Lilley, 2004 in: Klaarenbeek, 2020). Maps are a visual medium for the representation of a physical reality that can only be translated into a two-dimensional form with the help of concepts and interpretation. In the analysis and manufacture of the map, interpretation and coding are required in each of the following steps: (1) the observation by the map maker of the space, (2) the coding leading to the representation of the space on a two-dimensional form and (3) the decoding that takes place by the map reader. Then, working with a HisGIS adds two more steps, namely (4) a new coding of the maps and (5) a new interpretation by the reader (Renes, 2003; Klaarenbeek, 2020).

Water and port histories in Rotterdam, Hamburg and London

The continued importance of Rotterdam, Hamburg and London as port cities but with changing spatial and institutional constellations raises the question of port-city-related resilience and path dependencies (Hein & Laar, 2020; Hein & Schubert, 2020). Public and private actors have planned and administered the relationship of water and port in different ways in these three cities. The maps in Figure 31.2 show that the development paths of water, port and city spaces and the actors who shape them are not always aligned. In the case of Rotterdam, the port has been the heart of city development, growing from the historic center to the outskirts, with urban institutions trying to

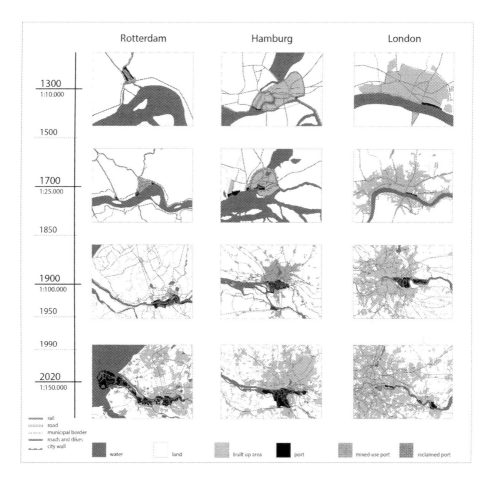

Figure 31.2 First draft for comparative geospatial mapping methodology, with the case study of London, Hamburg and Rotterdam.
Source: By Carola Hein, Yvonne van Mil, Blanka Borbely and Batuhan Ozaltun.

catch up with the expansion of the port. The case of the city-state Hamburg illustrates the development of long-term public leadership that has provided direction for the expanding port as well as for the growing city. The case of London is led by private investment, building and relocating a world-class port and administering it from the old city center, while local and national institutions only intervene to balance spatial or social short-comings of the private actors. This historical analysis and visualization points to the resilience of the port function, but each of these cities addresses challenges differently, creating specific development paths. Understanding these particular histories allows for a better understanding of sustainable future development.

The series of maps provides some insights into questions of spatial and institutional development of historical port cities in urban deltas, setting the stage for the identification of larger networks in which heritage structures exist. In 1300, the maps show fledgling cities controlled by dykes and dams and located near rivers for shipping, and as a place of crossing. The quays are part of the water structure in the city, where ships can load and unload safely. In 1700, access to shippable water for traditional industries and for trade became a key element for urban development, producing another set of related heritage structures. In Rotterdam and Hamburg, the ports expanded considerably through reclamation and the formation of new port islands in the Rivers Maas and Elbe. At the same time, they kept their mixed-use port area in the city center. By 1700, London was the most important port in Britain and many buildings and institutions documented that past. The institutions in charge of port and city are very diverse in the three case studies: Notably in Hamburg, port and city were largely under a single authority.

With industrialization in the late 19th century and the development of new forms of transport, private actors, port companies and some city governments created dedicated port areas separate from the urban spaces in all three cities. The port's facilities could not accommodate the rapid growth of ships. Specific patterns varied, but in every case, port spaces expanded dramatically and started to occupy land in the estuaries or rivers. In Rotterdam, the port expanded on both sides of the river and grew together with the smaller port areas of Delfshaven and Schiedam and developed as a transit port that served the needs of the German hinterland. During this time, the city had to adapt its port system in response to the New Waterway (Nieuwe Waterweg), constructed in 1872, which created a new connection to the sea. Hamburg also expanded the port on both sides of the river and the ports of Altona and Harburg that grew next to Hamburg would be integrated into the city state in 1937. Subsidized by the German Reich, the port was transformed into a transport zone, based on functional division and mono-functional districts. By 1900, the port of London was no longer an integral part of the city. London was still the most important port in the world, but it had already lost importance to Rotterdam and Hamburg, leaving behind heritage structures such as the Port of London Authority headquarters.

In the mid-20th century, containerization led to the abandonment of 19th-century innovations. Some of these sites have already become heritage sites—the Speicherstadt warehouse district in Hamburg being the most famous example. The maps of 2020 show land reclamation for port purposes. Old port areas became heritage sites. Separate master plans for the port and the city catered to their respective needs and interests. Maasvlakte 2 in Rotterdam provided for the needs of the growing port, the HafenCity development in Hamburg helped integrate former port areas into a multifunctional urban district. New interventions, such as the Thames Barrier in London, were erected as protection against the water. The water in the estuaries or rivers is channeled and controlled for shipping purposes, chemical industries, refineries and energy storage, while the city centers and old port areas are transformed to meet contemporary needs for living, working and leisure, with attempts made at the same time to safeguard their historical values. The comparative maps provide a foundation for the identification and analysis of interconnected maritime heritage and serves as a foundation for shared sustainable development.

Conclusions: toward an advanced research agenda

Planning for urbanized deltas with (potential) heritage sites is possible, but requires a clear understanding and necessity to be explored through developed methods. An understanding of long-term development can help build sustainable (port) cities and deltas. The historical perspective can facilitate future design beyond the fields of science, design and technology (Meyer, 2016). Historical geospatial mapping allows and enables detailed investigations from multidisciplinary perspectives. This first pilot study shows that a better understanding of historical processes in port cities and urban deltas can help identify particular processes and patterns of resilience relevant for the sustainable development of urban deltas and port cities. Our study explored how such a methodology (one of many) can provide a foundation to help reveal conflicts and consensus on water and heritage issues in urbanized deltas through a longitudinal perspective in order to address the challenges of water for livable historical cities. Our interest is to bring forth insight into interrelated issues of water, ports and historic cities, and the ways in which maritime practices have shaped practices and spaces in a networked way. Such a first step can lead to an advanced research agenda regarding heritage and sustainable development.

References

AADIPA-COAC (2020). PORT HERITAGE: Neighbourhoods with port/ports with town. Retrieved from: http://www.arqpla.com/curspatrimoni/index.php?option=com_content&view=article&id=38&Itemid=155&lang=en (Accessed on 10 December 2020).

Bosa, M. S. (Ed.) (2014). *Atlantic ports and the first globalisation, c. 1850–1930.* Cambridge Imperial and Post-Colonial Studies Series. Hampshire: Palgrave Macmillan. https://doi.org/10.1057/9781137327987.

Broeze, F. (Ed.) (1989). *Brides of the sea: Port cities of Asia from the 16th–20th centuries.* Honolulu; Kensington: University of Hawai'i Press; New South Wales University Press.

Broeze, F. (Ed.) (1997). *Gateways of Asia: Port cities of Asia in the 13th–20th centuries.* London: Kegan Paul International.

Copernicus (2016). CORINE land cover. Version 18.5. [dataset] Retrieved from: https://land.copernicus.eu/pan-european/corine-land-cover/lcc-2006-2012/view (Accessed on 10 December 2020).

Crul, M. R. J., Scholten, P., & Laar, P. van de (Eds.) (2019). *Coming to terms with superdiversity: The case of Rotterdam.* Cham, Switzerland: SpringerOpen.

de León, J. C. V., Bogardi, J., Dannenmann, S., & Basher, R. (2006). Early warning systems in the context of disaster risk management. *Entwicklung & Ländlicher Raum*, 2, 23–25. Retrieved from: https://www.eird.org/cd/indm/documentos/46fad12d0a62e5.38742613.pdf?id=281 (Accessed on 15 December 2020).

Eurogeographics (2017). EuroGlobalMap. Version 10.0. [dataset] Retrieved from: https://eurogeographics.org/maps-for-europe/open-data. (Accessed on 10 December 2020).

Global Administrative Boundaries (2018). GADM data. Version 36–2018. [dataset] Retrieved from: https://gadm.org/download_world.html. (Accessed on 21 December 2020).

Harley, J. B. (1968). The evaluation of early maps: Towards a methodology. *Imago Mundi*, 22: 62–74.

Hein, C. (Ed.) (2011). *Port cities: Dynamic landscapes and global networks.* New York: Routledge.

Hein, C. (2012). Modern cities: Interactions: Port cities. In P. Clark (Ed.). *Oxford handbook on cities in history* (pp. 809–827). Oxford and New York: Oxford University Press.

Hein, C. (2016). Port cities and urban waterfronts: How localized planning ignores water as a connector. *WIREs Water*, 3(May/June 2016), 419–38.

Hein, C. (2019). The port cityscape: Spatial and institutional approaches to port city relationships. *PORTUSplus*, 8. Retrieved from: https://portusplus.org/index.php/pp/article/view/1. (Accessed on 14 December 2020).

Hein, C., & Van Mil, Y. (2019). Towards a comparative spatial analysis for port city regions based on historical geo-spatial mapping. *PORTUSplus*, 8. Retrieved from: https://portusplus.org/index.php/pp/article/view/189 (Accessed on 10 December 2020).

Hein, C., &Van Mil, Y. (2020). Mapping as gap-finder: Geddes, tyrwhitt, and the comparative spatial analysis of port city regions. *Urban Planning*, 5(2), 152–166. https://doi.org/10.17645/up.v5i2.2803.

Hein, C., & Schubert, D. (2020). Resilience and path dependence: A comparative study of the port cities of London, Hamburg, and Philadelphia. *Journal of Urban History* 47(2), 389–419. https://doi.org/10.1177/0096144220925098.

Hein, C., & Mager, T. (2021). Port cities between global networks and local transformations Retrieved from: https://gahtc.org/pages/port-cities-between-global-networks-and-local-transformations (Accessed on 12 December 2020).

Hein, C., & Van de Laar, P. (2020). The separation of ports from cities: The case of Rotterdam. In A. Carpenter & R. Lozano (Eds.), *European port cities in transition*. Cham, Switzerland: Springer.

Hein, C, Luning, S & Van de Laar P. (2021a). Port CityCultures, Values, or Maritime Mindsets, Part 1: How to define and assess what makes port cities special, https://cpcl.unibo.it/issue/view/990.

Hein, C, Luning, S & Van de Laar P. (2021b). Port CityCultures, Values, or Maritime Mindsets, Part 2: Studying and Shaping Cultures in Port City Territories, https://cpcl.unibo.it/issue/view/1032.

Historic England (2020). Coastal, marine and maritime heritage. Retrieved from: https://historicengland.org.uk/research/current/discover-and-understand/coastal-and-marine/ (Accessed on 12 December 2020).

Klaarenbeek, R. (2020), *De herverkavelde stad. Kartografie van het naleven van stadskloosters in de Belgische steden Brussel, Antwerpen en Brugge (1773/1796–1860)* (Doctoral dissertation). Leuven: KU Leuven.

Konvitz, J. W. (1978). *Cities and the sea: Port city planning in Early Modern Europe.* Baltimore, MD: John Hopkins University Press.

Lee, R., & Lee, W. R. (1998). The socio-economic and demographic characteristics of port cities: A typology for comparative analysis? *Urban History*, 25(2), 147–172. Retrieved from: http://www.jstor.org/stable/44614055 (Accessed on 8 December 2020).

Lilley, K. (2004). Cities of God? Medieval urban forms and their Christian symbolism. *Transactions of the Institute of British Geographers*, 29(3), 296–313.

Martino, A., Fatiguso, F., & De Tommasi, G. (2015). Morphological and typological assessment of architectural heritage in historical Mediterranean ports. *STREMAH 2015–14th international conference on studies, repairs and maintenance of heritage architecture at a Coruña, Spain. Volume: structural studies, repairs and maintenance of heritage architecture XIV*, 15–28.

Meyer, H. (2016). Making urbanizing deltas more resilient by design. In C. Hein (Ed.), *International planning history society proceedings, 17th IPHS conference, history-urbanism-resilience*, 17–21 July 2016, V.03, TU Delft, TU Delft Open.

Miller, M. B. (2012). *Europe and the maritime world: A twentieth century history*. Cambridge: Cambridge University Press.

Miller, N. (2003). *Mapping the city: The language and culture of cartography in the renaissance*. London: Continuum.

OpenStreetMap contributors. (2020). OpenStreetMap. Retrieved from: https://planet.osm.org. https://www.openstreetmap.org (Accessed on 20 February 2020).

Interreg Italy - Croatia, Remember. (2020). Retrieved from: https://www.italy-croatia.eu/web/remember/about-the-project (Accessed on 12 January 2021).

Renes, H. (2003). Oude kaarten als bron voor onderzoek naar de geschiedenis van het landschap. *Caert-Thresoor, 23/3*, 87–94.

Pagés Sánchez, J. M., & Daamen, T. A. (2020). Using heritage to develop sustainable port–city relationships: Lisbon's shift from object-based to landscape approaches. In C. Hein (Ed.), *Adaptive strategies for water heritage* (pp. 383–399). Cham, Switzerland: Springer. https://doi.org/10.1007/978-3-030-00268-8_20.

Song, J. (2018). The rise of historic preservation in the urban land use planning system in Tokyo: Towards heritage-led regeneration. *National University of Singapore. The 18th International Planning History Society Conference*, Panel 16, p. 276 of 1343. Retrieved from: https://journals.open.tudelft.nl/iphs/issue/view/Yokohama/Download%20PDF (Accessed on 12 December 2020).

Urgent Safeguarding List (2020). Retrieved from: https://ich.unesco.org/en/dive&display=-threat (Accessed on 2 December 2020).

World Heritage Marine Programme (2020). Retrieved from: https://whc.unesco.org/en/marine-programme/ (Accessed on 10 December 2020).

32 Heritage data science

Scott Allan Orr

Introduction

'The data revolution' (Kitchen, 2014) is rapidly changing nearly every aspect of society. The use of data and data-driven approaches underpins infrastructure, healthcare, education, and economic decision-making. These approaches, regardless of specific application, hold the promise of identifying and processing patterns and trends beyond the capabilities of traditional approaches humans use to processing information to inform decision-making.

The heritage sector is not exempt from these developments. Decision-making increasingly relies on scientific activity to provide evidence, which is typically underpinned by robust experimental procedures and data. A notable area of earlier development is digital humanities (Schreibman et al., 2008), a field at the intersection of digital technologies and the humanities tradition (Terras, 2011). 'The data revolution' is further exacerbated within the heritage sector by the vast heritage assets that are 'born digital', with no physical analogue or manifestation (Palfrey & Gasser, 2011), which have accelerated the need to address the sustainability and management challenges posed by these resources.

While the generation and use of data is widespread and well-established, the concept of 'data science' provides a novel framework in which to understand its creation and use. In casual use, the terms 'statistics' and 'data science' may be used interchangeably. The astute observer might identify a distinction of scale or scope, specifying that something becomes data science when the data are 'big'. Or, they might believe that machine learning must be involved to be considered data science. These distinctions are artificial: statistical science and data science can be based on nearly any scale of data and may or may not include machine learning as part of their toolkit. These misperceptions of data science are rooted in the origins of data science within mathematical communities and primarily emphasising data analysis (Tukey, 1962). Both observers in the previous example have omitted two crucial things: context and critical reasoning.

Theory

Data science is a transdisciplinary field that incorporates several relevant bodies of knowledge and disciplinary approaches, including but not limited to statistics, informatics and communications technology (ICT), management, and sociology (Cao, 2017). These approaches are used in combination on the basis of three interrelated components: data, domain (context), and thinking (innovation). The heritage domain, with its unique and

DOI: 10.4324/9781003038955-39

pressing challenges and great diversity of value frameworks and disciplinary epistemologies, gives rise to the emerging field of *heritage data science* (Albuerne et al., 2018).

The data pipeline

The data pipeline is a set of activities that enable data-driven decision-making:

- Conceptualisation, including design, planning, and stakeholder engagement
- Acquisition, including methodological design, data collection, and documentation/recording
- Processing, including data cleaning, manipulation, synthesis, and conversion
- Analysis, including statistical summaries and algorithmic processes (e.g. machine learning)
- Visualisation and interpretation, typically including graphical representation of analysis, written descriptions, and discussion
- Curation and long-term management, including data documentation and storage

Although the components of the pipeline are conceptualised as a sequential procedure, they are strongly interrelated. For example, the nature of acquisition (how, when, where, and by whom) has significant implications for interpreting the analysis. If the analysis is ignorant of the methods, context, or design of data acquisition, there is a risk that a limitation or bias may not be identified.

The components of the data pipeline are most effectively implemented when iterated. The action of processing, analysing, visualising, and interpreting data rarely addresses all relevant aspects of the scenario. At each stage of the data pipeline or a data-driven project, it should be reassessed whether further work 'up the pipeline' is required, forming several interwoven feedback loops.

Heritage as a domain

Understanding the domain of a data science project is essential to formulating and undertaking data-driven decision-making. The domain provides context to any data science approach: it determines the subject of investigation and identifies the relevant challenges and questions to be addressed with data science. Within heritage data science, the data pipeline sits within the heritage domain: it is informed by heritage challenge(s) and aims to produce output that is relevant to the priorities of heritage stakeholders.

The 'subject' of heritage data science may be an individual heritage typology, or a set of heritage typologies. Heritage data science is not limited to the either the immaterial or material, as it transcends the false dichotomies of cultural/natural and tangible/intangible heritage (Fredheim & Khalaf, 2016). Thus, heritage data science is undertaken on material and/or immaterial culture to which a society ascribes value (Vecco, 2010). Data science methods are particularly adept at handling large sets of information, especially those that contain conflicting information and uncertainty and/or are based on subjectivity, perception, or belief. Thus, heritage data science is particularly useful for addressing complex challenges within the heritage domain.

The diverse interests of stakeholders in the heritage domain necessitate a wide range of activities within heritage data science. Building on an established framework in heritage science (NHSF, 2018), these activities can be broadly classified into

- Interpretation: furthering understanding of heritage
- Engagement: enabling and enhancing access to heritage
- Management: informing the stewardship of heritage, including but not limited to storage and maintenance

One of the grand challenges for heritage data science is to reconcile that the conceptualisation and implementation of a project are often rooted in several disciplinary epistemologies. As a transdisciplinary field, heritage data science incorporates methods and frameworks from several different traditions of study. This is exacerbated by the diversity of participants and applications of heritage data science. Some of these challenges are practical and can be addressed with the implementation of practical steps. For example, a widespread challenge to transdisciplinary working is a lack of shared terminology, or understanding of terminologies used by different disciplines (Tress at al., 2007). These challenges can be addressed by exercises, such as that proposed by the computer scientist and noted internet pioneer Jim Gray, in which the task to formulate *20 Questions* addressing a societal challenge assists in the process of normalising understanding of terminology and priorities (Gray, 2004). Other challenges, such as contrasts between disciplines in how claims are proved or substantiated, remain long-standing open challenges.

Thinking

Data science 'thinking' (Cao, 2018) embodies the distinction between data science different from existing developments in statistics and information science. Data science thinking emphasises the role of human intelligence by involving human intuition, belief, expectations, evaluation, and expertise into decision-making processes. In addition to the ability to collect, process, and analyse vast quantities of data, which can, in some cases, be routine tasks or trivial processes, heritage data scientists must also be imaginative and employ qualitative and critical reasoning.

Heritage data science thinking must critically assess the objective of study against the priorities of stakeholders. Heritage data science should actively address an open challenge within the engagement with, interpretation, or management of heritage. The nature of how this is undertaken depends on the context within the heritage domain and the relevant stakeholders, but should broadly seek to positively enhance the value of heritage and the benefits derived from it. These can be represented by one or several objectives, including (in no specific order; Table 32.1).

Table 32.1 Objectives assessed by heritage data science thinking against the priorities of stakeholders

Optimise	Economise	Reason
Automate	Imagine	Explain
Communicate	Calculate	Evaluate
Innovate	Critique	Prove
Demonstrate	Reduce	Predict

The appropriate objective(s) being determined from stakeholder priorities.

A framework

Heritage data science sits at the intersection of the heritage domain, critical and imaginative thinking, and data (Figure 32.1).

Figure 32.1 A framework for heritage data science, combining data, thinking, and the heritage domain.

Crucially, the data pipeline is integrated within this framework. This is fundamentally different than the model in which a challenge is generated within the heritage domain which inputs into the project, at which point the data pipeline is implemented by those with the relevant technical expertise. Although this may produce implications for the interpretation, engagement, and management of heritage, it is limited by its linearity. As a transdisciplinary field, the heritage domain (and relevant stakeholders) should be involved throughout the duration of a heritage data science project as a realisation of participatory science.

Heritage data

Heritage data is an inclusive term that includes data *as* heritage and data *about* heritage (Albuerne et al., 2018). The unified term is useful to identify common challenges and opportunities for heritage data: it is rooted in established concepts within sustainable heritage of value, significance, integrity, ethics, and authenticity; it imposes requirements for longevity; and it is the subject and/or output of transdisciplinary work.

Data as heritage

The UNESCO *Charter on the preservation of digital heritage* (2003) recognises digital assets as heritage that have lasting value and significance and should therefore be safeguarded under the same premises as other forms of heritage. The charter identifies a diverse range of types of digital materials that can be heritage, including texts, databases, still and moving images, audio, graphics, software, and web pages, among a wide and growing range of formats.

Data *about* heritage

There is an ever-increasing amount of data related to the provenance, conservation, management, and interpretation of heritage. These data can be qualitative and quantitative, scientific, or humanistic, come from diverse sources, and serve several different purposes. The unifying concept for these data is that they exist to produce, promote, and sustain value for heritage; thus, data about heritage are embedded in concepts of authenticity, ethics, integrity, and values, just as heritage to which they are relevant are embedded.

Sustainability of heritage data

Heritage data require the same diligence as other forms of heritage with regard to sustainable management. This is especially true as heritage data are often ephemeral, requiring purposeful maintenance and management to be retained. Accessing heritage data in proprietary formats, specifically the challenges associated with the maintenance of software required to access them, remains a significant challenge in heritage data science. This challenge is exacerbated when the proprietary format has in itself embedded heritage value, such as legacy video game platforms (see Eklund et al., 2019).

A heritage data science paradigm

Data science is argued to be an *exploratory* mode of science (Hey, 2009): a 'fourth paradigm' of undertaking scientific enquiry is proposed that is fundamentally different from its three predecessors that are, respectively, rooted in empirical evidence (observation), the scientific method (hypothesis testing), and computational science (scaling up analysis based on computation). In the fourth paradigm, data are big, abundant, and rich. In contrast to previous approaches to science, data are not collected based on a hypothesis; in the fourth paradigm, data are collected first, from which patterns, insights, and information are extracted. Collect first, ask questions second.

The 2006 House of Lords Science and Technology inquiry into science and heritage (House of Lords Science and Technology Committee, 2006) was grounded in the concept of heritage science as an applied field. The inquiry also acknowledged that 'basic and applied research…are inextricably intertwined' (ibid., p. 24). This lack of clarity on the nature of scientific activity within the remit of heritage science is a gap in the philosophical and theoretical canon of the field with implications for heritage data science: How can an applied field be exploratory?

Both basic research and applied research contribute to the aims of heritage science. Basic or 'pure' research is 'dedicated to managing and increasing knowledge of general validity' (Roll-Hansen, 2009, p. 3). Roll-Hansen distinguishes applied science as the area of intersection between science and politics: 'It depends highly on advanced scientific knowledge and methods but is dedicated to the solution of practical economic, social and political problems rather than the further development of such knowledge and methods' (ibid). The distinction between basic and applied research has been summarised as a dichotomy between fact and value (Proctor, 1991). Value is at the core of understanding materials and change within heritage science (Douglas-Jones et al., 2016). This would seem to suggest that heritage science should be classified as an applied science. However, a more enlightened perspective provided by Putnam appreciates the *distinction*, or 'entanglement' between fact and value, rather than a strict dichotomy (2002). Through the lens of current approaches within heritage science, this is demonstrated by damage functions, in which the value component is often decoupled from the dose–response function (Strlič et al., 2013). This demonstrates the integrated role of basic and applied approaches within heritage science, supporting their inclusion within its remit.

Placing data-driven approaches at the core of heritage science strengthens the argument for heritage data science to be, at least in part, a fundamental science. However, this discrepancy demonstrates the limited capacity in classifying heritage research as either basis or applied.

Models for classifying and organising scientific activity

Several models attempt to capture research activity more holistically than a dichotomy between basic and applied research which may be applicable for heritage data science.

The New Production of Knowledge proposes two ways of undertaking research which each correspond to a different kind of knowledge: Mode 1 represents the traditional academic and discipline-oriented research and knowledge; Mode 2 'operates within the context of application' and is 'transdisciplinary rather than mono- or multidisciplinary' (Gibbons et al., 1994, p. vii). This latter aspect resonates with the ideology and practice of integrating several disciplines to address heritage science challenges. Similarly, Mode 2 is organisationally transient as it does not have a stable hierarchy and is more reflexive and accountable to society. Mode 2, 'the new production of scientific knowledge', thus erases the distinction between basic and applied research.

The quadrant model of scientific research was proposed by Stokes (2011). Basic research and applied research are presented in the context of considerations of use and the quest for fundamental understanding. Using Pasteur's work in microbiology as an example, Stokes describes it as 'use-inspired' (or 'purpose-driven') basic research, demonstrating that basic research can be designed and undertaken in consideration of use. It is therefore 'applied', as it is informed by social and economic drivers, and 'basic', since it contributes more broadly to scientific understanding. This is contrasted by the work of Bohr and Edison, whose primarily contributions were in understanding atomic structure and quantum theory motivated by the need to develop devices for power generation and communication. 'Pasteur's quadrant' demonstrates that interactions between theoretical and practical problems can be highly productive (Roll-Hansen, 2009). Consideration of use is an important element of heritage science research; within the quadrant model, use-inspired basic research and applied research can thus be considered valid components of heritage science. Additionally, as identified by the House of Lords inquiry, the complexity of 'basic science underpinning conservation' (House of Lords

Science and Technology Committee, 2006, p. 80) is equally important and thus should be included within the remit of heritage data science.

Dudley (2013) proposed a three pillars model, in which basic research, use-inspired research, and development and industry share boundaries. The model includes a funding axis, since the question of the amount of funding allocated to each area is unavoidable. This model was developed based on concerns

> that the quadrant model minimizes the interface between fundamental research and industrial development, giving the misleading impression that research performed in Pasteur's quadrant has the greatest impact on industry. This erroneous impression has given rise to the paradigm of use-inspired research that dominates current thinking.
>
> (ibid., p. 339)

This model is similar to classifications within research and development (R&D) used by the OECD (Frascati, 2015): basic research, applied research, and experimental development. Experimental development is defined as 'systematic work, drawing on existing knowledge gained from research and practical experience and producing additional knowledge, which is directed to producing new products or processes or to existing products or processes' (ibid, p. 45).

A model for heritage data science

Models that separate basic and applied research as distinct entities are not appropriate within a heritage science context, since a dichotomous model cannot capture the complex and diverse nature of scientific activities relevant to heritage data science. Although the OECD and Dudley models also acknowledge the role of industry and development within the research landscape, they both depend on distinguishing between the arbitrary distinction between applied and basic research.

Heritage data science should adopt the quadrant model within the context of Mode 2 as set out in *The New Production of Scientific Knowledge*. This acknowledges that heritage data science can be basic (fundamental or purpose-motivated) and applied, operates within the context of application, and involves several disciplines. The inclusion of use-inspired basic research emphasises that heritage science can produce new knowledge that is relevant for the wider scientific community. This contextualised model could be further developed by introducing a third dimension: impact potential. Rather than a binary classification of 'yes' or 'no', a continuous relative scale might be more suited. While impact is notoriously difficult to measure (and quantify) in science at large (Ravenscroft et al., 2017), it is specifically challenging within the heritage domain (Dillon et al., 2014; Katrakazis et al., 2018). This model would acknowledge the diverse range of potential impacts and their respective timelines. For example:

- Basic research might reveal a previously unknown and imminent threat that would then dictate immediate research priorities in research areas with consideration of use;
- Use-inspired basic research could demonstrate the potential of a novel technology to be applied within a heritage context, although it might not be developed commercially (or become commercially viable) for several years;

- Applied research might produce an innovative management framework that addresses a heritage-specific need; this could be implemented within a heritage organisation in a relatively short time frame.

These examples (and subsequent extensions) demonstrate an important aspect of this model: different kinds of research activity can inform the others. The unexpected imminent threat feeds into subsequent use-inspired basic research and applied research. A novel technique developed might then require applied research to develop it and hone its operation and design to heritage applications. Adapting the quadrant model to heritage science, while including a more diverse range of scientific activities within its remit, encourages further interaction between the modes of research. Acknowledging the contribution of basic research within heritage data science enforces its transdisciplinary identity and reinforces its exploratory nature.

Opportunities and challenges

There are innumerable active areas of research and application within heritage data science: far too many to discuss in-depth or even list herein. Some key emerging areas include climate resilience, aggregated scientific analysis, critical heritage studies, as digital documentation, heritage in crisis, digital heritage, open and linked data, and citizen science and crowd-sourced approaches. Across these areas, opportunities are enabled by key developments in areas such as open and linked data, data standards, citizen and crowd-sourced science, critical communication, and Bayesian and fuzzy approaches. These areas are active within both research and practice and rapidly developing in capability and scope.

While these emerging themes within data science hold the promise of transforming the heritage sector, there remain several challenges to its successful implementation. Some of these challenges are universal to data science, while others are specific to the characteristics of the heritage sector.

Data quality

The colloquial phrase about data quality is 'garbage in–garbage out', acknowledging that poor-quality data lead to unreliable output. To be of good quality, data need to be fit for purpose within its intended use(s), such as decision-making and planning (Redman, 2008); high-quality is a true and accurate representation of the real-world entity; it should represent with limited bias. These present particular challenges for heritage data.

Heritage data, especially data *about* heritage, may have implicit bias or have caveats with implications for outputs. Heritage data science draws on a diverse range of data sources: many of which exist to produce data for their own purposes, for which the heritage application(s) are secondary. For example, heritage data science makes frequent use of climate data typically collected for regional and large-scale meteorological and climate monitoring (e.g. Orr et al., 2018; Brimblecombe et al., 2020). One of the challenges in using these datasets within a heritage context is determining whether they are suitably representative of the localised environment relevant to a heritage context. Similarly, these data are often collected at time intervals that do not represent the timescales of heritage phenomena. With a move towards open and FAIR data within and beyond the heritage sector (data that are findable, accessible, interoperable, and reusable), these types of challenges will become more prevalent. Thus, heritage data scientists must determine whether these data are sufficiently accurate and precise for the task, or whether it is necessary to produce a dataset with suitable spatial and temporal coverage.

The landscape of heritage data can be discriminatory. As chronicled in academic literature (Hoffmann, 2019) as well as popular reading (O'Neil, 2016), algorithmic decision-making implemented with the best intentions can be discriminatory. These often incorporate proxy data: data that are accessible or seemingly insensitive that indirectly represent an important factor within the context as a substitute for data that are difficult to collect or sensitive. These data may be correlated with other factors such as race, gender, and sexuality, or other sensitive characteristics, many of which have legislative protection in several regions.

Heritage data can be rooted in legacies that do not represent contemporary perspectives and discourse. A significant amount of heritage data has been curated by large cultural organisations or so-called memory institutions that aim to preserve, contextualise, and communicate canonical elements of culture, historical narrative, and collective memory. These institutions, and the social memory they have produced, are to varying extents rooted in West centrism, colonialism, and the world views of the social and moral values of the upper-class elite at the helm of their establishment and operation (West, 2010; Smith & Waterton, 2012). This manifests in the curation, management, and dissemination of heritage, which is increasingly embodied in the relevant heritage data. Thus, heritage data may represent particular narratives and themes, downplaying the importance of diverse and often conflicting social histories and perspectives. The implication for heritage data science is that any work undertaken with these data must contextualise output within this context, and strive to supplement both data and discussion accordingly. Citizen and crowd-sourced methods hold particular promise to supplement existing mainstream narratives and generate data: a successful example is *Pride of Place: England's LGBTQ Heritage* (Historic England, 2020). This initiative acknowledged that LGBTQ histories are embedded in England's built, cultural, and natural landscapes, but was severely underrepresented in heritage documentation. The initiative uses a map-based crowd-sourced approach to produce a geolocated dataset representing LGBTQ stories and places. In the absence of suitable supplementary data sources, heritage data science projects should transparently discuss potential bias and demonstrate an understanding of its culturally embedded nature.

Data integrity

Data integrity ensures consistency and accuracy of data over its entire lifetime, which is crucial to the successful implementation of data-driven decision-making. The challenges posed to data integrity within heritage data science are related to both physical assets, digital assets, and assets that have both physical and digital representations.

Physical heritage assets are an important source of heritage data. Beyond their metadata, they can provide new insight when made the subject of scientific investigation or perspective surveys. However, these data are fixed in time at their instance of collection, representing a snapshot. In reality, physical heritage assets are dynamic and ever-changing, as are perceptions of their physicality and value. Thus, data representing the state of a heritage collection may accurately represent its conditions, but this may be invalidated by subsequent changes of assets. Heritage data science needs to demonstrate awareness of the static nature of many heritage data and develop flexible approaches that respond to physical change and collection demographics, as well as emerging expectations of physical heritage assets and attitudes towards heritage.

Born digital assets equally undergo inevitable change, but their formats pose more significant challenges to long-term sustainability. Prone to natural bit-rot (a digital analogue of physical material decay; see Cerf, 2011), digital materials require regular checks and maintenance to be implemented as part of their preservation. The interfaces used to interrogate heritage data change; thus, data that were retrievable and useful at the time of acquisition may prove challenging to access as platforms and software become obsolete. Therefore, programmes of conversion and updating of data must be embedded into the long-term management of born digital assets to ensure they can be retrieved and accessed. For example, archived materials from early pioneers who incorporated CAD (computer aided design) and 3D software into architectural practice pose challenges. While the bespoke files hold a rich array of information pertaining to design process, curation and conversion may result in these becoming inaccessible. The remaining material after conversion may simply be a 2D rendering of the models, or presentation material relating to the models. However, how to reconcile conversion and maintenance with the potential to lose heritage value remains an open challenge within heritage data science.

By nature, sustainable heritage must consider long time horizons, often seeking data that can adequately represent the dynamic nature of heritage and its context over them. Within a data science paradigm, most *forward* considerations (e.g. forecasting and modelling) are born digital. However, there are a significant number of records reflecting the past that require digitisation in order to be incorporated into data science approaches. One example is the 'Data Rescue' initiative as part of the ACRE project run by the UK Met Office (http://www.met-acre.net). This initiative seeks to develop a comprehensive model of historical climate back to the early 19th century based on paper-based archives, such as the diaries of explorers and 'gentleman' scientists, reports, private archives, the records of port authorities, and ship logs books. A significant amount of resources are required to identify, digitise, and process these types of data, although natural language processing and citizen science can make important contributions.

More broadly, digitised assets may not be as rich as physical assets. Although they may include important information including essential metadata and text, contextual information may be lost. For example, watermarks have been shown to be an essential component of historical investigation of early printed documents (Calì, 2018). Depending on the digitisation procedure and storage format, important characteristics of the asset may not be included in a digitised record.

Transdisciplinarity and value-creating science

Heritage data science is by its nature culturally embedded. Resultantly, those involved (researchers, partner organisations, the public as a stakeholder) may hold vastly different views on how a heritage data science project should be undertaken. From the outset of a heritage data science project, care should be taken to identify the diverse array of stakeholders. Input should be solicited throughout the project's lifetime, from conceptualisation, acquisition, analysis, to dissemination and long-term curation.

By nature of studying heritage, heritage data science produces and enriches heritage value. Therefore, it is paramount to demonstrate awareness of what might be considered a perpetuating cycle of data science. Generating data and developing data-driven insights from it facilitates opportunities for further heritage data science within the area. Thus, heritage data science, and the topics it addresses, risks perpetuating a positive-reinforced feedback cycle in which heritage typologies and challenges studied become further studied, at the risk of others being neglected. Heritage data science should seek to identify cross-cutting expertise and methods that hold promise to address several heritage typologies and incorporate several overlapping heritage value frameworks and stakeholder perspectives.

Digital literacy

As a transdisciplinary field, heritage data science requires equal command of data and data-driven methods, the heritage domain, and imaginative thinking. This contrasts a multidisciplinary of interdisciplinary mode of working, in which those with expertise within the heritage domain work alongside those with relevant data skills. This poses a significant challenge to the training of heritage data scientists and upskilling of those working in the heritage sector.

Specialised training is required to produce transdisciplinary heritage data scientists who are equipped to address sustainability challenges within the heritage sector. Until recently, very few academic courses incorporated elements of heritage science data analysis, visualisation, use and reuse, digitisation, and data science (Albuerne et al., 2018). However, university-level courses are increasingly combining a deep understanding of the heritage domain with state-of-the-art computer science. One recent initiative is the MSc *Data Science for Cultural Heritage* at UCL, https://www.ucl.ac.uk/bartlett/heritage/study/data-science-cultural-heritage-msc, that address this gap by producing researchers and data scientists who are equipped to understand the complexities of working in the heritage domain, as well as more broadly in other challenging domains. Similarly, programmes focusing on conservation and restoration and heritage are increasingly embedding digital and statistical skills into their training, such as a two-week intensive module within the MSc in Conservation and Restoration of Cultural Heritage at the University of Amsterdam and plans to shortly integrate data science into the curriculum in conservation–restoration at the University of Antwerp.

More broadly, the heritage sector is realising that the future of work is going to be increasingly data-driven. Initiatives such as the *Mapping the Museum Digital Skills Ecosystem* project (Barnes et al., 2018) and the Heritage Alliance's *Heritage Digital* programme identify gaps and targeted training to upskill the heritage workforce. There is an ongoing need for flexible and accessible training that recognises the unique challenges posed to data science within the heritage sector.

Conclusions

Heritage data science is an emerging transdisciplinary field that informs heritage engagement with heritage and its interpretation and long-term management using data-driven approaches. It is also highly innovative and produces new knowledge and furthers understanding of society and the environment. Heritage data science requires in-depth comprehension of the complexities of the heritage domain and an awareness of science as culturally embedded enquiry, expertise in all aspects of the data pipeline, and an appreciation for the role of human intelligence and critical reasoning in research and decision-making. There are several open challenges within heritage data science: data quality and integrity especially with a focus on equality, diversity, and inclusion; a lack of frameworks for enabling transdisciplinary data-driven science in a heritage context; and limited educational pathways for both educating data scientists to work within the heritage domain and upskilling heritage professionals in preparation for the imminent digital transformation within the sector.

References

Albuerne, A., Grau-Bove, J., & Strlic, M. (2018). The role of heritage data science in digital heritage. In *EuroMed 2018: Digital heritage. Progress in cultural heritage: Documentation, preservation, and protection* (pp. 616–622). Cham: Springer. https://doi.org/10.1007/978-3-030-01762-0_54.

Barnes, S. A., Kispeter, E., Eikhof, D. R., & Parry, R. (2018). *Mapping the museum digital skills ecosystem - phase one report*. Leicester: University of Leicester. https://doi.org/10.29311/2018.01.

Brimblecombe, P., Hayashi, M., & Futagami, Y. (2020). Mapping climate change, natural hazards and Tokyo's built heritage. *Atmosphere*, *11*(7), 680. https://doi.org/10.3390/atmos11070680.

Calì, C. (2018). The importance of a project to enhance the watermarks of the Codex Atlanticus by Leonardo Da Vinci. In *5th international multidisciplinary scientific conference on social sciences and arts SGEM 2018* (pp. 333–338). Retrieved from: https://re.public.polimi.it/handle/11311/1116346.

Cao, L. (2017). Data science: challenges and directions. *Communications of the ACM*, *60*(8), 59–68. https://doi.org/10.1145/3015456.

Cao, L. (2018). Data science thinking. In *Data science thinking* (pp. 59–90). Cham: Springer. https://doi.org/10.1007/978-3-319-95092-1_3.

Cerf, V. G. (2011). Avoiding "bit rot": Long-term preservation of digital information [point of view]. *Proceedings of the IEEE*, *99*(6), 915–916. https://doi.org/10.1109/JPROC.2011.2124190.

Dillon, C., Bell, N., Fouseki, K., Laurenson, P., Thompson, A., & Strlič, M. (2014). Mind the gap: rigour and relevance in collaborative heritage science research. *Heritage Science*, *2*(1), 11. https://doi.org/10.1186/2050-7445-2-11.

Douglas-Jones, R., Hughes, J. J., Jones, S., & Yarrow, T. (2016). Science, value and material decay in the conservation of historic environments. *Journal of Cultural Heritage*, *21*, 823–833. https://doi.org/10.1016/j.culher.2016.03.007.

Dudley, J. M. (2013). Defending basic research. *Nature Photonics*, *7*(5), 338–339. https://doi.org/10.1038/nphoton.2013.105.

Eklund, L., Sjöblom, B., & Prax, P. (2019). Lost in translation: Video games becoming cultural heritage? *Cultural Sociology*, *13*(4), 444–460. https://doi.org/10.1177/1749975519852501.

Frascati, M. (2015). *Guidelines for collecting and reporting data on research and experimental development*. 7th ed. Paris: OECD Publishing.

Fredheim, L. H., & Khalaf, M. (2016). The significance of values: heritage value typologies re-examined. *International Journal of Heritage Studies*, *22*(6), 466–481. https://doi.org/10.1080/13527258.2016.1171247.

Gibbons, M. (Ed.) (1994). *The new production of knowledge: The dynamics of science and research in contemporary societies*. Atlanta, GA: Sage.

Gray, J. (2004). Online Science: the 20 questions approach. *Scientific Data Symposium*, Redmond, WA, 25 May 2004. Retrieved from: http://jimgray.azurewebsites.net/talks/SciData.ppt.

Hey, T. (Ed.) (2009). *The fourth paradigm: Data-intensive scientific discovery*. Redmond, WA: Microsoft Research.

Historic England (2020). *Pride of place: England's LGBTQ heritage*. Retrieved from: https://historicengland.org.uk/research/inclusive-heritage/lgbtq-heritage-project/ (Accessed on 24 August 2020).

Hoffmann, A. L. (2019). Where fairness fails: data, algorithms, and the limits of antidiscrimination discourse. *Information, Communication & Society*, *22*(7), 900–915. https://doi.org/10.1080/1369118X.2019.1573912.

House of Lords Science and Technology Committee (2006). *9th Report of Session 2005–06: Science and Heritage, Report with Evidence.* HL Paper 256.

Katrakazis, T., Heritage, A., Dillon, C., Juvan, P., & Golfomitsou, S. (2018). Enhancing research impact in heritage conservation. *Studies in Conservation, 63*(8), 450–465. https://doi.org/10.1080/00393630.2018.1491719.

Kitchin, R. (2014). *The data revolution: Big data, open data, data infrastructures and their consequences.* Atlanta, GA: Sage.

NHSF (2018). Strategic Framework for Heritage Science in the UK, 2018–2023. National Heritage Science Forum. Retrieved from: http://www.heritagescienceforum.org.uk/what-we-do/strategic-framework.

O'Neil, C. (2016). *Weapons of math destruction: How big data increases inequality and threatens democracy.* New York: Crown Publishing Group.

Orr, S. A., Young, M., Stelfox, D., Curran, J., & Viles, H. (2018). Wind-driven rain and future risk to built heritage in the United Kingdom: Novel metrics for characterising rain spells. *Science of the Total Environment, 640,* 1098–1111. https://doi.org/10.1016/j.scitotenv.2018.05.354.

Palfrey, J. G., & Gasser, U. (2011). *Born digital: Understanding the first generation of digital natives.* New York: Basic Books.

Proctor, R. (1991). *Value-free science?: Purity and power in modern knowledge.* Cambridge, MA: Harvard University Press.

Putnam, H. (2002). *The collapse of the fact/value dichotomy and other essays.* Cambridge, MA: Harvard University Press.

Ravenscroft, J., Liakata, M., Clare, A., & Duma, D. (2017). Measuring scientific impact beyond academia: An assessment of existing impact metrics and proposed improvements. *PloS one, 12*(3), e0173152. https://doi.org/10.1371/journal.pone.0173152.

Redman, T. C. (2008). *Data driven: Profiting from your most important business asset.* Cambridge, MA: Harvard Business Press.

Roll-Hansen, N. (2009). *Why the distinction between basic (theoretical) and applied (practical) research is important in the politics of science.* London School of Economics and Political Science, Contingency and Dissent in Science Project.

Schreibman, S., Siemens, R., & Unsworth, J. (Eds.) (2008). *A companion to digital humanities.* Hoboken, NJ: John Wiley & Sons.

Smith, L., & Waterton, E. (2012). Constrained by commonsense: The authorized heritage discourse in contemporary debates. In R. Skeates, C. McDavid, & J. Carman (Eds.), *The Oxford handbook of public archaeology* (pp. 153–171). Oxford: Oxford University Press.

Stokes, D. E. (2011). *Pasteur's quadrant: Basic science and technological innovation.* Wasington, DC: Brookings Institution Press.

Strlič, M., Thickett, D., Taylor, J., & Cassar, M. (2013). Damage functions in heritage science. *Studies in Conservation, 58*(2), 80–87. https://doi.org/10.1179/2047058412Y.0000000073.

Terras, M. (2011). Quantifying digital humanities (PDF). *UCL Centre for Digital Humanities.* Retrieved 24 August, 2020. Retrieved from: http://www.ucl.ac.uk/infostudies/melissa-terras/DigitalHumanitiesInfographic.pdf.

Tress, G., Tress, B., & Fry, G. (2007). Analysis of the barriers to integration in landscape research projects. *Land Use Policy, 24*(2), 374–385. https://doi.org/10.1016/J.LANDUSEPOL.2006.05.001.

Tukey, J. W. (1962). The future of data analysis. *The Annals of Mathematical Statistics, 33*(1), 1–67. https://doi.org/10.1214/aoms/1177704711.

UNESCO (2003). Charter on the preservation of digital heritage. Retrieved from: http://portal.unesco.org/en/ev.php-URL_ID=17721&URL_DO=DO_TOPIC&URL_SECTION=201.html.

Vecco, M. (2010). A definition of cultural heritage: From the tangible to the intangible. *Journal of Cultural Heritage, 11*(3), 321–324. https://doi.org/10.1016/j.culher.2010.01.006.

West, S. (2010). Heritage and class. In R. Harrison (Ed.), *Understanding the politics of heritage. Understanding global heritage* (pp. 270–303). Manchester and Milton Keynes: Manchester University Press and The Open University.

33 Capturing heritage significance

A critical analysis of economics-based methods

Mina Dragouni

Introduction

Following the general inception of sustainability, the idea of 'sustainable heritage', as conceived within mainstream economic theory of neoclassical economics and the prevailing political paradigm of neoliberalism, can be used to describe a reasonable compromise between protection and economic exploitation of natural and cultural landscapes, sites, monuments, artefacts or practices that have been inherited from the past to the present. From an economics standpoint, the sustainability hypothesis within the heritage realm is particularly rich in scope, encompassing various economic and financial considerations, such as the operation of healthy organisations, the provision and consumption of heritage services and the cost-effective and prudent use of resources employed in heritage goods production and distribution. When seen through the lenses of political economy, further questions of sustainable heritage emerge, revolving around the economic vibrancy of the sector, as reflected, for instance, in its contribution to GDP or employment, or its financial dependence on state patronage.

As an economist would suggest, heritage and its contents, both material and intangible, form an assemblage of goods and services that have economic qualities. Similar to other economic goods, their 'production' in the present becomes possible through a chain of processes (e.g. research, interpretation, conservation, curation), which necessitate the mobilisation of capital and other resources, such as labour, technology, knowledge and expertise. In this light, heritage goods are economic goods, participating in ongoing economic activities, decisions and results, irrespective of their value extending beyond the sphere of the economy. At the same time, heritage can be defined itself as a resource, of which the value is extracted and metabolised into other goods and services to serve our needs as society. In doing so, the economic importance of heritage is not confined to the sector but rather integrated into productive processes materialised in various branches of the economic system (e.g. creative industries, recreation, tourism), where it generates numerous indirect and induced economic effects.

In recent years, empirical enquiry into the economic impacts and value of protecting and engaging with the past and its remains has proliferated, while economics-based techniques to 'capture' and monetise heritage benefits have found increasing support within national culture departments and multi-lateral policy institutions, such as the United Nations, ICOMOS and the World Bank (see indicatively, Rypkema & Cheong, 2011; Licciardi & Amirtahmasebi, 2012; UNESCO/UNDP, 2013; Crossick & Kaszynska, 2016). Despite growing interest in the topic, the evaluation of heritage economic and broader societal significance remains a central concern for heritage specialists, economic analysts and policymakers (Throsby, 2019). Motivated by the relevance and criticality of the topic, the aim of this chapter is to provide an overview of the most popular economics-based approaches and evaluation tools used currently in the literature for assessing the direct and indirect contribution of heritage to economy and society.

The chapter engages with both academic research and formal evaluations of heritage programmes, as both contribute to our understanding of the problem area, allowing us to grasp how heritage significance is conceptualised and construed in economics-based studies. Our focus here on said approaches does not mean to imply that alternative scientific paths and methodological designs used in other social sciences or humanities disciplines, such as ethnography, grounded theory and phenomenology, are of less quality and worth when exploring the topic. However, making sense of and deconstructing these particular approaches is of paramount importance since public policy and institutions in western-type liberal state economies seem to increasingly favour evaluation exegeses of heritage significance that are subjected to monetisation using statistical and econometric frameworks.

To provide a picture of the ideological substratum of this occurrence, the chapter first explains the texture of current policy landscape in countries such as the UK, as this has been shaped by neoliberal reforms from the 1980s and onwards, the politics of heritage 'worth' and the rationale for translating the qualitative, non-consumptive elements of heritage engagement and participation into monetary values. It then moves on to frame theoretically the most common economics-based approaches to assessing heritage impacts and value, before reviewing critically the current state of the art. As the chapter analyses, existing work on the topic compiles a heritage significance 'toolbox' that features a variety of quantitative techniques, including hedonic pricing, stated and revealed preference methods along with the most recent wellbeing valuations. The strengths, weaknesses and complementarity of these tools are discussed, leading to conclusions regarding existing limitations and future research avenues.

Heritage economic significance: Why does it matter so much?

Heritage preservation and enhancement generate a plethora of direct, indirect and induced effects on the economy. As a multilayered sector embracing cultural activities related among others to museums, historic buildings, archaeological places and monuments, it fosters consumption and employment across various flows of 'back-of-house' functions (e.g. research, documentation, conservation) and 'front-of-house' services (e.g. curatorial, administrative, outreach). Parallel to these, heritage stimulates economic activity indirectly in the sectors of its supply chain, such as scientific research, education, architecture and engineering, to mention but a few. Furthermore, although heritage-related goods and services are typically not commercially traded as commodities, there are various markets peripheral to the sector (e.g. recreation, tourism, real estate, creative industries) which capitalise on their uniqueness and value (Harvey, 2012), thus multiplying the positive influences (induced effects) that they exercise on current economic processes.

Although heritage policy was always imbued with the notion of 'value', the quest for measuring heritage significance is relatively recent and, to a certain extent, it can be regarded as the offshoot of a new political economy, which, on the one hand, pushes for public sector minimisation and reform in the name of greater efficiency, while, on the other hand, nurtures an international policy landscape that promotes heritage as a mechanism for economic development (De Cesari, 2020). Although at global level, heritage remains largely a 'state project', in recent decades, neoliberal economic ideas have gradually nested in the political agendas of both developed and developing countries, generating progressively a public policy discourse that is better aligned to the market's logic. This new discourse extends to the heritage sector in a twofold way.

Firstly, similar to other state-provided goods, it is argued that the public financing of heritage protection and enhancement must contribute to economic development (Harvey, 2012), mainstreaming the definition of heritage resources as 'assets' (see, for instance, UK's National Planning Framework, par. 184) – a financial term which implies fixation with their growth potential. Furthermore, the strategic positioning of heritage as a driver for economic development pushes for the instrumentalisation of related activities in contributing to multiple policy areas and goals (Crossick & Kaszynska, 2016) while assigning greater weight to sectorial economic objectives in order to tackle fiscal stress in the Global North (e.g. the crisis-hit economy of Greece) or achieve sustainable development in the Global South, mainly through tourism income (De Cesari, 2020).

Secondly, neoliberal economic policy as a proponent of reduced state control and austerity have led to tightening governmental budgets, which in turn render the expression of heritage significance in monetary terms critical for policymakers that need to decide wisely on how to allocate limited public resources while maintaining fiscal accountability. In a time of weakening welfare-state politics, costs generated through provisions to heritage, which is a capital- and labour-intensive sector, could become an obvious burden for governmental budgets, while benefits, often subtle and non-monetary, remain hard to capture (CHCfE Consortium, 2015). Consequently, funds allocation to heritage increasingly calls for justification and 'hard' evidence of investment returns (Girard & Gravagnuolo, 2017), often rationalised as necessary for securing policy transparency in the management of public finances (Nijkamp, 2012). In this climate, expectations of delivering direct and spillover economy-related benefits, such as GDP rise, skills development and urban regeneration (Cebr, 2019), are embedded in justifications for financing the heritage sector in both policy and academic discourse.

Based on the aforementioned, one can see why heritage significance has come to matter so much and why defining its economic footprint is now a major concern for related policy. Demonstrating the economic benefits of monuments, sites and landscapes of heritage interest in monetary terms or evidencing that related investments maximise societal gains are gradually crystallising as prerequisites for justifying public spending on conservation, restoration, preservation and other heritage management tasks that keep us connected to the past. In addition, the generation of monetary value information is considered almost vital for choosing or prioritising between investment options, by comparing benefits and costs to achieve the highest economic payoff. This is exemplified in the 'Green Book', Britain's technical guide on evaluating public expenditure decisions, where it is stated that the outcomes of cultural services, although normally not assigned with market values, 'should be identified as far as possible and proportionately quantified and monetised' (HM Treasury, 2018, p. 62). As this rationale increasingly permeates policy texts and public financing, it necessitates the development of sophisticated solution methods to express heritage effects on society and economy numerically.

This discourse is not surprising, as ontologically, economics see the world as a universe of economic choices, namely trade-off decisions in allocating finite resources, such as time and money. In view of current politico-economic landscape and sustainability considerations, the key question as to how limited resources can be managed in an optimal way retains its relevance and impetus. The adoption of heritage-related projects entails the sacrifice of (public) capital and other resources that cannot be employed elsewhere. Thus, policymakers need to make informed decisions on the basis of opportunity costs (i.e. opportunities to create value through alternative activities) and choose those options that can deliver supposedly maximum gains to society. As is analysed in this chapter, there are specific branches of economic science, of which the methods are employed to deal with such questions. For example, welfare economics are used to explore whether the value gained from investing public capital in a certain heritage-based scheme is higher that the value lost from competing economic activities (Atkinson & Stiglitz, 2015). These economic appraisals assess the actual or potential effects and trade-offs of policy options, providing information to governments on how they can best improve social welfare (HM Treasury, 2018).

An economic analysis of heritage: key concepts and hypotheses

As mentioned previously, the idea of value(s) has long permeated heritage management and policy, understood mainly as a mosaic of characteristics and positive qualities, such as artistic, historical and educational. In neoclassical economics theory, said qualities are treated as a collective of elements that converge in a general notion of 'benefit' extracted during consumption of related goods and services. This benefit is believed to motivate an individual to accept the cost of acquiring the good or service at a certain value (price) that reflects derived benefits and thus economic value monetarily. This implies that in mainstream economics, value is understood and conceptually framed within the microeconomic notion of 'utility', defined as the level of one's satisfaction drawn from consuming a good or service (Begg et al., 2014).

Yet, contrary to 'ordinary' goods, a considerable number of heritage goods and services do not operate in regular markets. Rather, their supply to audiences, visitors and users often occurs at a price that neither reflects the actual benefits derived from their consumption nor the full costs involved in their provision (Bakhshi et al., 2015). This means that their economic value and thus significance remain concealed. As we will examine later on this chapter, the scope of many economic evaluation techniques is to extract some measurable information of heritage significance that translates the qualitative traits flowing from the consumption of heritage goods into quantitative expressions of value. We will see that estimating the monetary value of those goods does not necessitate their being sold and bought in the market as this can be extracted in other ways, such as exploring how much money people are willing to pay in order to enjoy them.

In heritage economics, we commonly distinguish the economic value of heritage goods as 'use' and 'passive-use' value (Throsby, 2019). A visit to a historic site entails the consumption of an economic good (service), where 'use value' may describe the pleasure of experiencing the site, namely the benefit extracted from satisfying a visitor's needs and desires (e.g. for knowledge and education). Visitation, as an economic decision, and its value (derived utility) would be normally reflected in the price of entry ticket, as the amount of money that is considered equivalent to enjoying that service. However, there are many occasions where the use values are not expressed in monetary terms through direct consumption as enjoyment does not always involve a financial transaction (e.g. free entrance at exhibitions in UK National Museums). In these cases, specific tools are deployed in order to capture this expression indirectly. Additional complexities also arise from the existence of 'passive use' values (i.e. existence/altruistic, option and bequest) that again abstain from market processes. The local residents of Athens may not visit the Acropolis Hill regularly (or ever) but they still appreciate its existence (e.g. through visual contact), the value it has for other people and society, and they may wish to maintain the visiting option for themselves or future residents.

In defining the economic value of heritage, it is therefore important to consider that the provision of related goods forms a field which consists of typically 'non-market' goods. This implies that although heritage goods are economic goods, in many cases, they exhibit certain specificities that make their economic valuation particularly complex. A key point of such differentiation and complexity is the fact that heritage is not always part of the market economy, at least directly. A vast amount of heritage resources are not tradable commodities, meaning that there is no specific market where these can be exchanged. Moreover, the provision of heritage goods and services often creates positive externalities to third parties and economic activities outside the heritage sphere. For instance, the organisation of a carnival festivity or a traditional music festival is likely to increase traffic in various local tourism businesses, increasing demand and consumption of non-heritage goods and services. Thus, to account for heritage significance, we need to consider both use and passive-use values along with positive spillovers to other sectors of the economy (i.e. indirect/induced effects).

Another key concept that flows from these specificities is the identification of heritage goods as 'public', which is assigned with specific meanings by economists. Based on economic theory, all economic goods form into certain categories on the basis of two criteria: rivalry and excludability. Public goods are defined as goods that are non-rival; i.e., they can be enjoyed by everyone, and non-excludable; i.e., it is hard to exclude someone from enjoying them. Based on these criteria, heritage goods can be considered on many occasions as public goods (Dragouni et al., 2018). This is because heritage consumption is normally not competitive, as, for example, a visit to a heritage site, monument or arts exhibition does not 'deplete' the good as long as visitation complies to capacity standards, whereas on many occasions, heritage enclosures are particularly hard (e.g. in historic cities) and/or economically inefficient (Stiglitz & Rosengard, 2015). Welfare economics, a particular branch of economic theory, is often applied to the heritage sector, as it deals with government intervention intended to the provision of public goods.

In terms of methodological design, there are two main approaches to translating heritage significance. These are economic impact studies and economic valuation approaches. The former examines the economic benefits and costs of heritage on (secondary) markets, whereas the latter seek to articulate the value of primary economic and intangible gains of heritage, estimating the utility derived directly or indirectly to individuals who engage with the sector. Techniques from mainstream economics employed in the field comprise a collection of quantitative research tools that allow some sort of value or impacts measurement through the systematic collection and analysis of numerical data, such as statistics. In the next section, we analyse the most popular of these methods as presented in the literature, drawing a key distinction between market-based and non-market-based approaches.

Capturing heritage significance: state of the art

In this section, we will describe analytically the most popular methods for assessing heritage significance through economics. Before doing so, it is important to distinguish economic analysis, which is interested in broader economic and societal outcomes, from pure financial analysis, which is mostly suitable for assessing the economic performance and viability of individual organisations and institutions. In particular, financial analysis is 'thinner' than economic analysis in the sense that it considers solely direct inputs and outputs in the course of an economic venture, such as revenue income from ticket sales and expenditure on staff salaries and wages. In contrast, economic analysis is more useful as a tool for cultural policy, as it takes into account some wider economic and societal influences of heritage. For instance, cost–benefit analysis moves beyond financial analysis to consider social benefits and costs. It measures the positive and negative outcomes of investing in a heritage good by considering its direct effects on users and non-users, externalities and passive-use values. Thus, it can be used as a tool for appraising specific policies or projects, by juxtaposing expected gains with potential losses for the various parties involved and inform policy with regard to capital expenditure options (de Rus, 2020).

As suggested earlier, there are presently two main economic analysis approaches to evaluating heritage significance, assessing impacts and value, respectively. The first strand of the literature that focuses on heritage impacts measures the effects of heritage on local economies through specific economic indicators, such as income, visitation, expenditure, salary levels and number of businesses, to mention but a few. Furthermore, market-based approaches to capturing impact, apart from sectorial information and in order to address complexities deriving from heritage goods and services that operate in irregular markets, may choose to draw data on peripheral areas of economic activity with the view to attribute market fluctuations to heritage as spillover effects of its symbolic agency (e.g. hedonic pricing). In contrast, the second strand of the literature that explores heritage value(s) aims to generate new (speculative) market information for heritage goods within the sector by translating their non-consumptive traits into monetary units (e.g. contingent valuation method). Such economic valuation tools are defined as 'non-market-based' techniques because they seek to translate non-market heritage gains into monetary units. All these are examined in more detail in the following paragraphs.

Market-based approaches

To begin with, one of the most standard tools for estimating the use value of tradeable goods is the 'market price method', whereby economic significance is extracted by observing supply and demand behaviour. The said method is applicable only when heritage goods and services in question operate in a commercial environment. When this condition holds, then the market price method can be applied to observe changes in consumer or producer surpluses (e.g. by estimating the difference between total revenues and total production costs before and after an investment or change to the quality of that good). Unfortunately, in the heritage realm, the process of evaluation is often less straightforward and much more complicated because market data are often unavailable (e.g. when heritage goods are non-commercial). Yet, even when available, observed data may not reflect all productive uses of a heritage good as a resource (King et al., 2000). Even here, widely accepted economics-based techniques, such as the 'productivity method', are not easily applicable to the field, as heritage quality does not affect productivity and costs of other marketed goods directly (e.g. tourist packages). Consequently, economists normally draw on alternative market-based methods to extract significance estimations indirectly, using the 'hedonic pricing' and the 'travel cost method'.

Both the hedonic pricing and travel cost method are described in economics as 're-vealed preferences' techniques. This terminology implies that since heritage goods and services do not operate in regular markets, the preferences of the public towards them, which would be otherwise known through the typical market price and supply–demand mechanisms, remain masked, and thus, researchers need to find ways in order to 'reveal' them. In both cases, the elicitation of individuals' preferences takes place indirectly, either by observing behaviour in proxy markets under the hypothesis that heritage goods would influence the prices of some market goods (hedonic pricing) or by observing economic decisions of consumers (travel cost method).

More specifically, the hedonic pricing technique can be employed to calculate the use value of heritage by observing market price variations which could be directly affected by heritage-related attributes. The overarching idea behind this method is that the market price of any good (e.g. residential property) expresses its various characteristics and attributes that together form its economic value. In particular, according to economic theory, the price of a good reflects, on the one hand, the characteristics of the good itself; for the residential property, these could be its architectural design, size, condition and so on – and, on the other hand, additional factors related to its surroundings (e.g. access to amenities, safety, pollution). With this in mind, heritage is considered as one such 'extra' attribute, related to the surrounding environment of the marketed good.

Hedonic pricing models can be applied to reveal information with regard to behaviour of related markets, examining their responsiveness to heritage policy (e.g. price variations), exploring how the presence or the quality of heritage non-market goods influences market prices, thus estimating their economic use value. Indicatively, Lazrak et al. (2014) employed the method in the designated historic landscape of Zaanstad in the Netherlands to explore the impact of heritage 'labelling' on real estate values of nearby private properties. They compared data prior to and after the designation to observe whether houses located within the conservation area had gained a price premium that could be attributed to this particular policy decision. Similarly, Franco and Macdonald (2018), using secondary geo-coded real estate market data, explored how historic architecture affected residential property values in Lisbon, Portugal, considering proximity, concentration and location factors.

Another avenue for estimating economic use values through the lens of consumer demand is the travel cost method. This allows to generate information of the benefits and costs associated to heritage sites, especially those located in remote areas. More specifically, compliant to the general economics' view of the world as a universe of choices and trade-offs, the method is based upon the proposition that a visit to a monument or site entails both time and travel costs, which can be translated as 'price' of access (King et al., 2000). In the heritage field, the application of the technique draws on data from survey participants – usually visitors to the area of interest, who provide information about the characteristics of their visit (e.g. length of trip, on-site expenses), their demographic profile (e.g. place of residence, income) and other details that could be fed into the model, such as experiences with the site (e.g. quality of services) or substitute goods/trips. These data are processed with the view to relate decision to visit and the variables determining such choice, providing the demand function for the heritage site of interest and its respective economic benefit (consumer surplus) for the average visitor.

For example, Brida et al. (2012) employed a travel cost model to explore visitor attractiveness of the South Tyrol's Museum of Archaeology in Bolzano and its capacity to serve as the city's tourist pull factor. They conducted on-site interviews with travellers to identify the determinants of repeat visitation, probing for travel, accommodation, subsistence and shopping expenses, along with visitors' socioeconomic status and personal preferences. As another option, Voltaire et al. (2017) applied the simplest version of the technique, the 'zonal travel cost method' to examine the economic value of Mont Saint Michel in France. Instead of collecting primary information, this alternative version draws on secondary macro-data (i.e. total visits per zones surrounding the site) to calculate visitation rates and extract consumer surplus. In applying this approach, the researchers sought to estimate the recreational benefits of the World Heritage Site as a means to defend preservation and management costs.

In brief, we observe that both of these empirical techniques present several advantages as they extract information by observing actual market or consumers' behaviour. Hedonic pricing is relatively inexpensive as it is based on secondary data, whereas the travel cost method can employ on-site surveys to collect data from large samples of visitors. However, admittedly, these techniques are limited by underplaying non-consumptive attributes of heritage and their economic value for non-users. For these reasons, researchers may opt for non-market-based methodological designs to answer the enigma of heritage significance.

Non-market-based approaches

Contrary to revealed preference tools, 'contingent valuation' and 'choice modelling' are coined as 'stated preference' techniques, because instead of observing actual (revealed) behaviour, they document individuals' stated preferences on the basis of some hypothetical scenario. They can be employed to estimate welfare gains and losses by eliciting citizens' willingness to pay for a positive change or their willingness to accept a negative change, either directly (contingent valuation) or indirectly (choice modelling). Compared to the revealed preference methods analysed earlier, stated preference approaches have the advantage of providing estimations of use and passive-use values, which remain unobserved by market-based tools. Moreover, they can be used for ex ante evaluations aiming to explore heritage value towards future changes through relevant scenarios.

More specifically, the rationale behind the contingent valuation method is to directly ask research participants to assign monetary values to the heritage goods in question, instead of inferring values from actual economic behaviour. To do so, it surveys individuals based on hypothetical scenarios/markets in order to elicit their willingness to pay for a particular attribute or good. To elicit the value that society or a certain community attribute to accessing a heritage public good in the present (users/visitors) or to securing its future existence (non-users/non-visitors), they rely on some sort of hypothetical scenario of voluntary donation or tax deduction, related for instance to a restoration/conservation project (e.g. for listed buildings or historic assemblies) or the introduction/increase of entry fees (e.g. for museums and heritage sites). As per design, data from audience/visitors and control samples (non-visitors) are collected to obtain use and passive-use values, respectively.

Related applications include the work of Lampi and Orth (2009), who used the method to examine the value of free entry policy to state-funded museums in Sweden and its effects on diversifying audiences. They requested visitors at the Museum of World Culture to state what exchange value (i.e. ticket price) they would be willing to pay for enjoying museum services, if free entrance was abolished. By the same token and considering the sector's limited financial resources, Sharifi-Tehrani et al. (2013)

employed the method to examine the potential of a dual pricing strategy for foreign and domestic visitors to the National Museum of Iran, collecting willingness-to-pay data with the view to set the maximum possible entrance price. Beyond museums, Lawton et al. (2018) employed contingent valuation to estimate the economic value and benefits associated with the conservation of historic cityscapes and cathedrals in four historic cities in England. They collected data through an online survey instrument that presented users and non-users with a scenario whereby local monuments were about to experience some large-scale damages, caused by extreme weather events and poor maintenance, which in turn would necessitate lengthy repair works and interrupt their public enjoyment. To avoid the scenario's materialisation, respondents were asked to state their willingness to make a donation for preventative works that would reduce the levels of risk and damage. Willingness to pay values were then elicited, expressed as the amounts allocated to said cause through the hypothetical voluntary payment mechanism.

Choice modelling is another tool to extract heritage monetary value by examining individual preferences of audiences and user communities in relation to specific characteristics of heritage management and policy decisions. Although choice modelling can be used in both revealed preferences or stated preferences protocols, it features mostly as a willingness-to-pay instrument in heritage-related studies, also coined in the literature as 'discrete choice experiment' (Thorsby, 2019). In this method, research participants are presented with various hypothetical scenarios of policy initiatives, each featuring a different set of attributes or different levels of same attributes related to an intervention (e.g. conservation project) or quality improvement (e.g. greater accessibility) together with an attribute which is expressed monetarily to reflect the cost for achieving scenario outcomes. Participants are requested to choose their preferable scenario, enabling the calculation of the monetary value they are willing to pay for presumably maximising their utility.

A case in point is the work of Jaffry and Apostolakis (2011), who employed a stated preferences choice-modelling protocol to assess visitors' likings over different management decisions at the British Museum of London, using voluntary contributions as a proxy for price and translating utility into monetary units. With the help of econometrics, they analysed visitors' survey data to observe utility changes and their monetary equivalent for attributes related to service operation (e.g. extended opening hours, staffing), facilities (e.g. ICT provisions) and content (e.g. more temporary exhibitions, community work). In another example dealing with natural heritage, Abrina and Bennett (2020) surveyed Filipino households regarding their willingness to pay for coral reef restoration. Using scenarios that featured different levels of fish abundance, fish species and coral cover as attributes that could be achieved through a monthly surcharge on their electricity bills, they asked respondents to choose between alternative futures of their coral reef heritage.

As implied earlier, a key advantage of non-market-based approaches is appraising economic significance not only across audiences and users but also communities and groups who may not engage with heritage at the time of the study but who, nonetheless, appreciate its existence and preservation. Furthermore, non-market-based methods can be used to capture future or potential effects as they do not rely on actual market data. Nevertheless, the main drawbacks of stated preferences techniques lie in their hypothetical nature (response bias) and the fact that individuals cannot always calibrate the costs of heritage management or its (re)production in the present nor they can predict the level of utility they would be able to derive or sacrifice from a future development.

Moreover, they may overstate and/or understate a monetary value based on their socio-economic condition (e.g. household income), political beliefs (e.g. their stance towards government, institutions and/or tax policy) and other factors that are hard to control.

Hybrid methods and emerging approaches

The extant literature also features methodological designs which generate information by employing a mixture of the above tools or by combining traditional with new techniques. Most notably, the previous decade has seen the emergence of the 'wellbeing' concept in the heritage policy agenda. Similar to previous techniques, wellbeing valuations serve as instruments for assessing welfare alterations owning to heritage and related policy initiatives; however, in this case, these are observed through individuals' subjective judgements of own happiness (subjective wellbeing) together with statistical data mining that aims to establish correlation to heritage engagement (Crossick, & Kaszynska, 2016). Therefore, the aim of subjective wellbeing valuation is not to measure economic value through willingness to pay but rather by validating positive causality between heritage quality goods and life satisfaction.

In a study commissioned by English Heritage, Fujiwara et al. (2014) applied the wellbeing valuation approach to the UK sector in order to assess the direct impacts of heritage visitation on life satisfaction across various user groups, as defined by age, socioeconomic profile, health and parental status. Regression analysis of survey data reported a significantly positive relationship between the two variables, suggesting that those who had visited at least one heritage site over the past year exhibited higher life satisfaction than control groups. Based on these results, the researchers provided use value estimations for different types of heritage sites (e.g. archaeological sites, historic places of worship), as the annual amount of money that would compensate current visitors for retaining the same level of wellbeing, if engaging with heritage was impossible. Likewise, in an AHRC-funded study, Bakhshi et al. (2015) adopted a mixed-methods contingent and wellbeing valuation approach to assess the direct effects of heritage on individuals who engage with the work of the Natural History Museum and Tate Liverpool Gallery. In these case studies, estimates of heritage goods' use values were extracted through hypothetical entry fee or donation scenarios (contingent valuation), while they were also expressed as money sums equivalent to the impact of heritage consumption on wellbeing. A hybrid application of the two methods was also performed, where museum visitors were asked to state the level of money compensation that would allow their life satisfaction to remain stable in the event of the museums' closure.

As with any mixed-method approach, hybrid methodologies can benefit from the strengths and lessen the weaknesses of the tools they employ (Dragouni, 2017). However, the wellbeing approach cannot be used to predict future welfare changes or examine passive-use values. It is characteristic that the studies reviewed above focus on user groups, i.e. on individuals that actually pay visits to heritage sites, providing no evidence for those that do not formally engage with the sector. In addition, as with state-preference techniques, wellbeing self-reports are subject to imperfect information and contextual influences. Indicatively, Fujiwara et al. (2014) intended to control for confounding economic and social factors, but admitted that it was difficult to do so thoroughly. It is also important to note that subjective wellbeing is theoretically based on the criterion of 'Pareto efficiency', which by nature focuses on individuals instead of society as a whole, failing to account for issues of inequality and disparities of accrued benefits between different social groups (Stiglitz & Rosengard, 2015). Rather, the focus of the method remains on idiosyncratic appraisals of life happiness instead of macroeconomic indicators that reflect a quality standard of living. Although it is perhaps naïve to assume that major issues, such as economic deprivation, could be ameliorated through heritage engagement alone, the recent popularity of the method does not come as a surprise as reducing a political economy problem to a matter of personal satisfaction is perfectly aligned to new market instrumentalism (Dragouni, 2020).

Concluding remarks and future research avenues

Witnessing increasing policy pressures and academic interest in extracting heritage monetary expressions of value and impact, this chapter reviews the economics-based 'toolbox' for capturing its significance for hosting economies and communities. As analysed in this chapter, there are various economics-based methods that are used exclusively or complementarily in various evaluation frameworks, such as cost–benefit analyses or satellite accounting methodologies, to provide estimations that can inform strategies towards sustainable heritage futures. Yet, in spite of growing attention to the topic, the evaluation of heritage socioeconomic significance remains a central problem question for those engaging in the field.

As it was shown, capturing heritage significance from an economics standpoint can be performed by exploring the direct impacts of the sector and/or the indirect impacts of related goods and services on secondary markets. Alternatively, the economic value of heritage goods can be estimated by observing changes revealed in peripheral markets, whereby heritage serves as an 'attribute' of non-heritage goods, by asking individuals to state their intentions for assuming the costs of preserving or giving up a heritage resource in a hypothetical setting, or even by attributing changes of personal wellbeing levels to heritage engagement. As the use of these approaches remains relatively new to the heritage realm, methodological refinement and enrichment is necessary before we can securely say that the economic significance of heritage is fully grasped through such methods. As it was already discussed, to be comprehensive, appraisals of heritage significance need to consider multiple levels of contribution to economic systems along with spillover effects on a plethora of markets that derive benefits from the (re)production of heritage. At the same time, quantitative and metrics tools need to be combined with qualitative approaches that will harness ethnography, discourse analysis and emotion to co-produce data and interpretation of significance with heritage users and citizens in order to apprehend heritage value-making processes more holistically.

Moving beyond methodological concerns, one might review critically the idea of heritage significance, its inception and evolution in current policy landscape. Shaped through a synergy between neoliberal politics and neoclassical economics tools, current discourse naturalises the question of its monetary appreciation. The whole task of economic evaluation, as rationale and process, is imbued with structural issues that need to be deconstructed so that alternative but equally legitimate approaches can emerge outside the canon. In their study exploring the connection between heritage and well-being, Fujiwara et al. (2014) provide an apt illustration of current policy fixation with neoliberal reasoning and argumentation, when stating that the allocation of 'scarce public resources' necessitates our weighing heritage benefits against budgetary costs, both expressed as monetary values, so that we can judge, as experts working from top to bottom, 'whether the intervention is worthwhile' (p. 4). Above all, a central argument of this chapter is that we need to treat such claims critically, not merely because public resources scarcity, nowadays presented as an axiom, remains in truth a political choice but also because it is crucial to interrogate the grounds of evidence upon which heritage worthiness is ultimately determined.

As demonstrated by our analysis, the value of heritage through the prism of mainstream economics is epistemologically founded upon the concepts of (individual) preferences and (individual) choices. On this basis, heritage value is narrowly defined as a monetary amount that one is willing to pay to enjoy or preserve in real life or hypothetically. This definition of value is grounded in the utility hypothesis, reflecting the benefits derived from a type of hedonistic and invariable consumption of market commodities. This view might seem particularly limited when considering that our interactions with and benefits derived from a monument, a piece of art or some other cultural text occur through Daedalian social processes of continuous motion. None of the values that are ascribed to heritage are static or intrinsic but rather place and time specific as (re)determined by a given society (Marx, 1990 [1867]). This implies that any assessment of heritage significance is historically situated in a specific system of ideas and beliefs, which at each point defines the texture of value and allocates it to the various heritage goods in question.

Thus, the issue of heritage evaluation is both a methodological and an epistemological problem; one that requires us to challenge the ways we seek to gain knowledge of social reality. Within current paradigm, where neoclassical economics serve as the blueprint, heritage value (i.e. utility) is increasingly understood as a trait or a collection of traits that require affirmation on the grounds of their capacity to satisfy individual (not social/collective) needs re-labelled as 'public welfare', but essentially following the same principles of preference satisfaction and utility maximisation. Inevitably, monolithic economics-based appraisals of heritage significance that are grounded in mainstream neoclassical and microeconomic theory reduce heritage values to their monetary expression while collateral gains that cannot be monetised (or even understood) are left aside.

Considering these limitations, it makes sense to suggest that theoretical and empirical approaches to heritage significance could be informed by alternative traditions of thought in economics. For instance, the dialectical approach to value in Marxist economics could serve as a useful analytical framework to address related questions and understand the social dynamics of value-making and heritage production processes. For Marx, use value nests in anything that satisfies a human need, whereas value is a social representation of use value in an economic system where use values are exchanged as commodities (i.e. monetary exchange values). Contrary to neoclassical utility theory that emphasises individual needs, the notion of use value here describes qualities generally regarded as useful in a given society. With this in mind, one could adapt Marxist theory of exploitation to analyse the appropriation of heritage resources in contemporary society, where similarly to labour, heritage maybe positioned as a source of surplus value and a means of value accumulation. In this light, exchange values (prices in real or artificial heritage markets) would not necessarily represent heritage use values accurately.

Overall, conceptual and methodological choices determine the nature and quality of information that can be gained and lost when investigating heritage significance. This suggests that future research needs to focus more on the processes and less on fragmented outcomes of heritage-making as a better way to prove its relevance and significance to society. This requires a departure from top-down approaches towards the development of more 'bottom-up' methodologies, which would be better attuned to the recent 'social turn' marked in the discipline (Lekakis & Dragouni, 2020). In parallel, economic interpretations of heritage significance, currently rooted almost exclusively in ideas and notions of mainstream neoclassical thought, could explore and experiment with heterodox economic theories and interpretations of value production processes as social relations and social acts of heritage production, with the view to inform and enrich their theoretical framing and analysis.

References

Abrina, T. A. S., & Bennett, J. (2020). Using choice modelling to estimate the non-market benefits of coral reef restoration in the Philippines. *Ocean & Coastal Management*, 185, 105039.

Atkinson, A. B., & Stiglitz, J. E. (2015). *Lectures on public economics: Updated edition*. Princeton, NJ: Princeton University Press.

Bakhshi, H., Fujiwara, D., Lawton, R., Mourato, S., & Dolan, P. (2015). *Measuring economic value in cultural institutions*. London: Arts and Humanities Research Council.

Begg, D., Vernasca, G., Fischer, S., & Dornbusch, R. (2014). *Economics*. 11th ed. London: McGraw-Hill Education.

Brida, J. G., Meleddu, M., & Pulina, M. (2012). Understanding urban tourism attractiveness: The case of the Archaeological Ötzi Museum in Bolzano. *Journal of Travel Research*, 51(6), 730–741.

Cebr (2019). *The heritage sector in England and its impact on the economy*. London: Centre for Economics & Business Research.

Crossick, G., & Kaszynska, P. (2016). *Understanding the value of arts & culture*. Swindon: UK Arts & Humanities Research Council.

Cultural Heritage Counts for Europe Consortium (2015). *Cultural heritage counts for Europe*. Krakow: International Cultural Centre.

De Cesari, C. (2020). Heritage beyond the nation-state? Nongovernmental organizations, changing cultural policies, and the discourse of heritage as development. *Current Anthropology*, 61(1), 30–56.

de Rus, G. (2020). *Introduction to cost-benefit analysis: Looking for reasonable shortcuts*. Cheltenham: Edward Elgar Publishing.

Dragouni, M. (2017). *Sustainable heritage tourism: Towards a community-led approach*. Doctoral thesis. UCL (University College London). Retrieved from: https://discovery.ucl.ac.uk/id/eprint/10025949/ (Accessed on 02 December 2020).

Dragouni, M. (2020). An economic analysis of heritage within the commons framework: Reflections on theory, methodology and social imagery. In S. Lekakis (Ed.), *Cultural heritage in the realm of the commons* (pp. 45–65). London: Ubiquity Press.

Dragouni, M., Fouseki, K., & Georgantzis, N. (2018). Community participation in heritage tourism planning: Is it too much to ask? *Journal of Sustainable Tourism*, 26(5), 759–781.

Franco, S. F., & Macdonald, J. L. (2018). The effects of cultural heritage on residential property values: Evidence from Lisbon, Portugal. *Regional Science and Urban Economics*, 70, 35–56.

Fujiwara, D., Cornwall, T., & Dolan, P. (2014). *Heritage and wellbeing*. London: English Heritage.

Girard, L. G., & Gravagnuolo, A. (2017). Circular economy and cultural heritage/landscape regeneration. Circular business, financing and governance models for a competitive Europe. *BDC. Bollettino Del Centro Calza Bini*, 17(1), 35–52.

Harvey, D. (2012). *Rebel cities: From the right to the city to the urban revolution*. London: Verso.

HM Treasury (2018). *The green book: Central government guidance on appraisal & evaluation*. London: The National Archives.

Jaffry, S., & Apostolakis, A. (2011). Evaluating individual preferences for the British Museum. *Journal of Cultural Economics*, 35(1), 49–75.

King, D. M., Mazzotta, M., & Markowitz, K. (2000). *Ecosystem valuation website*. US Department of Agriculture (Natural Resources Conservation Service), National Oceanographic and Atmospheric Administration. Retrieved from: http://www.ecosystemvaluation.org (Accessed on 03 December 2020).

Lampi, E., & Orth, M. (2009). Who visits the museums? A comparison between stated preferences and observed effects of entrance fees. *Kyklos*, 62(1), 85–102.

Lawton, R., Fujiwara, D., Mourato, S., Bakhshi, H., Lagarde, A., & Davies, J. (2018). *The economic value of heritage: A benefit transfer study*. London: Nesta.

Lazrak, F., Nijkamp, P., Rietveld, P., & Rouwendal, J. (2014). The market value of cultural heritage in urban areas: An application of spatial hedonic pricing. *Journal of Geographical Systems*, 16(1), 89–114.

Lekakis, S., & Dragouni, M. (2020). Heritage in the making: Rural heritage and its mnemeiosis at Naxos island, Greece. *Journal of Rural Studies*, 77, 84–92.

Licciardi, G., & Amirtahmasebi, R. (Eds.) (2012). *The economics of uniqueness*. Washington DC: The World Bank.

Marx, K. (1990) [1867]. *Capital volume 1*. London: Penguin.

Nijkamp, P. (2012). Economic valuation of cultural heritage. In G. Licciardi & R. Amirtahmasebi (Eds.), *The economics of uniqueness* (pp. 75–106). Washington, DC: The World Bank.

Rypkema, D., & Cheong, C. (2011). Measurements and indicators of heritage as development. *ICOMOS 17th General Assembly, 2011-11-27/2011-12-02* (pp. 755–762). Paris, France.

Sharifi-Tehrani, M., Verbic, M., & Chung, J. Y. (2013). An analysis of adopting dual pricing for museums: The case of the national museum of Iran. *Annals of Tourism Research*, 43, 58–80.

Stiglitz, J. E., & Rosengard, J. K. (2015). *Economics of the public sector*. 4th ed. New York: W.W. Norton Company Inc.

Throsby, D. (2019). Heritage economics: Coming to terms with value and valuation. In E. Avrami, S. Macdonald, R. Mason, & D. Myers (Eds.), *Values in heritage management* (pp. 199–209). Los Angeles, CA: The Getty Conservation Institute.

Voltaire, L., Lévi, L., Alban, F., & Boncoeur, J. (2017). Valuing cultural world heritage sites: an application of the travel cost method to Mont-Saint-Michel. *Applied Economics*, 49(16), 1593–1605.

UNESCO/United Nations Development Programme (2013). *Creative economy report: Widening local development pathways*. Retrieved from: http://www.unesco.org/culture/pdf/creative-economy-report-2013.pdf (Accessed on 15 July 2020).

34 A rambling field role for the heritage practitioner

A means to come to more socially sustainable heritage (re-) development projects

Gisèle Gantois

Introduction

Within the last decades, scholars from different disciplines increasingly recognise that cultural heritage has become an important actor and a key issue impacting contemporary societal change, thereby addressing its societal meaning at a universal level. However, at the same time, more attention is paid to the different forms of its local significance. This also changed the way in which heritage is conceived and tackled. The framing of its conditions is revised, leading to a shift from the domination of the historical characteristics towards the actual significance of heritage sites. Extinct remains of urban and rural cultures no longer exclusively dominate cultural heritage with its tangible and intangible expressions. Rather, heritage is considered to be organically integrated into the everyday life of local communities, whilst also having a use-value: it is lived in. (Waterton & Watson, 2011; Gravari-Barbas, 2014; Sarkki et al., 2019). This contributes to a different heritage process by favouring a more nuanced and socially sustainable re-use of historical places. In that sense, rather than merely considering heritage as a subject of change, the sustainability discourse consists of which changes heritage can generate and how it can act as an important driver towards the creation of an inclusive society (Fouseki et al., 2020, pp. 1–15). To "reinvent the field of (public) heritage as a revival of peoples' sense of belonging and participation in a living cultural community, not just the preservation or safeguarding of symbolic heritage elements" as stated by ICOMOS' former president Araoz (2011, pp. 55–60), it is indispensable to engage more of the voices of people who live with the heritage sites.

Unfortunately, wellbeing related to the interaction with heritage places is still to a great extent undervalued in development plans, which most often concentrate on a direct monetary return. The result is that a continually widening range of heritage has been transformed from an object of cultural significance, essentially embodied in the lives of local communities, to a marketable commodity. Additionally, in heritage practice, the focus has long been static, emphasising the building's history, its materiality and the technical aspects of conservation and restoration.

This chapter aims at demonstrating that with a different approach in architectural pedagogies and heritage practices we can come to a deeper understanding of a heritage site and its meaning today.

The here-presented three-step methodology of *Interactive Walking* (Gantois, 2019) departs from the traditional focus on the built or unbuilt dichotomy.

Step 1: Interactive Journeys of this novel method combines a rambling field role with chronologically sketching the observations and experiences. Hereby all of the heritage practitioner's sensory and sensual capacities are addressed, which, ultimately, turns observational walking into sensorial walking. A dynamic view is added to the static conception of heritage with its literal layers of archaeological remains whereby the attention shifts from exclusively focusing on the historical artefact as an isolated object towards considering its relationship with its (lived) context. This results in a reformulation of the concept of the value of heritage in which historical characteristics, intrinsic qualities and contemporary significance are inextricably linked. In this step, a decisive place is attributed to unplanned encounters. It unravels to which elements the attachment to heritage places is related by highlighting the underlying meaning of experience, "where experiences contain both the outward appearance and inward consciousness based on memory and image" (Creswell, 2007, p. 59).

In *Step 2: Spatial Narratives*, the practitioner's individual experiences with the heritage places are represented in artistic drawings and tactile models to be presented and discussed with the local community in exhibitions and debates on location.

In *Step 3: Cartes Parlantes* [Unveiling Maps], the individual experiences enriched with new narratives are then confronted and combined with existing expert knowledge and data. These counter maps and models already carry the germ of possibility in them.

Interactive Walking challenges the traditional material-focused view, nudging it towards a more anthropological perspective by directing attention primarily at the voiceless – that is, the "people overlooked as authorities capable of adjudicating their own sense of heritage" (Waterton & Smith, 2010, p. 10). Their voices are critical for understanding the way in which the studied heritage site is appropriated within daily life in a slowly grown and meaning-generating process that is difficult to define for outsiders and not always understandable solely through its visible expressions. These *nameless actors* are no longer considered to be simply users of the heritage place or passive observers; neither are they elements to be simply observed or educated. Instead, they are individuals who interact with each other and with the heritage place as a vital part of the community.

This method was first developed on a small scale within my architectural and restoration practice and has been further developed and tested together with my students of the International Master Studio at the faculty of Architecture (KU Leuven, Belgium) in three major real-life cases situated in Flanders: the post-industrial landscape of the Rupel region, more specifically the protected hamlet of Noeveren (academic year 2015–2016, 2018–2019 and 2020–2021), the peri-urban territory of the Brabantse Kouters (academic year 2016–2017) and the post-traumatic town of Ypres (academic year 2017–2018 and 2019–2020). These exercises were a fundamental part of the research *Tracing the deep significance of built heritage through encounters with undisclosed protagonists* (Gantois, 2019).[1]

A renewed vision on sustainability towards social sustainability

The conditions of the chosen cases can be linked to Meta conditions of other territories in Europe and beyond as worldwide some decisive transformations have occurred over the past several decades, which have also had an impact on the heritage sector and changed the way that policymakers view heritage. What Hartog (2015, pp. 115–116) calls the changing perspective on historicity, re-attributes a central role to heritage after an era of optimist futurism in which many historical sites were demolished.

In the mid-1970s, environmental protection and the preservation of monuments, objects, ways of life, landscapes, and animal species became major concerns in the context of increasing globalisation. This was translated in UNESCO's charters already in 1972 with the 'Convention Concerning the Protection of the World Cultural and Natural Heritage' codifying the idea of *world heritage* while relating heritage to its territory. Heritage became a major concern closely linked to environmental issues, recognising and including the tangible, the intangible and all living creatures.

Different official reports such as the Brundtland Report (1987) and the UNESCO Rio Declaration (1992) underscored this vision.

The Faro Convention Framework on the Value of Cultural Heritage for Society recognised already in 2005 "the need to put people and human values at the centre of an enlarged and cross-disciplinary concept of cultural heritage" (p. 6).

Unfortunately, in Flanders, as in other regions in Europe, the economic recession that has engulfed us since 2008, provoked a cutback of culture budgets and a significant reduction in traditional subventions. Owing to the need for alternative funding, and due to the challenge of urban pressure and post-industrial processes, as in other parts of the world, empty religious buildings, numerous former industrial sites and historical city centres and quarters have been impacted by increasing commercialisation. This process disrupts heritage from what it essentially is – namely, a long-term developed communal inheritance (Caserta & Russo, 2002, pp. 245–260; Mason, 2008, pp. 303–318; Porter & Shaw, 2013; Silberman, 2014, pp. 433–445). This concept denotes a social relationship beyond capitalisation and marketisation (Harvey, 2012, p. 73). The neglect of this distinctive aspect of heritage as a social relationship makes people claim the right-to-heritage when 'their' heritage is affected by new management plans or when they are excluded from the decision-making procedures (Emelianoff & Carballo, 2002, pp. 49–57; Gracia, 2003; Harvey, 2012, p. 73; Gantois, 2020, pp. 166–184). This process is marked by a growing movement of community empowerment within a renewed vision of sustainability towards social sustainability. This implies a move towards a vision of heritage that integrates alterity. In order for this view of heritage to play such a role, it is essential to adopt participatory approaches.

The Burra Charter (2013) explicitly draws attention to the importance of community involvement to enable the exploration of the cultural significance of all types of places, including natural, indigenous and historic sites with cultural values. The Cultural Heritage Counts for Europe report (Europa Nostra, 2015) stresses the multiple benefits of heritage and its equally important economic as well as social, cultural and environmental facets to come to the sustainable development of a heritage site. Moreover, the research entitled *Places that make us* (National Trust, 2017), updated in 2019 as *Why Places matter to people* (National Trust, 2019), demonstrates the strong emotional connection between people and places.

Today, the emphasis on public rights and responsibilities as seen in the formulation of heritage policy replaces the traditional, conventional top-down approach, which views the heritage practitioner as an omniscient expert, by a more interdisciplinary, community-based approach, long considered unthinkable to heritage specialists (e.g. Petzet, 2009: pp. 9–15 and p. 34).

According to the Association of Critical Heritage Studies' (ACHS) past President Lucie K. Morisset's comments on the ACHS website: "Collaboration is the cornerstone of this process, promoting a meeting of fields as different as material culture, political ecology, and local development".[2] These fields include anthropology, archaeology, architecture (history of), art history, arts & aesthetics, conservation ethnography, ethnomusicology, gender studies, geography, sociology and religious studies – to name a few.

While the social sciences have clearly been paying attention to the interactions between communities and built heritage and the site's territory, heritage practitioners still lack this approach. The previously mentioned heritage-related charters, conventions and research reports underscore the necessity of reflecting on cultural value, but they do not provide clear practice-oriented tools for detecting, unveiling and mapping the present societal significance. Heritage practitioners have at their disposal a range of well-known and effective methods for managing monuments and sites and addressing their financial accountability, but they are not trained to engage with local communities. Additionally, a heritage practitioner is usually an outsider in the heritage landscapes and buildings that he studies. As Rebecca Solnit (2005) formulates it, one is "illiterate in the landscapes' language, which is the colloquial expression of the place itself" (p. 10).

A rambling field role for the heritage practitioner

Neither the material focus on a single building, nor a solely theoretical (historical) study from above can provide sufficient insight into how things work. In other words: human behaviour or interaction with the heritage location cannot be understood without being on site and studying the context as part of people's daily lives and activities. This implies that, in order to gain insight into the heritage places, to observe its inhabitants and to express one's and others' understanding of it, a heritage practitioner must join those whose heritage s/he is working on. Therefore, the here presented three-step methodology of *Interactive Walking* starts from the idea that the best way to (re)-discover places and to understand people's deepest unspoken emotions, lived memories, traces and associations, "We have to take time to step across the roads, to visit the places of which the inhabitants tell" (Ingold & Vergunst, 2008, pp. 1–19).

The use of walking as a very fruitful research method is not new (see among others, Careri, 2002; Ingold, 2007; Ingold & Vergunst, 2008; Pierce & Lawhon, 2015; Springgay & Truman, 2019). Solnit's (2000) magnificent *Wanderlust* traces the long history of walking and, more particularly, its place in literature, thereby referring to a variety of authors who explore their worlds through strolling, ranging from Aristotle, Rousseau, Thoreau and Baudelaire to Wordsworth and Walter Benjamin. When dealing with architectural theory, many scholars have often referred to each of the aforementioned authors, as well as many others. There is indeed a compelling similarity between the way that a wandering writer uses the reading activity of walking to understand the (urban) landscape and the strategy used in *Interactive Walking* to detect possible relationships and different forms of significance.

The first step of the methodology, called *Interactive Journeys*, entails serendipitous wandering while graphically chronicling the observations and experiences in little jot booklets. While the region in which the heritage site is situated is pre-defined, the exact framing of the investigation area is not. It is delimitated through the act of walking itself, the efforts one wants to make (*How far will I go? How long will I walk?*) and through one's findings and the development of interest fields on location. In contrast to the traditional way of doing a survey, this first step is taken without preliminary investigations of written sources about the heritage place. This is intended to begin the process as open-mindedly as possible. To become thoroughly immersed in the site to investigate, the exploration on foot takes place over a considerable period of time as the pedestrian motion goes as slowly as experience demands. Each booklet represents one day of walking, or a one-day walk, with one or different starting points, a middle and an end.

The sketch booklet is folded from a conventional piece of paper, which guarantees the accessibility of the tool to all, because it is affordable and easy to create. The tiny A7 size is deliberately chosen because it is easy to take with as it fits in the palm of one's hand or in the pocket of one's jacket or trousers. By using a booklet rather than a simple paper, one is stimulated to take notes and sketch in chronological order, page by page, which makes it easier to recall the walk later on. In subsequent booklets, pencils, watercolours and pastels are used to convey more of the textures and atmospheres observed on location. At the end of a day of investigations, the booklet is unfolded to its initial A4 format to redraw on the reverse side, the *line of perambulation*. Once the practitioner has indicated his memory places, he can move along them in either direction, backwards or forwards as if he could *walk* through his memory by *re-reading* the landscapes he walked.

The series of jot booklets with the recordings of the observations and experiences serve as mnemonic devices, a means of memorising quantities of what would, otherwise, remain seemingly rote information. This construction of a personal *Memory Palace* (Yates, 2014/1966) brings to mind Warburg's *Mnemosyne Atlas* from a procedural point of view of wandering and associating. (Michaud, 2007) These booklets finally form a personal diary and are the basis of step 2: *Spatial Narratives*.

The initial outlines in the sketch booklets represent the individually chosen landmarks as isolated objects in order to create a memory of the investigation area. Walk after walk, the practitioner's observations overcome this first object-focus, and shift towards a context-locus viewpoint with expanded realms that appeal to the eye. Drawing intensifies our gaze, a gaze that regains depth through walking (Berger, 1972). As Zuccoli (2019) states: In this way, we can expand our action, reflection and participation.

Each time new journeys are depicted and traced, one's knowledge and understanding of the heritage place is strengthened, and the landmarks are linked step-by-step to images of real-life experiences at specific locations. Heritage structures are then no longer considered to be static, finished artefacts with a well-defined geographical surface; rather, they are active nodes participating in a lively process. Through the observation of events linked to heritage places, the attention is directed to the human figures, the other people involved.

To obtain better insight into why specific places are viewed as being significant and to determine one's and others' relationship to those places, in step 2: *Spatial Narratives*, the same routes are re-walked again. Some of the patterns of the jot booklets that use walking and sketching as a chronological *reading* and *writing* activity of the cultural landscape are retained. However, it surpasses the first step in that it trains the sensorial memorisation in a more conscious way as a means of exploring the changing cultural (urban) landscape and its behavioural impact.

The cultural landscape is now re-imagined. Unexpected insights and views are expressed from memory in distorted two-dimensional and three-dimensional representations. The artistic maps and models alternate between large and small scale and bird's eye and worm's eye views. They function as reflective tools for local residents to test and discuss the experience-based data, memories and imaginings collected while walking by the practitioner. The interactions and discussions through aesthetic engagement, involving 'active participation in the appreciative process, sometimes by overt physical action but always by creative perceptual involvement' (Berleant, 2013), shed light on how local people experience their living environment when constructing their own memory overview.

However, when wanting to communicate to outsiders, people who do not know the heritage sites, it is essential to transcend the individual experiences. Therefore, in step 3, *Cartes Parlantes*, they are combined and confronted with existing or newly recorded data whereby commonly accepted maps are recast as a series of re-cartographies. The results can be purely graphic, as well as physical models eventually supported by written notes.

Instead of being a 'neutral' chart, which only maps physical features, generally simplifying reality by slicing and cutting what naturally belongs together, they include the relationships between different historical layers, scales, topography, geography, ecology, temporal elements in the (urban) landscape and spatial configurations (Figure 34.1). But, most of all, they bring to light the (social) fabric with the people's everyday human practices and their interactions with heritage places. They open one up to the *implicit,* the invisible, the intangible and the empirically unverifiable whereby specific fields of interest are highlighted. Due to the focus on relational issues as crucial factors in the production of inherited space, the *Cartes Parlantes* are explicitly narrative. They have the agency to 'discover new worlds within the past and present; they inaugurate new grounds upon the hidden traces of a living context and actualize those potentials' (Corner, 1999, p. 213 as cited by Cattoor, 2014, p. 51).

The social potential of interactive walking

In that sense, the creative work of sketching the observations and experiences in *Interactive Journeys*, creating *Spatial Narratives* and composing *Cartes Parlantes* intensifies and reorients the practitioner's understanding of the heritage place. However, the methodology is practised 'less for its indexical, purely analytical or representational qualities, than for its *meaning generating* qualities and its ability to *shift our perception*' (Cattoor, 2014, p. 42). The significance of heritage cannot easily or clearly be defined, but the consecutive steps of *Interactive Walking* constantly increase in intensity.

This intensity and slowness, which, at first, is experienced as an inconvenience of the survey strategy, becomes an asset. It brings the practitioner into a liminal state, a state of *possibility* whereby his/her limited thoughts about the place are transcended. Similar to the way in which a writer does not know everything upfront about his story, the joy of discovery arises from the survey itself.

While walking and chronologically sketching the observations in a jot booklet, places are enriched with written notes, memories of other places, points of rest and encounters and inspiring perspectives and sceneries. However, they can also elucidate diversion, parallel ways and dead-end alleys, as would subplots and scenarios in a novel. In the beginning, what is plot and what is subplot, is always unexpected. This only becomes clear at the end of the survey. The practitioner might have had a sense of where the survey is going, but nothing comes alive until s/he walks and starts entering the lines on the white sheet of paper. This is one of the basic modes of walking, walking in search of something, the nature of which is totally unknown before the walk begins. Thus, the heritage practitioner starts having revelations and sees things that would never have been seen at the level of a predefined *synopsis*, as if s/he was on *pilgrimage*. By using a context-related narrative approach, curiosity is triggered, and the focus shifts towards thinking through walking and drawing rather than attempting to directly resolve or structure a pre-defined problem. As Rebecca Solnit (2000) puts it, "Moving on foot seems to make it easier to move in time, the mind wanders from plans to recollections to observations" (p. 5).

A rambling field role for the heritage practitioner 517

Figure 34.1 Spatial Narratives – Entering the Memory Palace – Noeveren – Wherever the feet go the eyes will follow soon. Clay model representing the experienced space of a former claypit as part of the cuesta in Noeveren along the river Rupel.
Source: Maria Carolina Lucas, Master Thesis – Inhabiting water, p. 25. Student International Master, Faculty of Architecture, KU Leuven – Academic Year 2018–2019.[3]

The newly developed method of *Interactive Walking* discloses new ways of observing by making the practitioner see things differently and more in detail than s/he would if s/he was only taking pictures or using cartographical techniques from the start of the survey. This is because 'a walker does not skip over much, sees things close up, and makes her vulnerable and accessible to local people and places' (Solnit, 2000, p. 128).

The students in architecture witnessed that the method of walking and drawing in jot booklets encouraged them to be more attentive to what they experienced and to observe more carefully. *Spatial Narratives* enhanced their ability to physically, mentally and visually engage with the cultural landscape, of which built heritage is an integrated part. When the students delved deeper into their specific field of interest by bringing together existing expert knowledge and their new findings based on their journeys and the journeys, stories and lives of others, in *Cartes Parlantes*, they were better able to identify possible project proposals.

Additionally, in a time in which selfies and snapshots are commonplace, some passers-by are drawn to the rare image of a stranger who sketches their living environment. The additional fact that this happens in a tiny booklet evokes their curiosity and invites them to take a closer look. This moment of intimacy increases the local people's eagerness to spontaneously tell stories about their lives linked to the heritage place. The journeys then become interactive in the sense that while sharing a meaningful experience with the people encountered on the walk, the researcher is being open to the potential changes that this may bring about. The jot booklet – at first an imaginary representation of the observer's lived reality – turns into an important communication tool that stimulates unprompted encounters. Although the "respondents", the people that are encountered while walking, are not explicitly requested to tell stories, the circumstances in which they feel comfortable talking about their value system, concerns, motivations, culture and the activities they hold are created.

It is specifically the idea of going out for a walk to have encounters with local people, anonymous "characters eager to tell their tale" (Solnit, 2000, p. 129) that turns *Interactive Walking* into a research method with societal potential. It can be considered as the union of a scenic walk for investigating the cultural landscape and a serendipitous stroll for the sake of occasionally meeting local people and having unplanned interactions with them. Conducting informal conversations as open interviews oftentimes proves to be more adequate to obtain deeper insight into the investigated heritage area than directly asking questions and this for two important reasons: Firstly, the people involved are often extremely suspicious about the practitioners' intentions because they relate them to the planning of new building developments; thus, they will not always be totally open when engaged in conversations. Secondly, when directly asked for it, it is not always evident for people to recognise the qualities inherent in their daily life routine; the fragile significance of it is often taken for granted.

In that sense, the methodology of *Interactive Walking* has many of the characteristics of narrative research and ethnographic fieldwork, now combined with the skills of the heritage practitioner such as the ability to draw by hand and to work with models. By framing the investigations in terms of narratives, it is possible to see different, and sometimes contradictory, layers of meaning, to bring them into useful dialogue with each other and to understand more about individual and social change. Thereby, the interest does not lie so much in the individual stories of the participants, although sometimes, they are collected individually; rather it lies in the 'network of data of narrative meaning' (Tamboukou, 2013, pp. 88–108). What they transfer is more a transcript of a bigger story – *multiple stories*. Consequently, when wanting to detect different forms of the deep significance of heritage places through narratives, there will be *multiple realities* or *multiple interpretations* – not just one. Different scenarios can be highlighted, and alternatives can be put forward. *Interactive Walking* is, therefore, "both a means and an end – to compose and to be" (Solnit, 2000, p.107).

Entering the other's memory palace

The strategy of both strolling around in order to collect and represent narratives, and the shape in which this happens, largely borrows from the mnemonic technique of constructing a *memory palace* as an intense visualisation technique, masterfully described by Frances Yates in *The Art of Memory* (2014/1966). The chronological structure of the jot booklets constitutes a continuous *chain* of representation of heritage places, discovered on the walking track. However, the thumbnail sketches are not only used to memorise a place (creating new memories) or relate to what was known before (associating).

If one wants to remember the encounter with a local actor and his activities in relationship with the heritage places, this entails imprinting a representation of his image enriched with accessories on definite *loci*, just as a writer, would (Yates, 2014/1966, p. 27). These images –as *imagines agentes* – which are "active, sharply denned, unusual, and which have the power of speedily encountering and penetrating the psyche" (Yates, 2014/1966, p. 17) are an aid to train the acute visual memory most practitioners in architecture already have. Repeated practice of the method intensifies the observational and drawing skills. However, as Dewey stated, it is important to carefully select "[...] the kind of present experiences that live fruitfully and creatively in subsequent experiences..." (Dewey, 1938/1963, pp. 15–16).

By re-thinking the heritage space, it is now possible to recall a related physical act, such as coming across, looking for shelter from rain and a place to sit, peering at the landscape with the view framed by a historical cross window, for example, thinking and associating with the place. Visiting, capturing and graphically recording the heritage places that people talk about off-the-record (explicit) or that they portray in their actions or make visual through the traces they leave (implicit), helps to gain a better understanding of how these places are rooted in the daily life of people. This enables the heritage practitioner to detect whether the studied heritage site still generates a cultural and social identity (Cohen, 1985, pp. 20–98) and still holds meaning for local communities.

By *re-reading* the cultural landscape through *Spatial Narratives* and *Cartes Parlantes,* a practitioner analyses how heritage structures (and other landscape objects) transform into images, sensations, dreams and emotions in an attempt to capture the intimacy and complexity of the links that unite people to their everyday environment. In fact, it is copying the way local people traverse their familiar environment through recourse to a mental map established through trial and error, by experience and by reading the signposts displayed in the environment, especially the built heritage structures with their archetypical forms, that make them easy to remember. This technique also refers back to times when there were no maps to guide ordinary people. "You made your way by sight, by memory, by history, by advice, by direction – and by luck" (Defoe & Wall, 2003, introduction).

A heritage site is not only important within the spatial fabric as a relic for its historical understanding or experience, in terms of the community's daily use and familiarity with the site. It can serve as a threshold of a collective memory of shared ideas of the past, but which are constantly renewed in the present with an opening towards the future. This process is supported by a myriad of personal memories, experiences and wishes. By interacting and associating with the site, individuals attribute different meanings or use to it through a slow but complex process of appropriation. People pass knowledge about the place down to each other, from one generation to another, preserving some characteristics of each memory revealed in the stories they tell and retell.

Sometimes only traces of this remain, but even if no mark is left, people remember what happened as if they *see* it inscribed in their collective memory. For example, for some residents, the landscapes of the past – although now invisible – still occupy the role of signposts over the present and real landscapes. At that moment, it is as though their thoughts and stories were fixed objects in the cultural landscape. In that sense, exploring the land by walking is the ideal way to share the experiences local actors have with "their" heritage sites in an attempt to enter their *Memory Palace*.

However, for the practitioner, as for any newcomer, the simple sight of a new place can trigger memories of other places. Through association, s/he can develop new narratives for the place. Another advantage of being an uninformed observer in the studied heritage landscape lies in the (re)-discovery of the qualities of the usual things that historical studies might neglect and that even natives, risk overlooking because they are an integral part of their daily routine.

In reverse, in the different case studies, it was experienced that when presenting the tactile drawings and models of step 2: *Spatial Narratives*, representing the lived environment of which heritage is an integrated part in exhibitions and debates on location, they appeal to the local actors and stimulate them to re-visit their own mental overview of the place and share their memories in new narratives. This awakens their awareness of the elements of the heritage landscapes, long-since forgotten, and their associations with their personal experiences. This helps them to formulate what they could not state in words before. They feel involved as these results are reformulated and reworked after their feedback is implanted in the subsequent research step of *Cartes Parlantes* and the further project development. The proposed method provides the people with an instrument that is profoundly documented in catalogues and reports and that not only shows the eventual conflicts of existing planning proposals but also alternatives based on grounded arguments so that they can communicate and negotiate with decision-makers about what is important to them. Moreover, policymakers have a well-founded basis for outlining detailed policy guidelines for example for new management plans.[4]

Heritage as processes of growth

The *Interactive Walking* methodology transcends the standardised ways of working when analysing a site in which objects are classified within the conventional cartography, where attention is nearly exclusively oriented towards the visually *explicit*, the physical matter in the form of spatial objects, such as heritage structures. This consists of masses and territories with enclosures or boundaries that determine the (urban) landscape. The in-between spaces appear to be empty and fixed. However, on foot, one travels the whole landscape, rather than just separate heritage sites. When walking, one's attention is not focused on the boundary lines of parcels that determine ownership, dividing the land into pieces; it is focused on the inner streets, paths and tracks that interconnect and to which locals retain rights (Solnit, 2000, p. 162).

In doing so, it becomes clear that, in reality, the territory of a historical building does not necessarily have clear boundaries. Rather, historical sites can be considered to be receptacles for people and events, endlessly moved, exchanged, replaced and forgotten. The cultural landscape might be viewed through the lens of the economy, families, religion or politics. Ultimately, all these aspects are intertwined and constitute a single reality. Thus, a subdivision of parcels historically fades into the background of authentic human experiences because these are omnipresent (Oles, 2008, p. 59). Moreover, even if physical walls or hedges sometimes surround these heritage places, mental barriers do not impede them. They became collective by their use, local significance or representation. By transcending an object-focused approach, it is possible to detect formal and informal uses of the place, which turn the existing historical mesh into a strong social fabric, a place of attachment, thereby facilitating fortuitous encounters.

Unfortunately, new development often neglects this fragile fabric and reframes the collectively used space by explicitly delimiting the private plots due to different landscaping or by defining the boundaries of their private territory using hedges and fences, obstructing an ancient right-of-way which grew over time. Interrupting the daily trajectories of people also means distorting their memories, lives and stories which alienates them from their living environment.

In reverse, exploring the periphery of the heritage place makes room for true interest in the lives and experiences of others. This more dynamic perspective on heritage highlights how heritage comes to be what it is by engaging with the experiences of those whose daily life is entangled with it as people's basic experiences with built heritage are active, not passive. The interactions and associations with heritage sites can, therefore, be defined by the verb-essence of the (architectural/spatial) experience (Pallasmaa, 2012, p. 68): attach, remember, refer, meet, enjoy, reject and hate. These verbal place specifications or connotations are transcultural and timeless. They may induce empathy or even repulsion, but a wandering practitioner can easily identify with these human activities as "Both spatially and socially, walking has the widest span; at the same time, it addresses daily life with all its universally recognisable human characteristics in a more in-depth way" (Solnit, 2000, pp. 97–101).

Once the ability is developed to fully experience with all one's senses *from inside*, it becomes possible to engage in the fascinating dialogue with the heritage of both the local resident and the newcomer, and the focus on heritage as an artefact, a finished *project or thing* can be transcended. As Ingold (2013, p. 12) notes, by bringing them back from their *passive* to their *active* materiality in a socially sustainable way, "we (can) rescue them from the cul-de-sac into which they had been cast and restore them to the currents of life". From that moment on, heritage places can be considered to be *processes of growth*.

Conclusions

Although additional research is needed to fine-tune some aspects of the proposed method, the different case studies evinced that *Interactive Walking* has undoubtedly important societal potential. The creative work of chronically sketching perceptions through mapping *from the ground* in *Interactive Journeys* not only enables memorising the place but also deepens the knowledge about it through unexpected encounters.

Spatial Narratives add a processual or narrative element – a story of how values happen for others, to what they are related and why and how they change. Recasting the observations and experiences in *Cartes Parlantes* makes things clearer and deepens the practitioner's understanding of the interactions between the different actors and the existing heritage landscape as a social narrative in relation to the (changing) identity of space.

Interactive Walking can be equated to sensemaking, which as Vygotsky and Kozulin (1934/1986, p. 86) refers to becoming familiar with what is unfamiliar. In that sense, new research avenues could be explored more specifically on the interaction with heritage and how this impacts the way people experience their environment. More research is needed on the deeper effects of associating with heritage places for newcomers, for example for refugees, whose journeys, lives and narratives have been interrupted dramatically. Could the method of *Interactive Walking* enhance the possibility of picking up the thread of one's life?

Heritage, and its links to the past, can of course spark moments of longing and remembrance, referring back to an idealised past that one might remember or to a past that one might never have experienced but still feels drawn to. Within the method of *Interactive Walking,* heritage is not just fixed in the past, but it is a means to trace an arch from past to present. It facilitates the detection and implementation of multiple narratives, old and new, registered through encounters with different people, without distinction. It embodies the successive movements, memories, experiences and dreams that are related to it, which highlight the multi-layered sense of the studied heritage place.

The essential concept of the newly developed methodology of *Interactive Walking* is that the initial problem statement is approached indirectly while following a non-linear path characterised by serendipity, meaning the discovery of an end that was not sought. Consequently, possibilities for redevelopment arise from the survey itself.

In the case of pseudo-serendipity, the heritage site and the community involved are forced to accept the new functions with predefined programs and plans, *like a rabbit pushed into a magic hat* (whether it fits or not). In that case, heritage development entails the maintenance of a rather superficial visual presence, which people, who deal with these heritage places on a regular basis, experience as something very invasive.

Spatial planning of a heritage site must of course go hand- in-hand with ecological issues, such as water and green management, and its related eco-systems and biotopes. However, from the perspective of societal sustainability today, the future of heritage and the will to take care of it is directly related to people's attachment to and/or interaction with it. After all, how long a historical building or site lasts depends on the value that generation after generation attributes to it. Therefore, it is fundamental to ensure that a heritage site remains in its local societal strata; otherwise, the restoration project will not be socially accepted.

The rambling field role is not only an integral part of the proposed method; it can also provide an answer to the multiplicity of roles of the heritage practitioner. It enables him/her to take different positions, distant or involved, without ceding his/her decision-making authority and expertise concerning the restoration, conservation or reuse of the heritage site while, simultaneously, engaging with the local actors, both known and undisclosed protagonists and members of multiple communities. The strategy reframes the idea of participation because it shifts the perspective from the conventional top-down approach to a more community-based approach. The argument that the essential features of the method, which are time and slowness, might represent a breakpoint in the applicability for architectural firms due to efficiency/economic reasons is countered by the fact that time is gained once the technical elaboration of the project begins as later protests against management plans, which often slow down further development of the project are prevented. However, a detailed examination is needed to calculate the long-term economic, societal and cultural benefits of the method in planning processes.

The new insights gained through the application of this methodology of routes, observations and stories can provide an answer to the question of how heritage sites can be continued in a socially sustainable way. It is, therefore, also future-oriented and becomes an inherent part of the planning process. Thus, *Interactive Walking* is a very effective research-by-design methodology with a high probability of success to arrive at more socially sustainable (re-) development plans for heritage sites in which local people are more explicitly represented.

Endnotes

1 This research is based on 35 years of experience as an architect specialised in the restoration of community heritage situated in peri-urban and post-industrial areas. It was only in 2012 that I decided to search for a way to transfer the knowledge I gathered on indirect participatory practices through the development of the novel three-step survey methodology of Interactive Walking to inject it into practice, practice based research and education at the Faculty and Department of Architecture KU Leuven in Belgium (where I teach in the design studios since 1991) (http://architectuur.kuleuven.be) and since 2020 also at the Raymond Lemaire International Centre for Conservation (RLICC), Master of Science in Conservation of Monuments and Sites (http://set.kuleuven.be/rlicc).
2 ACHS - https://www.criticalheritagestudies.org/history
3 The illustration is extracted from the work of Maria Carolina Lucas, a student of the master studio at the faculty of Architecture at KU Leuven of which the author was the academic promoter (International Master in the Architecture program at KU Leuven, Belgium. The academic year 2018–2019).

   In each of the case studies, the participating students offered their informed consent. The studio assignments derived from the cases are part of the classroom discussions and all students choose to participate. The students were not remunerated for their participation. They received an evaluation on the results of their architectural projects but not on the survey part, which was based on the developed method and served as the study material for the research in Gantois, G., 2019. *Tracing the deep significance of built heritage through the encounter of undisclosed protagonists.* Belgium: KU Leuven, unpublished.
4 This was assessed with exhibitions on location, different debates, a two-volume book drawn from the students' individual reflection papers and a "vision note" (Gantois, 2018) which summarized all the gathered knowledge both collected together with the students and from the participation project I See I See What You Can't See.

   The first exhibition took place in 2015, organized by the Province of Antwerp. This exhibition showed the first investigations in the Rupel region of 8 students in the framework of the professional seminar *Van Put tot Ieders Nut - Natuurlijke structuren en ruimtelijke ontwikkelingen in ontginningslandschappen en industriële sites. [From (clay)pit to a Collective Use - Natural structures and spatial developments in mining landscapes and industrial sites].*

   The second exhibition was organized on location in Noeveren (Boom) in a former machine hall of a brick manufactory and was facilitated by the local action group 'Noeveren Bezorgd' [Noeveren Worried] and was entitled *The Cuesta of the Rupel Region – New Challenges for its Cultural Heritage*. It presented the Interactive Journeys and Spatial Narratives of 8 master students to the local residents and local politicians. In 2016, a third exhibition was organized whereby the complete work of the students including *Cartes Parlantes* and project proposals were confronted with the work of artists, which increased the aesthetic engagement of the local residents. The exhibition was called *Noeveren Brandt [Noeveren burns]*.

   In 2017 a lecture and debate were organized by the local action group Noeveren Bezorgd in which local actors were invited to participate. I was the main speaker. Members of the local action group and local politicians were members of the discussion panel. The main theme of this evening was *De Cuesta van de Rupelstreek. Een fragiel? cultuurlandschap. [The Cuesta of the Rupel region. A fragile? Cultural landscape.]*

   In 2018, a participation project was launched in Noeveren in which local inhabitants were invited to walk and note a memory, a current event and a wish for the future on different locations on their walk.

   This participation project was called *Ik Zie Ik Zie wat Jij niet Ziet [I See I See What You Can't See]* followed by a political debate. The whole project was facilitated by the local action group Noeveren Bezorgd. The whole trajectory was published first in a two-volume book *The Cuesta of the Rupel Region. New challenges for its Cultural Heritage (VOL. I &*

VOL II) (Gantois et al., 2015) funded by the KU Leuven http://issuu.com/rupel/docs/cuesta_volume_1/1; http://issuu.com/rupel/docs/cuesta_volume_2/1 and the report *Ik Zie Ik Zie Wat Jij Niet Ziet. Visienota voor de wijk Noeveren* [*I See I See What You Can't See*, Vision note for the hamlet of Noeveren] (Gantois, 2018) funded by local action group Noeveren Bezorgd. Today, the experiences and knowledge gained from the case study of the post-industrial landscape of the Rupel region (Noeveren-Boom) form the basis for drawing up the heritage management plan and the residential environment plan for the hamlet of Noeveren, for which the municipality of Boom has appointed me as supporting expert.

References

Araoz, G. F. (2011). Preserving heritage places under a new paradigm. *Journal of Cultural Heritage Management and Sustainable Development, 1*(1), 55–60. https://doi.org/10.1108/20441261111129933.

Berger, J. (1972). *Ways of seeing*. London: British Broadcasting Corporation and Penguin Books.

Berleant, A. (2013). What is aesthetic engagement? *Contemporary Aesthetics, 11*(5), Retrieved from: https://digitalcommons.risd.edu/liberalarts_contempaesthetics/vol11/iss1/5?utm_source=digitalcommons.risd.edu%2Fliberalarts_contempaesthetics%2Fvol11%2Fiss1%2F5&utm_medium=PDF&utm_campaign=PDFCoverPages (Accessed on 14 May 2021).

Brundtland, G. (1987). *Report of the world commission on environment and development: Our common future*. United Nations General Assembly document A/42/427.

Careri, F. (2002). *Walkscapes: Walking as an aesthetic practice*. Ames, IA: Culicidae Architectural Press.

Caserta, S., & Russo, A. P. (2002). More means worse: Asymmetric information, spatial displacement and sustainable heritage tourism. *Journal of Cultural Economics 26*(4), 245–260. https://doi.org/10.1023/A:1019905923457.

Cattoor, B. (2014). *Atlas as design, designing atlases. Two cartographic explorations of implicit urbanisms in Southwest Flanders* (Unpublished PhD thesis). Leuven: KU Leuven.

Cohen, A. (1985). *The symbolic construction of community*. London: Routledge.

Corner, J. (1999). The agency of mapping: Speculation, Critique and Invention. *Mappings*, London: Reaction Books, 213–252.

Creswell, J. (2007). *Qualitative Inquiry and research design: Choosing among five approaches*. Thousand Oaks, CA: SAGE Publications, 59.

Defoe, D. (Author), & Wall, C. (Ed.) (2003). *A journal of the plague year*. London: Penguin Classics.

Dewey, J. (1938/1963). *Experience and education*. New York: Kappa Delta.

Emelianoff, C., & Carballo, C. (2002). La liquidation du patrimoine, ou la rentabilité du temps qui passe. *Les Annales de la Recherche Urbaine, 92*(1), 49–57.

Europa Nostra. (2015). *Cultural heritage counts for Europe report*. Retrieved from: https://www.europanostra.org/our-work/policy/cultural-heritage-counts-europe/ (Accessed on 13 December 2020).

Faro Convention. (2005). *Framework convention on the value of cultural heritage for society,* Council of Europe, 6. Retrieved from: https://www.coe.int/en/web/culture-and-heritage/faro-convention (Accessed on 13 December 2020).

Fouseki, K., Swensen, G., & Guttormsen, T. S. (2020). Heritage and sustainable urban transformations – A 'deep cities' approach. In K. Fouseki, S. Guttormsen, G. Swensen (Eds.), *Heritage and sustainable urban transformations deep cities* (pp. 1–15). Oxon and New York: Routledge.

Gantois, G. (2018). *Visienota omtrent de wijk Noeveren* (Unpublished report). Asse.

Gantois, G. (2019). *Tracing the deep significance of built heritage through encounters with undisclosed protagonists* (Unpublished PhD thesis). Leuven: KU Leuven.

Gantois, G. (2020). Small-scale heritage: The canary in the coal mine. In K. Fouseki, S. Guttormsen, & G. Swensen (Eds.), *Heritage and sustainable urban transformations, deep cities* (pp. 166–184). London: Routledge.

Gantois, G., Schoonjans, Y., & Scheerlinck, K. (2015). *The Cuesta of the Rupel Region, New challenges for its Cultural Heritage*. Boom/Gent – Belgium: KU Leuven. Retrieved from: http://issuu.com/rupel/docs/cuesta_volume_1/1; http://issuu.com/rupel/docs/cuesta_volume_2/1 (Accessed on 21 September 2016).

Gracia, J. J. E. (2003). *Old wine in new skins, the role of tradition in communication, knowledge, and group identity*. Milwaukee: Marquette University Press.

Gravari-Barbas, M. (2014). *New challenges for cultural heritage: Synthesis of the final report*. France: Université Paris, Agence Nationale de la Recherche.

Hartog, F. (2015). *Regimes of historicity*. New York: Columbia.

Harvey, D. (2012). *Rebel cities: From the right to the city to the urban revolution*. London-New York: Verso, 67–73. Retrieved from: http://www.abahlali.org/files/Harvey_Rebel_cities.pdf (Accessed on 21 January 2019).

ICOMOS (2013). *Burra charter, charter for the conservation of places of cultural significance*. Australia. Retrieved from: https://australia.icomos.org/wp-content/uploads/The-Burra-Charter-2013-Adopted-31.10.2013.pdf (Accessed on 21 January 2019).

Ingold, T. (2007). Up, across and along. *Lines: A brief history*. London: Routledge.

Ingold, T. (2013). *Making anthropology, archaeology, art and architecture*. London and New York: Routledge.

Ingold, T., & Lee Vergunst, J. (2008). Introduction. In T., Ingold, J., Lee Vergunst (Eds.), *Ways of walking: Ethnography and practice on foot* (pp. 1–19). London and New York: Routledge.

Lucas, M. C. (2019). *Inhabiting water. The natural elements and their involvement within the cultural heritage landscape. Case Study Noeveren*. (unpublished paper). Ghent: KU Leuven.

Mason, R. (2008). Be interested and beware: Joining economic valuation and heritage conservation. *International Journal of Heritage Studies*, *14*(4), 303–318. https://doi.org/10.1080/13527250802155810.

Michaud, P. (2007). *Aby Warburg and the image in motion*. New York: Zone Books.

National Trust. (2017). *Places that make us*. Retrieved from: https://nt.global.ssl.fastly.net/documents/places-that-make-us-research-report.pdf (Accessed on 15 April 2019).

National Trust. (2019). *Why places matter to people*. Retrieved from: https://nt.global.ssl.fastly.net/documents/places-matter-research-report.pdf (Accessed on 15 April 2019).

Oles, B. T. (2008). *Recovering the wall: Enclosure, ethics and the American landscape* (Unpublished PhD thesis). Massachusetts: Massachusetts Institute of Technology. Retrieved from: http://hdl.handle.net/1721.1/45437 (Accessed on 20 September 2018).

Pallasmaa, J. (2012). *The eyes of the skin. Architecture and the senses*. West Sussex: John Wiley & Sons Ltd.

Petzet, M. (2009). *International principles of preservation. Monuments and Sites series, Volume XX*. Berlin: Bäßler.

Pierce, J., & Lawhon, M. (2015). Walking as a method: Towards methodological forthrightness and comparability in urban geographical research. *The Professional Geographer*, *67*(4), 1–8. https://doi.org/10.1080/00330124.2015.1059401.

Porter, L., & Shaw, K. (2013). *Whose urban renaissance? An international comparison of urban regeneration strategies*. London: Routledge.

Sarkki, S., Ficko, A., Miller, D., Barlagne, C., Melnykovych, M., Mikko, J., Soloviy, I., & Nijnik, M. (2019). Human values as catalysts and consequences of social innovations. *Forest Policy and Economics*, *104*, 33–44. https://doi.org/10.1016/j.forpol.2019.03.006.

Silberman, N. A. (2014). Changing visions of heritage value. What role should the experts play? *Patrimoine Culturel Immateriel*. *36*(1), 433–445.

Solnit, R. (2000). *Wanderlust: A history of walking*. New York: Penquin Books.

Solnit, R. (2005). *A field guide to getting lost*. New York: Penquin Books.

Springgay, S., & Truman, S. E. (2019). *Walking methodologies in a more-than-human world: WalkingLab*. London: Routledge.

Tamboukou, M. (2013). A Foucauldian approach to narratives. In M. Andrews, C. Squire, & M. Tamboukou (Eds.), *Doing narrative research* (pp. 88–108). 2nd ed. Los Angeles, London, New Delhi, Singapore and Washington, DC: SAGE.

UNESCO. (1972). *World heritage convention*. Paris. Retrieved from: https://whc.unesco.org/en/conventiontext/ (Accessed on 15 March 2019).

UNESCO. (1992). *The Rio declaration on environment and development*. Rio de Janeiro: United Nations. Retrieved from: https://www.un.org/en/development/desa/population/migration/generalassembly/docs/globalcompact/A_CONF.151_26_Vol.I_Declaration.pdf (Accessed on 15 March 2019).

Vygotsky, L., & Kozulin, A. (1934/1986). *Thought and language*. Translation newly revised and edited by Kozulin Alex. Cambridge, MA: MIT Press, 86.

Waterton, E., & Smith, L. (2010). The recognition and misrecognition of community heritage. *International Journal of Heritage Studies*, 16(1–2). https://doi.org/10.1080/13527250903441671 (Accessed on 11 July 2019).

Waterton, E., & Watson, S. (Eds.) (2011). *Heritage and community development: Collaboration or contestation?* London: Routledge.

Yates, F. A. (2014/1966). *The Art of memory*. London: The Bodley Head. Retrieved from: http://www.alzhup.com/Reta/Docs/ArtOfMemory.pdf (Accessed on 17 October 2019).

Zuccoli, F. (2019). Intensifying our gaze in order to expand our action, reflection, and participation. *IMGJOURNAL*, 1(1), 354–365. https://doi.org/10.6092/issn.2724-2463/11081.

35 Teaching futures literacy for the heritage sector

Cornelius Holtorf

Introduction: acknowledging cultural diversity in time

The United Nation's Agenda 2030 has put planning and sustainable development on the agenda even of the realms of culture and heritage. That was not self-evident. Human culture and cultural heritage reflect in parts a history of expansion and exploitation, hardly able to illustrate what we today generally mean by sustainable development. But today, sustainability is nevertheless significant insofar as it is desirable that a variety of societal benefits of culture and heritage are sustained over long time periods. These benefits relate partly to the general ability of heritage to illustrate the significance of change in human history, which is a subject I discussed elsewhere (Holtorf, 2018). The benefits of heritage also relate to specific social and cultural values and uses of heritage that are highly context-specific and subject to change over time (De la Torre, 2013), which is addressed in the present paper. Taking both kinds of benefits together, the most important question of conservation is not how much heritage of any one period may or may not survive intact into the future but what story and (future) legacy of the past, as manifested in heritage, will come to benefit future generations the most.

Addressing the future benefits of heritage requires an important change in perspective. The 1994 Nara Document on Authenticity (ICOMOS, 1994) emphasized the significance of cultural diversity in appreciating heritage values and authenticity in different contexts. As the Document (in Article 6) alludes to, we must realize and understand the implications of the insight that this diversity applies not only spatially within the contemporary world but also temporally across different periods. The history of heritage thought over the past two and a half centuries illustrates how its values and indeed benefits have been changing. Originally growing out of Romantic Nationalism, the idea and notion of heritage have subsequently been linked to various educational ideals, with economic growth based on the "heritage industry" and cultural tourism growing in importance in recent decades and approaches to well-being through heritage emerging now (Lowenthal, 2015). In other words, the concept of heritage has been used and valued differently over the past few centuries. Heritage has also been managed differently over time in relation to changing perceptions of its significance and benefits in society (Table 35.1). Understanding and managing heritage thus need to acknowledge and accommodate cultural diversity, not only depending on *where* you are located in space but also *when* you are located in time.

Given this mutability of the values and benefits of heritage in society, we cannot assume that contemporary appreciations of heritage are inherent or timeless and will be shared by future generations. Indeed, heritage is not going to be of much use to future

Table 35.1 A necessarily flawed sketch of some changing practices and perceptions of the values and uses of heritage in European society over the past few centuries. The point is not accuracy in the details but to illustrate change over time

Heritage dimension	Pre-1700?	18th century	1st half 19th century	2nd half 19th century	1st half 20th century	2nd half 20th century	1st half 21st century
Identity	–	Folk culture	National identity			Cultural diversity	Identity politics
Learning	Curiosity		Private study	Bildung	Formal teaching		Storytelling
Visiting		Chance	Grand tour		Cultural tourism	Mass tourism	Overtourism
Quality of life	Re-use	Exploration	Discovery		Consumption		Engagement & well-being
Management focus	–	Fabric				Values	Uses
Protection	–		Describing	Listing	Legal preservation		Managing change

generations at all if it does not provide benefits to them. In that case, it would become an "unwanted gift" from the past that may be discarded swiftly (May 2021). That is the fundamental reason why there is a need for the heritage sector to engage systematically with the needs of future generations. In the words of Paolo Ceccarelli (2017, p. 6), "to correctly conserve the past we need to anticipate the future". The challenge of sustaining heritage over long time periods is closely linked to the challenge of providing credible benefits of heritage to future societies. Curiously, this challenge has until recently not been given much attention (but see e. g. Cameron, 2010; Hodder, 2010; Zetterström-Sharp, 2015; Ceccarelli, 2017; Merritt 2021; Högberg & Holtorf, 2021; Holtorf & Högberg, 2021a; Heritage 2022).

Arguably, the values and benefits of heritage in future society fall into the three different categories of anticipatory action discussed by Ben Anderson (2010). They may help preserve a valued way of life through *precautionary* prevention of what is known to threaten it, e. g. by supporting strategies to mitigate climate change. They may modify ways of life by *pre-empting* possible threats as they emerge, e. g. by extensively promoting human rights in the light of increasing cultural, economic and social tensions in the world. Or they may help *prepare* for a valued way of life under different circumstances, e. g. by enhancing cultural resilience. Identifying and working for alternative futures is to create the chance for better future lives than people may otherwise have had. These three anticipatory strategies concern the roles of heritage in managing the relations between present and future societies and they operate, therefore, in the domain of heritage futures, as defined by Holtorf and Högberg (2021a, p. 2).

The present chapter will discuss futures thinking in the heritage sector and how it can be taught to professionals. After a short exploration of how much we can actually know about the future, I will introduce the concept of "futures literacy" and discuss its significance for the heritage sector. To date, as I will show, explicit considerations of the future and future benefits of heritage are rare. I am, therefore, going on to argue that there is a need for more capacity-building in futures literacy for the heritage sector.

After reviewing a number of initiatives that have taken place to date, I will present promising training methods of teaching futures literacy to heritage professionals. In my conclusion, I will present important lessons for improving futures literacy in the heritage sector and thus helping ensure that benefits of heritage in society, although subject to change, can be sustained over long time periods.

What can we know about the future?

The future will not be like the present, and the longer we look ahead the more it will differ from the world we know today. Having said that, we do not know exactly and in detail how the future will differ from the present: it is inherently uncertain. Uncertainty means that the future is not entirely predetermined, and it can, therefore, not be predicted with any certainty. But the uncertainty of the future also means that human actions in the present matter more because they can influence the future (Harrison et al., 2020b, Part IV).

Although details about the future are unpredictable, futures are not formed entirely by chance. At a large enough scale, there are a number of discernible trends which we can try to understand. Paul Graves-Brown (2020) discussed this recently with reference to a text by H. G. Wells who made the following observation about piles of sand:

> Angular pit-sand has grains of the most varied shapes. Examined microscopically, you will find all sorts of angles and outlines and variations. Before you look you can say of no particular grain what its outline will be. And if you shoot a load of such sand from a cart you cannot foretell with any certainty where any particular grain will be in the heap that you make; but you can tell—you can tell pretty definitely—the form of the heap as a whole.

In other words, although the location and shape of each sand grain remain unpredictable, the size and shape of the entire heap do not vary to the same extent and can be foreseen with surprising accuracy. Something similar applies to human societies. The specific appearance and behaviour of each individual in specific situations cannot be predicted but the overall number, location and behaviour of large groups of individuals may be foreseeable and is subject to planning. The same pattern that in relation to the past makes detailed reconstructions futile and many reliable generalisations look banal makes detailed forecasts impossible and general anticipations trustworthy.

The heritage sector, too, can benefit from what may be called futures literacy (Miller, 2018; UNESCO, n.d.). Futures literacy is the ability to imagine and anticipate multiple possible futures that differ from the present. Futures literacy is relational; it never refers to the future as such but always to the future of something (e.g. a given heritage site or the notion of heritage as such). Recognising the significance of general trends and various possible developments in relation to these trends, futures literacy provides all of us with an understanding and appreciation of alternatives for action that will influence what is going to happen between now and then, especially when applied at the smaller scale. Futures literacy is thus liberating people to the extent that it removes the tight shackles of "presentism": the disabling and unquestioned assumption that the status quo is timeless.

This is not the place for a detailed analysis of changing conditions and discernible trends for the future. Initially, we will need to decide though which future we mean: next year is going to be more similar to this year than ten years, a hundred years or an even longer time ahead. Most often in society, we tend to consider the time of our children and grandchildren, that is the next 30–50 years. If assuming that specific timeframe, we can, with some confidence, expect certain demographic developments relating to population sizes in various world regions, patterns of urbanisation and migration, and people's increasing life expectancies and improved health. Climate change and its impact will increasingly make their mark. We will have to face the consequences of a variety of social divisions and cultural tensions around the world. We can also assert the continued importance of globalisation and technical innovations, not the least regarding applications of artificial intelligence and ever more advanced tools for digital communication. All this means that when we recognise that conditions in the world around us keep changing, heritage and the benefits it provides as well as how these benefits may be realised in society cannot by default be expected to remain the same, and we have to realise that they may need to change as well (see also Harvey & Perry, 2015; Holtorf, 2020). Therefore, the future matters to heritage, and it should be taken more seriously not only by heritage studies but also by heritage conservation and management.

The significance of understanding the future for the heritage sector

Up until now, the future has very rarely attracted detailed discussion in order to inform heritage management (but see Sandford, 2019; Harrison et al., 2020b; Holtorf and Bolin 2022; and various contributions to Holtorf & Högberg, 2021b). It is equally rare that heritage experts have acquired capacity in futures thinking in order to make better decisions on heritage and conservation in the present. This is very surprising given the ubiquity in heritage policy and legislation around the world of the claim that we need to preserve the remains of the past and safeguard our heritage for the benefit of future generations (Högberg et al., 2017). Critical observers like Dirk Spennemann (2007) have long suspected that such claims are nothing but empty clichés.

The desire to transmit the heritage to future generations is often treated as entirely self-evident and unproblematic as the future benefits of heritage are taken for granted. For example, topical expert groups that work for risk preparedness and disaster risk reduction hardly ever mention any risks that may result from the heritage itself (Jigyasu, 2013). Queries about actual future benefits likely to result from policies of conservation are usually deemed to be unnecessary (but see Willems, 2012; Karl, 2015, 2018). As Sarah May (2021) suggests, future generations tend to be imagined as if they were small children for whom we need to create the best future conditions possible. We expect them eventually to be grateful to us, although they may not yet realise the value of the heritage for which we make sacrifices today. In fact, as we all know, future generations will be mature adults, perfectly capable of making their own decisions and judgements, often disagreeing with well-intended but poorly informed decisions of their parents' generation.

Even the emerging discourse about heritage and sustainable development hardly considers anything else than known benefits of heritage in our present, addressing the challenges and aspirations of our own time and hoping that any positive outcomes will somehow carry forward into the future (e.g., Auclair & Fairclough, 2015; Albert et al., 2017). The World Heritage and Sustainable Development Policy of the United Nations Educational, Scientific and Cultural Organization (UNESCO, 2015) calls for more attention to heritage contributing to sustainable development (Larsen & Logan, 2018). It is, however, not clear yet how exactly the implementation of this policy will lead to improved futures thinking in heritage management.

Explicit discussions of the future in the literature about heritage are usually about ensuring continuity of present provisions or hoping to realise contemporary aspirations for a better world rather than preparing for anticipated future change. There are a number of studies addressing "the future of heritage" (e.g. Silberman & Liuzza, 2007), but they indulge in varieties of "presentism", discussing merely the author's various own aspirations for the future.

On the occasion of the 50th anniversary of the US National Historic Preservation Act in 2016, the US branch of the International Council of Monuments and Sites (US ICOMOS, 2016) invited submissions of short essays on the question "What can and should US preservation law and federal programs look like for the next 50 years?". The eight submissions published on the internet describe ideas that are already working in practice and make sense to us today. Although "lessons for US Historic Preservation in its second 50 years" were promised, visions of how historic preservation might have to change in a very different world 50 years ahead are not offered and maybe they were never expected either, as suggested by the associated poster series that looks backwards rather than forward while employing the slogan "Our Legacy, Our Future" (Figure 35.1).

Even at the highest international level, a potentially changed significance of heritage in the future is seldom considered. The UNESCO *Declaration on the Responsibilities of the Present Generations Towards Future Generations* (1997) asserts the importance of protecting the needs and interests of future generations. Despite the declaration explicitly referring to heritage in Articles 7 and 8, it is little known among heritage experts, even within UNESCO. The possible significance of appropriate uses of heritage to respond to the declaration's concern for "the fate of future generations in the face of the vital challenges of the next millennium" has never been given much attention in the heritage sector.

When heritage institutions commission foresight studies, their timeframe tends to be short, their aim is to support the status quo in heritage policy, and their content is usually limited to describing what is already the case. For example, Historic England's study on *Facing the Future: Foresight and the Historic Environment* (2015) anticipates threats and opportunities for the historic environment. Many of the trends described are relevant in the long term, but all the report considers is how the sector can maintain its present ambitions in managing the historic environment under changing circumstances. There is no analysis of how future generations might relate to and be affected by heritage in different ways than we are today, requiring new ambitions for managing what today we call the historic environment.

The most informative study of this kind is the Swedish National Heritage Board's study *Towards Future Heritage Management* (RAÄ, 2006). Its aim was to determine which trends would likely have an impact on the uses of heritage in society and how managing the cultural heritage up until 2015 would have to change to meet these trends more successfully. Much of the content remains relevant for the future even at the time of writing (a decade and a half after the study was published). The report identified, among others, the following areas of significance for future cultural heritage management (RAÄ, 2006, pp. 16–19):

1   A more heterogeneous society will emerge, with a greater diversity of understandings of society and its values.
2   Cityscapes and public spaces are transforming rapidly. The countryside is increasingly used for providing experiences and renewable energy rather than food.

Figure 35.1 Will the heritage preserved from the past be a valuable legacy in the future?
Source: © Preservation50, http://preservation50.org.

3   Increasing commercial uses of the cultural heritage risk decay, over-exploitation and idealisation, while re-appreciating authenticity.
4   Experts working in the public sector need to relate better to a growing variety of widely accessible views about heritage.
5   Demand for easily available and quality-assured knowledge about the past will grow.
6   The public heritage sector will need to adapt to increased regional variation and internationalisation, additional actors in a more heterogeneous society, growing emphasis on interactive engagements, and reduced public resources.

In order to develop the capability of understanding the significance of a variety of emerging and changing trends and of dealing with the anticipated challenges arising from them, it is helpful to improve futures literacy in the heritage sector.

Training in futures literacy

Continuous capacity-building in the heritage sector is important, and there is a long tradition of training and education, even at the global level (Logan & Wijesuriya, 2016). Training in futures literacy is a new area of competence, but as I indicated, it is of particular significance for managing heritage. In the following section, I will briefly review previous initiatives of training in futures literacy in the heritage sector.

The most notable examples of building capacity in futures literacy that I have come across in the heritage sector are the following initiatives carried out since 2009 in Belgium, France and, most recently, by my colleagues and me in Sweden. All of these initiatives had a different main aim than training. They focused primarily on anticipating futures in the context of planning in relation to the heritage sector.

Between 2009 and 2011, the Flemish support centre for cultural heritage FARO ran the Prisma project which involved 12 scenario workshops conducted all across Flanders in Belgium. In these workshops, invited local heritage experts explored the prospects of the heritage sector ten years ahead, in 2020. The project did not, however, consider in detail how heritage might contribute in new ways to Flemish society in 2020. Instead, the idea was to identify the most important societal trends, discuss any resulting opportunities and threats, and formulate possible strategies for the heritage sector to exploit opportunities and keep threats at bay. The project discussed a number of previously identified issues of perceived increasing significance, including "individualization or growing sense of community", "technopolis or back to basics", "diversity or uniformity" and "more or less free time" (Caals & Van Leeuwen, 2010). Unfortunately, the workshops did not all reach the desired outcomes as many participants had difficulty in considering the future playfully and with an open mind, to let go of the present, and to imagine their own expert practice differently (Vander Stichele, pers. comm.).

A French initiative focused on future research agendas in the heritage sector. As part of a larger project supported by the European Commission's Joint Programming Initiative on Cultural Heritage and Global Change (2011–2014), several activities were organized in Paris in 2012 with the goal of developing a strategic research agenda for cultural heritage for the next 20 years. This included a two-day futures literacy scenarios workshop (Rhisiart, 2018), a literary study of relevant drivers of change (JPI, 2013a), and a Delphi Study exploring the views of almost 100 heritage experts (JPI, 2013b). The results provide a rich collection of ideas and information that may still be able to inspire and stimulate reflection about discernible trends for future benefits and risks relating to cultural heritage. For example, the workshop concluded with two interesting suggestions for specific areas to be addressed in future cultural heritage research (Rhisiart, 2018, p. 117):

- Empowerment – how can cultural heritage research support empowerment and democratisation within society?
- Co-creation – how can policy be designed in a way that genuinely uses the knowledge and capacity distributed in society?

In 2012, the European Foresight Platform, a network building program supported by the European Commission, organized in Brussels, Belgium, a one-day workshop on the future of cultural heritage with a total of 13 participants (Koers et al., 2013). The aim of this small gathering was to identify trends and drivers of change that may impact cultural heritage in Europe, in order to support strategic thinking in the heritage sector concerning the creation, management, preservation, promotion, use and funding of heritage in the coming decades. The resulting report contains an outline of several relevant trends and developments in society, technology, economy, ecology, and politics and a discussion of their potential significance and some relevant implications for cultural heritage. Among others, the participants realised the significance over the coming decades of an ageing population. They reasoned that this will not only mean that the elderly will visit heritage sites more often, but they also acknowledged that the elderly nowadays may have different preferences in culture and arts than the elderly of the future will have (Koers et al., 2013, p. 6).

In 2014, a two-day Foresight Workshop was held in Paris by the PA.TER.MONDI consortium and supported by the French National Research Agency (ANR). The event brought together 60 researchers and heritage experts to take a prospective look at cultural heritage some 15 years ahead, in order to identify future research needs. The workshop involved an exploration of five alternative scenarios for heritage and led to a list of emerging issues in need of more research, with three particular challenges concerning the need to deal better with present and future diversity, emerging non-Western approaches and a possible future absence of heritage altogether (Gravari-Barbas, 2014).

Since 2019, we have been running eight Future Workshops at the UNESCO Chair on Heritage Futures for a total of ca 100 professionals and politicians managing cultural heritage sites in and around Kalmar in Southeast Sweden. The workshops were held on invitation after news of their availability and usefulness had spread by word of mouth. The workshops extended over a half-day and involved three steps: the participants (1) documented first how they perceived the future several decades ahead, (2) identified then major challenges they anticipated and (3) finally developed strategies to address these challenges using heritage. Subsequently, more work was still needed for each participating group to identify realistic goals and practical agendas for the heritage sites concerned. The feedback we collected confirmed that the workshops succeeded in raising a new set of issues and inspiring long-term agendas in managing heritage. For example, in the light of workshops we held with a project group developing cultural heritage tourism in the Kalmar-Öland region, participants commented that "the workshop emphasized the need for longer future planning", that it is "very useful to think further ahead than we commonly do and even to think with a larger and broader perspective" and that "all felt that they ought to think more about the future in planning their work". Similarly, after our workshop with the Steering Group of Kalmar Castle, participants stated that "engaging with foresight affected my ability to take more perspectives into account when we plan our work" and that "the workshop made me open for new ideas of how we could use the castle in the future" (all my translations).

Taken together, my discussion of these initiatives illustrates that the number of examples of building capacity in futures literacy related to heritage has been small and their focus has often been mostly on consequences of anticipated futures for continuing to manage heritage along existing lines. To date, there has been little interest in appreciating the kind of changing values and uses of heritage in a society that might call for very different approaches and priorities in managing heritage for the future. Although all the various workshops aimed primarily at improving the knowledge base for planning for the heritage sector, they succeeded also in training several hundred participating heritage professionals in dimensions of futures literacy.

How to teach futures literacy for the heritage sector

Futures literacy is an emerging area of education and professional capacity-building (see, e.g., Miller, 2018; Inayatullah, 2020; UNESCO, n.d.). It is about learning to anticipate multiple possible futures and each one's potential significance in relation to the present, not about predicting what will happen or determining specific strategies of action. Training methods in futures literacy should not be confused (although it may not always be easy) with methods of futures research and strategic foresight that form a large field with a wide range of available approaches such as extrapolation, probability analysis, survey research such as the Delphi method, monitoring, scenario-design, varying simulations and serious games, as well as relevant forms of social activism (Bell, 2003, pp. 239–317; Gidley, 2017; Heinonen et al., 2017; Andersson, 2018).

Enhanced futures literacy among heritage professionals is a precondition for successful planning related to specific heritage.[1] In Table 35.2, I have assembled some didactic methods and techniques for teaching futures literacy that have previously been used and are applicable to capacity-building in the heritage sector. My list is not exhaustive but a sample of what has been tried and tested and/or was discussed in relevant topical literature. A large variety of alternative suggestions for methods that can successfully engage audiences in relevant questions but have not been used directly in relation to futures literacy is available elsewhere (e.g. Slocum, 2003; Kessels et al., 2009; GO-Science, 2017; Clark, 2019). All methods have different strengths and weaknesses, and they work with audiences of different sizes and require different amounts of time. Which tool to choose in each circumstance will require careful advanced deliberation.

Concluding discussion: lessons for improving futures literacy in the heritage sector

Futures literacy is important for advancing sustainability in the context of heritage. It is, therefore, likely going to be of increasing importance in the research and practice of heritage studies and heritage management during the coming decades.

Based on experiences discussed in the literature and on our own practical experiences with capacity-building initiatives at the UNESCO Chair on Heritage Futures, there are several important insights to be drawn from previous attempts at engaging with the future in the heritage sector.

Firstly, improving futures literacy in the heritage sector involves not only a need to learn something new but also a need to unlearn something else. Many heritage professionals believe that the significance of heritage is timeless (not realising its temporal

Table 35.2 Some didactic methods and techniques for teaching futures literacy in the heritage sector

Didactic method	Idea	Audience	Strengths	Weaknesses
Card game "The Thing from the Future" (Candy, 2018; Harrison et al., 2020a, pp. 17–18; see also Tsing & Pollman, 2005); A variation was developed by Holtorf	Story telling about imaginary sites and artefacts from the future prompted by cards; a variation involves collective story telling with players extending a joint story.	A card game for three to five people; the collective variation works with up to eight people and takes ca 30 min per round	Provokes creative futures thinking that is easy, concrete, and very enjoyable; can break the ice in groups of all sorts of participants	Highly speculative; original version is somewhat convoluted
Performed speculative fiction, developed by Laura Watts (Munawar & Rydén, 2019, pp. 14–15; see also Edwards et al., 2016)	Engaged listening, can be combined with role play	Any attentive audience; role play requires a small group	Allows immersion into a scenario of the future	Ambitious writing requires advanced language skills; risk of one suggestive idea dominating the impact
Time travel to the future (Norlin, 2019)	Collective role-play into the future informed by a given scenario	Groups of adults or school classes	Allows immersion into a scenario of the future and stimulates important reflections in the present	Suspension of disbelief requires careful preparation of scenario and props
"Time Capsule Retrieval Service" (Willshire, 2017)	Imagined future time capsules in 2030, temporarily returning to us, containing six objects relating to four scenarios for currently emerging technologies	School class produces time capsules; conference participants interpret them.	Focus on possible future impacts of technologies encouraging backcasting; avoids misrepresenting or uncritically celebrating specific technologies	Extensive preparations; possible ambiguities of the time capsule content
Interactive video games (Koenitz, 2019)	Future scenarios simulating alternative strategies addressing complex societal challenges for the future	Game players	Educates players about varying consequences of specific actions	Dependent on access to digital games; learning effect relies on quality of data

Didactic method	Idea	Audience	Strengths	Weaknesses
Futures Workshop, developed by Holtorf and Högberg (see earlier, pp. 534–5)	Two groups describe in parallel a future and anticipated challenges; then they develop strategies to address the challenges of the other group	Two groups of 4–7 participants each; requires ca 3.5 hours	Makes participants reconsider the impact of their work on the future	Does not work well with larger groups or less time
Futures Literacy Laboratories, part of UNESCO Futures Literacy Programme (Miller, 2018, Rhisiart, 2018)	A learning process of (1) revealing own anticipatory assumptions, (2) reframing by making strange imaginary assumptions, (3) rethinking own assumptions after reflection	Intensive group workshop of adults or school children, over several days	Can deeply affect perceptions of what participants anticipate, influencing how they think and act today	The process can be somewhat confusing and unclear
Socratic Dialogue, developed by Bill Wei (Munawar & Rydén, 2019, pp. 10–11; see also Kessels et al., 2009, pp. 36–51)	Understanding everybody's judgment on a specific professional issue through formalised questioning	Several groups of six to eight participants work in parallel for ca 90 mins	Makes all participants speak and think about alternative viewpoints in professional practice	Does not work well for large groups and in plenaries

"boundedness") and that they are thus already addressing the future by advocating contemporary technologies of conservation (not realising the underlying "presentism"). It is important to make professionals aware of their unquestioned assumptions.

The very idea and social benefits of heritage – and thus its significance – have been changing extensively over the past few centuries, and there is no reason to assume that it will stay stable from now on (Table 35.1). Indeed, for many millennia, heritage was not necessarily recognized and valued in any specific ways at all. For large parts of the human past, tangible heritage was at best appreciated for the physical properties of its building material. Although it is useful to acknowledge the way in which the future has previously been evoked in the heritage sector, it is also important to realize that we should expect the significance of heritage in society to keep changing. Anything we do for the benefits of future generations, e.g., preserving tangible heritage, will need to take on board that the future values and uses of heritage are not likely to be the same as today, for better or worse. Appreciating alternative and emerging futures means to rethink and redevelop the stories we tell ourselves and thus reframing all we do (Inayatullah, 2020). What do we really mean today when we make sense of the heritage sector by reminding everybody that the heritage must be safeguarded for the benefit of future generations? Familiar futures may have come to stand in the way of desirable futures. In other words, existing practices of heritage management may have to be discontinued: new decisions could have expiry dates, require review in regular intervals, or could be forced to consider anticipated interests of futures generations (Kamijo et al., 2017; Högberg & Holtorf, 2021, pp. 266–267).

Secondly, futures literacy in the heritage sector will relativise and change current priorities. Some heritage professionals believe that addressing the possible needs of future generations is not as important as addressing the various pressing needs of present generations (cf. Henderson, 2020). This position is sometimes framed in terms of the future as a concern for affluent people and/or yet another form of Western hegemony but both claims are misconceived (Holtorf and Bolin, 2022).

The ongoing discussion in the society of the need to adopt practices of sustainable development implies a need to adopt long-term futures thinking. Taking the needs of future generations seriously is neither a luxury nor a Western idiosyncrasy but rather a sign of mature and responsible politics. This does not imply an ability to predict any specified future in great detail but rather a need to plan at a societal level in the light of more general trends and changing conditions that can be discerned. It requires a process of continuous learning about the future which must include the perspectives of all relevant stakeholders. There is a global shift towards sustainability thinking that we currently see affecting all realms of society and politics. Even the heritage sector ought to fit into a sustainability framework that appreciates the needs of future generations and develops an appropriate policy framework. We cannot just continue "business as usual", insisting that heritage inherently constitutes long-term thinking, when its future uses and benefits cannot in fact be taken for granted at all. That is why futures literacy is an important skill and future workshops developing concrete scenarios can be very helpful for understanding and contemplating alternative futures. Practical decisions of heritage management could subsequently be made in the light of several different scenarios, maximising likely benefits.

Thirdly, teaching and applying futures literacy in the heritage sector will require innovative ways of working. Some of the training methods I mentioned in Table 35.2 are rather playful in character and take the form of fun games or role-play. Even some of the more formal workshops adopt creative formats of dialogue and discussion. Such creative practices are necessary in order to broaden existing perspectives and break established routines of thought, paving the way for enhanced futures literacy. Although these methods can work well, some set-ups may be confusing, and user experiences must be taken into account.

Some professionals (such as those who were participating in the Belgium workshop) may prefer to address the future on the basis of factual knowledge and expect to receive confirmation not only for their existing future strategies but also for their existing methods of collaborative work. It is, therefore, important that creative approaches are framed in the right kind of informal setting, e.g., in the setting of an extended lunch break. In a second step, futures literacy will have to be applied to concrete professional challenges and may then adopt slightly more traditional formats. One promising approach to representing the interests of future generations in present-day decision-making involves role-play with special advocates representing the presumed interests of future generations (Otten, 2018). It has been demonstrated that decision-makers are influenced towards long-term thinking by the sheer presence of representatives of future generations (Kamijo et al., 2017).

Future research and other initiatives will need to not only extend the capability of the heritage sector to conceive multiple alternative futures and the way they each frame and impact heritage but also improve approaches and methods of capacity-building and training in futures literacy. Behind much of this work lies the necessity for the heritage sector to embrace the present and future change (Holtorf, 2018). Appreciating change is a precondition for heritage being able to sustain the provision of – albeit changing – benefits to future societies.

Acknowledgments

As always, I am grateful to Sarah May and Anders Högberg for stimulating my thoughts on heritage and the future in various ways. The text contains a few passages based on text first published in Holtorf & Högberg, 2021a. I also wish to acknowledge the very inspiring future sessions run by Kate Clark, Bill Wei and Laura Watts at our workshop in Amsterdam in 2019. Thanks to Alexander Vander Stichele for talking to me about the workshops in Flanders and to Annalisa Bolin, Kalliopi Fouseki and Anders Högberg for comments on a penultimate draft. We are all in this together.

Note

1 At the time of writing a short course on "New social futures from heritage" has been advertised by UCL to run for the first time in November 2020.

References

Albert, M.-T., Bandarin, F., & Pereira Roders, A. (Eds.) (2017). *Going beyond – Perceptions of sustainability in heritage studies, no. 2*. Cham: Springer.

Anderson, B. (2010). Preemption, precaution, preparedness: Anticipatory action and future geographies. *Progress in Human Geography, 34*(6), 777–798. https://doi.org/10.1177/0309132510362600.

Andersson, J. (2018). *The future of the world. Futurology, futurists, and the struggle for the post-cold war imagination*. Oxford: Oxford University Press.

Auclair, E., & Fairclough, G. (Eds.) (2015). *Theory and practice in heritage and sustainability*. London and New York: Routledge.

Bell, W. (2003). *Foundations of future studies. Human science for a new era*, vol. 1: *History, purposes, and knowledge*. 2nd ed. London/New York: Routledge.

Caals, A., & Van Leeuwen, J. (2010). *Anno 2020!? Trends, evoluties en uitdagingen voor een nieuw decennium*. Brussels: FARO. Retrieved from: https://faro.be/sites/default/files/bijlagen/e-documenten/trendboekje.pdf (Accessed on 30 September 2020).

Cameron, C. (2010). The unnatural history of heritage: What's the future for the past? *Journal of Heritage Tourism, 5*(3), 203–218. https://doi.org/10.1080/1743873X.2010.505289.

Candy, S. (2018). Gaming futures literacy. the thing from the future. In R. Miller (Ed.), *Transforming the future. Anticipation in the 21st century* (pp. 233–246). Paris: UNESCO and London and New York: Routledge.

Ceccarelli, P. (2017). Past is not a frozen concept: Considerations about heritage conservation in a fast changing world. *Built Heritage, 1*(3), 1–12. https://doi.org/10.1186/BF03545671.

Clark, K. (2019). *Playing with the past. Exploring values in heritage practice*. New York and Oxford: Berghahn.

De la Torre, M. (2013). Values and heritage conservation. *Heritage & Society, 6*(2), 155–166. https://doi.org/10.1179/2159032X13Z.00000000011.

Edwards, L., Maxell, D., Pillatt, T., & Downing, N. (2016). Beebots-a-lula, Where's my honey?: Design fictions and beeking. *NordiCHI'16: Proceedings of the 9th Nordic Conference on*

Human-Computer Interaction. Retrieved from: https://eprints.lancs.ac.uk/id/eprint/82527/1/beebots_author_final_version.pdf (Accessed on 21 September 2020).

Gidley, J. M. (2017). *The future. A very short introduction.* Oxford: Oxford University Press.

GO-Science (Government Office for Science) (2017). *The futures toolkit. Tools for futures thinking and foresight across UK government.* Edition 1.0. London: Government Office for Science. Retrieved from: https://www.gov.uk/government/publications/futures-toolkit-for-policy-makers-and-analysts (Accessed on 24 March 2021).

Gravari-Barbas, M. (2014). Foresight workshop: New challenges for cultural heritage. Synthesis of the final report. Paris. Retrieved from: https://www.pantheonsorbonne.fr/fileadmin/Colloques_IREST/ARP_Synthèse_mise_en_page_anglais.pdf (Accessed on 7 August 2020).

Graves-Brown, P. (2020). *Warranty.* [Blog post]. Retrieved from: http://slightlymuddy.com/kuriosum/?p=637 (Accessed on 7 August 2020).

Harrison, R., DeSilvey, C., Holtorf, C., & Macdonald, S. (2020a). 'For ever, for everyone…' In R. Harrison, C., DeSilvey, C., Holtorf, S., Macdonald, N., Bartolini, E., Breithoff, H., Fredheim, A., Lyons, S., May, J., Morgan, & S., Penrose (Eds.), *Heritage futures. Comparative approaches to natural and cultural heritage practices* (pp. 3–19). London: UCL Press.

Harrison, R., DeSilvey, C., Holtorf, C., Macdonald, S., Bartolini, N., Breithoff, E., Fredheim, H., Lyons, A., May, S., Morgan, J., & Penrose, S. (2020b). *Heritage futures. Comparative approaches to natural and cultural heritage practices.* London: UCL Press. Openly accessible at https://www.uclpress.co.uk/products/125035.

Harvey, D. C., & Perry, J. (2015). Heritage and climate change. The future is not the past. In D. C. Harvey & J. Perry (Eds.), *The future of heritage as climates change. Loss adaptation and creativity* (pp. 3–21). London and New York: Routledge.

Heinonen, S., Kuusi, O., & Salminen, H. (Eds.) (2017). *How do we explore our futures? Methods of futures research.* Acta Futura Fennica 10. Turku: Finnish Society for Futures Studies.

Henderson, J. (2020). Beyond lifetimes: Who do we exclude when we keep things for the future? *Journal of the Institute of Conservation,* 43(3), 195–212. https://doi.org/10.1080/19455224.2020.1810729.

Heritage, A. (2022). Bridging the past with the future: heritage and foresight. Rome: ICCROM. Retrieved from: https://www.iccrom.org/news/bridging-past-future-heritage-and-foresight (Accessed on 24 March 2022).

Historic England (2015). Facing the future: Foresight and the historic environment. Retrieved from: https://historicengland.org.uk/images-books/publications/facing-the-future/ (Accessed on 26 March 2016).

Hodder, I. (2010). Cultural heritage rights: From ownership and descent to justice and well-being. *Anthropological Quarterly,* 83(4), 861–882. https://doi.org/10.2307/40890842.

Högberg, A., & Holtorf, C. (2021). Final reflections: The future of heritage. In C. Holtorf & A. Högberg (Eds.), *Cultural heritage and the future* (pp. 264–269). London and New York: Routledge.

Högberg, A., Holtorf, C., May, S., & Wollentz, G. (2017). No future in archaeological heritage management? *World Archaeology,* 49(5), 639–647. https://doi.org/10.1080/00438243.2017.1406398.

Holtorf, C. (2018). Embracing change: How cultural resilience is increased through cultural heritage. *World Archaeology,* 50(4), 639–650. https://doi.org/10.1080/00438243.2018.1510340.

Holtorf, C. (2020). An archaeology for the future: From developing contract archaeology to imagining post-corona archaeology. *Post-Classical Archaeologies,* 10, 57–72. Retrieved from: http://www.postclassical.it/PCA_Vol.10_files/PCA10_Holtorf.pdf.

Holtorf, C., & Högberg, A. (2021a). Introduction: Cultural heritage as a futuristic field. In C. Holtorf & A. Högberg (Eds.), *Cultural heritage and the future* (pp. 1–28). London and New York: Routledge.

Holtorf, C., & Högberg, A. (Eds.) (2021b). *Cultural heritage and the future.* London and New York: Routledge.

Holtorf, C., & Bolin, A. (2022). Heritage futures: A conversation. *Journal of Cultural Heritage Management and Sustainable Development*, ahead of print. Retrieved from: https://doi.org/10.1108/JCHMSD-09-2021-0156.

ICOMOS (International Council of Monuments and Sites) (1994). *The Nara document on authenticity*. Retrieved from: https://www.icomos.org/charters/nara-e.pdf (Accessed on 30 September 2020).

Inayatullah, S. (2020). *Co-creating educational futures: Contradictions between the emerging future and the walled past*. Education Research and Foresight Working Paper 27. Paris: UNESCO. Retrieved from: https://unesdoc.unesco.org/ark:/48223/pf0000373581/PDF/373581eng.pdf.multi (Accessed on 22 September 2020).

Jigyasu, R. (2013). *Heritage and resilience: Issues and opportunities for reducing disaster risks*. Retrieved from: https://whc.unesco.org/document/122923 (Accessed on 26 October 2020).

JPI (2013a). *Foresight study and technological capability report 1: Report on drivers of change and the future of cultural heritage. Joint programming initiative on cultural heritage and global change*. Retrieved from: http://jpi-ch.eu/wp-content/uploads/D2.4-Part-1-Report-on-Drivers-of-Change-and-the-Future-of-Cultural-Heritage.pdf (Accessed on 7 August 2020).

JPI (2013b). *Foresight study and technological capability report 2: Real-time delphi study on the future of cultural heritage. Joint programming initiative on cultural heritage and global change*. Retrieved from: http://jpi-ch.eu/wp-content/uploads/JHEP_D2.4_Part2.pdf (Accessed on 7 August 2020).

Kamijo, Y., Komiya, A., Mifune, N., & Saijo, T. (2017). Negotiating with the future: Incorporating imaginary future generations into negotiations. *Sustainability Science*, *12*(3), 409–420. https://doi.org/10.1007/s11625-016-0419-8.

Karl, R. (2015). Every sherd is sacred. Compulsive hoarding in archaeology. In G. Sayeh, D. Henson & Y. F. Willumsen (Eds.), *Managing the archaeological heritage: Public archaeology in Europe* (pp. 24–37). Kristiansand: Vest-Agder-Museet.

Karl, R. (2018). Against Retention in situ. How to best preserve archaeology for 'future generations'? Archäologische Denkmalpflege. [Blog post]. Retrieved from: https://archdenk.blogspot.se/2018/02/against-retention-in-situ.html (Accessed on 15 March 2018).

Kessels, J., Boers, E., & Mostert, P. (2009). *Free space. Field guide to conversations*. Amsterdam: Boom.

Koenitz, H. (2019). Playful utopias. Sandboxes for the future. In B. Beil, G. S. Freyermuth & H. C. Schmidt (Eds.), *Playing Utopia. Futures in digital games* (pp. 253–265). Bielefeld: transcript.

Koers, W., van der Giessen, A., van Weelden, M., & Becker, J. (Eds.) (2013). *Future of cultural heritage – Impact of external developments. Paper incorporating the outcomes of the 18 Dec 2012 European Policy Workshop*, Brussels. Vienna. European Foresight Platform. Retrieved from: http://www.foresightfordevelopment.org/sobipro/download-file/46-979/54 (Accessed on 12 September 2020).

Larsen, P. B., & Logan, W. (Eds.) (2018). *World heritage and sustainable development. New directions in world heritage management*. London and New York: Routledge.

Logan, W., & Wijesuriya, G. (2016). The new heritage studies and education, training, and capacity-building. In W. Logan, M. Nic Craith & U. Kockel (Eds.), *A companion to heritage studies* (pp. 557–573). Chichester: Wiley Blackwell.

Lowenthal, D. (2015). *The past is a foreign country – revisited*. Revised 2nd ed. Cambridge: Cambridge University Press.

May, S. (2021). Heritage, thrift and our children's children. In C. Holtorf & A. Högberg (Eds.), *Cultural heritage and the future* (pp. 46–58). London and New York: Routledge.

Merritt, E. (2021). *TrendsWatch: Navigating a Disrupted Future*. Arlington: American Alliance of Museums Center for the Future of Museums. Retrieved from https://www.aam-us.org/programs/center-for-the-future-of-museums/trendswatch-navigating-a-disrupted-future/ (Accessed on 24 March 2022).

Miller, R. (2018). Futures Literacy Laboratories (FLL) in practice. An overview of key design and implementation issues. In: R. Miller (Ed.), *Transforming the future. Anticipation in the 21st century* (pp. 95–109). Paris: UNESCO; London and New York: Routledge. Retrieved from: https://www.taylorfrancis.com/books/e/9781351048002 (Accessed on 10 Nov 2020).

Munawar, N., & Rydén, H. (2019). *Thinking and planning the future in heritage management. Conference Report, ICOMOS University Forum, Amsterdam, 11–14 June 2019*. Retrieved from: https://issuu.com/lnu12/docs/conference_report_amsterdam_11-14_june_2019 (Accessed on 22 September 2020).

Norlin, A. (2019). *Att resa till framtiden. Report*. Kalmar: Kalmar County Museum. Retrieved from: http://urn.kb.se/resolve?urn=urn:nbn:se:lnu:diva-81727 (Accessed on 26 September 2020).

Otten, M. (2018). *Strong external representation of future generations: Legitimate and effective* (Unpublished Masters Thesis.) Department of Philosophy, University of Leiden. Retrieved from: http://hdl.handle.net/1887/65949 (Accessed on 21 September 20).

RAÄ (Riksantikvarieämbetet) (2006). *Towards future heritage management. The Swedish national heritage board's environmental scanning report*. Stockholm: Riksantikvarieämbetet. Retrieved from: http://www.raa.se/publicerat/9789172094581.pdf (Accessed on 10 November 2020).

Rhisiart, M. (2018). Cultural heritage research and the future. In R. Miller (Ed.), *Transforming the future. Anticipation in the 21st century* (pp. 110–118). Paris: UNESCO; London and New York: Routledge. Retrieved from: https://www.taylorfrancis.com/books/e/9781351048002 (Accessed on 10 November 2020).

Sandford, R. (2019). Thinking with heritage. Past and present in lived futures. *Futures*, 111, 71–80.

Silberman, N., & Liuzza, C. (2007). *The future of heritage. Changing visions, attitudes and contexts in the 21st century*. Interpreting the Past, vol. V(I). Brussels: Ename Center for Public Archaeology and Heritage Interpretation.

Slocum, N. (2003). *Participatory methods toolkit. A practitioner's manual*. Brussels: King Baudouin Foundation & Flemish Institute for Science and Technology Assessment. Retrieved from: http://cris.unu.edu/participatory-methods-toolkit-practitioners-manual (Accessed on 21 September 2020).

Spennemann, D. H. R. (2007). The futurist stance of historical societies: An analysis of position statements. *International Journal of Arts Management*, 9(2), 4–15.

Tsing, A., & Pollman, E. (2005). Global futures: The game. In D. Rosenberg & S. Harding (Eds.), *Histories of the future* (pp. 107–122). Durham, NC: Duke University Press.

UNESCO (1997). *Declaration on the responsibilities of the present generations towards future generations*. Paris: UNESCO General Conference.

UNESCO (2015). *World heritage and sustainable development policy*. Paris: UNESCO. Retrieved from: https://whc.unesco.org/document/139747 (Accessed on 2 October 2019).

UNESCO (n.d.). *Futures literacy: A skill for the 21st century*. Retrieved from: https://en.unesco.org/themes/futures-literacy (Accessed on 28 September 2020).

US ICOMOS (2016). *With a world of heritage so rich. Lessons from across the globe for us historic preservation in its second 50 years*. Retrieved from: http://www.usicomos.org/about/wwhsr/ (Accessed on 28 September 2020).

Willems, W. J. H. (2012). Problems with preservation in situ. In C. Bakels & H. Kamermans (Eds.), *The end of our fifth decade* (pp. 1–8). Leiden: Leiden University.

Willshire, J. V. (2017). *The time capsule retrieval service*. Retrieved from: https://www.smithery.com/2017/07/23/the-time-capsule-retrieval-service/ (Accessed on 21 September 2020).

Zetterström-Sharp, J. (2015). Heritage as future-making: Aspiration and common destiny in Sierra Leone. *International Journal of Heritage Studies*, 21(6), 609–627.

Index

Note: **Bold** page numbers refer to tables; *italic* page numbers refer to figures and page numbers followed by "n" denote endnotes.

Aalto, Alvar 392
Abbott, A. 367
Abid, S. K. 214, 220
Abrina, T. A. S. 505
absolute humidity (AH) 278
Addi, L. 375
aestheticism 400
aesthetic value, of historic monuments 47
Aga Khan Historic Cairo project *376*
agent of change: defined 19; heritage functions as 19–21
agricultural heritage 348–359; Asian rice fields 351–353; challenges 351–353; cooperation with protection systems 356; cultural heritage 355; cultural landscapes 350–351; European vineyards 351–354; heritage tourism as valuation of cultural capital 355–356; historical and intellectual development of sustainability 348–350; intellectual property rights 356–358; multi-designation and systemic protection 354–355; 1972 Convention 350–351; sustainable protection of 350–351; United Nations context 348–350
agroecology 339, 341n11
Ahmad, I. R. 467
Ahtisaari, Martti 184
Ahtisaari Plan 184, 186
Al-Azhar Mosque 372, *373*
Aleppo *(Halab)* 229–231; heritage values and local community in 234–237
Aleppo Project 233
alimentary heritage 328
Allmendinger, P. 412, 413, 415
All That's Left collective 253
Al Qaeda 204–205

alternative dispute resolution (ADR) methods 172, 179n2
Altieri, M. A. 356
Anderson, Ben 528
Anglo-American Empires 196
Anthropology of Food 340n3
Antiquities Law Number 55 200
Apostolakis, A. 505
Appelbaum, Barbara 421–422
Arab-Islamic civilisation 199
Arab nationalism 197
architectural conservation 421–424; ethical questions in 423–424; ethics and morality in 422–423; *vis-à-vis* ethics 421–422
The Architectural Review 394
archive: defined 19; heritage functions as 19–21
Aristotle 148
The Art of Memory (Yates) 519
Arts and Humanities Research Council 2
Ashley-Smith, Jonathan 45
Ashworth, Gregory 242, 382–383
Asian rice fields 351–353
Atarés Castle, Havana, Cuba 284–286
Athens Charter 308
Athens Charter for the Restoration of Historic Monuments 406–407, 421
Australia Charter on Places of Cultural Significance 216
Australia ICOMOS Charter for the Conservation of Places of Cultural Significance (Burra Charter) 407
authorised heritage discourse (AHD) 30, 88–89, 92, 368–369, 405, 410
Autoridad del Centro Histórico de la Ciudad de México 310
Avrami, E. 308

awareness-raising, and heritage conservation 142–143
axial coding 461
The Aztec Empire 310

Ba'ath rule in Iraq 198–200
Bakhshi, H. 506
Bali Cultural Landscape 72–73; case study analysis 73–74; management strategies 76–77; Outstanding Universal Value vs. community's version of values 74–76; and stakeholders' discrepancy 80–81; World Heritage status, local community's perspectives 78–79
Bandarin, Francesco 411
Barthel-Bouchier, D. 7
Battle of Aleppo (2012–2016) 229, 232
Bayat, Asef 254
belief: and indigenous heritage 63–64; in Nigeria 63–64
'bell jar' approach 30, 31, 35, 38; and ruins 36
Benjamin, W. 397
Bennett, J. 505
Berg, F. 308
Berger, John 397
Berger, S. 92
Bevan, R. 232
big data on heritage 475–477
BigPicnic project 324, **329**, 330–331
bio-deterioration 268–269
Black Lives Matter 416–417
Bloom, Paul 251
Bonazza, A. 267, 269
Botanic Garden Meise (BGM) 332
Bouchenaki, M. 220
boundary approach, urban heritage 365–378; defined 367; global south 375–378; in heritage discourse 368–375; overview 365–366; in social sciences 366–368
Bradford, K. 46
Bregazzi, H. 243
Brida, J. G. 503
Brimblecombe, P. 268
Brokerhof, A. W. 45, 46, 48
Brundtland Report 1, 58, 181
Build Back Better (BBB) 227
Bulow, A. E. 45
Burra Charter 22–23, 221, 274
The Burra Charter 421, 453, 513
business improvement districts (BIDs) 413
bygone past 371

Canadian Conservation Institute 45
Cannon, Alice 46, 48, 51, 53
Canter, D. 227
capacity-building, and heritage conservation 143–145
Carnic High Route 245–248
Cartes Parlantes 512, 516, 518–521, 523n4
case study(ies) 282–286; Atarés Castle, Havana, Cuba 284–286; decision-making of residents of heritage buildings 460–462; degradation of plastic museum artefacts 465–467; Golden Lane Estate, London, UK 282–284, *283*; Golden Mile Complex, Singapore *429*, 429–432; municipalities of Tixcacalcupul, Chichimila, and Sacalum in Yucatán, México 140–141; systems modelling for preventive collection management 463–465; Tan Si Chong Su (Po Chiak Keng), Singapore *426*, 426–428
Cassar, May 1, 394
causal loop diagram 461, **461**, 466–467
Ceccarelli, Paolo 528
Centre for Sustainable Heritage 1
Cergy Paris University (CYU) 359
Chamberlin, Powell and Bon (CP&B) architecture studio 282
change management: and heritage 30–40; overview 30–31; *see also* heritage management
change theory (systems theory) 33
Chapel of the Nativity of the Virgin case study: analysis, risk assessment 53; methodology, risk assessment 49–50
character witness 110–111
Chichimila in Yucatán, México 140–141
choice modelling 504–505
Christianity 155, 161, 214, 370
Chung, H.-L. 387
circular economy 439–441, 444–446; defined 452; initiatives 446
"citizen-oriented form of diplomacy" 170
citizen science (CS): correct methodology for data 118–119; data collection controls 120–121; defined 115; and ornithology 117; projects future 121–123; sustainable heritage conservation in 115–123; types of participation 116
Citizen Science Alliance 116
'citizen scientist' 115
City Museum of Stockholm 299
civic relationships, sustaining 146–147
Clapham, C. S. 374

Clark, Kate: *Playing with the Past* 27
classic engineering optimisation problem 294
Cleveland Museum of Art 175
climate analysis 277–278
climate-based daylight modelling (CBDM) 281
climate change 1, 7–8; bio-deterioration 268–269; carbonate stone, recession of 267; cryoclastism 268; cultural heritage 264–266; as driving factor for damage of built heritage 266–267; European research framework 264–265; methodology 265–266; overview 263–264; salt crystallization 267–268; sustainable heritage and 263–269; thermoclastism 269
Climate for Culture and Efficient Energy for EU Cultural Heritage (3ENCULT) 307
The Closing Circle 438
coding groups **313**
Cold War 245
collage, as heritage concept 32
Colombian Coffee Landscapes 358
Colston, Edward 417
Comite Europeen de Normalisation (CEN) 307
Commander of the US Central Command in the Middle East (CENTCOM) 200
community 6; -based monitoring (CBM) projects 117; involvement in heritage management 127–129; in regeneration 448–449; of relationships 453
complex social value (CSV) 439–440, 443, 449
computational fluid dynamic (CFD) simulation 278
conflict: cultural heritage in 184–186; and inscription on World Heritage List 184–186; management 451–452; new layers of cultural heritage during 232–234
Congres Internationaux d'Architecture Moderne (CIAM) 392
connective infrastructure 441, 452
conservation 27n2; correlation between sustainability and *163*; heritage 20, 21–22; in nature management 32; repair 406
conservation-planning: neoliberalisation of 412–415; policy approaches **416**; *see also* urban conservation
Convention Concerning the Protection of the World Cultural and Natural Heritage 513
Convention for the Protection of the Architectural Heritage of Europe 407
Convention for the Safeguarding of the Intangible Cultural Heritage 407

Convention on the Value of Cultural Heritage for Society 407–408
Corbusier, Le 392
CORINE Land Cover 478
Council of Cultural Affairs 389n1
Council of Europe 185
Covent Garden, London 408, 415
Covid-19 pandemic 222, 234
Creswell, J. W. 74
critical discourse analysis (CDA) 74
Cross-Cultural Foundation of Uganda (CCFU) 338
cross-disciplinary heritage research 2
cryoclastism 268
crystallization 267–268
The Cuesta of the Rupel Region – New Challenges for its Cultural Heritage 523n4
culinary tourism 329
cultural assets 172–173
cultural dimension, of food 330–331
cultural diplomacy 170–171; defined 170; and exhibitions 175; goals of 170; and ICR 176, 178–179; interconnection of museums with 171; museums in 171–172; and public diplomacy 170–171; and soft power 170
cultural diversity 527–529
cultural heritage 355, 395; climate change 264–266; in conflict 184–186; and inscription on World Heritage List 184–186; management 531–533; new layers, during conflict 232–234; religious 447–448; sustaining, in post-conflict Syria 226–237
Cultural Heritage Administration of Korea (CHA) 129–131
cultural landscapes 359n3
cultural property, and international cultural relations 169–179
Cultural Property Implementation Act (CPIA), US 175
Cultural Property Risk Analysis Model (CPRAM) 45
cultural relations 170–171; international 169–179
cultural sustainability 4
cultural value 61, 93, 220, 305–306
Culture for Development Indicators Suite 326
curating decay 30

Daamen, T. A. 472
Dadaocheng, Taipei City 383–384
dark heritage 6

data: about heritage 488; collection 475–477, 478–479; as heritage 488; integrity 492–493; pipeline 485; quality 491–492; science, heritage (*see* heritage data science); visualization 477, 478–479

data collection controls: and bias 120; and citizen science (CS) 120–121; and measurement error 120

Data Science for Cultural Heritage 494

datawheel methodology 476

daylight autonomy (DA) 281

Declaration of Nyeleni 338

deeper spatial understanding 473–475

The Degradation of Complex Modern Polymeric Objects in Heritage Collections: A System Dynamics Approach 465

Delphi method 535

Demossier, M. 327

deontological ethics 422, 424–425, 428

descriptive ethics 422–424, 427–428, 431

Design Partnership 429

DeSilvey, C. 31

developmental change 33–35, *34*; organizational principle 35; Roskilde Cathedral, Denmark 33–35, *34*

Dewey, John 519

difficult heritage: overview 106; perspective of the witness 107–108; and social sustainability 104–106; of traumatic pasts 106

'digital archaeologist' 118

'digital historian' 118

Digital Humanities at Delft University of Technology 475

digital literacy 494

diplomacy: cultural (*see* cultural diplomacy); public 170–171

dissolution, as heritage concept 32

Dissonant heritage: the management of past as a resource in conflict 6

dissonant nature of heritage 3, 6–7

diversity: cultural 527–529; and indigenous heritage 63–64; in Nigeria 63–64

'Dolomitenfreunde' association (Friends of the Austrian Alps) 246

Douglas, M. 46

dry bulb temperature (DBT) 278, 280

Dudley, J. M. 490

Duran-Casablancas, C. 464

Duru, M. 334

Dutch Belvedere Programme 408

dynamic approach, to sustainable heritage 2

dynamic hypothesis 461–462, 468

Earl, J. 272

eBird 117

economic: adding, value to local products 158–160; creating accommodations utilising old Japanese-style houses 157–158; heritage significance 498–501

economic beliefs: indigenous heritage 66–67; in Nigeria 66–67

economic sustainability 1, 44, 175, 355, 399

edible chronotope 333

edible souvenir 333

education, and sustainable community 156–157

Eisenstadt, S. 374

Elrasoul, R. 220

emotionalization: of cultural life 40; of social life 40

EN 16883 Conservation of cultural heritage 296

energy: behaviour 308; renewable production of 275; use and age, correlation 298–299

energy efficiency 316–318; for EU Historic Districts' Sustainability (EFFESUS) 306–307; and heritage buildings 309; renovations for 305; and thermal comfort in Mexico 309–311

energy efficiency in historic buildings 290–301; as an optimisation problem 294–295; background 291–293; in building stocks 298; case study 297–298; combining top-down with bottom-up analysis 299–300; dilemma about 293; energy use and age, correlation 298–299; heritage values and energy correlation 299; overview 290–291; policies for 292; in single buildings 295–296; as social practice 293–294; systematic approach 296–297

engaged preservation 25–26

Engelhardt, E. **324**

Enlightenment Europe movement 405

environment: and heritage 7–9; and society 7–9

environmental design strategies: case studies 282–286; climate analysis 277–278; for heritage 272–287; innovative 275–276; instrumental features 277; methodologies for 277; passive and low-energy space conditioning 276–277; simple changes 276; site analysis 278–279; smart zoning 276; state of art of 274–275; thermal performance analysis 279–280; visual environment 280–282

environmental impact assessment (EIA) 454

environmental (climate and pollution) parameters 265–266

environmental sustainability 8, 359n2
Eriksson, P. 309
essential values 24
ethics: deontological 422, 424–425, 428; descriptive 422–424, 427–428, 431; meta-ethics 424; normative 422–423
ethics in conservation, Singapore 421–433; architectural conservation 421–424; case study 426–432; dilemma 424–433
EU FP6 Project 264
EU Historic Districts' Sustainability (EFFESUS) 306–307, 309
EuroGlobalMap 478
European Agenda for Culture 263
European Commission 185, 449; European Foresight Platform 534; Horizon 2020 Programme 330; Joint Programming Initiative on Cultural Heritage and Global Change 533
European Foresight Platform 534
European Landscape Convention (Florence Convention) 407
European Union's Horizon 2020 Research and Innovation Programme 323
European vineyards 351–354, 359n7
evaluation framework for heritage conservation as social process 142–147
evolution, of indigenous heritage 61–62
expert witness 109–110
'extreme citizen science' 116
extrinsic values 47
eyewitnesses 108–109

Facing the Future: Foresight and the Historic Environment 531
'fake news' 107
Faneuil Hall Marketplace in Boston 415
Faro Convention Framework on the Value of Cultural Heritage for Society 513
Feilden, Bernard 421
Fetterman, D. M. 74
First International Congress of Architects and Technicians of Historic Monuments 406
food: memories, stories and social context of eating 332–333; -related activities 336; security 324, **324**, 340n1; sovereignty **324**
Food 2030 340n6
Food and Agriculture Organization of the United Nations (FAO) 356
food heritage 323–339; case study 330; cultural and social dimensions of food 330–331; emergence of 326–328; food memories, stories and social context of eating 332–333; heritage sustainability discourses 324–325; intangible cultural heritage 325–326; methodology 330; migration and 334–336; overview 323–324; sustainability and 328–330; sustainable development 325–326; traditional foodways 331–332
foodways 324, 331–332
Forsyth, M. 273
Fouseki, K. 65, 306, 309, 313, 320
FP7 Climate for Culture Project 264
Franco, S. F. 503
Frankena, W. 428
Freedom Seder 253
French National Research Agency (ANR) 535
French vineyards 353
Fujiwara, D. 506–507

Gachanga, Timothy 249
Gallou, E. 95
Garside, Paul 45, 46, 51
gender equality 144, 148
geographical indication (GI) system 356–358
George Town, Penang 426
Georgian Society 406
geospatial mapping 474–475, 477, 478, *479*, 481
Getty Conservation Institute 21, 24, 27n6
Getty–Italy agreement 173–174
Getty Museum 172–178
Giddens, Anthony 31
Giedion, Sigfried 392
Gillespie, A. 382
Glišić, Milan 187
Global Architectural History Teaching Collaborative (GAHTC) 473
globalisation: and indigenous heritage 64–65; and Nigeria 64–65; and political sustainability 182–183; and world heritage 182–183
Globally Important Agricultural Heritage Systems (GIAHS) 328, 356
Golden Lane Estate, London, UK 282–284, *283*
Golden Mile Complex, Singapore *429*, 429–432
Gomez-Bolea, A. 268
Gongju City 126–127
Graham, B. 409
Gray, Jim 486
Green Building Council (GBC) Historic Building 274–275

greenness of heritage 7
Griesemer, J. R. 367
Grossi, C. M. 268
Gulf War (1991) 199, 201

H2020 Project 264–265
Hamlyn, S. 46
Harald Bluetooth, king of Denmark 33
Harley, J. B. 478
Harrison, R. 59, 182
Hartog, F. 512
Haughton, G. 413
Hebron: Al-Shuhada Street *252*; cosmopolitan identity of 251; Freedom Seder 253; Hebron Protocol 253; Hebron Rehabilitation Committee 253; Jews in 256n18; justice weaving in 251–253; living heritage of 253; resistance groups in 253; uncoordinated activism in 254
Hebron Freedom Fund 253
hedonic pricing 502–503
Heemschut Heritage Association 406
Heller, A. 371
heritage 272–287; buildings and energy efficiency 309; care, responsibility for 144; and change management 30–40; climate change nexus 325; conservation 272, 316–318; dark 6; defined 22; as a domain 485–486; and environment 7–9; environmental problem 272–274; evolution of indigenous 61–62; functions as archive and agent 19–21; innovative 275–276; intangible 218–222; negative 6; participatory 5–6; and peacebuilding 241–256; practice in kenya and elsewhere 248–251; as processes of growth 520–521; regeneration schemes 275; as 'social glue' 138; and society 7–9; stocks 332; sustainability discourses 324–325; tangible 218–222; tourism 355–356; use and imagination 382–383; use and place vision 385–387; *see also* environmental design strategies; *specific entries*
Heritage Canada 406
heritage conservation 407–408; awareness-raising 142–143; capacity-building 143–145; case studies 140–141; conclusion and future research 149; data collection 141–142; evaluation framework for 142–147; exploring, as a social process 140; methodology 141; municipality of Chichimila, México 140–141; municipality of Sacalum, México 140–141; municipality of Tixcacalcupul, México 140–141; overview 138; participatory methodologies 139; planning 91; reinforcement of partnership 145–146; as social process 138–150; sustainable heritage and social sustainability 139; sustaining civic relationships 146–147
heritage data, defined 10
heritage data science 484–495, *487*; data about heritage 488; data as heritage 488; data integrity 492–493; data pipeline 485; data quality 491–492; digital literacy 494; framework 487; heritage as a domain 485–486; heritage data 487, *487*; model for 490–491; opportunities and challenges 491; paradigm 488–489; scientific activity, classifying and organising 489–490; sustainability of 488; theory 484–485; thinking 486–487; transdisciplinarity 493–494; value-creating science 493–494
heritage impact assessment (HIA) 454
heritage-led urban regeneration 11
heritage management 4–5, 91, 126; community involvement in 127–129; cultural 19; and cultural heritage 30; defined 3, 59; developmental change 33–35, *34*; holistic approach to 61; in Nigeria 58–68; transformational change 36–38, *37*; transitional approach to 36; transitional change 35–36
heritage practitioner 511–523; field role 514–516; heritage as processes of growth 520–521; other's memory 519–520; social potential of interactive walking 516–518; sustainability towards social sustainability 512–514
heritage protection life cycle 91, *91*
Heritage Resilience Against CLimate Events on Site (HERACLES) 264–265
heritage science 115
heritage significance 497–509; capturing 501–502; economic 498–501; emerging approaches 506–507; hybrid methods 506–507; market-based approaches 502–504; non-market-based approaches 504–506
heritage values 4, 24, 314–315; academic research on 47; assigned to historic buildings 315; change in, right to know about 131–133; conservation of 307–308; defined 308; domains of change as 33; and local community in Aleppo 234–237; in Nigeria 58–68; no right to be involved in reviving 133–135; protection of 131; and risk 4–5

heritagization process 38
Hidden Christians 155–156, 158–161
Hidden Christian Sites (Nagasaki Region) 155–156
high dynamic range (HDR) photography 282
HisGIS 478–479
historical consciousness 371
historic authenticity 272
historic building information modelling (HBIM) 275
historic buildings: climate responsiveness 273; defined 290; energy efficiency in 290–301; HBIM 275; image and value of 273; interior environments in 276; preservation of 292; restoration of 272–274; site analysis 278–279; smart zoning 276
Historic Center of Mexico City (CDMX) 311
Historic England 305, 307, 473; *Facing the Future: Foresight and the Historic Environment* 531
Historic England Heritage at Risk Register 46
historic masonry 273
historic urban: environments 405; fabric 406
Historic Urban Landscape (HUL) 405, 411, 441–444; approaches 412, 415, **416**, 442, 444, 453, 472; economic model 443; focus on tangible and intangible values of cultural heritage 442; landscape-based approach 443
historic value of historic monuments 47
Hitler, Adolf 425
Hogberg, A. 530
Holtorf, C. 530
House of Lords Science and Technology 488
human imagination 381
Husayn ibn Ali ibn Abi Talib 214
Hussein, Saddam 199–200, 203, 215
HVAC systems 275, 276
'hybrid' model 439

Ibrahim Maghazi Mosque 372
Ik Zie Ik Zie wat Jij niet Ziet (I See I See What You Can't See) 523n4
illuminance (lux) 281
image-driven media and resilience 392–402; heritage 397–398; knowledge 399–401; literature context 394–395; photography 397–398; Pinterest architect 395–396; rediscovering images as research tool 398–399; social media 392–394; ubiquitous formatting of social 396
indigenous communities: defined 59–60; in Nigeria 59–60

indigenous heritage: challenges 63–67; and colonialism 62–63; conflicting values between experts and communities 66; diversity and belief 63–64; economic beliefs 66–67; evolution of 61–62; framework for managing 62–63; and globalization 64–65; and Islamic jihad 62–63; lack of documentation 65; management of 60–61; and Western ideas 64–65; *see also* heritage
indoor luminous environment 281–282
Ingold, T. 521
Institute for Sustainable Heritage (ISH) 1
instrumental values 24
intangible cultural heritage (ICH) 323, **324**, 325–326, 328, 332, 335, **335**
intangible heritage 218–222, 326
intellectual property rights 356–358
Interactive Journeys method 511–512, 514, 516, 521, 523n24
interactive walking 10, 523n1; social potential of 516–518; three-step methodology 511–512
inter-community dynamics: and conservation 186; and political sustainability 186; and world heritage sites 186
International Alliance for the Protection of Heritage in Conflict Areas (ALIPH) 207
International Architecture Exhibit 472
International Art Exhibition 472
International Centre for the Study of the Preservation and Restoration of Cultural Property (ICCROM) 74, 90
International Civilian Office (ICO) 184
International Council on Monuments and Sites (ICOMOS) 406, 407; Burra Charter for Places of Cultural Significance (Australia) 47; Heritage at Risk Programme 46
international cultural relations (ICR) 169–179; and cultural diplomacy 170–171, 176, 178–179; and cultural property disputes 169–179; cultural relations 170–171; developing through restitution agreements 176–178; Getty–Italy agreement 173–174; museums in cultural diplomacy realm 171–172; negotiated agreements 174–176; overview 169; restitution of cultural assets as a tool 172–173; soft power 170–171
internationalisation, of cultural heritage 207–208
International National Trusts Organization (INTO) 332

International Restoration Charter (the Venice Charter) 406, 407
Interreg Central Europe Project 265
inter-spatial research methods 10
inter-temporal research methods 10
interviews: analysis, risk assessment 51–53; methodology, risk assessment 49
intrinsic values 47, 440–441, 443, 447–449
Iraq: Ba'ath rule in 198–200; cultural heritage, future of 208–209; early Iraqi state and cultural heritage 197–198; independence in 1932 198; internationalisation of cultural heritage 207–208; invasion and early occupation of 201–202; nationalism in 197; new political order, establishing 203–206; planning the Iraq war 200–201; political ruptures and cultural heritage of 196–210; State Board of Antiquities and Heritage (SBAH) 218
Iraq cultural heritage: early Iraqi state and 197–198; future of 208–209; internationalisation of 207–208; *see also* cultural heritage
Iraqi Institute for the Conservation of Antiquities and Heritage (IICAH) 207
Iraq–Iran war (1980–1988) 199
Iraq Museum 200
Iraq National Library and Archives (INLA) 201
Iraq Reconstruction and Management Office (IRMO) 202, 203
Iraq War 200–201
Islamic holy cities 215–218
Italian Mediterranean diet 331
Ito, N. 220

Jackson, M. 243
Jaffry, S. 505
Janssen, J. 408
Judaism 370
Judt, Tony 255

Kakure Kirishitan 161
Kalman, H. 422
Kaslegard, A. S. 36
Kenya: heritage-based peacebuilding in 250; heritage practice in 248–251
Kenyan Community Peace Museums and Heritage Foundation (CPMHF) 248–249
Keynes, Maynard 451
Khoury, Elias 233
knowledge: loss of heritage 143; preventive conservation 144; production 143
Kocher, S. 220

Kok Hui Chan, Jeffrey 9
Koohafkan, P. 356
Koppen–Geiger climate classification 268
Kosovo 185–186; independence of 184; March 2004 riots in 185; post-war period in 186; and post-war political status 183–184; WH sites in 186
Kozulin, A. 521

Labrador, A. M. 2
Lagerqvist, M. 414
Lagueux, Maurice 421
Lamont, M. 366
Lampi, E. 504
landscape: as attractive force 450–451; and utility 451–452
land-use planning 423
Larkham, P. 408
Lawton, R. 505
lay witness 111–112
Lazrak, F. 503
Leadership in Energy and Environmental Design (LEED) 274–275
Learner, T. 465
Leger, Fernand 392
Levi, D. 220
life cycle assessment (LCA) 445
life cycle cost (LCC) 295, 297, 299
Linguistics Association of Nigeria (LAN) 65
Lipfert function 267
literacy teaching 527–539; cultural diversity 527–529; didactic methods and techniques 535–539, **536–537**; knowledge about future 529–533; training in 533–535
'living memory' 108
living standards of buildings 292
Liwieratos, K. 59
local heritage 143
local products, adding value to 158–160
Lopes, M. A. R. 308
Lovink, Geert 396
low-energy space conditioning 276–277
Lucas, Maria Carolina 523n3

Macdonald, J. L. 503
MacGinty, R. 247
MacRae, G. 80
Madden, O. 465
Madinah 214, 216–217
Makkah 214, 216–218, 220, 222
al Maliki, Nouri 205
Mandela, Nelson 66

Manning, Martha: *Walking Distance* 246
Manning, Robert: *Walking Distance* 246
Mapping the Museum Digital Skills Ecosystem project 494
al-Mansur, Abbasid Caliph 199
market-based approaches 502–504
Martinez Cobo, Jose R. 60
Marx, K. 366
Matsuura, Koichiro 188
May, Sarah 530
McCabe, S. 269
McLuhan, Marshall 394
Meaning Theory (Williams) 233
mean radiant temperature (MRT) 280
Megoran, Nick 243
Memories from Aleppo 233
Memory Palace 515
Meskell, L. 189
meta-ethics 424; *see also specific entries*
Mexico City case study 305–320; changes to condition of buildings over time 315–316; conservation of heritage values 307–308; energy efficiency and thermal comfort 309–311; heritage buildings and energy efficiency 309; heritage values 314–315; methods and materials 311–314; overview 305–307; thermal comfort, energy efficiency and heritage conservation 316–318; thermal comfort and energy behaviour 308
Michalski, Stefan 45, 48
microclimatic site analysis 279
Middle East and North Africa (MENA) region 226
migrant acquisition 160–161
Mind the Gap research project 2
mission heliographique 392
Mnemosyne Atlas (Warburg) 394–395, 515
'The Modern Cult of Monuments, its Nature and Development' (Riegl) 47
Moholy-Nagy, Laszlo 392
Molnar, V. 366
Monument Monitor project 118
Morisset, Lucie K. 513
Morris, William 406
multi-criteria decision problem 294
museums in cultural diplomacy realm 171–172
Mutisya, Munuve 249

Najaf, Iraq 213–222; heritage conservation and Islamic holy cities 215–218; overview 213–215; Shrine of Imam Ali in 214, 219; tangible and intangible heritage 218–222; and urban cultural heritage conservation 215–218
Najaf Charter 219
Nanoscale Informal Science Education Network (NISE) 330
Naples National Archaeological Museum 174
Napoleonic Wars 245
Nara Document on Authenticity (ICOMOS) 216, 221, 407, 421, 527
Narrative of Syrian Heritage 232
National Archaeological Museum of Florence 174
National Commission for Museums and Monuments (NCMM) 63
National Heritage Board 427
National Heritage List for England (NHLE) 282
nationalism: Arab 197; Iraqi 197
National Lottery Heritage Fund 21
National Trust for Historic Preservation in the United States 406
Nazism 425
Ndoro, W. 332
Nebuchadnezzar II, king of Babylon 199
negative heritage 6
Negussie, E. 414
neoliberalisation 416; of conservation-planning 412–415, **416**
neo-Nazis 425
neo-patrimonialism 374
Netherlands 406
The New Production of Knowledge 489
The New Production of Scientific Knowledge 490
News on Live Maps 232
New Urban Agenda 441
Nigeria: contemporary challenges and implications in 60–62; framework for managing indigenous heritage 62–63; heritage management in 58–68; heritage values in 58–68; indigenous communities in 59–60; and indigenous heritage challenges 63–67; taboos 62
9/11 terrorist attacks 170
1972 Convention 349, 350–351, 358; *see also* World Heritage Convention
Noeveren Brandt [Noeveren burns] 523n4
non-market-based approaches 504–506
Norgaard, Bjorn 33, *34*
normative ethics 422–423
Nye, Joseph 170
Nyhavn, Copenhagen *409*, 415

Oers, Ron van 411
Office of Reconstruction and Humanitarian Assistance (ORHA) 200–202
O'Hagan, J. 93
Olagoke, O. A. 66
open coding process 460–461
Orth, M. 504
Osman, Mohammad Maliki 429
Ostrom, E. 338
Osun-Osogbo grove 61, 62, 66
Our creative Diversity 348
Outstanding Universal Value (OUV) 67, 72, 349, 472; of Bali Cultural Landscape 74–76

pacific heritage 3, 6–7
Pages Sánchez, J. M. 472
Palace of Westminster case study: analysis, risk assessment 54; methodology, risk assessment 50–51
Palestinian Liberation Organisation 253
palimpsest 32
Pareto efficiency 506
participation: public 190; as spatial practice 140
participatory approaches, and heritage research 3
participatory cultural heritage conservation 149; case studies 140–141; heritage conservation as social process 140; methodologies 139; social impacts of 138–150; sustainable heritage and social sustainability 139
participatory heritage 5–6; and reciprocal theoretical framework 93–95; and social sustainability 89–93; for social well-being 87–98; and sustainable management 89–93; systemic perspective 95–97, 97
'participatory science' 116
partnership: accountability 145; expertise 145; factors affecting involvement in decision-making 146; governance 146; money 145; principal actors 146; reinforcement of 145–146; routes of interactions 146; services and law prohibition 145; time 145–146
passive space conditioning 276–277
'passive use' values 500–501
Patriarchate of Peja/Peć Monastery *187*, 187
Pauperio, E. 45
peacebuilding: challenges 242–243; drawing strings together 254–256; heritage and 241–256; heritage practice in kenya 248–251; hopeful future 247–248; justice weaving in Hebron 251–253; peace efforts 246–247; possibilities of starting somewhere else 243–245; setting the scene 246; social non-movement, emergence of 253–254; sustainable practices, finding and identifying 254–256
Peck, J. 414
Pedersoli Jr, J. L. 48
Pendlebury, J. 92, 413, 414
people-centred approaches, and heritage 90
Pereira, N. 45
Philadelphia 18–19; and cultural heritage management 19; and preservation 18–19; and sustainable development 19
photogenia 401
photography 397–398
Pieterse, Nederveen 182
Pinterest architect 395–396
Places that make us (National Trust) 513
plastic museum artefacts 465–467
Playing with the Past (Clark) 27
Png, Ivan 429
political sustainability 7; conservation 186; and globalisation 182–183; and heritage 11; and inter-community dynamics 186; and world heritage 181–192; world heritage sites 186; *see also* sustainability
port city resilience 471–481; big data on heritage 475–477; case studies 477–479; data collected 475–477, 478–479; data sharing, dissemination, pilot studies 477; data visualization 477, 478–479; mapping for deeper spatial understanding 473–475; in Rotterdam, Hamburg and London 477–478, 479–481
post-conflict Syria: Aleppo *(Halab)* 229–231; cultural heritage during conflict 232–234; heritage values and local community in Aleppo 234–237; overview 226–228; sustaining cultural heritage in 226–237
power balance 88
preservation: engaged 25–26; in nature management 32; and Philadelphia 18–19
Preservation of Monuments Board 427
preventive collection management 463–465
Pride of Place: England's LGBTQ Heritage 492
primary value 440; *see also* intrinsic values
Privett, Helen 51
Probst, P. 64
productivity method 502
profound transformation 325
ProteCHt2save 265

proust effect 333
public diplomacy 170–171
public welfare 508
Punter, J. 227

quality of life improvement 161–162
Queen Margrethe II of Denmark 33
Queensway Shopping Centre in Singapore 431
'QuiskScan' method 45, 48, 52
Quist, H. O. 61

Radin, M. J. 370
rational-legal authority 374
reciprocal theoretical framework: conceptualise participation under 95; and participatory heritage 93–95
reciprocity 94–95
Recommendation on the Historic Urban Landscape 407–408
Redaelli, E. 408
rediscovering images as research tool 398–399
Reeve, A. 383
relative humidity (RH) 277
religious cultural heritage 447–448
renewable production of energy 275
resilience *see* port city resilience
responsible research and innovation (RRI) 324
responsive approach, to sustainable heritage 2, 8
restitution: of cultural assets 172–173; developing ICR through agreements 176–178
Rice Terraces of the Philippine Cordilleras 154
Richmond, O. P. 247
Riegl, A. 22–23; 'The Modern Cult of Monuments, its Nature and Development' 47
Riesch, H. 368
rigorous airtightness 273
risk, and heritage values 4–5
risk assessment 5; analysis 51–54; challenges of 46; defined 45; methodology 45, 49–51; and subjectivity 46; and value 48–49
Rogerson, Cordelia 51
Roll-Hansen, N. 489
Romao, X. 45
Roskilde Cathedral, Denmark 33–35, *34*
Rotterdam, Hamburg and London 477–478, 479–481
Royal Botanic Garden Edinburgh (RBGE) 336, 340n8
ruins: and 'bell jar' 36; defined 35
Ruskin, John 405–406

Sabri, R. 66
Sacalum (Yucatán, México) 140–141
Safeguarding cultural heritage through Technical and Organisational Resources Management (STORM) 265
Salahadeen, liberator of Jerusalem 199
salt crystallization 267–268
SCAPE Trust 119
scientific activity, classifying and organising 489–490
scientific process 115
Scott, M. 413–414
Scottish Coastal Heritage at Risk Project (SCHARP) 117, 119
Second Congress of Architects and Specialists of Historic Buildings in Venice 406
self: -efficacy 144–145; journey of 248–251
Semes, S. W. 424
Sempuku Kirishitan 161
Sen, Amartya 149
Sendai Framework for Disaster Risk Reduction 2015–2030 263
Sennett, R. 433
Serlio, Sebastiano 396
Sharifi-Tehrani, M. 504–505
Shipley, R. 383
Shove, E. 147
Shrine of Imam Ali, in Najaf 214, 219
Silberman, N. A. 2
Silman, R. 432
Simon, Herbert 27n5
Singapore Heritage Society 430
site analysis 278–279
Six Day War (1967) 251
smart zoning 276
Smith, B. J. 268, 269
Smith, J. 409, 410
Smith, L. 368–369, 374, 409
social dimension, of food 330–331
social ecology 339, 340n10
social impact assessment (SIA) 454
social media 392–394
social non-movement 253–254
social sciences 366–368
social sustainability: and difficult heritage 104–106; inclusive 335; overview 104; and participatory heritage 89–93; sustainability towards 512–514; and sustainable heritage 139
social values 23–24
social well-being: defined 87; participatory heritage for 87–98

societal values 24, 26, 348
society: and environment 7–9; and heritage 7–9
Society for the Protection of Ancient Buildings (SPAB) 406
Society for the Protection of Ancient Buildings Manifesto 421
socio-economic development 263
socio-spatial methodology 474
soft power 170–171
solar panels 273
solar radiation data 278
Sollis, J. 37
Solnit, Rebecca 514, 517
Sontag, Susan 392
Spatial Narratives 512, 515–516, *517*, 518–521, 523n4
Spennemann, Dirk 530
The Stalker 32, 40n1
Star, S. L. 367
Statement of Outstanding Universal Value (SOUV) 74–76
Steering Group of Kalmar Castle 535
Stig Sorensen, M. L. 244
Stokes, D. E. 489
St. Pancras Railway Station in London 423
stratigraphy 32
strong sustainability 38
Subak landscape 72, *73*, 77, 79, 81; *see also* Bali Cultural Landscape
Subak Pakerisan Watershed *73*, 76, 78, 80–81
subjectivity, and risk assessment 46
Sukur Cultural Landscape (SCL) 62, 65
surface recession 267
surface temperatures 280
sustainability 328–330; as concept for change as heritage 31–32; concept of 273; correlation between conservation and *163*; cultural 4; defined 31; efforts in historic buildings 274; environmental 8; and perspective of witness 107–108; social 139; strong 38; weak 38; *see also specific types*
sustainable community: accommodation creation and Japanese-style houses 157–158; adding value to local products 158–160; analysis 156–157; education 156–157; introduction and context 154–155; migrant acquisition 160–161; quality of life improvement 161–162; sustainable heritage through 154–164; theory and methodology 155–156
sustainable development (SD) 325–326; as concept for change as heritage 31–32; defined 58; dimensions of 349; and Philadelphia 19
Sustainable Development Goals (SDGs) 1, 138, 181, 227, 325
sustainable heritage 1–2, 11; conservation, citizen science 115–123; dynamic approach 2; -led transformation 9–10; responsive approach 2; and social sustainability 139; through sustainable community 154–164; and values-centered conservation theory 20–21
sustainable management, and participatory heritage 89–93
sustainable regeneration 441–444, 446–447
sustainable renovation 292
sustainable tourism 263, 355
sustaining: civic relationships 146–147; cultural heritage in post-conflict Syria 226–237
sustaining heritage: defined 17; and sustaining values 17–18
sustaining heritage places 381–389; heritage-led urban regeneration initiatives 383–384; overview 381–382; urban imaginaries, heritage use and place vision 385–387; use and imagination in urban regeneration 382–383
sustaining values: and sustaining heritage 17–18; values-centered conservation theory 17
Sutton, D. 334
Svalbard Global Seed Vault 327
Swedish National Heritage Board 531
Swedish Planning and Building Act 297
Sweyn II of Denmark 33
Sylvester, C. 178
Syrian Heritage Archive 232
system dynamics 459–469; decision-making of residents of heritage buildings 460–462; degradation of plastic museum artefacts 465–467; systems modelling for preventive collection management 463–465

tangible heritage 218–222
Tan Khuan Seng 426
Tan Si Chong Su (Po Chiak Keng), Singapore *426*, 426–428
Tarkovsky, Andrei 32, 40n1
Taylor, K. 60
teaching futures literacy *see* literacy teaching
team-based inquiry (TBI) 324, 330
techno-economic optimisation 298
Temple Bar in Dublin 415

Text Catalogue 400
Thelen, D. 367
thermal comfort 316–318; and energy behaviour 308; in Mexico 309–311
thermal performance analysis 279–280
thermal transmittance (U-value) 275
thermoclastism 269
Therond, D. 90
Till, Jeremy 399
Tixcacalcupul (Yucatán, México) 140–141
Tooro Botanical Gardens 332
Too Young to Die 430
Torre, Marta de la 21, 27n6, 47
total economic value (TEV) 440
Towards Future Heritage Management 531
Traditional-Modern System (TMS) 63
training in literacy teaching 533–535; *see also* literacy teaching
transdisciplinarity 493–494
transfer of development rights (TDR) 474
transformational change 36–38, *37*; defined 36; and heritage management 36–38; and heritage value 37; and 'ruinification' 37
transitional change 35–36; defined 35; prison in downtown Oslo, Norway 35–36; ruins of the Kaiser Wilhelm Memorial Church 35
triple cultural heritage 61

UCL Centre for Extreme Citizen Science (UCL ExCite) 116
UNIDROIT Convention on Stolen or Illegally Exported Cultural Objects 172
United Nations (UN) 1, 59, 181; Agenda 2030 441, 527; *Post-2015 Report* 215; Resolution 1244 (Security Council) 183–184
United Nations Declaration on the Rights of Indigenous Peoples (UNDRIP) 66
United Nations Development Programme 348
United Nations Educational, Scientific and Cultural Organization (UNESCO) 30, 185, 308, 310, 326, 328–329, 332, 334–335, 348; Chair on Heritage Futures 534–535; *Charter on the preservation of digital heritage* 488; Convention Concerning the Protection of the World Cultural and Natural Heritage 513; Convention on the Means of Prohibiting and Preventing the Illicit Import, Export and Transfer of Ownership of Cultural Property 172; Creative Cities Network (UCNN) 329; Culture in City Reconstruction and Recovery, CURE 2018 227; *Declaration on the Responsibilities of the Present Generations Towards Future Generations* 531; Outstanding Universal Value 67; 'Revive the Spirit of Mosul' 208; World Heritage 425–426, 471; World Heritage List 154
United Nations Interim Administration Mission deployed in Kosovo (UNMIK) 184–185
United Nations Sustainable Development Goals (UN SDG) 349
United States: Coalition Provisional Authority 218; Cultural Property Implementation Act (CPIA) 175
universal myth 371
UN Post-conflict Recovery 231
unreflected universality 371
urban conservation 9, 405–408; and Islamic holy cities 215–218; *see also* conservation-planning
urban heritage *see* boundary approach, urban heritage
urban imaginaries 382, 385–387
Urban Redevelopment Authority 427, 431
urban regeneration: in Dadaocheng, Taipei City 383–384; heritage use and imagination in 382–383
Urban Regeneration Station (URS) project 384
urban values-centred regeneration 438–454; circular economy 439–441, 444–446; community in the regeneration 448–449; conflict management 451–452; historic urban landscape 441–444; landscape and utility 451–452; landscape as an 'attractive force' 450–451; religious cultural heritage 447–448; sustainable regeneration 441–444, 446–447; value-centred approach 439–441
useful daylight illuminance (UDI) 281

value(s) 3; adding, to local products 158–160; defined 58; discourse 21; extrinsic 47; and heritage management 46–48; importance for heritage 17; intrinsic 47, 440–441, 443, 447–449; and knowledge 409–412; and Nigeria 60–61; and risk assessment 48–49; in risk assessment at heritage sites 44–55; typologies 23
value-centred approach 439–441, 453
value-creating science 493–494
'value pie' 48
values-centered conservation theory (VCC) 17, 19; adapting 25–27; criticism of 24–25; defined 22; engaged preservation 25–26; and heritage conservation 20, 21–22; ideas in

22–23; and modern heritage conservation practices 21; shortcomings of 24–25; and sustainable heritage 20–21
Vandesande, A. 139
Veldpaus, L. 413, 414
Venice Charter 421
Viejo-Rose, D. 191, 244
Vinci, Leonardo da 285
Viollet-le-Duc, Eugène 405
Visby (Sweden) *291,* 299
visual environment 280–282
Voltaire, L. 504
'vulnerable value' 48
Vygotsky, L. 521

Waanyi women 59
Wagner, C. 246
Walking Distance (Manning and Manning) 246
Waller, Robert 45, 48, 51–52
Warburg, Aby 401; *Mnemosyne Atlas* 394–395, 515
Ward, K. 413
Waterton, E. 92, 410
Watson, S. 92
weak sustainability 38
Webb, A. L. 309
Weber, M. 366
Web GIS tool 265
Wells, J. C. 221, 370
Wenger, Susanne 64
Western ideas: and indigenous heritage 64–65; and Nigeria 64–65
wet bulb temperature (WBT) 280
While, A. 410
Why Places matter to people 513
Wildavsky, A. 46
Williams, D. R.: *Meaning Theory* 233
Winter, Tim 244
Wintersteiner, W. 246
witness: character witness 110–111; expert witness 109–110; eyewitnesses 108–109; lay witness 111–112; moral responsibility of 107; perspective and sustainability 107–108; and social media 107
Wohlmuther, C. 246
Wolfe, P. 251
work and transferable skills 144–145
World Commission on Culture and Development 348
World Culture Reports 348
World Health Organization (WHO) 87
world heritage: conservation and inter-community dynamics 186; cultural heritage in conflict 184–186; and globalisation 182–183; Kosovo and post-war political status 183–184; overview 181; Patriarchate of Peja/Peć Monastery 187; and political sustainability 182–183, 186; and World Heritage List 184–186; World heritage sites 186
World Heritage Committee (WHC) 72; on Bali Cultural Landscape 74–76; Subak attributes from 75
World Heritage Convention (WHC) 58, 161, 349, 407
World Heritage List: conditions of integrity 216; cultural heritage in conflict and inscription on 184–186; and cultural landscapes 351
World Heritage sites 164; and conservation 186; and inter-community dynamics 186; political sustainability 186
World War I 243, 245
World War II 198, 234

Yarrow, T. 309
Yates, Frances: *The Art of Memory* 519
Yeo Kang Shua 9
Yohanis, Y. G. 308
York Conservation Trust 414
Youth Against Settlement 253

Zittoun, T. 382
Zooniverse 116, 119, 121, 122